PROTEIN TRAFFICKING IN NEURONS

PROTEIN TRAFFICKING IN NEURONS

Editor

ANDREW J. BEAN, Ph.D.

Department of Neurobiology and Anatomy
The University of Texas Health Science Center at Houston
Houston, Texas

AMSTERDAM • BOSTON • HEIDELBERG • LONDON
NEW YORK • OXFORD • PARIS • SAN DIEGO
SAN FRANCISCO • SINGAPORE • SYDNEY • TOKYO

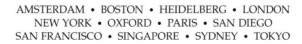

ELSEVIER

Academic Press is an imprint of Elsevier

Elsevier Academic Press
30 Corporate Drive, Suite 400, Burlington, MA 01803, USA
525 B Street, Suite 1900, San Diego, California 92101-4495, USA
84 Theobald's Road, London WC1X 8RR, UK

Cover image courtesy of Lorenzo Morales. Title page image courtesy of Sharon Pepper.

This book is printed on acid-free paper. ∞

Library of Congress Cataloging-in-Publication Data
Protein trafficking in neurons / editor, Andrew Bean.
        p. ; cm.
    Includes bibliographical references and index.
    ISBN-13: 978-0-12-369437-9 (hardback : alk. paper)
    ISBN-10: 0-12-369437-X (hardback : alk. paper)
    1. Proteins—Physiological transport. 2. Neurons. I. Bean, Andrew.
    [DNLM: 1. Protein Transport. 2. Neurons—metabolism.   QU 120
  P9675 2007]
  QP551.P76 2007
  572'.69—dc22

                                                            2006010166

British Library Cataloguing in Publication Data
A catalogue record for this book is available from the British Library

ISBN 13: 978-0-12-369437-9
ISBN 10: 0-12-369437-X

For all information on all Elsevier Academic Press publications visit our Web site at www.books.elsevier.com

Printed and bound by CPI Group (UK) Ltd, Croydon, CR0 4YY

Transferred to Digital Print 2012

This book is dedicated to A.R. Bean.

Contents

15. Exocytic Release of Glutamate from Astrocytes: Comparison to Neurons

W. Lee and V. Parpura

VII. PROTEIN TRAFFICKING AND NEURONAL DISEASE

16. Trafficking Defects in Huntington's Disease

E. Trushina and C.T. McMurray

17. Neuronal Protein Trafficking in Alzheimer's Disease and Niemann-Pick Type C Disease

A.M. Cataldo and R.A. Nixon

18. Trafficking of the Cellular Prion Protein and Its Role in Neurodegeneration

O. Chakrabarti and R.S. Hegde

Contributors

Numbers in parentheses indicate the pages on which the authors' contribution begin.

Andres Barria (203) Department of Physiology and Biophysics, University of Washington, Seattle, WA, USA

Angela I.M. Barth (45) Department of Molecular and Cellular Physiology, Stanford University School of Medicine, Stanford, CA, USA

Anne M. Cataldo (391) Departments of Psychiatry and Neuropathology, Harvard Medical School Laboratory for Molecular Neuropathology, McLean Hospital, Belmont, MA, USA

Oishee Chakrabarti (413) Cell Biology and Metabolism Branch, National Institute of Child Health and Human Development/NIH, Bethesda, MD, USA

Lu Chen (175) Department of Molecular and Cell Biology, University of California, Berkeley, CA, USA

Kathryn H. Condon (143) Department of Neurobiology, Duke University Medical Center, Durham, NC, USA

Robert H. Edwards (225) Departments of Neurology and Physiology, UCSF School of Medicine, San Francisco, CA, USA

Michael D. Ehlers (143) HHMI, Department of Neurobiology, Duke University Medical Center, Durham, NC, USA

Robert T. Fremeau, Jr. (225) Department of Neuroscience, Amgen, Inc., Thousand Oaks, CA, USA

Leora Gollan (63) Department of Physiology and Cellular Biophysics College of Physicians and Surgeons, Center for Neurobiology and Behavior, Columbia University, New York, NY, USA

Volker Haucke (125) Department of Membrane Biochemistry, Institute for Chemistry-Biochemistry, Freie Universität Berlin, Berlin, Germany

Ramanujan S. Hegde (413) Cell Biology and Metabolism Branch, National Institute of Child Health and Human Development/NIH, Bethesda, MD, USA

Erika L.F. Holzbaur (29) School of Medicine, University of Pennsylvania, Philadelphia, PA, USA

Jürgen Klingauf (125) Department of Membrane Biophysics, Max-Planck-Institute for Biophysical Chemistry, Göttingen, Germany

William Lee (329) Department of Cell Biology and Neuroscience, University of California, Riverside, CA, USA

Bita Maghsoodi (175) Department of Molecular and Cell Biology, University of California, Berkeley, CA, USA

T.F.J. Martin (305) Department of Biochemistry, University of Wisconsin-Madison, Madison, WI, USA

C.T. McMurray (369) Molecular Neuroscience Program, Mayo Clinic, Rochester, Rochester, MN, USA

Hiroaki Misonou (243) Department of Pharmacology, School of Medicine University of California, Davis, CA, USA

W. James Nelson (45) Department of Molecular and Cellular Physiology, Stanford University School of Medicine, Stanford, CA, USA

Ralph A. Nixon (391) Departments of Psychiatry and Cell Biology, New York University School of Medicine Center for Dementia Research, Nathan Kline Institute, Orangeburg, NY, USA

Vladimir Parpura (329) Department of Cell Biology and Neuroscience, University of California, Riverside, CA, USA

Eran Perlson (29) School of Medicine, University of Pennsylvania, Philadelphia, PA, USA

Timothy A. Ryan (97) Department of Biochemistry, Weill Medical College of Cornell University, New York, NY, USA

Sethu Sankaranarayanan (97) Merck Research Labs, West Point, PA, USA

Peter Scheiffele (63) Department of Physiology and Cellular Biophysics College of Physicians and Surgeons, Center for Neurobiology and Behavior, Columbia University, New York, NY, USA

James S. Trimmer (243) Department of Pharmacology, School of Medicine, University of California, Davis, CA, USA

E. Trushina (369) Molecular Neuroscience Program, Mayo Clinic, Rochester, Rochester, MN, USA

Helene Vacher (243) Department of Pharmacology, School of Medicine University of California, Davis, CA, USA

Violet Votin (45) Department of Molecular and Cellular Physiology, Stanford University School of Medicine, Stanford, CA, USA

M. Neal Waxham (3) Department of Neurobiology and Anatomy, University of Texas Medical School at Houston, Houston, TX, USA

Charles Yeaman (271) Department of Anatomy and Cell Biology, University of Iowa Carver College of Medicine, Iowa City, IA, USA

Mei Zhen (75) Samuel Lunenfeld Research Institute, Mount Sinai Hospital, Toronto, Ontario, Canada

Acknowledgments

The rapid insight into the molecular mechanisms of protein trafficking gained since the early 1980s is a tribute to the many fine scientists throughout the world working on these issues. In addition to the authors, many people have contributed directly to this book. The idea for this monograph resulted from teaching graduate students at The Graduate School of Biomedical Sciences, University of Texas Health Science Center at Houston, and from discussions with Dr. Thomas A. Vida. Lorenzo Morales spent countless hours creating the cover image. Sharon Pepper provided editorial assistance, comic relief, and the image you'll find facing the title page. Yasmin Lotfi and Brandon Grossman read and commented on selected chapters. Dr. Johannes Menzel, Kirsten Funk, and Sarah Hajduk at Elsevier have been understanding and kind, and have provided guidance throughout the production process.

Acknowledgments

PROTEIN MOVEMENT

The roles played by passive diffusion and active transport in the motility of proteins, lipids, and organelles is the subject of the first section. Recent developments in optical methods that can be used to examine molecular mobility (Waxham), and the role of the cytoskeleton and attendant motor proteins on vesicle and organelle movement (Perlson and Holzbaur) address cell biological issues whose importance for neuronal function is discussed. Cytoskeletal remodeling is required for neuronal morphogenesis and the development and outgrowth of neurites. A discussion of cytoskeleton-dependent remodeling and trafficking (Votin et al.) related to synapse development serves to segue the discussion to that of synapse development in Section 2.

Please note that the previous printing included colour figures throughout the book.

The colour figures are now only available on the Companion Website:

http://www.elsevierdirect.com/companion.jsp?ISBN=9780123694379

1

Molecular Mobility in Cells Examined with Optical Methods

M. NEAL WAXHAM

The efficient delivery of cellular constituents to their proper location is fundamental to all aspects of cell biology and is of particular interest to neuroscientists, in part, due to the unique and complex architecture of neurons. Compartments near the end of dendrites and, in particular axons, can be far from the cell soma; these large distances (axons can be as long as one meter) present a difficult problem for delivery of macromolecules. Diffusion due to the random walk of molecules is extremely efficient over short (μm) distances, but movement of molecules over larger distances requires active transport and expenditure of energy. A significant proportion of this book is devoted to describing the process of protein trafficking through the use of vesicle movement and active transport. The purpose of this chapter is to address the issue of protein mobility from the point of view of nondirected random walk of molecules. Whether one is interested in receptors and ion channels movement in the membrane, the range of action of a given second messenger, or the activation and movement

TABLE 1.1. **Physical properties of a 100 kDa protein.**

Property	Value	Comment
Mass	166×10^{-24} kg	Mass of one mole/Avogadro's constant
Density	1.38×10^3 kg/m^3	1.38 times the density of water
Volume	120 nm^3	Mass/density
Radius	3 nm	Assuming a spherical shape
Drag Coefficient (in water @ 20°C)	60 pN.s/m	From Stoke's Law
Diffusion Coefficient (in water @ 20°C)	67 μm^2/s	From the Stoke's-Einstein relationship
Average Speed	8.6 m/s	From the Equipartion principle

Modified from Howard (2001).

of transcriptional factors to the nucleus, all require an understanding of the basic principles of translational diffusion. Diffusion of molecules in well-mixed dilute environments is relatively straightforward, and follows a set of basic well-established principles. However, the interior of cells presents a much more complex environment to a moving molecule. Macromolecular crowding, viscosity, physical barriers, and specific and/or nonspecific binding each can influence the distance a given molecule can travel per unit time. The goal of this chapter is to provide a brief description of each of the factors that influence translational diffusion relevant to cell biologists. A discussion of the present techniques used to quantify translational diffusion also is presented, along with some examples of how each of these techniques has helped advance our understanding of molecular mobility in the cellular setting.

I. BROWNIAN MOTION AND THE FUNDAMENTALS OF DIFFUSION

An appreciation for molecular movement requires a rudimentary understanding of the forces sensed by molecules. For objects of the size relevant to cell biology, the most important forces include mechanical (viscous), thermal (collisional), and chemical. One reason that other forces can

be ignored is that protein molecules, and cells for that matter, have little inertia in relation to the large viscous forces from the environment. Gravity for instance has little effect on objects the size of proteins. There are several excellent books that discuss the basic principles and mathematics of diffusion (Berg 1993; Crank 1956). The following brief discussion borrows from the book by Howard (2001), to which the reader is referred for a more complete treatment of these concepts.

A. Viscosity and Collisional Forces

Viscosity is of fundamental importance to diffusion. As a molecule moves through a stationary liquid, the drag force sensed is related to the molecule's velocity, and a drag coefficient that takes into account the size and shape of the object and the viscosity of the liquid. A simple formula representing this relationship is:

$$F = v \cdot 6\pi\eta r, \qquad (1\text{-}1)$$

where v equals velocity, η equals viscosity, and r is the hydrodynamic radius of the molecule.[1]

Thus, the force sensed by a molecule increases as the velocity, viscosity, or size of the molecule increases. An approximate viscous force sensed by an average size protein of 100 kDa is approximately 480 pN (picoNewtons; see Table 1.1 for additional

---

[1]Viscosity is also fundamentally important to the process of protein folding. A description of "microviscosity" is not developed in this chapter.

physical properties of a 100 kDa protein; adapted from Howard 2001). For reference, the viscosity (in cP (centipoises) at 20°C) of water is 1, acetone is 0.32, a 50% solution of Ficoll 400 is 600, and glycerol is 1408. Cell cytoplasm is thought to have a viscosity of approximately 1–3, and the viscosity of the membrane bilayer is approximately 50–100 (discussed later in this chapter).

Whereas viscous forces retard movement, collisional (thermal) forces drive molecular movement. A protein in solution moves because of the immense number of collisions experienced from water molecules. Since these collisions are not directed, the movement of the protein molecule is random, and this is the basic premise of the random walk taken by molecules observed as Brownian motion. When two objects collide (e.g., a water molecule and a protein), the force produced is related to the rate of change of momentum and momentum, in turn, is related to the velocity and mass of the molecule. Water molecules travel at significant velocity (600 m/s), but their momentum is small due to their small mass. Their number, however, is enormous (water has a molar concentration of 55.35 mol/L), producing a massive number of collisions per unit time. This randomly directed force is also called the thermal force, and for a 100 kDa protein is calculated to be approximately 500 pN. The thermal force is significant enough to balance the viscous force noted earlier. However, as Howard (2001) notes, even with the relative large instantaneous speed of a 100 kDa protein (8.6 m/s), the average distance that the protein moves before its direction is randomized by collisions is only ~0.024 nm. This is an almost imperceptibly small step size relative to the size of the protein (~3 nm in diameter).

In total, it is thermal/collisional events that provide the force for moving molecules, but the viscosity of the solution and size of the molecule undergoing collisions that retard movement. The movement of proteins through solution is said to be *over-*

damped. Overdamped relates the fact that inertial forces are very small in relation to the viscous forces. An important consequence of this statement is that the drag force sensed by a protein is proportional to its velocity, and vice versa, since the inertial forces due to their small mass can be ignored. Ignoring inertial forces is a necessary prerequisite for applying Stoke's law to describe the forces associated with protein motion in cytoplasm or membranes.

The velocity of a molecule in three dimensions can be calculated by the relationship:

$$v_{rms} = \sqrt{\frac{3kT}{m}}, \qquad (1-2)$$

where v_{rms} is the root mean square velocity, k is the Boltzmann's constant (1.381 × 10^{-23} J/K), T is temperature (in Kelvin), and m is the mass of the molecule.

B. Diffusion

Diffusion, and factors that alter diffusion, is the central concept of this chapter. Diffusion is the random motion of an object due to collisions with other particles and is characterized by rapid abrupt changes in direction. Einstein is credited with relating diffusion to the Brownian motion observed at the macro and microscopic levels. This is formalized in the Einstein relation:

$$D = \frac{kT}{6\pi\eta r}, \qquad (1-3)$$

where D is the diffusion coefficient, k is Boltzmann's constant, T is temperature, η is viscosity, and r is the radius of gyration of the particle. Since the radius of a sphere is proportional to the cube root of its mass, we can get a reasonable approximation of D based on the molecular weight of the molecule of interest. This relationship also highlights that significant differences in molecular weight have only modest impacts on D due to the inverse cube root relationship. This equation permits simple

TABLE 1.2.   Distance/time relationship for one-dimensional diffusion of different sized objects in water.

Object	Distance Traveled			
	1 μm	100 μm	10 mm	1 m
K⁺	0.25 ms	2.5 s	2.5×10^4 s (7 hours)	2.5×10^8 s (8 years)
Protein (3 nm radius)	5 ms	50 s	5×10^5 s (6 days)	5×10^9 s (150 years)
Organelle (0.5 μm radius)	1 s	10^4 s (3 hr)	10^8 s (3 years)	10^{12} s (30 million years)

Modified from Howard (2001).

relationships to be identified. The diffusion coefficient is inversely proportional to the size and viscosity of the medium, meaning the larger the molecule or the more viscous the solution, the smaller the diffusion coefficient.

Diffusion can also be considered from the perspective of distance traveled over time. The mean squared displacement of a particle increases in proportion to time (see Berg (1993) for more discussion). For diffusion

$$\langle x^2 \rangle = nDt, \qquad (1\text{-}4)$$

where $\langle x \rangle$ represents the ensemble average of particle displacement; $n = 2$, 4, or 6, for one-, two-, or three-dimensional diffusion, respectively; D is the diffusion coefficient; and t is time. From this formula, we see that displacement, x, is proportional to the square root of time, so in order for a particle to wander twice as far, it takes four times as long.

Knowing the diffusion coefficient makes it similarly possible to calculate how long it takes for a molecule to travel a given distance. This relationship can be formalized (for three dimensions) as:

$$t = \frac{x^2}{6D}, \qquad (1\text{-}5)$$

where x is displacement, D is the diffusion coefficient of the molecule, and t is time.

It is clear that diffusion is an extremely efficient process for moving molecules over short distances. However, as a consequence of the square root of time relationship, the time cost for traveling distances greater than a few μm can become prohibitive. Some examples are shown in Table 1.2 (modified from Howard 2001). As discussed later in this chapter, if we consider that the cytoplasm presents additional barriers to diffusion, it becomes obvious that the random walk of proteins is insufficient to deliver molecules efficiently at distances more than a few μm.

II. A VIEW OF CYTOPLASM AND MEMBRANE FROM THE SINGLE MOLECULE PERSPECTIVE

The previous discussion of diffusion has assumed the objects are moving in a homogeneous, uncrowded environment. This is clearly not the situation when considering molecular movement in cellular cytoplasm or membranes. These complexities present a daunting challenge to the study of protein mobility in living cells, but significant progress has been made. Advances in electron microscopy are providing important details concerning the spatial geometry of macromolecules within cells. Similar advances in light and fluorescence microscopy and spectroscopy are providing quantitative experimental data on the non-idealities of diffusion in cytoplasm and cellular membranes.

A. Properties of Cytoplasm—Organelles and Cytoskeleton

Remarkable advances have been made in the last decade on tomographic reconstruction of cellular cytoplasm using the electron microscope (EM). Figure 1.1 shows an example of such work. This is a tomographic reconstruction of an insulin-secreting mammalian cell line (HIT-T15) and the area shown is centered on the Golgi apparatus (Marsh et al. 2001). One is immediately struck by the density and geometric complexities presented by the intracellular organelles. For perspective, the bright green tubes represent microtubules, which have an approximate diameter of 25 nm. It is evident that no two regions within the cytoplasm are identical. Also critical is the recognition that these macromolecular structures themselves are not static. The EM tomogram presented in Figure 1.1 represents only those molecules of large enough size and density to be reliably reconstructed (ribosomes of ~25 nm diameter are the smallest structures visualized in this reconstruction). Within the intraorganellar spaces are cytoplasms rich with proteins and metabolites.

We can take EM tomograms and other data and make reasonable approximations for the volumes and surface areas of different cellular components (reviewed by Luby-Phelps 2000). As examples, the total surface area occupied by mitochondria of a typical cell is on the order of hundreds to thousands of μm^2, and the endoplasmic reticulum in secretory cells can present as much as 30,000 μm^2 of surface area. In total, intracellular membrane surface area is estimated to be approximately 100,000 μm^2 cell, an order of magnitude larger than the surface area of the plasma membrane.

The cytoplasm is rich in cytoskeletal elements. Actin, tubulin, and intermediate filaments are typically the most abundant proteins present in cells. Each of these proteins can assemble into polymers with unique structure and functional properties. The levels of actin are estimated to be approximately 4 mg/ml in cells, half being in the polymerized F-actin state, and the remainder in the soluble G-actin state. F-actin has a diameter of approximately 8 nm and exists in a dynamic state of assembly and disassembly. Tubulin (α- and β-tubulin) is the fundamental component of microtubules that have an approximate diameter of 25 nm. Intermediate filaments are approximately 10 nm in diameter and are more stable than either F-actin or microtubules. Intermediate filaments can be assembled from different monomeric components, including vimentin and neurofilaments. Neurofilaments are the fundamental elements that provide axons with their remarkable structural resilience. The surface area occupied by the array of cytoskeletal elements also presents significant obstacles between which cytoplasmic molecules must navigate. Ultrastructural data suggests that the lattice of cytoskeletal elements, including but not exclusive of the three main elements discussed earlier, present as much as 70,000–90,000 μm^2 of surface area (Luby-Phelps 2000). Obviously, these values vary significantly depending on the particular cell type under investigation. The state of cellular differentiation and cell division are also significant factors in determining the content of organellar and cytoskeletal components.

B. Properties of Cytoplasm—Water, Protein, and Other Soluble Constituents

The space outside of organelles and cytoskeleton is filled with fluid whose composition has a major potential impact on the translational and rotational diffusion of molecules. Water composes approximately 70 percent of this space. Experimental approaches to examine water properties inside cells (nuclear magnetic resonance

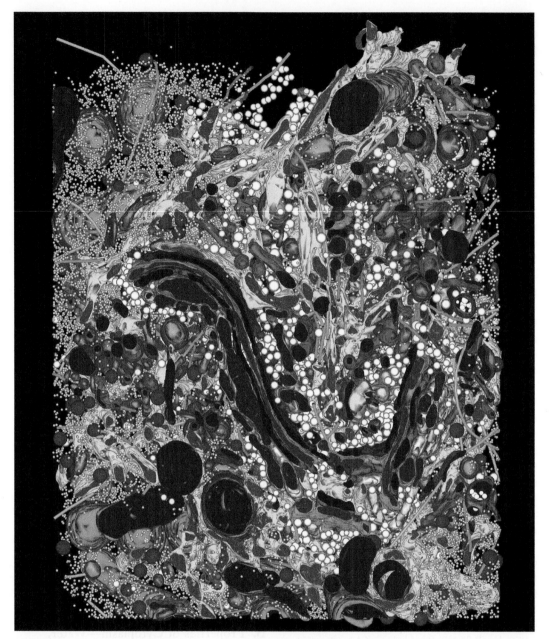

FIGURE 1.1.   A 3D reconstruction of the Golgi region in an insulin-secreting, mammalian cell. Three serial 400-nm-thick sections cut from a high pressure frozen, freeze-substituted, and plastic-embedded HIT-T15 cell were reconstructed by dual axis EM tomography. The software package IMOD was used to model all visible objects within the resulting reconstructed volume ($3.1 \times 3.2 \times 1.2\,\mu m^3$). The Golgi complex with seven cisternae (C1–C7) is at the center. The color coding is as follows: C1, light blue; C2, pink; C3, cherry red; C4, green; C5, dark blue; C6, gold; C7, bright red. The Golgi is displayed in the context of all surrounding organelles, vesicles, ribosomes, and microtubules: endoplasmic reticulum (ER), yellow; membrane-bound ribosomes, blue; free ribosomes, orange; microtubules, bright green; dense core vesicles, bright blue; clathrin-negative vesicles, white; clathrin-positive compartments and vesicles, bright red; clathrin- negative compartments and vesicles, purple; mitochondria, dark green.

Image courtesy of Dr. Brad Marsh, Institute for Molecular Bioscience, The University of Queensland, Australia. Originally published in the Inaugural Article: Organellar relationships in the Golgi region of the pancreatic beta cell line, HIT-T15, visualized by high resolution electron tomography. *Proc. Natl. Acad. Sci. USA* (2001) 98; 2399–2406.

and quasielastic neutron scattering) are somewhat difficult to interpret, but suggest that the overall rotational mobility of intracellular water is reduced about twofold. Importantly, these studies highlight the concept that bulk water has distinctly different properties than water associated with macromolecules or other intracellular surfaces and the twofold difference in rotational mobility is the sum of both these water compartments. The ordering of water near surfaces reduces its chemical activity, which can impact chemical reactions. As pointed out by Luby-Phelps (2000), an alternative way of thinking about this issue is that water will be more concentrated (and ordered) around hydrophilic surfaces than around hydrophobic surfaces.

An additional, and perhaps more meaningful method for probing the aqueous environment of the cytoplasm is to study the rotational mobility of labeled molecules. Rotational mobility is dictated largely by solvent viscosity and thus provides this fundamental parameter necessary for understanding the translational mobility of objects. Nuclear magnetic resonance of labeled or endogenous proteins, spin-labels, and fluorescent labels all have been used to examine the viscosity of the aqueous intracellular environment (reviewed in Luby-Phelps 2000). A distillation of this data is that the viscosity of the aqueous intracellular environment does not appear significantly different from bulk water with, at most, a two- to threefold increase in reported viscosity. Interestingly, there is also little variation in viscosity throughout the cell including the nucleus. These data indicate that intracellular viscosity is similar to water, and that viscosity *per se* produces only a modest impact on intracellular diffusion, at least for molecules the size of small fluorescent dyes.

To apply the Einstein relationship to analyze particle behavior, we assume that interactions between diffusing particles can be neglected. Even though this assumption can be satisfied rather easily *in vitro*, this assumption cannot be made when analyzing the diffusive behavior of particles inside cells. As noted earlier, the cellular cytoplasm is a complex and dynamic matrix of organelles that will present significant surface area to diffusing molecules. In addition, the fluid-filled space surrounding organelles is not diluted. The protein content of cells is estimated to be between 17 and 35 percent by weight (Luby-Phelps 2000; Minton 2001). This concentration of macromolecules results in significant probabilities for protein-protein collisions and also produces a volume exclusion effect, termed macromolecular crowding. By assessing the translational mobility of molecules of increasing size at increasing concentrations of Ficoll-70 (an inert molecule used to produce macromolecular crowding), it was determined that macromolecular crowding produces significant slowing of translational diffusion (Dauty and Verkman 2004). Surprisingly, the impact of the size of the diffusing molecule was relatively insensitive to the effects of crowding, and these authors concluded that significantly hindered diffusion of larger (>500 kDa) molecules must be due to immobile obstacles in addition to the slowed diffusion due to crowding. Others have shown similar slowing of diffusion in the presence of crowding obstacles but have found that molecules <500 kDa also suffer hindered diffusion (Weiss et al. 2004).

C. Properties of Membranes—Lipids and Integral Membrane Proteins

Cellular membranes are two-dimensional structures approximately 60–100 nm in thickness that are composed of lipids and proteins (see Figure 1.2). Estimates are that approximately 40 percent of the plasma membrane is lipids, and the other 60 percent is protein (Choquet and Triller 2003). The Sanger-Nicholson fluid-mosaic model for membrane structure remains an excellent starting point for discussion (modifications will be introduced later in

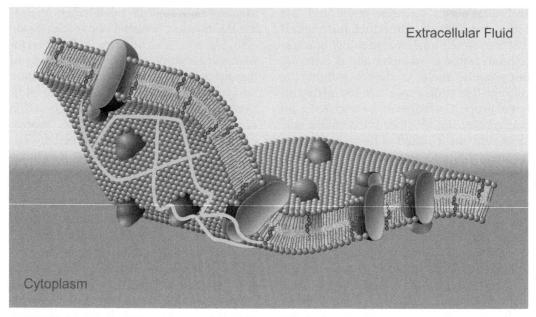

FIGURE 1.2. **Model of the plasma membrane.** The lipid bilayer is folded and cut away to reveal general features of the membrane architecture. Transmembrane proteins are shown in blue floating in the viscous environment of the lipid bilayer shown in gold. Cholesterol is shown as magenta structures embedded in the membrane. Cytoskeletal elements, shown as yellow lines on the cytoplasmic face, lie closely apposed to the membrane and can alter the translational mobility of membrane proteins by serving as anchors or corrals.

TABLE 1.3. **Lipid composition from membranes (percent by weight).**

Membrane	Phospholipids	Cholesterol	Glycolipids	Cholesterol esters and other
Plasma	57	15	6	22
Golgi	57	9	0	34
ER	85	5	0	10
Inner Mito	92	0	0	8
Nuclear	85	5	0	10

Modified from Lehninger (1982).

the chapter). Membranes are fluid structures at normal physiological temperatures with the lipids being held together by numerous cooperative noncovalent interactions. Due to the hydrophobic nature of the hydrocarbon tails, and the associated attractive van der Walls forces, lipids spontaneously form closed structures, but at the same time, maintain a reasonable degree of fluidity.

Diffusion of molecules in the cytoplasm provides three degrees of freedom in the x,

y, and z axes. Diffusion in membranes is constrained by the loss of movement in the z dimension. At a theoretical level, this would enhance the mean squared displacement of a particle for a given time interval (see Eq. 1-4). The lipid bilayer, however, is significantly different in its composition than the cytoplasm and is significantly different in composition between different cellular organelles (see Table 1.3; from Lehninger 1982). The concentration of phospholipids in the bilayer is very high, resulting in an increase in membrane viscosity that is sensed by a moving particle. The viscosity of the membrane has been likened to that of olive oil, some 50 to 100 times that of water (Stryer 1988). The membrane viscosity plays a dominant role in the translational mobility of proteins in the membrane. This can lead to significantly slowed translational mobility relative to what would be seen of the same sized molecule in solution.

Membrane fluidity (viscosity) can also be controlled by the composition of the fatty acyl chains of the lipids. Increased fluidity is produced by acyl chains with shorter lengths and decreased degrees of saturation. Membrane fluidity in eukaryotic cells is also regulated through the concentration of cholesterol in the bilayer. Cholesterol intercalates into the lipid membranes and, in doing so, increases lipid disorder and results in increased fluidity. Cholesterol at higher concentrations, can also have the opposite effect by preventing free motion of the fatty acyl chains, thus decreasing fluidity and forming microdomains of certain types of lipids.

The concept of lipid rafts highlights the heterogeneous environment of the bilayer, where lipids themselves can form small microdomains (<250–300 nm) that influence the lateral mobility of membrane-bound constituents. Lipid rafts are described as detergent insoluble membrane domains constructed largely of cholesterol and sphingolipids that contain saturated fatty acid chains that permit tight packing, leading to a liquid ordered phase. This ordering influences translational diffusion of the resident lipids, but also influences the mobility of lipids in the immediate domain surrounding the raft as the raft itself produces a barrier, albeit mobile, to translational diffusion. Lipid rafts have many proposed functions including roles in endocytosis, internalization of toxins and viruses, calcium homeostasis, and protein sorting (Zajchowski and Robbins 2002). However, one of their best studied roles

is in fostering the association (or potentially excluding) of signaling molecules into effective complexes. A number of membrane-associated proteins contain glycosylphosphatidylinositol- (GPI-) linked moieties that facilitate their association with lipid rafts, restricting their translational mobility and facilitating the probability of interactions. Some of the best studied GPI-anchored proteins are those involved in growth factor signaling; the epidermal growth factor (EGF) receptor and platelet-derived growth factor receptor each are enriched in lipid rafts. EGF receptor activation, tyrosine kinase phosphorylation, and recruitment of adapter proteins all occur in the confines of lipid rafts. The Src family of protein kinases is also GPI-linked and concentrates into rafts. Lipid rafts themselves also seem to be dynamic structures, assembling and disassembling as needed.

Typical diffusion coefficients for lipids in the plasma membrane are in the range of 10^{-9} to $10^{-8}\,cm^2/s$, whereas values in synthetic bilayers are nearly an order of magnitude higher (10^{-8} to $10^{-7}\,cm^2/s$) as measured by FPR (fluorescence photobleaching recovery). Clearly there are factors affecting lipid diffusion in the cellular environment not represented in synthetic bilayers. Constraints to consider are the packing of unique lipid domains, high density of membrane-bound protein components, and the interactions of the cytoplasmic face of the membrane with cytoskeletal elements that help maintain the shape and integrity of the plasma membrane.

III. DIFFUSION AND MOBILITY OF PROTEINS IN CELLS STUDIED WITH BIOPHYSICAL TECHNIQUES

A. Diffusion of Proteins within Cytoplasm

Proteins vary widely in size, and size is inversely proportional to the rate of

TABLE 1.4. Comparison of diffusion coefficients from *in vitro* and *in situ* fpr measurements.

Protein	Radius (nm)	D_s (in solution)	D_c (in cytoplasm)	D_c/D_s	% mobile
Calmodulin	2.1	102	<4	0.039	81
GFP	2.5	87	27	0.31	82
BSA	3.2	67	6.8	0.1	77
Creatine kinase	3.3	65	<4.5	0.07	50–80
Enolase	3.8	56	13.5	0.24	100
IgG	4.7	46	6.7	0.15	54

D = diffusion coefficients ($\mu m^2/s$); modified from Luby-Phelps (2000).

diffusion if one considers proteins as chemically inert spheres. This is obviously not the case. Proteins present complex chemical surfaces and can deviate significantly in shape from simple spheres. Protein shape is somewhat less important since measurement of the hydrodynamic radius of a protein in solution accounts for nonspherical shapes and can be used as an accurate parameter in calculating a diffusion coefficient through the Einstein relationship. The chemical surface of a protein, however, can fundamentally change the diffusive behavior of a protein inside cells. Sites for binding to mobile or immobile elements have the potential to significantly decrease the apparent mobility (quantified as a decreased diffusion coefficient) of a protein. Tabulation (see Table 1.4) of some examples compiled by Luby-Phelps (2000) highlights this point and provides a sense of the magnitude of differences one detects in diffusion coefficients.

Translational diffusion in cytoplasm and membranes is most conveniently studied using fluorescence techniques. In particular, fluorescence photobleaching recovery (FPR), also known as fluorescence recovery after photobleaching (FRAP), and fluorescence correlation spectroscopy (FCS) have been applied widely to quantify diffusion in different cellular compartments. The two techniques have inherent advantages and disadvantages. An experimental set-up to accomplish such experiments is shown in Figure 1.3. This particular set-up uses mul-

tiphoton excitation to illuminate the fluorescent specimens.

A typical FPR experiment is accomplished by first generating an image of a cell and then using the image to target the laser beam to a location for analysis. A short, high-intensity laser pulse is used to irreversibly photobleach a population of the labeled molecules (typically shaped as a spot) and the recovery of new labeled molecules in the focal volume is followed over time (see Box 1.1). The fitting functions used to extract the recovery time (τ_d) are highly system- and parameter-dependent, but in the simplest case, follow a single exponential shape described by

$$F(t) = \frac{F_o + F_\infty(t/\tau_d)}{1 + (t/\tau_d)}, \qquad (1\text{-}6)$$

where F_o is fluorescence before the bleach pulse and F_∞ is the fluorescence at infinite time.

The diffusion coefficient can be extracted from the recovery time by the relationship

$$\tau_d = \frac{\omega_{xy}^2 \gamma}{4D}, \qquad (1\text{-}7)$$

where τ_d is the time constant of recovery, ω_{xy} is the radius of the bleach spot, and γ is a correction factor for the amount of bleaching (Lippincott-Schwartz et al. 2001). Another important parameter that can be derived from FPR curves is the percentage of an immobile fraction, if one is present. A simple equation describing this relationship is

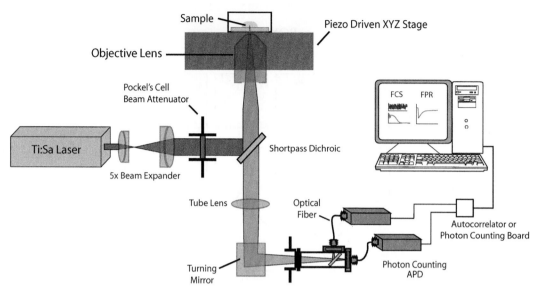

FIGURE 1.3. **Two-photon FPR, FCS, and FCCS set-up.** A titanium: sapphire laser provides two-photon excitation, and the beam is expanded and reflected onto the back focal aperture of a high NA objective lens to form a diffraction limited spot. Photons emitted from the fluorescent sample are collected by the objective and directed to avalanche photodiode detectors (APDs). By scanning the sample using a piezo driven stage, one can produce a "target" image on the screen to position the sample in a desired subcellular location for subsequent spectroscopic measurements. For FCS, detected photons are autocorrelated on a commercially available board with the resulting output being displayed on a computer screen in real time. For FPR measurements, a Pockels cell is used for rapid beam modulation to produce brief, high-intensity bleaching pulses. Photon counts acquired before and after the bleaching pulse are collected into user determined time bins and stored on the computer for subsequent analysis. FPR and FCS use only one of the two detectors. FCCS utilizes both detectors, and the signal is cross-correlated.

$$M_f = \frac{F_\infty - F_o}{F_i - F_o}, \qquad (1\text{-}8)$$

where F_∞ is the fluorescence at infinite time, F_i is the prebleach fluorescence, and F_o is the fluorescence immediately after the bleaching pulse. For more on the fitting of photobleaching recovery curves, see Verkman (2002); Lippincott-Schwartz et al. (2001); Weiss (2004).

A general conclusion that can be drawn from such FPR data is that the experimentally measured diffusion coefficient of a protein in cytoplasm does not correlate well with the radius of gyration. Additionally, for many of these proteins the fluorescence recovery was incomplete, indicating that a significant, but variable, fraction of the fluorescent protein was bound to an immobile element (see Verkman (2002) for discussion of experimental limitations). One interesting exception is the widely used genetically encoded fluorescent tag, green fluorescent protein (GFP). GFP diffusion in cytoplasm is slowed approximately two- to fourfold relative to water (see Table 1.5; from Lippincott-Schwartz et al. (2001)). This two- to fourfold slowing of diffusion is typical of other "inert" tracer molecules like Ficolls or dextrans analyzed by FPR. GFP thus serves as an ideal inert tag for fluorescently labeling proteins because of its apparent absence of binding to cytoplasmic proteins or organelles. The two- to fourfold slowing of translational diffusion can be ascribed to three potential reasons: increased viscosity, binding, and collisions with intracellular molecules/barriers. However, collisional effects appear the most likely mechanism for the slowed diffusion (Verkman 2002).

BOX 1.1

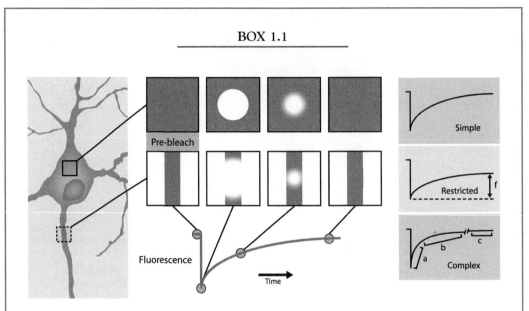

Fluorescence photobleaching recovery (FPR) is a technique used to quantify lateral diffusion of fluorescent molecule in cytosol (3D diffusion), membranes (2D diffusion), and other neuronal compartments. The panel on the left shows a rendering of a neuron where an area in the soma and an area in one of the processes were chosen for analysis. FPR is accomplished by collecting a baseline of fluorescence before applying a rapid, high-intensity laser pulse to produce a well-defined area of irreversibly bleached fluorophores. One then follows the time course of fluorophore recovery into the focal volume. Recovery is proportional to the diffusion coefficient of the fluorophores under investigation. This is diagrammatically shown in the middle panels. The top four panels demonstrate a spot photobleach in the cytoplasm where lateral diffusion repopulates fluorescence in the area that was photobleached. The recovery, schematized on the graph at the right, demonstrates simple diffusion in the soma. In some cases, the dimensions of the compartment under investigation may fall within the focal volume element as in neuronal dendrites or axons. Such an experiment is shown in the middle four panels where incomplete fluorescence recovery is portrayed. The graph shown in the right middle panel demonstrates evidence of this immobile fraction where fluorescence does not return to prebleach levels. By fitting the recovery curves (see text) one can calculate a decay constant that is directly proportional to the diffusion coefficient of the fluorophore. FPR curves from cellular measurements often require more complex fitting than simple single exponential models (right top panel) and also may fail to completely recover (right middle panel), indicating immobile molecules and revealing the complexities of translational diffusion inside cells.

These effects have been analyzed by comparison of fluorescently labeled tracer molecules in water to those in concentrated solutions of dextrans of Ficolls (Luby-Phelps 2000). Such FRP experiments indicate that diffusion in cytoplasm resembles diffusion in solutions of 12–13 percent Dextran or Ficoll.

As the size of macromolecules increases, additional factors influence their

TABLE 1.5. Diffusion of GFP and
GFP-fusion proteins determined by FPR.

Molecule	D ($\mu m^2/s$)
GFP in water	87
GFP in cytoplasm	25
GFP in the ER lumen	5–10
GFP in the mitochondrial matrix	20–30
ER Membrane	
GFP-VSV G-protein	0.45
GFP-signal recognition particle	0.26
Golgi Membrane	
GFP-galactosyltransferase	0.54
Nucleoplasm	
GFP-fibrillarin	0.53
GFP-ERCC1/XPF	15
Plasma Membrane	
GFP-cadherin	0.03–0.04

Modified from Lippincott-Schwartz (2001).

translational mobility. In particular, molecules >500 kDa exhibit significantly decreased rates of diffusion (Verkman 2002; Luby-Phelps 2000). The physical basis for this decreased diffusion is hypothesized to be a sieve-like effect as large macromolecules navigate through the intracellular cytoskeletal matrix. The size threshold for this sieving behavior is a matter of some debate (Luby-Phelps et al. 1986; Seksek et al. 1997; Lukacs et al. 2000), but a safe conclusion is that molecules larger than 500 kDa will suffer hindered diffusion. This sieving is attributed to a meshwork formed by cytoskeletal elements. As the cytoskeleton is neither homogeneous nor static in its structure, hindered diffusion due to sieving effects will be quite variable.

Obviously, significant sieving would occur for even the smallest intracellular organelles such as synaptic or small transport vesicles (50 nm in diameter). Single-particle tracking of 80 nm diameter fluorescent microspheres in fibroblast cytoplasm revealed diffusion coefficients of 2.6–4×10^{-11} cm^2/s (Luby-Phelps 2000). This is 500- to 1000-fold slower than diffusion in solution. Similar values were determined

for secretory vesicles expressing GFP fusion proteins (Burke et al. 1997), chromaffin secretory granules (Steyer et al. 1997), and particularly slow diffusion ($D = 5 \times 10^{-13}$ cm^2/s) was noted of synaptic vesicles in hippocampal neurons (Jordan et al. 2005). These experiments reveal the fundamental need for directional trafficking and active transport to efficiently move even the smallest organelles through the cytoplasm.

GFP can be fused to a variety of proteins that target the reporter to specific organelles. FPR studies of GFP in the mitochondrion revealed a three- to fourfold slowing of diffusion relative to water similar to what has been discovered for the mobility of GFP in the cytoplasm (Verkman 2002). This indicates that the aqueous phase of the mitochondrial matrix is similar to that in cytoplasm. A similar strategy was used to target GFP to the endoplasmic reticulum (ER), and FPR was used to examine diffusion in the ER lumen. In this organelle, diffusion was slowed nine- to 18-fold relative to water, indicating that the ER lumen presents an environment to GFP that hinders its translational mobility. The convoluted lumen of these small organelles confounds the application of simple models for diffusion. Applying the appropriate physical model is a necessary component for the proper interpretation of the FPR data (see Verkman (2002) for further discussion).

Fluorescence correlation spectroscopy (FCS) is a complementary method to FPR for quantifying translational diffusion *in vitro* and in living cells. FCS has found significant applications in cell biology since the mid 1990s, and although the number of papers is growing quickly, specific results from FCS lag behind those for FPR. The fundamental principles of FCS are described in Box 1.2. FCS compares the photon intensity profile produced from fluorophores entering and exiting the focal volumes as time progresses. The fluctuations become more dissimilar at increasing times leading to decay in the autocorrelation function. From this simple scenario, one can deduce that

BOX 1.2

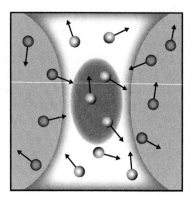

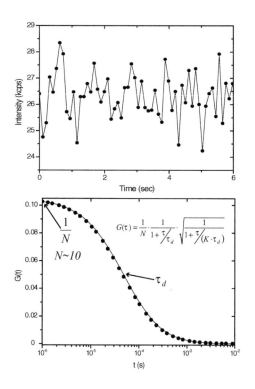

$$G(\tau) = \frac{1}{N} \cdot \frac{1}{1 + \tau/\tau_d} \cdot \sqrt{\frac{1}{1 + \tau/(K \cdot \tau_d)}}$$

$\frac{1}{N}$

$N \sim 10$

τ_d

Fluorescence correlation spectroscopy (FCS) is a sensitive technique capable of resolving fluorescent properties at the single molecule level. For a typical set-up for FCS experiments consult Figure 1.3. Using multiphoton excitation, one obtains a well-described focal volume element as shown in the top left panel. The green oval represents the approximate shape of the multiphoton illuminated volume element and the lighter hourglass outline is the approximate shape of a typical single photon illumination volume. The most common form of FCS analysis involves autocorrelation of the photon counts (a typical intensity trace is shown in the top right panel) to produce the curve shown in the bottom right panel. Two important parameters are derived from fits of this data. The first parameter is the ability to calculate the number of molecules in the focal volume and is inversely related to

$G(0)$; $1/N$ from the analytical fitting function. With knowledge of the illuminated volume element and the number of molecules, one can calculate the concentration of fluorophore. The second parameter, translational mobility, is derived from the decay of the autocorrelation curve and depends on the transit time of the molecules in the focal volume. A simple relationship (see the text) relates this τ_d to the diffusion coefficient of the fluorophore.

fewer numbers of molecules (and therefore fewer photons) in the observation volume will result in a more dissimilar intensity profile. As the concentration of fluorophore increases, the amplitude (G(0) on the y axis) decreases. FCS thus lends itself to situations of low fluorophore number, which is often more relevant in the cellular setting. By introducing or expressing fewer fluorescently tagged molecules, the experimental system is less perturbed from its normal equilibrium. More concretely, if one adds a significant excess of fluorescent molecules over those endogenously present, binding sites might become saturated, leading to the appearance of free diffusion of the labeled ligand when a significant fraction actually is bound at the normal concentration of ligand. This is of fundamental importance when one wishes to quantify translational mobility in the complex intracellular setting.

In FCS, molecules that traverse the volume more quickly (that have faster diffusion coefficients) produce an autocorrelation function that decays at an earlier time point. The simplest analytic function used to fit the autocorrelation data is:

$$G(\tau) = \frac{1}{N} \cdot \frac{1}{1 + \tau/\tau_d} \cdot \sqrt{\frac{1}{1 + \tau/(K \cdot \tau_d)}}, \quad (1\text{-}9)$$

The parameters determined by such fitting are τ_d the half-time for decay, N is the number of molecules, and K describes the shape of the focal volume (ω_z/ω_{xy}). τ_d is related to the diffusion coefficient of the molecule through the relationship:

$$\tau_d = \frac{\omega_{xy}^2}{4D}, \quad (1\text{-}10)$$

As noted previously, the diffusion coefficient of a molecule is inversely related to the radius of gyration (or approximately the cube root of its molecular weight), and distinguishing different bound species of a molecule inside cells with autocorrelation analysis alone has useful but limited application.

FCS can also provide a direct assessment of the number of molecules in the focal volume. Knowing the molecule number and the average fluorescence intensity, one can calculate how bright a particular molecule is, termed molecular brightness. Molecular brightness is a parameter that can be used to analyze complex mixtures of molecules and analysis tools such as the fluorescence intensity distribution analysis (Palo et al. 2002) or the photon counting histogram (Chen et al. 1999b) have capitalized on this information to extract additional biologically relevant information (for review see Elson 2001). Since oligomerization of proteins is a fundamentally important process in cell biology, analyzing how the molecular brightness of molecules is altered by a given stimulus can reveal the dynamics of protein complex formation.

FCS can be accomplished with single photon excitation through commercially available instruments based on standard confocal microscopes (Confocor 2, Carl Zeiss Inc.). This instrument provides efficient transitions between imaging a cellular preparation and collecting FCS data at specific cellular loci. This is a very powerful combination that permits assessing diffusion, and other photophysical processes accessible with FCS, in specific subcellular compartments. There are potential drawbacks with FCS however, and the limitations and cautions have been well reviewed (Haustein and Schwille 2003). Some of these include the data collection time needed to produce an accurate autocorrelation curve, which can require tens of seconds of stable photon intensity and is sometimes problematic in the heterogeneous cellular environment. Photobleaching and autocorrelating background cellular fluorescence can produce artifact laden data. Multiphoton excitation is also useful for FCS measurements. Some adjustments must be made in equations describing the fitting function, and relating τ_d to the diffusion coefficient (Schwille et al. 1999) but otherwise the data collection and analysis is similar to one-photon FCS.

An excellent example of the unique application of FCS comes from the analysis of the biochemical cascade affecting the flagellar motor in *Escherichia coli* responsible for tumbling behavior (Cluzel et al. 2000). FCS measurements were made of GFP-tagged Che-Y in single *E. coli*, which permitted an accurate estimate of the concentration of the protein and how it related to the tumbling behavior. By analyzing single cells both at the biochemical and behavioral level, a much steeper activation curve was derived than possible by measuring across populations of bacteria using more conventional biochemical means. The exceptional capability of FCS to quantify the amount of protein present in a single *E. coli* provided unique insight into this biochemical process.

FCS also was used to examine the transport properties of tubulin and creatine kinase in neuronal axons (Terada et al. 2000). The results showed that it was possible to distinguish unique diffusion of these two proteins in complexes undergoing active transport within the axon. Similarly, FCS was used to examine diffusion of an inert tetramethylrhodamine-10 kDa Dextran tracer in neuronal dendrites (Gennerich and Schild 2002). New models appropriate for fitting FCS data collected from the restricted geometry of dendrites were developed and were used to show that diffusion along the dendrite was slowed one- to twofold relative to diffusion of the same tracer in neuronal soma. However, diffusion across the dendrite was slowed some 90-fold. The extensive cytoskeletal network in dendrites was suggested to be the probable barrier to diffusion across the dendrite.

B. Diffusion of Proteins in Membranes

Diffusion of proteins in membranes is much slower than in cytoplasm. Like the cytoplasm, the membrane also presents a heterogeneous and crowded environment in which the protein moves. This can involve collisions with other proteins, partitioning into distinct lipid domains and collisions with cytoskeletal elements that underlie the plasmalemma. FPR analysis of GFP-tagged receptors has been widely investigated and diffusion coefficients of 5×10^{-10} cm^2/s are typical, with some exceptions. For example, translational diffusion of rhodopsin in rod photoreceptors seems more rapid (3.5–4×10^{-9} cm^2/s; Poo and Cone 1974). This study also reported that the viscosity of the rod photoreceptor membrane based on the translational mobility of rhodopsin was approximately 1 P, some 100 times that of water (1 cP).

A theoretical treatment of membrane diffusion reveals that protein oligomerization plays only a minimal role in altering translational mobility. The mathematical formula relating size to diffusion in membranes is referred to as the Saffman-Delbruck equation (Saffman and Delbruck 1975) and is:

$$D = cT \ln\left[\left(\frac{k}{\eta a}\right) - 0.5772\right], \quad (1\text{-}11)$$

where D is the translational diffusion coefficient, c and k are constants accounting for the aqueous phase viscosity and membrane thickness, T is the absolute temperature, η is the viscosity of the membrane, and a and h are the radius and height of the transmembrane segment of the protein. Monomer to tetramer transition only increases the translational diffusion rate by 1.1-fold. Increases to 100-mers produce a tenfold decrease in diffusion rate (Kusumi et al. 2005), assuming a radius of the transmembrane segment of 0.5 nm. Portions of transmembrane proteins that extend into the extracellular space and cytoplasm have minimal effect on diffusion, since they are moving in a medium some 50- to 100-fold less viscous than the transmembrane segment(s) within the bilayer. Based on this argument, the diffusion coefficients of many membrane-bound proteins collapse into a relatively small range.

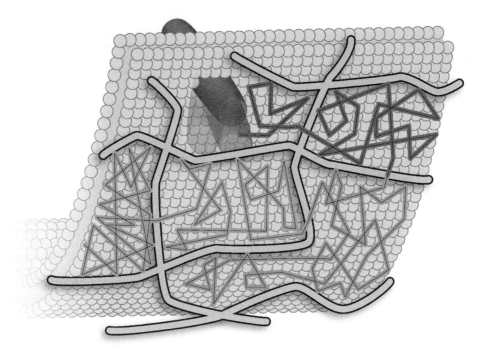

FIGURE 1.4. Hop-diffusion model of membrane protein movement. This rendering shows the cytoplasmic face of the membrane where cytoskeletal elements (yellow structures) form a submembranous lattice of "corrals." A typical integral membrane protein is shown in blue in the top left corral. The trajectories of proteins measured with single particle tracking techniques are schematized as different colored lines undergoing random walks within each corral. Translational diffusion within a corral is typical of diffusion in synthetic bilayers, but proteins must "hop" the cytoskeletal fences to move from one corral to the next within the membrane. This barrier restricts diffusion and produces significant slowing of translational diffusion on a macroscopic scale. Figure redrawn from Fujiwara et al. (2002), *J. Cell. Biol.*

Diffusion of proteins in cellular membranes is slowed some five- to 50-fold relative to their mobility in pure lipid membranes. Recent advances in high-speed data capture and single particle imaging have provided a plausible explanation for these observations (see Figure 1.4). Cellular membranes have a well-developed cytoskeletal matrix lying just beneath the plasmalemmal surface. This matrix produces barriers to the translational mobility of lipids and membrane associated proteins. The mechanistic impact projected onto receptor mobility is that diffusion rates are typical of those found in pure lipid bilayers within the confines or "corrals" formed by the matrix. However, the corrals hinder the diffusion of the receptors on larger spatial scales. To sample greater distances, the receptors must hop these fences and a "hop-diffusion" model was put forth to explain the macroscopic behavior of receptor mobility. Recent experimental data has directly supported this model (Kusumi et al. 2005). By tracking the trajectories of single particles (single particle tracking or SPT) at extremely high temporal resolution (25 µsec sampling rate), it was possible to reconstruct the random walk of particles in the membrane (see example in Figure 1.4). Individual particle motion was found to be restricted to domains within the membrane, but at certain points the particle would hop to a new domain where it would again sample the environment through diffusion similar to its rate in pure lipid membranes.

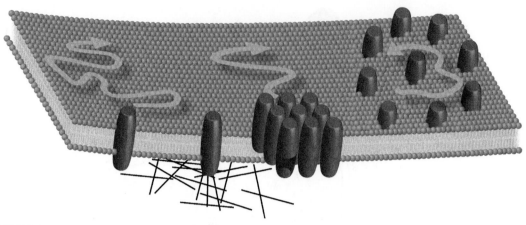

FIGURE 1.5. **Mechanisms that can slow membrane protein diffusion.** A cut-away view of a membrane is rendered in gray and integral membrane proteins are rendered in either blue or red. The unhindered diffusion of a membrane protein is shown on the far left. Next to it is a membrane protein that is immobile due to strong interactions with the underlying cytoskeleton. Oligomerization of a protein or lipid (as in a lipid raft) or through protein-protein interactions slows translational diffusion. Illustrated on the far right is hindered diffusion due to collisions either with other membrane bound proteins or via cytoskeletal corrals (see Figure 1.4). Reproduced in a modified form with permission from Nature Reviews Molecular Cell Biology (Lippincott-Schwartz et al., 2001) copyright (2001) Macmillan Magazines Ltd.

A summary of the types of interactions that an integral protein might encounter that would hinder its diffusion is shown in Figure 1.5 (modified from Lippincott-Schwartz et al. 2001).

The behavior of the AMPA and NMDA subtypes of glutamate receptors in neuronal membranes also has been examined with single particle tracking. The diffusion coefficients of AMPA receptors (median of $0.01\,\mu m^2/s$) was approximately four times faster than NMDA receptors (median of $0.0023\,\mu m^2/s$) in extrasynaptic areas, and at synapses, the diffusion coefficients were similar (0.028 and $0.021\,\mu m^2/s$, respectively). Interestingly, the percentage of the two receptors that were mobile in either synaptic or nonsynaptic sites were similar, and the percentage of mobile receptors could be affected by various stimuli (see Groc et al. (2004) for details and review by Choquet and Triller 2003).

FCS has found a unique application in the study of membrane protein oligomerization by capitalizing on changes in molecular brightness and translational mobility

as monomers form dimers or higher order structures. The dynamics of light-induced rhodopsin oligomerization were assessed using FCS (Kahya et al. 2002). The translational mobility of rhodopsin reconstituted in lipid bilayers decreased in a time-dependent manner (time course of minutes) from $1.0\,\mu m^2/s$ to $0.2\,\mu m^2/s$ upon exposure to light. The cell surface expression and oligomerization state of serotonin (5-HT3) receptors also was examined in neurons using FCS (Pick et al. 2003). It is also possible using FCS to analyze binding constants for fluorescent ligands to their receptors (see Gosch and Rigler (2005) for discussion). The interaction of ligands with GABA receptors on neuronal membranes is one example of the application of FCS in the measurement of binding constants (Meissner and Haberlein 2003).

It is possible to examine the rate of transport of membrane-bound molecules from one intracellular compartment to the next using fluorescence techniques. FPR has been used to examine both anterograde and retrograde transport rates of proteins between the ER and the Golgi. By using GFP-tagged transport proteins and photobleaching either the pool resident in the ER or the Golgi, one can then follow the recovery of the fluorescence from one pool to the next. This experiment establishes the cycling time between individual compartments and how different proteins are trafficked within and between intracellular organelles and the plasma membrane (for review see Lippincott-Schwartz et al. (2001) and references therein). Many variations on this theme can be envisioned to study the kinetics of various transport processes.

C. Protein-Protein Interactions

Protein-protein interactions are fundamental to all aspects of cell biology, and two techniques have been developed to assess such intracellular interactions. Fluorescence Resonance Energy Transfer (FRET) is a powerful technique that permits an assessment of the distance between two fluorophores, and has been widely applied to assess protein-protein interactions in the cellular setting. To obtain strong FRET signals, the two fluorophores must be in close enough proximity (<100 Å) that they can be said to be bound to each other (see Box 1.3 for details on the basic principles of FRET). Genetically encoded fluorescent proteins are valuable for these experiments. The excitation and emission spectra of CFP and YFP make them a good donor/acceptor pair in the design of FRET experiments. If the two fluorophores are in close enough proximity, excitation of CFP leads to resonance energy transfer to YFP with subsequence emission in the YFP channel. In addition, the CFP signal is decreased proportionally to the amount of energy lost in the transfer. Although genetically encoded fluorescent proteins have unique advantages, many other standard organic fluorophore pairs have been successfully used for FRET studies.

FRET studies can be accomplished in a standard fluorimeter; however, when coupled with fluorescence microscopy, one can obtain spatial and temporal information about activation of a given signaling pathway or where and when two proteins interact within a cell. A popular use of such FRET pairs is in the design of reporter molecules for various second messengers. A domain placed between a chimera of CFP and YFP that can bind to a second messenger molecule and alter the proximity of the fluorophores to each other can be used to report changes in concentration of that second messenger. These include biosensors for Ca^{2+}, cAMP, cGMP, and protease activity among others (Lippincott-Schwartz et al. 2001). FRET is also an attractive tool to examine the oligomerization state of membrane receptors and has been successfully applied to study the EGF receptors, and β_2-adrenergic receptors among others. FRET also was used to examine the dynamics of protein kinase interactions with anchoring and substrate proteins and to examine the dynamics of the formation of the SNARE complex necessary for exocytotic release of vesicles (see Lippincott-Schwartz et al. (2001) for other examples).

Another method growing in popularity to assess protein-protein interactions *in vitro* and in cells is fluorescence cross-correlation spectroscopy (FCCS). FCCS is a direct extension of FCS described earlier; the principle is described in Box 1.4. Unlike FRET that places proximity constraints on the two fluorophores, FCCS assesses whether two fluorophores are behaving as one molecule as they traverse the focal volume. When two molecules are not bound to each other, their photon intensity profiles differ. When they move together as one, the intensity profiles are identical, and the magnitude of the cross-correlation increases. Several recent reviews are available that describe the advantages and disadvantages of FCCS and that cover the practical aspects of data analysis (Zipfel and Webb 2001; Gosch and Rigler 2005; Bacia and Schwille 2003; Chen et al. 1999a). One advantage of FCCS is that the two

BOX 1.3

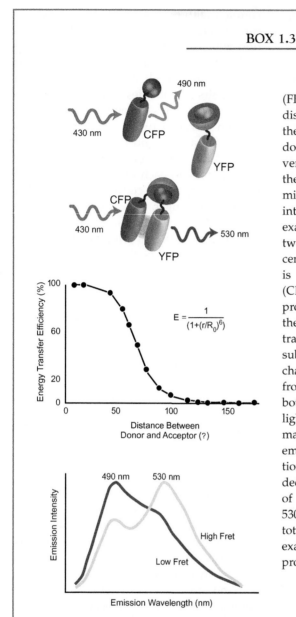

$$E = \frac{1}{(1+(r/R_0)^6)}$$

Fluorescence resonance energy transfer (FRET) is a tool used for determining the distance between two fluorophores. FRET is the nonradiative transfer of energy from donor to acceptor molecules and is inversely proportional to the sixth power of the distance, as shown on the plot in the middle panel. As such, protein-protein interactions inside cells can be assessed by examining whether FRET occurs between two proteins tagged with different fluorescent molecules. In the example, one protein is tagged with cyan fluorescent protein (CFP) and the other with yellow fluorescent protein (YFP). When the distance is small, the CFP molecule is excited and energy transfer can occur to the YFP molecule with subsequent emission detected in the YFP channel. The emission spectrum resulting from such an interaction is shown in the bottom panel. When excited with 430 nm light, in low FRET conditions there is a maximum reported at the peak of the CFP emission profile. Under high FRET conditions, the 490 nm emission peak of CFP is decreased (this is quenching due to loss of energy to the YFP molecule) while the 530 nm emission peak of YFP is increased. In total, FRET provides a methodology to examine the real-time dynamics of protein-protein interactions in living cells.

fluorophores could be too far apart for a successful FRET experiment. For example, two proteins may interact, but if the two fluorophores used to assess the interaction are not in close enough proximity (or are at unusual geometries relative to each other), no FRET will occur and a false negative will be reported. FCCS is not constrained in this manner; the distance between the two probes is not relevant. In fact, FRET is undesirable in cross-correlation measurements (see Bacia and Schwille (2003) for more discussion. Our laboratory has recently established that FCCS can be success-

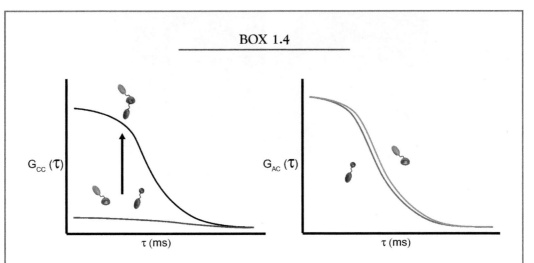

BOX 1.4

Fluorescence cross-correlation spectroscopy (FCCS) is a direct extension of fluorescence correlation spectroscopy. The apparatus for making FCCS measurements is identical to that for FCS measurements (see Figure 1.3), with the exception that the sample contains two spectrally distinct fluorophores whose signal is distinguished with independent detectors. The cross-correlated signal shown in the left panel is a direct indication of the concomitant movement of the two fluorophores as they transit through the focal volume. When diffusing separately (blue curve), the cross correlation signal is low; when diffusing together (bound; black curve) the cross correlation amplitude is high. The right panel represents data that would be generated from analyzing autocorrelation of the same set of proteins in the unbound states. Therefore, cross-correlation can offer a significant capacity to detect protein-protein interactions.

fully implemented to examine the Ca^{2+}-dependent interaction of calmodulin and Ca^{2+}/calmodulin-dependent protein kinase II in living cells (see Figure 1.6 and Kim et al. 2004; 2005).

An extension of this technique is that one could theoretically monitor two proteins that do not interact with each other, but that both interact with a third protein (a scaffold protein or linker protein of some kind) or are assembled in a higher order complex. FCCS would be an ideal approach to detect such a ternary (or potentially even higher order) interaction. A similar application would be assessing the interaction of two proteins across the plasma membrane where distances between the fluorophores

become prohibitive for FRET type studies. In fact, the time-resolved dynamics of association of a membrane bound receptor (the IgE receptor) with a signaling molecule (Lyn kinase) recently was examined using FCCS in live cells (Larson et al. 2005). Additionally, binding interactions can be assessed using coincidence analysis, which can bring the time required for measurement to the millisecond level (Heinze et al. 2002), dramatically increasing the possibility of evaluating binding interactions driven by physiological stimuli. The proof-of-principle experiment has been performed showing that FCCS coincidence analysis can be extended to three interacting molecules, each labeled with a

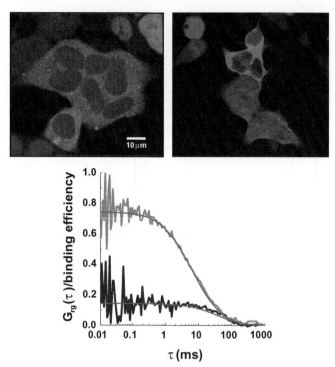

FIGURE 1.6.   **Example of FCCS measurements in live cells.** HEK293 cells were transfected with EGFP-tagged CaM-kinase II (green in the images) and then Alexa-633-CaM (red in the images) was introduced using electroporation. The confocal images show the morphology and distribution of the labels under high Ca^{2+} (top left panel) and low Ca^{2+} (top right panel) conditions. FCCS measurements were made using an experimental system similar to that shown in Figure 1.3. The bottom panel shows the cross-correlation curves of the red and green signals after elevating intracellular Ca^{2+} levels (light blue curve) followed by EGTA treatment to remove Ca^{2+} (purple curve). The results demonstrate the Ca^{2+} mediated binding and unbinding of CaM to CaM-kinase II in live cells. Figure taken from Kim et al. (2004), PNAS.

different colored fluorophore (Heinze et al. 2003).

Another interesting application of FCCS is in monitoring the sorting of cargo through the endocytic pathway (Bacia et al. 2002). By labeling the two subunits of cholera toxin with spectrally separable fluorophores (Cy2 and Cy5 in this case), it was possible to examine where in the endocytic pathway the toxin subunits dissociated. Confocal microscopy was used to image and place the focal spot where FCCS was to be accomplished. The results showed that the subunits do not completely dissociate until the cholera toxin has been delivered to the Golgi apparatus. Interestingly, the toxins could be labeled with either Cy2 alone or Cy5 alone and cross-correlation was still detected due to the two proteins moving together within the same vesicle. This result sets the precedent that cross-correlation could be used to assess a variety of questions concerning the temporal and spatial dynamics of cargo transport in live cells.

IV. CONCLUSIONS AND FUTURE DIRECTIONS

Over the last 15 years, the development of sophisticated fluorescent imaging and spectroscopy techniques has led to a significantly deeper understanding of molecular and organellar movement inside living cells. There is reason to believe these advancements will continue at a similar pace. Developments in microscope design, like the use of multiphoton excitation (Denk et al. 1990) and total-internal reflection microscopy (Axelrod et al. 1983), and the development of new (4π; see Hell et al. 2004) microscopy approaches, continue to increase the resolution of imaging and spectroscopy information. The efficiency of photon collection has advanced significantly and is not an area where much more improvement would seem possible; however development of new fluorescent probes will provide important new possibilities. The application of genetically encoded fluorescent proteins (GFP and the broad family of its derivatives) in the early 1990s provided an explosion of new applications and information relative to protein and organellar movement, and there remains a wealth of information to be gained. Further developments in fluorescent probe design will also continue to enhance the ability to visualize and quantify protein movement.

One of the most significant limitations for interpreting the wealth of data presently available is inadequate understanding of the cellular environment from a single molecule level perspective. We now have techniques (e.g., FCS and single particle tracking) that can analyze the behavior of single particles on the surface and inside of living cells, instead of having to rely on techniques that evaluate the average behavior of large ensembles of molecules. As noted, studies on the movement of lipids and proteins in the plasma membrane have advanced new models that resolve some long-standing issues in the field. To fully appreciate this type of data we must accept that heterogeneity is both expected and important. Cells are complex and dynamic, and at first principle one would have to assume no two measurements should be

exactly the same. Unfortunately, the molecular mechanisms that produce heterogeneity in intracellular diffusion are numerous and presently provide a significant barrier to a clear understanding of these processes. Development of accurate physical and mathematical models is needed, to guide interpretation of the complex data from intracellular measurements. A final step will be to merge data across disciplines. This will involve consolidating structural data, both cellular and high resolution single particle data, translational mobility data, protein-protein interaction data, and biochemical and enzymatic data. Developing a platform to meet this challenge will require computational approaches so that once established, specific cellular processes of interest can be evaluated as they evolve in both time and space.

Acknowledgements

I would like to acknowledge the long-standing collaborative efforts between Dr. Petra Schwille, Dr. Warren Zipfel, and our group to develop MPFCR, MPFCS, and MPFCCS for studying temporal and spatial signal transduction pathways in nerve cells. Particularly noteworthy are the contributions of Dr. Sally Kim, Dr. Katrin Heinze, Dr. Vijay Iyer, and Ms. Molly Rossow. I also thank Dr. Hugo Sanabria and Dr. Yoshihisa Kubota for their ongoing efforts and stimulating discussions in the development of these and other techniques for understanding the complexities of intracellular diffusion. I would also like to thank Dr. Sally Kim, Dr. Hugo Sanabria, Dr. Yoshihisa Kubota, Dr. Florence Noel, and Dr. Andrew Bean for critical review of the present manuscript and Mr. Lorenzo Morales for his contributions in producing the original graphics. Finally, financial support from the Texas Higher Education Coordinating Board, the Human Frontiers in Science Program, the W.M. Keck Center for the Neurobiology of Learning and Memory, and the National Institute of Health is gratefully acknowledged.

References

Axelrod, D., Thompson, N.L., Burghardt, T.P. (1983). Total internal reflection fluorescent microscopy. *J. Microsc.* **129 Pt 1**, 19–28.

Bacia, K., Majoul, I.V., Schwille, P. (2002). Probing the endocytic pathway in live cells using dual-color fluorescence cross-correlation analysis. *Biophys. J.* **83**, 1184–1193.

Bacia, K., Schwille, P. (2003). A dynamic view of cellular processes by in vivo fluorescence auto- and cross-correlation spectroscopy. *Methods* **29**, 74–85.

Berg, H.C. (1993). Random walks in biology. Princeton University Press, Princeton.

Burke, N.V., Han, W., Li, D., Takimoto, K., Watkins, S.C., Levitan, E.S. (1997). Neuronal peptide release is limited by secretory granule mobility. *Neuron* **19**, 1095–1102.

Chen, Y., Muller, J.D., Berland, K.M., Gratton, E. (1999a). Fluorescence fluctuation spectroscopy. *Methods* **19**, 234–252.

Chen, Y., Muller, J.D., So, P.T., Gratton, E. (1999b). The photon counting histogram in fluorescence fluctuation spectroscopy. *Biophys. J.* **77**, 553–567.

Choquet, D., Triller, A. (2003). The role of receptor diffusion in the organization of the postsynaptic membrane. *Nat. Rev. Neurosci.* **4**, 251–265.

Cluzel, P., Surette, M., Leibler, S. (2000). An ultrasensitive bacterial motor revealed by monitoring signaling proteins in single cells. *Science* **287**, 1652–1655.

Crank, J. (1956). The mathematics of diffusion. Clarendon Press, Oxford.

Dauty, E., Verkman, A.S. (2004). Molecular crowding reduces to a similar extent the diffusion of small solutes and macromolecules: Measurement by fluorescence correlation spectroscopy. *J. Mol. Recognit.* **17**, 441–447.

Denk, W., Strickler, J.H., Webb, W.W. (1990). Two-photon laser scanning fluorescence microscopy. *Science* **248**, 73–76.

Elson, E.L. (2001). Fluorescence correlation spectroscopy measures molecular transport in cells. *Traffic.* **2**, 789–796.

Fujiwara, T., Ritchie, K., Murakoshi, H., Jacobson, K., Kusumi, A. (2002). Phospholipids undergo hop diffusion in compartmentalized cell membrane. *J. Cell Biol.* **157**, 1071–1081.

Gennerich, A., Schild, D. (2002). Anisotropic diffusion in mitral cell dendrites revealed by fluorescence correlation spectroscopy. *Biophys. J.* **83**, 510–522.

Gosch, M., Rigler, R. (2005). Fluorescence correlation spectroscopy of molecular motions and kinetics. *Adv. Drug Deliv. Rev.* **57**, 169–190.

Groc, L., Heine, M., Cognet, L., Brickley, K., Stephenson, F.A., Lounis, B., Choquet, D. (2004). Differential activity-dependent regulation of the lateral mobilities of AMPA and NMDA receptors. *Nat. Neurosci.* **7**, 695–696.

Haustein, E., Schwille, P. (2003). Ultrasensitive investigations of biological systems by fluorescence correlation spectroscopy. *Methods* **29**, 153–166.

Heinze, K.G., Jahnz, M., Schwille, P. (2003). Triple fluorescence coincidence analysis (TFCA): Direct analysis of ternary complex formation. *Biophys. J.* **84**, 472a–473a.

Heinze, K.G., Rarbach, M., Jahnz, M., Schwille, P. (2002). Two-photon fluorescence coincidence analysis: Rapid measurements of enzyme kinetics. *Biophys. J.* **83**, 1671–1681.

Hell, S.W., Dyba, M., Jakobs, S. (2004). Concepts for nanoscale resolution in fluorescence microscopy. *Curr. Opin. Neurobiol.* **14**, 599–609.

Howard, J. (2001). Mechanics of motor protein and the cytoskeleton. Sunderland: Sinauer Associates, Inc.

Jordan, R., Lemke, E.A., Klingauf, J. (2005). Visualization of synaptic vesicle movement in intact synaptic boutons using fluorescence fluctuation spectroscopy. *Biophys. J.* **89**, 2091–2102.

Kahya, N., Wiersma, D.A., Poolman, B., Hoekstra, D. (2002). Spatial organization of bacteriorhodopsin in model membranes. Light-induced mobility changes. *J. Biol. Chem.* **277**, 39304–39311.

Kim, S.A., Heinze, K.G., Bacia, K., Waxham, M.N., Schwille, P. (2005). Two-photon cross-correlation analysis of intracellular reactions with variable stoichiometry. *Biophys. J.* **88**, 4319–4336.

Kim, S.A., Heinze, K.G., Waxham, M.N., Schwille, P. (2004). Intracellular calmodulin availability accessed with two-photon cross-correlation. *Proc. Natl. Acad. Sci. U S A* **101**, 105–110.

Kusumi, A., Nakada, C., Ritchie, K., Murase, K., Suzuki, K., Murakoshi, H. et al. (2005). Paradigm shift of the plasma membrane concept from the two-dimensional continuum fluid to the partitioned fluid: High-speed single-molecule tracking of membrane molecules. *Annu. Rev. Biophys. Biomol. Struct.* **34**, 351–378.

Larson, D.R., Gosse, J.A., Holowka, D.A., Baird, B.A., Webb, W.W. (2005). Temporally resolved interactions between antigen-stimulated ige receptors and lyn kinase on living cells. *J. Cell Biol.* **171**, 527–536.

Lehninger, A.L. (1982). Principles of biochemistry. New York: Worth Publishers, Inc.

Lippincott-Schwartz, J., Snapp, E., Kenworthy, A. (2001). Studying protein dynamics in living cells. *Nat. Rev. Mol. Cell Biol.* **2**, 444–456.

Luby-Phelps, K. (2000). Cytoarchitecture and physical properties of cytoplasm: volume, viscosity, diffusion, intracellular surface area. *In* Microcompartmentation and phase separation in cytoplasm,

H. Walter, D.E. Brooks, P.A. Srere, eds., 189–221. Academic Press, London.

Luby-Phelps, K., Taylor, D.L., Lanni, F. (1986). Probing the structure of cytoplasm. *J. Cell Biol.* **102**, 2015–2022.

Lukacs, G.L., Haggie, P., Seksek, O., Lechardeur, D., Freedman, N., Verkman, A.S. (2000). Size-dependent DNA Mobility in cytoplasm and nucleus. *J. Biol. Chem.* **275**, 1625–1629.

Marsh, B.J., Mastronarde, D.N., Buttle, K.F., Howell, K.E., Mcintosh, J.R. (2001). Organellar relationships in the Golgi region of the pancreatic beta cell line, HIT-T15, visualized by high resolution electron tomography. *Proc. Natl. Acad. Sci. U S A* **98**, 2399–2406.

Meissner, O., Haberlein, H. (2003). Lateral mobility and specific binding to GABA(A) receptors on hippocampal neurons monitored by fluorescence correlation spectroscopy. *Biochemistry* **42**, 1667–1672.

Minton, A.P. (2001). The influence of macromolecular crowding and macromolecular confinement on biochemical reactions in physiological media. *J. Biol. Chem.* **276**, 10577–10580.

Palo, K., Brand, L., Eggeling, C., Jager, S., Kask, P., Gall, K. (2002). Fluorescence intensity and lifetime distribution analysis: Toward higher accuracy in fluorescence fluctuation spectroscopy. *Biophys. J.* **83**, 605–618.

Pick, H., Preuss, A.K., Mayer, M., Wohland, T., Hovius, R., Vogel, H. (2003). Monitoring expression and clustering of the ionotropic 5HT3 receptor in plasma membranes of live biological cells. *Biochemistry* **42**, 877–884.

Poo, M., Cone, R.A. (1974). Lateral diffusion of rhodopsin in the photoreceptor membrane. *Nature* **247**, 438–441.

Saffman, P.G., Delbruck, M. (1975). Brownian motion in biological membranes. *Proc. Natl. Acad. Sci. U S A* **72**, 3111–3113.

Schwille, P., Haupts, U., Maiti, S., Webb, W.W. (1999). Molecular dynamics in living cells observed by fluorescence correlation spectroscopy with one- and two-photon excitation. *Biophys. J.* **77**, 2251–2265.

Seksek, O., Biwersi, J., Verkman, A.S. (1997). Translational diffusion of macromolecule-sized solutes in cytoplasm and nucleus. *J. Cell Biol.* **138**, 131–142.

Steyer, J.A., Horstmann, H., Almers, W. (1997). Transport, docking and exocytosis of single secretory granules in live chromaffin cells. *Nature* **388**, 474–478.

Stryer, L. (1988). Biochemistry. W.H. Freeman and Co., New York.

Terada, S., Kinjo, M., Hirokawa, N. (2000). Oligomeric tubulin in large transporting complex is transported via kinesin in squid giant axons. *Cell* **103**, 141–155.

Verkman, A.S. (2002). Solute and macromolecule diffusion in cellular aqueous compartments. *Trends in Biochem. Sci.* **27**, 27–33.

Weiss, M. (2004). Challenges and artifacts in quantitative photobleaching experiments. *Traffic* **5**, 662–671.

Weiss, M., Elsner, M., Kartberg, F., Nilsson, T. (2004). Anomalous subdiffusion is a measure for cytoplasmic crowding in living cells. *Biophys. J.* **87**, 3518–3524.

Zajchowski, L.D., Robbins, S.M. (2002). Lipid rafts and little caves. Compartmentalized signaling in membrane microdomains. *Eur. J. Biochem.* **269**, 737–752.

Zipfel, W.R., Webb, W.W. (2001). In vivo diffusion measurements using multiphoton excitation fluorescence photobleaching recovery and fluorescence correlation spectroscopy. *In* Methods in cellular imaging, A. Periasamy, ed., 216–235. Oxford University Press, Oxford.

2

The Role of Molecular Motors in Axonal Transport

ERAN PERLSON AND ERIKA L.F. HOLZBAUR

I. INTRODUCTION

A key milestone in the study of neuronal transport occurred about 20 years ago with the identification of the axoplasmic motors kinesin and dynein that generate movement along microtubules in opposite directions (Brady 1985; Vale, Schnapp et al. 1985; Vale, Reese et al. 1985; Paschal and Vallee 1987). Conventional kinesin (kinesin I) was identified as a novel force-generating pro-tein that moves toward the plus end of the microtubule. Cytoplasmic dynein, an isoform of the flagellar motor first purified as a force generating ATPase 40 years ago (Gibbons 1966), was identified as the motor protein for retrograde axonal transport. Since then, the function, structure, mechanism, and cargo interactions of these motor proteins have been studied in detail, providing new insights into axonal transport. These studies also tell us that there is much still to investigate and that many questions still remain unanswered.

Both cytoskeletal filaments and motor proteins are required for axonal transport. The cytoskeleton is a network of three types of protein filaments: microtubules, actin, and intermediate filaments. Microtubules are stiff hollow tubes 25 nm in diameter, built from 13 parallel protofilaments generated from the head-to-tail assembly of αβ-tubulin dimers. Microtubules have structural polarity; there are two distinct ends, a plus (fast growing) end and a minus (slow growing) end. Actin filaments (microfilaments) are helical polymers of the protein actin with a diameter of 5–9 nm. Like microtubules, actin filaments have polarity with distinct plus and minus ends. Axons also contain neurofilaments, a form of intermediate filaments. Neurofilaments assemble from three subunits (NF-H, NF-M, and NF-L) into a fibrous protein filament, 10 nm in diameter that forms a ropelike network. Intermediate filaments are distinct from both actin filaments and microtubules because they lack overall structural polarity and do not dynamically disassemble and reassemble within the cell. Unlike intermediate filaments, both tubulin and actin are highly conserved proteins, which may explain the conservation of the transport machinery through evolution.

In addition to their structural roles, both microtubules and actin filaments provide active tracks for directed motility within the cell. The polarized microtubule cytoskeleton provides long-range "rails" for efficient and directed movement of cargo. Actin filaments also serve as tracks for transport, although this is typically more dispersive transport over shorter distances. In contrast, there are no known motors that use neurofilaments as tracks. Neurofilaments are motile within the cell, apparently relying on the microtubule cytoskeleton for their movement.

Structurally, molecular motors consist of two functional parts: a motor domain that interacts with cytoskeletal filaments and

converts chemical energy into mechanical energy (energy from ATP hydrolysis is used for walking on the cytoskeletal filament); and a tail domain that interacts with cargo directly or through accessory chains or adaptors. Members of the kinesin superfamily and cytoplasmic dynein are the key motors in the axonal traffic machinery. Motor proteins from the myosin superfamily also play some role in axonal transport; however this movement is thought to be short range, such as the dispersal of cargo to the cell periphery. In axons, microtubules are oriented with plus ends (fast growing) directed toward the periphery and minus ends (slow growing) toward the cell body. Conventional kinesin is an anterograde axonal transporter that supplies new material to the axon and the presynaptic terminal. Cytoplasmic dynein is the retrograde transporter that moves organelles, vesicles, and signaling complexes to the cell body. In dendrites, microtubules are of mixed polarity so it is possible to find both motors moving in both directions.

II. KINESIN

Forty-five members of the kinesin superfamily, known as KIFs, have been identified in mice and humans. All KIFs have a conserved globular motor domain that binds both nucleotide and microtubules. KIFs are separated into three major classes according to where the motor domain is within the polypeptide sequence: N-terminal motor domain (N-kinesin), C-terminal motor domain (C-kinesin), or Middle motor domain (M-kinesin). Most kinesins expressed in neurons are in the N-terminal class (Hirokawa and Takemura 2005).

Conventional kinesin, kinesin-I, is a hetero-tetramer composed of two heavy chains (kif5A, kif5B, or kif5C) and two light chains (KLC1, KLC2). Motility assays have shown that kinesin-I is an efficient cellular motor due to coordination of its two motor domains. In vitro, kinesin-I can take hundreds of 8nm steps along the microtubule with a load of up to 6pN before detaching (Coppin et al. 1996). The two motor domains function in a hand-over-hand mechanism, in which the rear head moves forward while the front head stays bound to the microtubules. This kind of walking causes the stalk to rotate 180 degrees every step. Single-molecule experiments have supported the hand-over-hand model (Yildiz et al. 2004). However, it is not yet clear if it is a symmetric or asymmetric walk. Models for this stepping have been reviewed recently (Yildiz and Selvin 2005).

Outside of the motor domain, the sequences of the KIFs are divergent, with distinct cargo binding domains. This diversity may explain how members of the kinesin superfamily can transport different cargoes in the cell. Many members of the kinesin superfamily have been identified as motors for specific cellular cargos. For example, KIF17 transports NMDA-receptor associated vesicles in dendrites (Setou et al. 2000) and KIF3 transports fodrin-containing vesicles in axons (Takeda et al. 2000).

Most kinesins move toward the plus end of microtubules, which takes them from the cell body toward the nerve terminal. However, KIFC seems to move toward the minus end transporting vesicles in dendrites (Saito et al. 1997). The KIF2 family (M-kinesin) is unique in having both plus-end-directed motor activity and microtubule-depolymerizing activity. In a KIF2A -/- mouse, microtubule depolymerizing activity in the growth cones of mice was decreased, suggesting that KIF2 may control axon collateral extension (Homma et al. 2003).

The cargo specificity of kinesin is thought to depend on the diversity encoded by different kinesin genes; however, interactions with adaptor/scaffolding proteins also have been shown to mediate specificity; for example,

- Adaptor protein 1 (AP1), which links KIF13 to M6PR (Nakagawa et al. 2000)

- JIP scaffolding proteins from the JNK signaling pathway, which link kinesin to APOER2 (Verhey et al. 2001) and APP (Inomata et al. 2003)
- Glutamate receptor-interacting protein 1 (GRIP1), which links kinesin to the AMPA receptor (Setou et al. 2002)
- Tripartite scaffolding protein complex containing LIN10, LIN2, and LIN7, which mediates the interaction between KIF17 and NMDAR (Setou et al. 2000)

(For details, see Hirokawa and Takemura 2005.)

To insure selective transport, proteins and mRNA are transported in various organelles and protein complexes that are uniquely recognized by different adaptors. Differential-binding sites might control kinesin cargo destination. Cargo that binds to kinesin light chain may travel to axons, whereas binding to the kinesin tail will result in transport to dendrites (Setou et al. 2002; Kanai et al. 2004). There are still many open questions regarding motor-cargo regulation, motor activation, and complex formation and dissociation.

III. CYTOPLASMIC DYNEIN

Cytoplasmic dynein is the key motor for retrograde axonal transport. Dynein is a microtubule motor protein that uses the energy of ATP hydrolysis to move along microtubules. Dynein moves toward the minus end of the microtubule, which in axons is oriented toward the cell body. Dynein is a two-million Dalton complex composed of two heavy chains and multiple intermediate, light intermediate, and light chain subunits. The two heavy chains fold to form motor domains on a flexible stalk. The base of the complex functions primarily to bind to an associated multi-subunit protein complex named dynactin (Karki and Holzbaur 1995), and to cargo (Vallee et al. 2004). The motor domains are globular heads comprised of multiple AAA

(ATPases associated with various cellular activities) domain repeats, resulting in a ring structure that has close to heptad symmetry (Koonce and Samso 2004). The affinity of the dynein head for microtubules is regulated by nucleotide binding and hydrolysis at the P1 site in the first AAA repeat motif. The microtubule binding sites for dynein map to stalks that project from the globular head between the fourth and fifth AAA domains (Gee et al. 1997).

Because the AAA1 motif is relatively far from the microtubule binding site, this regulation is likely to involve a conformational change transduced around the AAA ring, or a change in the interaction of the stalk domain with the head, as suggested by EM images (Burgess et al. 2003). Kinetic studies indicate that ATP binding to the P1 site induces the release of the protein from microtubules, and following ATP hydrolysis and an accompanying conformational change, the enzyme rebinds to the microtubule and produces a 1.4 pN of force per ATP hydrolyzed (Holzbaur and Johnson 1989; Mallik et al. 2004). AAA1 is the primary site of ATP hydrolysis; AAA5–6 lack the P-loop consensus sequence and probably do not bind nucleotide. AAA2–4 each have a P-loop motif and may be capable of either binding or hydrolyzing ATP, potentially in a regulatory role. Mutation studies of the P-loop in AAA3 were found to disrupt the communication pathway between the hydrolytic site in AAA1 and the microtubule-binding domain such that ATP was hydrolyzed at AAA1 but the motor failed to be released from tight binding to microtubules by ATP (Silvanovich et al. 2003).

To date, no atomic-resolution structures of the dynein motor have been solved, so the mechanism of dynein movement remains poorly understood. Dynein might act like a winch (reviewed in Burgess and Knight 2004), where the head, rotating against the stem, would pull on the stalk. The stalk would act like a cable, capable of sustaining tension. Rotation implies that the linker subdomain of the stem moves across the head domain. This could change its interactions with the AAA modules, providing a means by which nucleotide binding to AAA modules 2–4 could be coupled to the powerstroke. Using an optical trap technique, Mallik et al. (2004) looked at single-molecule nanometry of dynein and showed that its step size decreases with increasing load. At high loads, cytoplasmic dynein takes 8 nm steps, at lower loads the steps are longer, and at zero load some steps are up to 32 nm long. A model has been proposed, in which the stepping distance per ATP is changed by the load-induced binding of ATP at secondary sites in the dynein head so that cytoplasmic dynein functions as a gear in response to load. To explain these properties, it is proposed that load induces ATP binding to AAA3, and possibly also to AAA2 and AAA4, and this produces a more compact structure of the head that is better able to transmit force and generate shorter step. Thus, the dynein motor may function very differently from kinesin and myosin.

Although dynein is morphologically uniform, it functions in a wide range of cellular processes. In addition to axonal retrograde transport, dynein is required for diverse functions, including ER to Golgi vesicular transport, neurofilament transport, mRNA localization, and mitotic spindle assembly. For its proper function in the cell, dynein requires another protein complex named dynactin.

IV. DYNACTIN

Dynactin is a large protein complex (1 MDa), composed of 11 distinct subunits with some present more than once per complex, making a total of more than 20 subunits per complex. Dynactin is required for most of the cellular functions of dynein *in vivo*. It is believed to act both to increase the efficiency of the dynein motor and to couple it to cargos. A short actin-like protein named Arp1 forms a filament at the complex base (Schafer et al. 1994). CapZ binds at one end and p62, p25, p27, and Arp11 bind at the other end, capping the Arp1 filament (Karki and Holzbaur 1999). p150[Glued], the largest subunit of dynactin, dimerizes to form a sidearm that projects from the Arp1 filament. Dynamitin (p50) forms a tetramer at the shoulder domain that docks the p150[Glued] sidearm to the Arp1 filament at the base (Schroer 2004). It is believed that the complex base is a cargo binding domain (interacting with spectrin, for example), whereas motors bind to the complex via p150[Glued] (Karki and Holzbaur 1995; Blangy et al. 1997; Deacon et al. 2003). p150[Glued] also binds directly to microtubules in a nucleotide independent manner via an N-terminal CAP-Gly domain (Waterman-Storer et al. 1995). Overexpression of dynamitin leads to the dissociation of the dynactin complex, effectively inhibiting dynein/dynactin function in the cell. Thus, dynamitin expression has been used as a tool to probe dynein/dynactin function in neurons (Waterman-Storer et al. 1997), especially because dynein and dynactin knockouts are lethal early in embryogenesis in both fly and mice.

V. MYOSIN

Myosins are actin-dependent molecular motors that use ATP energy for transport. Members of the myosin superfamily also play a role in axonal transport, however this movement on actin tracks is thought to be short range and mainly near the cell periphery. Dendritic spines or presynaptic terminals have few if any microtubules, but do contain actin filaments that may serve as tracks for myosin (Bridgman 2004). The myosin superfamily is composed of 20 structurally and functionally distinct classes of motors (Krendel and Mooseker 2005). As with kinesins, myosins appear to be rather specialized. In neurons, myosin II is involved in neuronal migration and growth cone motility. Myosin V is associated with synaptic vesicle proteins moving along microtubules (Prekeris and Terrian 1997; Bridgman 1999), and motor neurons from homozygous myosin Va null mice have slower retrograde transport compared with wild-type cells (Lalli et al. 2003). However, physiological studies on myosin V null mice have not revealed a role for the motor in neurotransmitter release or recycling (Miyata et al. 2000; Schnell and Nicoll 2001).

VI. TRANSPORT REGULATION AND CONTROL

The different motors that carry cargo in the cell must be regulated and work in concert to achieve accurate and efficient transport. How these different motors (kinesin, dynein, and myosin) cooperate, and whether there are specific mechanisms that activate or repress motor function, are questions that are not yet fully understood. Good candidates for motor activators or repressors are signaling proteins like protein kinases and G proteins. For example, rab GTPases regulate myosin V (Langford 2002) and dynein (Jordens et al. 2001; Short et al. 2002).

There is evidence that motors may interact or cooperate in cargo movement. Using fluorescence imaging with one nanometer accuracy (FIONA), Kural et al. (2005) analyzed organelle movement by kinesin and dynein. They found that dynein and kinesin do not work against each other during peroxisome transport *in vivo*. Rather, multiple kinesins or multiple dyneins work together, producing up to 10 times the speed observed *in vitro*. The mechanism by which this cooperativity occurs is not clear. Is there a switch that turns off one or both of the motors? Most evidence suggests that dynein activation and/or attachment is mediated by the dynactin complex (reviewed in Karki and Holzbaur 1999). Dynactin may enable dynein to participate efficiently in bidirectional transport, increasing its ability to stay "on" during minus end motion and keeping it "off" during plus end motion (Gross et al. 2002). However, Deacon et al. (2003) found that dynactin also may regulate transport by kinesin II, since both kinesin II and dynein bind to an overlapping site on the p150^{Glued} subunit of dynactin, suggesting that competition at that site might control the motor activities. Another protein that may play a role in the regulation of multiple motors is Klarsicht (klar). In the *Drosophila* mutant klar, but not in wild type flies, both plus and minus motors were engaged simultaneously (Welte et al. 1998), raising the possibility that klar might activate or repress the motors. However, no vertebrate homolog for klar has been identified. Another level of regulation may result from selective degradation of the motors. For example, kinesin moves out to the synapse but does not return (Hirokawa et al. 1991; Kondo et al. 1994; Yamazaki et al. 1995; Yang and Goldstein 1998). The regulation of motors in the cell is complex and further work is needed.

VII. AXONAL TRANSPORT

Peripheral neurons are highly polarized cells, with processes that extend many orders of magnitude longer than the diameters of the corresponding cell bodies. For example the axon length may be more than a thousand times the width of the cell body and the axoplasm may constitute more than 90 percent of the cell's volume (Howe and Mobley 2004).

Though both genetic and cellular studies have shown that kinesins are the primary motor for anterograde transport and dynein the primary motor for retrograde transport, there is also clear evidence that the function of these motors *in vivo* is interdependent. Studies on axonal transport have shown that inhibition of either kinesin or dynein/dynactin results in a bidirec-

tional block in transport (Brady et al. 1991; Waterman-Storer et al. 1997), and genetic analysis has shown that mutations in kinesin, dynein, and dynactin show strong dominant genetic interactions (Martin et al. 1999). Biochemical interactions have been noted between kinesin II and dynactin (Deacon et al. 2003), and kinesin I and dynein (Ligon et al. 2004), consistent with the possibility that direct interactions between oppositely directed motor proteins may modulate transport in the neuron.

An additional possibility is that perturbation of either motor may affect overall cellular pathways for protein synthesis and processing. For example, as dynein is required for normal ER-to-Golgi trafficking, disruption of dynein/dynactin function might perturb the synthesis of adaptors/scaffolding proteins required for the interaction between kinesin and its cargo, or of kinesin activator/repressors, thus leading to an indirect block in anterograde transport. Also, motor dysfunction is likely to have pleiotropic detrimental effects on neuronal function. The motor proteins move a variety of cargos on microtubule tracks, including membrane organelles, protein complexes, complexes of nucleic acids, signaling molecules, neuroprotective and repair molecules, and are required for the removal of misfolded or aggregated protein and vesicular and cytoskeleton components (Schliwa and Woehlke 2003) (Gunawardena and Goldstein 2004; Holzbaur 2005). (see Fig. 2.1)

VIII. PROTEIN DEGRADATION

One of the functions dynein might play in the neuron is a "garbage truck," in which the motor returns misfolded or aggregated proteins back to the cell body for degradation. Johnston et al. (2002) found dynein-dependent transport of misfolded protein into aggresomes and Burkhardt et al. (1997) found that dynein transports both

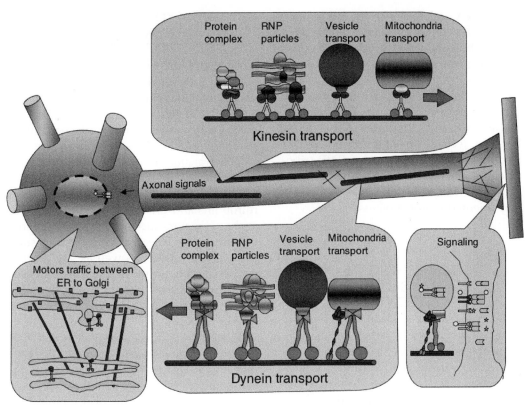

FIGURE 2.1. Axonal transport is driven by microtubule motor proteins. Microtubule motor proteins drive long distance transport in the axon. Kinesin and members of the kinesin superfamily drive the anterograde transport of proteins, RNP particles, vesicles, and mitochondria from cell body outward. Cytoplasmic dynein and its activator dynactin drive the retrograde transport of proteins, RNAs, vesicles, and mitochondria back to the cell body. The association of motors with cargo is mediated by an array of adaptor/scaffolding proteins, which provide specificity and may regulate axonal transport. Microtubule motors also mediate vesicle trafficking in the cell body, and participate in signaling events such as the transport of trophic factors.

vesicles to the lysosome and lysosomes along microtubules. Dynein has also been suggested to play a role in autophagy (Ravikumar et al. 2005).

IX. mRNA LOCALIZATION

The cell body classically has been considered the exclusive source of axonal proteins. However, significant evidence has accumulated recently to support the view that protein synthesis can occur in axons themselves, remote from the cell body (Piper and Holt 2004). mRNA and transla-

tional machinery were found to be in neurites (van Minnen and Syed 2001). One possibility is that the transport machinery is actively transporting both RNAs and ribosomes from the cell body into and along the axon. It is unclear how mRNAs are targeted, but there is evidence implicating the 3-foot and 5-foot untranslated regions of mRNAs in this process. RNA-binding proteins might be used as adaptors for mRNA localization. For example, β-actin mRNA is localized in neurite growth cones by using specific sequence within the 3'UTR termed zipcode. The zipcode binding protein (ZBP1) serves as an adaptor between the β-actin mRNA and the microtubule motors. Both the β-actin mRNA and ZBP1 were found to be colocalized in transport granules along microtubules (Micheva et al. 1998; Bassell and Singer 2001; Eom et al. 2003). Recently kinesin was found to transport large multisubunit complexes of proteins and mRNA in dendrites (Kanai 2004). Moreover, recent live cell imaging revealed the bidirectional movement of RNP (ribonucleoprotein particles) along the axon (Ling et al. 2004), likely the result of oppositely oriented motor proteins.

X. MITOCHONDRIA TRANSPORT

The intracellular distribution of mitochondria in neurons is important for providing an energy source close to where it is required for various cellular functions. Disruption of normal mitochondrial distribution may result in toxicity and perhaps neurodegenerative disease. In addition to the requirement for ATP production, it is believed that the mitochondria have the capacity to buffer Ca^{2+}. Mitochondria accumulate in active growth cones during neuronal development (Morris and Hollenbeck 1993), and in response to local energy changes or metabolic demand mitochondria may translocate from the cell body to the axon (Hollenbeck 1996). Mitochondria are transported anterogradely by KIF5 and retrogradely by dynein (Hollenbeck 1996; Ligon and Steward 2000; Nangaku et al. 1994; Tanaka et al. 1998). Syntabulin, a peripheral membrane-associated protein, links kinesin to mitochondria insuring axonal trafficking and proper mitochondria distribution in neurons (Su et al. 2004). In *Drosophila*, axonal transport of mitochondria depends on the novel protein Milton that might act as a linker between the mitochondria and kinesin (Stowers et al. 2002).

XI. AXON SIGNALING STRATEGY

The cell body of a neuron must receive accurate and timely information from the periphery to maintain neuronal viability and cell function, suggesting that specific mechanisms exist to transmit such information along the axon to the cell body.

The transfer of vital information may be mediated by several factors, including secreted protein factors such as neurotrophins and growth factors, membrane-anchored factors such as adhesion proteins, soluble membrane-permeant factors such as lipid-derivatives, as well as gas, electrical excitation, and ionic flux kinase-mediated signal transduction cascades (Campenot and MacInnis 2004; Howe and Mobley 2004).

Movement of these survival factors may occur in a passive or active manner. A signal may simply diffuse through the cytoplasm, eventually reaching the cell body and producing its effect. However, molecular crowding, as a result of the presence of organelles, cytoskeleton, and high protein concentrations, effectively restricts free diffusion of molecules, making this mechanism unlikely, even over short distances (Sodeik 2000) (see Chapter 1 for further discussion). Measured values of slow protein diffusion and rapid dephosphorylation showed that phospho-protein trafficking only by diffusion is not possible (Kholodenko 2003). Communication over long distances requires an active process. Other possible signal mechanisms may involve a regenerating wave that is amplified through signal transducers located along the axon length. For example, ionic flux changes, or amplified kinase signaling (Nguyen et al. 1998) would produce very low fidelity and specificity and a low signal to noise.

Another potential mechanism for signaling at a distance would involve a signaling factor that interacts with motor proteins and moves along the axon via the cyto-skeleton. This signaling factor would be protected from degradation in vesicle or a protein complex. It could be a signaling/cargo factor protected in vesicle or signaling/cargo factor protected in a protein complex (Delcroix et al. 2003; Howe and Mobley 2004). An example of this mechanism is the signaling of target-derived growth factors. Neurotrophins bind to their receptors, are internalized, and are retrogradely transported through the axon to the cell body to elicit local signal transduction cascades that modify or control cell survival and differentiation (Sofroniew et al. 2001). There is also evidence of a nonendosomal mechanism for retrograde signaling (MacInnis and Campenot 2002), in which protein adaptors bridge the signaling factors to the motor protein complex (Hanz et al. 2003; Cavalli et al. 2005; Perlson et al. 2005). In this regard, Trk receptor transport was shown to be blocked by using the dynein inhibitor (EHNA) (Reynolds et al. 1998) and by introduction of exogenous dynamitin into axons (Heerssen et al. 2004). Furthermore, Trk was found to bind directly to dynein light chain (Yano et al. 2001).

XII. SIGNALING ADAPTORS

It has become evident that axonal transport of molecules depends on scaffolding/adaptor proteins that are recognized specifically by molecular motors (Vale 2003; Gunawardena and Goldstein 2004). A variety of sorting proteins have been identified as scaffolds between cargo and molecular motors (Nakagawa et al. 2000; Setou et al. 2002). The receptor/signaling-adaptor-motor interactions represent a basic strategy for binding of cargo by molecular motors. It is thought that adaptor proteins that link receptors to motors determine the specificity of cargo-motor linkage. Defining the adaptor proteins that attach cargo to the

motor will provide a better understanding of how signals are conveyed. Examples of this adaptor protein strategy are numerous. Spectrin links the dynein complex to vesicles (Muresan et al. 2001). The JIP scaffolding proteins that assemble molecules in the JNK/MAP kinase signaling pathway interacts with kinesin and might serve as an adaptor linking kinesin to JNK signaling complexes and a transmembrane receptor (Bowman et al. 2000; Verhey et al. 2001; Yano and Chao 2004). Sunday driver (syd), a JNK scaffolding protein, interacts with dynactin after nerve injury, creating an axonal signaling complex (Cavalli et al. 2005). α and β importins link NLS survival signaling proteins to the dynein machinery after nerve injury (Hanz et al. 2003). Vimentin, a member of the type III intermediate filament family, interacts directly with the importin/dynein signaling complex and with phosphorylated Erks in injured nerve leading to survival (Perlson et al. 2005).

XIII. SURVIVAL SIGNALING

The neurotrophin family is the best studied of the retrograde survival signaling factors in neurons (review in Zweifel et al. 2005). Ligand binding induces the dimerization of neurotrophin receptor tyrosine-kinases, and triggers their internalization, autophosphorylation, and recruitment of downstream effectors, as well as activation of signal transduction cascades. Cytoplasmic dynein was found to transport the activated Trk receptor and to promote the survival of target-dependent neurons (Heerssen et al. 2004). The survival signals are believed to be transported to the cell body via signaling vesicles (signaling endosome model; Delcroix et al. 2003) as described earlier, or via protein complexes that move along microtubules, and contain the receptor, ligand, and a variety of signaling molecules like Rap1/ERK1/2, p38MAPK, ERK5, PI3K/Akt, and c-June/JNK (Ginty and Segal 2002; Howe and Mobley 2004; Yano and Chao 2004). However, the specific signals required to maintain the health of the motor neuron are not yet known.

XIV. DEATH SIGNALS

Axonal signaling factors also can initiate cell death. Recent work has identified a novel mechanism of retrograde signaling, in which caspase-8 interacts directly with the p150^{Glued} subunit of the dynactin complex, leading to apoptosis and neuron death (Carson et al. 2005). Caspase-8 activation and its retrograde transport are reduced in mice lacking the p75 subunit of the Trk receptor, suggesting the involvement of this receptor in the apoptosis process. Whereas mature neurotrophins bind with high affinity to Trk receptors and produce signals for cell survival, proneurotrophins bind with high affinity to the p75 subunit of the receptor (Lee et al. 2001), causing cell death. The p75 subunit can also bind to neurotrophins with low affinity, suggesting that proteolytic cleavage of proneurotrophins may regulate cell fate. The current model suggests that neurotrophin signaling is modulated by combinatorial regulation and differential signaling cascades involving the various Trks and the p75 receptor, leading to either survival or death (Kuruvilla et al. 2004).

XV. VIRUSES

Viruses use the cytoskeletal transport machinery for intracellular transport. Live cell imaging reveals that viruses move along microtubules in both directions, suggesting that viruses can recruit multiple motors. Using fluorescent tags, viruses have been shown to move along microtubules at speeds similar to those observed for endogenous cargos (Bearer et al. 2000;

Rietdorf et al. 2001; Seisenberger et al. 2001; Smith and Enquist 2002; Smith et al. 2004). The translocation of adenovirus, herpesvirus, and retroviral capsid are blocked by dynamitin overexpression, suggesting that these particles use the dynactin/dynein motor for transport (Suomalainen et al. 1999; Dohner et al. 2002). These viruses may have a viral receptor either for dynein or for dynactin, which allows them to engage the microtubule motor system for efficient transport. Recent studies have revealed several potential direct interactions between viral components and microtubule motors (reviewed in Dohner and Sodeik 2005). Thus far, only dynein subunits, and not dynactin, have been identified as binding domains for virus proteins. For example, herpes simplex virus capsid protein VP26 was found to interact with dynein light chain, suggesting a role for dynein in the retrograde transport of the virus (Douglas et al. 2004).

It is clear that viruses can highjack microtubule motors, but it is not yet clear if viral infection has a detrimental effect on the transport of endogenous cargos. Viral infection of the nervous system has the potential to affect transport efficiency or even to trigger retrograde signaling pathways. This in turn may lead to downstream stress or degeneration of the affected neurons. Links between viral infection of the nervous system and subsequent susceptibility to neurodegenerative disease currently are being examined.

XVI. NEURODEGENERATIVE DISEASE

Recent progress has shown that mutations in microtubule motor proteins can cause neurodegenerative disease (reviewed in Holzbaur 2004). The disease-causing mutations have been observed in the motors that power both anterograde and retrograde axonal transport. Mutations in members of the kinesin superfamily have been linked to specific forms of neuronal degeneration. Patients with Charcot-Marie-Tooth type 2A (CMT2A) neurodegenerative disease were found to have a mutation in neuronal kinesin KIF1B (Zhao et al. 2001). Consistent with this observation, KIF1B knockout mice die shortly after birth with multiple neurological defects, whereas the KIF1B heterozygotes have significant defects in protein transport two months after birth, with subsequent progressive muscle weakness and motor discoordination. Mutations in other members of the kinesin superfamily also have been linked to disease. For example, mutations in KIF5A (Reid et al. 2002) and in KIF21A (Yamada et al. 2003) have been observed in patients that develop axonal degeneration.

Transgenic mice with a targeted disruption of dynein/dynactin function due to overexpression of dynamitin develop late-onset progressive motor neuron degeneration and muscle atrophy (LaMonte et al. 2002). ENU-induced mutations in the cytoplasmic dynein heavy chain result in progressive motor neuron degeneration in Loa and Cra1 heterozygous mice (Hafezparast et al. 2003), whereas, when homozygous, both the Loa and Cra1 mutations are lethal. Finally, a missense mutation in the p150^{Glued} subunit of dynactin (G59S) was identified as the cause of late-onset progressive motor neuron degeneration and muscle atrophy in a North American family (Puls et al. 2003). The missense mutation destabilizes the polypeptide, leading to an enhanced tendency to aggregate specifically in motor neurons (Levy et al., manuscript submitted; Puls et al. 2005).

Although these neuropathies are caused by specific defects in genes encoding motor protein subunits, a number of studies have suggested that defects in axonal transport are linked more broadly to the pathogenic mechanisms observed in other neurodegenerative diseases (see Chapters 16–18). For example, defects in axonal transport have been observed in mouse models of the fatal neurodegenerative disease amyo-trophic lateral sclerosis (ALS). Inhibition of both anterograde and retrograde axonal transport have been observed in transgenic mice expressing mutant forms of SOD1 (Bruijn et al. 2004; Ligon et al. 2005). Surprisingly, analysis of the progeny of a SOD1^{G93A}/Loa cross that express both mutant SOD1 and mutant dynein showed rescue of transport rates as well as some amelioration of disease. Inhibition of axonal transport alone causes neurodegeneration and muscle atrophy; however, the coexpression of the mutant proteins somehow may alter the balance between survival and death signals, thus postponing, but not blocking, eventual cell death.

XVII. SUMMARY

In the two decades of research since the motors that drive axonal transport first were identified, it has become clear that the complex mechanisms involved can provide exquisite control of intracellular trafficking and signaling. A highly regulated balance between anterograde and retrograde transport may regulate whether survival or death signals will arrive at the cell body. This control may occur at multiple levels, including the relative expression of motors and adaptors/scaffolding proteins, and modulation of transport function. In the nerve periphery there is likely to be regulation at the level of activation in response to signal. And along the axon, the local concentration of the transport components may control the complex balance between association/disassociation of motor, adaptor, and cargo. Further studies should provide a more comprehensive understanding of the machinery involved in trafficking, signaling, and transport along the axon.

References

Azzouz, M., Ralph, G.S., Storkebaum, E., Walmsley, L.E., Mitrophanous, K.A., Kingsman, S.M. et al. (2004). VEGF delivery with retrogradely transported lentivector prolongs survival in a mouse ALS model. *Nature* **429 (6990)**, 413–417.

Bassell, G.J., Singer, R.H. (2001). Neuronal RNA localization and the cytoskeleton. *Results & Problems in Cell Diff.* **34**, 41–56.

Bearer, E.L., Breakefield, X.O., Schuback, D., Reese, T.S., LaVail, J.H. (2000). Retrograde axonal transport of herpes simplex virus: Evidence for a single mechanism and a role for tegument. *Proc. N. A. Sci. U S A* **97 (14)**, 8146–8150.

Blangy, A., Arnaud, L., Nigg, E.A. (1997). Phosphorylation by p34cdc2 protein kinase regulates binding of the kinesin-related motor HsEg5 to the dynactin subunit p150. *J. Biol. Chem.* **272 (31)**, 19418–19424.

Bowman, A.B., Kamal, A., Ritchings, B.W., Philp, A.V., McGrail, M., Gindhart, J.G., Goldstein, L.S. (2000). Kinesin-dependent axonal transport is mediated by the sunday driver (SYD) protein [see comment]. *Cell* **103 (4)**, 583–594.

Brady, S.T. (1985). A novel brain ATPase with properties expected for the fast axonal transport motor. *Nature* **317 (6032)**, 73–75.

Bridgman, P.C. (1999). Myosin Va movements in normal and dilute-lethal axons provide support for a dual filament motor complex. *J. Cell Biol.* **146 (5)**, 1045–1060.

Bridgman, P.C. (2004). Myosin-dependent transport in neurons. *J. Neurobiol.* **58 (2)**, 164–174.

Bruijn, L.I., Miller, T.M., Cleveland, D.W. (2004). Unraveling the mechanisms involved in motor neuron degeneration in ALS. *Ann. Rev. Neurosci.* **27**, 723–749.

Burgess, S.A., Knight, P.J. (2004). Is the dynein motor a winch? *Current Opinion in Structural Biology* **14 (2)**, 138–146.

Burgess, S.A., Walker, M.L., Sakakibara, H., Knight, P.J., Oiwa, K. (2003). Dynein structure and power stroke [see comment]. *Nature* **421 (6924)**, 715–718.

Burkhardt, J.K., Echeverri, C.J., Nilsson, T., Vallee, R.B. (1997). Overexpression of the dynamitin (p50) subunit of the dynactin complex disrupts dynein-dependent maintenance of membrane organelle distribution. *J. Cell Biol.* **139 (2)**, 469–484.

Campenot, R.B., MacInnis, B.L. (2004). Retrograde transport of neurotrophins: Fact and function. *J. Neurobiol.* **58 (2)**, 217–229.

Carson, C., Saleh, M., Fung, F.W., Nicholson, D.W., Roskams, A.J. (2005). Axonal dynactin p150glued transports caspase-8 to drive retrograde olfactory receptor neuron apoptosis. *J. Neurosci.* **25 (26)**, 6092–6104.

Cavalli, V., Kujala, P., Klumperman, J., Goldstein, L.S. (2005). Sunday driver links axonal transport to damage signaling. *J. Cell Biol.* **168 (5)**, 775–787.

Coppin, C.M., Finer, J.T., Spudich, J.A., Vale, R.D. (1996). Detection of sub-8-nm movements of kinesin by high-resolution optical-trap microscopy. *Proc. N. Acad. Sci U S A* **93 (5)**, 1913–1917.

Deacon, S.W., Serpinskaya, A.S., Vaughan, P.S., Lopez Fanarraga, M., Vernos, I., Vaughan, K.T., Gelfand, V.I. (2003). Dynactin is required for bidirectional organelle transport [see comment]. *J. Cell Biol.* **160 (3)**, 297–301.

Delcroix, J.D., Valletta, J.S., Wu, C., Hunt, S.J., Kowal, A.S., Mobley, W.C. (2003). NGF signaling in sensory neurons: Evidence that early endosomes carry NGF retrograde signals [see comment]. *Neuron* **39 (1)**, 69–84.

Dohner, K., Sodeik, B. (2005). The role of the cytoskeleton during viral infection. *Curr. Topics in Microbiol. Immunol.* **285**, 67–108.

Dohner, K., Wolfstein, A., Prank, U., Echeverri, C., Dujardin, D., Vallee, R., Sodeik, B. (2002). Function of dynein and dynactin in herpes simplex virus capsid transport. *Mol. Biol. Cell* **13 (8)**, 2795–2809.

Douglas, M.W., Diefenbach, R.J., Homa, F.L., Miranda-Saksena, M., Rixon, F.J., Vittone, V. et al. (2004). Herpes simplex virus type 1 capsid protein VP26 interacts with dynein light chains RP3 and Tctex1 and plays a role in retrograde cellular transport. *J. Biol. Chem.* **279 (27)**, 28522–28530.

Echeverri, C.J., Paschal, B.M., Vaughan, K.T., Vallee, R.B. (1996). Molecular characterization of the 50-kD subunit of dynactin reveals function for the complex in chromosome alignment and spindle organization during mitosis. *J. Cell Biol.* **132 (4)**, 617–633.

Eom, T., Antar, L.N., Singer, R.H., Bassell, G.J. (2003). Localization of a beta-actin messenger ribonucleoprotein complex with zipcode-binding protein modulates the density of dendritic filopodia and filopodial synapses. *J. Neurosci.* **23 (32)**, 10433–10444.

Gee, M.A., Heuser, J.E., Vallee, R.B. (1997). An extended microtubule-binding structure within the dynein motor domain. *Nature* **390 (6660)**, 636–639.

Gibbons, I.R. (1966). Studies on the adenosine triphosphatase activity of 14 S and 30 S dynein from cilia of Tetrahymena. *J. Biol. Chem.* **241 (23)**, 5590–5596.

Ginty, D.D., Segal, R.A. (2002). Retrograde neurotrophin signaling: Trk-ing along the axon. *Curr. Op. Neurobiol.* **12 (3)**, 268–274.

Gross, S.P., Welte, M.A., Block, S.M., Wieschaus, E.F. (2002). Coordination of opposite-polarity microtubule motors. *J. Cell Biol.* **156 (4)**, 715–724.

Gunawardena, S., Goldstein, L.S. (2004). Cargo-carrying motor vehicles on the neuronal highway: Transport pathways and neurodegenerative disease. *J. Neurobiol.* **58 (2)**, 258–271.

Hafezparast, M., Klocke, R., Ruhrberg, C., Marquardt, A., Ahmad-Annuar, A., Bowen, S. et al. (2003). Mutations in dynein link motor neuron degeneration to defects in retrograde transport. *Science* **300 (5620)**, 808–812.

Hanz, S., Perlson, E., Willis, D., Zheng, J.Q., Massarwa, R., Huerta, J.J. et al. (2003). Axoplasmic importins enable retrograde injury signaling in lesioned nerve. *Neuron* **40 (6)**, 1095–1104.

Heerssen, H.M., Pazyra, M.F., Segal, R.A. (2004). Dynein motors transport activated Trks to promote survival of target-dependent neurons. *Nat. Neurosci.* **7 (6)**, 596–604.

Hirokawa, N., Sato-Yoshitake, R., Kobayashi, N., Pfister, K.K., Bloom, G.S., Brady, S.T. (1991). Kinesin associates with anterogradely transported membranous organelles in vivo. *J. Cell Biol.* **114 (2)**, 295–302.

Hirokawa, N., Sato-Yoshitake, R., Yoshida, T., Kawashima, T. (1990). Brain dynein (MAP1C) localizes on both anterogradely and retrogradely transported membranous organelles in vivo. *J. Cell Biol.* **111 (3)**, 1027–1037.

Hirokawa, N., Takemura, R. (2005). Molecular motors and mechanisms of directional transport in neurons. *Nat. Rev. Neurosci.* **6 (3)**, 201–214.

Hollenbeck, P.J. (1996). The pattern and mechanism of mitochondrial transport in axons. *Frontiers in Biosci.* **1**, d91–102.

Holzbaur, E.L. (2004). Motor neurons rely on motor proteins. *Trends in Cell Biol.* **14 (5)**, 233–240.

Holzbaur, E.L., Johnson, K.A. (1989). Microtubules accelerate ADP release by dynein. *Biochem.* **28 (17)**, 7010–7016.

Homma, N., Takei, Y., Tanaka, Y., Nakata, T., Terada, S., Kikkawa, M. et al. (2003). Kinesin superfamily protein 2A (KIF2A) functions in suppression of collateral branch extension. *Cell* **114 (2)**, 229–239.

Howe, C.L., Mobley, W.C. (2004). Signaling endosome hypothesis: A cellular mechanism for long distance communication. *J. Neurobiol.* **58 (2)**, 207–216.

Inomata, H., Nakamura, Y., Hayakawa, A., Takata, H., Suzuki, T., Miyazawa, K., Kitamura, N. (2003). A scaffold protein JIP-1b enhances amyloid precursor protein phosphorylation by JNK and its association with kinesin light chain 1. *J. Biol. Chem.* **278 (25)**, 22946–22955.

Johnston, J.A., Illing, M.E., Kopito, R.R. (2002). Cytoplasmic dynein/dynactin mediates the assembly of aggresomes. *Cell Mot. & the Cytoskel.* **53 (1)**, 26–38.

Jordens, I., Fernandez-Borja, M., Marsman, M., Dusseljee, S., Janssen, L., Calafat, J. et al. (2001). The Rab7 effector protein RILP controls lysosomal transport by inducing the recruitment of dynein-dynactin motors. *Curr. Biol.* **11 (21)**, 1680–1685.

Kanai, Y., Dohmae, N., Hirokawa, N. (2004). Kinesin transports RNA: Isolation and characterization of an RNA-transporting granule [see comment]. *Neuron* **43 (4)**, 513–525.

Karki, S., Holzbaur, E.L. (1995). Affinity chromatography demonstrates a direct binding between cytoplasmic dynein and the dynactin complex. *J. Biol. Chem.* **270 (48)**, 28806–28811.

Karki, S., Holzbaur, E.L. (1999). Cytoplasmic dynein and dynactin in cell division and intracellular transport. *Curr. Opinion Cell Biol.* **11 (1)**, 45–53.

Kaspar, B.K., Llado, J., Sherkat, N., Rothstein, J.D., Gage, F.H. (2003). Retrograde viral delivery of IGF-1 prolongs survival in a mouse ALS model. *Science* **301 (5634)**, 839–842.

Kholodenko, B.N. (2003). Four-dimensional organization of protein kinase signaling cascades: The roles of diffusion, endocytosis and molecular motors. *J. Exp. Biol.* **206 (Pt 12)**, 2073–2082.

Kondo, S., Sato-Yoshitake, R., Noda, Y., Aizawa, H., Nakata, T., Matsuura, Y., Hirokawa, N. (1994). KIF3A is a new microtubule-based anterograde motor in the nerve axon. *J. Cell Biol.* **125 (5)**, 1095–1107.

Koonce, M.P., Samso, M. (2004). Of rings and levers: the dynein motor comes of age. *Trends in Cell Biol.* **14 (11)**, 612–619.

Krendel, M., Mooseker, M.S. (2005). Myosins: Tails (and heads) of functional diversity. *Physiol.* **20**, 239–251.

Kural, C., Kim, H., Syed, S., Goshima, G., Gelfand, V.I., Selvin, P.R. (2005). Kinesin and dynein move a peroxisome in vivo: A tug-of-war or coordinated movement? *Science* **308 (5727)**, 1469–1472.

Kuruvilla, R., Zweifel, L.S., Glebova, N.O., Lonze, B.E., Valdez, G., Ye, H., Ginty, D.D. (2004). A neurotrophin signaling cascade coordinates sympathetic neuron development through differential control of TrkA trafficking and retrograde signaling [see comment]. *Cell* **118 (2)**, 243–255.

Lalli, G., Gschmeissner, S., Schiavo, G. (2003). Myosin Va and microtubule-based motors are required for fast axonal retrograde transport of tetanus toxin in motor neurons. *J. Cell Sci.* **116 (Pt 22)**, 4639–4650.

LaMonte, B.H., Wallace, K.E., Holloway, B.A., Shelly, S.S., Ascano, J., Tokito, M. et al. (2002). Disruption of dynein/dynactin inhibits axonal transport in motor neurons causing late-onset progressive degeneration. *Neuron* **34 (5)**, 715–727.

Langford, G.M. (2002). Myosin-V, a versatile motor for short-range vesicle transport. *Traffic* **3 (12)**, 859–865.

Lee, R., Kermani, P., Teng, K.K., Hempstead, B.L. (2001). Regulation of cell survival by secreted proneurotrophins. *Science* **294 (5548)**, 1945–1948.

Levy, J.R., Holzbaur, E.L. (2006) Cytoplasmic dynein/dynactin function and dysfunction in motor neurons. *Int. J. Dev. Neurosci.* **24**, 103–111.

Ligon, L.A., LaMonte, B.H., Wallace, K.E., Weber, N., Kalb, R.G., Holzbaur, E.L. (2005). Mutant superoxide dismutase disrupts cytoplasmic dynein in motor neurons. *Neuroreport* **16 (6)**, 533–536.

Ligon, L.A., Steward, O. (2000). Role of microtubules and actin filaments in the movement of mitochondria in the axons and dendrites of cultured hippocampal neurons. *J. Comp. Neurol.* **427 (3)**, 351–361.

Ling, S.C., Fahrner, P.S., Greenough, W.T., Gelfand, V.I. (2004). Transport of Drosophila fragile X mental retardation protein-containing ribonucleoprotein granules by kinesin-1 and cytoplasmic dynein. *Proc. Natl. Acad. Sci. U S A* **101 (50)**, 17428–17433.

MacInnis, B.L., Campenot, R.B. (2002). Retrograde support of neuronal survival without retrograde transport of nerve growth factor [see comment]. *Science* **295 (5559)**, 1536–1539.

Mallik, R., Carter, B.C., Lex, S.A., King, S.J., Gross, S.P. (2004). Cytoplasmic dynein functions as a gear in response to load. *Nature* **427 (6975)**, 649–652.

Micheva, K.D., Vallee, A., Beaulieu, C., Herman, I.M., Leclerc, N. (1998). Beta-actin is confined to structures having high capacity of remodelling in developing and adult rat cerebellum. *Eur. J. Neurosci.* **10 (12)**, 3785–3798.

Miyata, M., Finch, E.A., Khiroug, L., Hashimoto, K., Hayasaka, S., Oda, S.I. et al. (2000). Local calcium release in dendritic spines required for long-term synaptic depression. *Neuron* **28 (1)**, 233–244.

Morris, R.L., Hollenbeck, P.J. (1993). The regulation of bidirectional mitochondrial transport is coordinated with axonal outgrowth. *J. Cell Sci.* **104 (Pt 3)**, 917–927.

Muresan, V., Stankewich, M.C., Steffen, W., Morrow, J.S., Holzbaur, E.L., Schnapp, B.J. (2001). Dynactin-dependent, dynein-driven vesicle transport in the absence of membrane proteins: A role for spectrin and acidic phospholipids. *Mol. Cell* **7 (1)**, 173–183.

Nakagawa, T., Setou, M., Seog, D., Ogasawara, K., Dohmae, N., Takio, K., Hirokawa, N. (2000). A novel motor, KIF13A, transports mannose-6-phosphate receptor to plasma membrane through direct interaction with AP-1 complex [see comment]. *Cell* **103 (4)**, 569–581.

Nangaku, M., Sato-Yoshitake, R., Okada, Y., Noda, Y., Takemura, R., Yamazaki, H., Hirokawa, N. (1994). KIF1B, a novel microtubule plus end-directed monomeric motor protein for transport of mitochondria. *Cell* **79 (7)**, 1209–1220.

Nguyen, Q.T., Parsadanian, A.S., Snider, WD., Lichtman, J.W. (1998). Hyperinnervation of neuromuscular junctions caused by GDNF overexpression in muscle. *Science* **279 (5357)**, 1725–1729.

Paschal, B.M., Vallee, R.B. (1987). Retrograde transport by the microtubule-associated protein MAP 1C. *Nature* **330 (6144)**, 181–183.

Perlson, E., Hanz, S., Ben-Yaakov, K., Segal-Ruder, Y., Seger, R., Fainzilber, M. (2005). Vimentin-dependent spatial translocation of an activated MAP kinase in injured nerve. *Neuron* **45 (5)**, 715–726.

Piper, M., Holt, C. (2004). RNA translation in axons. *Ann. Rev. Cell & Devel. Biol.* **20**, 505–523.

Prekeris, R., Terrian, D.M. (1997). Brain myosin V is a synaptic vesicle-associated motor protein: Evidence for a Ca2+-dependent interaction with the synaptobrevin-synaptophysin complex. *J. Cell Biol.* **137 (7)**, 1589–1601.

Puls, I., Jonnakuty, C., LaMonte, B.H., Holzbaur, E.L., Tokito, M., Mann, E. et al. (2003). Mutant dynactin in motor neuron disease. *Nat. Genet.* **33 (4)**, 455–456.

Ravikumar, B., Acevedo-Arozena, A., Imarisio, S., Berger, Z., Vacher, C., O'Kane, C.J. et al. (2005). Dynein mutations impair autophagic clearance of aggregate-prone proteins. *Nat. Genet.* **37 (7)**, 771–776.

Reid, E., Kloos, M., Ashley-Koch, A., Hughes, L., Bevan, S., Svenson, I.K. et al. (2002). A kinesin heavy chain (KIF5A) mutation in hereditary spastic paraplegia (SPG10) [see comment]. *Amer. J. Hum. Genetics* **71 (5)**, 1189–1194.

Reynolds, A.J., Bartlett, S.E., Hendry, I.A. (1998). Signaling events regulating the retrograde axonal transport of 125I-beta nerve growth factor in vivo. *Brain Research* **798 (1–2)**, 67–74.

Rietdorf, J., Ploubidou, A., Reckmann, I., Holmstrom, A., Frischknecht, F., Zettl, M. et al. (2001). Kinesin-dependent movement on microtubules precedes actin-based motility of vaccinia virus [see comment]. *Nat. Cell Biol.* **3 (11)**, 992–1000.

Saito, N., Okada, Y., Noda, Y., Kinoshita, Y., Kondo, S., Hirokawa, N. (1997). KIFC2 is a novel neuron-specific C-terminal type kinesin superfamily motor for dendritic transport of multivesicular body-like organelles. *Neuron* **18 (3)**, 425–438.

Schafer, D.A., Gill, S.R., Cooper, J.A., Heuser, J.E., Schroer, T.A. (1994). Ultrastructural analysis of the dynactin complex: an actin-related protein is a component of a filament that resembles F-actin. *J. Cell Biol.* **126 (2)**, 403–412.

Schliwa, M., Woehlke, G. (2003). Molecular motors [see comment]. *Nature* **422 (6933)**, 759–765.

Schnapp, B.J., Reese, T.S. (1989). Dynein is the motor for retrograde axonal transport of organelles. *Proc. Natl. Acad. Sci. U S A* **86 (5)**, 1548–1552.

Schnell, E., Nicoll, R.A. (2001). Hippocampal synaptic transmission and plasticity are preserved in myosin Va mutant mice. *J. Neurophys.* **85 (4)**, 1498–1501.

Schroer, T.A. (2004). Dynactin. *Ann. Rev. Cell & Devel. Biol.* **20**, 759–779.

Seisenberger, G., Ried, M.U., Endress, T., Buning, H., Hallek, M., Brauchle, C. (2001). Real-time single-molecule imaging of the infection pathway of an adeno-associated virus [see comment]. *Science* **294 (5548)**, 1929–1932.

Setou, M., Nakagawa, T., Seog, D.H., Hirokawa, N. (2000). Kinesin superfamily motor protein KIF17 and mLin-10 in NMDA receptor-containing vesicle transport. *Science* **288 (5472)**, 1796–1802.

Setou, M., Seog, D.H., Tanaka, Y., Kanai, Y., Takei, Y., Kawagishi, M., Hirokawa, N. (2002). Glutamate-receptor-interacting protein GRIP1 directly steers kinesin to dendrites. *Nature* **417 (6884)**, 83–87.

Short, B., Preisinger, C., Schaletzky, J., Kopajtich, R., Barr, F.A. (2002). The Rab6 GTPase regulates recruitment of the dynactin complex to Golgi membranes. *Curr. Biol.* **12 (20)**, 1792–1795.

Silvanovich, A., Li, M.G., Serr, M., Mische, S., Hays, T.S. (2003). The third P-loop domain in cytoplasmic dynein heavy chain is essential for dynein motor function and ATP-sensitive microtubule binding. *Mol. Biol. Cell* **14 (4)**, 1355–1365.

Smith, G.A., Enquist, L.W. (2002). Break ins and break outs: Viral interactions with the cytoskeleton of mammalian cells. *Ann. Rev. Cell & Devel. Biol.* **18**, 135–161.

Smith, G.A., Pomeranz, L., Gross, S.P., Enquist, L.W. (2004). Local modulation of plus-end transport targets herpesvirus entry and egress in sensory axons. *Proc. Natl. Acad. Sci. U S A* **101 (45)**, 16034–16039.

Sodeik, B. (2000). Mechanisms of viral transport in the cytoplasm. *Trends in Microbiol.* **8 (10)**, 465–472.

Sofroniew, M.V., Howe, C.L., Mobley, W.C. (2001). Nerve growth factor signaling, neuroprotection, and neural repair. *Ann. Rev. Neurosci.* **24**, 1217–1281.

Stowers, R.S., Megeath, L.J., Gorska-Andrzejak, J., Meinertzhagen, I.A., Schwarz, T.L. (2002). Axonal transport of mitochondria to synapses depends on Milton, a novel Drosophila protein. *Neuron* **36 (6)**, 1063–1077.

Su, Q., Cai, Q., Gerwin, C., Smith, C.L., Sheng, Z.H. (2004). Syntabulin is a microtubule-associated protein implicated in syntaxin transport in neurons [see comment]. *Nat. Cell Biol.* **6 (10)**, 941–953.

Suomalainen, M., Nakano, M.Y., Keller, S., Boucke, K., Stidwill, R.P., Greber, U.F. (1999). Microtubule-dependent plus and minus end-directed motilities are competing processes for nuclear targeting of adenovirus. *J. Cell Biol.* **144 (4)**, 657–672.

Takeda, S., Yamazaki, H., Seog, D.H., Kanai, Y., Terada, S., Hirokawa, N. (2000). Kinesin superfamily protein 3 (KIF3) motor transports fodrin-associating vesicles important for neurite building. *J. Cell Biol.* **148 (6)**, 1255–1265.

Tanaka, Y., Kanai, Y., Okada, Y., Nonaka, S., Takeda, S., Harada, A., Hirokawa, N. (1998). Targeted disruption of mouse conventional kinesin heavy chain, kif5B, results in abnormal perinuclear clustering of mitochondria. *Cell* **93 (7)**, 1147–1158.

Vale, R.D. (2003). The molecular motor toolbox for intracellular transport. *Cell* **112 (4)**, 467–480.

Vale, R.D., Reese, T.S., Sheetz, M.P. (1985). Identification of a novel force-generating protein, kinesin, involved in microtubule-based motility. *Cell* **42 (1)**, 39–50.

Vale, R.D., Schnapp, B.J., Mitchison, T., Steuer, E., Reese, T.S., Sheetz, M.P. (1985). Different axoplasmic proteins generate movement in opposite directions along microtubules in vitro. *Cell* **43 (3 Pt 2)**, 623–632.

Vallee, R.B., Williams, J.C., Varma, D., Barnhart, L.E. (2004). Dynein: An ancient motor protein involved in multiple modes of transport. *J. Neurobiol.* **58 (2)**, 189–200.

van Minnen, J., Syed, N.I. (2001). Local protein synthesis in invertebrate axons: From dogma to dilemma. *Res. & Prob. Cell Diff.* **34**, 175–196.

Verhey, K.J., Meyer, D., Deehan, R., Blenis, J., Schnapp, B.J., Rapoport, T.A., Margolis, B. (2001). Cargo of kinesin identified as JIP scaffolding proteins and associated signaling molecules [see comment]. *J. Cell Biol.* **152 (5)**, 959–970.

Waterman-Storer, C.M., Karki, S.B., Kuznetsov, S.A., Tabb, J.S., Weiss, D.G., Langford, G.M., Holzbaur, E.L. (1997). The interaction between cytoplasmic dynein and dynactin is required for fast axonal transport. *Proc. Natl. Acad. Sci. U S A* **94 (22)**, 12180–12185.

Welte, M.A., Gross, S.P., Postner, M., Block, S.M., Wieschaus, E.F. (1998). Developmental regulation of vesicle transport in Drosophila embryos: Forces and kinetics. *Cell* **92 (4)**, 547–557.

Yamada, K., Andrews, C., Chan, W.M., McKeown, C.A., Magli, A., de Berardinis, T. et al. (2003). Heterozygous mutations of the kinesin KIF21A in congenital fibrosis of the extraocular muscles type 1 (CFEOM1). *Nat. Genet.* **35 (4)**, 318–321.

Yamazaki, T., Selkoe, D.J., Koo, E.H. (1995). Trafficking of cell surface beta-amyloid precursor protein: Retrograde and transcytotic transport in cultured neurons. *J. Cell Biol.* **129 (2)**, 431–442.

Yang, Z., Goldstein, L.S. (1998). Characterization of the KIF3C neural kinesin-like motor from mouse. *Mol. Biol. Cell* **9 (2)**, 249–261.

Yano, H., Chao, M.V. (2004). Mechanisms of neurotrophin receptor vesicular transport. *J. Neurobiol.* **58 (2)**, 244–257.

Yano, H., Lee, F.S., Kong, H., Chuang, J., Arevalo, J., Perez, P. et al. (2001). Association of Trk neurotrophin receptors with components of the cytoplasmic dynein motor. *J. Neurosci.* **21 (3)**, RC125.

Yildiz, A., Selvin, P.R. (2005). Kinesin: Walking, crawling or sliding along? *Trends in Cell Biol.* **15 (2)**, 112–120.

Yildiz, A., Tomishige, M., Vale, R.D., Selvin, P.R. (2004). Kinesin walks hand-over-hand. *Science* **303 (5658)**, 676–678.

Zhao, C., Takita, J., Tanaka, Y., Setou, M., Nakagawa, T., Takeda, S. et al. (2001). Charcot-Marie-Tooth disease type 2A caused by mutation in a microtubule motor KIF1B beta [erratum appears in *Cell* 2001 **106 (1)**, 127]. *Cell* **105 (5)**, 587–597.

Zweifel, L.S., Kuruvilla, R., Ginty, D.D. (2005). Functions and mechanisms of retrograde neurotrophin signalling. *Nat. Rev. Neurosci.* **6 (8)**, 615–625.

3

Role of APC Complexes and the Microtubule Cytoskeleton in Neuronal Morphogenesis

VIOLET VOTIN, W. JAMES NELSON,
AND ANGELA I. M. BARTH

Neuronal morphogenesis requires cytoskeletal remodeling, and recent studies point to novel regulators, such as adenomatous polyposis coli protein (APC), that are involved in this process. In addition to functioning as a tumor suppressor and regulator of Wnt signaling, APC regulates microtubule dynamics and organization. APC accumulates at microtubule plus ends at the distal tips of growing neurites and axons. Targeting of APC to the ends of microtubules likely involves the microtubule motor kinesin II and cotransport of other proteins, such as β-catenin and the Par complex. APC regulates microtubule organization and dynamics at the tips of neurites during neuronal morphogenesis, and this function is modified by two other components of the Wnt signaling pathway, GSK3β and β-catenin. In the context of protein trafficking, the regulation of microtubule organization is particularly relevant for neurite outgrowth and synaptogenesis.

I. INTRODUCTION

The complex functions of eukaryotic cells depend on specialization of subcellular regions. The cytoskeleton provides the structure and transport tracks to generate and maintain regions of the cell with distinctive shapes and protein compositions. Neurons are the ultimate example of a cell functionally specialized by the cytoskeleton.

Neurons are specialized for communication by their plasma membrane extensions, termed neurites. A typical polarized neuron specifies two kinds of neurites: several dendrites for receiving signals from other cells, and one long axon for transmitting signals away from the cell body. Outgrowth and differentiation of neurites requires a dynamic cytoskeleton. We will focus on microtubules and microtubule-associated proteins in neuronal morphogenesis. Proteins transported along microtubules specifically to the distal tip of the neurite regulate cytoskeletal changes, which in turn affect protein transport.

Adenomatous polyposis coli (APC) is a microtubule-regulating protein that dynamically accumulates at neurite tips during neuronal polarization. Recent data suggest that APC and associated proteins are transported to growth cones along microtubules and affect outgrowth of early neurites and axons by modifying microtubule dynamics.

II. ROLE OF MICROTUBULES IN NEURONAL POLARIZATION

A. Neuronal Polarization

Hippocampal neurons *in situ* or in culture form one axon that is several times longer than the multiple dendrites (Bartlett and Banker 1984). Banker-type neuronal cultures allow the study of such neuronal polarization independent of direct contact with glial cells (Banker 1980; Goslin et al. 1998). In these cultures, neuronal morphogenesis proceeds through several stereotypical stages (Dotti et al. 1988) (see Figure 3.1). First, the neuron attaches and sends out thin lamellipodia that condense into cylindrical neurites. Subsequently, the neuron polarizes; that is, one of the neurites is distinguished as the future axon by morphology and protein composition. The growth rate of the future axon increases, while the growth rates of the future dendrites stall. Ultimately, only the axon develops a uniformly thin shaft terminating in a wide growth cone.

The composition of the axon differs from that of the cell body and dendrites. The axon contains fewer ribosomes, more synaptic vesicles, and certain cytoskeletal and membrane proteins (Craig and Banker 1994). Axons and early undifferentiated neurites orient all of their microtubule plus ends distally—that is, pointing away from the cell body. Whereas this orientation is maintained in the axon, the future dendrites elongate and establish dendritic morphology, including a microtubule array of mixed orientation (Baas et al. 1989).

Differences in shape and protein composition between axons and dendrites lead to differences in protein trafficking, and vice versa. Most axonal proteins are synthesized in the cell body (Hirokawa and Takemura 2005) and may travel long distances (up to several meters) to the axonal growth cone. Due to uniform microtubule orientation, a plus-end-directed microtubule motor would more consistently transport cargo toward the distal end of axons and early neurites compared to dendrites. Furthermore, differential trafficking of proteins may help initially specify, or later reinforce, neurite identity.

Actin dynamics and organization were previously thought to be sufficient to mediate neurite growth and axon specifica-

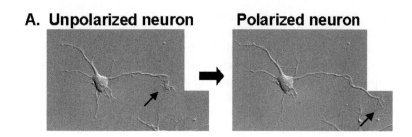

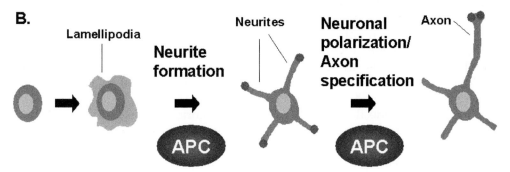

FIGURE 3.1. **Neuronal polarization involves APC re-distribution.** (A) Still images from time-lapse videomicroscopy of hippocampal neurons one day in culture. The first frame shows an unpolarized neuron, with the tip of the future axon marked by an arrow. The second frame shows the same neuron that has polarized after 3 hr 20 min. The axon (arrow) is significantly longer than the other neurites. (B) Schematic of the stages of neuronal morphogenesis. APC (red dots) redistributes from the tips of most neurites of the unpolarized neuron to the tip of the axon in the polarized neuron.

tion. During morphogenesis, the growth cone of the future axon is more dynamic and more susceptible to actin loss upon drug treatment than that of the other neurites (Bradke and Dotti 1999). Actin-depolymerizing drugs increase axon-like neurite growth and are locally sufficient to signal one neurite to elongate like an axon (Bradke and Dotti 1999).

Older studies emphasized, and more recent studies rediscovered, the important role of microtubules in neurite growth and axon specification. Three aspects of microtubules are involved: microtubule organization and dynamics, actin-microtubule coordination, and microtubule-associated proteins.

B. Organization and Dynamics of Microtubules

As described earlier, neurites of unpolarized neurons and axons of polarized neurons share a high growth rate and similar microtubule orientation. Orientation and dynamics of microtubules may be important for membrane addition and neurite growth at the distal neurite tip (Zakharenko and Popov 1998). Microtubule depolymerizing drugs inhibit neurite growth in PC12 cells (Keith 1990). Axon growth is sensitive to microtubule depolymerization at the distal axon, but not at the cell body (Yu and Baas 1995; Bamburg et al. 1986). Low doses of vinblastine that inhibit microtubule dynamics but not microtubule assembly, inhibiting growth cone motility, indicate that dynamic microtubules are required for persistent forward growth of the growth cone (Tanaka et al. 1995).

C. Actin-Microtubule Coordination

Microtubules respond to actin organization during both outgrowth of the axon and growth toward or away from guidance cues (Zhou and Cohan 2004; Dent and Gertler 2003). Actin depletion increases microtubule-microtubule interactions, which may be sufficient for the proper orientation and transport of axonal microtubules (Hasaka et al. 2004). When the bundles of actin, but not the actin meshwork, are disrupted on one side of the growth cone, microtubules extend into the peripheral domain of the other side where actin bundles are still intact, and the growth cone turns in that direction (Zhou et al. 2002). Thus, in the peripheral growth cone, actin bundles may act as tracks that promote microtubule stabilization or assembly forward in the direction of growth, although retrograde flow of actin also may transport microtubules rearward (Schaefer et al. 2002). Actin depolymerization increases the penetration of microtubules into the peripheral domain of the growth cone (Forscher and Smith 1988) and also increases axonal outgrowth (Bradke and Dotti 1999). Thus, the dense actin meshwork and retrograde flow of actin in the peripheral domain normally may act to restrict microtubule extension and thereby neurite extension. Destabilization of actin specifically in the axon or future axon could explain increased microtubule-dependent transport. Organelles, TGN-derived vesicles, and proteins localize exclusively to the axon in early polarized neurons, and exclusively to one neurite, presumably the future axon, in many unpolarized neurons (Bradke and Dotti 1997).

D. Microtubule-Associated Proteins

Several microtubule-associated proteins (MAPs) that affect microtubule dynamics and organization also affect neurite growth, including CRMP-2, tau and MAP1B. CRMP-2 promotes polymerization of microtubules (Fukata et al. 2002) and formation of axons (Inagaki et al. 2001). Tau bundles microtubules, affects axonal transport, and is required for the formation and maintenance of the axon (Kosik and Caceres 1991; Johnson and Stoothoff 2004; Caceres et al. 1991; Wagner et al. 1996). MAP1B expression increases resistance of microtubules to depolymerization (Goold et al. 1999), and MAP2/MAP1B double knockout mice have defects in microtubule bundling and neurite outgrowth in hippocampal neurons (Teng et al. 2001). Interestingly, the kinase GSK3β seems to regulate these proteins (Cole et al. 2004; Yoshimura et al. 2005; Wagner et al. 1996; Goold and Gordon-Weeks 2001; Trivedi et al. 2005). GSK3β also locally regulates the association of APC with microtubules in axon tips (see later).

APC is enriched at sites of active neurite growth: neurite tips in unpolarized neurons and axon tips but not stalled dendrite tips in polarizing neurons (Shi et al. 2004; Votin et al. 2005) (see Figure 3.1B). This localization suggests that APC plays a role in neurite outgrowth, including the increased growth that distinguishes an axon from dendrites. APC function also has been studied in cell types that do not distinguish axons and dendrites. For example, pheochromocytoma (PC12) cells form several equivalent neurites, and dorsal root ganglia (DRG) neurons form one long branched axon but no distinct dendrites (Zhou et al. 2004; Votin et al. 2005). Taken together, these data reveal how APC and associated proteins are transported to the neurite tip and affect neuronal morphogenesis.

III. TARGETING OF APC TO TIPS OF NEURITES

A. APC Associates with Microtubules

APC is a large, multifunctional protein. Its interaction with microtubules is particularly relevant to its intracellular trafficking and consequent effects on neuronal morphogenesis (see Figure 3.2). APC binds

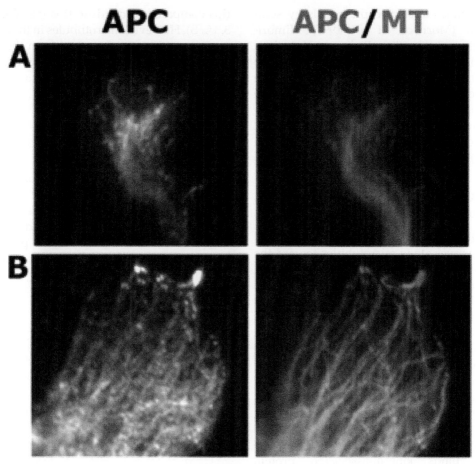

FIGURE 3.2. **APC localizes to microtubule ends at the tips of membrane extensions.** (A) The neurite tip of a hippocampal neuron three days in culture shows APC clusters (red) enriched at the ends of microtubules (green) by immunofluorescence, using an affinity-purified polyclonal antibody against the APC-2 fragment (Nathke et al. 1996). (B) The extension tip of an MDCK epithelial cell shows similar localization of APC to microtubule ends.

microtubules directly through a basic domain, as well as indirectly through EB1, a protein that binds the growing plus ends of microtubules (Askham et al. 2000; Munemitsu et al. 1994). Endogenous APC forms clusters at the plus ends of microtubules in neurons (Zhou et al. 2004; Shimomura et al. 2005; Votin et al. 2005) and in migrating and polarized epithelial cells (Nathke et al. 1996; Mogensen et al. 2002; Reilein and Nelson 2005). APC in distal axons resists actin depolymerization but not microtubule depolymerization (Shimomura et al. 2005). Microtubule-dependent localization of APC to ends of membrane extensions was confirmed by eventual loss of APC upon microtubule depolymerization in both neuroblasts and epithelial cells (Morrison et al. 1997; Nathke et al. 1996). Stabilization of microtubules by taxol increased APC at tips of neurite extensions in neuroblasts (Morrison et al. 1997; Nathke et al. 1996). APC clusters do not disassemble immediately after nocodazole treatment in neuroblasts or epithelial cells, and remain associated with basal epithelial membranes after disrupting cells by soni-cation (Morrison et al. 1997; Mimori-Kiyosue et al. 2000; Reilein and Nelson 2005). Therefore, APC clusters may mediate the interaction of microtubules with the cortical membrane and thereby play a role in microtubule dynamics and capture at the membrane (Reilein and Nelson 2005; Mimori-Kiyosue and Tsukita 2003) (see later).

B. APC Associates with Motor Proteins

GFP-tagged APC expressed in epithelial cells moves along microtubules toward the plus end (Mimori-Kiyosue et al. 2000). Amino-terminal APC, lacking microtubule- and EB1-binding sites, still accumulates at the distal ends of axons (Zhou et al. 2004). Thus, direct binding of APC to micro-tubules may not be necessary for its local-ization. Amino-terminal APC moves along microtubules toward the cell periphery in a manner dependent on ATP hydrolysis (Mimori-Kiyosue et al. 2000), suggesting that movement represents transport by motor proteins. In fact, amino-terminal APC binds kinesin-associated protein KAP3, which is part of the kinesin motor KIF3 (Jimbo et al. 2002; Takeda et al. 2000). At the tips of epithelial extensions, KAP3 is enriched in APC clusters, and the KAP3-binding armadillo domain of APC is required for accumulation of APC at clus-ters (Jimbo et al. 2002). Reduced expression of kinesin heavy chain prevents APC from being transported to the membrane in epithelial cells (Cui et al. 2002).

These studies suggest, it seems likely that APC is transported toward micro-tubule plus-ends through interaction with KAP3 and the plus-end-directed motor KIF3 that is required for neurite extension (Takeda et al. 2000). Since all axonal micro-tubules, but only about half of dendritic microtubules, orient their plus ends distally, kinesin-mediated transport could contri-bute to the accumulation of APC in axon tips compared to dendrite tips (see Figure 3.3A, B). However, microtubules in the axon are not continuous. Transport from the cell body to the axon tip may require switching cargo from a microtubule track to a con-necting actin track (Brown 1999).

C. APC May Cotransport Other Proteins

Several proteins require transport down the axon to reach their site of action dur-ing neuronal morphogenesis. Since APC is transported down the axon and forms complexes with some of these proteins (see next), APC may affect their motor-dependent transport.

β-Catenin

β-catenin may be transported to micro-tubule ends in a complex with APC and/or N-cadherin, a cell-cell adhesion protein that binds β-catenin. KAP3 (see later) is required for localization of β-catenin and N-cadherin to the plasma membrane of neuroepithelial cells, as well as appropriate patterning and proliferation control (Teng et al. 2005). Binding of APC to KAP3 is required for localization of APC to the tip of epithelial extensions (Jimbo et al. 2002). However, it is unclear whether β-catenin, N-cadherin and APC are transported in the same kinesin complex and if APC regulates trans-port of β-catenin and N-cadherin. β-catenin coimmunoprecipitates with APC, KAP3, and the motor subunits of KIF3 (KIF3A/B) in epithelial cells (Jimbo et al. 2002) and with KAP3 and KIF3B in embryonic brain (Teng et al. 2005). The structural role of β-catenin in neuronal morphogenesis is discussed later.

Par Complex

Recent evidence suggests that transport of APC to axon tips is important for neu-ronal polarity, in part because APC binds and transports Par3. Par3 is a polarity complex that includes the GTPase Cdc42,

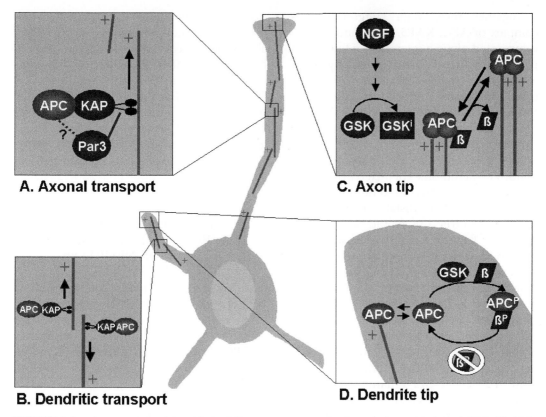

A. Axonal transport

C. Axon tip

B. Dendritic transport

D. Dendrite tip

FIGURE 3.3. **Axons and dendrites likely differ in transport and microtubule association of APC.** (**A**) APC associates with the motor KIF3 (black two-headed structure) by binding kinesin-association protein 3 (KAP). Par3 binds KIF3 and may interact with APC. We hypothesize that the complex is transported along microtubules (green lines) toward their plus ends (arrows), that is, distally in axons and both distally and proximally in dendrites (**B**). (**C**) At the axon tip, GSK3β (GSK) is phosphorylated inactivated (GSK^{iP}, rectangle) and thereby locally inactivated. (In DRG neurons, inactivation is downstream of extracellular NGF signals.) APC associates with microtubules and likely transiently with β-catenin (β). Levels of β-catenin in APC clusters may modulate APC function at tips of microtubules. (**D**) Our model suggests that active GSK3β at the dendrite tip, conversely, phosphorylates the small amount of APC transported to dendrite tips, preventing bundling of microtubules and growth of dendrites.

Par6, and atypical protein kinase C (aPKC), which mediates asymmetric morphogenesis and cell migration (Etienne-Manneville and Hall 2003b). Par3 colocalizes with APC during neuronal polarization; both are at the ends of most undifferentiated neurites in unpolarized hippocampal neurons and at the end of only the axon in polarized neurons (Shi et al. 2003; 2004). Furthermore, Par3 and APC overlap in puncta along the neurite that could represent transport packets of preassembled protein complex (Shi et al. 2004; Ahmari et al. 2000). As localization of APC to the tip of epithelial extensions depends on association with KAP3, localization of Par3 to the tip of neurite extensions depends on KIF3A (Nishimura et al. 2004; Shi et al. 2004; Jimbo et al. 2002). Par3 can be coimmunoprecipitated with APC and KIF3A from embryonic brain (Shi et al. 2004). The functional significance of this interaction is suggested by a dominant-negative APC mutant that causes defects in neuronal polarity and in Par3 transport out of the cell body in hippocampal neurons (Shi et al. 2004).

Together these data suggest that a complex of APC, KAP3, KIF3, and Par3 is transported along a microtubule toward its plus end, which is oriented distally in an axon. The axon-like orientation of microtubules in undifferentiated neurites could facilitate transport of APC and Par3 to the distal tip of neurites. This model also explains the lack of APC and Par3 enrichment in mature dendrites, because dendritic microtubules of mixed orientation would not efficiently concentrate kinesin-transported proteins distally. Thus, slower-growing dendrite tips would not undergo APC/Par/β-catenin-mediated cytoskeletal changes that result in neurite growth.

Components of the Par complex are required for neuronal polarity, although the mechanism is unclear. Disruption of Par3 localization to the neurite tip prevents axon specification (Nishimura et al. 2004; Shi et al. 2004; Shi et al. 2003). Recent evidence suggests that the interaction of Par3 with STEF, a Rac GEF (guanine nucleotide exchange factor) related to Tiam1 at the axon tip regulates actin dynamics required for axon outgrowth (Nishimura et al. 2005).

Synaptic Proteins

Synapse formation requires changes in neuronal shape and protein distribution. Although it is unknown if presynaptic and postsynaptic proteins are transported in neurites with APC, APC may cluster synaptic proteins at the plasma membrane. Both APC and β-catenin localize to synapses (Matsumine et al. 1996; Uchida et al. 1996). An APC mutant that cannot bind EB1 or the postsynaptic density protein PSD-93, reduces the surface localization of nicotinic acetylcholine receptors at postsynaptic clusters in ciliary ganglion neurons (Temburni et al. 2004). PSD-93 binds to acetylcholine receptors and stabilizes their clustering at synapses (Parker et al. 2004), suggesting a function for APC in synaptic stability. β-catenin is required for localiza-

tion of vesicles to presynaptic sites in hippocampal neurons (Bamji et al. 2003). This vesicle localization is independent of cadherin-mediated adhesion and Wnt signaling, and correlates with recruitment of PDZ proteins (Bamji et al. 2003). It will be interesting to determine whether binding of β-catenin to APC clusters, perhaps via alterations in microtubule function, may affect protein or vesicle recruitment required for synaptogenesis.

IV. ROLE OF APC IN MICROTUBULE DYNAMICS AT TIPS OF NEURITES

Aside from its potential role in cotransport of proteins involved in neuronal morphogenesis along microtubules, APC has direct effects on microtubules. Enrichment of APC at microtubule ends, initially in all neurite tips and later only in the axon tip (see the previous section), suggest a role in microtubule reorganization during growth of neurites and axons.

APC promotes microtubule growth and stability. Carboxy-terminal fragments of APC are sufficient to promote microtubule assembly *in vitro* (Munemitsu et al. 1994). APC bundles microtubules *in vitro* and prevents microtubule depolymerization *in vitro* and in APC-overexpressing epithelial cells (Zumbrunn et al. 2001).

Does APC mediate the microtubule reorganization necessary for neurite growth? In PC12 cells, nerve growth factor (NGF) induces neurite outgrowth, increases resistance of microtubules to the drug colchicine (Black et al. 1986; Black and Greene 1982), and increases APC protein levels (Dobashi et al. 1996). APC is required for NGF-mediated neurite outgrowth in PC12 cells (Dobashi et al. 2000). In DRG neurons, inhibition of NGF signaling dissociates APC from growth cone microtubules and decreases microtubule density in the distal axon (Zhou et al. 2004). Disrupting the binding of endogenous APC to micro-

A. Absence of Wnt

B. Presence of Wnt

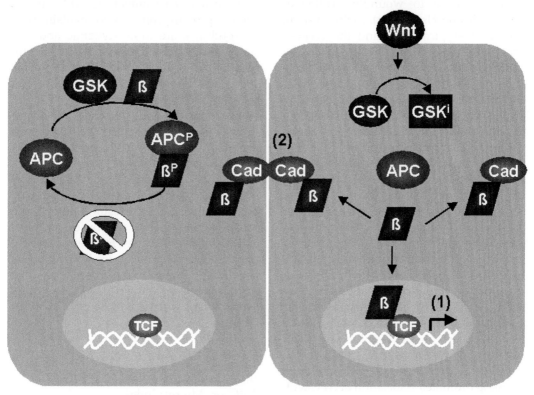

FIGURE 3.4. **Wnt mediates proliferation and differentiation in neural precursors.** (A) GSK3β activity leads to phosphorylation and degradation of β-catenin, except β-catenin is bound to cadherin. (B) Wnt signaling blocks GSK3β activity and β-catenin degradation. Accumulated β-catenin (1) turns on transcription through interaction with the transcription factor Tcf, which is bound to DNA and (2) strengthens cell-cell adhesion by binding cadherin at the lateral membrane. Transcriptional role effects are important for proliferation and neuronal differentiation of neural precursors, and adhesion role effects may be important for branching in neurons. For simplicity, we left out many important Wnt signaling pathway components, such as axin.

tubules, by expression of mutant APC or EB1 constructs, also inhibits axon growth (Zhou et al. 2004). These data suggest that localization of APC to microtubule ends is required for microtubule stability during neurite outgrowth mediated by NGF.

Interestingly, NGF and Wnt signaling pathways may intersect at APC and GSK3β (see Figures 3.3 and 3.4). NGF locally inhibits GSK3β at the axon tip (Zhou et al. 2004). The canonical Wnt signaling pathway, which is important in development and in cancer (Logan and Nusse 2004), also inhibits GSK3β. Inactive GSK3β cannot phosphorylate APC or β-catenin, and thereby modulates the cytoskeletal effects of APC in neuronal morphogenesis, as described next.

A. GSK3β Regulates the Effects of APC on Microtubules

Association of APC with microtubules is regulated by GSK3β. Phosphorylation of APC by PKA and then GSK3β prevents APC from bundling microtubules *in vitro*

(Zumbrunn et al. 2001). Regulation of APC has been studied in astrocytes that polarize in response to wounding (Etienne-Manneville and Hall 2003a). In this system, activity of atypical PKC and the GTPase Cdc42 are required to locally inactivate GSK3β by phosphorylation at the leading edge of astrocyte extensions. In DRG neurons, NGF and downstream signaling locally inactivate GSK3β in the axonal growth cone (Zhou et al. 2004). Localized inactivation of GSK3β, leading to hypophosphorylation of APC, is necessary for the localization of APC clusters to microtubule ends in both the leading edge of astrocytes, and in the axonal growth cone of DRG neurons (Zhou et al. 2004; Etienne-Manneville and Hall 2003a).

APC may therefore bind and capture microtubules specifically at the tips of neurites due to regulation by GSK3β activity (see Figure 3.3C, D). Axon outgrowth correlates with enrichment of inactive GSK3β in the axon tip compared to the axon shaft or dendrite tip (Zhou et al. 2004). In the axonal growth cone, localized inactivation of GSK3β promotes association of APC with microtubules and may therefore restrict APC-mediated microtubule capture to the growth cone (Zhou et al. 2004). Global GSK3β inhibition by drugs seems to inappropriately increase the sites marked for microtubule capture and thereby membrane outgrowth, both in the tips of future dendrites and along the axon shaft, further emphasizing the importance of localized inhibition of GSK3β in the axon tip. Unregulated GSK3β inhibition prevents axon specification in hippocampal neurons (Shi et al. 2004), potentially because microtubule capture occurs at every neurite tip, so that no one neurite grows persistently and becomes the axon. Moreover, global GSK3β inhibition in DRG neurons causes the normally unbranched axon shaft to form many tubulin-containing extensions, with APC concentrated at their bases, in DRG neurons (Zhou et al. 2004). This is a further suggestion that microtubules captured by

APC at the cell periphery continue to grow and exhibit dynamic instability. The normal enrichment of APC in the axonal growth cone could explain why microtubules are more dynamic and less stable in the axonal growth cone as compared to the axonal shaft (Bamburg et al. 1986; Tanaka et al. 1995; Ferreira and Caceres 1989; Shea 1999).

The capture and bundling of microtubules by hypophosphorylated APC may promote plus-end-distal orientation of microtubules in axons and early neurites. Alternatively, accumulation of APC at early neurite tips and at axon tips compared to dendrite tips may be a result, rather than a cause, of the higher proportion of plus-end-distal microtubules. Furthermore, APC may mediate a positive feedback loop by stabilizing microtubule orientation that further increases delivery of APC and other proteins and vesicles to the tips of growing neurites.

B. β-Catenin Regulates the Effects of APC on Microtubules

Canonical Wnt signaling involves inhibition of GSK3β that is normally in a complex with APC and β-catenin. Subsequent hypophosphorylation of β-catenin prevents its degradation by the ubiquitin-proteasome pathway. The resulting accumulation of β-catenin then mediates cell-cell adhesion and transcription via the Tcf family of transcription factors (Logan and Nusse 2004).

In early development, β-catenin-mediated transcription regulates neuronal differentiation (see Figure 3.4). APC mutant embryonic stem cells that cannot downregulate β-catenin have defects in differentiation of the three germ layers and in the expression of neuronal markers (Kielman et al. 2002). Hippocampal development in mice depends on β-catenin-mediated transcription through Tcf/LEF proteins (Galceran et al. 2000). Wnt, through stabilization of β-catenin, promotes proliferation of early-stage neuroepithelial precursor cells and neuronal differentiation of late-stage cortical neural precursor cells (Chenn and Walsh 2002; Hirabayashi et al. 2004).

In differentiated neurons, Wnt and β-catenin have structural roles. Wnt signaling affects the axonal cytoskeleton and acts as an attractive guidance cue over a short time scale, suggesting that its major role is related to cytoskeletal rather than transcriptional activity (Salinas 1999; Lyuksyutova et al. 2003). β-catenin mutations cause mistargeting of axons in *Xenopus* retinal ganglion cells *in vivo* (Elul et al. 2003). β-catenin also influences axonal and dendrite branching; effects that may depend on binding to cell-cell adhesion proteins or to PDZ proteins, not on transcriptional regulation (Yu and Malenka 2003; 2004; Elul et al. 2003).

Recent data indicate a role of β-catenin in regulating the structural APC function at neurite tips (Votin et al. 2005). A stabilized β-catenin mutant that is not phosphorylated by GSK3β accumulates in APC clusters at neurite tips and inhibits neurite initiation and growth in PC12 cells. Inhibition of neurite growth is independent of Tcf-mediated transcription and cell-cell contact. Instead, β-catenin is likely affecting how APC clusters regulate microtubules.

In support of a role for reduced β-catenin turnover in the regulation of APC clusters, cortical APC clusters are thought to be sites of β-catenin degradation. Proteasome inhibitors increase the amount of β-catenin in APC clusters at extension tips (Jimbo et al. 2002). Stabilized β-catenin mutants accumulate in APC clusters at the tip of epithelial extensions and inhibit cell extension during epithelial tubulogenesis (Barth et al. 1997; Pollack et al. 1997). In this regard, when reduction of kinesin heavy chain prevents APC from localizing to the plasma membrane, total and nuclear β-catenin levels increase (Cui et al. 2002).

C. APC May Affect Actin-Microtubule Coordination

Effects of APC on microtubules may affect the coordination of actin and microtubule networks. Actin filaments terminate near APC clusters in epithelial cells (Iizuka-Kogo et al. 2005) and coalign with microtubules that are tipped with APC in neurons (Zhou et al. 2004). APC and Asef, a Rac-specific GEF also called ARHGEF4, both are highly expressed in developing brain (Bhat et al. 1994; Akong et al. 2002; Yoshizawa et al. 2003), and APC stimulates Asef-mediated actin reorganization and epithelial morphogenesis (Kawasaki et al. 2000). In polarized migration of non-neuronal cells, APC interacts with the actin-crosslinking protein IQGAP1 and is required for actin meshwork formation (Watanabe et al. 2004). APC is also involved in microtubule capture and stabilization through CLIP-170 and EB1, two microtubule plus-end binding proteins (Wen et al. 2004; Watanabe et al. 2004). It is unclear whether APC mediates actin/microtubule coordination in neurons, although effects on actin may be mediated through APC interaction with the Par complex (see the previous section).

V. SUMMARY

Transport of APC and associated proteins to neurite tips is essential for neurite outgrowth. APC at neurite and axon tips plays a key role in microtubule reorganization necessary for growth. Microtubule association with APC, and likely microtubule capture by APC, depend on localized inactivation of GSK3β at axon tips. Wnt and NGF signaling pathways may further overlap via β-catenin, which exhibits a novel structural function in regulating APC clusters, in addition to being downregulated by APC in canonical Wnt signaling. Understanding the mechanism by which APC influences cytoskeleton-dependent processes that regulate axon guidance will likely reveal novel information about the protein trafficking events underlying synapse development.

References

Ahmari, S.E., Buchanan, J., Smith, S.J. (2000). Assembly of presynaptic active zones from cytoplasmic transport packets. *Nat. Neurosci.* **3**, 445–451.

Akong, K., McCartney, B.M., Peifer, M. (2002). Drosophila APC2 and APC1 have overlapping roles in the larval brain despite their distinct intracellular localizations. *Dev. Biol.* **250**, 71–90.

Askham, J.M., Moncur, P., Markham, A.F., Morrison, E.E. (2000). Regulation and function of the interaction between the APC tumour suppressor protein and EB1. *Oncogene* **19**, 1950–1958.

Baas, P., Black, M., Banker, G. (1989). Changes in microtubule polarity orientation during the development of hippocampal neurons in culture. *J. Cell Biol.* **109**, 3085–3094.

Bamburg, J.R., Bray, D., Chapman, K. (1986). Assembly of microtubules at the tip of growing axons. *Nature* **321**, 788–790.

Bamji, S.X., Shimazu, K., Kimes, N., Huelsken, J., Birchmeier, W., Lu, B., Reichardt, L.F. (2003). Role of β-catenin in synaptic vesicle localization and presynaptic assembly. *Neuron* **40**, 719–731.

Banker, G.A. (1980). Trophic interactions between astroglial cells and hippocampal neurons in culture. *Science* **209**, 809–810.

Barth, A.I., Pollack, A.L., Altschuler, Y., Mostov, K.E., Nelson, W.J. (1997). NH2-terminal deletion of beta-catenin results in stable colocalization of mutant beta-catenin with adenomatous polyposis coli protein and altered MDCK cell adhesion. *J. Cell Biol.* **136**, 693–706.

Bartlett, W.P., Banker, G.A. (1984). An electron microscopic study of the development of axons and dendrites by hippocampal neurons in culture. I. Cells which develop without intercellular contacts. *J. Neurosci.* **4**, 1944–1953.

Bhat, R.V., Baraban, J.M., Johnson, R.C., Eipper, B.A., Mains, R.E. (1994). High levels of expression of the tumor suppressor gene APC during development of the rat central nervous system. *J. Neurosci.* **14**, 3059–3071.

Black, M.M., Greene, L.A. (1982). Changes in the colchicine susceptibility of microtubules associated with neurite outgrowth: Studies with nerve growth factor-responsive PC12 pheochromocytoma cells. *J. Cell Biol.* **95**, 379–386.

Black, M.M., Aletta, J.M., Greene, L.A. (1986). Regulation of microtubule composition and stability during

nerve growth factor-promoted neurite outgrowth. *J. Cell Biol.* **103**, 545–557.

Bradke, F., Dotti, C.G. (1997). Neuronal polarity: Vectorial cytoplasmic flow precedes axon formation. *Neuron* **19**, 1175–1186.

Bradke, F., Dotti, C.G. (1999). The role of local actin instability in axon formation. *Science* **283**, 1931–1934.

Brown, S.S. (1999). Cooperation between microtubule- and actin-based motor proteins. *Annu. Rev. Cell Dev. Biol.* **15**, 63–80.

Caceres, A., Potrebic, S., Kosik, K.S. (1991). The effect of tau antisense oligonucleotides on neurite formation of cultured cerebellar macroneurons. *J. Neurosci.* **11**, 1515–1523.

Chenn, A., Walsh, C.A. (2002). Regulation of cerebral cortical size by control of cell cycle exit in neural precursors. *Science* **297**, 365–369.

Cole, A.R., Knebel, A., Morrice, N.A., Robertson, L.A., Irving, A.J., Connolly, C.N., Sutherland, C. (2004). GSK-3 phosphorylation of the Alzheimer epitope within collapsin response mediator proteins regulates axon elongation in primary neurons. *J. Biol. Chem.* **279**, 50176–50180.

Craig, A.M., Banker, G. (1994). Neuronal polarity. *Annu. Rev. Neurosci.* **17**, 267–310.

Cui, H., Dong, M., Sadhu, D.N., Rosenberg, D.W. (2002). Suppression of kinesin expression disrupts adenomatous polyposis coli (APC) localization and affects beta-catenin turnover in young adult mouse colon (YAMC) epithelial cells. *Exp. Cell Res.* **280**, 12–23.

Dent, E.W., Gertler, F.B. (2003). Cytoskeletal dynamics and transport in growth cone motility and axon guidance. *Neuron* **40**, 209–227.

Dobashi, Y., Bhattacharjee, R.N., Toyoshima, K., Akiyama, T. (1996). Upregulation of the APC gene product during neuronal differentiation of rat pheochromocytoma PC12 cells. *Biochem. Biophys. Res. Commun.* **224**, 479–483.

Dobashi, Y., Katayama, K., Kawai, M., Akiyama, T., Kameya, T. (2000). APC protein is required for initiation of neuronal differentiation in rat pheochromocytoma PC12 cells. *Biochem. Biophys. Res. Commun.* **279**, 685–691.

Dotti, C., Sullivan, C., Banker, G. (1988). The establishment of polarity by hippocampal neurons in culture. *J. Neurosci.* **8**, 1454–1468.

Elul, T.M., Kimes, N.E., Kohwi, M., Reichardt, L.F. (2003). N- and C-terminal domains of beta-catenin, respectively, are required to initiate and shape axon arbors of retinal ganglion cells in vivo. *J. Neurosci.* **23**, 6567–6575.

Etienne-Manneville, S., Hall, A. (2003a). Cdc42 regulates GSK-3beta and adenomatous polyposis coli to control cell polarity. *Nature* **421**, 753–756.

Etienne-Manneville, S., Hall, A. (2003b). Cell polarity: Par6, aPKC and cytoskeletal crosstalk. *Curr. Opin. Cell Biol.* **15**, 67–72.

Ferreira, A., Caceres, A. (1989). The expression of acetylated microtubules during axonal and dendritic growth in cerebellar macroneurons which develop in vitro. *Brain Res. Dev. Brain Res.* **49**, 205–213.

Forscher, P., Smith, S. (1988). Actions of cytochalasins on the organization of actin filaments and microtubules in a neuronal growth cone. *J. Cell Biol.* **107**, 1505–1516.

Fukata, Y., Itoh, T.J., Kimura, T., Mâenager, C., Nishimura, T., Shiromizu, T. et al. (2002). CRMP-2 binds to tubulin heterodimers to promote microtubule assembly. *Nat. Cell Biol.* **4 (8)**, 583–591.

Galceran, J., Miyashita-Lin, E.M., Devaney, E., Rubenstein, J.L., Grosschedl, R. (2000). Hippocampus development and generation of dentate gyrus granule cells is regulated by lef1. *Development* **127**, 469–482.

Goold, R.G., Owen, R., Gordon-Weeks, P.R. (1999). Glycogen synthase kinase 3β phosphorylation of microtubule-associated protein 1b regulates the stability of microtubules in growth cones. *J. Cell Sci.* **112**, 3373–3384.

Goold, R.G., Gordon-Weeks, P.R. (2001). Micro-tubule-associated protein 1b phosphorylation by glycogen synthase kinase 3β is induced during PC12 cell differentiation. *J. Cell Sci.* **114**, 4273–4284.

Goslin, K., Asmussen, H., Banker, G. (1998). Rat hippocampal neurons in low-density culture. *In* Culturing nerve cells, G. Banker, Goslin, K, eds., 339–370. The MIT Press, Cambridge, MA.

Hasaka, T.P., Myers, K.A., Baas, P.W. (2004). Role of actin filaments in the axonal transport of microtubules. *J. Neurosci.* **24**, 11291–11301.

Hirabayashi, Y., Itoh, Y., Tabata, H., Nakajima, K., Akiyama, T., Masuyama, N., Gotoh, Y. (2004). The Wnt/beta-catenin pathway directs neuronal differentiation of cortical neural precursor cells. *Development* **131**, 2791–2801.

Hirokawa, N., Takemura, R. (2005). Molecular motors and mechanisms of directional transport in neurons. *Nat. Rev. Neurosci.* **6**, 201–214.

Iizuka-Kogo, A., Shimomura, A., Senda, T. (2005). Colocalization of APC and DLG at the tips of cellular protrusions in cultured epithelial cells and its dependency on cytoskeletons. *Histochem. Cell Biol.* **123**, 67–73.

Inagaki, N., Chihara, K., Arimura, N., Mâenager, C., Kawano, Y., Matsuo, N. et al. (2001). CRMP-2 induces axons in cultured hippocampal neurons. *Nat. Neurosci.* **4**, 781–782.

Jimbo, T., Kawasaki, Y., Koyama, R., Sato, R., Takada, S., Haraguchi, K., Akiyama, T. (2002). Identification of a link between the tumour suppressor APC and the kinesin superfamily. *Nat. Cell Biol.* **4**, 323–327.

Johnson, G.V.W., Stoothoff, W.H. (2004). Tau phosphorylation in neuronal cell function and dysfunction. *J. Cell Sci.* **117**, 5721–5729.

Kawasaki, Y., Senda, T., Ishidate, T., Koyama, R., Morishita, T., Iwayama, Y. et al. (2000). ASEF, a link between the tumor suppressor APC and G-protein signaling. *Science* **289**, 1194–1197.

Keith, C.H. (1990). Neurite elongation is blocked if microtubule polymerization is inhibited in PC12 cells. *Cell Motil. Cytoskeleton* **17**, 95–105.

Kielman, M.F., Rindapaa, M., Gaspar, C., van Poppel, N., Breukel, C., van Leeuwen, S. et al. (2002). APC modulates embryonic stem-cell differentiation by controlling the dosage of beta-catenin signaling. *Nat. Genet.* **32**, 594–605.

Kosik, K.S., Caceres, A. (1991). Tau protein and the establishment of an axonal morphology. *J. Cell Sci. Suppl.* **15**, 69–74.

Logan, C.Y., Nusse, R. (2004). The Wnt signaling pathway in development and disease. *Annu. Rev. Cell Dev. Biol.* **20**, 781–810.

Lyuksyutova, A.I., Lu, C.-C., Milanesio, N., King, L.A., Guo, N., Wang, Y. et al. (2003). Anterior-posterior guidance of commissural axons by Wnt-frizzled signaling. *Science* **302**, 1984–1988.

Matsumine, A., Ogai, A., Senda, T., Okumura, N., Satoh, K., Baeg, G.H. et al. (1996). Binding of APC to the human homolog of the drosophila discs large tumor suppressor protein. *Science* **272**, 1020–1023.

Mimori-Kiyosue, Y., Shiina, N., Tsukita, S. (2000). Adenomatous polyposis coli (APC) protein moves along microtubules and concentrates at their growing ends in epithelial cells. *J. Cell Biol.* **148**, 505–518.

Mimori-Kiyosue, Y., Tsukita, S. (2003). "Search-and-capture" of microtubules through plus-end-binding proteins (+TIPS). *J. Biochem. (Tokyo)* **134**, 321–326.

Mogensen, M.M., Tucker, J.B., Mackie, J.B., Prescott, A.R., Nathke, I.S. (2002). The adenomatous polyposis coli protein unambiguously localizes to microtubule plus ends and is involved in establishing parallel arrays of microtubule bundles in highly polarized epithelial cells. *Journal of Cell Biology* **157**, 1041–1048.

Morrison, E.E., Askham, J.M., Clissold, P., Markham, A.F., Meredith, D.M. (1997). The cellular distribution of the adenomatous polyposis coli tumour suppressor protein in neuroblastoma cells is regulated by microtubule dynamics. *Neuroscience* **81**, 553–563.

Munemitsu, S., Souza, B., Muller, O., Albert, I., Rubinfeld, B., Polakis, P. (1994). The APC gene product associates with microtubules *in vivo* and promotes their assembly *in vitro*. *Cancer Res.* **54**, 3676–3681.

Nathke, I.S., Adams, C.L., Polakis, P., Sellin, J.H., Nelson, W.J. (1996). The adenomatous polyposis coli tumor suppressor protein localizes to plasma membrane sites involved in active cell migration. *J. Cell Biol.* **134**, 165–179.

Nishimura, T., Kato, K., Yamaguchi, T., Fukata, Y., Ohno, S., Kaibuchi, K. (2004). Role of the Par-3-KIF3 complex in the establishment of neuronal polarity. *Nat. Cell Biol.* **6**, 328–334.

Nishimura, T., Yamaguchi, T., Kato, K., Yoshizawa, M., Nabeshima, Y., Ohno, S. et al. (2005). Par-6-Par-3 mediates Cdc42-induced rac activation through the Rac GEFs STEF/Tiam1. *Nat. Cell Biol.* **7**, 270–277.

Parker, M.J., Zhao, S., Bredt, D.S., Sanes, J.R., Feng, G. (2004). PSD93 regulates synaptic stability at neuronal cholinergic synapses. *J. Neurosci.* **24**, 378–388.

Pollack, A.L., Barth, A.I.M., Altschuler, Y., Nelson, W.J., Mostov, K.E. (1997). Dynamics of beta-catenin interactions with APC protein regulate epithelial tubulogenesis. *J. Cell Biol.* **137**, 1651–1662.

Reilein, A., Nelson, W.J. (2005). APC is a component of an organizing template for cortical microtubule networks. *Nat. Cell Biol.* **7**, 463–473.

Salinas, P.C. (1999). Wnt factors in axonal remodelling and synaptogenesis. *Biochem. Soc. Symp.* **65**, 101–109.

Schaefer, A.W., Kabir, N., Forscher, P. (2002). Filopodia and actin A guide the assembly and transport of two populations of microtubules with unique dynamic parameters in neuronal growth cones. *J. Cell Biol.* **158**, 139–152.

Shea, T.B. (1999). Selective stabilization of microtubules within the proximal region of developing axonal neurites. *Brain Res. Bull.* **48**, 255–261.

Shi, S.-H., Jan, L.Y., Jan, Y.-N. (2003). Hippocampal neuronal polarity specified by spatially localized mPar3/mPar6 and PI 3-kinase activity. *Cell* **112**, 63–75.

Shi, S.-H., Cheng, T., Jan, L.Y., Jan, Y.N. (2004). APC and GSK-3β are involved in mPar3 targeting to the nascent axon and establishment of neuronal polarity. *Curr. Biol.* **14**, 2025–2032.

Shimomura, A., Kohu, K., Akiyama, T., Senda, T. (2005). Subcellular localization of the tumor suppressor protein APC in developing cultured neurons. *Neurosci. Lett.* **375**, 81–86.

Takeda, S., Yamazaki, H., Seog, D.-H., Kanai, Y., Terada, S., Hirokawa, N. (2000). Kinesin superfamily protein 3 (KIF3) motor transports fodrin-associating vesicles important for neurite building. *J. Cell Biol.* **148**, 1255–1266.

Tanaka, E., Ho, T., Kirschner, M.W. (1995). The role of microtubule dynamics in growth cone motility and axonal growth. *J. Cell Biol.* **128 (1–2)**, 139–155.

Temburni, M.K., Rosenberg, M.M., Pathak, N., McConnell, R., Jacob, M.H. (2004). Neuronal nicotinic synapse assembly requires the adenomatous polyposis coli tumor suppressor protein. *J. Neurosci.* **24**, 6776–6784.

Teng, J., Takei, Y., Harada, A., Nakata, T., Chen, J., Hirokawa, N. (2001). Synergistic effects of MAP2 and MAP1b knockout in neuronal migration, dendritic outgrowth, and microtubule organization. *J. Cell Biol.* **155**, 65–76.

Teng, J., Rai, T., Tanaka, Y., Takei, Y., Nakata, T., Hirasawa, M. et al. (2005). The KIF3 motor transports N-cadherin and organizes the developing neuroepithelium. *Nat. Cell Biol.* **7**, 474–482.

Trivedi, N., Marsh, P., Goold, R.G., Wood-Kaczmar, A., Gordon-Weeks, P.R. (2005). Glycogen synthase kinase-3{beta} phosphorylation of MAP1b at Ser1260 and Thr1265 is spatially restricted to growing axons. *J. Cell Sci.* **118**, 993–1005.

Uchida, N., Honjo, Y., Johnson, K.R., Wheelock, M.J., Takeichi, M. (1996). The catenin/cadherin adhesion system is localized in synaptic junctions bordering transmitter release zones. *J. Cell Biol.* **135**, 767–779.

Votin, V., Nelson, W.J., Barth, A.M. (2005). Role of APC and beta-catenin in neurite outgrowth. Submitted.

Wagner, U., Utton, M., Gallo, J.M., Miller, C.C. (1996). Cellular phosphorylation of tau by GSK-3 beta influences tau binding to microtubules and microtubule organization. *J. Cell Sci.* **109**, 1537–1543.

Watanabe, T., Wang, S., Noritake, J., Sato, K., Fukata, M., Takefuji, M. et al. (2004). Interaction with IQGAP1 links APC to rac1, cdc42, and actin filaments during cell polarization and migration. *Dev. Cell* **7**, 871–883.

Wen, Y., Eng, C.H., Schmoranzer, J., Cabrera-Poch, N., Morris, E.J.S., Chen, M. et al. (2004). EB1 and APC bind to mDia to stabilize microtubules downstream of Rho and promote cell migration. *Nat. Cell Biol.* **6**, 820–830.

Yoshimura, T., Kawano, Y., Arimura, N., Kawabata, S., Kikuchi, A., Kaibuchi, K. (2005). GSK-3β regulates phosphorylation of CRMP-2 and neuronal polarity. *Cell* **120**, 137–149.

Yoshizawa, M., Sone, M., Matsuo, N., Nagase, T., Ohara, O., Nabeshima, Y., Hoshino, M. (2003). Dynamic and coordinated expression profile of dbl-family guanine nucleotide exchange factors in the developing mouse brain. *Gene Expr. Patterns* **3**, 375–381.

Yu, W., Baas, P.W. (1995). The growth of the axon is not dependent upon net microtubule assembly at its distal tip. *J. Neurosci.* **15**, 6827–6833.

Yu, X., Malenka, R.C. (2003). β-catenin is critical for dendritic morphogenesis. *Nat. Neurosci.* **6**, 1169–1177.

Yu, X., Malenka, R.C. (2004). Multiple functions for the cadherin/catenin complex during neuronal development. *Neuropharmacology* **47**, 779–786.

Zakharenko, S., Popov, S. (1998). Dynamics of axonal microtubules regulate the topology of new membrane insertion into the growing neurites. *J. Cell Biol.* **143**, 1077–1086.

Zhou, F.Q., Cohan, C.S. (2004). How actin filaments and microtubules steer growth cones to their targets. *J. Neurobiol.* **58**, 84–91.

Zhou, F.Q., Waterman-Storer, C.M., Cohan, C.S. (2002). Focal loss of actin bundles causes microtubule redistribution and growth cone turning. *J. Cell Biol.* **157**, 839–849.

Zhou, F.Q., Zhou, J., Dedhar, S., Wu, Y.H., Snider, W.D. (2004). NGF-induced axon growth is mediated by localized inactivation of GSK-3beta and functions of the microtubule plus end binding protein APC. *Neuron* **42**, 897–912.

Zumbrunn, J., Kinoshita, K., Hyman, A.A., Nathke, I.S. (2001). Binding of the adenomatous polyposis coli protein to microtubules increases microtubule stability and is regulated by GSK3 beta phosphorylation. *Current Biology* **11**, 44–49.

II

SYNAPTIC DEVELOPMENT

The formation of synapses is a critical event in the development of neuronal circuits and is the subject of the second section. Following axon outgrowth and target finding, the growth cone makes tremendous morphological and functional alterations as it differentiates into a presynaptic terminal. Mechanisms underlying the organization of signaling and adhesion molecules in synaptic development (Gollan and Scheiffele) as well as the regulation of two key aspects of presynaptic terminal differentiation, active zone assembly and presynaptic growth, will be discussed (Zhen).

4

Assembly of Synapses in the Vertebrate Central Nervous System

LEORA GOLLAN AND PETER SCHEIFFELE

The formation of synapses during development of the nervous system requires a series of cell-cell signaling events. Coordinating the trafficking of synaptic proteins and the structural assembly of pre- and postsynaptic membrane domains represents a challenge to cell biologists. This chapter focuses on the molecular pathways that contribute to synaptogenesis in the vertebrate central nervous system. Recent studies have led to the identification of three general mechanisms that underlie the organization of signaling and adhesion molecules in synaptic development: the activation of signaling cascades by secreted growth factors; the nucleation of cytoplasmic scaffolds by adhesion molecules; and the direct aggregation of synaptic ion channels by extracellular protein-protein interactions. In combination, these diverse mechanisms are thought to direct the structural and functional differentiation of synapses during the development of the central nervous system.

I. INTRODUCTION

The formation of synapses is a critical event in the development of neuronal circuits. The appropriate expression, trafficking, and assembly of protein complexes play a central role in synapse formation. Protein complexes mediate signaling and adhesion events that transform axo-dendritic contacts into pre- and postsynaptic membrane specializations and thereby orchestrate synaptic differentiation.

The formation of synaptic specializations occurs with remarkable precision. Organizing the structure of a single synaptic contact requires the coordinate assembly of presynaptic exocytic machinery and postsynaptic neurotransmitter receiving units at a highly specialized cell-cell junction. It is important that synaptic junctions remain stable over long time periods and at the same time, synapses retain the ability to rapidly alter their structure and function in response to neuronal activity.

Beyond the organization of individual synapses, synaptic formation is controlled with respect to synapse number and cellular specificity. Mechanisms exist that control the establishment of synapses in the appropriate number with the appropriate target cells. Alterations of target specificity or the formation of excessive or insufficient numbers of connections results in dysfunction of the neuronal circuit.

Much of what is known about the molecular mechanisms of synapse formation has been learned using invertebrate systems, such as *Drosophila* and *C. elegans* or using the vertebrate neuromuscular junction as a model (for review, see Sanes and Lichtman (1999) and Jin (2002)). In this chapter, we will focus on CNS synaptogenesis in vertebrates. Although current knowledge in this field is still quite preliminary, recent studies have revealed some interesting candidate molecules that may contribute to this process. We will first review the cell biological principles of synaptic organization and then discuss some of the more recent insights into the molecular mechanisms of synapse formation in the CNS.

II. CELL BIOLOGY OF CNS SYNAPSES

A. Synapses Are Specialized Cell Junctions

Like other cell-cell junctions, such as epithelial tight junctions and adherens junctions, synapses are close appositions of two cellular membranes that are linked through adhesion molecules. In fact, many molecular components of synapses and epithelial junctions are common, such as the homophilic adhesion molecules (e.g., cadherins and nectins) and cytoplasmic scaffolding molecules of the MAGUK family.

Scaffolding proteins at epithelial junctions and synapses tether ion channels at the specific plasma membrane domains. For example, the Lin2 / Lin7 / Lin10 complex (also termed CASK / Veli / Mint) has been implicated in basolateral concentration of cell surface receptors in epithelia as well as the concentration of ion channels in the pre- and postsynaptic membranes of neurons (Butz et al. 1998; Kaech et al. 1998; Rongo et al. 1998). Furthermore, the polarized delivery of proteins to the lateral epithelial membrane and the synaptic junction share common cytoplasmic machinery. One example that received significant attention is the exocyst complex, a multiprotein complex involved in basolateral delivery of transmembrane proteins in epithelial cells and the transport of NMDA-receptors to the postsynaptic membrane in hippocampal neurons (Grindstaff et al. 1998; Sans et al. 2003) (see Chapter 13). Therefore, similar mechanisms for protein transport and localization govern the concentration of adhesion molecules and ion channels at synaptic and nonsynaptic junctions of neuronal and nonneuronal cells, respectively.

Much of what is known of the epithelial system can be applied to the understanding of cell biological principles of protein recruitment to synaptic junctions; however, there are also important differences between the two systems. First, epithelial junctions are symmetric whereas synapses are highly asymmetric. This asymmetry is fundamental for the ability of synapses to mediate directional flow of information. Presynaptic structures contain specialized sites for rapid vesicle fusion and neurotransmitter release whereas postsynaptic membranes contain the appropriate neurotransmitter receptors that control activity of the postsynaptic cell. Asymmetry is thought to be achieved by polarized sorting of ion channels to axon and dendrites (Craig and Banker 1994) and also by neuron-specific molecules that mediate heterophilic adhesion and thereby provide directionality for cell-cell interactions.

Another important difference between epithelial and synaptic junctions is their size. The size of synaptic adhesion sites is tightly regulated and single synapses are significantly smaller than the large junctional complexes that wrap around a cell within an epithelial monolayer. The geometry of axons and dendrites may contribute to the significantly smaller synaptic adhesions, but the lateral growth of synapses also appears to be controlled by neuronal signaling and by the presence of glial processes that ensheath the synaptic junction.

B. Conversion of Afferent Axons into Synaptic Terminals

One fruitful approach to understanding the mechanisms of synaptic differentiation has been to chronicle the morphological rearrangements of afferent axons during synapse formation. In the developing nervous system, many axons extend over long distances as single unbranched fibers. Upon arrival in the target area, the axonal growth cones undergo a series of transformations that represent a prelude to synapse formation. Axon advancement slows, and growth cones often spread and enlarge sig-

nificantly (Mason et al. 1997; Yoshihara et al. 1997; Jontes et al. 2000). These alterations in growth cone shape suggest that signals from the target area increase the adhesiveness of the axonal membrane with the substratum and that the dynamics of the axonal cytoskeleton are altered. Cell-cell contacts between axons and their target cells are established either directly from the growth cone or through filopodial protrusions emanating from the dendrite (Fiala et al. 1998; Niell et al. 2004). These early contacts transform into axonal varicosities, which are filled with neurotransmitter containing vesicles and resemble immature presynaptic terminals. Simultaneously, axons tend to form branches that frequently emerge from the sites of axonal varicosities (Alsina et al. 2001; Hua and Smith 2004). These observations indicate that there are multiple signals derived from the target area that transform the ingrowing afferent: a stop-growth signal, a signal that promotes axonal branching, and a differentiation signal that induces the formation of presynaptic varicosities in axons.

One remarkable observation from *in vitro* studies has been that the transformation of a growing axon into a presynaptic terminal occurs with a very fast time course. When target cells were manipulated into contact with their presynaptic partners, accumulation of synaptic vesicles and functional transmission could be detected within minutes of contact (Xie and Poo 1986; Evers et al. 1989; Dai and Peng 1993). These early studies led to several important conclusions: (1) axons and target cells contain the machinery required for synaptic transmission *before* synapses are formed, and (2) axon-target interactions provide an instructive signal for the assembly of functional synaptic terminals. Although these studies were conducted with *Xenopus* motor neurons and muscle cells, more recent *in vitro* time lapse imaging studies conducted with cultured vertebrate neurons confirmed a rapid time course for central synapse assembly (Friedman et al. 2000; Zhai et al.

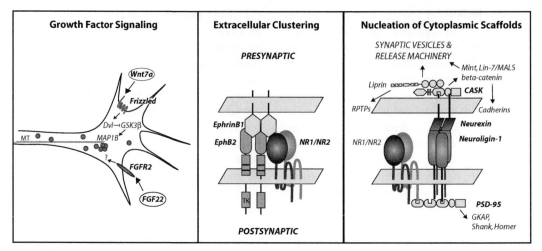

FIGURE 4.1. Three general mechanisms that contribute to the differentiation and functional organization of synaptic membranes:

(**Left Panel**) *Retrograde growth factor signaling.* Wnt and FGF growth factors have been characterized in the vertebrate system (Hall et al. 2000; Umemori et al. 2004). Both factors are secreted from target cells and interact with axonal receptors. Binding of Wnt7a to the 7-TM receptor Frizzled activates Disheveled (Dvl), which inhibits GSK3β-mediated phosphorylation of microtubule-associated protein 1B (MAP-1B). This signaling pathway ultimately regulates microtubule stability. The mechanism of action of FGF22 is less clear, but it was shown that it is dependent on FGFR2 in the afferent.

(**Middle Panel**) *Extracellular clustering of synaptic membrane proteins.* NMDA-receptors containing NR1/NR2 subunits are recruited to synaptic contacts through formation of an extracellular tri-partite complex with presynaptic EphrinB1 and postsynaptic EphB2 (Dalva et al. 2000). Similarly, postsynaptic AMPA-receptor clustering is induced by the protein NARP (not shown), which is released from axons (O'Brien et al. 1999). Both examples use direct extracellular interactions to concentrate ion channels at synapses.

(**Right panel**) *Nucleation of cytoplasmic scaffolds by trans-synaptic interactions.* Pre- and postsynaptic scaffolds are assembled around a trans-synaptic complex formed between postsynaptic neuroligin-1 and presynaptic neurexin (Dean et al. 2003; Graf et al. 2004; Chih et al. 2005). Through PDZ-domain and other protein-protein interactions, additional scaffolding proteins are recruited, which in turn can immobilize other synaptic transmembrane proteins such as cadherins or receptor-protein tyrosine phosphatases (RPTPs) (Serra-Pages et al. 1998; Perego et al. 2000; Olsen et al. 2005). Similarly, expanding postsynaptic scaffolds can recruit postsynaptic NMDA-receptors (NR1/NR2) (Kim and Sheng 2004). The interactions of multiple different trans-synaptic adhesion systems is likely integrated at the level of common scaffolding proteins.

2001). The critical molecular signals that induce the differentiation of pre- and postsynaptic domains upon cell-cell contact are still not understood, but recent studies have revealed a number of interesting candidate signaling systems that are discussed later.

III. MOLECULAR MECHANISMS OF SYNAPSE FORMATION

One simple marker for the transformation of a growth cone into a nerve terminal is the clustering of synaptic vesicles. Several *in vitro* studies have used vesicle clustering to identify target-derived signaling molecules, as well as cell adhesion molecules that might contribute to the induction of synaptic differentiation (reviewed in Scheiffele (2003) and Waites et al. (2005)). From this work, three different mechanisms have emerged that can promote synapse formation (see Figure 4.1):

(1) The activation of cell surface signaling receptors through target-derived growth factors.

(2) The recruitment of ion channels to incipient synaptic connections through extracellular protein-protein interactions.

(3) The nucleation of pre- and post-synaptic scaffolds at the cytoplasmic tails of trans-synaptic adhesion molecules.

Examples of these three mechanisms are discussed next.

A. Induction of Presynaptic Differentiation by Target-derived Growth Factors

The morphological transformation of growth cones in the target area is in part dependent on signaling molecules that are secreted from target cells. In the cerebellum, Wnt7a is secreted by the granule cells and induces axonal branching and recruitment of presynaptic proteins in mossy fiber axons, the presynaptic partner of granule cells (Hall et al. 2000). This effect of Wnt is mediated by its receptor Frizzled, which activates Disheveled to inhibit GSK3β (see Figure 4.1). Inhibition of GSK3β has at least two cellular consequences: It prevents β-catenin degradation and thereby results in activation of β-catenin-TCF/LEF target genes; and it alters the phosphorylation state of the microtubule-associated protein MAP1B, which results in increased microtubule stability (see Figure 4.1) (Krylova et al. 2000; Ciani et al. 2004). Time-lapse imaging studies revealed that growth cones respond to direct application of Wnt growth factors within 10 to 20 minutes. This rapid time course suggests that the initial axonal remodeling induced by Wnt growth factors occurs by direct modulation of the cytoskeleton and not by transcriptional regulation.

A second growth factor that is released from target cells and that promotes vesicle clustering in pontine axons is a subset of fibroblast growth factors (FGFs). Early studies by Peng and colleagues implicated FGF2 as a synaptogenic molecule that could induce the clustering of synaptic vesicle proteins in cultured neurons (Dai and Peng 1995). Subsequent, more extensive studies revealed that FGF2 is a potent inducer of presynaptic differentiation in cerebellar mossy fiber axons *in vitro* and *in vivo* (Umemori et al. 2004). This activity requires expression of FGF receptor 2 in the presynaptic cell. However, the cell biological mechanisms by which FGF signaling controls vesicle accumulation require further investigation.

Even though Wnts and FGFs are both secreted molecules it remains to be determined whether these factors act locally or as diffusible factors. Both growth factors are closely associated with the extracellular matrix and cellular membranes and therefore may act as cell-cell contact dependent signals rather than affecting ingrowing afferents over longer distances.

B. Recruitment of Ion Channels by Extracellular Domain Interactions

Considering that the precise alignment of pre- and postsynaptic machinery is one of the hallmarks of synaptic junctions, it is interesting that studies have revealed that extracellular domains of ion channels are targets for synaptic differentiation signals. Such interactions are perfectly positioned to tether channel proteins in direct contact with the opposing synaptic membrane. The first such example was the postsynaptic AMPA-type glutamate receptor that binds to a protein called NARP (neuronal activity regulated pentraxin) through its extracellular domain. NARP is released from axons and forms oligomers that promote clustering of AMPA-receptors and subsequent recruitment of cytoplasmic AMPA-receptor binding proteins (O'Brien et al. 1999; 2002). Protein localization studies and loss-of-function analyses have revealed that NARP acts only at glutamatergic shaft synapses, not at glutamatergic synapses formed on dendritic spines (Mi et al. 2002). However, a protein with analogous function might act at dendritic spines, since the extracellular portion of the AMPA-receptor subunit GluR2 has been shown to contribute to spine morphogenesis through an as yet unknown extracellular ligand (Passafaro et al. 2003).

A second example of extracellular interactions that play a role in the recruitment of neurotransmitter receptors to developing synapses was described for the NMDA-type glutamate receptor (see Figure 4.1). The extracellular domain of the subunit NR1 was shown to form a tri-partite complex with the receptor tyrosine kinase EphB2 and its ligand EphrinB (Dalva et al. 2000; Takasu et al. 2002). Application of an oligomeric form of recombinant EphrinB resulted in the formation of postsynaptic specializations containing NMDA-receptors and scaffolding molecules. This suggests that presynaptic EphrinBs might provide an anterograde signal that contributes to the organization of postsynaptic structures. Knock-out studies confirmed a critical role for EphB receptors in postsynaptic differentiation, since mice lacking multiple EphB receptors have reduced levels of postsynaptic NMDA-receptors and have dendritic spines with aberrant morphology (Grunwald et al. 2001; Henderson et al. 2001; Henkemeyer et al. 2003). Earlier in development during axon guidance, EphB receptor/EphrinB complexes mediate retrograde signaling into the EphrinB expressing cell (Cowan and Henkemeyer 2002). Analogous functions for presynaptic Ephrins at the synapse might provide a mechanism for bidirectional signaling that coordinates the differentiation of the pre- and postsynaptic specializations.

Finally, work at the neuromuscular junction suggests that extracellular interactions may also contribute to the recruitment of presynaptic ion channels. The pore-forming $\alpha 1$ subunits of Cav2.1 and Cav2.2, the P/Q- and N-type voltage-gated calcium channels have been shown to directly interact with the extracellular matrix protein laminin-β_2

(Nishimune et al. 2004). This interaction is required for the normal assembly of presynaptic active zones and the recruitment of synaptic vesicles to motor neuron terminals. This represents an additional example of how extracellular interactions with a synaptic ion channel promote channel recruitment or retention in the synaptic membrane. Further work will be required to determine whether a similar mechanism mediates voltage-gated calcium channel recruitment at neuron-neuron synapses.

C. Nucleation of Synaptic Scaffolds by Adhesion Molecules

The mechanical stability of synapses and the precise apposition of pre- and postsynaptic membrane domains implies that adhesion molecules might bridge the synaptic cleft and thereby keep both sides of the synapse in register. Ultrastructural studies revealed the presence of different sites of cell-cell adhesion at the synapse— including the linkage across the synaptic cleft, as well as at puncta adherentia that flank the synaptic cleft. A third site of cellular adhesion at central synapses are the end feet of glial cells that tightly enclose the synaptic junctions.

Most attention in recent years has been focused on neuronal adhesion molecules that directly bridge the synaptic cleft and that might instruct the assembly of synaptic structures. Several such adhesion molecules have been identified. The synaptogenic activity of these adhesive factors is thought to rely on the nucleation of cytoplasmic scaffolds at the pre- and postsynaptic side, which subsequently facilitate the recruitment of other cytoplasmic and cell surface molecules to an incipient synaptic connection.

In vitro experiments with cultured neurons revealed that exposing axonal or dendritic processes to specific adhesive triggers is sufficient to induce pre- and postsynaptic structures. One extensively studied adhesion system with synapse-inducing function is the neuroligin-neurexin complex. Neuroligins and neurexins are heterophilic adhesion proteins that form a trans-synaptic complex (Ichtchenko et al. 1995; Nguyen and Südhof 1997; Song et al. 1999). Cytoplasmic sequences in neuroligin-1 target this protein exclusively to the synaptic sites in dendrites and ensure exclusion from the axons (Dresbach et al. 2004; Iida et al. 2004; Rosales et al. 2005). This strict polarity of neuroligin-1 provides this heterophilic adhesion system with the required directionality for the assembly of asymmetric synaptic structures. Contact of axons with neuroligin-1 ectopically expressed in a nonneuronal cell induces clustering of axonal neurexins and the cytoplasmic scaffolding molecule CASK (Scheiffele et al. 2000; Dean et al. 2003). These neuroligin-induced axonal contacts then mature into functional presynaptic release sites, including active zone components and clusters of synaptic vesicles. Conversely, contact of dendrites with neurexin-1 expressing cells leads to the formation of neuroligin clusters and postsynaptic structures, such as the accumulation of the scaffolding protein PSD95 and NMDA receptors (Graf et al. 2004). These experiments suggest that the neuroligin-neurexin adhesion system can nucleate pre- and postsynaptic structures. To a large extent these structures are thought to assemble around scaffolding components that bind to the cytoplasmic tails of neuroligins and neurexins (see Figure 4.1). For example, postsynaptic neuroligin-1 binds to PSD95, which in turn can recruit NMDA-receptors and other scaffolding components such as GKAP, Homer, and Shank (Irie et al. 1997; Meyer et al. 2004; Prange et al. 2004; Chih et al. 2005).

Interestingly, the neuroligin-2 isoform, which differs in its cytoplasmic sequences from neuroligin-1, is found primarily at GABAergic synapses. Extracellular aggregation of neuroligin-2 results in not only the recruitment of PSD95, but also of gephrin, a scaffolding protein specific to inhibitory synapses (Graf et al. 2004; Varoqueaux et al. 2004). Selective interaction of excitatory and inhibitory postsynaptic scaffolding proteins with different neuroligin cytoplasmic tail sequences therefore might contribute to the nucleation of specific postsynaptic scaffolds. However, interactions between extracellular domains that may contribute to this process have not been ruled out.

Similar experiments revealed synaptogenic activities for the cell adhesion molecule SynCAM, a homophilic Ig-domain protein (Biederer et al. 2002). Interestingly, SynCAM and neuroligin-1 differ in their cytoplasmic interactions with scaffolding proteins. These differences may explain the differential recruitment of NMDA and AMPA-type glutamate receptors to neuroligin-1- and SynCAM-induced synapses, respectively (Sara et al. 2005). Neuroligin-1 recruits PSD95 and NMDA-receptors through the type I PDZ-binding motif in neuroligin, whereas SynCAM contains a type II PDZ-binding motif that contributes to the recruitment of functional AMPA-receptors to synapses. The importance of these selective scaffolding interactions was confirmed in experiments showing that a chimaeric protein, which includes neuroligin-1 containing the cytoplasmic sequences of SynCAM, is capable of stimulating AMPA receptor recruitment (Sara et al. 2005).

Another presumptive cell adhesion molecule that has been specifically linked to the synaptic recruitment of AMPA receptors through interaction with scaffolding molecules is Dasm1 (Shi et al. 2004). However, the extracellular binding partners of this protein are currently unknown.

It is likely that several adhesion molecules that act through mechanisms similar to the neuroligin-neurexin complex and SynCAM will cooperate in the nucleation of synaptic scaffolds and the subsequent recruitment of neurotransmitter receptors. Candidate molecules include the nectins, a family of homophilic Ig-domain proteins that are similar to SynCAM (Satoh-

Horikawa et al. 2000; Mizoguchi et al. 2002). The integration of multiple adhesive signals likely occurs at the level of the cytoplasmic scaffold. There is substantial crosstalk between different scaffolding components; for example, the CASK/Mint/MALS complex that binds to neurexins is linked to beta-catenin, the cytoplasmic binding partner of cadherins (see Figure 4.1; Hata et al. 1996; Perego et al. 2000; Bamji et al. 2003). Such crosstalk might be employed for the sequential recruitment of different adhesive factors to a growing adhesion site as has been observed for cadherins and nectins in epithelial cells (Takahashi et al. 1999; Tachibana et al. 2000).

IV. SYNAPTIC SPECIFICITY AND MOLECULAR DIVERSITY OF CELL SURFACE RECEPTORS

The studies just discussed have provided insights into the cell biological mechanisms that might contribute to the assembly and maturation of synaptic structures. Considering that a single neuron can receive as many as 10,000 synapses from several different presynaptic partners, the question arises as to how selective connections with different synaptic partners are specified. Studies on developing motor neurons revealed that neurons are specified with regard to their connectivity pattern before the cells extend axons (Jessell 2000; Landmesser 2001). A large part of this specification is encoded by combinations of transcription factors that are expressed in subsets of neuronal populations. However, the cell surface molecules that impose specificity on the process of synapse formation have not been identified yet.

Cell adhesion molecules represent excellent candidate molecules for directing the specificity of synaptic connectivity. Neurofascin, a member of the L1-family of adhesion molecules, was shown to determine sites of synapse formation of basket cell axons on the axon initial segment of Purkinje cells (Ango et al. 2004). Another class of adhesion molecules that has been proposed to underlie selective synaptic connectivity in the CNS are the cadherins. Different isoforms of type I cadherins are selectively localized to subsets of synapses where they mediate homophilic adhesion between pre- and postsynaptic partners (Fannon and Colman 1996; Huntley and Benson 1999). Interestingly, afferent and target cells often express the same cadherin isoforms, suggesting that homophilic cadherin-mediated adhesion may contribute to axon pathfinding (Takeichi 1988; Obst-Pernberg and Redies 1999; Shapiro and Colman 1999).

Considering the diversity of neuronal cell types engaged in selective cell-cell interactions, it has been hypothesized that large families of adhesion molecules might generate an adhesive code that specifies the identity of individual neurons and their synaptic interactions. In the vertebrate CNS, two gene families have received particular attention due to the large number of variants that they encode: protocadherins and neurexins (Serafini 1999). Neurexins were the first family of highly polymorphic neuronal cell surface receptors to be identified (Ushkaryov et al. 1992). More than 1000 neurexin variants are generated from three neurexin genes through transcription from alternative promoters and alternative splicing (Missler et al. 1998a). Importantly, the diversity of neurexins is restricted to their extracellular domains, whereas the cytoplasmic tails are highly conserved between isoforms. Binding of neuroligins, neurexin's partners in heterophilic adhesion, as well as binding of other ligands to the extracellular domain of neurexin, is neurexin isoform-specific (Ichtchenko et al. 1995; Missler et al. 1998b; Sugita et al. 2001; Boucard et al. 2005). Thus, selective interactions through the extracellular domains of adhesion molecules are coupled to a common intracellular pathway at synapses. Neurexin isoforms show differential expression in various neuronal populations (Püschel and Betz 1995; Ullrich et al. 1995), but it is unclear how many variants are expressed in a single neuron and how many neurexin variants coexist at individual synapses.

The second family of highly diverse adhesion molecules are the protocadherins. Three clusters of genes, termed alpha, beta, and gamma, encode 58 protocadherin genes that differ in their extracellular domains but have related intracellular sequences (Kohmura et al. 1998; Wu and Maniatis 1999; 2000). Gamma protocadherins are localized to neuronal synapses where they may mediate homophilic adhesion, although heterophilic interactions with other ligands have been suggested (Phillips et al. 2003; Frank et al. 2005). Deletion of the entire gamma protocadherin gene cluster results in postnatal lethality and loss of synapses, demonstrating that this gene family is critically important for synapse formation and/or maintenance (Wang et al. 2002; Weiner et al. 2005). One remarkable finding made in the analysis of alpha protocadherin isoform expression was that subsets of neurons of the same cell type differed in their protocadherin expression profiles (Kohmura et al. 1998). Moreover, expression of variable exons in single cells was monoallelic (Esumi et al. 2005). Therefore, the molecular diversity of protocadherin cell surface receptors reveals a significant heterogeneity of neuronal populations that might underlie selective cell-cell interactions. Though further studies are required to clarify the distribution and function of individual protocadherin isoforms, it is intriguing to speculate that functionally different subsets of neuronal populations might be recruited into specific neuronal circuits based on their protocadherin repertoire.

V. CONCLUDING REMARKS

In recent years, several exciting advances have led to the identification of a number of molecular mechanisms that can contribute to synapse assembly. Although this review focused exclusively on neuron-derived synaptic factors, it should be pointed out that a role for glial-derived factors as synaptogenic signals is also emerging (Pfrieger and Barres 1997; Murai et al. 2003; Christopherson et al. 2005). The challenge for the near future will be to identify how the various signaling mechanisms are employed and regulated during synapse formation and plasticity *in vivo*. Synapses may rely on different cell signaling systems, or many systems may cooperate at single synapses. One limiting factor for such analysis has been the availability of specific labeling techniques for identified synapses *in vivo*. However, recent advances in cell-specific transgenic approaches and the increase in characterized markers has begun to overcome these issues (Leighton et al. 2001; Gong et al. 2003). Understanding the hierarchy of cellular signaling that determines the appropriate wiring diagram, with respect to the integration of different molecular signals, as well as the interplay of cell intrinsic programs and the activity-dependent synaptic competition, will be a future challenge.

References

Alsina, B., Vu, T., Cohen-Cory, S. (2001). Visualizing synapse formation in arborizing optic axons in vivo: Dynamics and modulation by BDNF. *Nat. Neurosci.* **4**, 1093–1101.

Ango, F., di Cristo, G., Higashiyama, H., Bennett, V., Wu, P., Huang, Z.J. (2004). Ankyrin-based subcellular gradient of neurofascin, an immunoglobulin family protein, directs GABAergic innervation at purkinje axon initial segment. *Cell* **119**, 257–272.

Bamji, S.X., Shimazu, K., Kimes, N., Huelsken, J., Birchmeier, W., Lu, B., Reichardt, L.F. (2003). Role of beta-catenin in synaptic vesicle localization and presynaptic assembly. *Neuron* **40**, 719–731.

Biederer, T., Sara, Y., Mozhayeva, M., Atasoy, D., Liu, X., Kavalali, E.T., Südhof, T.C. (2002). SynCAM, a synaptic adhesion molecule that drives synapse assembly. *Science* **297**, 1525–1531.

Boucard, A.A., Chubykin, A.A., Comoletti, D., Taylor, P., Südhof, T.C. (2005). A splice code for trans-synaptic cell adhesion mediated by binding of neuroligin 1 to alpha- and beta-neurexins. *Neuron* **48**, 229–236.

Butz, S., Okamoto, M., Südhof, T.C. (1998). A tripartite protein complex with the potential to couple synaptic vesicle exocytosis to cell adhesion in brain. *Cell* **94**, 773–782.

Chih, B., Engelman, H., Scheiffele, P. (2005). Control of excitatory and inhibitory synapse formation by neuroligins. *Science* **307**, 1324–1328.

Christopherson, K.S., Ullian, E.M., Stokes, C.C., Mullowney, C.E., Hell, J.W., Agah, A. et al. (2005). Thrombospondins are astrocyte-secreted proteins that promote CNS synaptogenesis. *Cell* **120**, 421–433.

Ciani, L., Krylova, O., Smalley, M.J., Dale, T.C., Salinas, P.C. (2004). A divergent canonical WNT-signaling pathway regulates microtubule dynamics: disheveled signals locally to stabilize microtubules. *J. Cell Biol.* **164**, 243–253.

Cowan, C.A., Henkemeyer, M. (2002). Ephrins in reverse, park and drive. *Trends Cell Biol.* **12**, 339–346.

Craig, A.M., Banker, G. (1994). Neuronal polarity. *Annu. Rev. Neurosci.* **17**, 267–310.

Dai, Z., Peng, H.B. (1993). Elevation in presynaptic Ca^{2+} level accompanying initial nerve-muscle contact in tissue culture. *Neuron* **10**, 827–837.

Dai, Z., Peng, H.B. (1995). Presynaptic differentiation induced in cultured neurons by local application of basic fibroblast growth factor. *J. Neurosci.* **15**, 5466–5475.

Dalva, M.B., Takasu, M.A., Lin, M.Z., Shamah, S.M., Hu, L., Gale, N.W., Greenberg, M.E. (2000). EphB receptors interact with NMDA receptors and regulate excitatory synapse formation. *Cell* **103**, 945–956.

Dean, C., Scholl, F.G., Choih, J., DeMaria, S., Berger, J., Isacoff, E., Scheiffele, P. (2003). Neurexin mediates the assembly of presynaptic terminals. *Nat. Neurosci.* **6**, 708–716.

Dresbach, T., Neeb, A., Meyer, G., Gundelfinger, E.D., Brose, N. (2004). Synaptic targeting of neuroligin is independent of neurexin and SAP90/PSD95 binding. *Mol. Cell. Neurosci.* **27**, 227–235.

Esumi, S., Kakazu, N., Taguchi, Y., Hirayama, T., Sasaki, A., Hirabayashi, T. et al. (2005). Monoallelic yet combinatorial expression of variable exons of the protocadherin-alpha gene cluster in single neurons. *Nat. Genet.* **37**, 171–176.

Evers, J., Laser, M., Sun, Y.A., Xie, Z.P., Poo, M.M. (1989). Studies of nerve-muscle interactions in Xenopus cell culture: Analysis of early synaptic currents. *J. Neurosci.* **9**, 1523–1539.

Fannon, A.M., Colman, D.R. (1996). A model for central synaptic junctional complex formation based on the differential adhesive specificities of the cadherins. *Neuron* **17**, 423–434.

Fiala, J.C., Feinberg, M., Popov, V., Harris, K.M. (1998). Synaptogenesis via dendritic filopodia in developing hippocampal area CA1. *J. Neurosci.* **18**, 8900–8911.

Frank, M., Ebert, M., Shan, W., Phillips, G.R., Arndt, K., Colman, D.R., Kemler, R. (2005). Differential expression of individual gamma-protocadherins during mouse brain development. *Mol. Cell. Neurosci.* **29**, 603–616.

Friedman, H.V., Bresler, T., Garner, C.C., Ziv, N.E. (2000). Assembly of new individual excitatory synapses: Time course and temporal order of synaptic molecule recruitment. *Neuron* **27**, 57–69.

Gong, S., Zheng, C., Doughty, M.L., Losos, K., Didkovsky, N., Schambra, U.B. et al. (2003). A gene expression atlas of the central nervous system based on bacterial artificial chromosomes. *Nature* **425**, 917–925.

Graf, E.R., Zhang, X., Jin, S.X., Linhoff, M.W., Craig, A.M. (2004). Neurexins induce differentiation of GABA and glutamate postsynaptic specializations via neuroligins. *Cell* **119**, 1013–1026.

Grindstaff, K.K., Yeaman, C., Anandasabapathy, N., Hsu, S.C., Rodriguez-Boulan, E., Scheller, R.H., Nelson, W.J. (1998). Sec6/8 complex is recruited to cell-cell contacts and specifies transport vesicle delivery to the basal-lateral membrane in epithelial cells. *Cell* **93**, 731–740.

Grunwald, I.C., Korte, M., Wolfer, D., Wilkinson, G.A., Unsicker, K., Lipp, H.P. et al. (2001). Kinase-independent requirement of EphB2 receptors in hippocampal synaptic plasticity. *Neuron* **32**, 1027–1040.

Hall, A.C., Lucas, F.R., Salinas, P.C. (2000). Axonal remodeling and synaptic differentiation in the cerebellum is regulated by WNT-7a signaling. *Cell* **100**, 525–535.

Hata, Y., Butz, S., Südhof, T.C. (1996). CASK: A novel dlg/PSD95 homolog with an N-terminal calmodulin-dependent protein kinase domain identified by interaction with neurexins. *J. Neurosci.* **16**, 2488–2494.

Henderson, J.T., Georgiou, J., Jia, Z., Robertson, J., Elowe, S., Roder, J.C., Pawson, T. (2001). The receptor tyrosine kinase EphB2 regulates NMDA-dependent synaptic function. *Neuron* **32**, 1041–1056.

Henkemeyer, M., Itkis, O.S., Ngo, M., Hickmott, P.W., Ethell, I.M. (2003). Multiple EphB receptor tyrosine kinases shape dendritic spines in the hippocampus. *J. Cell Biol.* **163**, 1313–1326.

Hua, J.Y., Smith, S.J. (2004). Neural activity and the dynamics of central nervous system development. *Nat. Neurosci.* **7**, 327–332.

Huntley, G.W., Benson, D.L. (1999). Neural (N)-cadherin at developing thalamocortical synapses provides an adhesion mechanism for the formation of somatopically organized connections. *J. Comp. Neurol.* **407**, 453–471.

Ichtchenko, K., Hata, Y., Nguyen, T., Ullrich, B., Missler, M., Moomaw, C., Südhof, T.C. (1995). Neuroligin 1: A splice site-specific ligand for beta-neurexins. *Cell* **81**, 435–443.

Iida, J., Hirabayashi, S., Sato, Y., Hata, Y. (2004). Synaptic scaffolding molecule is involved in the synaptic clustering of neuroligin. *Mol. Cell. Neurosci.* **27**, 497–508.

Irie, M., Hata, Y., Takeuchi, M., Ichtchenko, K., Toyoda, A., Hirao, K. et al. (1997). Binding of neuroligins to PSD-95. *Science* **277**, 1511–1515.

Jessell, T.M. (2000). Neuronal specification in the spinal cord: Inductive signals and transcriptional codes. *Nat. Rev. Genet.* **1**, 20–29.

Jin, Y. (2002). Synaptogenesis: Insights from worm and fly. *Curr. Opin. Neurobiol.* **12**, 71–79.

Jontes, J.D., Buchanan, J., Smith, S.J. (2000). Growth cone and dendrite dynamics in zebrafish embryos: early events in synaptogenesis imaged in vivo. *Nat. Neurosci.* **3**, 231–237.

Kaech, S.M., Whitfield, C.W., Kim, S.K. (1998). The LIN-2/LIN-7/LIN-10 complex mediates basolateral membrane localization of the *C. elegans* EGF receptor LET-23 in vulval epithelial cells. *Cell* **94**, 761–771.

Kim, E., Sheng, M. (2004). PDZ domain proteins of synapses. *Nat. Rev. Neurosci.* **5**, 771–781.

Kohmura, N., Senzaki, K., Hamada, S., Kai, N., Yasuda, R., Watanabe, M. et al. (1998). Diversity revealed by a novel family of cadherins expressed in neurons at a synaptic complex. *Neuron* **20**, 1137–1151.

Krylova, O., Messenger, M.J., Salinas, P.C. (2000). Dishevelled-1 regulates microtubule stability: A new function mediated by glycogen synthase kinase-3beta. *J. Cell Biol.* **151**, 83–94.

Landmesser, L.T. (2001). The acquisition of motoneuron subtype identity and motor circuit formation. *Int. J. Dev. Neurosci.* **19**, 175–182.

Leighton, P.A., Mitchell, K.J., Goodrich, L.V., Lu, X., Pinson, K., Scherz, P. et al. (2001). Defining brain wiring patterns and mechanisms through gene trapping in mice. *Nature* **410**, 174–179.

Mason, C.A., Morrison, M.E., Ward, M.S., Zhang, Q., Baird, D.H. (1997). Axon-target interactions in the developing cerebellum. *Perspect. Dev. Neurobiol.* **5**, 69–82.

Meyer, G., Varoqueaux, F., Neeb, A., Oschlies, M., Brose, N. (2004). The complexity of PDZ domain-mediated interactions at glutamatergic synapses: A case study on neuroligin. *Neuropharmacology* **47**, 724–733.

Mi, R., Tang, X., Sutter, R., Xu, D., Worley, P., O'Brien, R.J. (2002). Differing mechanisms for glutamate receptor aggregation on dendritic spines and shafts in cultured hippocampal neurons. *J. Neurosci.* **22**, 7606–7616.

Missler, M., Fernandez-Chacon, R., Südhof, T.C. (1998a). The making of neurexins. *J. Neurochem.* **71**, 1339–1347.

Missler, M., Hammer, R.E., Südhof, T.C. (1998b). Neurexophilin binding to alpha-neurexins. A single LNS domain functions as an independently folding ligand-binding unit. *J. Biol. Chem.* **273**, 34716–34723.

Mizoguchi, A., Nakanishi, H., Kimura, K., Matsubara, K., Ozaki-Kuroda, K., Katata, T. et al. (2002). Nectin: An adhesion molecule involved in formation of synapses. *J. Cell Biol.* **156**, 555–565.

Murai, K.K., Nguyen, L.N., Irie, F., Yamaguchi, Y., Pasquale, E.B. (2003). Control of hippocampal dendritic spine morphology through ephrin-A3/EphA4 signaling. *Nat. Neurosci.* **6**, 153–160.

Nguyen, T., Südhof, T.C. (1997). Binding properties of neuroligin 1 and neurexin 1beta reveal function as heterophilic cell adhesion molecules. *J. Biol. Chem.* **272**, 26032–26039.

Niell, C.M., Meyer, M.P., Smith, S.J. (2004). In vivo imaging of synapse formation on a growing dendritic arbor. *Nat. Neurosci.* **7**, 254–260.

Nishimune, H., Sanes, J.R., Carlson, S.S. (2004). A synaptic laminin-calcium channel interaction organizes active zones in motor nerve terminals. *Nature* **432**, 580–587.

O'Brien, R., Xu, D., Mi, R., Tang, X., Hopf, C., Worley, P. (2002). Synaptically targeted narp plays an essential role in the aggregation of AMPA receptors at excitatory synapses in cultured spinal neurons. *J. Neurosci.* **22**, 4487–4498.

O'Brien, R.J., Xu, D., Petralia, R.S., Steward, O., Huganir, R.L., Worley, P. (1999). Synaptic clustering of AMPA receptors by the extracellular immediate-early gene product Narp. *Neuron* **23**, 309–323.

Obst-Pernberg, K., Redies, C. (1999). Cadherins and synaptic specificity. *J. Neurosci. Res.* **58**, 130–138.

Olsen, O., Moore, K.A., Fukata, M., Kazuta, T., Trinidad, J.C., Kauer, F.W. et al. (2005). Neurotransmitter release regulated by a MALS-liprin-alpha presynaptic complex. *J. Cell Biol.* **170**, 1127–1134.

Passafaro, M., Nakagawa, T., Sala, C., Sheng, M. (2003). Induction of dendritic spines by an extracellular domain of AMPA receptor subunit GluR2. *Nature* **424**, 677–681.

Perego, C., Vanoni, C., Massari, S., Longhi, R., Pietrini, G. (2000). Mammalian LIN-7 PDZ proteins associate with beta-catenin at the cell-cell junctions of epithelia and neurons. *EMBO J.* **19**, 3978–3989.

Pfrieger, F.W., Barres, B.A. (1997). Synaptic efficacy enhanced by glial cells in vitro. *Science* **277**, 1684–1687.

Phillips, G.R., Tanaka, H., Frank, M., Elste, A., Fidler, L., Benson, D.L., Colman, D.R. (2003). Gamma-protocadherins are targeted to subsets of synapses and intracellular organelles in neurons. *J. Neurosci.* **23**, 5096–5104.

Prange, O., Wong, T.P., Gerrow, K., Wang, Y.T., El-Husseini, A. (2004). A balance between excitatory and inhibitory synapses is controlled by PSD-95 and neuroligin. *Proc. Natl. Acad. Sci. U.S.A.* **101**, 13915–13920.

Püschel, A., Betz, H. (1995). Neurexins are differentially expressed in the embryonic nervous system of mice. *J. Neurosci.* **15**, 2849–2856.

Rongo, C., Whitfield, C.W., Rodal, A., Kim, S.K., Kaplan, J.M. (1998). LIN-10 is a shared component of the polarized protein localization pathways in neurons and epithelia. *Cell* **94**, 751–759.

Rosales, C.R., Osborne, K.D., Zuccarino, G.V., Scheiffele, P., Silverman, M.A. (2005). A cytoplasmic motif targets neuroligin-1 exclusively to dendrites of cultured hippocampal neurons. *Eur. J. Neurosci.* **22**, 2381–2386.

Sanes, J.R., Lichtman, J.W. (1999). Development of the vertebrate neuromuscular junction. *Annu. Rev. Neurosci.* **22**, 389–442.

Sans, N., Prybylowski, K., Petralia, R.S., Chang, K., Wang, Y.X., Racca, C. et al. (2003). NMDA receptor trafficking through an interaction between PDZ proteins and the exocyst complex. *Nat. Cell Biol.* **5**, 520–530.

Sara, Y., Biederer, T., Atasoy, D., Chubykin, A., Mozhayeva, M.G., Südhof, T.C., Kavalali, E.T. (2005). Selective capability of SynCAM and neuroligin for functional synapse assembly. *J. Neurosci.* **25**, 260–270.

Satoh-Horikawa, K., Nakanishi, H., Takahashi, K., Miyahara, M., Nishimura, M., Tachibana, K. et al. (2000). Nectin-3, a new member of immunoglobulin-like cell adhesion molecules that shows homophilic and heterophilic cell-cell adhesion activities. *J. Biol. Chem.* **275**, 10291–10299.

Scheiffele, P. (2003). Cell-cell signaling during synapse formation in the CNS. *Annu. Rev. Neurosci.* **26**, 485–508.

Scheiffele, P., Fan, J., Choih, J., Fetter, R., Serafini, T. (2000). Neuroligin expressed in nonneuronal cells triggers presynaptic development in contacting axons. *Cell* **101**, 657–669.

Serafini, T. (1999). Finding a partner in a crowd: neuronal diversity and synaptogenesis. *Cell* **98**, 133–136.

Serra-Pages, C., Medley, Q.G., Tang, M., Hart, A., Streuli, M. (1998). Liprins, a family of LAR transmembrane protein-tyrosine phosphatase-interacting proteins. *J. Biol. Chem.* **273**, 15611–15620.

Shapiro, L., Colman, D.R. (1999). The diversity of cadherins and implications for a synaptic adhesive code in the CNS. *Neuron* **23**, 427–430.

Shi, S.H., Cheng, T., Jan, L.Y., Jan, Y.N. (2004). The immunoglobulin family member dendrite arborization and synapse maturation 1 (Dasm1) controls excitatory synapse maturation. *Proc. Natl. Acad. Sci. U.S.A.* **101**, 13346–13351.

Song, J.Y., Ichtchenko, K., Südhof, T.C., Brose, N. (1999). Neuroligin 1 is a postsynaptic cell-adhesion molecule of excitatory synapses. *Proc. Natl. Acad. Sci. U.S.A.* **96**, 1100–1105.

Sugita, S., Saito, F., Tang, J., Satz, J., Campbell, K., Südhof, T.C. (2001). A stoichiometric complex of neurexins and dystroglycan in brain. *J. Cell Biol.* **154**, 435–445.

Tachibana, K., Nakanishi, H., Mandai, K., Ozaki, K., Ikeda, W., Yamamoto, Y. et al. (2000). Two cell adhesion molecules, nectin and cadherin, interact through their cytoplasmic domain-associated proteins. *J. Cell Biol.* **150**, 1161–1176.

Takahashi, K., Nakanishi, H., Miyahara, M., Mandai, K., Satoh, K., Satoh, A. et al. (1999). Nectin/PRR: An immunoglobulin-like cell adhesion molecule recruited to cadherin-based adherens junctions through interaction with Afadin, a PDZ domain-containing protein. *J. Cell Biol.* **145**, 539–549.

Takasu, M.A., Dalva, M.B., Zigmond, R.E., Greenberg, M.E. (2002). Modulation of NMDA receptor-dependent calcium influx and gene expression through EphB receptors. *Science* **295**, 491–495.

Takeichi, M. (1988). The cadherins: Cell-cell adhesion molecules controlling animal morphogenesis. *Development* **102**, 639–655.

Ullrich, B., Ushkaryov, Y.A., Südhof, T.C. (1995). Cartography of neurexins: More than 1000 isoforms generated by alternative splicing and expressed in distinct subsets of neurons. *Neuron* **14**, 497–507.

Umemori, H., Linhoff, M.W., Ornitz, D.M., Sanes, J.R. (2004). FGF22 and its close relatives are presynaptic organizing molecules in the mammalian brain. *Cell* **118**, 257–270.

Ushkaryov, Y.A., Petrenko, A.G., Geppert, M., Südhof, T.C. (1992). Neurexins: Synaptic cell surface proteins related to the alpha-latrotoxin receptor and laminin. *Science* **257**, 50–56.

Varoqueaux, F., Jamain, S., Brose, N. (2004). Neuroligin 2 is exclusively localized to inhibitory synapses. *Eur. J. Cell Biol.* **83**, 449–456.

Waites, C.L., Craig, A.M., Garner, C.C. (2005). Mechanisms of vertebrate synaptogenesis. *Annu. Rev. Neurosci.* **28**, 251–274.

Wang, X., Weiner, J.A., Levi, S., Craig, A.M., Bradley, A., Sanes, J.R. (2002). Gamma protocadherins are required for survival of spinal interneurons. *Neuron* **36**, 843–854.

Weiner, J.A., Wang, X., Tapia, J.C., Sanes, J.R. (2005). Gamma protocadherins are required for synaptic development in the spinal cord. *Proc. Natl. Acad. Sci. U.S.A.* **102**, 8–14.

Wu, Q., Maniatis, T. (1999). A striking organization of a large family of human neural cadherin-like cell adhesion genes. *Cell* **97**, 779–790.

Wu, Q., Maniatis, T. (2000). Large exons encoding multiple ectodomains are a characteristic feature of protocadherin genes. *Proc. Natl. Acad. Sci. U.S.A.* **97**, 3124–3129.

Xie, Z.P., Poo, M.M. (1986). Initial events in the formation of neuromuscular synapse: Rapid induction of acetylcholine release from embryonic neuron. *Proc. Natl. Acad. Sci. U.S.A.* **83**, 7069–7073.

Yoshihara, M., Rheuben, M.B., Kidokoro, Y. (1997). Transition from growth cone to functional motor nerve terminal in Drosophila embryos. *J. Neurosci.* **17**, 8408–8426.

Zhai, R.G., Vardinon-Friedman, H., Cases-Langhoff, C., Becker, B., Gundelfinger, E.D., Ziv, N.E., Garner, C.C. (2001). Assembling the presynaptic active zone: A characterization of an active one precursor vesicle. *Neuron* **29**, 131–143.

5

Presynaptic Terminal Differentiation

MEI ZHEN

Chemical synapses mediate a majority of the communication in the nervous system (Katz 1969). Synapse formation, therefore, is essential for the establishment of neural circuits that are responsible for physical and cognitive functions. Identifying molecular mechanisms that underlie synapse development will provide key insight, not only into normal development, but also the potential cause for the onset of mental illness (Couteaux and Pecot-Dechavassine 1970; Katz 1969).

Protein components of chemical synapses have been identified in both vertebrate and invertebrate nervous systems.

Although each model system has advantages in the experimental approaches that can be used (e.g., biochemistry in vertebrates versus genetics in invertebrates), many common and conserved components have been discovered. This convergence provides a strong basis for guided cross-examination of the functional conservation between components of vertebrate and invertebrate synapses. It also serves as a tool to help unravel common molecular mechanisms that govern synapse development, and in addition, allows a comparison of the difference between vertebrate and invertebrate synapse development and function. This chapter will focus on mechanisms that regulate two aspects of presynaptic terminal differentiation in vertebrates and invertebrates: *active zone assembly* and *presynaptic growth*. It will also provide a future perspective on how to bridge the knowledge obtained from vertebrate and invertebrate synapses by various experimental approaches.

I. MORPHOLOGY OF THE PRESYNAPTIC TERMINI OF VERTEBRATE AND INVERTEBRATE SYNAPSES

The ultrastructure of chemical synapses displays common basic features despite the morphological variations among animal species. In mature chemical synapses, dense populations of synaptic vesicle pools accumulate at presynaptic termini. These synaptic vesicles cluster around electron-dense regions that are tightly associated with the plasma membrane. These specialized structures, called presynaptic active zones, facilitate synaptic vesicles tethering and fusion at the plasma membrane to release neurotransmitters (Couteaux and Fessard 1975; Harlow et al. 2001; Rosenmund et al. 2003; Zhai and Bellen 2004). The presynaptic active zones are aligned precisely with the postsynaptic density (PSD), a morphological feature that promotes the efficacy and fidelity of synaptic transmission. Endocytosis and recycling of synaptic vesicles are postulated to take place at regions flanking the presynaptic active zones.

A major morphological variation for different types of synapses in various organisms is the shape and size of presynaptic active zones (Zhai and Bellen 2004), which ranges from train-track dense lines in mature frog neuromuscular junctions (NMJs) (Harlow et al. 2001; Heuser et al. 1974) to matrix/web-like short filamentous electron patches in mammalian cerebellum and hippocampal neurons (Akert et al.

1971; Landis et al. 1988; Pfenninger et al. 1972; Phillips et al. 2001). However, all presynaptic active zones can be dissected into common structural components: the plasma membrane region where the vesicle fusion occurs and the electron-dense cytomatrix that associates tightly with the plasma membranes and extends into the cytoplasm. This region is proposed to facilitate the tethering and docking of synaptic vesicles. The shape and size of the electron dense cytomatrix affects how synaptic vesicles organize around the active zones. In mammalian CNS neurons, synaptic vesicles fit snuggly among the "dents" in the web-like electron-dense matrix (Landis et al. 1988; Phillips et al. 2001). At *Drosophila* neuromuscular junctions, where active zones contain a T-bar shape extension, vesicles are clustered around the T-bar platform (Atwood et al. 1993). At frog neuromuscular junctions, two parallel rows of synaptic vesicles are docked along "train track"-like structures at the active zone (Harlow et al. 2001; Heuser et al. 1974).

II. PROTEIN COMPONENTS OF PRESYNAPTIC TERMINI

Extensive biochemical and genetic analysis of presynaptic proteins in vertebrates and invertebrates has established that various protein complexes function in regulating the exocytosis and endocytosis of synaptic vesicles (Sudhof 2004). Recent forward genetic screens and mutant analysis using invertebrate models are rapidly expanding the list of presynaptic proteins that regulate presynaptic differentiation (Broadie and Richmond 2002; Jin 2002; Marques 2005; Packard et al. 2003; Schaefer and Nonet 2001). The wealth of proteins invites further analysis of the isolated molecular mechanisms by which these proteins act, including their potential physical interactions with the established presynaptic machinery. It is important to point out that the components of some protein complexes are somewhat promiscuous and variable under different studying conditions, and we have focused on those proteins/complexes relevant to synaptogenesis and exocytosis in this chapter. Detailed discussion of protein machinery that regulates the recycling of synaptic vesicles can be found in Chapters 6, 7, and 14.

III. SNAREs

The core SNARE (soluble N-ethylmaleimide-sensitive component attachment protein receptor) complex is composed of 18–32 kDa membrane-associated proteins, including a vesicle SNARE synaptobrevin (VAMP) and two target SNAREs syntaxin and synaptosomal-associated protein-25 (SNAP-25). These SNARE proteins form a bundle of four α-helices at the synaptic terminal, and this complex formation is sufficient for membrane fusion, suggesting that this complex provides the driving force for the fusion between synaptic vesicles and plasma membrane at the active zones (Chen and Scheller 2001; Jahn et al. 2003; Jahn and Sudhof 1999) (see Chapter 6).

A. SNARE Regulating Proteins

Proteins that interact with the SNAREs and modulate SNARE complex formation include Synaptotagmin, Munc18/UNC-18, and Munc13/UNC-13. The vesicle membrane protein Synaptotagmin binds calcium, and physically interacts with syntaxin, SNAP-25, and calcium channels. It functions as a Ca^{2+} sensor that controls the SNARE complex assembly in the presence of calcium (additional reviews by Chapman 2002; Koh and Bellen 2003; Marek and Davis 2002; Maximov and Sudhof 2005; Tokuoka and Goda 2003). Munc13/UNC-13 promotes the SNARE complex formation through its physical interaction with syntaxin, which leads to a conformational change in syntaxin in favor of its assembly

into the SNARE complex (Aravamudan et al. 1999; Augustin et al. 1999; Brose et al. 2000; Richmond et al. 1999; 2001). Munc18/Unc-18, a member of the Sec1p protein family, binds syntaxin (Hata et al. 1993) as well as the Mint-Cask-Neurexin scaffolding complex (see later). Munc18 is required for the docking of dense-core vesicles in mouse chromaffin cells and synaptic vesicles at *Caenorhabditis elegans* synapses, as well as the fusion of synaptic vesicles of the mouse CNS (Verhage et al. 2000; Voets et al. 2001; Weimer et al. 2003). How Munc18/Unc-18 regulates SNARE complex during exocytosis is still under debate (Rizo and Sudhof 2002; Weimer and Richmond 2005).

B. Scaffolding Proteins

Scaffolding proteins form the organizing cytomatrix in the presynaptic terminal. One proposed complex contains Piccolo, Bassoon, RIM/UNC-10, CAST/ERC, and Liprin-α/SYD-2 (Ko et al. 2003; Zamorano and Garner 2001). This complex is tightly associated with the electron-dense regions of the active zones by immunoelectron microscopy studies, and is proposed to form the cytomatrix structure at the active zone. This cytomatrix creates an extremely tight and organized protein network at the active zone region. The MAL complex is composed of Veli/Lin-7, CASK/Lin-2, and Mint/Lin-10 (Butz et al. 1998; Kaech et al. 1998) and is thought to be the base of another scaffold. Both protein complexes have been reported to interact with proteins that regulate SNAREs, and may therefore organize the distribution of endocytosis and exocytosis machineries. For example, RIM/UNC-10 binds Munc13/UNC-13 (Schoch et al. 2002), and Mint-1 associates with Munc18-1 (Okamoto and Sudhof 1997). The MAL complex interacts with liprin-α/SYD-2 through CASK (Olsen et al. 2005), providing a potential molecular bridge for further intermolecular interactions among these protein complexes.

Genetic studies in invertebrate synapses, in particular *Drosophila* and *C. elegans*, have begun to reveal many new presynaptic proteins that control the growth and/or stabilization of presynaptic structures. Several cell surface signaling molecules such as the BMP molecule Gbb and TGF-β receptor Wit (Aberle et al. 2002; Marques et al. 2002; McCabe et al. 2003), the LAR-type receptor tyrosine phosphotases (Ackley et al. 2005; Dunah et al. 2005; Kaufmann et al. 2002), and the Wnt/Wg receptor Dfz2 (Packard et al. 2002), as well as various intracellular signaling molecules such as Ser/Thr kinase SAD-1/SAD-A/SAD-B (Crump et al. 2001; Kishi et al. 2005), GSK-3/Sgg kinase (Franco et al. 2004), LIM1 kinase (Eaton and Davis 2005), and p38 MAP kinase cascades (Nakata et al. 2005); scaffolding/adaptor proteins UNC-16/JSAP1/JIP3 (Byrd et al. 2001) and Nervous Wreck (Coyle et al. 2004); microtubule-binding protein Futsch (Hummel et al. 2000; Roos et al. 2000); and putative regulators for protein turnover such as late endosome-lysosomal protein Spinster (Sweeney and Davis 2002) and ubiquitin-ligases RPM-1/HIW (Schaefer et al. 2000; Wan et al. 2000; Zhen et al. 2000), FSN-1 (Liao et al. 2004), and APC complex (van Roessel et al. 2004) have been suggested to regulate various aspects of presynaptic development. For example, the size of presynaptic termini and the number and structures of active zones (TGF-β, Wnt and RPTP signaling, Spinster and E3 ligases), as well as vesicle transport and localization (SAD-1 and JSAP) may be controlled by these proteins. Although these proteins have been detected at presynaptic termini, their ultrastructural localization remains to be determined. Interestingly, some of these proteins, such as RPM-1, Nervous Wreck, and DLK, are enriched at presynaptic regions excluded from the active zones and synaptic vesicles (Coyle et al. 2004; Nakata et al. 2005; Wan et al. 2000; Wu et al. 2005; Zhen et al. 2000). FSN-1, although present broadly along the nerve processes, is also excluded from the active zones and associ-ated synaptic vesicles at the presynaptic terminal (Liao et al. 2004). This subpresynaptic region, termed loosely as *peri-active zones*, may include a domain at the presynaptic termini that regulate the development of synaptic structures (Sone et al. 2000). Defining the ultrastructural localization of these proteins and the physical and genetic interactions between these proteins and the existing synaptic protein complex will provide insights into the function of these proteins.

C. Developmental Regulation of Presynaptic Differentiation

Identifying signals that induce the differentiation and assembly of presynaptic structures and the molecular machinery that regulates this process, resulting in the maturation, elimination, and remodeling of nascent synapses, has expanded our understanding of synapse development. For the remainder of the chapter, we will focus on recent advances in two aspects of presynaptic differentiation: the assembly of active zones and regulation of synaptic growth.

The Assembly of Presynaptic Active Zones

The presynaptic active zone is essential for synaptic functions as it serves as release sites for synaptic vesicles. In cultured hippocampal neurons, the accumulation of active zone proteins at the synapse can be detected 30 minutes after the initial synaptic contact, preceding that of vesicles and postsynaptic proteins, suggesting that active zone formation also may play a vital role in stabilizing synaptic contacts prior to stable synapse formation (Friedman et al. 2000). Active zone assembly is a dynamic process regulated both by activity and development. In tetrad synapses of the *Drosophila* visual system, light stimulation induces changes in the number of presynaptic active zones (Brandstatter et al. 1999). Long-term strengthening or potentiation of *Drosophila* NMJs and mammalian hippocampal neurons results in the expansion of existing active zones, or the addition of new active zones (Reiff et al. 2002; Weeks et al. 2000). Ultrastructural analysis from *Drosophila* and *Sarcophaga* also has suggested that the density of active zones is maintained during development (Meinertzhagen et al. 1998). Moreover, *C. elegans* GABAergic NMJs incorporate additional active zone components as they mature, implying an increase in active zone size and/or number (Yeh et al. 2005).

A limited number of presynaptic proteins are localized exclusively at the active zones (Dresbach et al. 2001; Garner et al. 2000; Zhen and Jin 2004). Most active zone proteins, such as P/Q and N-type Ca^{2+} channels (Kawasaki et al. 2004; Robitaille et al. 1993; Wachman et al. 2004), syntaxin (Bennett et al. 1992), RIM (Castillo et al. 2002; Koushika et al. 2001; Schoch et al. 2002), and Munc-13 (Aravamudan et al. 1999; Augustin et al. 1999; Richmond et al. 1999) are components or regulators for SNARE complex formation during exocytosis. Scaffolding proteins Bassoon, Piccolo, CAST/ELK, and Liprin-α/SYD-2 likely form a highly interactive protein complex-based matrix that connects these vesicle release machineries with the basic cytoskeleton networks composed of actin, tubulin, myosin, and spectrin. They likely facilitate exocytosis (Dick et al. 2003; Fenster et al. 2000; Ohtsuka et al. 2002) through their interactions with the SNARE complex, modulating proteins such as RIM and Munc-13.

Despite the interactions among various protein components of the active zone and their role in exocytosis, *in vivo* studies do not suggest a role for these interactions in active zone assembly. In contrast to the postsynaptic assembly, whereby the sequential recruitment of postsynaptic components through scaffolding proteins appears important (Bresler et al. 2004; Kim and Sheng 2004), deletion of most active zone proteins does not significantly alter active zone morphology or the incorpora-

tion of other active zone proteins *in vivo*. Mouse *Rim* and *Munc13-1* mutants display normal active zone morphology (Schoch et al. 2002). Bassoon knockout mice display normal active zone morphology in hippocampus and cerebellum (Altrock et al. 2003). In the retina, some photoreceptor synapses display an abnormal structure, where the electron-dense ribbon extension from the active zone fails to be anchored (Dick et al. 2003). However, the assembly of the electron-dense regions remains unaffected and the Bassoon-interacting proteins, RIM and Piccolo, remain localized to active zones.

One possible explanation for these observations is that there is functional compensation by related mouse homologous genes. Genetic analysis in *C. elegans* and *Drosophila* synapses, where most of these genes are present as single genes, also does not support a role in active zone assembly. *C. elegans unc-10* (RIM homologue) and *unc-13* mutants, as well as *Drosophila unc-13* mutants display normal morphology of active zones (Aravamudan et al. 1999; Koushika et al. 2001; Weimer et al. 2003). *C. elegans elk/cast* null mutants display no behavioral or synaptic transmission defects and the localization of other known active zone proteins (such as the CAST/ELK, UNC-13 and SYD-2 protein in *unc-10* mutants, and *vice versa*) remains unaffected (Deken et al. 2005), suggesting normal active zone assembly. SYD-2/liprin-α are the only active zone proteins that regulate the morphology of active zones in *Drosophila* and *C. elegans* (Kaufmann et al. 2002; Yeh et al. 2005; Zhen and Jin 1999). Vertebrate liprin-α physically can interact with active zone proteins CAST, RIM, and CASK (Ohtsuka et al. 2002; Olsen et al. 2005; Schoch et al. 2002). Loss of SYD-2/liprin-α function in *C. elegans* and *Drosophila* leads to abnormalities in active zone size, shape, and electron density (Kaufmann et al. 2002; Zhen and Jin 1999). Despite the morphological changes, UNC-10 and CAST, its interacting proteins, were localized properly at the active zone in *syd-2* mutants (Deken et al. 2005). Taken together with the synaptic transmission defects that are consistently observed in these mutants, these active zone protein components are most likely to be involved in synaptic transmission (e.g., tethering vesicles and promoting exocytosis), rather than recruiting other active zone proteins and establishing presynaptic active zones.

So what regulates the assembly of the active zone? A combination of biochemical analysis and imaging studies suggest that in cultured hippocampal neurons, most identified active zone proteins are pre-assembled and packaged in 80 nm dense-core vesicles that are delivered to the nascent synaptic sites (Ahmari et al. 2000; Shapira et al. 2003; Zhai et al. 2001). This unitary assembly model suggests a quantal deposition of active zone packets upon fusion of these specialized vesicles with the plasma membrane during synaptogenesis. Green fluorescent protein (GFP)-tagged Bassoon was used to visualize the transport of these active zone packets and active zone development in cultures of hippocampal neurons (Bresler et al. 2004; Shapira et al. 2003). Intriguingly, in early embryonic neurons, GFP-Bassoon packets are highly motile and dynamic, displaying high-speed retro- and anterotransport along the axons. These fluorescent packets become more stationary during the culturing of the early embryonic neurons. The fast dynamics of active zone protein packets suggests a mechanism of rapid active zone assembly in juvenile neurons undergoing active synaptogenesis and activity-dependent active zone modulation. A focus for future studies is to identify genes that regulate the biogenesis, packaging, and fusion of these active zone vesicles.

Recent development in vital active zone specific fluorescent markers holds great potential not only to visualize active zone development, but also to uncover the molecular mechanisms for active zone assembly. GFP-tagged Bassoon greatly facilitated the analysis of active zone assembly in vertebrate cultured neurons. GFP tagging of Cacophony, a calcium channel $\alpha 1$ subunit, allows the visualization of active zones at *Drosophila* NMJs (Kawasaki et al. 2004). GFP-tagged SYD-2/liprin-α, a marker that labels the presynaptic active zones in *C. elegans*, not only has allowed a direct visualization of active zone changes at *C. elegans* GABAergic NMJs during development, but also has led to the first genetic screen for mutations that affect the active zone (Yeh et al. 2005). Although the scale of the screen was small, 12 genetic loci that affect the distribution and morphology of the SYD-2 active zone marker were recovered. Most of the recovered mutants display no obvious axon outgrowth defects, suggesting that they likely represent genes specific for synapse development. The most intriguing mutants recovered from this screen are those preferentially affecting the active zone marker without significantly altering the synaptic vesicle pool morphology, as they likely represent genes that regulate recruitment of specific active zone components and active zone formation. Uncovering the molecular identity of these invertebrate genetic loci, followed by functional validation in vertebrates, will help to reveal key genetic pathways that control active zone development (Jin 2002).

Regulation of Presynaptic Growth— Involvement of Multiple Signaling Pathways

Genetic analysis of invertebrate and vertebrate synapses has implicated multiple signaling molecules in regulating the initiation and differentiation of presynaptic termini. In this section we discuss the roles of TGF-β and WNT signaling cascades, Lar-type receptor tyrosine phosphatases, and multiple intracellular kinases.

D. TGF-β Signaling

The transforming growth factor-β TGF-β superfamily proteins regulate many aspects of nervous system development including neuronal proliferation, differentiation, and apoptosis (Angley et al. 2003; Brionne et al. 2003; Farkas et al. 2003; Gomes et al. 2005; Lein et al. 2002; Rios et al. 2004). The role of TGF-β/BMP signaling at synaptic differentiation was first suggested by the expression of several BMPs and their receptors in specific neuronal populations and at NMJs (Packard et al. 2003; Shi and Massague 2003). Chronic exposure of cultured sympathetic neurons to some BMPs promotes dendrite growth (Charytoniuk et al. 2000; McLennan and Koishi 1994). BMP4 induces the upregulation of synaptic proteins neurexin and synaptotagmin in sympathetic neurons (Horbinski et al. 2002).

In the canonical TGF-β signaling pathway, TGF-β and the related bone morphogenetic proteins (BMPs) activate two receptor Ser/Thr kinases, type I and type II receptors. Activation of these receptors results in the phosphorylation of a family of cytoplasmic proteins called R-SMADs, which then associate with Co-SMADs. The complex translocates into the nucleus, and subsequently activates genes responsive to TGF-β. Noncanonical pathways where TGF-β activates small GTPases, MAPK (Erk, JNK, and p38) kinase pathways, and the cytoskeletal regulator LIM kinase, independently of the SMAD proteins, have also been described (Derynck and Zhang 2003; Feng and Derynck 2005).

The first direct demonstration of specific *in vivo* requirement of TGF-β/BMP signaling in presynaptic development and synaptic growth came from studies at *Drosophila* NMJs (Marques 2005; McCabe et al. 2004; Packard et al. 2003). Mutations in a *Drosophila* BMP-7 homologue *Gbb*, the type I TGF-β receptors, *Tkv* and *Sax*, a type II receptor *Wit*, as well as the intracellular signal transducers R-SMAD, *Mad* and Co-Smad *Med* cause similar abnormal phenotypes at *Drosophila* NMJs: smaller synapses with aberrant presynaptic morphology including the dissociation of pre- and postsynaptic membrane and floating active zones (Aberle et al. 2002; Marques et al.

2002; McCabe et al. 2004). The total number of synapses and active zones also is reduced during larval development. In particular, the type II receptor *Wit* is shown to be required at the presynaptic termini to restore the normal synapse size, and to allow the anchoring of electron dense T-bars at the active zone regions at the plasma membrane and the tight association of pre- and postsynaptic membrane. Its ligand, *Gbb,* is secreted from muscle fibers and functions as a retrograde signal to promote the development of normal presynaptic morphology. These results suggest that TGF-β/BMP genes form a signaling cascade that regulates presynaptic growth in *Drosophila.*

In addition to its effects on presynaptic morphology, *Gbb* also is required for synaptic transmission. Interestingly, *Gbb* is expressed by neurons, and its neuronal expression is required to restore the normal synaptic transmission at *Drosophila* NMJ (McCabe et al. 2003). Thus, neuronal *Gbb* may activate separate signaling cascades in motoneurons to control synaptic function. An alternative explanation came from a study suggesting that motoneuron-expressed *Gbb* is required for synaptic transmission between motoneurons and their innervating interneurons (Baines 2004). Thus, *Gbb* secreted from motoneurons may also function as a retrograde signal to induce presynaptic development of innervating interneurons. The failure in the interneuron synapses indirectly results in the functional silencing of a morphologically normal motoneuron presynaptic terminal. Another demonstrated role for *Gbb/Wit* retrograde signaling in the central nervous system is their requirement to express neuropeptide FMRFa in a subset of neuroendocrine Tv neurons (Allan et al. 2003; McCabe et al. 2003). Whereas *Wit* is required in Tv neurons for FMRFa expression, *Gbb* is provided by the neurohemal organ that receives the peptides.

TGF-β signaling in motoneurons also regulates NMJ stability in *Drosophila* (Eaton and Davis 2005). Mutations in genes that define the canonical *gbb/wit* retrograde TGF-β signaling pathway, *gbb*, *wit*, *tkv*, *mad*, and *medea* all lead to increased retraction events at the NMJs. However, *tkv*, *mad*, and *medea* mutants display less severe defects for synapse retraction when compared to *gbb* and *wit* mutants, suggesting that TGF-β also signals through a SMAD-independent pathway to regulate synapse stabilization. Eaton et al. suggests that this additional synapse-stabilization function is mediated through a cytoplasmic Ser/Thr kinase LIM1 kinase that is both present and required presynaptically to stabilize synapses (Eaton and Davis 2005).

LIMK1 physically interacts with a C-terminal region of the *Wit* receptor that is not required for SMAD-dependent TGF-β signaling (Eaton and Davis 2005; Foletta et al. 2003). *Drosophila* LIMK1 previously has been implicated in the growth cone motility during axon outgrowth of the central nervous system through phosphorylating and inactivating an actin depolymerizing factor, cofilin (Endo et al. 2003; Ng and Luo 2004). At the NMJs, however, the expression pattern of LIMK1 does not support any strong association with synaptic actin, and the requirement of LIMK1 in synapse stabilization does not seem to involve cofilin, suggesting a currently undefined molecular pathway through which LIMK1 stabilizes synapses (Eaton and Davis 2005).

Genetic evidence strongly suggests that retrograde TGF-β signaling mediated by a specific BMP family ligand regulates both synaptic development and synaptic transmission at *Drosophila* NMJs, and in the central nervous system. However, TGF-β signaling cascades that regulate synapse development have yet to be identified in other nervous systems. The human genome encodes more than 29 TGF-β superfamily proteins but only 5 type I receptors and 7 type II receptors, suggesting that specific ligand activation likely plays a major role in achieving signaling specificity in mammals. Identifying TGF-β molecules that localize at developing or mature synapses would be essential to reveal vertebrate candidates that regulate synapse development. The *C. elegans* genome encodes four TGF-β superfamily proteins. Two BMP family ligands, DAF-7 and DBL-1, both expressed in neurons, regulate the developmental programming and body size, respectively, through the canonical SMAD-dependent signaling cascades (Patterson and Padgett 2000). A third ligand, UNC-129, is secreted by muscles to guide the dorsal migration of motoneuron axons (Colavita et al. 1998). The function of the fourth BMP-like molecule, TIG-4, is unknown. Mutations in *C. elegans* type I and II receptors, as well as the R-SMAD and Co-SMAD proteins, do not result in significant defects in axon guidance or locomotion, suggesting that UNC-129 signals through a different pathway. It therefore remains possible that any *C. elegans* or mammalian TGF-β ligands involved in regulating synapse development may signal through SMAD-independent pathways.

E. WNT Signaling

WNTs are secreted proteins that regulate many aspects of development, including cell fate determination, cell polarity, cell migration, pattern formation, axon morphology, and synapse differentiation in multiple organisms. WNTs activate many different signaling cascades. In the canonical Wnt signaling pathway, WNT activates a Wnt receptor Frizzled and a coreceptor Arrow/LRP5/6, which leads to the membrane recruitment and activation of a cytoplasmic signal transducer Disheveled. Disheveled in turn inactivates a GSK-3β/APC/Axin complex that normally keeps the cytoplasmic β-catenin at low levels. The inhibition of GSK-3β kinases leads to the stabilization and accumulation of cytoplasmic β-catenin, and its subsequent translocation to the nucleus to activate TCF-mediated transcription. Several β-catenin-independent Wnt signaling pathways have been described. In most cases, they require Frizzled receptors and Disheveled, but signal through other intracellular signaling cascades, including Rho family GTPases, heterotrimeric G proteins, Ca^{2+}/PKC, and JNK (Adler 2002; Jones 1993; Katanaev et al. 2005; Logan and Nusse 2004; McEwen and Peifer 2000; Veeman et al. 2003). Identification of a new class of WNT receptors, the Derailed/Lin-18/RyK family atypical receptor kinase in *Drosophila, C. elegans*, and mammals, revealed additional signaling effectors for WNT signaling (Inoue et al. 2004; Lu et al. 2004; Yoshikawa et al. 2003). Studies on Wnt-mediated regulation in synaptogenesis appear to be defining novel mechanisms through which Wnts signal (see later).

In the vertebrate nervous system, secreted Wnt proteins function as retrograde signals to promote presynaptic differentiation (Ciani and Salinas 2005). In the developing mouse cerebellum, Wnt7A is secreted by granule cell (GC) neurons to induce axon remodeling and synaptic vesicle protein synapsin I accumulation in cultured mossy fibers, a presynaptic partner to GC neurons. Consistent with the inductive role of Wnt7A in presynaptic remodeling, a transient reduction of the total number of synapses is observed in the Wnt-7a knock-out mouse cerebellum (Hall et al. 2000). In the spinal cord, Wnt3 is expressed in motoneurons and can induce axon branching and synapsin I (a presynaptic protein) clustering in the contacting sensory neurons in spinal cord transplants (Krylova et al. 2002). Interestingly, retrograde WNT signaling likely regulates presynaptic differentiation through a set of intracellular events different from other WNT pathways. In neuron cultures, the retrograde induction of presynaptic differentiation is activated through Dishevelled/DVL-1 and GSK-3β, but is independent of β-catenin or transcription (Ciani et al. 2004; Krylova et al. 2000). In this divergent pathway, DVL-1 activation stabilizes microtubules locally by inhibiting GSK-3β-

dependent phosphorylation of the microtubule-binding protein MAP-1B (Ciani et al. 2004).

At *Drosophila* glutamatergic NMJs, the loss of function of either Wg (a *Drosophila* Wnt molecule) or a Wg receptor Dfz2 leads to a reduction of synapse number and abnormal pre- and postsynaptic morphology in type I boutons, including the lack of presynaptic active zones and the absence of postsynaptic specialization (SSR). Both Wg and Dfz2 are expressed by neurons and muscles. The neuronal Wg, secreted from the presynaptic termini, is essential for the development of presynaptic active zones and postsynaptic SSR. Dfz2 also is required in neurons for proper presynaptic development at *Drosophila* NMJs (Packard et al. 2002). Unlike Wg and Dfz2, GSK-3β/Sgg is concentrated preferentially at presynaptic termini (Franco et al. 2004; Packard et al. 2002). Sgg is required in motoneurons to regulate synapse numbers and bouton size, likely by affecting the phosphorylation of the microtubule-binding protein Futsch (a *Drosophila* homologue of MAP-1B) (Franco et al. 2004). The expression of dominant negative GSK-3β in retinal ganglion cells alters their branching patterns and the size of presynaptic boutons in Zebrafish (Tokuoka et al. 2002). Wnt therefore may induce presynaptic differentiation through a conserved Frizzle/Disheveled/GSK-3β cascade that controls microtubule dynamics at developing synapses.

The presence of Dfz2, but not GSK-3β, at the postsynaptic termini in the *Drosophila* NMJs suggests that the defective postsynaptic development in *Wg* and *Dfz2* mutants is either a secondary defect due to the lack of presynaptic innervation, or that WNT regulates postsynaptic differentiation through different transducers. Recent studies support the later hypothesis, and reveal direct roles of Wnts in postsynaptic differentiation in both *Drosophila* and mammals through different signaling pathways. At *Drosophila* NMJs, Wg regulates postsynaptic development through a novel, noncanonical pathway where the cleavage and nuclear import of the C-terminal region of the postsynaptic Frizzled-2 protein affects transcriptional regulation (Mathew et al. 2005). In mouse hippocampal neurons, Wnt-7b promotes dendritic development through Disheveled-mediated activation of Rac and JNK signaling cascades (Rosso et al. 2005).

Given the presence of multiple WNT molecules and their large number of transduction pathways, it would not be surprising for different types of neurons and synapses to use WNT signaling for various aspects of synapse development. Finally, cross-talk between TGF-β and WNT signaling regulates development and tumorogenesis (Attisano and Labbe 2004; Theisen et al. 1996; Yu et al. 1996), suggesting that this interaction may affect synaptic differentiation.

F. LAR RPTP Signaling

Leukocyte antigen-related (LAR)-type receptor tyrosine phosphatases are cell adhesion/signaling molecules that regulate axon guidance and neurite outgrowth. Although the role of *Drosophila* LAR in axon guidance is well established (Desai et al. 1996; Krueger et al. 1996; Maurel-Zaffran et al. 2001), its involvement in synapse formation and function has just begun to be revealed. Liprin-α, the mammalian homologue of SYD-2 family protein, was identified as a specific interactor for the intracellular domain of the LAR-family RPTPs, and Liprin-α recruits LAR to focal adhesions in mammalian cell cultures (Pulido et al. 1995; Serra-Pages et al. 1995). The discovery of SYD-2, the *C. elegans* Liprin-α homologue as a presynaptic protein affecting active zone morphology has implicated the LAR RPTP pathway in synapse differentiation (Zhen and Jin 1999). Studies have since confirmed both the presence and function of LAR RPTP and Liprin-α family proteins at synapses in *C. elegans*, *Drosophila*, and mouse (Dunah et al. 2005;

Kaufmann et al. 2002; Wyszynski et al. 2002; Yeh et al. 2005; Zhen and Jin 1999).

At *Drosophila* NMJs, whereas DLar is present only along the axons, DLiprin-α protein is both pre- and postsynaptic (Kaufmann et al. 2002). Consistent with a role of Liprin-α in active zone morphology at *C. elegans* NMJs, both DLar and DLiprin-α are required to promote synaptic growth, and to regulate the size and shape of presynaptic active zones (Kaufmann et al. 2002). Furthermore DLiprin-α is required for DLar function in regulating synapse morphology, suggesting that the physical interaction between these two proteins is important for their functions, and it is the presynaptic DLiprin-α that regulates active zone development (Kaufmann et al. 2002). At *C. elegans* NMJs, both SYD-2 and PTP-3 (*C. elegans* LAR) are present only at the presynaptic termini (Ackley et al. 2005; Yeh et al. 2005; Zhen and Jin 1999). *C. elegans* LAR has two isoforms, PTP-3A and PTP-3B, that play independent roles in synapse formation and axon guidance (Ackley et al. 2005). The long isoform PTP-3A appears to be the major player in synapse formation. PTP-3A localizes exclusively in presynaptic termini, and functions in a genetic pathway with *syd-2* to restrict the size of synapses (and likely active zones). Furthermore, PTP-3A colocalizes with *C. elegans* active zone proteins SYD-2 and UNC-10/RIM, and the synaptic localization of PTP-3A depends on SYD-2, further supporting the hypothesis that Liprin-α and Lar RPTP function together to control synapse development (Ackley et al. 2005).

How RPTPs mediate signaling remains a mystery. Although many extracellular matrix proteins and secretion factors have been found to interact with the extracellular domains of RPTPs, their physiological relevance as *in vivo* RPTP ligands remains to be tested. *In vitro* studies suggest that mammalian LAR can bind a laminin-nidogen complex. Mutations in *C. elegans* Nid-1/nidogen lead to synapse morphology defects similar to those in *ptp-3A* and *syd-2* mutants and altered/mislocalization of PTP-3A and SYD-2 proteins at the synapse (Ackley et al. 2005), suggesting that Laminin-Nidogen is an *in vivo* ligand for LAR during synaptogenesis. A recent study in *Drosophila* showed that the heparan sulfate proteoglycan Syndecan (Sdc) is an *in vivo* ligand for DLar in motoneuron axon guidance, but the incomplete penetrance of the *Sdc* mutant phenotype suggests that Syndecan is not the only ligand for DLar (Fox and Zinn 2005). Since *C. elegans* PTP3A and PTP3B differ only in their extracellular domains, it will be interesting to determine whether different extracellular ligands are used to activate Lar signaling for axon guidance and synapse differentiation.

The downstream signal transducers for RPTPs are not well understood. Aside from Liprin-α, other proteins that interact with the cytoplasmic domains of LAR include a tyrosine kinase Abl and its substrate Ena (Wills et al. 1999a,b), a guanine exchange factor Trio (Debant et al. 1996), and a β-catenin/cadherin cell adhesion complex (Kypta et al. 1996). In *Drosophila*, Abl kinase and DLar phosphotase are proposed to balance the phosphorylation dynamics of Ena, a putative regulator for actin polymerization during motoneuron axon guidance (Gertler et al. 1996; Lanier et al. 1999; Wills et al. 1999a,b). Trio also collaborates with DLar, Abl, and Ena to regulate axon guidance in embryonic motoneurons and photoreceptor cells (Bateman et al. 2000; Liebl et al. 2000; Maurel-Zaffran et al. 2001). It would be interesting to examine if LAR and Liprin-α regulate synapse differentiation through Trio, Abl, and Ena.

In mammalian hippocampal neurons, both LAR and Liprin-α are present at pre- and postsynaptic termini of glutamatergic synapses, and are required for postsynaptic development and function (Dunah et al. 2005; Wyszynski et al. 2002). LAR, and the interactions between LAR, Liprin-α, and a postsynaptic scaffolding protein GRIP, are required for the recruitment of the β-catenin/cadherin complex to the dendritic spines. Since the interaction between β-catenin and cadherin is regulated by tyrosine phosphorylation (Ozawa and Kemler 1998), LAR likely affects postsynaptic assembly and function through dephosphorylating β-catenin. These studies focused on a postsynaptic role of LAR and Liprin-α, and it is presently unknown whether LAR and Liprin-α also affect presynaptic development. Given the complex functional interaction among different classes of RPTPs during axon outgrowth (Desai et al. 1997; Sun et al. 2001), the roles of other RPTPs at synapse differentiation deserve further investigation.

Intracellular Regulation of Presynaptic Differentiation: Ubiquitin-Proteasome System

Since nascent synapses receive multiple inductive signals, it is important for the developing synaptic termini to integrate and regulate the activity of multiple signaling cascades to ensure the development of synapses with the desired morphology and strength. Ubiquitination, a process by which the addition of ubiquitin to a protein leads to its degradation, processing, or relocation, has emerged as a key regulatory mechanism for presynaptic differentiation (for reviews, see Chisholm and Jin 2005; DiAntonio and Hicke 2004; Teng and Tang 2005; Yi and Ehlers 2005). The specificity of protein substrate targeting for ubiquitination is achieved largely through different classes of E3 ubiquitin ligases that mediate interactions with specific protein targets (Glickman and Ciechanover 2002; Hicke and Dunn 2003). The two best-studied E3 ubiquitin ligases are multimeric protein complexes known as the anaphase promoting complex (APC) and Skp-Cullin and F-box-containing complex (SCF), both of which were first discovered because of their requirement for yeast cell cycle progression (Glickman and Ciechanover 2002; Hicke and Dunn 2003; Peters 2002; Tyers and Jorgensen 2000). Genetic analysis in *C. elegans* and *Drosophila* nervous systems has revealed roles for specific ubiquitin ligases in both pre- and postsynaptic development.

The importance of the ubiquitination system in the formation, maintenance, and remodeling of synapses was first suggested by studies where proteasome inhibition resulted in an increase in synaptic strength and the number of synaptic contacts in *Drosophila* and Aplysia, respectively (Speese et al. 2003; Zhao et al. 2003). Short-term treatments with proteasome inhibitors in flies led to an acute increase in the level of Dunc-13, a presynaptic protein that enhances synaptic vesicle release and is ubiquitinated *in vivo* (Aravamudan and Broadie 2003; Speese et al. 2003). Antagonizing the ubiquitination pathway in neurons by expression of the fly deubiquitin enzymes *FAF* or UBP2 results in synaptic overgrowth and dysfunction (DiAntonio et al. 2001). Several postsynaptic scaffolding proteins are ubiquitinated (Ehlers 2003). The first direct evidence of the ubiquitination system for regulating intracellular signaling pathways during presynaptic differentiation came from the discovery of a conserved protein family that includes vertebrate Pam/Phr1, *Drosophila* Highwire and *C. elegans* RPM-1 (Burgess et al. 2004; Schaefer et al. 2000; Wan et al. 2000; Zhen et al. 2000).

These large proteins consist of conserved protein modules, including a Ran-type guanine exchange factor domain, a large internal region unique to the Pam/HIW/RPM-1 family proteins and multiple Ring fingers, a hallmark motif for E3 ubiquitin ligases (Joazeiro and Weissman 2000). All three proteins are expressed in the nervous system, and RPM-1 is restricted at the periactive zones of synapses (Burgess et al. 2004; Murthy et al. 2004; Schaefer et al. 2000; Wu et al. 2005; Zhen et al. 2000). *Drosophila hiw* mutants are characterized by an overgrowth of synapses, with a twofold increase of synaptic bouton numbers but with smaller bouton size, suggesting that HIW negatively regulates synaptic differentiation (Wan et al. 2000; Wu et al. 2005).

Loss of RPM-1 function in *C. elegans* causes more complex defects in the development of presynaptic structures. Some GABAergic NMJs in *rpm-1* mutants have larger synapses that contain increased numbers of active zones. Other regions that normally possess GABAergic NMJs completely lack synaptic development (Zhen et al. 2000). Mechanosensory neurons fail to develop stable synapses, suggesting the possible dual roles for *rpm-1* in both the restricting, or initiation/maintenance of the synapses (Schaefer et al. 2000). Both HIW and RPM-1 are required in presynaptic motoneurons for normal synapse morphology (Schaefer et al. 2000; Wu et al. 2005; Zhen et al. 2000).

RPM-1/HIW family proteins possess E3 ubiquitin ligase activity. *hiw* mutants have genetic interactions with *Faf*, a deubiquitination enzyme that antagonizes the ubiquitination pathway (DiAntonio et al. 2001; Huang et al. 1995). Whereas *hiw* is synthetically lethal with overexpressed *faf*, the *hiw* synapse phenotype can be rescued by loss of function of *faf*. This genetic interaction is consistent with opposite activities of HIW and FAF. A *C. elegans* neuron-specific F-box protein FSN-1 has been shown to interact physically with RPM-1, and recruits Cullin and Skp to form a protein complex reminiscent of a SCF-like E3 ubiquitin ligase complex that regulate yeast cell cycle (Liao et al. 2004). *C. elegans fsn-1* mutants are phenotypically similar, although synaptic morphology defects with *rpm-1* are less severe, and *fsn-1* functions in the *rpm-1* genetic pathway to regulate presynaptic differentiation. Moreover, genetic analysis suggests that the phenotype of *hiw, rpm-1*, and *fsn-1* mutants depends on several signaling cascades, and the levels of some of these molecules are negatively regulated by HIW/RPM-1 and FSN-1 (Liao et al. 2004; McCabe et al. 2004; Murthy et al. 2004; Nakata et al. 2005).

The *Drosophila Hiw* likely regulates synaptic growth through *Wit*-mediated TGF-β signaling (McCabe et al. 2004). The loss of function of the TGF-β type II receptor *Wit* and the co-SAMD *Med* restored the number of synaptic boutons in *hiw* mutants, but not other defects such as the bouton size or reduced synaptic transmission. Overexpression of constitutively activated TGF-β type I receptor *Tkv* in *hiw* mutants results in further increases in synaptic bouton numbers. Furthermore, a direct physical interaction between Hiw and MAD proteins, consistent with a hypothesis that *Hiw* directly targets components of the TGF-β signaling pathway for ubiquitination (McCabe et al. 2004). Only the defect in synaptic bouton number in *hiw* mutants is dependent on TGF-β signaling, suggesting that Hiw regulates additional signaling cascades. Alternatively, TGF-β signaling regulates other aspects of synaptic differentiation through different and currently unknown mechanisms.

A recent study on mammalian PAM/Phr-1 suggests that endogenous PAM/Phr-1 physically associates with the TSC1/TSC2 protein complex in the embryonic rat brain, and colocalizes with TSC1/TSC2 in cultured primary cortical neurons (Murthy et al. 2004). The TSC1/TSC2 complex restricts cell growth/size by inhibiting Tor kinase signaling, and the activity of TSC1/TSC2 complex is regulated by Akt and PI3-kinase in both *Drosophila* and mammals (Garami et al. 2003; Inoki et al. 2002; Long et al. 2005; Potter et al. 2002; Stocker et al. 2003). Overexpression of dTSC1/dTSC2 complex in *Drosophila* eye leads to a reduction in the eye size. This phenotype is reduced further when *hiw* activity is eliminated, supporting a potential role of HIW/PAM in negatively regulating TSC1/TSC2 activity in the CNS (Murthy et al. 2004). Whether TSC1/TSC2 complex plays a role during synaptogenesis in the CNS remains to be determined.

In *C. elegans*, *rpm-1* negatively regulates the activity of a specific MAP kinase cascade, consisting of DLK-1 (MAPKKK), MKK-4 (MAPKK), and a P38 family MAPK PMK-3. Loss of function of these kinases rescues the synapse morphology defects

of *rpm-1* mutants. DLK-1 localization is restricted at the periactive zone region, where RPM-1 resides. The levels of DLK-1 protein are increased in *rpm-1* mutants, and RPM-1 can mediate ubiquitination of DLK-1, suggesting that DLK-1 may be a target for RPM-1-mediated ubiquitination and degradation (Nakata et al. 2005). The synaptic defect in *fsn-1* mutants is also dependent on the activity of DLK-1/MKK-4/PMK-3, suggesting that it may participate in RPM-1-mediated down-regulation of this signaling cascade (Liao and Zhen, unpublished). *fsn-1* also negatively regulates the signaling mediated through a receptor tyrosine kinase ALK since *alk* mutants rescue the synapse morphology defects of *fsn-1* mutants, and ALK protein level increases in *fsn-1* mutants (Liao et al. 2004).

The existence of multiple binding partners of HIW, PAM, RPM-1, and FSN-1, as well as their genetic interactions with distinct signaling pathways, are consistent with the hypothesis that RPM-1/HIW/PAM and FSN-1 participate in different protein complexes to target multiple protein substrates for ubiquitinated-mediated proteolysis. Furthermore, the putative guanine-exchange factor motif that is present in this protein family, and indispensable for its function during synapse differentiation (Zhen et al. 2000), may also allow the RPM-1/HIW family proteins to couple intracellular signaling with protein turnover.

Complementary studies in *Drosophila* and *C. elegans* recently have shown that another well-known E3 ubiquitin ligase complex, APC/C, also plays a direct role at synapses to regulate their morphology and function (Juo and Kaplan 2004; van Roessel et al. 2004). Loss of APC/C function at *Drosophila* neuromuscular junctions leads to synapse overgrowth (an increase of bouton numbers without altering the size of individual boutons) and an increase in synaptic transmission at glutamergic NMJs. Not only does the APC/C complex regulate synapse differentiation directly after terminal cell division, the core subunits are concentrated at *Drosophila* mature NMJs, suggesting a direct role of APC/C complex in regulating synapse development. Intriguingly, the increase in the bouton number can be reverted to normal only when APC/C function is restored presynaptically, whereas the synaptic function defects require the postsynaptic expression of APC/C. This suggests that APC functions on different substrates in the pre- and postsynaptic terminal to control synapse differentiation. In this regard, the presynaptic protein Liprin-α/SYD-2, containing multiple conserved APC destruction box motifs, was identified as either a key substrate or a downstream effector through which APC regulates the bouton numbers. Loss-of-function mutations in DLiprin-α restore the synapse overgrowth defects of fly APC mutants; furthermore, the level of presynaptic Liprin-α increases in APC backgrounds. However, the role of APC in regulating the efficacy of synaptic transmission appears independent of Liprin-α. Instead, APC regulates the abundance of specific glutamate receptors at postsynaptic termini in both *Drosophila* and *C. elegans* glutamergic synapses. Therefore the APC/C complex also targets multiple, distinctive synaptic proteins to regulate synapse development and synaptic function.

We have only begun to understand the roles of a small fraction of E3 ubiquitin ligases. In addition to RPM-1/HIW/Phr, APC and SCF complexes, other E3 ubiquitin-ligases have been, and very likely will continue to be found present at synapses and regulate synapse assembly and function (Araki and Milbrandt 2003; Dreier et al. 2005; Ehlers 2003; Pak and Sheng 2003; Staropoli et al. 2003). Identifying the targets of these ligases not only will help define their roles in synapse differentiation, but also provide important tools for identifying key regulators of synapse differentiation that are likely missed by loss-of-function studies due to functional redundancies. For an example, in *C. elegans*, *dlk-1*, *mkk-4*, *pmk-3*, and *alk* mutants have near wild-type synapse morphology. It is only in *rpm-1* and *fsn-1* mutant backgrounds that their requirement for synapse development are revealed. Thus, other components in signaling pathways regulated by *rpm-1/hiw*, *fsn-1*, and APC complexes may be identified in future genetic screens.

IV. FUTURE PERSPECTIVES

Understanding the mechanisms underlying presynaptic differentiation has been accelerated by the convergence of information obtained from vertebrate and invertebrate synapses. Many key players in synapse function were identified via biochemical purification of presynaptic protein complexes from vertebrate brain tissues. The development of invertebrate synapses as a genetically accessible model for synapse development has been facilitated by live-imaging tools that allow direct visualization of synapse development and has resulted in an increase in discovery of genetic regulators for presynaptic differentiation. It is worth noting that most, if not all, characterized invertebrate presynaptic regulators are members of conserved protein families, inviting functional analysis of these proteins in vertebrates. The functions of these genes can now be studied and compared in different organisms due to the development of many tools that allow detailed morphological and physiological analyses of invertebrate and vertebrate synapses. A recent *C. elegans* genome-wide screen on genes required for synaptic function demonstrated the power of a genetic model that may benefit studies in other animals (Sieburth et al. 2005). Using a systematic RNAi-mediated functional knockdown of the entire open reading frames encoded in the *C. elegans* genome, a total of 185 genes that regulate synaptic function, 132 of which had not been characterized previously, and at least 24 of these newly identified proteins showed expression patterns in the nervous system (Sieburth et al. 2005). Functional analysis of these genes in *C. elegans* and other organisms should greatly accelerate our understanding of the genetic controls for the development and function of synapses.

References

Aberle, H., Haghighi, A.P., Fetter, R.D., McCabe, B.D., Magalhaes, T.R., Goodman, C.S. (2002). Wishful thinking encodes a BMP type II receptor that regulates synaptic growth in Drosophila. *Neuron* **33**, 545–558.

Ackley, B.D., Harrington, R.J., Hudson, M.L., Williams, L., Kenyon, C.J., Chisholm, A.D., Jin, Y. (2005). The two isoforms of the Caenorhabditis elegans leukocyte-common antigen related receptor tyrosine phosphatase PTP-3 function independently in axon guidance and synapse formation. *J. Neurosci.* **25**, 7517–7528.

Adler, P.N. (2002). Planar signaling and morphogenesis in Drosophila. *Dev. Cell* **2**, 525–535.

Ahmari, S.E., Buchanan, J., Smith, S.J. (2000). Assembly of presynaptic active zones from cytoplasmic transport packets. *Nat. Neurosci.* **3**, 445–451.

Akert, K., Moor, H., Pfenninger, K. (1971). Synaptic fine structure. *Adv. Cytopharmacol.* **1**, 273–290.

Allan, D.W., St Pierre, S.E., Miguel-Aliaga, I., Thor, S. (2003). Specification of neuropeptide cell identity by the integration of retrograde BMP signaling and a combinatorial transcription factor code. *Cell* **113**, 73–86.

Altrock, W.D., tom Dieck, S., Sokolov, M., Meyer, A.C., Sigler, A., Brakebusch, C. et al. (2003). Functional inactivation of a fraction of excitatory synapses in mice deficient for the active zone protein bassoon. *Neuron* **37**, 787–800.

Angley, C., Kumar, M., Dinsio, K.J., Hall, A.K., Siegel, R.E. (2003). Signaling by bone morphogenetic proteins and Smad1 modulates the postnatal differentiation of cerebellar cells. *J. Neurosci.* **23**, 260–268.

Araki, T., Milbrandt, J. (2003). ZNRF proteins constitute a family of presynaptic E3 ubiquitin ligases. *J. Neurosci.* **23**, 9385–9394.

Aravamudan, B., Broadie, K. (2003). Synaptic Drosophila UNC-13 is regulated by antagonistic G-protein pathways via a proteasome-dependent degradation mechanism. *J. Neurobiol.* **54**, 417–438.

Aravamudan, B., Fergestad, T., Davis, W.S., Rodesch, C.K., Broadie, K. (1999). Drosophila UNC-13 is essential for synaptic transmission. *Nat. Neurosci.* **2**, 965–971.

Attisano, L., Labbe, E. (2004). TGFbeta and Wnt pathway cross-talk. *Cancer Metastasis Rev.* **23**, 53–61.

Atwood, H.L., Govind, C.K., Wu, C.F. (1993). Differential ultrastructure of synaptic terminals on ventral longitudinal abdominal muscles in Drosophila larvae. *J. Neurobiol.* **24**, 1008–1024.

Augustin, I., Rosenmund, C., Sudhof, T.C., Brose, N. (1999). Munc13-1 is essential for fusion competence of glutamatergic synaptic vesicles. *Nature* **400**, 457–461.

Baines, R.A. (2004). Synaptic strengthening mediated by bone morphogenetic protein-dependent retrograde signaling in the Drosophila CNS. *J. Neurosci.* **24**, 6904–6911.

Bateman, J., Shu, H., Van Vactor, D. (2000). The guanine nucleotide exchange factor trio mediates axonal development in the Drosophila embryo. *Neuron* **26**, 93–106.

Bennett, M.K., Calakos, N., Scheller, R.H. (1992). Syntaxin: A synaptic protein implicated in docking of synaptic vesicles at presynaptic active zones. *Science* **257**, 255–259.

Brandstatter, J.H., Fletcher, E.L., Garner, C.C., Gundelfinger, E.D., Wassle, H. (1999). Differential expression of the presynaptic cytomatrix protein bassoon among ribbon synapses in the mammalian retina. *Eur. J. Neurosci.* **11**, 3683–3693.

Bresler, T., Shapira, M., Boeckers, T., Dresbach, T., Futter, M., Garner, C.C. et al. (2004). Postsynaptic density assembly is fundamentally different from presynaptic active zone assembly. *J. Neurosci.* **24**, 1507–1520.

Brionne, T.C., Tesseur, I., Masliah, E., Wyss-Coray, T. (2003). Loss of TGF-beta 1 leads to increased neuronal cell death and microgliosis in mouse brain. *Neuron* **40**, 1133–1145.

Broadie, K.S., Richmond, J.E. (2002). Establishing and sculpting the synapse in Drosophila and C. elegans. *Curr. Opin. Neurobiol.* **12**, 491–498.

Brose, N., Rosenmund, C., Rettig, J. (2000). Regulation of transmitter release by Unc-13 and its homologues. *Curr. Opin. Neurobiol.* **10**, 303–311.

Burgess, R.W., Peterson, K.A., Johnson, M.J., Roix, J.J., Welsh, I.C., O'Brien, T.P. (2004). Evidence for a conserved function in synapse formation reveals Phr1 as a candidate gene for respiratory failure in newborn mice. *Mol. Cell Biol.* **24**, 1096–1105.

Butz, S., Okamoto, M., Sudhof, T.C. (1998). A tripartite protein complex with the potential to couple synaptic vesicle exocytosis to cell adhesion in brain. *Cell* **94**, 773–782.

Byrd, D.T., Kawasaki, M., Walcoff, M., Hisamoto, N., Matsumoto, K., Jin, Y. (2001). UNC-16, a JNK-signaling scaffold protein, regulates vesicle transport in C. elegans. *Neuron* **32**, 787–800.

Castillo, P.E., Schoch, S., Schmitz, F., Sudhof, T.C., Malenka, R.C. (2002). RIM1alpha is required for presynaptic long-term potentiation. *Nature* **415**, 327–330.

Chapman, E.R. (2002). Synaptotagmin: A Ca(2+) sensor that triggers exocytosis? *Nat. Rev. Mol. Cell Biol.* **3**, 498–508.

Charytoniuk, D.A., Traiffort, E., Pinard, E., Issertial, O., Seylaz, J., Ruat, M. (2000). Distribution of bone morphogenetic protein and bone morphogenetic protein receptor transcripts in the rodent nervous system and up-regulation of bone morphogenetic protein receptor type II in hippocampal dentate gyrus in a rat model of global cerebral ischemia. *Neuroscience* **100**, 33–43.

Chen, Y.A., Scheller, R.H. (2001). SNARE-mediated membrane fusion. *Nat. Rev. Mol. Cell Biol.* **2**, 98–106.

Chisholm, A.D., Jin, Y. (2005). Neuronal differentiation in C. elegans. *Curr. Opin. Cell Biol.* **17**, 682–689.

Ciani, L., Krylova, O., Smalley, M.J., Dale, T.C., Salinas, P.C. (2004). A divergent canonical WNT-signaling pathway regulates microtubule dynamics: Disheveled signals locally to stabilize microtubules. *J. Cell Biol.* **164**, 243–253.

Ciani, L., Salinas, P.C. (2005). WNTs in the vertebrate nervous system: From patterning to neuronal connectivity. *Nat. Rev. Neurosci.* **6**, 351–362.

Colavita, A., Krishna, S., Zheng, H., Padgett, R.W., Culotti, J.G. (1998). Pioneer axon guidance by UNC-129, a C. elegans TGF-beta. *Science* **281**, 706–709.

Couteaux, R., Fessard, M.A. (1975). [Differentiation factors of active zones of presynaptic membranes]. *C. R. Acad. Sci. Hebd. Seances Acad. Sci. D.* **280**, 299–301.

Couteaux, R., Pecot-Dechavassine, M. (1970). [Synaptic vesicles and pouches at the level of "active zones" of the neuromuscular junction]. *C. R. Acad. Sci. Hebd. Seances Acad. Sci. D.* **271**, 2346–2349.

Coyle, I.P., Koh, Y.H., Lee, W.C., Slind, J., Fergestad, T., Littleton, J.T., Ganetzky, B. (2004). Nervous wreck, an SH3 adaptor protein that interacts with Wsp, regulates synaptic growth in Drosophila. *Neuron* **41**, 521–534.

Crump, J.G., Zhen, M., Jin, Y., Bargmann, C.I. (2001). The SAD-1 kinase regulates presynaptic vesicle clustering and axon termination. *Neuron* **29**, 115–129.

Debant, A., Serra-Pages, C., Seipel, K., O'Brien, S., Tang, M., Park, S.H., Streuli, M. (1996). The multidomain protein Trio binds the LAR transmembrane tyrosine phosphatase, contains a protein kinase domain, and has separate rac-specific and rho-specific guanine nucleotide exchange factor domains. *Proc. Natl. Acad. Sci. U S A* **93**, 5466–5471.

Deken, S.L., Vincent, R., Hadwiger, G., Liu, Q., Wang, Z.W., Nonet, M.L. (2005). Redundant localization mechanisms of RIM and ELKS in Caenorhabditis elegans. *J. Neurosci.* **25**, 5975–5983.

Derynck, R., Zhang, Y.E. (2003). Smad-dependent and Smad-independent pathways in TGF-beta family signalling. *Nature* **425**, 577–584.

Desai, C.J., Gindhart, J.G., Jr., Goldstein, L.S., Zinn, K. (1996). Receptor tyrosine phosphatases are required for motor axon guidance in the Drosophila embryo. *Cell* **84**, 599–609.

Desai, C.J., Krueger, N.X., Saito, H., Zinn, K. (1997). Competition and cooperation among receptor tyrosine phosphatases control motoneuron growth cone guidance in Drosophila. *Development* **124**, 1941–1952.

DiAntonio, A., Haghighi, A.P., Portman, S.L., Lee, J.D., Amaranto, A.M., Goodman, C.S. (2001). Ubiquitination-dependent mechanisms regulate synaptic growth and function. *Nature* **412**, 449–452.

DiAntonio, A., Hicke, L. (2004). Ubiquitin-dependent regulation of the synapse. *Annu. Rev. Neurosci.* **27**, 223–246.

Dick, O., tom Dieck, S., Altrock, W.D., Ammermuller, J., Weiler, R., Garner, C.C. et al. (2003). The presynaptic active zone protein bassoon is essential for photoreceptor ribbon synapse formation in the retina. *Neuron* **37**, 775–786.

Dreier, L., Burbea, M., Kaplan, J.M. (2005). LIN-23-mediated degradation of beta-catenin regulates the abundance of GLR-1 glutamate receptors in the ventral nerve cord of C. elegans. *Neuron* **46**, 51–64.

Dresbach, T., Qualmann, B., Kessels, M.M., Garner, C.C., Gundelfinger, E.D. (2001). The presynaptic cytomatrix of brain synapses. *Cell. Mol. Life Sci.* **58**, 94–116.

Dunah, A.W., Hueske, E., Wyszynski, M., Hoogenraad, C.C., Jaworski, J., Pak, D.T. et al. (2005). LAR receptor protein tyrosine phosphatases in the development and maintenance of excitatory synapses. *Nat. Neurosci.* **8**, 458–467.

Eaton, B.A., Davis, G.W. (2005). LIM Kinase-1 controls synaptic stability downstream of the type II BMP receptor. *Neuron* **47**, 695–708.

Ehlers, M.D. (2003). Activity level controls postsynaptic composition and signaling via the ubiquitin-proteasome system. *Nat. Neurosci.* **6**, 231–242.

Endo, M., Ohashi, K., Sasaki, Y., Goshima, Y., Niwa, R., Uemura, T., Mizuno, K. (2003). Control of growth cone motility and morphology by LIM kinase and Slingshot via phosphorylation and dephosphorylation of cofilin. *J. Neurosci.* **23**, 2527–2537.

Farkas, L.M., Dunker, N., Roussa, E., Unsicker, K., Krieglstein, K. (2003). Transforming growth factor-beta(s) are essential for the development of midbrain dopaminergic neurons in vitro and in vivo. *J. Neurosci.* **23**, 5178–5186.

Feng, X.H., Derynck, R. (2005). Specificity and versatility in TGF-signaling through smads. *Annu. Rev. Cell Dev. Biol.* **21**, 659–693.

Fenster, S.D., Chung, W.J., Zhai, R., Cases-Langhoff, C., Voss, B., Garner, A.M. et al. (2000). Piccolo, a presynaptic zinc finger protein structurally related to bassoon. *Neuron* **25**, 203–214.

Foletta, V.C., Lim, M.A., Soosairajah, J., Kelly, A.P., Stanley, E.G., Shannon, M. et al. (2003). Direct signaling by the BMP type II receptor via the cytoskeletal regulator LIMK1. *J. Cell Biol.* **162**, 1089–1098.

Fox, A.N., Zinn, K. (2005). The heparan sulfate proteoglycan syndecan is an in vivo ligand for the Drosophila LAR receptor tyrosine phosphatase. *Curr. Biol.* **15**, 1701–1711.

Franco, B., Bogdanik, L., Bobinnec, Y., Debec, A., Bockaert, J., Parmentier, M.L., Grau, Y. (2004). Shaggy, the homolog of glycogen synthase kinase 3, controls neuromuscular junction growth in Drosophila. *J. Neurosci.* **24**, 6573–6577.

Friedman, H.V., Bresler, T., Garner, C.C., Ziv, N.E. (2000). Assembly of new individual excitatory synapses: time course and temporal order of synaptic molecule recruitment. *Neuron* **27**, 57–69.

Garami, A., Zwartkruis, F.J., Nobukuni, T., Joaquin, M., Roccio, M., Stocker, H. et al. (2003). Insulin activation of Rheb, a mediator of mTOR/S6K/4E-BP signaling, is inhibited by TSC1 and 2. *Mol. Cell* **11**, 1457–1466.

Garner, C.C., Kindler, S., Gundelfinger, E.D. (2000). Molecular determinants of presynaptic active zones. *Curr. Opin. Neurobiol.* **10**, 321–327.

Gertler, F.B., Niebuhr, K., Reinhard, M., Wehland, J., Soriano, P. (1996). Mena, a relative of VASP and Drosophila Enabled, is implicated in the control of microfilament dynamics. *Cell* **87**, 227–239.

Glickman, M.H., Ciechanover, A. (2002). The ubiquitin-proteasome proteolytic pathway: Destruction for the sake of construction. *Physiol. Rev.* **82**, 373–428.

Gomes, F.C., Sousa Vde, O., Romao, L. (2005). Emerging roles for TGF-beta1 in nervous system development. *Int. J. Dev. Neurosci.* **23**, 413–424.

Hall, A.C., Lucas, F.R., Salinas, P.C. (2000). Axonal remodeling and synaptic differentiation in the cerebellum is regulated by WNT-7a signaling [see comments]. *Cell* **100**, 525–535.

Harlow, M.L., Ress, D., Stoschek, A., Marshall, R.M., McMahan, U.J. (2001). The architecture of active zone material at the frog's neuromuscular junction. *Nature* **409**, 479–484.

Hata, Y., Slaughter, C.A., Sudhof, T.C. (1993). Synaptic vesicle fusion complex contains unc-18 homologue bound to syntaxin. *Nature* **366**, 347–351.

Heuser, J.E., Reese, T.S., Landis, D.M. (1974). Functional changes in frog neuromuscular junctions studied with freeze-fracture. *J. Neurocytol.* **3**, 109–131.

Hicke, L., Dunn, R. (2003). Regulation of membrane protein transport by ubiquitin and ubiquitin-binding proteins. *Annu. Rev. Cell Dev. Biol.* **19**, 141–172.

Horbinski, C., Stachowiak, E.K., Chandrasekaran, V., Miuzukoshi, E., Higgins, D., Stachowiak, M.K. (2002). Bone morphogenetic protein-7 stimulates initial dendritic growth in sympathetic neurons through an intracellular fibroblast growth factor signaling pathway. *J. Neurochem.* **80**, 54–63.

Huang, Y., Baker, R.T., Fischer-Vize, J.A. (1995). Control of cell fate by a deubiquitinating enzyme encoded by the fat facets gene. *Science* **270**, 1828–1831.

Hummel, T., Krukkert, K., Roos, J., Davis, G., Klambt, C. (2000). Drosophila Futsch/22C10 is a MAP1B-like protein required for dendritic and axonal development. *Neuron* **26**, 357–370.

Inoki, K., Li, Y., Zhu, T., Wu, J., Guan, K.L. (2002). TSC2 is phosphorylated and inhibited by Akt and suppresses mTOR signalling. *Nat. Cell Biol.* **4**, 648–657.

Inoue, T., Oz, H.S., Wiland, D., Gharib, S., Deshpande, R., Hill, R.J., Katz, W.S., Sternberg, P.W. (2004). C. elegans LIN-18 is a Ryk ortholog and functions in parallel to LIN-17/Frizzled in Wnt signaling. *Cell* **118**, 795–806.

Jahn, R., Lang, T., Sudhof, T.C. (2003). Membrane fusion. *Cell* **112**, 519–533.

Jahn, R., Sudhof, T.C. (1999). Membrane fusion and exocytosis. *Annu. Rev. Biochem.* **68**, 863–911.

Jin, Y. (2002). Synaptogenesis: insights from worm and fly. *Curr. Opin. Neurobiol.* **12**, 71–79.

Joazeiro, C.A., Weissman, A.M. (2000). RING finger proteins: Mediators of ubiquitin ligase activity. *Cell* **102**, 549–552.

Jones, E.G. (1993). GABAergic neurons and their role in cortical plasticity in primates. *Cereb. Cortex* **3**, 361–372.

Juo, P., Kaplan, J.M. (2004). The anaphase-promoting complex regulates the abundance of GLR-1 glutamate receptors in the ventral nerve cord of C. elegans. *Curr. Biol.* **14**, 2057–2062.

Kaech, S.M., Whitfield, C.W., Kim, S.K. (1998). The LIN-2/LIN-7/LIN-10 complex mediates basolateral membrane localization of the C. elegans EGF receptor LET-23 in vulval epithelial cells. *Cell* **94**, 761–771.

Katanaev, V.L., Ponzielli, R., Semeriva, M., Tomlinson, A. (2005). Trimeric G protein-dependent frizzled signaling in Drosophila. *Cell* **120**, 111–122.

Katz, B. (1969). The release of neural transmitter substances. Liverpool University Press, Liverpool.

Kaufmann, N., DeProto, J., Ranjan, R., Wan, H., Van Vactor, D. (2002). Drosophila liprin-alpha and the receptor phosphatase Dlar control synapse morphogenesis. *Neuron* **34**, 27–38.

Kawasaki, F., Zou, B., Xu, X., Ordway, R.W. (2004). Active zone localization of presynaptic calcium channels encoded by the cacophony locus of Drosophila. *J. Neurosci.* **24**, 282–285.

Kim, E., Sheng, M. (2004). PDZ domain proteins of synapses. *Nat. Rev. Neurosci.* **5**, 771–781.

Kishi, M., Pan, Y.A., Crump, J.G., Sanes, J.R. (2005). Mammalian SAD kinases are required for neuronal polarization. *Science* **307**, 929–932.

Ko, J., Na, M., Kim, S., Lee, J.R., Kim, E. (2003). Interaction of the ERC family of RIM-binding proteins with the liprin-alpha family of multidomain proteins. *J. Biol. Chem.* **278**, 42377–42385.

Koh, T.W., Bellen, H.J. (2003). Synaptotagmin I, a Ca2+ sensor for neurotransmitter release. *Trends Neurosci.* **26**, 413–422.

Koushika, S.P., Richmond, J.E., Hadwiger, G., Weimer, R.M., Jorgensen, E.M., Nonet, M.L. (2001). A postdocking role for active zone protein Rim. *Nat. Neurosci.* **4**, 997–1005.

Krueger, N.X., Van Vactor, D., Wan, H.I., Gelbart, W.M., Goodman, C.S., Saito, H. (1996). The transmembrane tyrosine phosphatase DLAR controls motor axon guidance in Drosophila. *Cell* **84**, 611–622.

Krylova, O., Herreros, J., Cleverley, K.E., Ehler, E., Henriquez, J.P., Hughes, S.M., Salinas, P.C. (2002). WNT-3, expressed by motoneurons, regulates terminal arborization of neurotrophin-3-responsive spinal sensory neurons. *Neuron* **35**, 1043–1056.

Krylova, O., Messenger, M.J., Salinas, P.C. (2000). Dishevelled-1 regulates microtubule stability: A new function mediated by glycogen synthase kinase-3beta. *J. Cell Biol.* **151**, 83–94.

Kypta, R.M., Su, H., Reichardt, L.F. (1996). Association between a transmembrane protein tyrosine phosphatase and the cadherin-catenin complex. *J. Cell Biol.* **134**, 1519–1529.

Landis, D.M., Hall, A.K., Weinstein, L.A., Reese, T.S. (1988). The organization of cytoplasm at the presynaptic active zone of a central nervous system synapse. *Neuron* **1**, 201–209.

Lanier, L.M., Gates, M.A., Witke, W., Menzies, A.S., Wehman, A.M., Macklis, J.D. et al. (1999). Mena is required for neurulation and commissure formation. *Neuron* **22**, 313–325.

Lein, P.J., Beck, H.N., Chandrasekaran, V., Gallagher, P.J., Chen, H.L., Lin, Y. et al. (2002). Glia induce dendritic growth in cultured sympathetic neurons by modulating the balance between bone morphogenetic proteins (BMPs) and BMP antagonists. *J. Neurosci.* **22**, 10377–10387.

Liao, E.H., Hung, W., Abrams, B., Zhen, M. (2004). An SCF-like ubiquitin ligase complex that controls presynaptic differentiation. *Nature* **430**, 345–350.

Liebl, E.C., Forsthoefel, D.J., Franco, L.S., Sample, S.H., Hess, J.E., Cowger, J.A. et al. (2000). Dosage-sensitive, reciprocal genetic interactions between the Abl tyrosine kinase and the putative GEF trio reveal trio's role in axon pathfinding. *Neuron* **26**, 107–118.

Logan, C.Y., Nusse, R. (2004). The Wnt signaling pathway in development and disease. *Annu. Rev. Cell Dev. Biol.* **20**, 781–810.

Long, X., Lin, Y., Ortiz-Vega, S., Yonezawa, K., Avruch, J. (2005). Rheb binds and regulates the mTOR kinase. *Curr. Biol.* **15**, 702–713.

Lu, W., Yamamoto, V., Ortega, B., Baltimore, D. (2004). Mammalian Ryk is a Wnt coreceptor required for stimulation of neurite outgrowth. *Cell* **119**, 97–108.

Marek, K.W., Davis, G.W. (2002). Transgenically encoded protein photoinactivation (FlAsH-FALI): Acute inactivation of synaptotagmin I. *Neuron* **36**, 805–813.

Marques, G. (2005). Morphogens and synaptogenesis in Drosophila. *J. Neurobiol.* **64**, 417–434.

Marques, G., Bao, H., Haerry, T.E., Shimell, M.J., Duchek, P., Zhang, B., O'Connor, M.B. (2002). The Drosophila BMP type II receptor Wishful Thinking regulates neuromuscular synapse morphology and function. *Neuron* **33**, 529–543.

Mathew, D., Ataman, B., Chen, J., Zhang, Y., Cumberledge, S., Budnik, V. (2005). Wingless signaling at synapses is through cleavage and nuclear import of receptor DFrizzled2. *Science* **310**, 1344–1347.

Maurel-Zaffran, C., Suzuki, T., Gahmon, G., Treisman, J.E., Dickson, B.J. (2001). Cell-autonomous and -nonautonomous functions of LAR in R7 photoreceptor axon targeting. *Neuron* **32**, 225–235.

Maximov, A., Sudhof, T.C. (2005). Autonomous function of synaptotagmin 1 in triggering synchronous release independent of asynchronous release. *Neuron* **48**, 547–554.

McCabe, B.D., Hom, S., Aberle, H., Fetter, R.D., Marques, G., Haerry, T.E. et al. (2004). Highwire regulates presynaptic BMP signaling essential for synaptic growth. *Neuron* **41**, 891–905.

McCabe, B.D., Marques, G., Haghighi, A.P., Fetter, R.D., Crotty, M.L., Haerry, T.E. et al. (2003). The BMP homolog Gbb provides a retrograde signal that regulates synaptic growth at the Drosophila neuromuscular junction. *Neuron* **39**, 241–254.

McEwen, D.G., Peifer, M. (2000). Wnt signaling: Moving in a new direction. *Curr. Biol.* **10**, R562–564.

McLennan, I.S., Koishi, K. (1994). Transforming growth factor-beta-2 (TGF-beta 2) is associated with mature rat neuromuscular junctions. *Neurosci. Lett.* **177**, 151–154.

Meinertzhagen, I.A., Govind, C.K., Stewart, B.A., Carter, J.M., Atwood, H.L. (1998). Regulated spacing of synapses and presynaptic active zones at larval neuromuscular junctions in different genotypes of the flies Drosophila and Sarcophaga. *J. Comp. Neurol.* **393**, 482–492.

Murthy, V., Han, S., Beauchamp, R.L., Smith, N., Haddad, L.A., Ito, N., Ramesh, V. (2004). Pam and its ortholog highwire interact with and may negatively regulate the TSC1.TSC2 complex. *J. Biol. Chem.* **279**, 1351–1358.

Nakata, K., Abrams, B., Grill, B., Goncharov, A., Huang, X., Chisholm, A.D., Jin, Y. (2005). Regulation of a DLK-1 and p38 MAP kinase pathway by the ubiquitin ligase RPM-1 is required for presynaptic development. *Cell* **120**, 407–420.

Ng, J., Luo, L. (2004). Rho GTPases regulate axon growth through convergent and divergent signaling pathways. *Neuron* **44**, 779–793.

Ohtsuka, T., Takao-Rikitsu, E., Inoue, E., Inoue, M., Takeuchi, M., Matsubara, K. et al. (2002). Cast: A novel protein of the cytomatrix at the active zone of synapses that forms a ternary complex with RIM1 and munc13-1. *J. Cell. Biol.* **158**, 577–590.

Okamoto, M., Sudhof, T.C. (1997). Mints, Munc18-interacting proteins in synaptic vesicle exocytosis. *J. Biol. Chem.* **272**, 31459–31464.

Olsen, O., Moore, K.A., Fukata, M., Kazuta, T., Trinidad, J.C., Kauer, F.W. et al. (2005). Neurotransmitter release regulated by a MALS-liprin-alpha presynaptic complex. *J. Cell Biol.* **170**, 1127–1134.

Ozawa, M., Kemler, R. (1998). The membrane-proximal region of the E-cadherin cytoplasmic domain prevents dimerization and negatively regulates adhesion activity. *J. Cell Biol.* **142**, 1605–1613.

Packard, M., Koo, E.S., Gorczyca, M., Sharpe, J., Cumberledge, S., Budnik, V. (2002). The Drosophila Wnt, wingless, provides an essential signal for pre- and postsynaptic differentiation. *Cell* **111**, 319–330.

Packard, M., Mathew, D., Budnik, V. (2003). Wnts and TGF beta in synaptogenesis: old friends signalling at new places. *Nat. Rev. Neurosci.* **4**, 113–120.

Pak, D.T., Sheng, M. (2003). Targeted protein degradation and synapse remodeling by an inducible protein kinase. *Science* **302**, 1368–1373.

Patterson, G.I., Padgett, R.W. (2000). TGF beta-related pathways. Roles in Caenorhabditis elegans development. *Trends Genet.* **16**, 27–33.

Pfenninger, K., Akert, K., Moor, H., Sandri, C. (1972). The fine structure of freeze-fractured presynaptic membranes. *J. Neurocytol.* **1**, 129–149.

Phillips, G.R., Huang, J.K., Wang, Y., Tanaka, H., Shapiro, L., Zhang, W. et al. (2001). The presynaptic particle web: Ultrastructure, composition, dissolution, and reconstitution. *Neuron* **32**, 63–77.

Potter, C.J., Pedraza, L.G., Xu, T. (2002). Akt regulates growth by directly phosphorylating Tsc2. *Nat. Cell Biol.* **4**, 658–665.

Pulido, R., Serra-Pages, C., Tang, M., Streuli, M. (1995). The LAR/PTP delta/PTP sigma subfamily of trans-membrane protein-tyrosine-phosphatases: Multiple human LAR, PTP delta, and PTP sigma isoforms are expressed in a tissue-specific manner and associate with the LAR-interacting protein LIP.1. *Proc. Natl. Acad. Sci. U S A* **92**, 11686–11690.

Reiff, D.F., Thiel, P.R., Schuster, C.M. (2002). Differential regulation of active zone density during long-term strengthening of Drosophila neuromuscular junctions. *J. Neurosci.* **22**, 9399–9409.

Richmond, J.E., Davis, W.S., Jorgensen, E.M. (1999). UNC-13 is required for synaptic vesicle fusion in C. elegans. *Nat. Neurosci.* **2**, 959–964.

Richmond, J.E., Weimer, R.M., Jorgensen, E.M. (2001). An open form of syntaxin bypasses the requirement for UNC-13 in vesicle priming. *Nature* **412**, 338–341.

Rios, I., Alvarez-Rodriguez, R., Marti, E., Pons, S. (2004). Bmp2 antagonizes sonic hedgehog-mediated proliferation of cerebellar granule neurones through Smad5 signalling. *Development* **131**, 3159–3168.

Rizo, J., Sudhof, T.C. (2002). Snares and Munc18 in synaptic vesicle fusion. *Nat. Rev. Neurosci.* **3**, 641–653.

Robitaille, R., Adler, E.M., Charlton, M.P. (1993). Calcium channels and calcium-gated potassium channels at the frog neuromuscular junction. *J. Physiol. Paris* **87**, 15–24.

Roos, J., Hummel, T., Ng, N., Klambt, C., Davis, G.W. (2000). Drosophila Futsch regulates synaptic micro-tubule organization and is necessary for synaptic growth. *Neuron* **26**, 371–382.

Rosenmund, C., Rettig, J., Brose, N. (2003). Molecular mechanisms of active zone function. *Curr. Opin. Neurobiol.* **13**, 509–519.

Rosso, S.B., Sussman, D., Wynshaw-Boris, A., Salinas, P.C. (2005). Wnt signaling through Dishevelled, Rac and JNK regulates dendritic development. *Nat. Neurosci.* **8**, 34–42.

Schaefer, A.M., Hadwiger, G.D., Nonet, M.L. (2000). Rpm-1, a conserved neuronal gene that regulates targeting and synaptogenesis in C. elegans. *Neuron* **26**, 345–356.

Schaefer, A.M., Nonet, M.L. (2001). Cellular and molecular insights into presynaptic assembly. *Curr. Opin. Neurobiol.* **11**, 127–134.

Schoch, S., Castillo, P. E., Jo, T., Mukherjee, K., Geppert, M., Wang, Y. et al. (2002). RIM1 alpha forms a protein scaffold for regulating neurotransmitter release at the active zone. *Nature* **415**, 321–326.

Serra-Pages, C., Kedersha, N.L., Fazikas, L., Medley, Q., Debant, A., Streuli, M. (1995). The LAR transmembrane protein tyrosine phosphatase and a coiled-coil LAR-interacting protein co-localize at focal adhesions. *Embo. J.* **14**, 2827–2838.

Shapira, M., Zhai, R. G., Dresbach, T., Bresler, T., Torres, V.I., Gundelfinger, E.D. et al. (2003). Unitary assembly of presynaptic active zones from Piccolo-Bassoon transport vesicles. *Neuron* **38**, 237–252.

Shi, Y., Massague, J. (2003). Mechanisms of TGF-beta signaling from cell membrane to the nucleus. *Cell* **113**, 685–700.

Sieburth, D., Ch'ng, Q., Dybbs, M., Tavazoie, M., Kennedy, S., Wang, D. et al. (2005). Systematic analysis of genes required for synapse structure and function. *Nature* **436**, 510–517.

Sone, M., Suzuki, E., Hoshino, M., Hou, D., Kuromi, H., Fukata, M. et al. (2000). Synaptic development is controlled in the periactive zones of Drosophila synapses. *Development* **127**, 4157–4168.

Speese, S.D., Trotta, N., Rodesch, C.K., Aravamudan, B., Broadie, K. (2003). The ubiquitin proteasome system acutely regulates presynaptic protein turnover and synaptic efficacy. *Curr. Biol.* **13**, 899–910.

Staropoli, J.F., McDermott, C., Martinat, C., Schulman, B., Demireva, E., Abeliovich, A. (2003). Parkin is a component of an SCF-like ubiquitin ligase complex and protects postmitotic neurons from kainate excitotoxicity. *Neuron* **37**, 735–749.

Stocker, H., Radimerski, T., Schindelholz, B., Wittwer, F., Belawat, P., Daram, P. et al. (2003). Rheb is an essential regulator of S6K in controlling cell growth in Drosophila. *Nat. Cell Biol.* **5**, 559–565.

Sudhof, T.C. (2004). The synaptic vesicle cycle. *Annu. Rev. Neurosci.* **27**, 509–547.

Sun, Q., Schindelholz, B., Knirr, M., Schmid, A., Zinn, K. (2001). Complex genetic interactions among four receptor tyrosine phosphatases regulate axon guidance in Drosophila. *Mol. Cell Neurosci.* **17**, 274–291.

Sweeney, S.T., Davis, G.W. (2002). Unrestricted synaptic growth in spinster: A late endosomal protein implicated in TGF-beta-mediated synaptic growth regulation. *Neuron* **36**, 403–416.

Teng, F.Y., Tang, B.L. (2005). APC/C regulation of axonal growth and synaptic functions in postmitotic neurons: The Liprin-alpha connection. *Cell. Mol. Life Sci.* **62**, 1571–1578.

Theisen, H., Haerry, T.E., O'Connor, M.B., Marsh, J.L. (1996). Developmental territories created by mutual antagonism between Wingless and Decapentaplegic. *Development* **122**, 3939–3948.

Tokuoka, H., Goda, Y. (2003). Synaptotagmin in Ca2+ dependent exocytosis: Dynamic action in a flash. *Neuron* **38**, 521–524.

Tokuoka, H., Yoshida, T., Matsuda, N., Mishina, M. (2002). Regulation by glycogen synthase kinase-3beta of the arborization field and maturation of retinotectal projection in zebrafish. *J. Neurosci.* **22**, 10324–10332.

van Roessel, P., Elliott, D.A., Robinson, I.M., Prokop, A., Brand, A.H. (2004). Independent regulation of synaptic size and activity by the anaphase-promoting complex. *Cell* **119**, 707–718.

Veeman, M.T., Axelrod, J.D., Moon, R.T. (2003). A second canon. Functions and mechanisms of beta-catenin-independent Wnt signaling. *Dev. Cell* **5**, 367–377.

Verhage, M., Maia, A.S., Plomp, J.J., Brussaard, A.B., Heeroma, J.H., Vermeer, H. et al. (2000). Synaptic assembly of the brain in the absence of neurotransmitter secretion. *Science* **287**, 864–869.

Voets, T., Toonen, R.F., Brian, E.C., de Wit, H., Moser, T., Rettig, J. et al. (2001). Munc18-1 promotes large dense-core vesicle docking. *Neuron* **31**, 581–591.

Wachman, E.S., Poage, R.E., Stiles, J.R., Farkas, D.L., Meriney, S.D. (2004). Spatial distribution of calcium entry evoked by single action potentials within the presynaptic active zone. *J. Neurosci.* **24**, 2877–2885.

Wan, H.I., DiAntonio, A., Fetter, R.D., Bergstrom, K., Strauss, R., Goodman, C.S. (2000). Highwire regulates synaptic growth in Drosophila. *Neuron* **26**, 313–329.

Weeks, A.C., Ivanco, T.L., Leboutillier, J.C., Racine, R.J., Petit, T.L. (2000). Sequential changes in the synaptic structural profile following long-term potentiation in the rat dentate gyrus. II. Induction/early maintenance phase. *Synapse* **36**, 286–296.

Weimer, R.M., Richmond, J.E. (2005). Synaptic vesicle docking: A putative role for the Munc18/Sec1 protein family. *Curr. Top. Dev. Biol.* **65**, 83–113.

Weimer, R.M., Richmond, J.E., Davis, W.S., Hadwiger, G., Nonet, M.L., Jorgensen, E.M. (2003). Defects in synaptic vesicle docking in unc-18 mutants. *Nat. Neurosci.* **6**, 1023–1030.

Wills, Z., Bateman, J., Korey, C.A., Comer, A., Van Vactor, D. (1999a). The tyrosine kinase Abl and its substrate enabled collaborate with the receptor phosphatase Dlar to control motor axon guidance. *Neuron* **22**, 301–312.

Wills, Z., Marr, L., Zinn, K., Goodman, C.S., Van Vactor, D. (1999b). Profilin and the Abl tyrosine kinase are required for motor axon outgrowth in the Drosophila embryo. *Neuron* **22**, 291–299.

Wu, C., Wairkar, Y.P., Collins, C.A., DiAntonio, A. (2005). Highwire function at the Drosophila neuromuscular junction: spatial, structural, and temporal requirements. *J. Neurosci.* **25**, 9557–9566.

Wyszynski, M., Kim, E., Dunah, A.W., Passafaro, M., Valtschanoff, J.G., Serra-Pages, C. et al. (2002). Interaction between GRIP and liprin-alpha/SYD2 is required for AMPA receptor targeting. *Neuron* **34**, 39–52.

Yeh, E., Kawano, T., Weimer, R.M., Bessereau, J.L., Zhen, M. (2005). Identification of genes involved in synaptogenesis using a fluorescent active zone marker in Caenorhabditis elegans. *J. Neurosci.* **25**, 3833–3841.

Yi, J.J., Ehlers, M.D. (2005). Ubiquitin and protein turnover in synapse function. *Neuron* **47**, 629–632.

Yoshikawa, S., McKinnon, R.D., Kokel, M., Thomas, J.B. (2003). Wnt-mediated axon guidance via the Drosophila Derailed receptor. *Nature* **422**, 583–588.

Yu, K., Sturtevant, M.A., Biehs, B., Francois, V., Padgett, R.W., Blackman, R.K., Bier, E. (1996). The Drosophila decapentaplegic and short gastrulation genes function antagonistically during adult wing vein development. *Development* **122**, 4033–4044.

Zamorano, P.L., Garner, C.C. (2001). Unwebbing the presynaptic web. *Neuron* **32**, 3–6.

Zhai, R.G., Bellen, H.J. (2004). The architecture of the active zone in the presynaptic nerve terminal. *Physiology (Bethesda)* **19**, 262–270.

Zhai, R.G., Vardinon-Friedman, H., Cases-Langhoff, C., Becker, B., Gundelfinger, E.D., Ziv, N.E., Garner, C.C. (2001). Assembling the presynaptic active zone: a characterization of an active one precursor vesicle. *Neuron* **29**, 131–143.

Zhao, Y., Hegde, A.N., Martin, K.C. (2003). The ubiquitin proteasome system functions as an inhibitory constraint on synaptic strengthening. *Curr. Biol.* **13**, 887–898.

Zhen, M., Huang, X., Bamber, B., Jin, Y. (2000). Regulation of presynaptic terminal organization by C. elegans RPM-1, a putative guanine nucleotide exchanger with a RING-H2 finger domain. *Neuron* **26**, 331–343.

Zhen, M., Jin, Y. (1999). The liprin protein SYD-2 regulates the differentiation of presynaptic termini in C. elegans. *Nature* **401**, 371–375.

Zhen, M., Jin, Y. (2004). Presynaptic terminal differentiation: Transport and assembly. *Curr. Opin. Neurobiol.* **14**, 280–287.

SECTION

III

EXO-/ENDOCYTOSIS

Once pre- and postsynaptic elements have differentiated and formed a functional synapse, the fidelity and efficiency of synaptic transmission depends on the precise alignment and function of pre- and postsynaptic membranes as well as their corresponding release mechanisms and postsynaptic receptors. Physiological methods for estimating presynaptic function coupled with biochemical/molecular investigation of the molecular machinery underlying the synaptic vesicle cycle (Sankaranarayanan and Ryan; Haucke and Klingauf) have provided a window into the mechanisms underlying neurotransmission.

EXO-/ENDOCYTOSIS

6

Neuronal Exocytosis

SETHU SANKARANARAYANAN AND TIMOTHY A. RYAN

Synaptic transmission is mediated by repeated cycles of exocytosis followed by endocytosis of synaptic vesicles at nerve terminals. This chapter reviews neuronal exocytosis and the recent advances in the molecular and cell biological understanding of this process.

I. INTRODUCTION

Neurotransmission is the fundamental process that drives information transfer between neurons and their targets. It regulates both excitatory and inhibitory functions in the central nervous system (CNS), underlies sensory processing, and regulates autonomic and motor outputs in species ranging from small invertebrates to highly evolved mammals. Modulation of synaptic transmission is believed to drive cognitive processes such as learning and memory (Kandel et al. 2000).

Neurotransmission occurs at specialized regions between neurons and their targets, called the synapse (see Figure 6.1). The synapse is a highly specialized contact between a presynaptic and a postsynaptic cell, built to transmit information with high fidelity. This information transfer is mediated via exocytosis of neurotransmitters from synaptic vesicles that activate

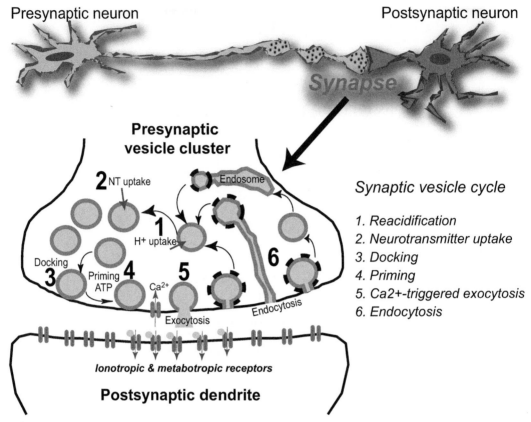

FIGURE 6.1. The synaptic vesicle cycle. Neurotransmission occurs at specialized sites of contact between neurons called the synapse. The synapse is the site of contact between the presynaptic terminal and the post-synaptic dendrite or cell body. Synaptic vesicles in the nerve terminal undergo a synchronized cycle of events that underlie neurotransmission. Synaptic vesicles are filled with neurotransmitter following reacidification of vesicles. Once filled with neurotransmitter the vesicles dock at the active zone, a specialized region of plasma membrane in the presynaptic terminal opposing the synaptic cleft. The vesicles then undergo an ATP-dependent priming step, which makes them fusion competent. Arrival of an action potential at the nerve terminal leads to rapid increase in cytosolic calcium that triggers exocytosis. Following fusion, the vesicles are retrieved by endocytosis. Endocytosis is predominantly mediated by clathrin-mediated process and can generate new vesicles either directly at the plasma membrane or at the tips of deep plasma membrane invaginations or via an intermediate endosomal compartment.

postsynaptic receptors. Repeated cycles of fusion and retrieval of synaptic vesicles allows efficient neurotransmission.

A. The Synaptic Vesicle Cycle

Activity-dependent exocytosis and endocytosis of synaptic vesicles underlies all neurotransmission in the CNS. An action potential that invades the nerve terminal results in depolarization of the membrane leading to calcium influx, via voltage-dependent calcium channels. Rapid fusion of predocked vesicles causes neurotransmitter release into the synaptic cleft where it may activate postsynaptic receptors. Rapid buffering and clearance mechanisms lead to recovery of intracellular calcium and allow the synapse to return to its resting state. Following exocytosis of vesicles, membrane components are retrieved from the plasma membrane via endocytosis. The internalized vesicles are then refilled with neurotransmitter, mobilized, and reprimed to populate the readily releasable pool of vesicles (see Figure 6.1). The process of synaptic vesicle cycling has been extensively reviewed (Murthy and De Camilli 2003; Sudhof 2004). Endocytosis occurs predominantly via clathrin-mediated mechanism directly from the plasma membrane, via deep membrane invaginations, or via intermediate endosomal structures. Synaptic vesicle endocytosis is driven by protein-protein interactions involving adaptor proteins and clathrin coats (reviewed in Slepnev and De Camilli 2000; Brodin et al. 2000). Synaptic vesicles are locally recycled, ensuring high fidelity of synaptic transmission and overcoming the spatial and energy constraints brought about by the large distances between the soma and synaptic terminal of neurons.

The synaptic vesicle cycle is mediated via a coordinated set of trafficking events that occur at the nerve terminal and the postsynaptic neuron. In the last 20 years, innovations in molecular, biochemical, genetic, and physiological methods to study synaptic transmission have led to a dramatic increase in our understanding of the molecular players that mediate synaptic vesicle recycling and have helped to precisely delineate the kinetic aspects of the cycle. Genetic and functional studies in *Drosophila* and *C. elegans* have provided novel insights into proteins regulating synaptic vesicle recycling (reviewed in Stimson and Ramaswami 1999; Harris et al. 2001; Richmond and Broadie 2002; Kidokoro 2003). Although protein-protein interactions are thought to mediate specificity and selectivity of recycling events, there is growing consensus that local changes in membrane lipids can regulate both membrane curvature and protein interactions with the membrane during vesicle fusion and retrieval (Wenk and De Camilli 2003). This chapter will focus on the various aspects of neuronal exocytosis with particular emphasis on small synaptic vesicle recycling in the CNS.

B. Historical Perspective and Methods Used to Analyze Neurosecretion

Synaptic transmission initially was studied in great detail using electrophysiological recordings in the neuromuscular junction. The neuromuscular synapse was readily accessible, and provided easily obtained stable recordings that allowed detailed kinetic measurements (Katz 1966). Characterization of miniature end plate potentials in the neuromuscular synapse revealed that neurotransmitter was stored in packets or quanta. Each *quanta* contained roughly similar concentrations of acetylcholine, and release events were found to be a simple multiple of the fundamental quanta (Katz 1971). Release of a single quanta led to a miniature synaptic currents, whereas exocytosis of a large number of quanta led to large postsynaptic currents that resulted in a muscle action potential (Katz and Miledi 1969). These studies also revealed that exocytosis is regulated by extracellular calcium (Katz and Miledi 1965) and that synaptic transmission can be modulated over a range of time scales (Katz and Miledi 1967; Magleby and Zengel 1976; Magleby 1979; Zucker and Regher 2002). Thus, postsynaptic recordings of electrical signals in muscle led to fundamental discoveries on the workings of a synapse and continue to guide our view of synaptic transmission. However, these methods provide only indirect information about presynaptic events underlying neurotransmitter release. Until recently it was not possible to probe directly the events at the synaptic terminal, and the quantal nature of transmitter release in CNS synapses was not clear.

Advent of new methods that enabled direct measurement of exocytosis and endocytosis of vesicles in neuroendocrine cells and presynaptic nerve terminals have led to a better understanding of the cell biology of presynaptic vesicle trafficking. Direct measurements of membrane capacitance from secretory cells (Neher and Marty

1982), amperometric signals arising from secreted catecholamines (Michael and Wightman 1999), and optical measurements of vesicle trafficking using fluorescent probes (Murthy 1999; Ryan and Reuter 2001; Rizzoli et al. 2003) have proved to be extremely useful for studying presynaptic events (see Figure 6.2).

In the late 1980s, the invention of the patch clamp method to record membrane electrical events led to many novel insights into the functioning of ion channels and synaptic transmission (Hamill et al. 1981). This method was optimized to measure membrane capacitance in secretory cells to track changes in membrane surface area brought about by exocytosis and endocytosis of secretory vesicles (Neher and Marty 1982) (see Figure 6.2A). These methods, in combination with intracellular calcium measurements, have resulted in a better

understanding of the kinetics of exocytosis and endocytosis in chromaffin cells (Heinemann et al. 1994), retinal bipolar terminals (von Gersdorff and Matthews 1994; 1997), and the giant presynaptic terminal from the calyx of Held (a relay nucleus in the auditory pathway) (Bollmann et al. 2000).

Electrochemical detection of catecholamines released from chromaffin cells has complemented capacitance measurements, resulting in a better understanding of the kinetics of exocytosis (Chow et al. 1996; Xu et al. 1998) (see Figure 6.2B). Capacitance measurements and electrochemical detection have been combined very elegantly to measure single vesicle fusion events and the associated release of neurotransmitter using cell-attached recordings from chromaffin cells (Albillos et al. 1997; Dernick et al. 2003) (see Figure

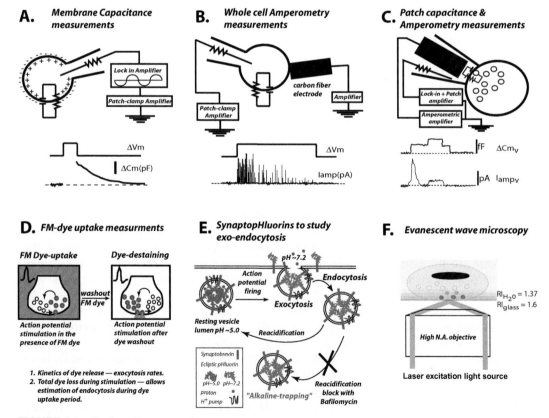

FIGURE 6.2.   For legend see opposite page.

6.2C). Such patch amperometric recording methods showed that opening of a fusion pore precedes the complete fusion of the granule and also revealed that some of the individual fusion events are transient.

Small synaptic terminals in the CNS measure about 0.5–1 μm in diameter and contain ~100–200 vesicles; terminals in the neuromuscular junction are much larger and contain ~10,000 vesicles. However, the

◄ _____

FIGURE 6.2. **Techniques to study exocytosis and endocytosis in neurons. A.** Membrane capacitance measurements allow the direct monitoring of the net membrane surface area of cells, which can be patch-clamped, with high temporal resolution. A net increase in membrane surface area during exocytosis leads to an increase in membrane capacitance (in the order of 1–100 pF); endocytosis leads to a decrease in membrane capacitance. This method has been used extensively in chromaffin cells, bipolar nerve terminals from retina, and more recently in the giant presynaptic terminal of the calyx of Held. **B.** Whole cell amperometry allows detection of catecholamines secreted from chromaffin cells and sympathetic neurons with high temporal resolution. A carbon fiber electrode is placed in close proximity to a cell. Oxidation of the secreted catecholamine at the carbon fiber electrode produces an oxidation current that is measured by an amplifier. This method has the sensitivity to detect individual vesicle fusion events (current amplitudes of ~1–2 pA). This method has been combined with capacitance measurements to determine how changes in membrane capacitance relate to secretion of catecholamines in chromaffin cells. **C.** Patch capacitance and amperometry is a combination of **A** and **B**, but used to measure single vesicle fusion events in a membrane patch using the cell-attached patch clamp method. This method has extremely high signal-to-noise and can measure capacitance changes arising from single vesicle fusion events (1–4 fF). A carbon fiber electrode is placed inside a patch pipet, and allows direct correlation of a capacitance event with secretion of catecholamine in chromaffin cells. Both a capacitance step and a spike of current in the amperometric trace accompany a complete fusion event. Transient fusion events are characterized by a step jump in capacitance followed by a step decrease in capacitance of similar amplitude. **D.** FM-dye uptake studies of presynaptic vesicle recycling. A pulse-chase protocol is used where vesicles are loaded with dye during a train of action potentials. Vesicles are first loaded with FM-dye by stimulating neurons in the presence of dye or by adding dye after a variable time delay after stimulation. Following dye wash out, exocytosis can be monitored with a second train of action potentials. During the loading phase, dye is taken up into newly endocytosing vesicles. The dye gets trapped within vesicles and does not passively diffuse out since it is positively charged. Following complete washout of dye from the outside, the terminals appear as fluorescent puncta. A second stimulation of the nerve terminals leads to dye release as vesicles undergo exocytosis. The time course of dye release during stimulation provides estimates of exocytosis. The amount of total dye that is released during the second stimulus provides an estimate of the amount of endocytosis that occurred during the loading phase. Performing multiple dye-loading experiments with various time delays after stimulation and then measuring uptake can provide an estimate of the time course of endocytosis. This method has been applied extensively to the frog neuromuscular junction and primary CNS neurons in culture and also in brain slices. **E.** Synaptic vesicle membrane protein recycling. A pH sensitive GFP molecule called ecliptic pHluorin fused to the lumenal domain of synaptobrevin can be used to study recycling of vesicles in terminals expressing this construct. Under resting conditions, when the vesicle interior is at pH of ~5.0, the pHluorin signal is completely quenched. During action potential firing the vesicles undergo exocytosis and the pHluorin moiety gets deprotonated leading to robust increase in fluorescence. Following end of the stimulus, the vesicles are endocytosed and the fluorescence recovers by reacidification of vesicles. The rise in fluorescence during stimulation represents the net balance of exocytosis and endocytosis. The fluorescence recovery kinetics, following end of stimulus, provide robust measure of endocytosis since reacidification is very rapid. Application of Bafilomycin results in a complete block of reacidification and allows quantification of all exocytic events. This alkaline trapping of vesicles allows one to count all exocytic events. This enables estimation of endocytosis during periods of stimulation by performing quantitative comparisons of fluorescence signals during stimulation with and without Bafilomycin. **F.** Evanescent Wave microscopy. This novel microscopy method makes use of the principle of total internal reflection of light when it travels from a medium of high refractive index like glass into a medium of low refractive index such as water. When the angle of incidence of the beam at the interface reaches a certain critical angle, it undergoes total internal reflection. However, not the entire beam is reflected at the interface and light penetrates into the second medium for a short distance, thus setting up an evanescent field. The depth of the evanescent field is determined by the numerical aperture of the objective and the difference in the refractive indices of the two media. This method allows one to illuminate ~100–200 nm beneath the plasma membrane with minimal excitation of fluorescence from deeper parts of the cell. It has been used to study exocytosis of dense core granules in chromaffin cells and small synaptic vesicles in bipolar terminals.

size of these terminals makes them difficult to patch clamp and record membrane capacitance changes. In the early 1990s, a novel imaging method was developed by Betz and colleagues to measure vesicle recycling in presynaptic terminals in the neuromuscular junction (reviewed in Cochilla et al. 1999). The styryl dye FM1-43 labels recycling vesicles in synaptic terminals when recycling is stimulated in the presence of the dye (see Figure 6.2D). The FM dyes are amphipathic molecules that intercalate into the outer leaflet of plasma membrane and are trapped in endocytosing vesicles. Following washout of dye in external medium, a subsequent train of action potentials allows visualization of vesicle exocytosis. Such pulse-chase experiments with FM dyes in hippocampal neurons have led to estimation of exocytic rates, the recycling vesicle pool size, and the time course of endocytosis in small synaptic terminals of the CNS (Ryan et al. 1993, 1996; Ryan and Smith 1995). Sensitive fluorescence measurements in these small synaptic terminals have demonstrated the quantal nature of vesicle cycle in the CNS (Ryan et al. 1997). Following endocytosis, vesicles become rapidly available for exocytosis and seem to maintain their identity during subsequent rounds of exocytosis (Murthy and Stevens 1998).

The advent of green fluorescent proteins led to the development of fusion constructs of vesicle or synaptic proteins to follow the activity-dependent cycling of these proteins at the synapse. A novel pH sensitive version of GFP (ecliptic pHluorin) was developed and fused to the lumenal domain of synaptobrevin, an integral membrane protein, in order to study vesicle recycling in synapses (Miesenbock et al. 1998). This fusion protein, called synaptopHluorin, allows real-time visualization of the trafficking of synaptic vesicles as they undergo cycles of exo- and endocytosis (Sankaranarayanan and Ryan 2000a). Ecliptic pHluorin has a pKa of ~7.1, therefore, when it is localized within a synaptic vesicle (pH of ~5.5)

the fluorescence is completely quenched (Sankaranarayanan et al. 2000b). During action potential firing, as the vesicles fuse with the plasma membrane the pHluorin moiety is externalized, leading to a large increase in fluorescence. The fluorescence signal recovers after endocytosis and reacidification of the vesicles. Since reacidification of vesicles occurs within seconds, the time course of fluorescence recovery over longer time scales is dictated by the time course of endocytosis of vesicles (Sankaranarayanan and Ryan 2000a). During action potential firing, the fluorescence signal represents the net balance of exo- and endocytosis. In order to estimate endocytic time course during action potential firing, a method called *alkaline trapping* was developed (Sankaranarayanan and Ryan 2001). Bafilomycin, an inhibitor of the vesicular proton pump, blocks reacidification of vesicles enabling measurement of all exocytic events that occur during a train of action potentials. Quantitative fluorescence difference signals in individual boutons, in the presence and absence of bafilomycin, allows estimation of endocytosis rates during stimulus trains.

Although light microscopic imaging of small synaptic terminals has provided novel insight into the process of synaptic vesicle recycling, the underlying processes involved in exocytosis and endocytosis of single vesicles are still below the limits of resolution of an optical microscope. These problems may be partially overcome by using the technique of evanescent wave microscopy or total internal reflection microscopy, which allows one to restrict the illumination to a narrow band near the bottom surface of a cell (Axelrod et al. 1992). In this technique, total internal reflection of a laser excitation beam at the glass-water interface allows a small evanescent wave to penetrate through the water. This evanescent wave decays exponentially with a space constant of 50 to 150 nm and allows visualization of fluorescent molecules and organelles in the subplasmalemmal region with high contrast (Steyer and Almers 2001). This method has been used extensively in chromaffin cells (Steyer et al. 1997) and also in larger bipolar terminals from the retina (Zenisek et al. 2000) to study the final steps in the process of exocytosis. At present it is difficult to use this method to study fusion events in synapses, as the postsynaptic membrane prevents access of the evanescent field.

C. Biogenesis and Organization of the Presynaptic Terminal

Neurons are polarized cells that have a cell body, numerous dendritic processes, and usually a single axon that innervates the target. The axon terminals, which form synapses, are characterized by a dense cluster of small synaptic vesicles that seem to maintain their identity over repeated cycles of exo-endocytosis. An important question that is still debated is the origin of synaptic vesicles at the synaptic terminal. Like all membranous compartments in a cell, synaptic vesicles originate in the cell body in the ER-Golgi biosynthetic pathway. The possible mechanisms by which synaptic vesicles initially populate the nerve terminal have been intensely debated, resulting in multiple hypotheses (Hannah et al. 1999; Roos and Kelly 2000).

One hypothesis is that each of the components of the nerve terminal is transported down the axon in individual packets. This would suggest that each component of the presynaptic terminal would carry a sorting peptide signal or a lipid-binding domain that will enable it to be preferentially trafficked to the nerve terminal. Such sorting signals would be similar to signals that regulate trafficking to the apical and basolateral membrane in epithelial cells (Rodriguez-Boulan 2005). However, such axonal sorting signals have not been identified thus far.

A second hypothesis is that a preassembled complex of synaptic vesicle proteins, endocytotic machinery, and large dense core vesicles is transported as a single unit or *prototerminal* along microtubules and form the presynaptic complex of proteins. This would suggest that a few proteins have sorting signals that target them toward the axonal terminal along microtubules, and other proteins get cotransported to the terminal by association. Recent studies in neurons in culture using a combination of live-cell imaging and ultrastructural analysis suggests that small organelles containing varied vesicular and tubulovesicular membrane structures are trafficked along the axon and form synapses at sites where the growth cone meets a dendritic spine (Ahmari et al. 2000). These organelles or puncta are much larger in size than small synaptic vesicles, and transported at speeds of ~0.5 μm/s along axons. Such organelles seem to undergo a form of immature activity-dependent recycling along axonal membranes as measured using electrophysiological methods and fluorescent dye-uptake studies (Zakharenko 1999; Kraszewski 1995). During this process of exo-endocytosis, small synaptic-like vesicles are generated along the axon via a Brefeldin-A sensitive mechanism. Brefeldin-A is a macrocyclic lactone from fungi, which disrupts ARF coat-dependent vesicle budding from endosomes. Such Brefeldin-A sensitive generation of vesicles is mediated via ARF and AP3 adaptor complex (Faundez et al. 1998). However, once the synaptic terminals mature, synaptic vesicles are generated by a Brefeldin-A insensitive, clathrin-coat and AP2-adaptor dependent mechanism (Shi et al. 1998). Interestingly, in mice lacking the neural adhesion molecule N-CAM, the Brefeldin-A sensitive recycling pathway is maintained within the nerve terminal in mature neuromuscular junctions but participates in ectopic vesicle cycling outside of the active zone (Polo-Parada et al. 2001). These data suggest the existence of a mechanism whereby a prefabricated synaptic vesicle cluster with all necessary presynaptic components is transported to the nerve terminal. Formation of a stable dendritic contact leads to establishment of a synaptic terminal with mature recycling process. How then is the mature presynaptic terminal and synaptic structures established?

During synaptogenesis, in addition to generating the vesicle cluster at the nerve terminal, the specializations that make up the presynaptic active zone are rapidly organized. Recent work has led to a better understanding of the molecular and cellular events leading to synapse formation and the timing of the establishment of the presynaptic and postsynaptic units of a synapse, in primary neuronal cultures (see Chapters 4 and 5 for addition discussion). These studies have revealed that scaffolding proteins such as Piccolo, Bassoon, and Rim are transported in dense-core vesicles of ~80 nm diameter along the axon (Shapira et al. 2003; Ziv and Garner 2004). It appears that arrival of these organelles at axonal filopodia and formation of a semistable interaction with the plasma-membrane signal the formation of a primordial presynaptic site. This site is then populated by synaptic vesicle clusters that mature to undergo activity-dependent exo- and endocytosis. There seems to be a critical temporal window of ~15–60 minutes during which this primordial presynaptic site is converted to a mature presynaptic site (Friedman et al. 2000).

These presynaptic events occur coincidentally with organization of cellular adhesion molecules such as N-cadherin, NCAM, which stabilize dynamic contacts and allow the activation of intracellular signalling cascades that induce pre- and postsynaptic differentiation. The protein interactions that occur at synaptic sites are coupled to intracellular signalling events and to intracellular scaffolding molecules that link with the presynaptic vesicle cluster (Scheiffele 2003). N-cadherin is a neuronal cell-adhesion molecule that interacts with catenins and the spectrin/actin cytoskeleton (Uchida et al. 1996; Bruses 2000; Bamji et al. 2003, 2005; Nishimura 2002). NCAM, a member of the

IgG superfamily of adhesion molecules, binds spectrin and helps to stabilize synaptic contacts, and appears to play a role in triggering final maturation of the recycling machinery (Polo-Parada et al. 2001). On the postsynaptic side, molecules such as neuroligin 1 and 2 (Scheiffele et al. 2000), and the immunoglobulin domain-containing protein SynCAM, serve to promote synaptic vesicle clustering along overlying axons (Biederer et al. 2002). SynCAM is a homophilic adhesion molecule; neuroligin, a heterotopic inducer of synaptic sites, induces the clustering of presynaptically expressed β-neurexin, which together lead to formation of functional active zones and synapses (Nam and Chen 2005; Sara et al. 2005a). Thus, the molecular events that lead to the formation of mature synapses are temporally synchronized in the pre- and postsynaptic sites and culminate in the organization of the presynaptic vesicle cluster, the active zone, and the postsynaptic densities (Chapters 4 and 5 directly address these issues).

D. The Presynaptic Active Zone

The active zone is a highly specialized region of the cytoplasm in the presynaptic nerve terminal that directly faces the synaptic cleft. This region of the nerve terminal holds a small subset of vesicles that are believed to undergo rapid exocytosis on arrival of a nerve impulse. In CNS synapses, the active zone is an electron-dense structure and contains hexagonal grid-like structures with synaptic vesicles embedded in the depressions of the grid (Akert et al. 1971). In contrast, the active zones of neuromuscular synapses have elongated structures such as ridges, beams, and pegs, containing vesicles lined up like beads on a string (Harlow et al. 2001). Synapses in sensory pathways, such as the bipolar rod cells in the retina and the inner hair cells in the auditory epithelium, which release transmitter continuously in response to graded changes in membrane potential, have structures called ribbons or dense bodies (Lenzi and von Gersdorff 2001). The ribbons and dense bodies have presynaptic specializations that are similar to those of the active zone of conventional synapses. The ribbons help to localize vesicles and may function like a conveyer belt to help deliver vesicles to the plasma membrane (Zenisek et al. 2000). Thus, the active zone complex mediates vesicle recruitment to the plasma membrane and directs exocytosis precisely onto postsynaptic receptors.

In the last five to 10 years, numerous candidate presynaptic scaffolding proteins have been identified (reviewed in Dresbach et al. 2001) (see Chapters 4 and 5 for additional information). These include members of the Rab3 interacting molecule (RIM) (RIM1α, 2α/β/γ, 3γ, and 4γ) (Wang et al. 1997; Wang and Sudhof 2003), Piccolo/aczonin (Fenster et al. 2000; Wang et al. 1999), Bassoon family (tom Dieck et al. 1998), Munc13 (Munc13-1, -2, and -3; Brose et al. 1995), ERC (ELKS/RIM/CAST) proteins including ERC1b and ERC2 (Wang et al. 2002; Ohtsuka et al. 2002), and α-Liprins (Schoch et al. 2002; Ko et al. 2003). Piccolo and Bassoon are large proteins, homologous to each other and are part of the cytomatrix involved in protein-protein interactions and signalling (tom Dieck et al. 1998). RIM shares homology with both Piccolo and Bassoon and was identified as an interactor of Rab3, a vesicle associated protein. RIM1α and RIM2α have three types of domains: an N-terminal zinc-finger domain that interacts with the synaptic vesicle protein Rab3 and the active zone protein Munc13-1; a central PDZ domain that binds the C-terminus of ERCs; and two C-terminal C2 domains, of which the second, so-called C2B domain binds to α-Liprins. RIM binding protein, RIM-BP, is an SH3-domain containing protein that binds to RIMs (Wang et al. 2000). Other components of the active zone contain the tripartite protein complex comprising Mint, CASK, and Veli (Butz et al. 1998). CASK is a member of the membrane-associated guanylate kinase family of proteins, which interacts with calcium channels, β-neurexins, mint, and Veli.

The assembly of the cytoskeletal matrix of the active zone (CAZ) seems to occur in a hierarchical fashion. Munc13-1 requires the interaction with the N-terminal zinc-finger domain of RIM1α/2α for it synaptic localization (Schoch et al. 2002; Dulubova et al. 2005). RIM1α/2α binds to the coiled-coil proteins ERC1a and ERC2 and with α-Liprins, but these proteins do not require RIM1α for their synaptic localization (Wang et al. 2002; Ko et al. 2003). The amino terminus of ERC2/Cast1, binds the coiled-coil domain of both Piccolo and Bassoon (Takao-Rikitsu 2004). Syntaxin, the target SNARE on the plasma membrane, binds to Munc13-1 (Betz et al. 1997). The synaptic vesicles in the cluster can thus bind to the active zone via binding of the vesicle SNARE synaptobrevin to syntaxin and SNAP-25 and binding of the vesicle Rab proteins Rab3A, 3B, 3C, and 3D to the N-terminal domain of RIM1α and RIM2α, which also binds to Munc13-1 (Wang et al. 1997, 2000; Betz et al. 2001; Schoch et al. 2002). These data indicate that multiple protein-protein interactions mediated by a number of proteins including RIM, Rab3, ERCs, α-Liprin, Piccolo, and Bassoon might be crucial in the molecular assembly of the CAZ.

Genetic ablation of Bassoon led to inactivation of a significant fraction of glutamatergic synapses in mice (Altrock et al. 2003). At these synapses, vesicles are clustered and docked in normal numbers but are unable to fuse. These data suggest that Bassoon is not essential for synapse formation but plays an essential role in secretion from a subset of glutamatergic synapses. In contrast, photoreceptor ribbons lacking Bassoon are not anchored to the presynaptic active zones. This results in impaired photoreceptor synaptic transmission and abnormal dendritic branching of neurons that are postsynaptic to photoreceptors

(Dick et al. 2003). Similarly, anchoring of inner hair cell ribbons is impaired in mice lacking the presynaptic scaffolding protein Bassoon, leading to a decrease in the presynaptic readily releasable vesicle pool, and impaired synchronous auditory signalling (Khimich et al. 2005). These results suggest that Bassoon is required for anchoring ribbons to the presynaptic release site and thereby regulates neurotransmitter release. In both worms and mice, genetic disruption of RIM function does not produce obvious morphological defects but leads to reduced neurotransmitter release (Koushika et al. 2001; Schoch et al. 2002) suggesting that RIM may not have an essential function in active zone organization.

The vesicle pool docked at the plasma membrane has been examined using electron microscopy and imaging techniques. In neurosecretory cells, granules seem to be docked at plasma membrane sites and multiple exocytic events have been measured at select locations on the membrane, suggestive of hotspots for vesicle docking and release (Steyer et al. 1997; Oheim et al. 1998). Some granules also seem to undergo reversible docking, suggesting that this process is dynamically regulated in chromaffin cells (Steyer et al. 1997) and in retinal bipolar terminals (Zenisek et al. 2000). At synaptic terminals, newly endocytosed vesicles can gain access to the docking sites within a few minutes, even in the absence of significant emptying of the docking sites due to exocytosis, further suggesting that the docked pool of vesicles is actively regulated over a timescale of minutes (Ryan et al. 1996; Murthy and Stevens 1999). In adrenal chromaffin cells, depletion of ATP leads to loss of the readily releasable pool of vesicles over a period of minutes, suggesting an energy-dependent step in maintaining the docked and primed pool of vesicles (Holz et al. 1989; Hay and Martin 1992; Parsons et al. 1995). Thus, although the molecular machinery of the presynaptic active zone has been identified and characterized, the precise interplay between the

components and the relationships of these proteins in mediating and regulating vesicle docking is still far from being completely understood. It remains unclear whether all docked vesicles actually participate in neurotransmitter release and recycling, as few experiments have demonstrated a complete depletion of the morphologically defined docked pool.

E. Synaptic Vesicles—Local Neurotransmitter Filling of Vesicles

Unlike secretion of proteins and neuropeptides that require transport of newly synthesized material from the Golgi apparatus, the local membrane recycling and synthesis of amines such as glutamate, aspartate, GABA, serotonin, acetylcholine, norepinephrine, and epinephrine enable rapid filling of neurotransmitter at the nerve terminal. Following exocytosis, vesicles are internalized and rapidly reacidified, which provides the appropriate driving force required to refill them with neurotransmitter.

Synaptic vesicles are small (20–25 nm radius) and abundant organelles whose function is to take up and release neurotransmitters. Purified vesicles have a protein:phospholipid ratio of 1:3 with an unremarkable lipid composition (40% phosphatidylcholine, 32% phosphatidylethanolamine, 12% phosphatidylserine, 5% phosphatidylinositol, 10% cholesterol, wt/wt; Benfenati et al. 1989). Synaptic vesicle proteins fall into two major classes: transport proteins involved in neurotransmitter uptake and trafficking proteins that participate in synaptic vesicle exo- and endocytosis and recycling. Transport proteins are composed of a vacuolar-type proton pump that generates the electrochemical gradient, which fuels neurotransmitter uptake and vesicular neurotransmitter transporters that mediate the actual uptake (reviewed by Fykse and Fonnum 1996; see Chapter 11). The trafficking proteins are complex and include intrinsic membrane proteins, proteins associated via posttranslational lipid modifications, and peripherally bound proteins.

Synaptic vesicles accumulate and store neurotransmitters at high concentrations by active transport, driven by a vacuolar proton pump whose activity establishes an electrochemical gradient across the vesicle membrane (Maycox et al. 1988; Tabb et al. 1992). The vesicle proton pump is the largest multiprotein complex (1 million Dalton) in the vesicle, consisting of at least 13 subunits (Arata et al. 2002). The proton pump is composed of a larger peripheral complex called V1, which mediates the ATPase activity, and an integral membrane complex called V0, which mediates proton translocation. Although the V0 subunit is thought to be primarily involved in proton transport, it has been suggested to have a direct role in mediating yeast vacuolar fusion as a fusion pore. The V0 subunit either alone, or in conjunction with the t-SNARE syntaxin, can mediate the fusion pore that is formed during vacuole fusion (Bayer et al. 2003). In *Drosophila*, mutants of the V0 component have decreased evoked release of neurotransmitter and an accumulation of vesicles, suggesting a role in the final stages of vesicle fusion (Heisinger et al. 2005).

Vesicular uptake is mediated by four distinct neurotransmitter uptake systems. Transporters use either a membrane potential gradient or a combination of membrane potential and pH to transport neurotransmitter (reviewed by Fykse and Fonnum 1996). Vesicular glutamate transporters (VGlut1-VGlut3) mediate uptake of glutamate using the membrane potential gradient for energy whereas vesicular GABA transporters use both membrane potential and the proton gradient to provide the driving force (Fremeau et al. 2004). Monoamines (catecholamines, histamine, and serotonin) are taken up into vesicles by two differentially expressed transporters (Erickson et al. 1992; Liu et al. 1992). GABA and glycine (McIntyre et al. 1997; Sagne et al. 1997) and acetylcholine (Roghani et al. 1994; Varoqui and Erickson 1997) each have single transporters. The four families of transporters are distantly related to each other, but differ mechanistically (see Chapter 11 for additional discussion).

Expression of a particular transporter type is a major determinant of the type of neurotransmitter used by that neuron. When VGlut1 was transfected into GABAergic neurons, their synapses became glutamatergic, in addition to secreting GABA (Takamori et al. 2000). Thus, simple expression of the glutamatergic transporter was enough to specify neurotransmitter type. VGlut3 is present in many neurons not previously considered glutamatergic (e.g., the cholinergic interneurons of the striatum; Fremeau et al. 2002, Schafer et al. 2002), suggesting that coexpression of glutamate transporters with other neurotransmitter transporters may allow some neurons to use multiple classical transmitters.

The amount of transmitter released per synaptic vesicle exocytic event is dependent on both the size and the concentration of neurotransmitter in each vesicle (reviewed in van der Kloot 1991; Sulzer and Pothos 2000). Variation in single quanta of release may be due to vesicles of different sizes (Bekkers et al. 1990; Bruns et al. 2000) that possess different total amounts of neurotransmitter, or to differences in absolute concentration of neurotransmitter per vesicle. In the calyx of Held synaptic terminals, glutamate transport into vesicles is enhanced by increasing the cytosolic glutamate concentration (Yamashita et al. 2003). Overexpression of the vesicular monoamine transporter increases the amount of release mediated by exocytosis of a single vesicle (Pothos et al. 2000), whereas hemizygous deletion of this transporter dramatically alters monoaminergic signalling in brain, suggesting that levels of transporter expression are physiologically regulated (Wang et al. 1997).

F. Protein Players in Synaptic Vesicle Exocytosis—The SNARE Hypothesis

The mechanism by which membrane fusion occurs has been a subject of intense investigation because of the universal nature of the phenomenon that drives trafficking between intracellular compartments and the plasma membrane. The debate over membrane fusion has focused on whether fusion is mediated by lipids, proteins, or both lipids and proteins (Lindau and Almers, 1995).

It is now well established that membrane fusion *in vivo* requires the specific interaction of the membrane proteins called *soluble N-ethylmaleimide sensitive fusion protein attachment receptors* (SNAREs) in the vesicle and target membranes. For synaptic vesicle fusion the vesicle or v-SNARE is synaptobrevin or VAMP (vesicle associated membrane protein), and the target or t-SNARE are the proteins syntaxin and SNAP-25 (synaptosomal-associated protein-25) (Sollner et al. 1993a). The SNARE hypothesis postulates that membrane fusion results from the interaction of specific vesicle and target SNAREs that bring their respective membranes into close apposition leading to fusion (Sollner et al. 1993b). This hypothesis originally was based on *in vitro* fusion studies using purified proteins (Sollner et al. 1993b) and on results of genetic studies in yeast, which led to the identification of the *Sec* family of yeast secretion mutants (Sanders and Schekman 1993). Before the formulation of the SNARE hypothesis it was well known that clostridial and botulinum toxins were potent neurotoxins and blocked neurotransmission. Subsequent work revealed that these toxins cleave various SNARE proteins and lead to the disruption of synaptic vesicle fusion (Schiavo et al. 1994).

Synaptic vesicle fusion is driven by four-helical bundle formation or zippering of four domains present in syntaxin, SNAP-25, and synaptobrevin. The R-SNARE motif from synaptobrevin, the Qa-SNARE motif of syntaxin, and the Qb and Qc motif in SNAP-25 mediate this core complex of SNAREs that drives vesicle fusion (Sutton et al. 1998; Hughson 1999; Chen et al. 2001). In purified liposome preparations, the SNAREs represent the minimal machinery required for membrane fusion, although the kinetics of the fusion event are extremely slow (Weber et al. 1998; Parlati et al. 1999). Other protein components in the vesicle membrane, presynaptic terminal, plasma membrane, and a calcium sensor are essential to regulate the speed of fusion that is critical in neurotransmission. During fusion the SNAREs are assembled into a helical bundle that is disassembled prior to subsequent rounds of fusion (Sollner et al. 1993c; Otto et al. 1997). The disassembly of SNARE complexes is mediated by the ATPase function of NSF or N-ethylmaleimide sensitive fusion protein (Wilson et al. 1992). Block of NSF action leads to progressive accumulation of fusion incompetent vesicles (Schweizer et al. 1998). Cleavage of SNAREs by toxins leads to abolition of vesicle fusion and a similar progressive accumulation of vesicles at nerve terminals (Xu et al. 1998; Sakaba et al. 2005).

Although there is general agreement that SNAREs are essential for vesicle membrane fusion, genetic ablation of certain SNAREs that normally are present at the synapse did not lead to an expected phenotype. Deletion of SNAREs such as synaptobrevin or SNAP-25 led to an abolishment of approximately 90% of synaptic vesicle fusion (Schoch et al. 2001). In these examples, calcium-dependent exocytosis was more severely compromised than spontaneous fusion and hypertonic solution-induced exocytosis. Similar results were found when SNAP-25 was knocked out in mice (Washbourne et al. 2002). These results suggest that although synchronized calcium-dependent exocytosis absolutely requires these specific SNAREs, alternate SNAREs may serve to carry out fusion at the synapse as well.

In contrast, when Munc18 was deleted in mice there was a complete loss of exocytosis, suggesting that other proteins besides SNAREs are essential for neuronal exocytosis (Verhage et al. 2000). Munc18 or Sec1 seems to control the final steps of fusion in yeast and in nerve terminals (Graham et al. 2004; Gallwitz and Jahn 2003). Munc18 is a cytosolic protein that binds to the N-terminal region of syntaxin and folds back onto the SNARE domain into a closed conformation (Misisura 2000). Syntaxin must dissociate from Munc18 before it can form the core complex with SNAP-25 and synaptobrevin (Dulubova 1999; 2003). This Munc18 interaction with syntaxin would serve to inhibit fusion of vesicles. However, Munc18 can bind to the assembled yeast SNARE complex and in *in vitro* studies (Carr et al. 1999; Scott et al. 2004). Other proteins such as Munc13 and RIM may allow syntaxin to transition from a closed to open state, which is necessary for SNARE complex assembly (Rizo and Sudhof 2002). In *C. elegans*, the loss of exocytosis due to Munc13 or RIM mutations can be overcome by expression of the open form of syntaxin (Richmond et al. 2001).

Other proteins that are thought to perform an essential function in SNARE assembly include Synaptophysins, Complexins, Tomosyn, and Amisyn. Synaptophysins are abundant synaptic vesicle membrane proteins that bind directly to synaptobrevin. Synaptobrevin binding to the SNARE complex and to synaptophysins are mutually exclusive, suggesting that synaptobrevin binding to synaptophysin may decrease core complex formation (Edelmann 1995). Complexins or Synaphins are small cytosolic proteins that can bind to assembled SNARE complexes. Complexins seem to interact in a groove in the C-terminus of the SNARE complex that is not present in partially assembled complexes (Pabst et al. 2000; 2002). In the squid giant synapse disruption of complexin-SNARE interactions prevent SNARE complex assembly and decrease transmitter release (Tokumaru et al. 2001). Mice lacking complexins have a severe decrease in calcium triggered neurotransmitter release (Reim et al. 2001). Tomosyn and Amisyn are soluble cytosolic proteins that have an R-SNARE domain similar to that of Synaptobrevin. Both proteins inhibit exocytosis by interfering with the ability of Synaptobrevin to form core-complexes (Hatsuzawa et al. 2003).

The nature of the fusion process in physiological systems is not completely understood due the difficulty in directly studying this process. Electrophysiological studies using Patch-amperometry have revealed a transient opening of a fusion pore preceding a small percentage of dense core fusion events in chromaffin cells (Albillos et al. 1997). In some instances the fusion pore opening is reversible and referred to as flickering fusion (Henkel and Almers 1996; Lindau and Alvarez de Toledo 2003). Under physiological calcium concentrations the frequency of these events is very low, but when external calcium is elevated these transient fusion pore openings become more frequent (Ales et al. 1999). Similar studies have not been possible in small synaptic terminals from neurons due to the inability to patch clamp nerve terminals. Recently, capacitance measurements in large CNS terminals such as the calyx of Held have provided novel insight into the kinetics of fast exocytosis and endocytosis (Sun and Wu 2001; Sun et al. 2002). The molecular nature of the vesicle retrieval process is not known at central synapses. A direct reversal of the fusion process at the release site has been proposed based on kinetic measurements of vesicle cycling in small synaptic terminals (Klingauf et al. 1998; Pyle et al. 2000). However imaging studies of small synaptic vesicle fusion in large bipolar terminals using FM-dyes have suggested that dye release from fusing vesicles is complete within 30 ms of stimulus onset, suggesting a fusion pore that allows lipid continuity between the vesicle and plasma membrane (Zenisek et al. 2002).

Although membrane lipid dyes are freely mobile during exocytosis, integral plasma membrane components and dense core granules seem to escape much more slowly based on evidence from dense core granule exocytosis (Perrias et al. 2004). These data suggest that the fusion pore likely is composed of both proteins and lipids and allows short amphipathic lipid dyes to escape rapidly from the vesicle membrane. As mentioned earlier, the V0 subunit of the proton pump may serve as a fusion pore based on studies of *Drosophila* mutants (Hiesinger et al. 2005). Other proteins such as synaptotagmin and the SNARE's syntaxin and SNAP-25 may also regulate fusion pore dynamics during calcium-triggered exocytosis (Wang et al. 2003; Bai et al. 2004).

G. Priming of Vesicles—What Tells a Vesicle When It Is Fusion Ready

Immediately prior to or following the docking of a vesicle with the plasma membrane, synaptic vesicles undergo a series of maturation steps, making them fusion competent. Collectively these steps are referred to as priming. Priming may involve protein-protein interactions; second messenger mediated events such as calcium signalling (von-Ruden and Neher 1993; Smith et al. 1998) or protein kinase activation (Gillis et al. 1996); or local lipid changes mediated by diacylglycerol (Rhee et al) or PIP2 (phosphatidyl-inositol 4,5, bisphosphate) (Eberhard et al. 1990; Hay and Martin 1993, 1995). In dense core granule exocytosis, an ATP-dependent priming step has been identified as essential for vesicles to become fusion-competent (Holz et al. 1989; Parsons et al. 1995). ATP-dependent small synaptic vesicle fusion in bipolar terminals also has been established (Heidelberger 1998; Heidelberger et al. 2002). This ATP-dependence could be mediated via NSF-dependent dissociation of SNARE complexes that allow them to form productive core complexes necessary for fusion. Alternatively, the ATP-dependence also could be due to essential phosphorylation reactions driven by protein (Gillis et al. 1996; Saitoh et al. 2001) or lipid kinases that make vesicles fusion competent (Dipaolo et al. 2004).

Presynaptic proteins such as Munc13, Rab3, and RIM may play an essential role in priming. Munc13, a presynaptic protein, initially was identified in an uncoordinated locomotion mutant screen in a *C. elegans*. Three isoforms of Munc13 (Munc13-1, Munc13-2, and Munc13-3) are present in mammals. Munc13 has an N-terminal C2 domain, a central C1/C2 tandem domain, and a C-terminal C2 domain (Brose et al. 1995). Munc13s are targets of the diacylglycerol (DAG) second messenger pathway. The C1 domain function of Munc13-1 is essential for DAG and phorbol ester binding that causes potentiation of neurotransmitter release in hippocampal neurons (Rhee et al. 2002). DAG, a membrane phospholipid produced in response to receptor activation, can regulate vesicle release via Munc13. Munc13 contains a calmodulin binding domain that mediates the calcium-dependent short-term synaptic enhancement that occurs during sustained action potential activity (Junge et al. 2004). Munc13s interact with the SNARE protein syntaxin (Betz et al. 1997) and promote SNARE complex formation and fusion competence of synaptic vesicles (Richmond et al. 2001). Genetic studies in mouse, fly, and nematode have established an essential role for Munc13 in synaptic vesicle priming and for maintaining the readily releasable pool of vesicles (Aravamudan et al. 1999; Augustin et al. 1999; Richmond et al. 1999). In Munc13-1 knockout mice, a small number of ultrastructurally normal glutamatergic synapses cannot release transmitter in response to action potentials, calcium-ionophore, or hypertonic sucrose solution (Augustin et al. 1999; Rosenmund et al. 2002). In Munc13-1/13-2 double knockout mice, glutamatergic synapses, but not GABAergic synapses, have severe deficits in evoked and spontaneous release events, but form normal numbers of synapses with typical ultrastructural features (Varoqueaux et al. 2002).

Rab proteins are small GTP binding cytosolic proteins, which associate peripherally with the synaptic vesicle membrane and seem to play a critical role in synaptic vesicle recycling. Rab3A is the most abundant Rab protein in the brain. Rab3A undergoes cycles of vesicle association and dissociation during synaptic vesicle exo- and endocytosis (Fischer von Mollard 1991). When Rab3A binds to GTP it associates with the vesicle membrane. During or after calcium-triggered synaptic vesicle fusion, the GTP is hydrolyzed to GDP and Rab-GDP dissociates from vesicles. Thus Rabs can play important roles in determining the progression of the vesicle membrane through the different steps in the vesicle cycle. In Rab3A knockout mice, short-term synaptic plasticity is diminished in hippocampal CA1 synapses without a change in the readily releasable pool of vesicles (Geppert et al. 1994a). In hippocampal mossy fiber synapses that exhibit protein kinase A (PKA)-dependent presynaptic form of LTP, Rab3A deletion leads to inhibition of LTP (Castillo et al. 1997). Since Rab3A is not a direct PKA substrate the LTP is not due to direct phosphorylation of Rab3A. The Rab3A effectors rabphilin and RIM1α have N-terminal zinc-finger domains that interact with all Rab3 isoforms, have a central PKA phosphorylation site, and have two C-terminal C2 binding sites (Yamaguchi et al. 1993; Li et al. 1994). Rabphilin also undergoes a calcium-dependent cycling on vesicles mediated via its interaction with Rab3. However deletion of rabphilin failed to produce an impairment in vesicle trafficking (Schluter et al. 1999).

Although RIM seems to be involved in active zone assembly, in both worms and mice, genetic disruption of RIM function does not produce obvious morphological defects but leads to reduced neurotransmitter release (Koushika et al. 2001; Schoch et al. 2002) and abolition of presynaptic LTP in the mossy fiber synapse (Castillo et al. 2002). The regulatory role of RIM on release may be mediated by its interaction with Munc13, a critical player in the priming of synaptic vesicle for exocytosis (Betz et al. 2001). In neuronal cultures from mice lacking RIM1α, a large reduction in the readily releasable pool of vesicles and altered short-term plasticity is observed. However, no effect on synapse formation, spontaneous release, overall Ca^{2+} sensitivity of release, or synaptic vesicle recycling has been observed (Calakos 2004). It has been suggested that RIM1α phosphorylation by PKA is essential for mossy fiber LTP (Lonart et al. 2003). Thus both the GTP-dependent binding of Rab3A to the N-terminal region of RIM1α and its phosphorylation might lead to a change in active zone structure that results in an increase in the number of release-ready vesicles or a change the efficiency of release.

The lipid metabolite PtdIns(4,5)P2 (phosphatidyl inositol bis-phosphate) is an essential priming factor in exocytosis from neuroendocrine cells (Eberhard et al. 1990; Hay and Martin 1993; Hay et al. 1995). However, its role in neuronal exocytosis was not well established until recently when a deletion mutant for phosphatidyl inositol-phosphate kinase type 1γ was analyzed (DiPaolo et al. 2004). The phosphorylation of PtdIns(4)P by PIP kinase type 1γ is the major pathway for the generation of PtdIns(4,5)P2. Decreased levels of PtdIns(4,5)P2 in the brain and an impairment in the depolarization-dependent synthesis of PtdIns(4,5)P2 in nerve terminals led to early postnatal lethality and synaptic defects in mice (Dipaolo et al. 2004). The synaptic defects include decreased frequency of miniature currents, enhanced synaptic depression, a smaller readily releasable pool of vesicles, delayed endocytosis, and slower recycling kinetics.

H. Calcium Regulation of Neurotransmitter Release

The ultimate trigger that mediates exocytosis following the arrival of an action potential in the nerve terminal is the localized increase in presynaptic calcium. Although vesicles fuse under resting conditions, leading to miniature synaptic currents in the postsynaptic cell, the probability of release increases dramatically following elevation of cytosolic calcium. The large concentration gradient between intracellular (~100 nM) and extracellular calcium (~2 mM) leads to a rapid rise in presynaptic calcium after the opening of voltage-dependent calcium channels. Primed and docked vesicles sense local calcium changes in the range of 10 to 100 μM and undergo rapid exocytosis of neurotransmitter into the synaptic cleft (Schneggenburger and Neher 2005).

Classical studies using the frog neuromuscular junction delineated the relationship between extracellular calcium concentration and transmitter release and suggested that the cooperative action of two or more calcium ions was necessary to drive exocytosis (Katz 1969). The effective calcium concentration required to trigger rapid fusion of vesicles has been directly determined via intracellular calcium measurements following flash photolysis of caged calcium in neuroendocrine cells and large synaptic terminals (Ellis-Davis 2003). Rapid and uniform elevation of intraterminal calcium concentrations have allowed determination of the calcium requirements for exocytosis. These experiments have revealed that the effective or half maximal calcium concentration required to trigger exocytic release is ~10–20 μM in neuroendocrine chromaffin cells (Heinemann et al. 1994), ~100–200 μM in retinal bipolar neurons (Heidelberger et al. 1994) and ~5–10 μM in the giant synapse of the calyx of Held—the cochlear cell relay nucleus in the midbrain (Bollmann et al. 2000; Bollmann and Sakmann 2005). The very steep relationship between cytosolic calcium concentration and vesicle fusion suggests that exocytosis is a highly cooperative process requiring binding of at least two to five calcium ions to a calcium sensor.

The calcium-dependence of exocytosis is likely mediated by calcium sensors located on the vesicle membrane or at the site of fusion on the plasma membrane. The primary candidate for a calcium sensor for exocytosis is synaptotagmin I, an integral vesicle membrane protein (Chapman 2002). Synaptotagmin I is a single transmembrane protein with a short intravesicular domain, and two cytosolic C2 domains (C2A and C2B) (Perin et al. 1990; Geppert et al. 1991). The C2A domain binds three calcium ions and the C2B domain binds two calcium ions, suggestive of cooperative binding of calcium (Ubach et al. 1998). Although the intrinsic affinity of the C2 domains for calcium is low, the apparent affinity increases approximately 1000-fold upon phospholipid binding (Fernandez-Chacon et al. 2001; Bai et al. 2004). Genetic ablation of synaptotagmin I is lethal in mice. In hippocampal cells derived from these mice, a selective loss of the fast calcium-triggered synchronous exocytosis without a deficit in asynchronous release was observed (Geppert et al. 1994b). Neurons derived from a different mouse with a single point mutation in the synaptotagmin I C2A domain resulting in a lower affinity for calcium displayed a change in the apparent affinity for calcium-triggered exocytosis (Fernandez-Chacon et al. 2001). These results suggest that interaction of the C2 domain with lipids and proteins in the plasma membrane catalyze SNARE-mediated membrane fusion in response to calcium (Bai and Chapman 2004). However, studies in *Drosophila* have shown that only the C2B, and not the C2A, domain is essential in calcium-triggered synchronous release (Robinson et al. 2002).

In addition to synaptotagmins 1 and 2, numerous other proteins with C2 domains have been identified on vesicles such as synaptotagmin 3, 6, 7 (Sudhof 2002), DOC2, rabphilin, as well as the active zone proteins such as RIM and Piccolo (Rizo and Sudhof 1998). These proteins could mediate the calcium-sensitivity of release. SNAP-25 deletion mutants also have altered calcium-triggered exocytosis and loss of the rapid phase of exocytosis in chromaffin cells, suggesting that SNAP-25 could mediate the calcium-dependence for exocytosis (Sorensen et al. 2003). In *Drosophila*, synaptobrevin, and syntaxin mutants have alterations in the cooperativity of calcium-dependent release (Stewart et al. 2000). Genetic ablation of Complexin 1 and 2 in mice also results in severe deficits in calcium-triggered exocytosis (Reim et al. 2001). Finally, cysteine string proteins (CSPs), when deleted in flies, led to a shift in calcium-dependence of exocytosis to higher calcium without a change in cooperativity (Dawson-Scully et al. 2000). Thus, there are a number of proteins that have subtle roles in regulating the calcium-dependent fusion of synaptic vesicles.

An important mechanism that contributes to the high fidelity of calcium-triggered neurotransmitter release is the close localization of presynaptic calcium channels to the fusion machinery. A high density of calcium channels of the P/Q type (Cav2.1) and N-type (Cav2.2) are located at the presynaptic nerve terminal (Robitaille et al. 1993). These channels directly interact with components of the SNARE complex at the nerve terminal (Sheng et al. 1998). The synprint or the synaptic interaction sites, located in the intracellular loops of calcium channels, binds to SNAP-25, syntaxin, and synaptotagmin in a calcium-dependent manner (optimal binding at $10–30\,\mu M$) (Sheng et al. 1997; Yokoyama 2005). Introduction of the synprint peptides into presynaptic neurons significantly reduced neurotransmitter release in frog neuromuscular junctions and cervical ganglion neurons (Mochida et al. 1996) possibly due to detachment of docked vesicles from calcium channels. These studies suggest

that presynaptic calcium channels not only provide the calcium signal required for fusion, but also contain structural elements that are essential for vesicle docking, priming, and fusion processes (Catterall 1999).

How is the calcium concentration near the release sites controlled? In addition to the opening of calcium channels, allowing calcium to enter the bouton, there is evidence that calcium-induced calcium release from ER stores (Emptage et al. 2001) can enhance the calcium concentrations at the terminal. Rapid buffering of calcium by cytosolic buffers and the activation of a sodium/calcium exchanger and calcium-pumps leads to rapid recovery of the calcium signal in the nerve terminal. Mitochondria also participate in calcium buffering in nerve terminals. Thus, the tight interplay between specific channels, transporters, and mitochondria is essential for a fine-tuned regulation of Ca^{2+} concentrations and transmitter release in individual boutons (Scotti et al. 1999).

Although it is believed that the fusion machinery and the underlying calcium-dependence of spontaneous fusion are similar to that of evoked release, recent studies indicate that the underlying mechanisms of spontaneous fusion may be different from that of evoked release (Sara et al. 2005b).

I. Synaptic Vesicle Pools and Their Regulation

Ultrastructurally, the total vesicle pool in hippocampal synapses is ~100–200 vesicles with about 10% docked at the plasma membrane (Harata et al. 2001; Schikorski and Stevens 2001). Synaptic vesicle pools have been defined functionally in a variety of ways, usually based on the kinetics of their use. It had long been assumed that the vesicles that fuse first by a brief train of action potentials correspond to the docked vesicle pool, or perhaps the first two rows of vesicles. However, recent experiments using photoconversion of FM 1-43 at the frog neuromuscular junction showed that the pool of vesicles preferentially released by a brief stimulus were distributed throughout many rows of vesicles beyond the active zone, whereas other docked vesicles were not mobilized to fuse (Rizzoli and Betz 2004). Thus, the preference for being the first-used vesicles during repetitive stimulation does not appear to be based on some privileged location with respect to the active zone, but must instead reflect a biochemical advantage. The term readily releasable pool usually has been used to define those vesicles that are used first during stimulation, and often has been thought of as equivalent to those that are released during a brief hypertonic shock (Rosenmund and Stevens 1996). However, hypertonic sucrose usually releases a much larger pool than those released by the first 20 to 40 AP in dissociated hippocampal neurons (Moulder and Mennerick 2005).

In cultured hippocampal neurons, maximal labeling of vesicles in terminals is observed after about 600 action potentials, suggesting that this stimulus number is equivalent to the total recycling pool in these synapses (Ryan and Smith 1995). All the vesicles that are labelled with FM 1-43 can be released with an exponential time course during subsequent trains of action potentials. Photoconversion of FM 1-43 after maximal labelling revealed that only a fraction (~25–40%) of all the vesicle had taken up dye, thus defining a third pool of nonrecycling vesicles (Harata et al. 2001). Similar conclusions have been reached with FM 1-43 staining in the calyx of Held (de Lange et al. 2003) as well as in hippocampal synapses using synaptopHluorin (that examines the fraction of vesicles that become alkaline during prolonged stimulation in the presence of bafilomycin) (Fernandez-Alfonso et al. 2001; Li et al. 2005). Strikingly, releasable vesicles are not only anatomically indistinguishable, but they are also distributed at random in the vesicle cluster (Harata et al. 2001; Micheva

and Smith 2005; de Lange et al. 2003). Whether nonreleasable vesicles can be made to participate in recycling, and what the role of this reluctant pool is remains unknown. This raises the question of how vesicle clusters are maintained over time and what regulates their ability to participate in the recycling pool in a mature synaptic terminal.

Synaptic vesicles are clustered at the terminal and show an abrupt drop off in density at the edge of the terminal (Heuser and Reese 1981). Based on ultrastructural studies, actin has been proposed to be a potential scaffold that holds vesicles within the cluster (Landis et al. 1988; Hirokawa et al. 1989). However many studies have shown that actin largely is excluded from the synaptic vesicle cluster and might play a role in preventing vesicles from escaping from the terminal (Dunaevsky and Connor 2000; Colicos 2001), or in directing vesicles to the cluster following retrieval (Shupliakov et al. 2002). Live cell imaging in hippocampal terminals showed that actin is localized largely to the periphery of the synaptic vesicle cluster (Sankaranarayanan et al. 2003). However, acute disruption of actin did not lead to significant impairment of presynaptic vesicle recycling in hippocampal terminals (Li and Murthy 2001; Morales et al. 2000; Sankaranarayanan et al. 2003). In contrast, in the large lamprey reticular-spinal synapse, actin tracks are believed to shuttle endocytosing vesicles back to vesicle cluster. In these *en passant* synapses, disruption of actin led to a dramatic loss of vesicles in the terminals, suggesting that in these synapses actin may play a different role (Shupliakov et al. 2002).

Synapsins are abundant synaptic proteins that bind vesicles and have been suggested to play an important role in regulating vesicle pool mobilization during action potential firing (reviewed in Hilfiker et al. 1999). Synapsins bind to each other, to vesicles, and to the actin cytoskeleton in a manner that is regulated by phosphorylation (Greengard et al. 1994). Synapsins also possess an ATP binding domain, suggesting a possible enzymatic role (Hosaka and Sudhof 1998; Hosaka et al. 1999). Synapsin knockdown leads to a reduction in the size of the vesicle cluster and a decrease in the reserve pool of vesicles (Rosahl et al. 1993, 1995; Pieribone et al. 1995; Ryan et al. 1996). Synapsins tagged with EGFP are localized in synaptic terminals. Synapsin Ia dissociates from synaptic vesicles, disperses into axons during action potential firing, and reclusters to synapses after the cessation of synaptic activity (Chi et al. 2001). Site-directed mutational analysis of synapsins suggested that phosphorylation is essential for synapsin dispersion during activity and regulates the frequency-dependent mobilization of vesicles (Chi et al. 2001; 2003). These data suggest that synapsins might play an essential role in clustering vesicles in a reversible, activity, and phosphorylation-dependent manner.

Vesicle clustering also can be mediated by scaffolding proteins in the terminal such as Piccolo or Bassoon. These scaffolding proteins have long flexible loops that can bind vesicles and cluster them within the nerve terminal (Zhang et al. 2000; Sanmarti-Vila 2000). Rab3, a peripheral vesicle membrane protein, also has been suggested to play a role in vesicle clustering. Rab3 deletion in *C. elegans* led to vesicle dispersion in synapses, but this phenomenon is not observed in mammalian synapses (Nonet et al. 1997). Deletion of α-synuclein, a protein linked to Parkinson's disease, in mice led to a reduction in the reserve pool of vesicles, suggesting that this protein may play a role in vesicle clustering (Cabin et al. 2002; Murphy et al. 2000).

In mature synapses, vesicle pool mobilization is regulated by phosphorylation events. Treatment with a broad-spectrum kinase inhibitor staurosporine decreased vesicle pool mobilization from the reserve to the releasable pool (Becherer et al. 2001) and slowed the recruitment of newly endocytosed vesicles into the cycling pool (Li and Murthy 2001). In contrast, a phosphatase inhibitor led to dispersion of vesicles in neuromuscular synapses (Betz and Henkel 1994). Myosin light chain kinase inhibitors led to a reduction in the size of recycling pool of vesicles but did not affect fusion of the readily releasable vesicle pool (Ryan 1999). Thus, phosphorylation-dependent mobilization of vesicles from the reserve pool may regulate the size of the total recycling pool of vesicles.

Although functional presynaptic vesicle pools are fairly stable during cycles of exo-endocytosis, these pools are not static and can be modulated rapidly by protein phosphorylation events. A large fraction of the vesicles in hippocampal synapses do not seem to recycle, suggesting a reserve that can be tapped into during presynaptic potentiation. The potential triggers and targets for mobilizing this reserve pool are yet unknown.

II. CONCLUSIONS

Neuronal exocytosis is the final step in a cycle that leads to information transfer across synapses. The vesicle cycle is highly regulated via essential vesicular and plasma membrane proteins that mediate the various steps including neurotransmitter loading into vesicles, docking, priming, calcium-sensing, exocytosis, and recycling. Although tremendous progress has been achieved in understanding the roles of individual proteins using elegant physiological, biochemical, and genetic methods, there is much that remains unknown regarding the sequence and kinetics of protein-protein interactions that drive vesicle recycling. Future developments in single molecule and organelle imaging methods and robust methods for depleting or increasing protein levels *in vivo* will lead to novel insights on synaptic vesicle recycling. Further understanding of the synaptic vesicle cycle will provide novel insight into the processes underlying learning and memory and cognitive disorders. Knowledge of the subcellular events that control neurotransmission will ultimately provide novel targets to impact cognitive disorders in humans.

References

Ahmari, S.E., Buchanan, J., Smith, S.J. (2000). Assembly of presynaptic active zones from cytoplasmic transport packets. *Nat. Neurosci.* **3**, 445–451.

Akert, K., Moor, H., Pfenninger, K. (1971). Synaptic fine structure. *Adv. Cytopharmacol.* **1**, 273–290.

Albillos, A., Dernick, G., Horstmann, H., Almers, W., Alvarez de Toledo, G., Lindau, M. (1997). The exocytotic event in chromaffin cells revealed by patch amperometry. *Nature* **389**, 509–512.

Ales, E., Tabares, L., Poyato, J.M., Valero, V., Lindau, M., Alvarez de Toledo, G. (1999). High calcium concentrations shift the mode of exocytosis to the kiss-and-run mechanism. *Nat. Cell Biol.* **1**, 40–44.

Altrock, W.D., tom Dieck, S., Sokolov, M., Meyer, A.C., Sigler, A., Brakebusch, C. et al. (2003). Functional inactivation of a fraction of excitatory synapses in mice deficient for the active zone protein bassoon. *Neuron* **37**, 787–800.

Arata, Y., Nishi, T., Kawasaki-Nishi, S., Shao, E., Wilkens, S., Forgac, M. (2002). Structure, subunit subunit function and regulation of the coated vesicle and yeast vacuolar (H+)-ATPases. *Biochim. Biophys. Acta* **1555**, 71–74.

Aravamudan, B., Fergestad, T., Davis, W.S., Rodesch, C.K., Broadie, K. (1999). Drosophila UNC-13 is essential for synaptic transmission. *Nat. Neurosci.* **2**, 965–971.

Augustin, I., Rosenmund, C., Sudhof, T.C., Brose, N. (1999). Munc13-1 is essential for fusion competence of glutamatergic synaptic vesicles. *Nature* **400**, 457–461.

Axelrod, D., Hellen, E.H., Fulbright, R. (1992). Total internal reflection fluorescence. *In* Topics in fluorescence spectroscopy, Vol. 3, J.R. Lakowicz, ed., 289–343. Plenum, New York, NY.

Bai, J., Wang, C.T., Richards, D.A., Jackson, M.B., Chapman, E.R. (2004). Fusion pore dynamics are regulated by synaptotagmin, t-SNARE interactions. *Neuron* **41**, 929–942.

Bai, J., Chapman, E.R. (2004). The C2 domains of synaptotagmin—Partners in exocytosis. *Trends Biochem. Sci.* **29**, 143–151.

Bai, J., Tucker, W.C., Chapman, E.R. (2004). PIP2 increases the speed of response of synaptotagmin and steers its membrane-penetration activity toward the plasma membrane. *Nat. Struct. Mol. Biol.* **11**, 36–44.

Bamji, S.X., Shimazu, K., Kimes, N., Huelsken, J., Birchmeier, W., Lu, B., Reichardt, L.F. (2003). Role of beta-catenin in synaptic vesicle localization and presynaptic assembly. *Neuron* **40**, 719–731.

Bamji, S.X. (2005). Cadherins: Actin with the cytoskeleton to form synapses. *Neuron* **47**, 175–178.

Bayer, M.J., Reese, C., Buhler, S., Peters, C., Mayer, A. (2003). Vacuole membrane fusion: V0 functions after trans-SNARE pairing and is coupled to the Ca2+-releasing channel. *J. Cell Biol.* **162**, 211–222.

Becherer, U., Guatimosim, C., Betz, W.J. (2001). Effects of staurosporine on exocytosis and endocytosis at frog motor nerve terminals. *J. Neurosci.* **21**, 782–787.

Bekkers, J.M., Richerson, G.B., Stevens, C.F. (1990). Origin of variability in quantal size in cultured hippocampal neurons and hippocampal slices. *Proc. Natl. Acad. Sci. U S A* **87**, 5359–5362.

Benfenati, F., Greengard, P., Brunner, J., Bahler, M. (1989). Electrostatic and hydrophobic interactions of synapsin I and synapsin I fragments with phospholipid bilayers. *J. Cell Biol.* **108**, 1851–1862.

Betz, W.J., Henkel, A.W. (1994). Okadaic acid disrupts clusters of synaptic vesicles in frog motor nerve terminals. *J. Cell Biol.* **124**, 843–854.

Betz, A., Okamoto, M., Benseler, F., Brose, N. (1997). Direct interaction of the rat unc-13 homologue Munc13-1 with the N terminus of syntaxin. *J. Biol. Chem.* **272**, 2520–2526.

Betz, A., Thakur, P., Junge, H.J., Ashery, U., Rhee, J.S., Scheuss, V. et al. (2001). Functional interaction of the active zone proteins Munc13-1 and RIM1 in synaptic vesicle priming. *Neuron* **30**, 183–196.

Biederer, T., Sara, Y., Mozhayeva, M., Atasoy, D., Liu, X., Kavalali, E.T., Sudhof, T.C. (2002). SynCAM, a synaptic adhesion molecule that drives synapse assembly. *Science* **297**, 1525–1531.

Bollmann, J.H., Sakmann, B., Borst, J.G. (2000). Calcium sensitivity of glutamate release in a calyx-type terminal. *Science* **289**, 953–957.

Bollmann, J.H., Sakmann, B. (2005). Control of synaptic strength and timing by the release-site Ca2+ signal. *Nat. Neurosci.* **8**, 426–434.

Brodin, L., Low, P., Shupliakov, O. (2000). Sequential steps in clathrin-mediated synaptic vesicle endocytosis. *Curr. Opin. Neurobiol.* **10**, 312–320.

Brose, N., Hofmann, K., Hata, Y., Sudhof, T.C. (1995). Mammalian homologues of C. elegans unc-13 gene define novel family of C2-domain proteins. *J. Biol. Chem.* **270**, 25273–25280.

Bruns, D., Riedel, D., Klingauf, J., Jahn, R. (2000). Quantal release of serotonin. *Neuron* **28**, 205–220.

Bruses, J.L. (2000). Cadherin-mediated adhesion at the interneuronal synapse. *Curr. Opin. Cell Biol.* **12**, 593–597.

Butz, S., Okamoto, M., Sudhof, T.C. (1998). A tripartite protein complex with the potential to couple synaptic vesicle exocytosis to cell adhesion in brain. *Cell* **94**, 773–782.

Cabin, D.E., Shimazu, K., Murphy, D., Cole, N.B., Gottschalk, W., McIlwain, K.L. et al. (2002). Synaptic vesicle depletion correlates with attenuated synaptic responses to prolonged repetitive stimulation in mice lacking alpha-synuclein. *J. Neurosci.* **22**, 8797–8807.

Calakos, N., Schoch, S., Sudhof, T.C., Malenka, R.C. (2004). Multiple roles for the active zone protein RIM1alpha in late stages of neurotransmitter release. *Neuron* **42**, 889–896.

Carr, C.M., Grote, E., Munson, M., Hughson, F.M., Novick, P.J. (1999). Sec1p binds to SNARE complexes and concentrates at sites of secretion. *J. Cell Biol.* **146**, 333–344.

Castillo, P.E., Janz, R., Sudhof, T.C., Tzounopoulos, T., Malenka, R.C., Nicoll, R.A. (1997). Rab3A is essential for mossy fibre long-term potentiation in the hippocampus. *Nature* **388**, 590–593.

Castillo, P.E., Schoch, S., Schmitz, F., Sudhof, T.C., Malenka, R.C. 2002. RIM1α is required for presynaptic long-term potentiation. *Nature* **415**, 327–330.

Catterall, W.A. (1999). Interactions of presynaptic Ca2+ channels and snare proteins in neurotransmitter release. *Ann. N.Y. Acad. Sci.* **868**, 144–159.

Chapman, E.R. (2002). Synaptotagmin: A Ca(2+) sensor that triggers exocytosis? *Nat. Rev. Mol. Cell Biol.* **3**, 498–508.

Chen, Y.A., Scales, S.J., Scheller, R.H. (2001). Sequential SNARE assembly underlies priming and triggering of exocytosis. *Neuron* **30**, 161–170.

Chi, P., Greengard, P., Ryan, T.A. (2001). Synapsin dispersion and reclustering during synaptic activity. *Nat. Neurosci.* **4**, 1187–1193.

Chi, P., Greengard, P., Ryan, T.A. (2003). Synaptic vesicle mobilization is regulated by distinct synapsin I phosphorylation pathways at different frequencies. *Neuron* **38**, 69–78.

Chow, R.H., Klingauf, J., Heinemann, C., Zucker, R.S., Neher, E. (1996). Mechanisms determining the time course of secretion in neuroendocrine cells. *Neuron* **16**, 369–376.

Cochilla, A.J., Angleson, J.K., Betz, W.J. (1999). Monitoring secretory membrane with FM1-43 fluorescence. *Annu. Rev. Neurosci.* **22**, 1–10.

Colicos, M.A., Collins, B.E., Sailor, M.J., Goda, Y. (2001). Remodeling of synaptic actin induced by photoconductive stimulation. *Cell* **107**, 605–616.

Dawson-Scully, K., Bronk, P., Atwood, H.L., Zinsmaier, K.E. (2000). Cysteine-string protein increases the calcium sensitivity of neurotransmitter exocytosis in Drosophila. *J. Neurosci.* **20**, 6039–6047.

de Lange, R.P., de Roos, A.D., Borst, J.G. (2003). Two modes of vesicle recycling in the rat calyx of Held. *J. Neurosci.* **23**, 10164–10173.

Dernick, G., Alvarez de Toledo, G., Lindau, M. (2003). Exocytosis of single chromaffin granules in cell-free inside-out membrane patches. *Nat. Cell Biol.* **5**, 358–362.

Dick, O., tom Dieck, S., Altrock, W.D., Ammermuller, J., Weiler, R., Garner, C.C. et al. (2003). The presynaptic active zone protein bassoon is essential for photoreceptor ribbon synapse formation in the retina. *Neuron* **37**, 775–786.

Di Paolo, G., Moskowitz, H.S., Gipson, K., Wenk, M.R., Voronov, S., Obayashi, M. et al. (2004). Impaired PtdIns(4,5)P2 synthesis in nerve terminals produces defects in synaptic vesicle trafficking. *Nature* **431**, 415–422.

Dresbach, T., Qualmann, B., Kessels, M.M., Garner, C.C., Gundelfinger, E.D. (2001). The presynaptic cytomatrix of brain synapses. *Cell. Mol. Life Sci.* **58**, 94–116.

Dulubova, I., Sugita, S., Hill, S., Hosaka, M., Fernandez, I., Sudhof, T.C., Rizo, J. (1999). A conformational switch in syntaxin during exocytosis: Role of munc18. *EMBO J.* **18**, 4372–4382.

Dulubova, I., Yamaguchi, T., Arac, D., Li, H., Huryeva, I., Min, S.W. et al. (2003). Convergence and divergence in the mechanism of SNARE binding by Sec1/Munc18-like proteins. *Proc. Natl. Acad. Sci. U S A* **100**, 32–37.

Dulubova, I., Lou, X., Lu, J., Huryeva, I., Alam, A., Schneggenburger, R. et al. (2005). A Munc13/RIM/Rab3 tripartite complex: From priming to plasticity? *EMBO J.* **24**, 2839–2850.

Dunaevsky, A., Connor, E.A. (2000). F-actin is concentrated in nonrelease domains at frog neuromuscular junctions. *J. Neurosci.* **20**, 6007–6012.

Eberhard, D.A., Cooper, C.L., Low, M.G., Holz, R.W. (1990). Evidence that the inositol phospholipids are necessary for exocytosis. Loss of inositol phospholipids and inhibition of secretion in permeabilized cells caused by a bacterial phospholipase C and removal of ATP. *Biochem. J.* **268**, 15–25.

Edelmann, L., Hanson, P.I., Chapman, E.R., Jahn, R. (1995). Synaptobrevin binding to synaptophysin: A potential mechanism for controlling the exocytotic fusion machine. *EMBO J.* **14**, 224–231.

Ellis-Davies, G.C. (2003). Development and application of caged calcium. *Methods Enzymol.* **360**, 226–238.

Emptage, N.J., Reid, C.A., Fine, A. (2001). Calcium stores in hippocampal synaptic boutons mediate short-term plasticity, store-operated Ca2+ entry, and spontaneous transmitter release. *Neuron* **29**, 197–208.

Erickson, J.D., Eiden, L.E., Hoffman, B.J. (1992). Expression cloning of a reserpine-sensitive vesicular monoamine transporter. *Proc. Natl. Acad. Sci. U S A* **89**, 10993–10997.

Faundez, V., Horng, J.T., Kelly, R.B. (1998). A function for the AP3 coat complex in synaptic vesicle formation from endosomes. *Cell* **93**, 423–432.

Fenster, S.D., Chung, W.J., Zhai, R., Cases-Langhoff, C., Voss, B. et al. (2000). Piccolo, a presynaptic zinc finger protein structurally related to Bassoon. *Neuron* **25**, 203–214.

Fernandez-Alfonso, T., Sankaranarayanan, S., Ryan, T.A. (2001). Barium converts non-recycling synaptic vesicles to recycling vesicles in CNS synapses. *Soc. Neurosci.* **386**, 14.

Fernandez-Alfonso, T., Ryan, T.A. (2004). The kinetics of synaptic vesicle pool depletion at CNS synaptic terminals. *Neuron* **41**, 943–953.

Fischer von Mollard, G., Sudhof, T.C., Jahn, R. (1991). A small GTP-binding protein (rab3A) dissociates from synaptic vesicles during exocytosis. *Nature* **349**, 79–81.

Forgac, M. (1999). Structure and properties of the clathrin-coated vesicle and yeast vacuolar V-ATPases. *J. Bioenerg. Biomembr.* **31**, 57–65.

Friedman, H.V., Bresler, T., Garner, C.C., Ziv, N.E. (2000). Assembly of new individual excitatory synapses: Time course and temporal order of synaptic molecule recruitment. *Neuron* **27**, 57–69.

Fremeau, R.T. Jr, Burman, J., Qureshi, T., Tran, C.H, Proctor, J., Johnson, J. et al. (2002). The identification of vesicular glutamate transporter 3 suggests novel modes of signaling by glutamate. *Proc. Natl. Acad. Sci. U S A* **99**, 14488–14493.

Fremeau, R.T. Jr, Voglmaier, S., Seal, R.P., Edwards, R.H. (2004). VGLUTs define subsets of excitatory neurons and suggest novel roles for glutamate. *Trends Neurosci.* **27**, 98–103.

Fykse, E.M., Fonnum, F. (1996). Amino acid neurotransmission: Dynamics of vesicular uptake. *Neurochem. Res.* **21**, 1053–1060.

Gallwitz, D., Jahn, R. (2003). The riddle of the Sec1/Munc-18 proteins—New twists added to their interactions with SNAREs. *Trends. Biochem. Sci.* **28**, 113–116.

Geppert, M., Archer, B.T. III, Sudhof, T.C. (1991). Synaptotagmin II: A novel differentially distributed form of synaptotagmin. *J. Biol. Chem.* **266**, 13548–13552.

Geppert, M., Bolshakov, V.Y., Siegelbaum, S.A., Takei, K., De Camilli, P., Sudhof, T.C. (1994a). The role of rab3A in neurotransmitter release. *Nature* **369**, 493–497.

Geppert, M., Goda, Y., Hammer, R.E., Li, C., Rosahl, T.W., Stevens, C.F., Sudhof, T.C. (1994b). Synaptotagmin I: A major Ca2+ sensor for transmitter release at a central synapse. *Cell* **79**, 717–727.

Gillis, K.D., Mossner, R., Neher, E. (1996). Protein kinase C enhances exocytosis from chromaffin cells by increasing the size of the readily releasable pool of secretory granules. *Neuron* **16**, 1209–1220.

Goda, Y., Stevens, C.F. (1994). Two components of transmitter release at a central synapse. *Proc. Natl. Acad. Sci. U S A* **91**, 12942–12946.

Graham, M.E., Barclay, J.W., Burgoyne, R.D. (2004). Syntaxin/Munc18 interactions in the late events during vesicle fusion and release in exocytosis. *J. Biol. Chem.* **279**, 32751–32760.

Greengard, P., Benfenati, F., Valtorta, F. (1994). Synapsin I, an actin-binding protein regulating synaptic vesicle traffic in the nerve terminal. *Adv. Second Messenger Phosphoprotein Res.* **29**, 31–45.

Hamill, O.P., Marty, A., Neher, E., Sakmann, B., Sigworth, F.J. (1981). Improved patch-clamp techniques for high-resolution current recording from cells and cell-free membrane patches. *Pflugers. Arch.* **391**, 85–100.

Hannah, M.J., Schmidt, A.A., Huttner, W.B. (1999). Synaptic vesicle biogenesis. *Annu. Rev. Cell Dev. Biol.* **15**, 733–798.

Harata, N., Pyle, J.L., Aravanis, A.M., Mozhayeva, M., Kavalali, E.T., Tsien, R.W. (2001). Limited numbers of recycling vesicles in small CNS nerve terminals: Implications for neural signalling and vesicular cycling. *Trends Neurosci.* **24**, 637–643.

Harlow, M.L., Ress, D., Stoschek, A., Marshall, R.M., McMahan, U.J. (2001). The architecture of active zone material at the frog's neuromuscular junction. *Nature* **409**, 479–484.

Harris, T.W., Schuske, K., Jorgensen, E.M. (2001). Studies of synaptic vesicle endocytosis in the nematode C. elegans. *Traffic* **2**, 597–605.

Hatsuzawa, K., Lang, T., Fasshauer, D., Bruns, D., Jahn, R. (2003). The R-SNARE motif of tomosyn forms SNARE core complexes with syntaxin 1 and SNAP-25 and down-regulates exocytosis. *J. Biol. Chem.* **278**, 31159–31166.

Hay, J.C., Martin, T.F. (1992). Resolution of regulated secretion into sequential MgATP-dependent and calcium-dependent stages mediated by distinct cytosolic proteins. *J. Cell Biol.* **119**, 139–151.

Hay, J.C., Martin, T.F. (1993). Phosphatidylinositol transfer protein required for ATP-dependent priming of Ca(2+)-activated secretion. *Nature* **366**, 572–575.

Hay, J.C., Fisette, P.L., Jenkins, G.H., Fukami, K., Takenawa, T., Anderson, R.A., Martin, T.F. (1995). ATP-dependent inositide phosphorylation required for Ca(2+)-activated secretion. *Nature* **374**, 173–177.

Heidelberger, R., Heinemann, C., Neher, E., Matthews, G. (1994). Calcium dependence of the rate of exocytosis in a synaptic terminal. *Nature* **371**, 513–515.

Heidelberger, R. (1998). Adenosine triphosphate and the late steps in calcium-dependent exocytosis at a ribbon synapse. *J. Gen. Physiol.* **111**, 225–241.

Heidelberger, R., Sterling, P., Matthews, G. (2002). Roles of ATP in depletion and replenishment of the releasable pool of synaptic vesicles. *J. Neurophysiol.* **88**, 98–106.

Heinemann, C., Chow, R.H., Neher, E., Zucker, R.S. (1994). Kinetics of the secretory response in bovine chromaffin cells following flash photolysis of caged Ca2+. *Biophys. J.* **67**, 2546–2557.

Henkel, A.W., Almers, W. (1996). Fast steps in exocytosis and endocytosis studied by capacitance measurements in endocrine cells. *Curr. Opin. Neurobiol.* **6**, 350–357.

Heuser, J.E., Reese, T.S. (1981). Structural changes after transmitter release at the frog neuromuscular junction. *J. Cell Biol.* **88**, 564–580.

Hibino, H., Pironkova, R., Onwumere, O., Vologod-skaia, M., Hudspeth, A.J., Lesage, F. (2002). RIM binding proteins (RBPs) couple Rab3-interacting molecules (RIMs) to voltage-gated Ca(2+) channels. *Neuron* **34**, 411–423.

Hiesinger, P.R., Fayyazuddin, A., Mehta, S.Q., Rosenmund, T., Schulze, K.L., Zhai, R.G. et al. (2005). The v-ATPase V0 subunit a1 is required for a late step in synaptic vesicle exocytosis in Drosophila. *Cell* **121(4)**, 607–620.

Hilfiker, S., Pieribone, V.A., Czernik, A.J., Kao, H.T., Augustine, G.J., Greengard, P. (1999). Synapsins as regulators of neurotransmitter release. *Philos. Trans. R. Soc. Lond. B. Biol. Sci.* **354**, 269–279.

Hirokawa, N., Sobue, K., Kanda, K., Harada, A., Yorifuji, H. (1989). The cytoskeletal architecture of the presynaptic terminal and molecular structure of synapsin 1. *J. Cell Biol.* **108**, 111–126.

Holz, R.W., Bittner, M.A., Peppers, S.C., Senter, R.A., Eberhard, D.A. (1989). MgATP-independent and MgATP-dependent exocytosis. Evidence that MgATP primes adrenal chromaffin cells to undergo exocytosis. *J. Biol. Chem.* **264**, 5412–5419.

Hosaka, M., Sudhof, T.C. (1998). Synapsins I and II are ATP-binding proteins with differential Ca2+ regulation. *J. Biol. Chem.* **273**, 1425–1429.

Hosaka, M., Hammer, R.E., Sudhof, T.C. (1999). A phospho-switch controls the dynamic association of synapsins with synaptic vesicles. *Neuron* **24(2)**, 377–387.

Hughson, F.M. (1999). Membrane fusion: Structure snared at last. *Curr. Biol.* **9**, R49–52.

Junge, H.J., Rhee, J.S., Jahn, O., Varoqueaux, F., Spiess, J., Waxham, M.N. et al. (2004). Calmodulin and Munc13 form a Ca2+ sensor/effector complex that controls short-term synaptic plasticity. *Cell* **118**, 389–401.

Kandel E.R., Schwartz J.H., Jessell T.M. (2000). Principles of neural science, 4e. McGraw-Hill Inc., New York, NY.

Katz, B., Miledi, R. (1965). The effect of calcium on acetylcholine release from motor nerve terminals. *Proc. R. Soc. Lond. B. Biol. Sci.* **161**, 496–503.

Katz, B. (1966). Nerve, muscle, and synapse. McGraw-Hill, New York, NY.

Katz, B., Miledi, R. (1967). Modification of transmitter release by electrical interference with motor nerve endings. *Proc. R. Soc. Lond. B. Biol. Sci.* **167(6)**, 1–7.

Katz, B., Miledi, R. (1969). Spontaneous and evoked activity of motor nerve endings in calcium Ringer. *J. Physiol.* **203**, 689–706.

Katz, B. (1971). Quantal mechanism of neural transmitter release. *Science* **173**, 123–126.

Khimich, D., Nouvian, R., Pujol, R., tom Dieck, S., Egner, A., Gundelfinger, E.D., Moser, T. (2005). Hair cell synaptic ribbons are essential for synchronous auditory signalling. *Nature* **434**, 889–894.

Kidokoro, Y. (2003). Roles of SNARE proteins and synaptotagmin I in synaptic transmission: Studies at the Drosophila neuromuscular synapse. *Neurosignals* **12**, 13–30.

Klingauf, J., Kavalali, E.T., Tsien, R.W. (1998). Kinetics and regulation of fast endocytosis at hippocampal synapses. *Nature* **394**, 581–585.

Ko, J., Na, M., Kim, S., Lee, J.R., Kim, E. (2003). Interaction of the ERC family of RIM-binding proteins with the liprin-α family of multidomain proteins. *J. Biol. Chem.* **278**, 42377–42385.

Koushika, S.P., Richmond, J.E., Hadwiger, G., Weimer, R.M., Jorgensen, E.M., Nonet, M.L. (2001). A postdocking role for active zone protein Rim. *Nat. Neurosci.* **4**, 997–1005.

Kraszewski, K., Mundigl, O., Daniell, L., Verderio, C., Matteoli, M., De Camilli, P. (1995). Synaptic vesicle dynamics in living cultured hippocampal neurons visualized with CY3-conjugated antibodies directed against the lumenal domain of synaptotagmin. *J. Neurosci.* **15**, 4328–4342.

Landis, D.M., Hall, A.K., Weinstein, L.A., Reese, T.S. (1988). The organization of cytoplasm at the presynaptic active zone of a central nervous system synapse. *Neuron* **1**, 201–209.

Lenzi, D., von Gersdorff, H. (2001). Structure suggests function: The case for synaptic ribbons as exocytotic nanomachines. *Bioessays* **23**, 831–840.

Li, C., Takei, K., Geppert, M., Daniell, L., Stenius, K., Chapman, E.R. et al. (1994). Synaptic targeting of rabphilin-3A, a synaptic vesicle Ca2+/phospholipid-binding protein, depends on rab3A/3C. *Neuron* **13**, 885–898.

Li, Z., Murthy, V.N. (2001). Visualizing postendocytic traffic of synaptic vesicles at hippocampal synapses. *Neuron* **31**, 593–605.

Li, Z., Burrone, J., Tyler, W.J., Hartman, K.N., Albeanu, D.F., Murthy, V.N. (2005). Synaptic vesicle recycling studied in transgenic mice expressing synaptopHluorin. *Proc. Natl. Acad. Sci. U S A* **102**, 6131–6136.

Lindau, M., Alvarez de Toledo, G. (2003). The fusion pore. *Biochim. Biophys. Acta.* **1641**, 167–173.

Lindau, M., Almers, W. (1995). Structure and function of fusion pores in exocytosis and ectoplasmic membrane fusion. *Curr. Opin. Cell Biol.* **7**, 509–517.

Liu, Y., Peter, D., Roghani, A., Schuldiner, S., Prive, G.G. et al. (1992). A cDNA that suppresses MPP+ toxicity encodes a vesicular amine transporter. *Cell* **70**, 539–551.

Lonart, G., Schoch, S., Kaeser, P.S., Larkin, C.J., Sudhof, T.C., Linden, D.J. (2003). Phosphorylation of RIM1α by PKA triggers presynaptic long-term potentiation at cerebellar parallel fiber synapses. *Cell* **115**, 49–60.

Magleby, K.L., Zengel, J.E. (1976). Long term changes in augmentation, potentiation, and depression of transmitter release as a function of repeated synaptic activity at the frog neuromuscular junction. *J. Physiol.* **257**, 471–494.

Magleby, K.L. (1979). Facilitation, augmentation, and potentiation of transmitter release. *Prog. Brain. Res.* **49**, 175–182.

McIntire, S.L., Reimer, R.J., Schuske, K., Edwards, R.H., Jorgensen, E.M. (1997). Identification and characterization of the vesicular GABA transporter. *Nature* **389**, 870–876.

Maycox, P.R., Deckwerth, T., Hell, J.W., Jahn, R. (1988). Glutamate uptake by brain synaptic vesicles. Energy dependence of transport and functional reconstitution in proteoliposomes. *J. Biol. Chem.* **263**, 15423–15428.

Michael, D.J., Wightman, R.M. (1999). Electrochemical monitoring of biogenic amine neurotransmission in real time. *J. Pharm. Biomed. Anal.* **19**, 33–46.

Micheva, K.D., Smith, S.J. (2005). Strong effects of subphysiological temperature on the function and plasticity of mammalian presynaptic terminals. *J. Neurosci.* **25**, 7481–7488.

Miesenbock, G., De Angelis, D.A., Rothman, J.E. (1998). Visualizing secretion and synaptic transmission with pH-sensitive green fluorescent proteins. *Nature* **394**, 192–195.

Misura, K.M., Scheller, R.H., Weis, W.I. (2000). Three-dimensional structure of the neuronal-Sec1-syntaxin 1a complex. *Nature* **404**, 355–362.

Mochida, S., Sheng, Z.H., Baker, C., Kobayashi, H., Catterall, W.A. (1996). Inhibition of neurotransmission by peptides containing the synaptic protein interaction site of N-type Ca2+ channels. *Neuron* **17**, 781–788.

Montecucco, C. (1998). Protein toxins and membrane transport. *Curr. Opin. Cell Biol.* **10**, 530–536.

Morales, M., Colicos, M.A., Goda, Y. (2000). Actin dependent regulation of neurotransmitter release at central synapses. *Neuron* **27**, 539–550.

Moulder, K.L., Mennerick, S. (2005). Reluctant vesicles contribute to the total readily releasable pool in glutamatergic hippocampal neurons. *J. Neurosci.* **25**, 3842–3850.

Murphy, D.D., Rueter, S.M., Trojanowski, J.Q., Lee, V.M. (2000). Synucleins are developmentally expressed, and alpha-synuclein regulates the size of the presynaptic vesicular pool in primary hippocampal neurons. *J. Neurosci.* **20**, 3214–3220.

Murthy, V.N., De Camilli, P. (2003). Cell biology of the presynaptic terminal. *Annu. Rev. Neurosci.* **26**, 701–728.

Murthy, V.N., Stevens, C.F. (1998). Synaptic vesicles retain their identity through the endocytic cycle. *Nature* **392**, 497–501.

Murthy, V.N. (1999). Optical detection of synaptic vesicle exocytosis and endocytosis. *Curr. Opin. Neurobiol.* **9**, 314–320.

Nam, C.I., Chen, L. (2005). Postsynaptic assembly induced by neurexin-neuroligin interaction and neurotransmitter. *Proc. Natl. Acad. Sci. U S A* **102**, 6137–6142.

Neher, E., Marty, A. (1982). Discrete changes of cell membrane capacitance observed under conditions of enhanced secretion in bovine adrenal chromaffin cells. *Proc. Natl. Acad. Sci. U S A* **79**, 6712–6716.

Nishimura, W., Yao, I., Iida, J., Tanaka, N., Hata, Y. (2002). Interaction of synaptic scaffolding molecule and β-catenin. *J. Neurosci.* **22**, 757–765.

Nonet, M.L., Staunton, J.E., Kilgard, M.P., Fergestad, T., Hartwieg, E., Horvitz, H.R. et al. (1997). Caenorhabditis elegans rab-3 mutant synapses exhibit impaired function and are partially depleted of vesicles. *J. Neurosci.* **17**, 8061–8073.

Oheim, M., Loerke, D., Stuhmer, W., Chow, R.H. (1998). The last few milliseconds in the life of a secretory granule. Docking, dynamics and fusion visualized by total internal reflection fluorescence microscopy (TIRFM). *Eur. Biophys. J.* **27**, 83–98.

Ohtsuka, T., Takao-Rikitsu, E., Inoue, E., Inoue, M., Takeuchi, M. (2002). Cast: A novel protein of the cytomatrix at the active zone of synapses that forms a ternary complex with RIM1 and munc13-1. *J. Cell Biol.* **158**, 577–590.

Otto, H., Hanson, P.I., Jahn, R. (1997). Assembly and disassembly of a ternary complex of synaptobrevin, syntaxin, and SNAP-25 in the membrane of synaptic vesicles. *Proc. Natl. Acad. Sci. U S A* **94**, 6197–6201.

Pabst, S., Hazzard, J.W., Antonin, W., Sudhof, T.C., Jahn, R., Rizo, J., Fasshauer, D. (2000). Selective interaction of complexin with the neuronal SNARE complex. Determination of the binding regions. *J. Biol. Chem.* **275**, 19808–19818.

Pabst, S., Margittai, M., Vainius, D., Langen, R., Jahn, R., Fasshauer, D. (2002). Rapid and selective binding to the synaptic SNARE complex suggests a modulatory role of complexins in neuroexocytosis. *J. Biol. Chem.* **277**, 7838–7848.

Parlati, F., Weber, T., McNew, J.A., Westermann, B., Sollner, T.H., Rothman, J.E. (1999). Rapid and efficient fusion of phospholipid vesicles by the alpha-helical core of a SNARE complex in the absence of an N-terminal regulatory domain. *Proc. Natl. Acad. Sci. U S A* **96**, 12565–12570.

Parsons, T.D., Coorssen, J.R., Horstmann, H., Almers, W. (1995). Docked granules, the exocytic burst, and the need for ATP hydrolysis in endocrine cells. *Neuron* **15**, 1085–1096.

Perin, M.S., Fried, V.A., Mignery, G.A., Jahn, R., Sudhof, T.C. (1990). Phospholipid binding by a synaptic vesicle protein homologous to the regulatory region of protein kinase C. *Nature* **345**, 260–263.

Perrais, D., Kleppe, I.C., Taraska, J.W., Almers, W. (2004). Recapture after exocytosis causes differential retention of protein in granules of bovine chromaffin cells. *J. Physiol.* **560**, 413–428.

Pieribone, V.A., Shupliakov, O., Brodin, L., Hilfiker-Rothenfluh, S., Czernik, A.J., Greengard, P. (1995). Distinct pools of synaptic vesicles in neurotransmitter release. *Nature* **375**, 493–497.

Polo-Parada, L., Bose, C.M., Landmesser, L.T. (2001). Alterations in transmission, vesicle dynamics, and transmitter release machinery at NCAM-deficient neuromuscular junctions. *Neuron* **32**, 815–828.

Pothos, E.N., Larsen, K.E., Krantz, D.E., Liu, Y., Haycock, J.W., Setlik, W. et al. (2000). Synaptic vesicle transporter expression regulates vesicle phenotype and quantal size. *J. Neurosci.* **20**, 7297–7306.

Pyle, J.L., Kavalali, E.T., Piedras-Renteria, E.S., Tsien, R.W. (2000). Rapid reuse of readily releasable pool vesicles at hippocampal synapses. *Neuron* **28**, 221–231.

Reim, K., Mansour, M., Varoqueaux, F., Mc-Mahon, H.T., Sudhof, T.C., Brose, N., Rosenmund, C. (2001). Complexins regulate a late step in Ca^{2+}-dependent neurotransmitter release. *Cell* **104**, 71–81.

Rhee, J.S., Betz, A., Pyott, S., Reim, K., Varoqueaux, F., Augustin, I. et al. (2002). Beta phorbol ester- and diacylglycerol-induced augmentation of transmitter release is mediated by Munc13s and not by PKCs. *Cell* **108**, 121–133.

Richards, D.A., Guatimosim, C., Betz, W.J. (2000). Two endocytic recycling routes selectively fill two vesicle pools in frog motor nerve terminals. *Neuron* **27**, 551–559.

Richards, D.A., Guatimosim, C., Rizzoli, S.O., Betz, W.J. (2003). Synaptic vesicle pools at the frog neuromuscular junction. *Neuron* **39**, 529–541.

Richmond, J.E., Davis, W.S., Jorgensen, E.M. (1999). UNC-13 is required for synaptic vesicle fusion in C. elegans. *Nat. Neurosci.* **2**, 959–964.

Richmond, J.E., Weimer, R.M., Jorgensen, E.M. (2001). An open form of syntaxin bypasses the requirement for UNC-13 in vesicle priming. *Nature* **412**, 338–341.

Richmond, J.E., Broadie, K.S. (2002). The synaptic vesicle cycle: Exocytosis and endocytosis in Drosophila and C. elegans. *Curr. Opin. Neurobiol.* **12**, 499–507.

Rizo, J., Sudhof, T.C. (1998). C2-domains, structure and function of a universal Ca^{2+}-binding domain. *J. Biol. Chem.* **273**, 15879–15882.

Rizo, J., Sudhof, T.C. (2002). Snares and Munc18 in synaptic vesicle fusion. *Nat. Rev. Neurosci.* **3**, 641–653.

Rizzoli, S.O., Richards, D.A., Betz, W.J. (2003). Monitoring synaptic vesicle recycling in frog motor nerve terminals with FM dyes. *J. Neurocytol.* **32**, 539–549.

Rizzoli, S.O., Betz, W.J. (2004). The structural organization of the readily releasable pool of synaptic vesicles. *Science* **303**, 2037–2039.

Rizzoli, S.O., Betz, W.J. (2005). Synaptic vesicle pools. *Nat. Rev. Neurosci.* **6**, 57–69.

Robinson, I.M., Ranjan, R., Schwarz, T.L. (2002). Synaptotagmins I and IV promote transmitter release independently of Ca^{2+} binding in the C(2)A domain. *Nature* **418**, 336–340.

Robitaille, R., Garcia, M.L., Kaczorowski, G.J., Charlton, M.P. (1993). Functional colocalization of calcium and calcium-gated potassium channels in control of transmitter release. *Neuron* **11**, 645–655.

Rodriguez-Boulan, E., Kreitzer, G., Musch, A. (2005). Organization of vesicular trafficking in epithelia. *Nat. Rev. Mol. Cell Biol.* **6**, 233–247.

Roghani, A., Feldman, J., Kohan, S.A., Shirzadi, A., Gundersen, C.B., Brecha, N., Edwards, R.H. (1994). Molecular cloning of a putative vesicular transporter for acetylcholine. *Proc. Natl. Acad. Sci. USA* **91**, 10620–10624.

Roos, J., Kelly, R.B. (2000). Preassembly and transport of nerve terminals: A new concept of axonal transport. *Nat. Neurosci.* **3(5)**, 415–417.

Rosahl, T.W., Geppert, M., Spillane, D., Herz, J., Hammer, R.E. et al. (1993). Short term synaptic plasticity is altered in mice lacking synapsin I. *Cell* **75**, 661–670.

Rosahl, T.W., Spillane, D., Missler, M., Herz, J., Selig, D.K. et al. (1995). Essential functions of synapsins I and II in synaptic vesicle regulation. *Nature* **375**, 488–493.

Rosenmund, C., Stevens, C.F. (1996). Definition of the readily releasable pool of vesicles at hippocampal synapses. *Neuron* **16**, 1197–1207.

Rosenmund, C., Sigler, A., Augustin, I., Reim, K., Brose, N., Rhee, J.S. (2002). Differential control of vesicle priming and short-term plasticity by Munc13 isoforms. *Neuron* **33**, 411–424.

Ryan, T.A., Reuter, H., Wendland, B., Schweizer, F.E., Tsien, R.W., Smith, S.J. (1993). The kinetics of synaptic vesicle recycling measured at single presynaptic boutons. *Neuron* **11**, 713–724.

Ryan, T.A., Smith, S.J. (1995). Vesicle pool mobilization during action potential firing at hippocampal synapses. *Neuron* **14**, 983–989.

Ryan, T.A., Smith, S.J., Reuter, H. (1996). The timing of synaptic vesicle endocytosis. *Proc. Natl. Acad. Sci. USA* **93**, 5567–5571.

Ryan, T.A., Li, L., Chin, L.S., Greengard, P., Smith, S.J. (1996). Synaptic vesicle recycling in synapsin I knock-out mice. *J. Cell Biol.* **134**, 1219–1227.

Ryan, T.A., Reuter, H., Smith, S.J. (1997). Optical detection of a quantal presynaptic membrane turnover. *Nature* **388**, 478–482.

Ryan, T.A. (1999). Inhibitors of myosin light chain kinase block synaptic vesicle pool mobilization during action potential firing. *J. Neurosci.* **19**, 1317–1323.

Sagne, C., El Mestikawy, S., Isambert, M.F., Hamon, M., Henry, J.P., Giros, B., Gasnier, B. (1997). Cloning of a functional vesicular GABA and glycine transporter by screening of genome databases. *FEBS Lett.* **417**, 177–183.

Saitoh, N., Hori, T., Takahashi, T. (2001). Activation of the epsilon isoform of protein kinase C in the mammalian nerve terminal. *Proc. Natl. Acad. Sci. USA* **98**, 14017–14021.

Sakaba, T., Stein, A., Jahn, R., Neher, E. (2005). Distinct kinetic changes in neurotransmitter release after SNARE protein cleavage. *Science* **309**, 491–494.

Sankaranarayanan, S., Ryan, T.A. (2000a). Real-time measurements of vesicle-SNARE recycling in synapses of the central nervous system. *Nat. Cell Biol.* **2**, 197–204.

Sankaranarayanan, S., De Angelis, D., Rothman, J.E., Ryan, T.A. (2000b). The use of pHluorins for optical measurements of presynaptic activity. *Biophys. J.* **79**, 2199–2208.

Sankaranarayanan, S., Ryan, T.A. (2001). Calcium accelerates endocytosis of vSNAREs at hippocampal synapses. *Nat. Neurosci.* **4**, 129–136.

Sankaranarayanan, S., Atluri, P.P., Ryan, T.A. (2003). Actin has a molecular scaffolding, not propulsive, role in presynaptic function. *Nat. Neurosci.* **6**, 127–135.

Sanmarti-Vila, L., tom Dieck, S., Richter, K., Altrock, W., Zhang, L., Volknandt, W. et al. (2000). Membrane association of presynaptic cytomatrix protein bassoon. *Biochem. Biophys. Res. Commun.* **275**, 43–46.

Sara, Y., Biederer, T., Atasoy, D., Chubykin, A., Mozhayeva, M.G., Sudhof, T.C., Kavalali, E.T. (2005a). Selective capability of SynCAM and neuroligin for functional synapse assembly. *J. Neurosci.* **25(1)**, 260–270.

Sara, Y., Virmani, T., Deak, F., Liu, X., Kavalali, E.T. (2005b). An isolated pool of vesicles recycles at rest and drives spontaneous neurotransmission. *Neuron* **45**, 563–573.

Schafer, M.K., Varoqui, H., Defamie, N., Weihe, E., Erickson, J.D. (2002). Molecular cloning and functional identification of mouse vesicular glutamate transporter 3 and its expression in subsets of novel excitatory neurons. *J. Biol. Chem.* **277**, 50734–50748.

Scheiffele, P., Fan, J., Choih, J., Fetter, R., Serafini, T. (2000). Neuroligin expressed in nonneuronal cells triggers presynaptic development in contacting axons. *Cell* **101**, 657–669.

Scheiffele, P. (2003). Cell–cell signaling during synapse formation in the CNS. *Annu. Rev. Neurosci.* **26**, 485–508.

Schekman, R. (1992). Genetic and biochemical analysis of vesicular traffic in yeast. *Curr. Opin. Cell Biol.* **4**, 587–592.

Schiavo, G., Rossetto, O., Montecucco, C. (1994). Clostridial neurotoxins as tools to investigate the molecular events of neurotransmitter release. *Semin. Cell Biol.* **5(4)**, 221–229.

Schikorski, T., Stevens, C.F. (2001). Morphological correlates of functionally defined synaptic vesicle populations. *Nat. Neurosci.* **4**, 391–395.

Schluter, O.M., Schnell, E., Verhage, M., Tzonopoulos, T., Nicoll, R.A., Janz, R. et al. (1999). Rabphilin knock-out mice reveal that rabphilin is not required for rab3 function in regulating neurotransmitter release. *J. Neurosci.* **19**, 5834–5846.

Schneggenburger, R., Neher, E. (2005). Presynaptic calcium and control of vesicle fusion. *Curr. Opin. Neurobiol.* **15**, 266–274.

Schoch, S., Deak, F., Konigstorfer, A., Mozhayeva, M., Sara, Y., Sudhof, T.C., Kavalali, E.T. (2001). SNARE function analyzed in synaptobrevin/VAMP knockout mice. *Science* **294**, 1117–1122.

Schoch, S., Castillo, P.E., Jo, T., Mukherjee, K., Geppert, M., Wang, Y. et al. (2002). RIM1alpha forms a protein scaffold for regulating neurotransmitter release at the active zone. *Nature* **415**, 321–326.

Schweizer, F.E., Dresbach, T., DeBello, W.M., O'Connor, V., Augustine, G.J., Betz, H. (1998). Regulation of neurotransmitter release kinetics by NSF. *Science* **279**, 1203–1206.

Scott, B.L., Van Komen, J.S., Irshad, H., Liu, S., Wilson, K.A., McNew, J.A. (2004). Sec1p directly stimulates SNARE-mediated membrane fusion in vitro. *J. Cell Biol.* **167**, 75–85.

Scotti, A.L., Chatton, J.Y., Reuter, H. (1999). Roles of Na(+)-Ca2+ exchange and of mitochondria in the regulation of presynaptic Ca2+ and spontaneous glutamate release. *Philos. Trans. R. Soc. Lond. B. Biol. Sci.* **354**, 357–364.

Shapira, M., Zhai, R.G., Dresbach, T., Bresler, T., Torres, V.I., Gundelfinger, E.D. et al. (2003). Unitary assembly of presynaptic active zones from Piccolo-Bassoon transport vesicles. *Neuron* **38**, 237–252.

Sheng, Z.H., Yokoyama, C.T., Catterall, W.A. (1997). Interaction of the synprint site of N-type Ca2+ channels with the C2B domain of synaptotagmin I. *Proc. Natl. Acad. Sci. U S A* **94**, 5405–5410.

Sheng, Z.H., Westenbroek, R.E., Catterall, W.A. (1998). Physical link and functional coupling of presynaptic calcium channels and the synaptic vesicle docking/fusion machinery. *J. Bioenerg. Biomembr.* **30**, 335–345.

Shi, G., Faundez, V., Roos, J., Dell'Angelica, E.C., Kelly, R.B. (1998). Neuroendocrine synaptic vesicles are formed in vitro by both clathrin-dependent and clathrin-independent pathways. *J. Cell Biol.* **143**, 947–955.

Shupliakov, O., Bloom, O., Gustafsson, J.S., Kjaerulff, O., Low, P., Tomilin, N. et al. (2002). Impaired recycling of synaptic vesicles after acute perturbation of the presynaptic actin cytoskeleton. *Proc. Natl. Acad. Sci. U S A* **99(14)**, 476–481.

Slepnev, V.I., De Camilli, P. (2000). Accessory factors in clathrin-dependent synaptic vesicle endocytosis. *Nat. Rev. Neurosci.* **1**, 161–172.

Smith, C., Moser, T., Xu, T., Neher, E. (1998). Cytosolic Ca2+ acts by two separate pathways to modulate the supply of release-competent vesicles in chromaffin cells. *Neuron* **20**, 1243–1253.

Sollner, T., Whiteheart, S.W., Brunner, M., Erdjument-Bromage, H., Geromanos, S., Tempst, P., Rothman, J.E. (1993a). SNAP receptors implicated in vesicle targeting and fusion. *Nature* **362**, 318–324.

Sollner, T., Bennett, M.K., Whiteheart, S.W., Scheller, R.H., Rothman, J.E. (1993b). A protein assembly-disassembly pathway in vitro that may correspond to sequential steps of synaptic vesicle docking, activation, and fusion. *Cell* **75**, 409–418.

Sollner, T., Bennett, M.K., Whiteheart, S.W., Scheller, R.H., Rothman, J.E. (1993c). A protein assembly-disassembly pathway in vitro that may correspond to sequential steps of synaptic vesicle docking, activation, and fusion. *Cell* **75**, 409–418.

Sorensen, J.B., Nagy, G., Varoqueaux, F., Nehring, R.B., Brose, N., Wilson, M.C., Neher, E. (2003). Differential control of the releasable vesicle pools by SNAP-25 splice variants and SNAP-23. *Cell* **114**, 75–86.

Stewart, B.A., Mohtashami, M., Trimble, W.S., Boulianne, G.L. (2000). SNARE proteins contribute to calcium cooperativity of synaptic transmission. *Proc. Natl. Acad. Sci. U S A* **97(13)**, 955–960.

Steyer, J.A., Horstmann, H., Almers, W. (1997). Transport, docking and exocytosis of single secretory granules in live chromaffin cells. *Nature* **388**, 474–478.

Steyer, J.A., Almers, W. (2001). A real-time view of life within 100 nm of the plasma membrane. *Nat. Rev. Mol. Cell Biol.* **2**, 268–275.

Stimson, D.T., Ramaswami, M. (1999). Vesicle recycling at the Drosophila neuromuscular junction. *Int. Rev. Neurobiol.* **43**, 163–189.

Sudhof, T.C. (2002). Synaptotagmins: Why so many? *J. Biol. Chem.* **277**, 7629–7632.

Sudhof, T.C. (2004). The synaptic vesicle cycle. *Annu. Rev. Neurosci.* **27**, 509–547.

Sulzer, D., Pothos, E.N. (2000). Regulation of quantal size by presynaptic mechanisms. *Rev. Neurosci.* **11**, 159–212.

Sun, J.Y., Wu, L.G. (2001). Fast kinetics of exocytosis revealed by simultaneous measurements of presynaptic capacitance and postsynaptic currents at a central synapse. *Neuron* **30**, 171–182.

Sun, J.Y., Wu, X.S., Wu, L.G. (2002). Single and multiple vesicle fusion induce different rates of endocytosis at a central synapse. *Nature* **417**, 555–559.

Sutton, R.B., Fasshauer, D., Jahn, R., Brunger, A.T. (1998). Crystal structure of a SNARE complex involved in synaptic exocytosis at 2.4 A resolution. *Nature* **395**, 347–353.

Tabb, J.S., Kish, P.E., Van Dyke, R., Ueda, T. (1992). Glutamate transport into synaptic vesicles. Roles of membrane potential pH gradient and intravesicular pH. *J. Biol. Chem.* **267**, 15412–15418.

Takao-Rikitsu, E., Mochida, S., Inoue, E., Deguchi-Tawarada, M., Inoue, M., Ohtsuka, T., Takai, Y. (2004). Physical and functional interaction of the active zone proteins, CAST, RIM1, and Bassoon, in neurotransmitter release. *J. Cell Biol.* **164**, 301–311.

Takamori, S., Rhee, J.S., Rosenmund, C., Jahn, R. (2000). Identification of a vesicular glutamate transporter that defines a glutamatergic phenotype in neurons. *Nature* **407**, 189–194.

Tokumaru, H., Umayahara, K., Pellegrini, L.L., Ishizuka, T., Saisu, H., Betz, H. et al. (2001). SNARE complex oligomerization by synaphin/complexin is essential for synaptic vesicle exocytosis. *Cell* **104**, 421–432.

tom Dieck, S., Sanmarti-Vila, L., Langnaese, K., Richter, K., Kindler, S. et al. (1998). Bassoon, a novel zinc-finger CAG/glutamine-repeat protein selectively localized at the active zone of presynaptic nerve terminals. *J. Cell Biol.* **142**, 499–509.

Ubach, J., Zhang, X., Shao, X., Sudhof, T.C., Rizo, J. (1998). Ca2+ binding to synaptotagmin: How many Ca2+ ions bind to the tip of a C2- domain? *EMBO J.* **17**, 3921–3930.

Uchida, N., Honjo, Y., Johnson, K.R., Wheelock, M.J., Takeichi, M. (1996). The catenin/cadherin adhesion system is localized in synaptic junctions bordering transmitter release zones. *J. Cell Biol.* **135**, 767–779.

Umeda, T., Okabe, S. (2001). Visualizing synapse formation and remodeling: Recent advances in real-time imaging of CNS synapses. *Neurosci. Res.* **40(4)**, 291–300.

van der Kloot, W. (1991). The regulation of quantal size. *Prog. Neurobiol.* **36**, 93–130.

Varoqui, H., Erickson, J.D. (1997). Vesicular neurotransmitter transporters. Potential sites for the regulation of synaptic function. *Mol. Neurobiol.* **15(2)**, 165–191.

Verhage, M., Maia, A.S., Plomp, J.J., Brussaard, A.B., Heeroma, J.H., Vermeer, H. et al. (2000). Synaptic assembly of the brain in the absence of neurotransmitter secretion. *Science* **287**, 864–869.

von Gersdorff, H., Borst, J.G. (2002). Short-term plasticity at the calyx of held. *Nat. Rev. Neurosci.* **3**, 53–64.

von Gersdorff, H., Matthews, G. (1994). Dynamics of synaptic vesicle fusion and membrane retrieval in synaptic terminals. *Nature* **367**, 735–739.

von Gersdorff, H., Matthews, G. (1997). Depletion and replenishment of vesicle pools at a ribbon-type synaptic terminal. *J. Neurosci.* **17**, 1919–1927.

von Gersdorff, H., Matthews, G. (1999). Electrophysiology of synaptic vesicle cycling. *Annu. Rev. Physiol.* **61**, 725–752.

von Ruden, L., Neher, E. (1993). A Ca-dependent early step in the release of catecholamines from adrenal chromaffin cells. *Science* **262**, 1061–1065.

Wang, C.T., Lu, J.C., Bai, J., Chang, P.Y., Martin, T.F., Chapman, E.R., Jackson, M.B. (2003). Different domains of synaptotagmin control the choice between kiss-and-run and full fusion. *Nature* **424**, 943–947.

Wang, Y., Okamoto, M., Schmitz, F., Hofman, K., Sudhof, T.C. (1997). RIM: A putative Rab3-effector in regulating synaptic vesicle fusion. *Nature* **388**, 593–598.

Wang, Y.M., Gainetdinov, R.R., Fumagalli, F., Xu, F., Jones, S.R., Bock, C.B. et al. (1999). Aczonin, a 550-kD putative scaffolding protein of presynaptic active zones, shares homology regions with Rim and Bassoon and binds profilin. *J. Cell Biol.* **147**, 151–162.

Wang, Y., Sugita, S., Sudhof, T.C. (2000). The RIM/NIM family of neuronal C2 domain proteins. Interactions with Rab3 and a new class of Src homology 3 domain proteins. *J. Biol. Chem.* **275**, 20033–20044.

Wang, Y., Liu, X., Biederer, T., Sudhof, T.C. (2002). A family of RIM-binding proteins regulated by alternative splicing: Implications for the genesis of synaptic active zones. *Proc. Natl. Acad. Sci. USA* **99**, 14464–14469.

Wang, Y., Sudhof, T.C. (2003). Genomic definition of RIM proteins: Evolutionary amplification of a family of synaptic regulatory proteins. *Genomics* **81**, 126–137.

Washbourne, P., Thompson, P.M., Carta, M., Costa, E.T., Mathews, J.R., Lopez-Bendito, G. et al. (2002). Genetic ablation of the t-SNARE SNAP-25 distinguishes mechanisms of neuroexocytosis. *Nat. Neurosci.* **5**, 19–26.

Weber, T., Zemelman, B.V., McNew, J.A., Westermann, B., Gmachl, M., Parlati, F. et al. (1998). SNAREpins: Minimal machinery for membrane fusion. *Cell* **92**, 759–772.

Wenk, M.R., De Camilli, P. (2004). Protein-lipid interactions and phosphoinositide metabolism in membrane traffic insights from vesicle recycling in nerve terminals. *Proc. Natl. Acad. Sci. U S A* **101**, 8262–8269.

Wightman, R.M., Caron, M.G. (1997). Knockout of the vesicular monoamine transporter 2 gene results in neonatal death and supersensitivity to cocaine and amphetamine. *Neuron* **19**, 1285–1296.

Wilson, D.W., Whiteheart, S.W., Wiedmann, M., Brunner, M., Rothman, J.E. (1992). A multisubunit particle implicated in membrane fusion. *J. Cell Biol.* **117**, 531–538.

Wu, L.G., Betz, W.J. (1998). Kinetics of synaptic depression and vesicle recycling after tetanic stimulation of frog motor nerve terminals. *Biophys. J.* **74**, 3003–3009.

Xu, T., Binz, T., Niemann, H., Neher, E. (1998). Multiple kinetic components of exocytosis distinguished by neurotoxin sensitivity. *Nat. Neurosci.* **1**, 192–200.

Xu, T., Rammner, B., Margittai, M., Artalejo, A.R., Neher, E., Jahn, R. (1999). Inhibition of SNARE complex assembly differentially affects kinetic components of exocytosis. *Cell* **99**, 713–722.

Yamaguchi, T., Shirataki, H., Kishida, S., Miyazaki, M., Nishikawa, J., Wada, K. et al. (1993). Two functionally different domains of rabphilin-3A, Rab3A p25/smg p25A-binding and phospholipid- and Ca(2+)-binding domains. *J. Biol. Chem.* **268**, 27164–27170.

Yamashita, T., Ishikawa, T., Takahashi, T. (2003). Developmental increase in vesicular glutamate content does not cause saturation of AMPA receptors at the calyx of held synapse. *J. Neurosci.* **23**, 3633–3638.

Yan, Q., Sun, W., McNew, J.A., Vida, T.A., Bean, A.J. (2004). Ca2+ and N-ethylmaleimide-sensitive factor differentially regulate disassembly of SNARE complexes on early endosomes. *J. Biol. Chem.* **279**, 18270–18276.

Yokoyama, C.T., Myers, S.J., Fu, J., Mockus, S.M., Scheuer, T., Catterall, W.A. (2005). Mechanism of SNARE protein binding and regulation of Cav2 channels by phosphorylation of the synaptic protein interaction site. *Mol. Cell Neurosci.* **28**, 1–17.

Zakharenko, S., Chang, S., O'Donoghue, M., Popov, S.V. (1999). Neurotransmitter secretion along growing nerve processes: Comparison with synaptic vesicle exocytosis. *J. Cell Biol.* **144**, 507–518.

Zenisek, D., Steyer, J.A., Almers, W. (2000). Transport, capture and exocytosis of single synaptic vesicles at active zones. *Nature* **406**, 849–854.

Zenisek, D., Steyer, J.A., Feldman, M.E., Almers, W. (2002). A membrane marker leaves synaptic vesicles in milliseconds after exocytosis in retinal bipolar cells. *Neuron* **35**, 1085–1097.

Zhang, L., Volknandt, W., Gundelfinger, E.D., Zimmermann, H. (2000). A comparison of synaptic protein localization in hippocampal mossy fiber terminals and neurosecretory endings of the neurohypophysis using the cryo-immunogold technique. *J. Neurocytol.* **29**, 19–30.

Ziv, N.E., Garner, C.C. (2004). Cellular and molecular mechanisms of presynaptic assembly. *Nat. Rev. Neurosci.* **5**, 385–399.

Zucker, R.S., Regehr, W.G. (2002). Short-term synaptic plasticity. *Annu. Rev. Physiol.* **64**, 355–405.

7

Endocytosis in Neurons

VOLKER HAUCKE AND JÜRGEN KLINGAUF

Neurons are communication specialists that contain elaborate machinery for the release and reception of chemical neurotransmitters. Endocytosis of pre- and postsynaptic membrane proteins as well as the internalization of postsynaptic receptors and ion channels contributes to the regulation of synapse structure and function. During the last few years we have witnessed enormous progress in our understanding of the mechanics and regulation of vesicle formation and the roles that specialized adaptor and accessory proteins play in this process. Moreover, ever more sophisticated molecular tools have allowed us to unravel many of the secrets of endocytic vesicle cycling in the nervous system.

I. ENDOCYTOSIS IN NEURONS

Neurons are highly polarized cells with axonal and somatodendritic membrane domains that serve to spatially segregate outgoing from incoming signals. During chemical neurotransmission, neurotransmitter molecules stored in the lumen of neurosecretory vesicles undergo calcium-regulated exocytosis by fusion with the presynaptic plasma membrane. Neurons can secrete a variety of nonpeptide as well as peptidergic transmitters via two types of secretory organelles, synaptic vesicles (SVs) and large dense-core vesicles, also termed secretory granules (SGs). Nonpeptide neurotransmitters such as acetylcholine (ACh), noradrenaline, glutamate, glycine, and GABA are released from SVs (Murthy and De Camilli 2003), whereas neuropeptides are stored in, and released from, SGs (Torrealba and Carrasco 2004). SGs are formed directly at the exit site of the trans-Golgi network and transported down the axon to their release sites (De Camilli et al. 2000). Following neuroexocytosis, neurotransmitter molecules bind to postsynaptic receptors leading to an electrical response in the postsynaptic neuron. Endocytosis has emerged as a basal mechanism for maintaining and regulating pools of presynaptic recycling vesicles, as well as exo-endocytic shuttling of plasma membrane receptors, ion channels, and transporters (see Chapters 8, 9, 11, and 12).

Since neurons frequently must sustain high rates of neurosecretion for prolonged periods, SVs undergo local endocytic recycling within the nerve terminal. Evidence for such presynaptic vesicle endocytosis was provided in the early 1970s with the introduction of extracellular tracers for neuronal endocytosis in electron microscopy. Elegant studies using the frog neuromuscular junction by Heuser and Reese (1973) demonstrated stimulation-dependent uptake of horseradish peroxidase into SVs, providing evidence for their endocytic recycling. The observation that clathrin-coated pits accumulate at or near the active zone of nerve terminals during periods of high exo-endocytic activity suggested an important role for clathrin-mediated endocytosis in the recycling pathway (Heuser 1989). Although the important role of clathrin and its partners in presynaptic vesicle cycling remains undisputed, several recent studies imply that a faster, more direct, "kiss-and-run" pathway of exo-endocytosis may prevail under milder stimulation conditions (see later for further details) (Galli and Haucke 2004).

In addition to its important role in SV recycling at nerve endings, clathrin-dependent internalization has been shown to modulate the number of postsynaptic neurotransmitter receptors in response to ligand binding or after eliciting intracellular signaling cascades (see Chapters 8–10). Ionotropic glutamate or $GABA_A$ receptors undergo both constitutive and regulated cycling between recycling endosomal compartments, the cell surface, and the postsynaptic density (Bredt and Nicoll 2003; Ehlers 2000; Sheng and Kim 2002). Under conditions mimicking long-term depression (LTD) in cultured neurons or hippocampal slices, postsynaptic neurotransmitter ion channels such as AMPA-type glutamate receptors become internalized in a clathrin- and dynamin-dependent manner, thereby downregulating the response of the postsynaptic neuron. Similar mechanisms may operate for other receptors both at synapses and within neuronal cell bodies. Much less is known about clathrin- and dynamin-independent endocytic mechanisms and their membrane cargo substrates.

II. THE CLATHRIN-DEPENDENT ENDOCYTIC MACHINERY

Although clathrin-coated vesicles (CCVs) are found ubiquitously in all eukaryotic cells and tissues, their components are particularly enriched in the brain (Galli and Haucke 2004; Murthy and De Camilli 2003). Clathrin, the heterotetrameric adaptor complex (AP-2), as well as monomeric adaptors and accessory proteins including epsin, eps15, AP180, HIP1/HIP1R, and amphiphysin, play an early role in coat formation (Conner and Schmid 2003; Takei and Haucke 2001). Recruitment of AP-2 to the plasma membrane is a cooperative, and presumably highly regulated process, that involves interactions with phosphoinositides, membrane cargo, and a variety of AP-2α binding partners (Galli and Haucke 2004; Takei and Haucke 2001). Although neurons use a clathrin-dependent endocytic mechanism similar to that used in nonneuronal cells, many components of the clathrin machinery are enriched in brain. In addition, endocytosis at synapses involves endocytic protein isoforms, including splice-variants of clathrin light chains and α_A-adaptin, AP180, auxilin, intersectin, and dynamin-I, a mechanochemical GTPase mediating vesicle fission (Hinshaw 2000; Murthy and De Camilli 2003; Szymkiewicz et al. 2004) (see Table 7.1). AP180/CALM, epsins, and HIP1/HIP1R (huntingtin interacting proteins) have been suggested to function as cargo adaptors, although direct interaction with cargo has not been demonstrated for any of them.

A. Mechanics of Clathrin-Coated Pit Formation

CCVs are formed by the coordinated assembly of clathrin triskelia, formed from three tightly linked heavy chains and their associated light chains on the plasma membrane. The recruitment and polymerization of the outer clathrin layer is assisted by mono- and heterotetrameric adaptor proteins that simultaneously bind to clathrin, membrane lipids, and in many cases to transmembrane cargo proteins (Ehrlich et al. 2004; Owen et al. 2004). Perhaps the most important clathrin adaptor is the heterotetrameric AP-2 complex comprised of two large subunits (α and $\beta2$), a medium subunit ($\mu2$), and a small subunit ($\sigma2$). The two large subunits together with $\sigma2$ and the amino-terminal domain of $\mu2$ (N-$\mu2$) form the trunk or core domain of AP-2 (Collins et al. 2002), which are joined by extended, flexible "hinges" to the appendage or ear domains of α- and $\beta2$-adaptins. Since AP-2 associates with clathrin, a variety of accessory endocytic proteins, phosphatidylinositol 4,5-bisphosphate [PI(4,5)P$_2$], and membrane cargo proteins, it has been postulated to serve as a main protein inter-

action hub during coated pit assembly (Praefcke et al. 2004). Many accessory proteins such as epsins, AP180/CALM, and amphiphysin also serve an adaptor function by linking clathrin assembly to membrane bud formation. These mono- or dimeric adaptors possess a folded lipid-binding domain (Gallop and McMahon 2005) linked to a more flexible portion of the protein harboring short clathrin- and AP-2-binding motifs, which may aid in stabilizing nascent coated pits during the assembly process (see Table 7.1). Transmembrane cargo proteins are recognized by adaptor proteins during CCP assembly, which bind to endocytic sorting motifs within their cytoplasmic tails. These motifs include tyrosine-based Yxxø (where ø is a bulky hydrophobic residue) and dileucine motifs, which bind directly to distinct sites within the AP-2 core domain (Höning et al. 2005). Yxxø motifs have been cocrystallized with the carboxy-terminal portion of the AP-2 μ-subunit (C-$\mu2$) to which they bind in an extended conformation (Owen et al. 1999). C-$\mu2$ also harbors a structurally unresolved binding site for basic internalization motifs found in a variety of multimeric membrane proteins including postsynaptic AMPA-receptors (Kastning et al., in preparation), GABA$_A$-receptors (Kittler et al. 2005), the α_{1b}-adrenergic receptor (Diviani et al. 2003), and members of the synaptotagmin family of calcium sensor proteins (Chapman et al. 1998; Grass et al. 2004; Haucke et al. 2000). Most, if not all, of these proteins undergo ligand- or stimulus-dependent internalization, suggesting that AP-2 binding signals may be preferentially used for regulated internalization.

Multiple lines of evidence suggest that the assembly of plasma membrane CCPs is dependent on the presence of PI(4,5)P$_2$ within the membrane. PI(4,5)P$_2$ is concentrated on the plasma membrane (De Matteis and Godi 2004). Elevated levels of PI(4,5)P$_2$ due to genetic inactivation of the inositol phosphatase synaptojanin 1 lead to perinatal lethality in mice and to the accumulation

TABLE 7.1.   Overview of endocytic coat and accessory proteins, their domain structure, binding partners, and function. Many of these factors exist as brain- or neuron-specific isoforms (see text for details). An excellent, regularly updated overview of endocytic accessory proteins can also be found at http://www2.mrc-lmb.cam.ac.uk/groups/hmm/adaptors/Table2.htm.

Coat/ accessory protein	Structural Features	Motif/ binding partners	Functions	Neuronal Isoforms
Clathrin HC & LCa/b	triskelia & lattice formation by CHC; regulatory role of LCa/b	HC binds accessory proteins via LLDLD or PWxxW clathrin box motifs; LCs bind HIP1, calmodulin, hsc70	endocytosis via formation of polyhedral lattice; pre-assembled at plasmalemma & synapses	neuron-specific light chains a and b
AP-2 (α, β2, μ2, σ2 subunits)	trunk & appendage domains connected via flexible hinge regions	Yxxø, [DE]xxxL[LI], and basic cargo motifs via μ2; membrane PI(4,5)P₂via α, μ2; accessory proteins via WVxF, DxF, and FxDxF appendage domain binding motifs; clathrin HC	plasma membrane endocytosis; connects clathrin coat to cargo sorting	neuron-specific long splice form of α_A-adaptin & enriched at synapses
Dynamin	large multidomain GTPase; self-oligomerizes on acidic membranes	GTP binding; PRD binds SH3 domain proteins; PH domain binds PI(4,5)P₂; self-assembly	vesicle scission on GTP hydrolysis	Dynamin I is brain-specific & enriched at synapses
Abp1	N-terminal ADFH domain, C-terminal SH3 domain	binds F-Actin (ADFH domain and charged region with consensus R/KXEEXR) & Synaptojanin1, Synapsin1, Dynamin1, and Piccolo via SH3 domain	links endocytosis with actin cytoskeleton, similar to HIP1R (both may interact)	
Amphiphysin	N-terminal BAR domain (dimerising, curvature-sensing/ inducing)	binds AP-2α and CHC via FxDxF, DxF, LLDLD, and PWxxW motifs; SH3 binds to PSRPNR in Dynamin I	induces or senses membrane curvature; coordinates clathrin & dynamin I recruitment	brain-specific splice variants of Amphiphysin 2 bind AP-2α and CHC
AP180/ CALM	N-terminal ANTH domain binds PI(4,5)P₂	binds AP-2α via FxDxF and DxF motifs, also binds CHC	recruits clathrin & promotes assembly on PI(4,5)P₂–rich membranes	AP180 is brain-specific & enriched at synapses
Auxilin	DnaJ domain	binds AP-2α and clathrin via DPW, WDW, and DLL motifs	CCV uncoating via stimulation of hsc70 ATPase & displacement of AP-2 adaptors	Auxilin 1 is brain-specific
Endophilins	N-terminal BAR domain (dimerising, curvature-sensing/ inducing); carboxy-terminal SH3 domain	SH3 binds PPxRP in dynamin I & PxRPP in Synaptojanin	senses/ induces curvature & recruits Synaptojanin	Endophilin 1 is brain-specific & enriched at synapses
Epsins	N-terminal ANTH domain binds PI(4,5)P₂ & bends membranes	binds AP-2α via DPW, and clathrin via clathrin box motifs; associates w/ Eps15 via NPF motifs	may coordinate membrane bending w/ clathrin/ AP-2 recruitment & internalization of cargo via UIM	
Eps15/ Eps15R	3 EH domains at amino-terminal end	binds APα via multiple DxF motifs; associates with NPF-containing proteins (i.e. epsin)	located at edges of coated pits & may thus provide a linker function; special role in internalization of ubiquitinated cargo via UIM	
HIP1/ HIP1R	N-terminal ANTH domain binds PI(4,5)P₂	binds clathrin via clathrin box & AP-2α via FxDxF and DxF	links endocytosis with actin cytoskeleton; required for AMPAR internalization	HIP1 enriched at postsynaptic side
Intersectin	multiple EH & SH3 domains plus Rho-GEF domain for Cdc42	binds NPFs via EH domains & PRDs via SH3 domains	link between Cdc42-mediated actin polymerization & endocytosis	
PIPK type I	PIP kinase domain that synthesizes PI(4,5)P₂	stimulated by interaction with Arf•GTP and talin (only γ isoform)	provides PI(4,5)P₂ required for initial stages of endocytosis; link with cell signaling	PIPK type Iγ is brain-specific & enriched at synapses
Stonins	C-terminal μ-homology domain binds synapto-tagmin	Stonin 2 interacts with AP-2α and μ2 via WVxFs; binds synaptotagmin via μ-homology domain	Stonin 2 facilitates synaptotagmin endocytosis	Stonin 2 is brain-specific & enriched in hippocampus
Synaptojanin	Inositol 5-phosphatase & SAC1 domains	binds AP-2α via FxDxF and WVxF motifs	PI(4,5)P₂ hydrolysis via phosphoinositide 5'- and 4' phosphatase activities; role in CCV uncoating	short 145 kDa isoform specifically expressed in mature neurons
Synaptotag-mins	C2 domains bind Ca²⁺, phospholipid, SNAREs, AP-2, and Stonin 2	basic motif implicated in binding to AP-2μ	may connect SV exo- and endocytosis by associating with AP-2 and stonin	Syt 1,2 are brain-specific
Syndapin	C-terminal SH3 domain binds N-WASP & Dynamin I		may link clathrin-mediated endocytosis with actin-based motility	

of CCVs within nerve endings due to decreased rates of presynaptic vesicle cycling (Cremona et al. 1999). Conversely, depletion of PI(4,5)P₂ by overexpression of a membrane-targeted 5-phosphatase domain (Krauss et al. 2003) or depletion of type I phosphatidylinositol 4-phosphate 5-kinase (PIPK) impairs receptor-mediated internalization of transferrin (Padron et al. 2003) or EGF due to mislocalization of AP-2 and plasma membrane pools of clathrin (Krauss et al. 2003).

A number of endocytic proteins including the α (Gaidarov and Keen 1999) and μ2 subunits of AP-2 (Rohde et al. 2002), AP180/CALM (Ford et al. 2001), HIP1/1R,

Dab2 (Mishra et al. 2002), epsin (Ford et al. 2002), and dynamin (Schmid et al. 1998) specifically bind to $PI(4,5)P_2$ (see Table 7.1). Structurally, binding is accomplished by a variety of different domains including a pleckstrin homology (PH) domain in the case of dynamin, ANTH, or ENTH domains within AP180/CALM and epsin, as well as surface-exposed basic patches within AP-2α and μ2. Several proteins related to the function of the actin cytoskeleton, such as WASP and profilin, that also affect clathrin-dependent endocytosis have been shown to associate with $PI(4,5)P_2$ (De Matteis and Godi 2004). In the case of AP-2, $PI(4,5)P_2$ may play an active role in stabilizing the open conformation of the complex, thus enabling cargo recognition by its μ2-subunit, facilitating the stable association of AP-2 with the membrane (Höning et al. 2005). Consistent with this scenario, clathrin/AP-2-coated pits have been shown to become stabilized in living cells upon encountering cargo receptors (Ehrlich et al. 2004). These data suggest that AP-2 recruitment and initiation of plasmalemmal CCPs is a highly cooperative process whereby stable membrane binding is achieved by a conformational change within AP-2μ subunit allowing for cargo recognition only in the context of a $PI(4,5)P_2$-rich membrane (Höning et al. 2005; Wenk et al. 2001). Stabilization of the active open AP-2 conformer at the presynaptic plasma membrane by binding to cargo proteins containing a Yxxø motif facilitates the interaction of AP-2μ with the synaptic vesicle protein synaptotagmin (Haucke et al. 2000), thus providing an additional means of positive cooperativity (Haucke and De Camilli 1999).

B. Clathrin-Coated Vesicles Bud from Endocytic Hot Spots

After the clathrin lattice is formed, endophilin, epsin, and amphiphysin are involved in membrane invagination and clathrin rearrangements (Conner and Schmid 2003; McMahon and Mills 2004;

Szymkiewicz et al. 2004; Takei and Haucke 2001). The GTPase dynamin is required for fisson of endocytic membrane vesicles (Hinshaw 2000) by mechanochemically constricting (pinchase) or expanding (poppase) the vesicle neck. Observation of clathrin-coated pit dynamics using evanescent wave microscopy indicates that during fission, dynamin recruitment to coated pits rapidly is followed by recruitment of actin (Merrifield 2004). Moreover perturbation of actin disrupts the endocytic reaction with accumulation of coated pits with wide necks (Shupliakov et al. 2002), suggesting a role for actin and dynamin-interacting accessory proteins in promoting constriction of the neck. In lamprey, snake (Teng and Wilkinson 2000), and fly neuromuscular synapses, the invagination of the membrane into pits occurs at distinct endocytic zones surrounding the active zones of exocytosis. FM1-43 photoconversion and serial section electron microscopy analysis revealed that labeled clathrin-coated endocytic vesicles are clustered near active zones consistent with local recycling vesicle pools at this synapse (see Figure 7.1).

One spatial cue, in addition to or coincident, with actin meshworks and presynaptic scaffolding proteins is $PI(4,5)P_2$, which is required for neurosecretory vesicle exocytosis (Gong et al. 2005; Milosevic et al. 2005) and endocytic recycling. Mice lacking the major brain-enriched type I PIPK (PIPK Iγ) have impaired neurotransmitter release as well as defects in the endocytic limb of the SV cycle (Di Paolo et al. 2004). Distinct "hot-spots of endocytosis" have also been described at the postsynaptic membrane (Blanpied et al. 2002; Jarousse and Kelly 2001) within dendritic spines (Racz et al. 2004) (see Figure 7.2).

III. MECHANISMS OF PRESYNAPTIC VESICLE CYCLING

Maintenance of synaptic transmission requires continuing retrieval and recycling

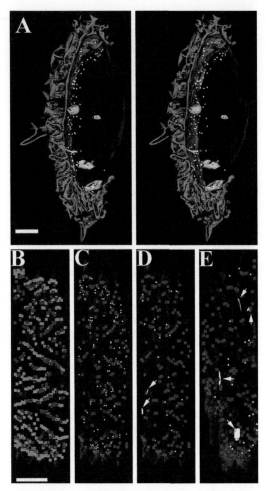

FIGURE 7.1. **Organization of the endocytic machinery near presynaptic active zones in snakes.** Endocytosis occurs near active zones (AZs) in snake motor boutons. **A.** Rendering of bouton portion from 31 serial EM sections. Stereo pair shows postjunctional folds of muscle fiber at left (blue). AZs lie on presynaptic membrane. Labeled vesicles (LVs) (white), shown as 50 nm spheres, were found predominantly near the presynaptic membrane. Also shown are four endosomes and two deep membrane invaginations (gray). **B, C.** Projections of fold centers (FCs; blue squares), AZs (red squares), and LVs (white spheres) onto a plane corresponding to that of the presynaptic membrane. **B.** AZs appeared near (or in direct apposition to) FCs. Some fold regions were not occupied by AZs. **C.** LVs were found clustered near AZs. **D, E.** Locations of coated pits (yellow spheres) and deep membrane invaginations (gray; arrows) relative to AZs in two rendered boutons. Bouton in **D** is the same as that in **A–C.** Scale bars: **A,** 1 μm; **B–E,** 1 μm. Reprinted with kind permission from Teng and Wilkinson 2000.

of SVs. Various modes of SV recycling have been proposed (see Figure 7.3). A fast "kiss-and-run" mechanism, where the vesicle connects only briefly with the plasma membrane without full membrane collapse (Aravanis et al. 2003; Ceccarelli et al. 1973; Galli and Haucke 2004; Gandhi and Stevens 2003; Klingauf et al. 1998; Koenig et al. 1998; Murthy and De Camilli 2003; Sun et al. 2002; Valtorta et al. 2001), perhaps without even leaving the plasma membrane ("kiss-and-stay") (Pyle et al. 2000). A slower retrieval pathway utilizing large infoldings and endosomes (Gad et al. 1998; Takei et al. 1996), or via clathrin-coated pits also exists (Galli and Haucke 2004; Heuser and Reese 1973; Murthy and De Camilli 2003; Pearse 1976; Shupliakov et al. 1997). Most of our understanding of the kinetics of stimulated endocytosis in synaptic boutons comes from electrophysiological membrane capacitance measurements of giant terminals like goldfish retinal bipolar cells (von Gersdorff and Matthews 1994), the calyx of Held (Sun et al. 2002), or from imaging experiments using fluorescent tracers. Vesicle cycling can be visualized by styryl dyes, such as the fluorescent vesicle marker FM 1-43 (Aravanis et al. 2003; Klingauf et al. 1998; Ryan et al. 1993, 1996; Wu and Betz 1996) or by imaging synaptopHluorin, a fusion construct between the SV protein synaptobrevin/VAMP-2 and a highly pH-sensitive GFP variant (Gandhi and Stevens 2003; Miesenbock et al. 1998; Sankaranarayanan and Ryan 2001) (see Chapter 6). However, these techniques may not unequivocally distinguish between different molecular mechanisms of membrane retrieval.

A. Kiss-and-Run Exo-Endocytosis

In hippocampal terminals a rapid kiss-and-run type of endocytosis has been proposed based on the observation of partial release of the slowly (~3 s) departitioning styryl dye FM 1-43 (Aravanis et al. 2003; Klingauf et al. 1998; Pyle et al. 2000) as well

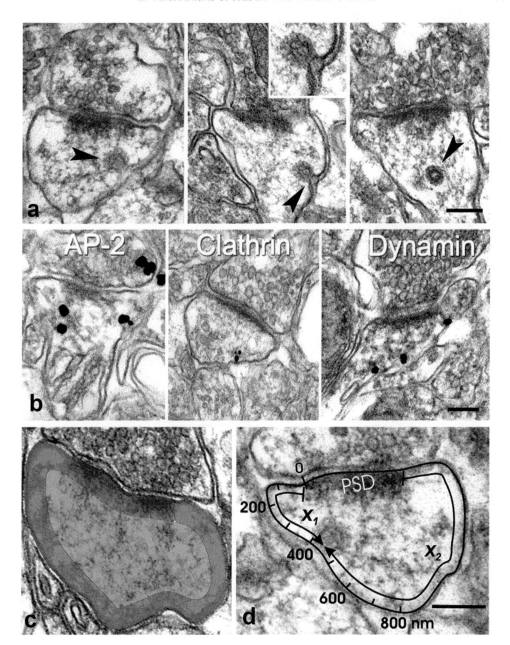

FIGURE 7.2. **Lateral organization of the endocytic machinery in dendritic spines.** Electron micrographs from CA1 hippocampus of adult rat. **(a)** Endocytosis in spines. Arrowheads point to coated pit assembly (left), vesicle scission (middle), and internal trafficking (right). **(b)** Silver-enhanced nanogold labeling of endocytic proteins. Nonsynaptic membrane was labeled even in spines that lacked identifiable coated structures. **(c)** Micrograph, color-coded to show spine core (green), postsynaptic shell (orange), and PSD region (blue). **(d)** Distances from each particle center to the closer (X_1) and farther (X_2) edges of the PSD were measured along the plasma membrane. Bars, 200 nm. Reprinted with kind permission from Racz et al. 2004; see http://www.nature.com/neuro/journal/v7/n9/abs/nn1303.html for further information.

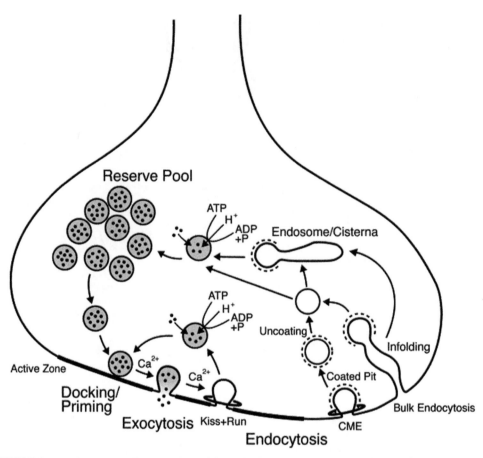

FIGURE 7.3. **Mechanisms of presynaptic vesicle cycling. Schematic depicting the various proposed modes of synaptic vesicle (SV) recycling.** A fast kiss-and-run mechanism, where the vesicle connects only briefly to the plasma membrane without full collapse *(kiss-and-run)*, a slow pathway via large infoldings and endosomes *(bulk endocytosis)*, and recovery of vesicle membrane by clathrin-coated pits (clathrin-mediated endocytosis, *CME*).

as based on the quantal and transient nature of fluorescence signals from single SVs labeled with synaptopHluorin (Gandhi and Stevens 2003). Pyle and colleagues (Pyle et al. 2000) used hypertonic solution to selectively fuse the readily releasable pool (RRP) of docked vesicles (Rosenmund and Stevens 1996) after labelling vesicles either with FM 1-43 or its more rapidly dissociating cousin FM 2-10. They discovered nonquantitative dye release in vesicles of the RRP for the slow dye FM 1-43, but not the faster FM 2-10. Thus, they proposed that vesicles of the RRP are preferentially recycled by a kiss-and-run mechanism and are rapidly reused. Stevens and Williams (2000)

found that styryl dyes are only partially released during a hypertonic challenge in hippocampal boutons by comparing fluorescence data and simultaneously recorded postsynaptic currents. Interestingly, no difference was observed between FM 1-43 and FM 2-10 in this study.

B. Bulk Retrieval

Another, much slower pathway that recycles SV membrane piecemeal via large infoldings and cisternae has been discovered at motor nerve terminals of frog (Miller and Heuser 1984; Richards et al. 2000), snake (Teng and Wilkinson 2000), endocrine cells

(Neher and Marty 1982; Rosenboom and Lindau 1994; Thomas et al. 1994), and may also be present in CNS synapses (de Lange et al. 2003; Gad et al. 1998; Holt et al. 2003; Paillart et al. 2003; Takei et al. 1996). The exact mechnism of bulk retrieval and its relationship to the clathrin pathway described next remains uncertain.

C. Clathrin-Mediated Vesicle Cycling

Following neurotransmitter release, clathrin-mediated endocytosis is thought to be the major pathway for SV recycling (Brodin et al. 2000; Brodsky et al. 2001; Heuser and Reese 1973). The first reconstruction of the time course of clathrin-mediated endocytosis was obtained from electron micrographs (Miller and Heuser 1984). Using a fusion construct of EGFP and either clathrin light (Gaidarov et al. 1999) or clathrin heavy chains (Damer and O'Halloran 2000), clathrin dynamics were visualized during constitutive endocytosis. The formation, fission, and movement of single CCVs has been imaged using evanescent field fluorescence microscopy (Merrifield et al. 2002). Blanpied et al. (2002) imaged clathrin-EGFP in hippocampal neurons to examine the kinetics of single coated pits in dendrites and spines. The time constants (tenths of seconds) measured for clathrin-mediated endocytosis in these imaging studies are surprisingly similar to the time course of membrane retrieval as reconstructed from electron micrographs of frog neuromuscular junctions (Miller and Heuser 1984). Mechanisms for more rapid vesicle reuse ($\tau \leq 10\,s$) may not require intermediate endosomal sorting (Murthy and Stevens 1998), as has been proposed for kiss-and-run or kiss-and-stay pathways (Aravanis et al. 2003; Fesce et al. 1994; Gandhi and Stevens 2003; Klingauf et al. 1998; Pyle et al. 2000; Sun et al. 2002). In the *Drosophila* neuromuscular junctions distinct vesicle sorting has been attributed to different endocytic routes and Ca^{2+} sources (Kuromi and Kidokoro 2002).

D. Presynaptic Vesicle Pools

Vesicles in presynaptic boutons are categorized depending on their release kinetics (Pieribone et al. 1995). Exo-endocytic cycling recruits only a fraction of available vesicles (Harata et al. 2001), the so-called "recycling pool" (RP), which typically contains about 30 to 45 vesicles. Part of the RP is called the readily releasable pool (RRP), and comprises vesicles (~5–10) available for immediate release as a result of stimulus onset, increase in local $[Ca^{2+}]_i$ (Schneggenburger and Neher 2000), or by a hypertonic pulse (Rosenmund and Stevens 1996; Stevens and Tsujimoto 1995). A recent study (Rizzoli and Betz 2004) suggests that, at least in frog NMJ, RRP vesicles are randomly distributed. However, an electron microscopic study has shown a correlation between RRP vesicles and those morphologically docked at the active zone (AZ) of cultured hippocampal neurons (Schikorski and Stevens 2001). Other recycling vesicles (the reserve pool) have a lower release probability, and the resting pool contains vesicles that are not recruited during a set stimulation period (Sudhof 2000). A future challenge will be to establish the molecular mechanisms that govern which vesicle recycling pathway is used in a given functional context.

IV. ENDOCYTOSIS OF POSTSYNAPTIC NEUROTRANSMITTER RECEPTORS

Over the past few years it has become clear that the strength of synaptic connections, in particular with respect to postsynaptic responses, is subject to plastic changes. At excitatory synapses, activation of glutamate receptors provides the primary depolarization in excitatory neurotransmission. Glutamate receptor-mediated postsynaptic currents are modulated by changes in their localization and surface expression. Glutamate receptor density thus appears to be carefully regulated by fine-tuning receptor synthesis, endosomal trafficking, and degradation (Bredt and Nicoll 2003; Sheng and Kim 2002). Similar findings have been reported for ionotropic GABA$_A$ (Kittler et al. 2000, 2005; van Rijnsoever et al. 2005) and glycine (Buttner et al. 2001; Rasmussen et al. 2002) receptors that regulate inhibitory neurotransmission in the nervous system and can also be internalized into endosomal compartments (more on this topic can be found in Chapters 8–10).

A. AMPA Receptor Internalization

AMPA receptors, heterotetramers composed of related GluR1-4 subunits, undergo dynamic redistribution in and out of the postsynaptic membrane. Most excitatory synapses form on dendritic spines, that emanate from the main shaft and usually bear a single synaptic contact at their heads. AMPA receptors, concentrated at the postsynaptic density (PSD) of dendritic spines, serve to propagate the signal (Sheng and Lee 2001) and are able to dynamically move into and out of the postsynaptic density by lateral diffusion (see Chapter 1). They may also undergo constitutive internalization (Man et al. 2000). Stimulation of glutamatergic synapses with AMPA, NMDA, or insulin has been shown to enhance AMPA receptor internalization by clathrin-mediated endocytosis (Beattie et al. 2000; Carroll et al. 1999; Lin et al. 2000; Luscher et al. 1999; Man et al. 2000). AMPA receptor internalization along the endocytic pathway correlates physiologically with activity-dependent long-term depression (LTD). Conversely, during long-term potentiation (LTP), a cellular model for learning and memory, an increase in the number of functional, cell-surface exposed AMPA receptors is observed at the postsynaptic membrane (Bredt and Nicoll 2003; Malinow and Malenka 2002). These receptors are thought to originate from an intracellular

reserve pool (Pickard et al. 2001) (see Figure 7.4). Endocytic removal of AMPA receptors occurs mostly from extrasynaptic sites (Ashby et al. 2004). This observation is consistent with the predominant localization of endocytic proteins including clathrin, AP-2, and dynamin lateral to the postsynaptic density (Racz et al. 2004) (see Figure 7.2).

The exact molecular mechanisms of the constitutive and regulated pathways for AMPA receptor internalization are not yet completely understood. Although all pathways are dependent on the GTPase dynamin, an accessory protein required for fission of both clathrin- and nonclathrin-coated vesicles, and its SH3 domain-containing binding partners (Man et al. 2000), they seem to be spatially segregated and differentially influenced by protein kinases (Ehlers 2000), phosphatases, and calcium ions (Beattie et al. 2000; Lin et al. 2000).

B. Endosomal Sorting of Internalized AMPA Receptors

Different stimuli differentially affect the subcellular localization of internalized receptors. AMPA receptors internalized via direct agonist stimulation (i.e., AMPA) colocalize with early endosomal markers such as early endosomal antigen 1 (EEA1), syntaxin 13, and internalized transferrin receptors. In contrast, AMPA receptors internalized via insulin- or NMDA-regulated signaling pathways, although initially present in EEA1-positive early endosomes, appear to segregate into distinct compartments, which may include late endosomes and lysosomes (Lee et al. 2004; also see Ehlers 2000). The mechanism underlying differential endosomal sorting remains unclear. Activated AMPA receptors colocalize with AP-2 (Carroll et al. 1999) and Eps15 (Man et al. 2000) in CCPs. Direct binding of the basic stretch within the cytoplasmic tail of the AMPA receptor, subunits GluR1-3, to the clathrin adaptor complex AP-2 is only required, however, for NMDA-

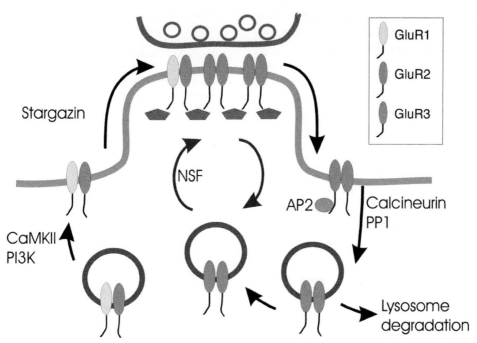

FIGURE 7.4. Model of AMPAR trafficking. AMPARs cycle between the postsynaptic membrane and intracellular compartments. Via a NSF-dependent mechanism, intracellular GluR2-GluR3 receptors exchange constantly with synaptic receptors. GluR1-GluR2 heteromeric receptors are retained in intracellular compartments but are delivered to the dendrite upon activation of NMDARs, CaMKII, and PI3K. Exocytosis of GluR1-GluR2 receptors occurs at extrasynaptic sites and is followed by lateral translocation into synapses. The activation of calcineurin and PP1 leads to the recruitment of the AP2 clathrin adaptor complex to AMPARs, resulting in endocytosis. Internalized AMPARs can be recycled to the surface or sorted to lysosomes for degradation. Circles in presynaptic terminal represent synaptic vesicles. Reprinted with kind permission from Sheng and Kim 2002.

induced AMPA receptor endocytosis (Lee et al. 2002), thus indicating that differential recognition modes at the cell surface may contribute to endosomal sorting. In nematodes GluR is subject to multiubiquitination, which may target glutamate receptors for internalization and late endosomal/lysosomal degradation (Burbea et al. 2002). Differential sorting of receptors recognized directly by endocytic adaptors or modified by ubiquitination has been observed in nonneuronal cells (e.g., internalized transferrin vs. epidermal growth factor receptors) (Dikic 2003; Hicke and Dunn 2003). Additionally, insulin-stimulated AMPA receptor internalization may be regulated by tyrosine phosphorylation (Ahmadian et al. 2004).

V. CONCLUSIONS

Endocytosis is critically important for both presynaptic and postsynaptic function. Synaptic vesicle cycling relies on endocytic retrieval of membrane and critically important regulatory proteins. Internalization and endocytic sorting controls the signaling mediated by postsynaptic membrane proteins such as receptors, ion channels, and transporters. There has been considerable progress toward understanding the mechanisms underlying endocytic trafficking, especially with regard to the initial internalization step. The role of sorting at later steps in the endocytic pathway cannot be underestimated and will likely be the subject of intensive study as mechanistic insight into protein sorting will clarify the rules that determine the ultimate fate of internalized proteins.

References

Ahmadian, G., Ju, W., Liu, L., Wyszynski, M., Lee, S.H., Dunah, A.W. et al. (2004). Tyrosine phosphorylation of GluR2 is required for insulin-stimulated AMPA receptor endocytosis and LTD. *Embo. J.* **23**, 1040–1050.

Aravanis, A.M., Pyle, J.L., Tsien, R.W. (2003). Single synaptic vesicles fusing transiently and successively without loss of identity. *Nature* **423**, 643–647.

Ashby, M.C., De La Rue, S.A., Ralph, G.S., Uney, J., Collingridge, G.L., Henley, J.M. (2004). Removal of AMPA receptors (AMPARs) from synapses is preceded by transient endocytosis of extrasynaptic AMPARs. *J. Neurosci.* **24**, 5172–5176.

Beattie, E.C., Carroll, R.C., Yu, X., Morishita, W., Yasuda, H., von Zastrow, M., Malenka, R.C. (2000). Regulation of AMPA receptor endocytosis by a signaling mechanism shared with LTD. *Nat. Neurosci.* **3**, 1291–1300.

Blanpied, T.A., Scott, D.B., Ehlers, M.D. (2002). Dynamics and regulation of clathrin coats at specialized endocytic zones of dendrites and spines. *Neuron* **36**, 435–449.

Bredt, D.S., Nicoll, R.A. (2003). AMPA receptor trafficking at excitatory synapses. *Neuron* **40**, 361–379.

Brodin, L., Low, P., Shupliakov, O. (2000). Sequential steps in clathrin-mediated synaptic vesicle endocytosis. *Curr. Opin. Neurobiol.* **10**, 312–320.

Brodsky, F.M., Chen, C.Y., Knuehl, C., Towler, M.C., Wakeham, D.E. (2001). Biological basket weaving: Formation and function of clathrin-coated vesicles. *Annu. Rev. Cell Dev. Biol.* **17**, 517–568.

Burbea, M., Dreier, L., Dittman, J.S., Grunwald, M.E., Kaplan, J.M. (2002). Ubiquitin and AP180 regulate the abundance of GLR-1 glutamate receptors at postsynaptic elements in C. elegans. *Neuron* **35**, 107–120.

Buttner, C., Sadtler, S., Leyendecker, A., Laube, B., Griffon, N., Betz, H., Schmalzing, G. (2001). Ubiquitination precedes internalization and proteolytic cleavage of plasma membrane-bound glycine receptors. *J. Biol. Chem.* **276**, 42978–42985.

Carroll, R.C., Lissin, D.V., von Zastrow, M., Nicoll, R.A., Malenka, R.C. (1999). Rapid redistribution of glutamate receptors contributes to long-term depression in hippocampal cultures. *Nat. Neurosci.* **2**, 454–460.

Ceccarelli, B., Hurlbut, W.P., Mauro, A. (1973). Turnover of transmitter and synaptic vesicles at the frog neuromuscular junction. *J. Cell Biol.* **57**, 499–524.

Chapman, E.R., Desai, R.C., Davis, A.F., Tornehl, C.K. (1998). Delineation of the oligomerization, AP-2

binding, and synprint binding region of the C2B domain of synaptotagmin. *J. Biol. Chem.* **273**, 32966–32972.

Colledge, M., Snyder, E.M., Crozier, R.A., Soderling, J.A., Jin, Y., Langeberg, L.K. et al. (2003). Ubiquitination regulates PSD-95 degradation and AMPA receptor surface expression. *Neuron* **40**, 595–607.

Collins, B.M., McCoy, A.J., Kent, H.M., Evans, P.R., Owen, D.J. (2002). Molecular architecture and functional model of the endocytic AP2 complex. *Cell* **109**, 523–535.

Conner, S.D., Schmid, S.L. (2003). Regulated portals of entry into the cell. *Nature* **422**, 37–44.

Cremona, O., Di Paolo, G., Wenk, M.R., Luthi, A., Kim, W.T., Takei, K. et al. (1999). Essential role of phosphoinositide metabolism in synaptic vesicle recycling. *Cell* **99**, 179–188.

Damer, C.K., O'Halloran, T.J. (2000). Spatially regulated recruitment of clathrin to the plasma membrane during capping and cell translocation. *Mol. Biol. Cell* **11**, 2151–2159.

De Camilli, P., Haucke, V., Takei, K., Mugnaini (2000). The structure of synapses. *In* Synapses. M.W. Cowan, T.C. Südhof, C.F. Stevens, eds., 89–133. The Johns Hopkins Univ Press, Baltimore.

de Lange, R.P., de Roos, A.D., Borst, J.G. (2003). Two modes of vesicle recycling in the rat calyx of Held. *J. Neurosci.* **23**, 10164–10173.

De Matteis, M.A., Godi, A. (2004). PI-loting membrane traffic. *Nat. Cell Biol.* **6**, 487–492.

Di Paolo, G., Moskowitz, H.S., Gipson, K., Wenk, M.R., Voronov, S., Obayashi, M. et al. (2004). Impaired PtdIns(4,5)P2 synthesis in nerve terminals produces defects in synaptic vesicle trafficking. *Nature* **431**, 415–422.

Dikic, I. (2003). Mechanisms controlling EGF receptor endocytosis and degradation. *Biochem. Soc. Trans.* **31**, 1178–1181.

Diviani, D., Lattion, A.L., Abuin, L., Staub, O., Cotecchia, S. (2003). The adaptor complex 2 directly interacts with the alpha 1B-adrenergic receptor and plays a role in receptor endocytosis. *J. Biol. Chem.* **278**, 19331–19340.

Ehlers, M.D. (2000). Reinsertion or degradation of AMPA receptors determined by activity-dependent endocytic sorting. *Neuron* **28**, 511–525.

Ehrlich, M., Boll, W., Van Oijen, A., Hariharan, R., Chandran, K., Nibert, M.L., Kirchhausen, T. (2004). Endocytosis by random initiation and stabilization of clathrin-coated pits. *Cell* **118**, 591–605.

Fesce, R., Grohovaz, F., Valtorta, F., Meldolesi, J. (1994). Neurotransmitter release: Fusion or "kiss-and-run"? *Trends Cell Biol.* **4**, 1–4.

Ford, M.G., Mills, I.G., Peter, B.J., Vallis, Y., Praefcke, G.J., Evans, P.R., McMahon, H.T. (2002). Curvature of clathrin-coated pits driven by epsin. *Nature* **419**, 361–366.

Ford, M.G., Pearse, B.M., Higgins, M.K., Vallis, Y., Owen, D.J., Gibson, A. et al. (2001). Simultaneous binding of PtdIns(4,5)P2 and clathrin by AP180 in the nucleation of clathrin lattices on membranes. *Science* **291**, 1051–1055.

Gad, H., Low, P., Zotova, E., Brodin, L., Shupliakov, O. (1998). Dissociation between Ca2+-triggered synaptic vesicle exocytosis and clathrin-mediated endocytosis at a central synapse. *Neuron* **21**, 607–616.

Gaidarov, I., Keen, J.H. (1999). Phosphoinositide-AP-2 interactions required for targeting to plasma membrane clathrin-coated pits. *J. Cell Biol.* **146**, 755–764.

Gaidarov, I., Santini, F., Warren, R.A., Keen, J.H. (1999). Spatial control of coated-pit dynamics in living cells. *Nat. Cell Biol.* **1**, 1–7.

Galli, T., Haucke, V. (2004). Cycling of synaptic vesicles: How far? How fast! *Sci. STKE.* **264**, re19.

Gallop, J.L., McMahon, H.T. (2005). BAR domains and membrane curvature: Bringing your curves to the BAR. *Biochem. Soc. Symp.* **72**, 223–231.

Gandhi, S.P., Stevens, C.F. (2003). Three modes of synaptic vesicular recycling revealed by single-vesicle imaging. *Nature* **423**, 607–613.

Gong, L.W., Di Paolo, G., Diaz, E., Cestra, G., Diaz, M.E., Lindau, M. et al. (2005). Phosphatidylinositol phosphate kinase type I gamma regulates dynamics of large dense-core vesicle fusion. *Proc. Natl. Acad. Sci. U S A* **102**, 5204–5209.

Grass, I., Thiel, S., Honing, S., Haucke, V. (2004). Recognition of a basic AP-2 binding motif within the C2B domain of synaptotagmin is dependent on multimerization. *J. Biol. Chem.* **279**, 54872–54880.

Haglund, K., Di Fiore, P.P., Dikic, I. (2003). Distinct monoubiquitin signals in receptor endocytosis. *Trends Biochem. Sci.* **28**, 598–603.

Harata, N., Pyle, J.L., Aravanis, A.M., Mozhayeva, M., Kavalali, E.T., Tsien, R.W. (2001). Limited numbers of recycling vesicles in small CNS nerve terminals: Implications for neural signaling and vesicular cycling. *Trends Neurosci.* **24**, 637–643.

Haucke, V., De Camilli, P. (1999). AP-2 recruitment to synaptotagmin stimulated by tyrosine-based endocytic motifs. *Science* **285**, 1268–1271.

Haucke, V., Wenk, M.R., Chapman, E.R., Farsad, K., De Camilli, P. (2000). Dual interaction of synaptotagmin with mu2- and alpha-adaptin facilitates clathrin-coated pit nucleation. *EMBO. J.* **19**, 6011–6019.

Heuser, J. (1989). The role of coated vesicles in recycling of synaptic vesicle membrane. *Cell Biol. Intl. Report.* **13**, 1063–1076.

Heuser, J.E., Reese, T.S. (1973). Evidence for recycling of synaptic vesicle membrane during transmitter release at the frog neuromuscular junction. *J. Cell Biol.* **57**, 315–344.

Hicke, L., Dunn, R. (2003). Regulation of membrane protein transport by ubiquitin and ubiquitin-binding proteins. *Annu. Rev. Cell. Dev. Biol.* **19**, 141–172.

Hinshaw, J.E. (2000). Dynamin and its role in membrane fission. *Annu. Rev. Cell Dev. Biol.* **16**, 483–519.

Holt, M., Cooke, A., Wu, M.M., Lagnado, L. (2003). Bulk membrane retrieval in the synaptic terminal of retinal bipolar cells. *J. Neurosci.* **23**, 1329–1339.

Höning, S., Ricotta, D., Krauss, M., Späte, K., Motley, A., Spolaore, B. et al. (2005). Phosphatidylinositol (4,5)-bisphosphate regulates sorting signal recognition by the clathrin-associated adaptor complex AP-2. *Mol. Cell.* **18**, 519–531.

Jarousse, N., Kelly, R.B. (2001). Endocytotic mechanisms in synapses. *Curr. Opin. Cell Biol.* **13**, 461–469.

Kittler, J.T., Chen, G., Höning, S., Bogdanov, Y.D., McAinsh, C., Arancibia, -.C., I.L. et al. (2005). Phospho-dependent binding of the clathrin AP2 adaptor complex to GABA_A receptors regulates the efficiacy of inhibitory synaptic transmission. *Proc. Natl. Acad. Sci. U S A* **102**, 1487–14876.

Kittler, J.T., Delmas, P., Jovanovic, J.N., Brown, D.A., Smart, T.G., Moss, S.J. (2000). Constitutive endocytosis of GABA_A receptors by an association with the adaptin AP2 complex modulates inhibitory synaptic currents in hippocampal neurons. *J. Neurosci.* **20**, 7972–7977.

Kittler, J.T., Thomas, P., Tretter, V., Bogdanov, Y.D., Haucke, V., Smart, T.G., Moss, S.J. (2004). Huntingtin-associated protein 1 regulates inhibitory synaptic transmission by modulating gamma-aminobutyric acid type A receptor trafficking. *Proc. Natl. Acad. Sci. U S A* **101**, 12736–12741.

Klingauf, J., Kavalali, E.T., Tsien, R.W. (1998). Kinetics and regulation of fast endocytosis at hippocampal synapses. *Nature* **394**, 581–585.

Koenig, J.H., Yamaoka, K., Ikeda, K. (1998). Omega images at the active zone may be endocytotic rather than exocytotic: implications for the vesicle hypothesis of transmitter release. *Proc. Natl. Acad. Sci. U S A* **95**, 12677–12682.

Krauss, M., Kinuta, M., Wenk, M.R., De Camilli, P., Takei, K., Haucke, V. (2003). ARF6 stimulates clathrin/AP-2 recruitment to synaptic membranes by activating phosphatidylinositol phosphate kinase type Igamma. *J. Cell Biol.* **162**, 113–124.

Kuromi, H., Kidokoro, Y. (2002). Selective replenishment of two vesicle pools depends on the source of Ca2+ at the Drosophila synapse. *Neuron* **35**, 333–343.

Lee, S.H., Liu, L., Wang, Y.T., Sheng, M. (2002). Clathrin adaptor AP2 and NSF interact with overlapping sites of GluR2 and play distinct roles in AMPA receptor trafficking and hippocampal LTD. *Neuron* **36**, 661–674.

Lee, S.H., Simonetta, A., Sheng, M. (2004). Subunit rules governing the sorting of internalized AMPA receptors in hippocampal neurons. *Neuron* **43**, 221–236.

Li, Y., Chin, L.S., Levey, A.I., Li, L. (2002). Huntingtin-associated protein 1 interacts with hepatocyte growth factor-regulated tyrosine kinase substrate and functions in endosomal trafficking. *J. Biol. Chem.* **277**, 28212–28221.

Lin, J.W., Ju, W., Foster, K., Lee, S.H., Ahmadian, G., Wyszynski, M. et al. (2000). Distinct molecular mechanisms and divergent endocytotic pathways of AMPA receptor internalization. *Nat. Neurosci.* **3**, 1282–1290.

Luscher, C., Xia, H., Beattie, E.C., Carroll, R.C., von Zastrow, M., Malenka, R.C., Nicoll, R.A. (1999). Role of AMPA receptor cycling in synaptic transmission and plasticity. *Neuron* **24**, 649–658.

Malinow, R., Malenka, R.C. (2002). AMPA receptor trafficking and synaptic plasticity. *Annu. Rev. Neurosci.* **25**, 103–126.

Man, H.Y., Lin, J.W., Ju, W.H., Ahmadian, G., Liu, L., Becker, L.E. et al. (2000). Regulation of AMPA receptor-mediated synaptic transmission by clathrin-dependent receptor internalization. *Neuron* **25**, 649–662.

McMahon, H.T., Mills, I.G. (2004). COP and clathrin-coated vesicle budding: Different pathways, common approaches. *Curr. Opin. Cell Biol.* **16**, 379–391.

Merrifield, C.J. (2004). Seeing is believing: Imaging actin dynamics at single sites of endocytosis. *Trends Cell Biol.* **14**, 352–358.

Merrifield, C.J., Feldman, M.E., Wan, L., Almers, W. (2002). Imaging actin and dynamin recruitment during invagination of single clathrin-coated pits. *Nat. Cell Biol.* **4**, 691–698.

Miesenbock, G., De Angelis, D.A., Rothman, J.E. (1998). Visualizing secretion and synaptic transmission with pH-sensitive green fluorescent proteins. *Nature* **394**, 192–195.

Miller, T.M., Heuser, J.E. (1984). Endocytosis of synaptic vesicle membrane at the frog neuromuscular junction. *J. Cell Biol.* **98**, 685–698.

Milosevic, I., Sorensen, J.B., Lang, T., Krauss, M., Nagy, G., Haucke, V. et al. (2005). Plasmalemmal phosphatidylinositol-4,5-bisphosphate level regulates the releasable vesicle pool size in chromaffin cells. *J. Neurosci.* **25**, 2557–2565.

Mishra, S.K., Keyel, P.A., Hawryluk, M.J., Agostinelli, N.R., Watkins, S.C., Traub, L.M. (2002). Disabled-2 exhibits the properties of a cargo-selective endocytic clathrin adaptor. *EMBO. J.* **21**, 4915–4926.

Murthy, V.N., De Camilli, P. (2003). Cell biology of the presynaptic terminal. *Annu. Rev. Neurosci.* **26**, 701–728.

Murthy, V.N., Stevens, C.F. (1998). Synaptic vesicles retain their identity through the endocytic cycle. *Nature* **392**, 497–501.

Neher, E., Marty, A. (1982). Discrete changes of cell membrane capacitance observed under conditions of enhanced secretion in bovine adrenal chromaffin cells. *Proc. Natl. Acad. Sci. U S A* **79**, 6712–6716.

Owen, D.J., Collins, B.M., Evans, P.R. (2004). Adaptors for clathrin coats: Structure and function. *Annu. Rev. Cell Dev. Biol.* **20**, 153–191.

Owen, D.J., Vallis, Y., Noble, M.E., Hunter, J.B., Dafforn, T.R., Evans, P.R., McMahon, H.T. (1999). A structural explanation for the binding of multiple ligands by the alpha-adaptin appendage domain. *Cell* **97**, 805–815.

Padron, D., Wang, Y.J., Yamamoto, M., Yin, H., Roth, M.G. (2003). Phosphatidylinositol phosphate 5-kinase Ibeta recruits AP-2 to the plasma membrane and regulates rates of constitutive endocytosis. *J. Cell Biol.* **162**, 693–701.

Paillart, C., Li, J., Matthews, G., Sterling, P. (2003). Endocytosis and vesicle recycling at a ribbon synapse. *J. Neurosci.* **23**, 4092–4099.

Pearse, B.M. (1976). Clathrin: A unique protein associated with intracellular transfer of membrane by coated vesicles. *Proc. Natl. Acad. Sci. U S A* **73**, 1255–1259.

Pickard, L., Noel, J., Duckworth, J.K., Fitzjohn, S.M., Henley, J.M., Collingridge, G.L., Molnar, E. (2001). Transient synaptic activation of NMDA receptors leads to the insertion of native AMPA receptors at hippocampal neuronal plasma membranes. *Neuropharmacology* **41**, 700–713.

Pieribone, V.A., Shupliakov, O., Brodin, L., Hilfiker-Rothenfluh, S., Czernik, A.J., Greengard, P. (1995). Distinct pools of synaptic vesicles in neurotransmitter release. *Nature* **375**, 493–497.

Praefcke, G.J., Ford, M.G., Schmid, E.M., Olesen, L.E., Gallop, J.L., Peak-Chew, S.Y. et al. (2004). Evolving nature of the AP2 alpha-appendage hub during clathrin-coated vesicle endocytosis. *EMBO. J.* **23**, 4371–4383.

Pyle, J.L., Kavalali, E.T., Piedras-Renteria, E.S., Tsien, R.W. (2000). Rapid reuse of readily releasable pool vesicles at hippocampal synapses. *Neuron* **28**, 221–231.

Racz, B., Blanpied, T.A., Ehlers, M.D., Weinberg, R.J. (2004). Lateral organization of endocytic machinery in dendritic spines. *Nat. Neurosci.* **7**, 917–918.

Raiborg, C., Rusten, T.E., Stenmark, H. (2003). Protein sorting into multivesicular endosomes. *Curr. Opin. Cell Biol.* **15**, 446–455.

Rasmussen, H., Rasmussen, T., Triller, A., Vannier, C. (2002). Strychnine-blocked glycine receptor is removed from synapses by a shift in insertion/degradation equilibrium. *Mol. Cell Neurosci.* **19**, 201–215.

Richards, D.A., Guatimosim, C., Betz, W.J. (2000). Two endocytic recycling routes selectively fill two vesicle pools in frog motor nerve terminals. *Neuron* **27**, 551–559.

Rizzoli, S.O., Betz, W.J. (2004). The structural organization of the readily releasable pool of synaptic vesicles. *Science* **303**, 2037–2039.

Rohde, G., Wenzel, D., Haucke, V. (2002). A phosphatidylinositol (4,5)-bisphosphate binding site within mu2-adaptin regulates clathrin-mediated endocytosis. *J. Cell Biol.* **158**, 209–214.

Rosenboom, H., Lindau, M. (1994). Exo-endocytosis and closing of the fission pore during endocytosis in single pituitary nerve terminals internally perfused with high calcium concentrations. *Proc. Natl. Acad. Sci. U S A* **91**, 5267–5271.

Rosenmund, C., Stevens, C.F. (1996). Definition of the readily releasable pool of vesicles at hippocampal synapses. *Neuron* **16**, 1197–1207.

Ryan, T.A., Reuter, H., Wendland, B., Schweizer, F.E., Tsien, R.W., Smith, S.J. (1993). The kinetics of synaptic vesicle recycling measured at single presynaptic boutons. *Neuron* **11**, 713–724.

Ryan, T.A., Smith, S.J., Reuter, H. (1996). The timing of synaptic vesicle endocytosis. *Proc. Natl. Acad. Sci. U S A* **93**, 5567–5571.

Sankaranarayanan, S., Ryan, T.A. (2001). Calcium accelerates endocytosis of vSNAREs at hippocampal synapses. *Nat. Neurosci.* **4**, 129–136.

Schikorski, T., Stevens, C.F. (2001). Morphological correlates of functionally defined synaptic vesicle populations. *Nat. Neurosci.* **4**, 391–395.

Schmid, S.L., McNiven, M.A., De Camilli, P. (1998). Dynamin and its partners: A progress report. *Curr. Opin. Cell Biol.* **10**, 504–512.

Schneggenburger, R., Neher, E. (2000). Intracellular calcium dependence of transmitter release rates at a fast central synapse. *Nature* **406**, 889–893.

Sheng, M., Kim, M.J. (2002). Postsynaptic signaling and plasticity mechanisms. *Science* **298**, 776–780.

Sheng, M., Lee, S.H. (2001). AMPA receptor trafficking and the control of synaptic transmission. *Cell* **105**, 825–828.

Shupliakov, O., Bloom, O., Gustafsson, J.S., Kjaerulff, O., Low, P., Tomilin, N. et al. (2002). Impaired recycling of synaptic vesicles after acute perturbation of the presynaptic actin cytoskeleton. *Proc. Natl. Acad. Sci. U S A* **99**, 14476–14481.

Shupliakov, O., Low, P., Grabs, D., Gad, H., Chen, H., David, C. et al. (1997). Synaptic vesicle endocytosis impaired by disruption of dynamin-SH3 domain interactions. *Science* **276**, 259–263.

Stevens, C.F., Tsujimoto, T. (1995). Estimates for the pool size of releasable quanta at a single central synapse and for the time required to refill the pool. *Proc. Natl. Acad. Sci. U S A* **92**, 846–849.

Stevens, C.F., Williams, J.H. (2000). "Kiss and run" exocytosis at hippocampal synapses. *Proc. Natl. Acad. Sci. U S A* **97**, 12828–12833.

Sudhof, T.C. (2000). The synaptic vesicle cycle revisited. *Neuron* **28**, 317–320.

Sun, J.Y., Wu, X.S., Wu, L.G. (2002). Single and multiple vesicle fusion induce different rates of endocytosis at a central synapse. *Nature* **417**, 555–559.

Szymkiewicz, I., Shupliakov, O., Dikic, I. (2004). Cargo- and compartment-selective endocytic scaffold proteins. *Biochem. J.* **383**, 1–11.

Takei, K., Haucke, V. (2001). Clathrin-mediated endocytosis: Membrane factors pull the trigger. *Trends Cell Biol.* **11**, 385–391.

Takei, K., Mundigl, O., Daniell, L., De Camilli, P. (1996). The synaptic vesicle cycle: A single vesicle budding step involving clathrin and dynamin. *J. Cell Biol.* **133**, 1237–1250.

Teng, H., Wilkinson, R.S. (2000). Clathrin-mediated endocytosis near active zones in snake motor boutons. *J. Neurosci.* **20**, 7986–7993.

Thomas, P., Lee, A.K., Wong, J.G., Almers, W. (1994). A triggered mechanism retrieves membrane in seconds after Ca(2+)-stimulated exocytosis in single pituitary cells. *J. Cell Biol.* **124**, 667–675.

Torrealba, F., Carrasco, M.A. (2004). A review on electron microscopy and neurotransmitter systems. *Brain Res. Brain Res. Rev.* **47**, 5–17.

Traub, L.M. (2003). Sorting it out: AP-2 and alternate clathrin adaptors in endocytic cargo selection. *J. Cell Biol.* **163**, 203–208.

Valtorta, F., Meldolesi, J., Fesce, R. (2001). Synaptic vesicles: Is kissing a matter of competence? *Trends Cell Biol.* **11**, 324–328.

van Rijnsoever, C., Sidler, C., Fritschy, J.M. (2005). Internalized GABA-receptor subunits are transferred to an intracellular pool associated with the postsynaptic density. *Eur. J. Neurosci.* **21**, 327–338.

Varoqueaux, F., Sigler, A., Rhee, J.S., Brose, N. Enk, C., Reim, K., Rosenmund, C. (2002). Total arrest of spontaneous and evoked synaptic transmission but normal synaptogenesis in the absence of Munc13-mediated vessicle priming. *Proc. Natl. Acad. Sci. U S A* **99**, 9037–9042.

von Gersdorff, H., Matthews, G. (1994). Dynamics of synaptic vesicle fusion and membrane retrieval in synaptic terminals. *Nature* **367**, 735–739.

Wenk, M.R., Pellegrini, L., Klenchin, V.A., Di Paolo, G., Chang, S., Daniell, L. et al. (2001). PIP kinase Igamma is the major PI(4,5)P(2) synthesizing enzyme at the synapse. *Neuron* **32**, 79–88.

Wu, L.G., Betz, W.J. (1996). Nerve activity but not intracellular calcium determines the time course of endocytosis at the frog neuromuscular junction. *Neuron* **17**, 769–779.

IV

RECEPTOR TRAFFICKING

The number of receptors and their residence time on the plasma membrane are critical determinants for the response of a neuron to extracellular cues, which can control synaptic plasticity, growth, and differentiation. The organization of the postsynaptic machinery (Condon and Ehlers) and how that machinery it utilized by different receptors is the subject of this section. Examination of the trafficking of glutamate receptor subtypes (Chen and Maghsoodi; Barria) provides an interesting example of similarities and differences in receptor trafficking mechanisms.

8

Postsynaptic Machinery for Receptor Trafficking

KATHRYN H. CONDON AND MICHAEL D. EHLERS

After extending axons, developing dendrites, and forming synapses, mature neurons are stably integrated into a network. However, the environment of the neuron continues to change and neural circuits continue to adapt. The trafficking of postsynaptic receptors is a primary means by which a mature neuron tunes its response to incoming signals.

This chapter will provide an overview of the machinery responsible for postsynaptic receptor trafficking. We discuss the various membrane domains and membrane compartments through which neurotransmitter receptors traffic during their lifetime—the endoplasmic reticulum, Golgi apparatus, endosomes, and the plasma membrane. Emphasis will be placed on how such membrane compartments, though present in other cell types, have evolved neuron-specific features that allow for dynamic trafficking to and from central nervous system synapses.

I. POSTSYNAPTIC MICROANATOMY: DENDRITES AND SPINES

Before we discuss the details of the machinery for receptor trafficking, we must first become familiar with the organization of the postsynaptic compartment of the neuron. Here we introduce postsynaptic anatomy and specializations of dendritic spines, the cytoskeleton, and the internal membrane compartments involved in receptor trafficking (see Figure 8.1), with emphasis on glutamatergic synapses. Detailed discussions of the roles and regulation of particular structures in receptor trafficking will follow in subsequent sections.

A. Synapses and the Postsynaptic Density

Asymmetric and Symmetric Synapses

In the mature neuron, 80 to 90% of synapses are asymmetric synapses, characterized by a thick postsynaptic density (PSD) on dendritic spines. These synapses are typically excitatory synapses that utilize glutamate as a neurotransmitter. Dendritic spines are mushroom-shaped membrane protrusions off the dendritic shaft that are typically less than 2 µm in length and contain a spine head about 1 µm in width. Longer protrusions lacking heads are classified as filopodia and often are not part of an active synapse. Spine morphology can

vary greatly among spines and within one spine over time. The head of the spine contains a specialized protein structure, the PSD, named for its dark, dense appearance in electron microscopic images of the synapse. The highly regulated and protein-rich organization of the PSD links postsynaptic receptors to scaffolding molecules, signaling molecules, and the actin cytoskeleton. This arrangement allows for changes in synaptic activity to be signaled through the spine to the rest of the neuron in a spatially and temporally regulated manner (DeFelipe et al. 2002; Nimchinsky et al. 2002).

Synapses onto the soma or dendritic shaft are typically symmetric synapses characterized by a thinner postsynaptic specialization. These are usually inhibitory synapses that utilize γ-amino butyric acid (GABA) as a neurotransmitter. In the symmetric synapse, the postsynaptic structures are less elaborate than dendritic spine synapses and the PSD is thin and less prominent (DeFelipe et al. 2002).

Neurotransmitter Receptors at Synapses

The characteristics of postsynaptic receptors vary greatly. Ionotropic glutamate receptors, particularly α-amino-3-hydroxy-5-methyl-4-isoxazolepropionic acid (AMPA) and N-methyl-D-aspartate (NMDA) receptors, are the primary receptor types found at the excitatory synapse. In response to presynaptic glutamate release and subsequent binding of glutamate to the receptor, these receptor channels open, allowing a flux of cations, thereby depolarizing the postsynaptic membrane. The magnitude of this ion flux and concomitant membrane depolarization is determined, in part, by the number of receptors present at the synapse (Malinow and Malenka 2002).

Attempts to determine the number of receptors at excitatory synapses using electophysiological techniques have estimated that the number of AMPA receptors at the hippocampal or Purkinje cell synapse ranges from zero to 230 receptors

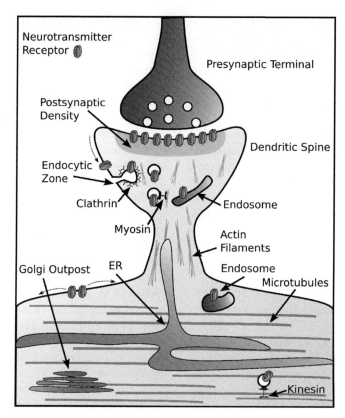

FIGURE 8.1. **Postsynaptic machinery for receptor trafficking.** The dendritic spine is a mushroom-shaped protrusion from the dendritic shaft of the postsynaptic neuron (yellow). Machinery for the neuronal secretory and endocytic pathways can be found within and near dendritic spines, permitting the movement of receptors in and out of the postsynaptic density, which sits directly apposed to the presynaptic terminal (blue). The endoplasmic reticulum (green) is a continuous structure throughout the soma and dendrites, and can extend into spines. Dendritic Golgi outposts (red) are discontinuous from the somatic Golgi and can be found in a subset of dendrites. The endocytic zone with clathrin-coated pits lies just lateral to the postsynaptic density. Endosomes (purple) are distributed in the dendrite and can be found in larger spines. Trafficking of vesicles and membrane compartments occurs by transport along microtubules (gray) in dendrites and actin filaments (orange) in dendritic spines, facilitated by kinesin and myosin molecular motors, respectively. Receptors also diffuse laterally in the plasma membrane between the postsynaptic density and extrasynaptic sites.

(Matsuzaki et al. 2001; Momiyama et al. 2003; Tanaka et al. 2005). Immunogold labeling and counting of receptors at adult hippocampal synapses, in contrast, has indicated the presence of just six to ten NMDA receptors per synapse (Racca et al. 2000), although this technique may underestimate receptor number. The receptor count at a synapse varies with the structure of the spine, with larger synapses containing more functional AMPA receptors (Matsuzaki et al. 2001). Smaller synapses (diameter less than

about 180 nm) can be completely devoid of AMPA receptors but still contain NMDA receptors (Takumi et al. 1999). Due to the absence of AMPA receptors and the voltage-dependent block of NMDA receptors by Mg^{2+}, these so-called "silent synapses" do not generate excitatory postsynaptic potentials during basal stimulation at typical resting membrane potentials. In contrast to AMPA receptors, NMDA receptor density is weakly correlated to synapse size, such that the AMPA-to-NMDA receptor ratio changes dramatically with the size of the synapse (Takumi et al. 1999; Racca et al. 2000). However, not all receptors respond to every stimulation. Nimchinsky et al. (2004) calculated that only three NMDA receptors and less than 10 AMPA receptors per synapse on CA1 hippocampal neurons open under low-frequency stimulation conditions. These numbers and the calculations by Matsuzaki et al. (2001) suggest that the postsynaptic receptors are not saturated by a single quantum of glutamate, allowing the synapse to respond to a greater dynamic range of stimulation.

B. Cytoskeleton: Morphology and Motion

The complex morphology of the neuron underscores the need for an organized route for trafficking, so that proteins and organelles can be delivered to the most distal segments of the neuron. As in other cell types, the neuronal cytoskeleton is composed of both microtubule and actin structures, which are critical for receptor trafficking and structural integrity. These two cytoskeletal elements play distinct roles in postsynaptic receptor trafficking, with the microtubule cytoskeleton typically providing long-range transport in the dendritic shaft and the actin cytoskeleton providing short-range transport of proteins in dendritic spines.

Microtubule Cytoskeleton

The microtubule cytoskeleton has a uniform orientation in the axon (plus-ends distal) and mixed polarity in the dendrites (Baas et al. 1988). As microtubule motors usually provide unidirectional transport along a microtubule, this difference in microtubule polarity permits different sets of molecular motors to associate with axons or dendrites. A detailed discussion of this transport and of microtubule motors can be found in Chapters 2 and 3.

Several studies have suggested that the transport of receptor-containing vesicles

to their appropriate dendritic destinations is mediated by interactions with kinesin superfamily protein (KIF) microtubule motors through adaptor proteins. The interaction between the Mint-CASK-Velis PDZ protein complex, the NMDA receptor subunit NR2B, and the dendrite-specific kinesin KIF17 is among the first examples of a link between postsynaptic receptor cargo and a microtubule motor protein (Setou et al. 2000). Furthermore, the activity of KIF17 is required for the proper synaptic targeting of NR2B (Guillaud et al. 2003). Similarly, glutamate receptor interacting protein 1 (GRIP1), an AMPA receptor-associated protein, binds directly to the heavy chain of the KIF5 kinesin motor (Setou et al. 2000). Disruption of this interaction reduces the amount of synaptic GRIP1 and GluR2, suggesting that the interaction between GRIP1 and KIF5 is required for proper delivery of AMPA receptors to synapses. At inhibitory synapses, gephyrin, a scaffold protein that binds tubulin, inhibitory glycine receptors, and $GABA_A$ receptors, links receptors to motor proteins through its interaction with the dynein light chains Dlc1 and Dlc2 (Fuhrmann et al. 2002). As components of the cytoplasmic dynein and myosin Va complexes, Dlc1 and Dlc2 may be involved in the proper targeting of $GABA_A$ receptors. However, this interaction does not appear to be essential for the synaptic localization of gephyrin.

Actin Cytoskeleton

As the dendritic spine does not contain microtubules, protein trafficking in the spine relies heavily on the actin cytoskeleton. Thus, not only are many actin-regulating proteins present in the spine, but many such proteins have been implicated in regulating receptor trafficking under constitutive and plasticity-inducing stimuli.

Transport of vesicles along the actin cytoskeleton is facilitated by the myosin family of motor proteins. Myosin classes I, II, V, and VI have all been found in neurons

(Bridgman 2004). Conventional myosins are protein complexes composed of two myosin heavy chains, one myosin regulatory light chain (RLC), and one myosin essential light chain (ELC). When assembled, this complex contains a motor domain that interacts with actin, a light chain binding region, coiled-coil regions that promote dimerization, and a globular tail region that binds to cargo. Class I myosins lack this coiled-coil region, do not dimerize, and so are unable to "walk" along actin filaments. All myosins, except for class VI myosins, travel toward the fast-growing plus or barbed ends of actin filaments; myosin VI travels toward the slow-growing minus or pointed ends of actin filaments.

Myosins II, V, and VI coordinate vesicle-mediated postsynaptic transport along actin filaments. In the myosin Va-deficient *dilute* mice, there are defects in the organization of the spine smooth endoplasmic reticulum of Purkinje neurons, suggesting that myosin Va contributes to spine organelle trafficking (Miyata et al. 2000). Myosin RLC, a component of the myosin II protein complex, directly interacts with the C-terminus of NR1 and NR2 (Amparan et al. 2005). Ca^{2+}/calmodulin interferes with this interaction between myosin RLC and NR1, suggesting a role for myosin II in calcium dependent remodeling of the postsynaptic density. With its backward movement on actin filaments and appearance at endocytic sites, myosin VI has been implicated in endocytosis. Disruption of myosin VI inhibits AMPA receptor internalization to the same degree as inhibiting clathrin-dependent endocytosis, yet in a manner specific for AMPA receptors (Osterweil et al. 2005). Given that myosin VI exists in a complex with the clathrin adaptor AP-2, AMPA receptors, and synapse-associated protein 97 (SAP97) (Osterweil et al. 2005), these findings suggest that myosin VI links internalized AMPA receptors to the actin cytoskeleton in the dendritic spine.

In addition to myosins, a large number of actin-binding proteins have been isolated from the PSD, and these actin-binding proteins interact with many postsynaptic proteins (Pak et al. 2001; Sala et al. 2001; Ehlers 2002; Carlisle and Kennedy 2005; Terry-Lorenzo et al. 2005). The interaction between actin-binding proteins and other PSD proteins regulates morphological changes of the dendritic spine in an activity dependent manner. As discussed earlier, the morphology of the dendritic spine directly correlates with its level of AMPA receptor expression, implying that these actin-binding proteins may regulate the strength of a synapse.

A direct link between spine morphology and an actin-regulatory protein is provided by the spine-associated Rap GTPase activating protein (GAP), SPAR, which has two actin interacting domains and RapGAP activity (Pak et al. 2001). SPAR binds the canonical postsynaptic PDZ scaffold protein PSD-95, and overexpression of SPAR reorganizes the actin cytoskeleton into aggregates. This remodeling is facilitated by serum-inducible kinase/polo-like kinase 2 (SNK/Plk2), which is regulated by plasticity-inducing stimuli (Pak and Sheng 2003). A second example is provided by two classic PSD proteins, Shank and Homer, that link the NMDA receptor to the actin cytoskeleton and cause massive spine growth when overexpressed (Sala et al. 2001). Finally, overexpression of the actin-binding proteins neurabin-I and spinophilin/neurabin-II in hippocampal neurons induces the formation of filopodia (Zito et al. 2004; Terry-Lorenzo et al. 2005), a process likely due to localized actin-bundling (Svitkina et al. 2003; Terry-Lorenzo et al. 2005). In addition, neurabin-I promotes the maturation of filopodia into dendritic spines through its interaction with protein phosphatase I (Terry-Lorenzo et al. 2005).

Interaction between the Microtubule and Actin Cytoskeletons

In order for a vesicle to be delivered to a spine, it must be transported along the microtubule cytoskeleton until it reaches the spine and then transferred to the actin cytoskeleton for short-range movement within the spine. To be efficient, there should be a direct interaction between the two transport systems. Evidence for this was demonstrated for the myosin Va motor, which interacts with KhcU, a microtubule-based motor (Huang et al. 1999). In myosin Va *dilute* mice, an accumulation of myosin Va-associated organelles was observed at microtubule ends in the axon, suggesting that their transport began on microtubules and ended abruptly at the microtubule-actin interface when myosin Va was absent (Bridgman 1999). Myosin V also interacts with the dynein light chain (Espindola et al. 2000; Naisbitt et al. 2000). These interactions support a model of dual microtubule/actin motors coordinating vesicle delivery to the synapse (Brown 1999).

C. Internal Membrane Compartments and the Neuronal Plasma Membrane

Neurons contain an extensive network of internal membrane compartments that function in protein trafficking. These include the endoplasmic reticulum (ER), Golgi apparatus, endosomal system, and the spine apparatus. In addition, the plasma membrane plays its own role in receptor trafficking. Of these, all but the still-mysterious spine apparatus will be considered in depth in subsequent sections.

The ER and Golgi apparatus are involved in the early stages of the neuronal secretory pathway. Receptors first are synthesized and processed in the ER and then are trafficked to the Golgi apparatus for further processing, sorting, and packaging for delivery to their final destination (Horton and Ehlers 2004). The endosomal system is composed of early endosomes, recycling endosomes, late endosomes, and lysosomes (Maxfield and McGraw 2004). This network of tubulo-vesicular structures receives internalized receptors from the plasma membrane and determines whether a receptor will be recycled or degraded. Though there is crosstalk between the exocytic and endocytic systems, the ER and Golgi apparatus primarily mediate the trafficking of newly synthesized receptors, whereas the endosomal system mediates the trafficking of preexisting receptors internalized from the plasma membrane.

The spine apparatus was first described by Gray in 1959 in an ultrastructural study of cortical synapses, where it appears as an electron dense stack of membranous cisternae. Thought to be a specialized domain of the smooth ER or Golgi-related membranes, the spine apparatus is found in large, mature spines with perforated PSDs (Spacek and Harris 1997). Mice deficient in synaptopodin, an actin/α-actinin binding protein, lack a spine apparatus and exhibit defects in long-term potentiation (LTP) induction and spatial learning (Deller et al. 2003). The spine apparatus may serve as a reservoir of neurotransmitter receptors, recognition molecules, channels, and pumps in mature synapses (Sytnyk et al. 2004). Consistent with this notion, colocalization of AMPA and NMDA receptors in the spine apparatus has been observed by immunogold labeling and electron microscopy (Racca et al. 2000). Though the function of the spine apparatus is not clear, these results suggest a role in membrane trafficking, calcium buffering, or synaptic plasticity.

The plasma membrane is also a site of receptor trafficking. Receptors can move from one location to another via lateral diffusion in the plasma membrane, allowing receptors to diffuse between the PSD and extrasynaptic regions or between sites of exocytosis and endocytosis (Triller and Choquet 2005) (see Chapter 1).

D. Dynamic Regulation of Receptor Trafficking at the Synapse

Receptor trafficking is a major mechanism for dynamically regulating synaptic strength. An early clue that regulation of receptor number and density contributes to synaptic plasticity came from the discovery of silent synapses (Isaac et al. 1995; Liao et al. 1995; Durand et al. 1996). The central observation was that some synapses contained functional NMDA receptors and no functional AMPA receptors. Upon LTP induction, however, these formerly silent synapses gained functional AMPA receptors. It is now clear that the addition of new AMPA receptors to a synapse accounts for the awakening of silent synapses and mediates the increase in synaptic strength after LTP induction at many glutamatergic synapses (Malenka and Bear 2004) providing a mechanism for postsynaptic modification of the synapse. Conversely, the endocytosis of AMPA receptors underlies long-term depression (LTD) (Carroll et al. 1999; Beattie et al. 2000; Wang and Linden 2000; Malenka and Bear 2004). Thorough discussions of AMPA receptor trafficking and subunit-specific NMDA receptor trafficking can be found in subsequent chapters.

II. TRAFFICKING OF NEW RECEPTORS: DENDRITIC SECRETORY ORGANELLES

A. Structure and Function of Dendritic Secretory Organelles

The secretory pathway is composed of the ER, the ER-Golgi intermediate compartment (ERGIC), the Golgi apparatus, and the trans-Golgi network (TGN). These large, membrane-bound structures are integral to receptor trafficking and necessary for protein synthesis, processing, sorting, and targeting to the plasma membrane. Proteins typically pass through these organelles in a sequential manner with transport from one compartment to the next provided by vesicular intermediates along the microtubule cytoskeleton.

Endoplasmic Reticulum and ERGIC

The ER is a continuous membrane-bound compartment that extends throughout the soma and dendrites (see Figure 8.2)

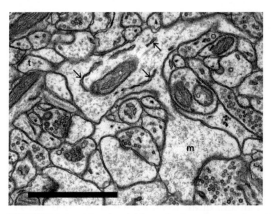 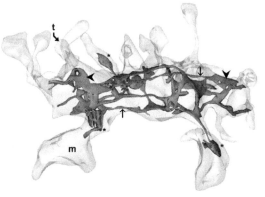

FIGURE 8.2. **The dendritic endoplasmic reticulum. Left:** In the electron micrograph, the ER (arrows) appears as thin cisternae at the base of a mushroom spine (m) and a thin spine (t). Scale bar: 1 μm. **Right:** 3D reconstruction of the ER in dendrites and spines reveals the continuous network formed by the ER in the dendrite and the occasional intrusion of the ER into dendritic spines (*). Larger flat compartments (arrowheads) connect with thin extensions (arrows) to form the ER network. Adapted from Cooney et al. (2002) with permission; copyright 2002 by the Society for Neuroscience.

(Broadwell and Cataldo 1983; Spacek and Harris 1997; Cooney et al. 2002). Regions of the ER are considered rough ER if ribosome-associated, and smooth ER if not. The smooth ER acts primarily as an internal calcium store, sequestering intracellular calcium by the action of the ER-associated Ca^{2+}-ATPase (Verkhratsky 2005). Activation of inositol triphosphate (IP3) receptors or ryanodine receptors releases calcium into the cytoplasm, triggering a variety of signaling pathways and altering synaptic properties (Finch and Augustine 1998; Verkhratsky 2005). Though smooth ER dominates in distal dendrites and dendritic spines, rough ER has been found in dendrites as well (Pierce et al. 2000). As integral membrane proteins, nascent receptor polypetides are synthesized by ribosomes bound to the membrane of the ER and translocated across the ER membrane to ensure proper topology. The translocon, formed by the Sec61 complex, is required for the stable association of ribosomes with the ER and functions as the pore for translocating nascent polypeptides across the ER membrane (White and von Heijne 2005). Immunogold labeling for Sec61α demonstrated that the rough ER extends throughout the dendrites, into spines, and sometimes near the postsynaptic density (Pierce et al. 2000, 2001). Once folded and assembled, receptors are trafficked from the ER to the Golgi via ERGIC, a sorting station that catches and returns proteins containing an ER retention/retrieval motif, including resident ER proteins (Vliet et al. 2003). Lipid biosynthesis, initial carbohydrate addition, and protein quality control are additional functions of the ER, whose regulation in dendrites remains poorly understood.

The Golgi Apparatus and the Trans-Golgi Network

The Golgi apparatus is composed of a stack of membrane-bound structures, or cisternae, that are categorized into three primary compartments: the *cis-*, *medial-*, and *trans*-Golgi. In secretory cells, these stacks are topologically oriented with the *cis* side facing the ER and the *trans* side facing the plasma membrane (Mogelsvang et al. 2004). As this orientation suggests, proteins are received from the ER by the *cis*-Golgi and sorted for final export by the *trans*-Golgi network. Membranous tubular structures connect the various stacks (Marsh et al. 2004; Trucco et al. 2004), and the Golgi apparatus appears continuous in the cell body and proximal dendrites (Gardiol et al. 1999; Horton and Ehlers 2003). The structure of the Golgi apparatus is determined by the microtubule cytoskeleton, and disruptions or rearrangements of microtubules cause Golgi dispersal and rearrangement (Horton and Ehlers 2004).

Though the somatic Golgi structure is similar to that of nonneuronal cell types, neurons possess a more extensive, dispersed network of Golgi elements for protein processing. Experiments exposing protein synthesis (Feig and Lipton 1993) and glycosylation (Torre and Steward 1996) in dendrites inspired the search for dendritic Golgi. Additionally, evidence for active ER exit sites throughout dendrites (Horton and Ehlers 2003; Aridor et al. 2004) is suggestive for the local presence of Golgi. This unusual organization of the Golgi apparatus in neurons has been elucidated by recent experiments that revealed the presence of discrete Golgi structures in the dendrites, termed Golgi outposts.

Initial evidence for dendritic Golgi arose from studies examining the glycosylation of newly synthesized proteins in dendrites, as well as immunolabeling for Golgi-specific proteins and visualization via light microscopy and electron microscopy (Torre and Steward 1996; Gardiol et al. 1999; Pierce et al. 2000, 2001). These studies found that protein markers for *cis-*, *medial-*, and *trans*-Golgi-labeled structures in a subpopulation of dendrites, including regions in dendrites far from the somatic Golgi (see Figure 8.3A–C). However, these studies could not deter-

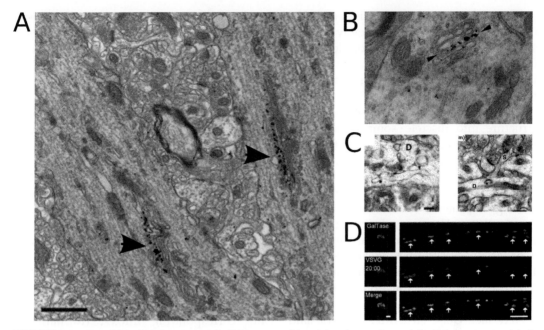

FIGURE 8.3. **Dendritic Golgi outposts. A–C**. Immunogold staining for markers of the Golgi apparatus label structures in the dendrites. **A**. GM130 labeled stacks of Golgi outposts (black arrowheads) in the apical dendrite (pink) of a CA1 hippocampal pyramidal neuron. Scale bar: 1 μm. Image courtesy of B. Rácz and R. J. Weinberg. **B**. CTR433-IR (black arrowheads) labeled membrane-bound structures in the dendrite of a ventromedial horn spinal cord neuron from an adult rat. Adapted from Gardiol et al. (1999) with permission; copyright 1999 by the Society for Neuroscience. **C**. Rab6 (red arrowheads) labeled cisternae in the dendrite (D) in a hippocampal neuron from an adult rat, including at the base of spine (S). Scale bar: 0.25 μm. Adapted from Pierce et al. (2001) with permission. **D**. Dendritic Golgi outposts engage in secretory trafficking. Structures in the soma (left) and discrete structures in the dendrite (right) of a cultured hippocampal neuron are labeled with GalTase-CFP (*top*), a marker for the Golgi apparatus. Twenty minutes after release from the ER, the secretory cargo VSVG-YFP (*middle*) merges with the GalTase-CFP positive structures in both the soma and dendrites (*bottom*), indicating that these structures are functional Golgi outposts. Scale bar: 5 μm. Adapted from Horton and Ehlers (2003) with permission; copyright 2003 by the Society for Neuroscience.

mine whether the occasional Golgi protein-positive structures observed in dendrites were functional Golgi structures. To address whether dendritic Golgi structures are functional, a recent study employed live-cell imaging techniques (Horton and Ehlers 2003). ER to Golgi traffic was visualized in live neurons by monitoring the movement of the yellow-fluorescent protein tagged temperature sensitive mutant of the vesicular stomatitis virus glycoprotein (VSVG-YFP), whose ER exit can be synchronized by a change in temperature from nonpermissive (39.5°C) to permissive (32°C). Upon leaving dendritic ER exit sites, VSVG-YFP carriers move bidirectionally in dendrites, and a significant fraction of these highly mobile carriers merge with highly stable and stationary structures located in the dendrites; the rest traffic to somatic Golgi (see Figure 8.3D) (Horton and Ehlers 2003). This capture by dendritic structures occurred over the same time range during which VSVG-YFP accumulates in the Golgi. Simultaneous visualization of Golgi markers confirmed that post-ER VSVG-YFP carriers merged with Golgi structures in the dendrites. As suggested by initial morphological studies, these Golgi markers labeled structures in the soma and immediately proximal dendrites, as well as discrete, discontinuous, distributed Golgi elements in distal dendrites (Horton and Ehlers 2003). Golgi outposts also contain newly synthesized brain-derived neurotrophic factor (BDNF), demonstrating that these structures serve both integral membrane proteins and secreted proteins.

The organization of neuronal Golgi structure into discrete somatic and dendritic networks likely facilitates the delivery of receptor proteins and other important cargo to their final destination. Dendritic Golgi outposts may enable processing of proteins synthesized locally in the dendrites. However, their distribution in dendrites does not intrinsically imply the synapse specificity that local processing is hypothesized to provide. In particular, Golgi outposts are not found near all synapses and, in fact, many dendrites completely lack Golgi structures (Horton and Ehlers 2003). Additionally, it is still unclear what determines whether an immature protein is delivered to the dendritic or somatic Golgi; that is, why only a subset of the VSVG-YFP carriers merge with the dendritic Golgi, whereas many others continue on the much longer path back to the soma (Horton and Ehlers 2003). Further experimentation is required to uncover the full role of the dendritic Golgi and the functional consequence of this unusual Golgi organization.

B. Local Dendritic Synthesis of Receptors

In neurons, certain mRNAs are transported into dendrites and local translation of these dendritic mRNAs occurs in response to synaptic activity and growth factor stimulation (Steward and Schuman 2001; Sutton and Schuman 2005). This phenomenon has been studied most thoroughly for mRNAs encoding cytoplasmic proteins, whose synthesis requires only mRNA, ribosomes, and associated soluble factors (e.g., charged tRNAs, initiation and elongation factors). However, the mRNAs for several receptor subunits, including

GluR1, NR1, and GlyRα also have been found in dendrites (Miyashiro et al. 1994; Racca et al. 1997). For mRNAs encoding integral membrane proteins (such as neurotransmitter receptors), the requirements for local synthesis are more complex as, in addition to mRNAs and ribosomes, production of integral membrane proteins requires all the organelles of the secretory pathway.

As discussed in the previous section, the necessary secretory trafficking machinery for receptor synthesis and processing can be found in some neuronal dendrites. However, this does not prove that local receptor synthesis in fact does occur. In support of local dendritic receptor synthesis, isolated dendrites can synthesize AMPA receptor subunits when transfected with the corresponding mRNA (Kacharmina et al. 2000). More recently, fluorescence-based detection of newly synthesized AMPA receptors demonstrated that transected dendrites can synthesize GluR1 and GluR2 subunits (Ju et al. 2004). Most importantly, these newly synthesized AMPA receptors traffic to the synapse and are incorporated into the postsynaptic membrane. Thus, the machinery for local synthesis and secretory trafficking of receptors is present and functional in at least some dendrites. Yet, it is worth pointing out that the majority of cargo carriers are emerging from the dendritic ER traffic all the way back to the cell body, and some dendrites lack detectable Golgi outposts (Horton and Ehlers 2003), emphasizing that local secretory trafficking may be the exception rather than the rule.

The ability to locally produce a receptor has profound implications for protein trafficking in neurons. Of considerable interest is the contribution to late-phase synaptic plasticity requiring new protein synthesis (Kelleher et al. 2004). Forms of synapse plasticity such as late-phase long-term potentiation (L-LTP) are input specific, suggesting that if dendritic synthesis of a neurotransmitter receptor is involved, the locally synthesized receptor would have to be delivered to the locally activated synapse. However, it is not yet known how far new receptors diffuse in internal membranes or how localized each step of the secretory pathway is in dendrites. Several studies have offered evidence that synaptic plasticity-inducing stimuli promote the local synthesis of proteins, and this remains an active area of inquiry (Steward and Schuman 2001; Sutton and Schuman 2005).

C. Receptor Processing in the ER

Quality Control in the ER

Once synthesized in the ER, receptors are subject to quality control. Export from the ER is carefully controlled by ER retention/retrieval motifs and ER associated degradation (ERAD). Cytoplasmic ER retention motifs (e.g., RXR motifs) have been identified in NMDA receptor subunits (Standley et al. 2000; Scott et al. 2001) and kainate receptor subunits (Ren et al. 2003a,b; Jaskolski et al. 2004). Residues in the pore-lining domain of AMPA receptor subunits also control the rate of ER export (Greger et al. 2002). Subunit assembly is thought to sterically mask exposed ER retention motifs (Zerangue et al. 1999) and thereby ensure that only properly assembled complexes are released from the ER. In some instances, ER retention motifs are altered so as to promote ER release. For example, dual phosphorylation of ER-retained NR1 overrides its RXR ER retention signal (Scott et al. 2003). This dual phosphorylation by cAMP-dependent protein kinase (PKA) and protein kinase C (PKC) may link the activity of signal transduction pathways with ER export in the dendritic secretory pathway.

Misfolded or defective proteins are targeted for destruction by ERAD (Meusser et al. 2005). Interestingly, the extracellular N-terminal domain of the NMDA receptor subunit NR1 is recognized by the F-box protein Fbx2 (Kato et al. 2005), a component of the Skp/cullin/Fbx (SCF) ubiquitin ligase complex that recognizes high-mannose glycans of polypeptides destined for ERAD (Yoshida et al. 2002). Expression of a dominant-negative mutant of Fbx2 increases NR1 levels and NMDA receptor-mediated currents in an activity-dependent manner (Kato et al. 2005), suggesting that activity-dependent regulation of NR1 targeting involves receptor retrotranslocation and ER-associated degradation.

Excessive production of secreted proteins overwhelms ER-associated chaperones and triggers a cellular program of gene transcription referred to as the unfolded protein response (UPR) (Hegde 2004; Schroeder and Kaufman 2005). Interestingly, glutamate receptor export in *C. elegans* is regulated by the UPR (Shim et al. 2004). Similarly, the export of the mammalian AMPA receptor subunit GluR1 from the ER can be increased either by upregulating ER chaperones or stargazin, giving stargazin a putative role as a nontraditional chaperone (Vandenberghe et al. 2005). Interaction with the UPR thus provides an additional mode of ER-mediated regulation to control receptor numbers.

ER Exit and ER Exit Sites

Release from the ER is often the rate-limiting step for the trafficking of new receptors to the plasma membrane. To be released from the ER, nascent secretory cargo is recruited to ER exit sites where COPII coats bud off vesicles departing the ER (Lee et al. 2004a). The COPII coat complex is comprised of Sar1, a small GTPase that initiates coat assembly, the cargo-binding Sec23/24 bimolecular complex, and Sec13/31, which together with Sec23/24 assemble into oligomeric coats (Lee et al. 2004a). The rate of export from the ER is determined in part by the ability of cargo to bind either directly or indirectly to Sec23/24 (Barlowe et al. 1994; Bi et al. 2002). Though the components of the COPII coat were originally ascertained from yeast studies, homologous COPII complexes are present and functional in neuronal dendrites (Horton and Ehlers 2003; Aridor et al. 2004). Direct association of postsynaptic receptors with COPII coats has been demonstrated for specific splice variants of the NMDA receptor subunit NR1 (Mu et al. 2003). Intriguingly, mRNA splicing of the exon encoding the COPII binding domain is regulated by neuronal activity, producing NMDA receptors that either contain or lack a di-valine motif that directly binds Sec23/24. Upon blockade of neuronal activity, NR1 splice variants containing the di-valine motif are predominant, leading to an enhanced rate of ER export and a corresponding increase in the steady-state abundance of postsynaptic NMDA receptors (Mu et al. 2003). Such activity-dependent mRNA splicing and ER export provides a means for homeostatically tuning the abundance of NMDA receptors at the synapse (Pérez-Otaño and Ehlers 2005).

D. Golgi Secretory Trafficking

As the final compartment for protein processing, the Golgi apparatus and TGN concentrate, modify, and sort departing cargo. Although the role of the Golgi in protein sorting is well-established for other cell types (Keller and Simons 1997; Donaldson and Lippincott-Schwartz 2000; Lippincott-Schwartz et al. 2000), mechanistic knowledge of cargo sorting in the neuronal Golgi is limited. Emerging evidence for protein sorting in the neuronal Golgi arises from studies of neuronal polarity, for which proteins are specifically delivered to axons or dendrites.

Initial studies of neuronal polarity focused on the transferrin receptor, as it is specifically localized to dendrites and excluded from axons. The transferrin receptor contains a C-terminal dendritic targeting sequence (West et al. 1997a,b) and appears to be initially delivered to dendrites, as vesicles containing the transferrin receptor are excluded from the axon (Burack et al. 2000). A model to explain this

suggests that the C-terminal targeting sequence may interact with a special microtubule motor whose action is restricted to dendrites (Burack et al. 2000). In contrast to the transferrin receptor, axonally targeted membrane proteins reside in a distinct set of carriers and appear to sort separately from transferrin receptors either by selective retention in axons or selective endocytosis from dendrites (Winckler et al. 1999; Sampo et al. 2003; Wisco et al. 2003). Such studies point to selective sorting of dendrite-destined cargo, perhaps in the Golgi, likely by selective association with cargo adaptors and coat proteins.

Other studies have contributed evidence in support of this model. Mutation of *unc-101*, a component of the *C. elegans* TGN-associated AP-1 clathrin adaptor complex, disrupts the polarized distribution of an odorant receptor so that it is expressed throughout the neuron, instead of being confined to the dendrite (Dwyer et al. 2001). AP-4, a clathrin adaptor associated with the Golgi apparatus, directly interacts with the ionotropic glutamate receptor δ2 and is thought to mediate its sorting in the Golgi (Yap et al. 2003). NMDA receptors form a complex with MAGUK PDZ-domain proteins and exocyst components early in secretory trafficking that may facilitate dendritic targeting (Sans et al. 2003). GABARAP, a GABA$_A$ receptor associated protein, may serve as a Golgi sorting factor for GABA$_A$ receptors (Kneussel et al. 2000). Similarly, gephyrin, a microtubule binding scaffold protein known to bind glycine receptors and GABA$_A$ receptors (Kneussel and Betz 2000), associates with Golgi-like membranes (Sur et al. 1995) and facilitates the synaptic insertion of glycine receptors (Hanus et al. 2004).

E. Delivery and Membrane Insertion of Receptors

For a postsynaptic receptor to be functional, it must be inserted into the plasma membrane and delivered into the synapse.

New insertion of glycine receptors initially was observed in cultured spinal neurons (Rosenberg et al. 2001). A thrombin-cleavable myc-tagged GlyR α1 construct was used to distinguish between glycine receptors that had visited the cell surface and those that had not. After removing the tag from receptors in the plasma membrane, the trafficking of newly synthesized glycine receptors was synchronized by the use of a temperature-dependent inhibition of Golgi exit. Newly inserted receptors were detected by surface labeling with an anti-myc antibody. Upon release of receptors from the Golgi, a synchronized delivery of glycine receptors appeared at the soma and near-proximal dendrites, but delivery to distal dendrites was unsynchronized and occurred much later. Glycine receptors delivered by intracellular paths were monitored by adding a second thrombin cleavage step after Golgi exit and labeling for intracellular receptors. Since no intracellular receptors were labeled, the authors concluded that the receptors were delivered from the soma to synaptic sites in distal dendrites through lateral diffusion in the plasma membrane (Rosenberg et al. 2001).

The pattern of AMPA receptor subunit insertion in cultured hippocampal neurons is dependent on the receptor subunit composition (Passafaro et al. 2001). Unlike the case for glycine receptors, newly inserted AMPA receptors appeared on the somatic and dendritic membranes with similar kinetics. When HA/thrombin-tagged GluR1 was expressed alone, weak nonsynaptic surface GluR1 appeared just three minutes after thrombin treatment and accumulation at synapses followed five minutes after thrombin treatment. This implies that GluR1 was delivered initially to the plasma membrane at extrasynaptic sites and subsequently trafficked to the synapse. In contrast, new HA/thrombin-tagged GluR2 subunits were inserted at sites closer to synapses. Co-expression of HA-tagged GluR1 and GluR2 resulted in an insertion profile mirroring that of HA-tagged GluR1 alone. These results indicate that AMPA receptor subunit composition has a profound effect on membrane insertion and synaptic delivery, with GluR1 receptors accumulating initially at extrasynaptic sites, whereas GluR2 receptors more rapidly reach the synaptic membrane following plasma membrane insertion.

Although the receptor insertion studies discussed earlier propose lateral diffusion as the final step following initial exocytosis for the delivery of the receptors to the synaptic membrane, neither study could fully distinguish between lateral diffusion and trafficking through the endosomal system. That is, though the glycine and GluR1-containing receptors are inserted extrasynaptically before they localize to the postsynaptic membrane, it is possible that extrasynaptic receptors are internalized and delivered to the synapse by endosomal trafficking and recycling. Alternatively, some combination of all three trafficking steps may contribute to their synaptic localization. A definitive determination of the trafficking itinerary awaits methods for tracking individual receptors from synthesis to synaptic localization.

III. ENDOCYTOSIS AND THE ENDOCYTIC ZONE

The endocytosis of postsynaptic receptors is an essential mechanism for maintaining receptors in a cycling pool and for rapidly depressing synaptic strength. The internalization of several postsynaptic receptors occurs constitutively and can also be regulated by external stimuli and synaptic activity.

In most cases studied to date, endocytosis of postsynaptic receptors occurs via clathrin coats. A thorough discussion of clathrin-mediated endocytosis can be found in Chapters 6 and 7. Briefly, clathrin adaptors (e.g., AP-2) bind transmembrane proteins and recruit clathrin to the membrane (Slepnev and de Camilli 2000). The assembly of the clathrin coat induces local invagination of the membrane, creating a clathrin-coated pit. This invaginated membrane is pinched off as a vesicle by the large GTPase dynamin, and the endocytotic vesicle travels in an actin-dependent manner to join internal compartments. Following clathrin-mediated endocytosis, cargo is delivered to early endosomes (also known as sorting endosomes), from where it is sorted to either recycling endosomes, late endosomes, or directly back to the plasma membrane (Maxfield and McGraw 2004). Trafficking between these different compartments allows internalized receptors to be sorted for either recycling or degradation (Ehlers 2000). These compartments and their contribution to receptor trafficking are discussed in detail in Chapter 7.

A. The Endocytic Zone

Studies of neuronal endocytosis have focused primarily on the presynaptic terminal, where recovery of vesicles after neurotransmitter release is integral to the efficiency of neurotransmission (Augustine et al. 1999; Slepnev and De Camilli 2000). Endocytosis occurs at locations in the membrane that are near to but distinct from sites of neurotransmitter release (Brodin et al. 2000; Miller and Heuser 1984; Teng et al. 1999). This organization presumably prevents the slower endocytic machinery from interfering with the faster exocytosis of neurotransmitter or from disrupting the local organization of the presynaptic active zone.

Endocytosis in the postsynaptic membrane has been visualized directly, starting with electron micrographs of clathrin coated vesicles and pits in dendritic spines (see Figure 8.4) (Spacek and Harris et al. 1997; Toni et al. 2001; Cooney et al. 2002; Petralia et al. 2003; Rácz et al. 2004). Some of these vesicles were shown to incorporate fluid-phase cargo introduced to the extracellular media (Cooney et al. 2002), demonstrating that these clathrin coated vesicles

were indeed products of endocytosis at the plasma membrane rather than originating from the secretory pathway.

Given the highly structured organization of the postsynaptic density, one could postulate the existence of a similar domain in the spine for the endocytosis of receptors. Recent studies have demonstrated that postsynaptic endocytosis indeed occurs in a specialized endocytic zone in the dendritic spine (Blanpied et al. 2002; Rácz et al. 2004). Spatial and temporal imaging of clathrin coats in dendritic spines revealed "hot spots" for endocytosis that repeatedly hosted endocytic sites (Blanpied et al. 2002). Closer examination demonstrated that these hot spots were stable endocytic zones adjacent to the postsynaptic density in dendritic spines (see Figure 8.4) (Blanpied et al. 2002). Quantitative analysis of the relationship between the postsynaptic density and markers for the clathrin coat by electron microscopy confirmed the presence of an endocytic zone lateral in the spine tangential to the postsynaptic density *in vivo* (Rácz et al. 2004). The formation of these endocytic zones paralleled the dynamic development of neurons, with clathrin coats in dendrites becoming increasingly stable as the neuron matured (Blanpied et al. 2002, 2003).

The capture and maintenance of a reliable pool of clathrin is critical for the formation of the endocytic zone. Indeed, photobleaching experiments indicate that clathrin pools are maintained locally, with little exchange of clathrin between dendritic spines (Blanpied et al. 2002). Individual vesicles are uncoated promptly after internalization, within $0.3\,\mu m$ of the coated pit on the plasma membrane (Blanpied et al. 2002).

The discovery of a stable postsynaptic endocytic zone highlights the importance of endocytosis in the dendritic spine. Given the rigid, stable organization of the postsynaptic density, it is difficult to envision enough space for membrane invagination and vesicle budding to occur without

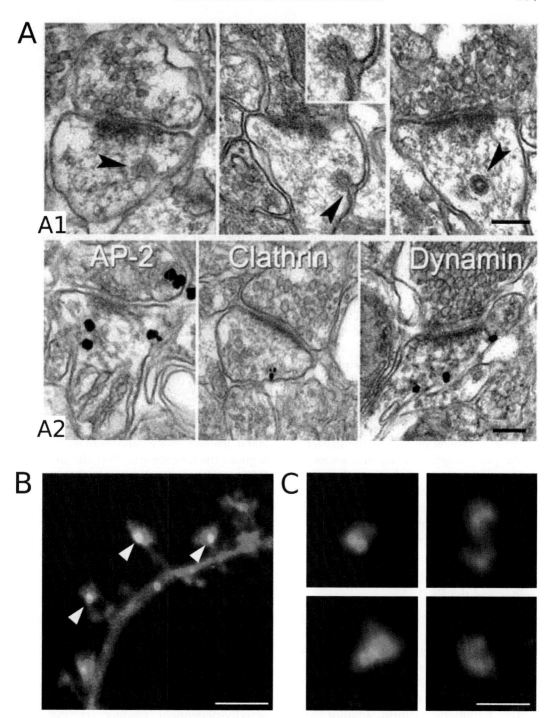

FIGURE 8.4.   **The endocytic zone at lateral domains of dendritic spines. A.** Endocytosis and the endocytic zone as revealed through electron microscopy and immunogold labeling. Endocytic clathrin-coated vesicles (**A1**) and proteins involved in clathrin-mediated endocytosis including AP-2, clathrin heavy chain, and dynamin (**A2**) can be found in the dendritic spine lateral to, but not within, the postsynaptic density. Scale bar: 200nm. Adapted from Rácz, et al. (2004) with permission. **B.** Clathrin-YFP (green) marks sites of endocytosis (arrowheads) in dendritic spines. Red: fluorescent cell fill by cyan fluorescent protein (CFP). Scale bar: 2μm. **C.** Clathrin-DsRed (red) and PSD-95-GFP (green) occupy nonoverlapping adjacent zones in the spine. Scale bar: 1μm. **B–C.** Adapted from Blanpied et al. (2002) with permission.

causing a large-scale disruption of the postsynaptic density. Regulated movement out of the postsynaptic density and into the endocytic zone also could serve as a means to induce specificity of cargo selection. AMPA receptors in the vicinity of clathrin coats show significantly less lateral mobility and enter a "confined" state (Borgdorff and Choquet 2002). It is not known whether the receptors undergo directed or channeled transport out of the postsynaptic density or simply diffuse in Brownian fashion to the endocytic zone. As in the presynaptic terminal, a dedicated domain for endocytosis may be critical for the efficiency of endocytosis. Alternately, a localized endocytic zone could facilitate local recycling and recapture of receptors, such that a dendritic spine need not share its pool of receptors with nearby synapses. The precise function of this intriguing spine endocytic zone remains to be determined.

Though the endocytic zone is the first identified membrane specialization in the lateral domain of spines, functional compartmentalization is likely a general feature of spine organization. Given the existence of the postsynaptic density and endocytic zones, one may postulate the existence of specified zones for other functions, such as exocytosis. The existence and molecular basis for such membrane zones in spines has yet to be delineated.

B. Endocytic Regulatory Proteins in Spines and the PSD

Although our current knowledge of the mechanisms for postsynaptic endocytosis is incomplete, many proteins with known roles in endocytosis, including AP-2, dynamin, cortactin, Rab5, Abp1, and Hip1R have been localized to dendritic spines.

The tetrameric clathrin adaptor AP-2 complex facilitates the interaction between endocytic cargo and the forming clathrin lattice (Kittler et al. 2000; Pearse et al. 2000; Lee et al. 2002). AP-2 subunits associate with several neurotransmitter receptor subunits, including GluR2 of the AMPA receptor and the GABA$_A$ receptor β and γ subunits (Kittler et al. 2000; Lee et al. 2002). Both of these receptors colocalize with AP-2 when in clathrin-coated pits (Carroll et al. 1999; Kittler et al. 2000; Lee et al. 2002).

Dynamins 1, 2, and 3 all have been found in the brain, and dynamins generally are thought to mediate fission of internalized vesicles from the plasma membrane (Gray et al. 2003). Though dynamin-1 is found primarily at presynaptic terminals, dynamin-2 is present at the PSD and interacts with actin, Shank, and cortactin binding protein 1 (CortBP1) (Okamoto et al. 2001). Overexpression of a dominant negative form of dynamin-2 inhibits the endocytosis of AMPA receptors following glutamate application (Carroll et al. 1999). Though dynamin-3 also is found in the postsynaptic density, its role in endocytosis is less clear (Gray et al. 2003). Dynamin 3 interacts with metabotropic glutamate receptor 5 (mGluR5) and Homer, but does not regulate the endocytosis of mGluR5 (Gray et al. 2003). Instead, splice variants of dynamin-3 regulate the formation of filopodia and dendritic spines through an interaction with cortactin and reorganization of the actin cytoskeleton (Gray et al. 2005).

Cortactin can be found in clathrin-coated pits, where it participates in clathrin-mediated endocytosis (Cao et al. 2003; Merrifield et al. 2005). Cortactin activates the Arp2/3 complex, which promotes actin filament assembly (Schafer et al. 2002). With its interactions with actin and dynamin-3, cortactin may bridge the Arp2/3 complex and the actin cytoskeleton (Hering and Sheng 2003). In neurons, cortactin is concentrated in dendritic spines and regulates dendritic spine morphogenesis in an activity-dependent manner (Hering and Sheng 2003; Rácz and Weinberg 2004).

Rab5 is a small GTPase known to associate with sorting endosomes in neurons and can be found at the endocytic zone (Brown et al. 2005). Signaling by Rab5 has been implicated in receptor endocytosis in neurons (Hoop et al. 1994; Kanaani et al. 2004; Brown et al. 2005) and in actin remodeling in nonneuronal cell types (Spaargaren and Bos 1999; Lanzetti et al. 2004). In particular, Rab5 is required for the activity-dependent endocytosis of AMPA receptors, and inhibiting Rab5 activity prevents long-term depression (Brown et al. 2005). This removal of AMPA receptors requires an interaction between Rab5 and the clathrin-dependent endocytic machinery (Brown et al. 2005).

Abp1, an F-actin binding protein, interacts with both dynamin and the actin cytoskeleton in dendritic spines (Kessels et al. 2001). Stimulation of NIH3T3 cells with growth factors leads to the recruitment of Abp1 at endocytic sites, and disruption of the Abp1-dynamin interaction inhibits transferrin receptor endocytosis in COS7 cells. As most of its binding partners (e.g., synapsin I, dynamin I, synaptojanin I) are either brain-specific or enriched in brain (Kessels et al. 2001), Abp1 may serve to link endocytic vesicles to the actin cytoskeleton.

Huntingtin-interacting protein 1R (Hip1R) localizes to vesicular structures near the PSD (Okano et al. 2003) and interacts directly with both actin and clathrin (Engqvist-Goldstein et al. 2001; Okano et al. 2003; Chen and Brodsky 2005; Legendre-Guillemin et al. 2005). Removing Hip1R via RNA interference from HeLa cells results in a stabilization of clathrin-coated vesicles and clathrin-coated pits with the actin cytoskeleton and corresponding defects in transferrin receptor endocytosis (Engqvist-Goldstein et al. 2004). These observations have suggested that Hip1R may regulate the interaction between the actin cytoskeleton and the endocytic machinery.

Many of these proteins link endocytosis with the actin cytoskeleton such that there is coordination between endocytosis and local reorganization of the actin cytoskeleton. Interestingly, the proteins that did not appear to directly affect endocytosis in the neuron—such as dynamin-3 and cortactin—had drastic effects on dendritic spine morphology. Disruption or stabilization of the actin cytoskeleton alters the turnover of clathrin coated pits at the cell surface in cultured hippocampal neurons, suggesting that the rate of endocytosis is determined by the balance of actin polymerization and depolymerization (Blanpied et al. 2002). Machinery for actin reorganization is recruited to sites of endocytosis before membrane scission in fibroblasts, and inhibiting actin polymerization blocks subsequent endocytic events (Merrifield et al. 2005). Hence, endocytosis relies on the reorganization of the actin cytoskeleton.

Besides encouraging the efficient movement of the internalized vesicle, linking actin and endocytosis allows for the localization and stabilization of the endocytic machinery at a particular patch of the plasma membrane. The actin cytoskeleton may provide a direct link between the postsynaptic density and the endocytic zone, allowing receptors to use the actin cytoskeleton as a pathway between their positions in the PSD and in the endocytic zone. Indeed, by providing a lattice for the organization of components within the dendritic spine, the actin cytoskeleton and associated proteins may be responsible for the localization of the endocytic zone to sites immediately adjacent to the PSD.

C. Receptor Endocytosis

Endocytosis of postsynaptic receptors (e.g., glutamate receptors) occurs constitutively and may be triggered by many different stimuli including ligand binding, protein ubiquitination, and various intracellular signaling cascades. The consequence of such endocytosis is typically a reduction in synaptic efficacy (Carroll et al. 2001). Other chapters in this book deal at length with the endocytosis of individual receptors (see Chapters 9 and 10); here we will introduce the general features and stimulus paradigms that induce postsynaptic receptor endocytosis.

Constitutive Endocytosis

Some receptors are constitutively removed from the plasma membrane. AMPA receptors, for example, have a high basal turnover rate, with half the receptors internalized in about eight to nine minutes (Ehlers 2000; Lin et al. 2000). Inhibiting endocytosis for 40 minutes can double AMPA receptor-mediated excitatory postsynaptic currents (Lüscher et al. 1999), indicating that endocytosis of AMPA receptors tightly regulates their synaptic expression. Internalization of AMPA receptors is clathrin-dependent (Man et al. 2000; Wang and Linden 2000) and involves an interaction of GluR2 subunits with the AP-2 clathrin adaptor (Lee et al. 2002). Although some studies have reported more robust endocytosis of GluR2- than GluR1-containing AMPA receptors (Lin et al. 2000; Lee et al. 2002), others have found nearly equivalent endocytosis of GluR1- and GluR2-containing AMPA receptors (Ehlers 2000; Park et al. 2004).

Compared to AMPA receptors, NMDA receptors are stable at the synapse, with 20 percent internalized in young neurons and just 5 percent in mature neurons after 30 minutes (Roche et al. 2001). Endocytosis of NMDA receptors is mediated by diverse motifs, including a YEKL motif in the C-terminus of the NR2B subunit of the NMDA receptor (Roche et al. 2001; Prybylowski et al. 2005), a YKKM motif in the C-terminus of NR2A (Lavezzari et al. 2004), and a conserved family of tyrosine-based motifs in the membrane proximal domains of both NR1 and NR2 subunits (Scott et al. 2004). The YEKL motif of NR2B interacts with the AP-2 complex, and this motif facilitates the constitutive removal of NR2B-containing receptors from the synapse in a manner negatively regulated by Fyn phosphorylation of the YEKL tyrosine residue (Prybylowski et al. 2005).

GABA$_A$ receptors are also constitutively internalized (Connolly et al. 1999; Kittler et al. 2000; Rijnsoever et al. 2004). This internalization is facilitated by a direct interaction between the GABA$_A$ receptor subunits β and γ with AP-2 (Kittler et al. 2000).

The constitutive endocytosis of receptors from the plasma membrane has several implications for the nature of the postsynapse. It highlights the dynamic nature of the postsynaptic membrane as there is constant turnover of AMPA receptors even in the basal state. This implies that the postsynaptic density is not so rigid as to prevent this fast removal of AMPA receptors. Also, constitutive cycling could provide a potential mechanism for rapid changes in the postsynapse in response to plasticity-inducing stimuli. Given the fast rate of turnover, inhibiting either endocytosis or exocytosis of AMPA receptors can quickly and robustly change the response of that synapse to synaptic activity (Lüscher et al. 1999).

Agonist-Induced Endocytosis

Exposure to agonist can also lead to receptor endocytosis. This phenomenon may allow a neuron to downregulate the surface expression of receptors in response to excess exposure to the activating ligand. G-protein coupled receptors (GPCRs) are regulated in this manner undergoing internalization and downregulation with exposure to agonist (Claing et al. 2002). Endocytosis of GPCRs relies on interactions with arrestins that recognize the activated receptors and interact with both clathrin and AP-2 (Claing et al. 2002). Neuronal GPCRs, such as dopamine receptors, share this mechanism. Exposure to dopamine induces the internalization of dopamine receptors (Sun et al. 2003) in a manner that requires specific interactions with different isoforms of dynamin (Kabbani et al. 2004). Agonist-induced endocytosis is not limited to GPCRs. Application of glutamate or AMPA induces the internalization of AMPA receptors but not NMDA receptors (Lissin et al. 1999). The structural and molecular basis of how ligand binding of AMPA receptors leads to endocytosis remains unclear.

Ubiquitin-Mediated Endocytosis

The ubiquitin-proteasome system is a recognized pathway for the destruction of proteins marked with a polyubiquitin chain (Yi and Ehlers 2005), though it also has specific roles in endocytosis. Mori et al. (1992) initially demonstrated that the platelet-derived growth factor (PDGF) receptor is ubiquitinated in the presence of a ligand that was known to induce its internalization. Hicke and Riezman (1996) then found in *S. cerevisiae* that binding of alpha factor to its receptor Ste2p promoted ubiquitination of Ste2p, which is required for ligand-stimulated endocytosis. Additional studies have shown that the monoubiquitination of receptor tyrosine kinases and G protein-coupled receptors at multiple sites is sufficient to induce internalization of the receptor (Shenoy et al. 2001; Haglund et al. 2003a,b; Hicke and Dunn 2003; Mosesson et al. 2003). The ubiquitin ligase complex for the epidermal growth factor receptor, Cbl-CIN85-endophilin, directs endocytosis through endophilins, which are known regulators of clathrin-coated vesicles (Soubeyran et al. 2002). In contrast to polyubiquitination that targets substrates to the proteasome for degradation, monoubiquitination is the typical ubiquitin-mediated signal that regulates receptor endocytosis and sorts through the endosomal system to the lysosome for degradation (Hicke and Dunn 2003; Haglund et al. 2003a,b; Mosesson et al. 2003; Yi and Ehlers 2005).

Ubiquitination mechanisms are presumed to exist for neurotransmitter receptors, though the extensiveness of its applicability is yet to be uncovered. The glycine receptor is marked with ubiquitin prior to its internalization and lysosomal mediated-degradation (Buttner et al. 2001). In *C. elegans*, the surface expression of the glutamate receptor GLR-1 also is regulated by ubiquitination in a manner dependent on AP-180, a clathrin adaptor protein (Burbea et al. 2002). Ubiquitination and subsequent proteasome-dependent degradation of substrates may also influence the endocytosis of receptors. Several studies have indicated that polyubiquitination and a properly functioning proteasome are required for the AMPA or NMDA-induced endocytosis of AMPA receptors, though the underlying mechanism is unclear (Colledge et al. 2003; Patrick et al. 2003; Bingol and Schuman 2004). Through these multiple mechanisms, ubiquitin may prove to be a powerful regulator of receptor endocytosis and the surface expression levels of receptors.

Activity-Induced Endocytosis

As mentioned earlier, receptor internalization can be constitutive, agonist-induced, or ubiquitin-mediated, though these mechanisms represent just a subset of those mediating receptor endocytosis. Synaptic activity in several forms also can regulate the surface expression of receptors through various signaling pathways, allowing the neuron to differentially regulate the number of receptors in the synapse depending on the types of receptors and stimuli.

The multiple mechanisms for activity-induced AMPA receptor endocytosis illustrate how different stimuli can activate different signaling pathways, yet converge to the same result of AMPA receptor internalization. LTD, for which low frequency stimulation leads to a prolonged decrease in the amplitude of excitatory postsynaptic currents, is mediated in large part by the clathrin-mediated endocytosis of AMPA receptors (Carroll et al. 1999; Beattie et al. 2000). NMDA receptor mediated LTD is dependent on NMDA receptors, calcium influx, calcineurin, and Rab5 (Carroll et al. 1999; Beattie et al. 2000; Ehlers 2000; Wang et al. 2000; Brown et al. 2005). However, activation of metabotropic glutamate receptors induces a different form of LTD, where NMDA receptors and AMPA receptors are internalized through a mechanism that requires local protein synthesis (Snyder et al. 2001). Finally, insulin exposure induces the rapid internalization of AMPA receptors (Lin et al. 2000).

IV. ENDOSOMES AND RECEPTOR RECYCLING

A. Endosomes: Early, Late, Recycling

Endosomes are internal membrane-bound structures that, in dendrites, can be categorized structurally as vesicular, multivesicular, or tubular (Cooney et al. 2002). Though endosomal compartments can be found throughout the soma and dendrites (Prekeris et al. 1998; Ehlers 2000), only about half of spines contain an endosomal structure, with very few spines containing a complete repertoire of smooth endoplasmic reticulum, endosomes, and small vesicles (see Figure 8.5) (Cooney et al. 2002). Dendritic endosomes typically are located within or at the base of dendritic spines. One sorting endosome serves, on average, about 20 spines, though larger, mushroom-shaped spines average more sorting endosomes per spine. Endosomes can either be highly mobile or stationary, and may move in anterograde or retrograde directions with speeds ranging between 0.2 to 0.5 µm per second (Prekeris et al. 1998).

Cargo remains for only a few minutes in early endosomes. In five to 10 minutes, early endosomes progressively acidify and mature into late endosomes. Membrane proteins targeted for degradation typically are tagged by monoubiquitination and recognized for endocytosis and sorting into the late endosomes through ubiquitin-interacting proteins such as epsins, Hrs, and GGAs (Golgi-localized, gamma ear-containing Arf-binding proteins) (Katzmann et al. 2002; Polo et al. 2002; Puertollano and Bonifacino 2004). Mono-ubiquitinated membrane proteins are rec-

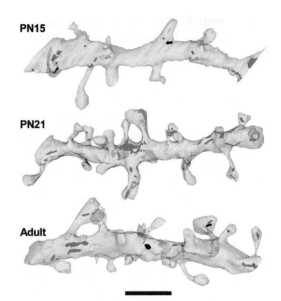

FIGURE 8.5. **Distribution of endosomes in dendrites.** Three-dimensional reconstruction of dendritic segments and endosomal transport machinery in hippocampal neurons from rats aged 15 days (*top*), 21 days (*middle*), and adult (*bottom*). Endocytic vesicles (yellow; coated pits, coated vesicles, and large vesicles), endosomes (red), small vesicles (blue), and amorphous vesicles (black) can be found within and at the base of spines, though endosomes are not found in all spines. Scale bar: 1 µm. Adapted from Cooney et al. (2002) with permission; copyright 2002, by the Society for Neuroscience.

ognized in the late endosomes by the multiprotein ESCRT complexes I, II, and III (endosomal sorting complex required for transport) (Katzmann et al. 2001, 2002; Babst et al. 2002a,b). These complexes sort and concentrate ubiquitinated cargo into vesicles that invaginate and bud from the late endosomal limiting membrane to form intralumenal multivesicular body (MVB) vesicles (Katzmann et al. 2001, 2002; Babst et al. 2002a,b). After fusion of MVB vesicles with lysosomes, resident hydrolases, proteases, and lipases break down lumenal material.

Many proteins do not remain in the early endosome while it matures into late endosomes and MVBs, but instead are returned to the plasma membrane or sorted into a subdomain of the early endosome that buds off tubules, giving rise to a recycling endosome compartment. Direct transport of receptors from early endosomes to the plasma membrane allows for rapid recycling with a $t_{1/2}$ of two to six minutes (Maxfield and McGraw 2004). Recycling endosomes are long-lived tubular organelles from which cargo transport to the plasma membrane is typically slower (~15–30 min) than direct transport out of the early endosome. Sequestration in the recycling endosome acts as a rate-limiting step for the delivery of recycling receptors to the membrane.

In dendrites, endosomal compartments regulate the availability of receptors for insertion in the synapse (Ehlers 2000; Lavezzari et al. 2004; Lee et al. 2004b; Park et al. 2004; Scott et al. 2004; Washbourne et al. 2004). The sorting of receptors for either recycling or degradation can be regulated in an activity-dependent manner, thus providing another mechanism for the activity-dependent remodeling of a synapse.

B. Receptor Recycling and Degradation

Receptors are cycled in and out of the synapse so that they may be inserted and removed from the plasma membrane multiple times. This recycling process allows tight control of the number of receptors at a synapse while limiting the need for rapid synthesis of new receptors. Whether receptor number is to be maintained, increased, or decreased, trafficking receptors through the endosomal system allows a synapse to rapidly alter its strength.

The first description of regulated endosomal sorting of a postsynaptic receptor came from studies in cortical and hippocampal neurons, where AMPA receptors (both GluR1-containing and GluR2/3-containing) are recycled in response to neuronal activity (Ehlers 2000). Activity accelerates ongoing endocytosis of AMPA receptors, whereas the rate of AMPA receptor internalization decreases with activity

blockade. Activation of NMDA receptors alone induces the endocytosis of AMPA receptors (Beattie et al. 2000; Ehlers 2000; Lin et al. 2000), which then traffic to recycling endosomes and are reinserted into the synapse (Ehlers 2000; Lee et al. 2004b). The sorting of AMPA receptors for recycling requires intracellular Ca^{2+}, the protein phosphatases PP1 and calcineurin, and PKA (Ehlers 2000). Ligand binding to AMPA receptors (in the absence of NMDA receptor activation) also triggers AMPA receptor endocytosis but, in this case, internalized receptors traffic to late endosomes.

The regulation of internalized NMDA receptors further illustrates the ability of cells to differentially traffic receptors. The NR2B and NR2A subunits of the NMDA receptor contain distinct endocytic motifs, each interacting with different adaptor proteins (Lavezzari et al. 2003, 2004). This allows these subunits to be independently trafficked. Indeed, although receptors containing both NR2A and NR2B initially are delivered to early endosomes, NR2A subsequently is trafficked to late endosomes and NR2B is delivered to recycling endosomes (Lavezzari et al. 2004; Scott et al. 2004). In addition, in the absence of overriding C-terminal trafficking signals, the conserved membrane proximal endocytic motifs of NR1 and NR2 subunits promote sorting of internalized receptors to late endosomes for degradation (Scott et al. 2004).

The fate of constitutively internalized receptors is dynamically determined. Much like AMPA receptors, kainate receptors undergo rapid constitutive endocytosis (Martin and Henley 2004), but it is the external signals that determine whether these receptors are recycled or degraded. Kainate stimulation induces degradation whereas NMDA receptor activation induces recycling (Martin and Henley 2004). The sorting of $GABA_A$ receptors, also constitutively internalized, is regulated by their association with huntingtin-associated protein 1 (HAP1) (Kittler et al. 2004). Coexpression of HAP1 and GABA receptor subunits increases miniature inhibitory postsynaptic current (mIPSC) frequency and GABA$_A$ receptor cell surface number by favoring GABA$_A$ receptor recycling over degradation (Kittler et al. 2004).

C. Functional Role of Receptor Recycling

Regulation of receptor number at a particular synapse through receptor recycling provides a rapid and effective means for altering synaptic strength in response to activity and local signals. The most celebrated example of such activity-dependent regulation is the insertion of AMPA receptors into the postsynaptic membrane during LTP (Malinow and Malenka 2002; Malenka and Bear 2004; Bredt and Nicoll 2003). Several compelling studies have demonstrated that LTP involves the delivery of AMPA receptors to the postsynaptic membrane (Shi et al. 1999; Hayashi et al. 2000; Zhu et al. 2002). However, until quite recently, the intracellular source of these mobilized AMPA receptors was unclear.

Recent studies indicate that the recycling of AMPA receptors is the critical trafficking step regulating AMPA receptor abundance at the postsynaptic membrane during LTP. Specifically, AMPA receptors inserted during LTP come from the recycling endosome (Park et al. 2004). Inhibiting membrane transport from recycling endosomes prevents NMDA receptor-dependent insertion of AMPA receptors into the dendritic plasma membrane and completely abolishes LTP. Inhibiting protein synthesis for four to six hours has no effect on NMDA receptor-dependent insertion of AMPA receptors, indicating that newly inserted AMPA receptors are not newly synthesized. Moreover, depletion of the cycling pool of AMPA receptors over a four-hour period in hippocampal neurons abolishes stimulus-dependent insertion, suggesting that, in this cell culture model of LTP, all AMPA receptors mobilized during LTP were present at the plasma membrane at some point in the four-hour period before the stimulus—that is, mobilized receptors are recycled receptors (Park et al. 2004).

How is NMDA receptor activation translated into enhanced AMPA receptor trafficking from recycling endosomes? Although much work remains to be done to answer this question, clues come from the fact that AMPA receptor recycling is stimulated by Ca^{2+} influx through NMDA receptors and requires PKA (Ehlers 2000). Thus, one could postulate the existence of a "calcium signal sensor" on recycling endosomes, the identity of which is not known. Such an endosomal calcium sensor would help link established upstream mechanisms of LTP (e.g., NMDA receptor activation, Ca^{2+} influx) to established downstream mechanisms of LTP (e.g., AMPA receptor insertion). Although hundreds of molecules have been implicated in LTP (Sanes and Lichtman 1999), only one organelle—the recycling endosome—has been found to supply mobilized AMPA receptors (Park et al. 2004), suggesting that recycling endosomes may provide a convergence point for diverse molecular mechanisms that mediate or modulate LTP.

How general are the membrane trafficking mechanisms activated by LTP-inducing stimuli? Much of the focus on protein trafficking during LTP has concentrated, for good reason, on the AMPA receptor, its subunit composition, and the protein binding partners of AMPA receptors that regulate trafficking (Malinow and Malenka 2002; Bredt and Nicoll 2003). However, LTP-inducing stimuli can promote generalized transport of completely unrelated cargo from the recycling endosome to the plasma membrane (Park et al. 2004). Thus, postsynaptic membrane trafficking mechanisms during LTP are not limited to AMPA receptors, and one possibility is that AMPA receptors are merely bystanders that get mobilized during LTP simply by virtue of their presence in recycling endosomes. Such a model predicts that additional molecular components are mobilized along with AMPA receptors during LTP (Park et al. 2004), perhaps as a "plasticity module" of molecules that alter diverse aspects of synapse composition, function, and structure.

The endosomal system provides both a source (via recycling) and sink (via sequestration and degradation) for receptors. Having a local pool of available receptors allows for rapid insertion into the synapse during LTP. Similarly, internalized receptors may be sequestered in recycling endosomes until they are needed at a later time or targeted to late endosomes for degradation to globally alter receptor levels for slower forms of synaptic plasticity (Turrigiano and Nelson 2004; Pérez-Otaño and Ehlers 2005).

V. THE EXTRASYNAPTIC PLASMA MEMBRANE AND LATERAL MOVEMENT OF RECEPTORS

A. Movement between Extrasynaptic and Synaptic Receptor Pools

Until now, this chapter has discussed endocytosis and exocytosis of receptors as the exclusive mechanisms for regulating the number of receptors at the synapse. Experiments monitoring the diffusion of receptors have shown that receptors also move laterally within the plasma membrane between extrasynaptic and synaptic areas (Triller and Choquet 2005). Thus, extrasynaptic receptors provide an additional source of receptors for incorporation into the synapse.

A functional exchange of receptors between synaptic and extrasynaptic pools was revealed by exploiting MK801, a quasi-irreversible open channel blocker of the NMDA receptor (Tovar and Westbrook 2002). Electrical stimulation of cultured autaptic hippocampal neurons to evoke synaptic glutamate release in the presence of MK801 irreversibly inhibited synaptic NMDA receptors. In contrast, both extrasynaptic and synaptic receptors were blocked by MK801 upon bath application of NMDA. Blockade of synaptic NMDA receptors initially abolished NMDA receptor-mediated excitatory postsynaptic currents (EPSCs), but this current recovered by 30 percent after 20 minutes (Tovar and Westbrook 2002). No recovery occurred when both extrasynaptic and synaptic receptors were blocked, implying that the exocytosis of new, intracellularly sequestered NMDA receptors was not the source of the recovery. Hence, the recovery was mediated by the rapid lateral movement of extrasynaptic NMDA receptors into the synapse (Tovar and Westbrook 2002).

Single particle tracking allows for the direct visualization of receptors moving into and out of the synapse. Quantum-dot labeled glycine receptors move between extrasynaptic, perisynaptic, and synaptic domains (Dahan et al. 2003). Diffusion rates were greatest for the extrasynaptic domains, whereas perisynaptic and synaptic domains contained both rapid and slow-moving receptors. This suggests that some factor anchored the slow-moving glycine receptors, whereas the nonanchored receptors could travel between domains with relative ease. A good candidate for such anchoring is gephryin, a cytoplasmic scaffold protein that links glycine receptors to microtubules. Similarly, NMDA and AMPA receptors have been observed diffusing rapidly between extrasynaptic and synaptic domains (Groc et al. 2004).

$GABA_A$ receptors also are incorporated into the synapse by lateral movement. A use-dependent "antagonist" of $GABA_A$ receptors was created by mutating the $\alpha 1$ subunit so that the addition of a sulfhydryl reagent would block the receptor through covalent binding but only when the receptor channel is open (Thomas et al. 2005). mIPSCs recorded from neurons were significantly inhibited by adding the sulfhydryl reagent, but the currents recovered after 10 minutes. This indicated that new receptors were rapidly trafficked into the synapse, as those at the synapse during the treatment were irreversibly inhibited. Inhibition of endocytosis and exocytosis indicated that GABA$_A$ receptor insertion from intracellular pools was relatively slow compared to the timescale of current recovery. Blockade of extrasynaptic and synaptic receptors through the bath application of GABA and the sulfhydryl reagent prevented the recovery of mIPSCs, suggesting that the synapse was resupplied with GABA$_A$ receptors by lateral diffusion from extrasynaptic sites (Thomas et al. 2005). For an in-depth discussion of diffusion mechanisms, see Chapter 1.

B. Activity-Dependent Regulation of Receptor Lateral Movement

Under basal conditions, extrasynaptic GluR2-containing AMPA receptors switch between rapid diffusion and stationary behavior with abrupt transitions between these two states (Borgdorff and Choquet 2002). As hippocampal neurons mature in culture, the stationary periods increase in frequency and duration, often in spatial correlation with sites of synaptic contact (Borgdorff and Choquet 2002). Extrasynaptic AMPA receptors have significantly greater mobility than extrasynaptic NMDA receptors, but within the synapse, the receptors have the same low mobility (Groc et al. 2004).

Changing the activity level alters the pattern of glutamate receptor diffusion. Increasing intracellular calcium concentration decreases mobility and lengthens the stationary periods of extrasynaptic AMPA receptors, causing a local accumulation of AMPA receptors (Borgdorff and Choquet 2002). In contrast, potassium chloride application increases AMPA receptor diffusion (Groc et al. 2004). Chronic (48 hours) treatment with tetrodotoxin decreases AMPA receptor diffusion, though an acute application has no effect (Groc et al. 2004). PKC activation increases NMDA receptor

and AMPA receptor mobility in both the extrasynaptic and synaptic pools (Groc et al. 2004). Thus, glutamate receptor activation and intracellular signaling regulate the mobility of synaptic and extrasynaptic receptors, providing an additional mode of activity-dependent receptor trafficking. Regulated lateral movement of glutamate receptors into and out of the synapse is an attractive mechanism for rapidly changing receptor number for synaptic plasticity.

C. Mechanisms for Lateral Movement of Receptors

The precise molecular machinery controlling receptor lateral mobility in different membrane domains remains unclear, but likely includes components of the PSD, the endocytic zone of dendritic spines, and transient associations with cytoskeletal anchoring proteins. In support of this notion, diffusing AMPA receptors enter a confined state when in proximity to membrane-associated clathrin coats (Borgdorff and Choquet 2002). In addition, disruption of either the actin or microtubule cytoskeleton alters the lateral mobility of mGluR5 (Serge et al. 2003). Single particle tracking of glycine receptors revealed that an association with gephryin significantly slows down glycine receptor movement, providing a diffusion trap (Meier et al. 2001). As presynaptic release of glycine triggers the local increase of cytoskeleton-tethered gephyrin (Kirsch and Betz 1998), the creation of this diffusion trap is regulated by activity.

Despite these insights into lateral movement of receptors near the synapse, the relative contributions of endocytosis/ exocytosis in the postsynaptic region and lateral movement from extrasynaptic regions to the synaptic insertion of receptors is unclear. Additionally, the mechanism for activity-dependent lateral movement of glutamate receptors is unknown. As most labels used for single molecule studies (e.g., quantum dots) are too large to be readily internalized, and it has not yet been possible to monitor lateral mobility immediately upon exocytosis, there remain important questions awaiting improvements in imaging technology and related techniques. Ultimately, it will be important to define the combined influences of extrasynaptic receptor pools and intracellular receptor pools in the activity-dependent incorporation and removal of receptors at the synapse.

VI. PERSPECTIVES

This chapter has presented the machinery for postsynaptic trafficking of receptors: the endoplasmic reticulum, Golgi apparatus, endosomal system, and the plasma membrane. Through these membrane compartments, receptors may travel throughout the dendritic structure or be maintained in close proximity to a particular dendrite spine. Inserting and removing receptors from the synapse allows the neuron to respond dynamically to its inputs, strengthening and weakening individual synapses as appropriate.

Though initial visualization of machinery for receptor trafficking in neurons dates back many years (e.g., Golgi 1898; Gray 1959), detailed, functional descriptions of such structures such as dendritic Golgi (e.g., Horton and Ehlers 2003) and neuronal endosomes (e.g., Prekaris et al. 1998; Ehlers 2000; Cooney et al. 2002; Park et al. 2004) have arisen just recently and are still incomplete. Indeed, the purpose of the spine apparatus remains elusive.

Given the intimate relationship between receptor trafficking and synaptic plasticity, many questions remain as to how ubiquitous cellular machinery is adapted for specialized use during neural development and synapse plasticity. For example, do dendritic Golgi confer local delivery of receptors for LTP at specific synapses or do they serve some other purpose? How does the machinery for endocytosis and sorting become localized to spines and does this matter for synapse-specific modification? More generally, it is unclear how much specificity exists in vesicular trafficking of receptors. For example, cargo could either be carefully selected through the recognition of entire signaling complexes or specific receptors may be selected and the receptors plus whichever proteins happen to be nearby are trafficked with the receptor.

Despite advances in following receptors at different stages of trafficking, receptors have not been tracked at the interface between endocytic and exocytic pathways. It would be of great interest to follow a particular receptor from the ER and Golgi to the plasma membrane and the synapse, and later from the synapse to endosomes.

Though one generally considers receptors moving through stable compartments, each of the ER, Golgi, endosomes, and plasma membrane are dynamic structures that are themselves being trafficked. Control mechanisms for organelle trafficking in dendrites remain fertile ground for future experimentation. Furthermore, despite the emphasis on plasticity, it is perhaps more surprising that synapse composition, synapse structure, and overall neuronal form are maintained in the face of the myriad processes of dynamic membrane trafficking.

This chapter described the postsynaptic machinery for receptor trafficking. In subsequent chapters, mechanisms for the trafficking of specific receptors will be discussed: the synaptic trafficking of AMPA receptors (see Chapter 9) and subunit-specific NMDA receptor trafficking (see Chapter 10).

Acknowledgements

We thank J. Hernandez, A. C. Horton, M. Park, L. Ehlers, J. J. Yi, M. Kennedy, and K. G. Condon for helpful comments. Work in the lab of MDE is supported by grants from the NIH and the Raymond and Beverly Sackler Foundation. MDE is an

Investigator of the Howard Hughes Medical Institute. KHC is supported by a Ruth K. Broad Foundation fellowship.

References

Amparan, D., Avram, D., Thomas, C.G., Lindahl, M.G., Yang, J., Bajaj, G., Ishmael, J.E. (2005). Direct interaction of myosin regulatory light chain with the NMDA receptor. *J. Neurochem.* **92**, 349–361.

Aridor, M., Guzik, A.K., Bielli, A., Fish, K.N. (2004). Endoplasmic reticulum export site formation and function in dendrites. *J. Neurosci.* **24**, 3770–3776.

Augustine, G.J., Burns, M.E., DeBello, W.M., Hilfiker, S., Morgan, J.R., Schweizer, F.E. et al. (1999). Proteins involved in synaptic vesicle trafficking. *J. Physiol.* **520 (Pt 1)**, 33–41.

Baas, P.W., Deitch, J.S., Black, M.M., Banker, G.A. (1988). Polarity orientation of microtubules in hippocampal neurons: Uniformity in the axon and nonuniformity in the dendrite. *Proc. Natl. Acad. Sci. U S A* **85**, 8335–8339.

Babst, M., Katzmann, D.J., Snyder, W.B., Wendland, B., Emr, S.D. (2002). Endosome-associated complex, ESCRT-II, recruits transport machinery for protein sorting at the multivesicular body. *Dev. Cell* **3**, 283–289.

Babst, M., Katzmann, D.J., Estepa-Sabal, E.J., Meerloo, T., Emr, S.D. (2002). Escrt-III: An endosome-associated heterooligomeric protein complex required for mvb sorting. *Dev. Cell* **3**, 271–282.

Barlowe, C., Orci, L., Yeung, T., Hosobuchi, M., Hamamoto, S., Salama, N. et al. (1994). COPII: A membrane coat formed by Sec proteins that drive vesicle budding from the endoplasmic reticulum. *Cell.* **77**, 895–907.

Beattie, E.C., Carroll, R.C., Yu, X., Morishita, W., Yasuda, H., von Zastrow, M., Malenka, R.C. (2000). Regulation of AMPA receptor endocytosis by a signaling mechanism shared with LTD. *Nat. Neurosci.* **3**, 1291–1300.

Bi, X., Corpina, R.A., Goldberg, J. (2002). Structure of the Sec23/24-Sar1 pre-budding complex of the COPII vesicle coat. *Nature* **419**, 271–277.

Bingol, B., Schuman, E.M. (2004). A proteasome-sensitive connection between PSD-95 and GluR1 endocytosis. *Neuropharmacology* **47**, 755–763.

Blanpied, T.A., Scott, D.B., Ehlers, M.D. (2002). Dynamics and regulation of clathrin coats at specialized endocytic zones of dendrites and spines. *Neuron* **36**, 435–449.

Blanpied, T.A., Scott, D.B., Ehlers, M.D. (2003). Age-related regulation of dendritic endocytosis associated with altered clathrin dynamics. *Neurobiol. Aging* **24**, 1095–1104.

Borgdorff, A.J., Choquet, D. (2002). Regulation of AMPA receptor lateral movements. *Nature* **417**, 649–653.

Bredt, D.S., Nicoll, R.A. (2003). AMPA receptor trafficking at excitatory synapses. *Neuron* **40**, 361–379.

Bridgman, P.C. (1999). Myosin Va movements in normal and dilute-lethal axons provide support for a dual filament motor complex. *J. Cell Biol.* **146**, 1045–1060.

Bridgman, P.C. (2004). Myosin-dependent transport in neurons. *J. Neurobiol.* **58**, 164–174.

Broadwell, R.D., Cataldo, A.M. (1983). The neuronal endoplasmic reticulum: Its cytochemistry and contribution to the endomembrane system. I. Cell bodies and dendrites. *J. Histochem. Cytochem.* **31**, 1077–1088.

Brodin, L., Löw, P., Shupliakov, O. (2000). Sequential steps in clathrin-mediated synaptic vesicle endocytosis. *Curr. Opin. Neurobiol.* **10**, 312–320.

Brown, S.S. (1999). Cooperation between microtubule- and actin-based motor proteins. *Annu. Rev. Cell Dev. Biol.* **15**, 63–80.

Brown, T.C., Tran, I.C., Backos, D.S., Esteban, J.A. (2005). NMDA receptor-dependent activation of the small GTPase Rab5 drives the removal of synaptic AMPA receptors during hippocampal LTD. *Neuron* **45**, 81–94.

Büttner, C., Sadtler, S., Leyendecker, A., Laube, B., Griffon, N., Betz, H., Schmalzing, G. (2001). Ubiquitination precedes internalization and proteolytic cleavage of plasma membrane-bound glycine receptors. *J. Biol. Chem.* **276**, 42978–42985.

Burack, M., Silverman, M., Banker, G. (2000). The role of selective transport in neuronal protein sorting. *Neuron* **26**, 465–472.

Burbea, M., Dreier, L., Dittman, J.S., Grunwald, M.E., Kaplan, J.M. (2002). Ubiquitin and AP180 regulate the abundance of GLR-1 glutamate receptors at postsynaptic elements in C. elegans. *Neuron* **35**, 107–120.

Cao, H., Orth, J.D., Chen, J., Weller, S.G., Heuser, J.E., McNiven, M.A. (2003). Cortactin is a component of clathrin-coated pits and participates in receptor-mediated endocytosis. *Mol. Cell Biol.* **23**, 2162–2170.

Carlisle, H.J., Kennedy, M.B. (2005). Spine architecture and synaptic plasticity. *Trends Neurosci.* **28**, 182–187.

Carroll, R.C., Lissin, D.V., von Zastrow, M., Nicoll, R.A., Malenka, R.C. (1999). Rapid redistribution of glutamate receptors contributes to long-term depression in hippocampal cultures. *Nat. Neurosci.* **2**, 454–460.

Carroll, R.C., Beattie, E.C., von Zastrow, M., Malenka, R.C. (2001). Role of AMPA receptor endocytosis in synaptic plasticity. *Nat. Rev. Neurosci.* **2**, 315–324.

Chen, C., Brodsky, F.M. (2005). Huntingtin-interacting protein 1 (Hip1) and Hip1-related protein (Hip1R) bind the conserved sequence of clathrin light chains and thereby influence clathrin assembly in vitro and actin distribution in vivo. *J. Biol. Chem.* **280**, 6109–6117.

Claing, A., Laporte, S.A., Caron, M.G., Lefkowitz, R.J. (2002). Endocytosis of G protein-coupled receptors:

Roles of G protein-coupled receptor kinases and beta-arrestin proteins. *Prog. Neurobiol.* **66**, 61–79.

Colledge, M., Snyder, E.M., Crozier, R.A., Soderling, J.A., Jin, Y., Langeberg, L.K. et al. (2003). Ubiquitination regulates PSD-95 degradation and AMPA receptor surface expression. *Neuron* **40**, 595–607.

Connolly, C.N., Kittler, J.T., Thomas, P., Uren, J.M., Brandon, N.J., Smart, T.G., Moss, S.J. (1999). Cell surface stability of gamma-aminobutyric acid type A receptors. Dependence on protein kinase C activity and subunit composition. *J. Biol. Chem.* **274**, 36565–36572.

Cooney, J.R., Hurlburt, J.L., Selig, D.K., Harris, K.M., Fiala, J.C. (2002). Endosomal compartments serve multiple hippocampal dendritic spines from a widespread rather than a local store of recycling membrane. *J. Neurosci.* **22**, 2215–2224.

Dahan, M., Lévi, S., Luccardini, C., Rostaing, P., Riveau, B., Triller, A. (2003). Diffusion dynamics of glycine receptors revealed by single-quantum dot tracking. *Science* **302**, 442–445.

DeFelipe, J., Alonso-Nanclares, L., Arellano, J.I. (2002). Microstructure of the neocortex: Comparative aspects. *J. Neurocytol.* **31**, 299–316.

Deller, T., Korte, M., Chabanis, S., Drakew, A., Schwegler, H., Stefani, G.G. et al. (2003). Synaptopodin-deficient mice lack a spine apparatus and show deficits in synaptic plasticity. *Proc. Natl. Acad. Sci. U S A* **100**, 10494–10499.

Donaldson, J.G., Lippincott-Schwartz, J. (2000). Sorting and signaling at the Golgi complex. *Cell* **101**, 693–696.

Durand, G.M., Kovalchuk, Y., Konnerth, A. (1996). Long-term potentiation and functional synapse induction in developing hippocampus. *Nature* **381**, 71–75.

Dwyer, N., Adler, C., Crump, J., L'Etoile, N., Bargmann, C. (2001). Polarized dendritic transport and the AP-1 mu1 clathrin adaptor UNC-101 localize odorant receptors to olfactory cilia. *Neuron* **31**, 277–287.

Ehlers, M.D. (2000). Reinsertion or degradation of AMPA receptors determined by activity-dependent endocytic sorting. *Neuron* **28**, 511–525.

Ehlers, M.D. (2002). Molecular morphogens for dendritic spines. *Trends Neurosci.* **25**, 64–67.

Engqvist-Goldstein, A., Warren, R., Kessels, M., Keen, J., Heuser, J., Drubin, D. (2001). The actin-binding protein Hip1R associates with clathrin during early stages of endocytosis and promotes clathrin assembly in vitro. *J. Cell Biol.* **154**, 1209–1223.

Engqvist-Goldstein, A.E.Y., Zhang, C.X., Carreno, S., Barroso, C., Heuser, J.E., Drubin, D.G. (2004). RNAi-mediated Hip1R silencing results in stable association between the endocytic machinery and the actin assembly machinery. *Mol. Biol. Cell* **15**, 1666–1679.

Espindola, F.S., Suter, D.M., Partata, L.B., Cao, T., Wolenski, J.S., Cheney, R.E. et al. (2000). The light chain composition of chicken brain myosin-Va: Calmodulin, myosin-II essential light chains, and 8-kDa dynein light chain/PIN. *Cell Motil. Cytoskeleton* **47**, 269–281.

Feig, S., Lipton, P. (1993). Pairing the cholinergic agonist carbachol with patterned Schaffer collateral stimulation initiates protein synthesis in hippocampal CA1 pyramidal cell dendrites via a muscarinic, NMDA-dependent mechanism. *J. Neurosci.* **13**, 1010–1021.

Finch, E.A., Augustine, G.J. (1998). Local calcium signalling by inositol-1,4,5-trisphosphate in Purkinje cell dendrites. *Nature* **396**, 753–756.

Fuhrmann, J.C., Kins, S., Rostaing, P., Far, O.E., Kirsch, J., Sheng, M. et al. (2002). Gephyrin interacts with Dynein light chains 1 and 2, components of motor protein complexes. *J. Neurosci.* **22**, 5393–5402.

Gardiol, A., Racca, C., Triller, A. (1999). Dendritic and postsynaptic protein synthetic machinery. *J. Neurosci.* **19**, 168–179.

Golgi, C. (1989). On the structure of nerve cells. 1898. *J. Microsc.* **155 (Pt 1)**, 3–7.

Gray, E.G. (1959). Electron microscopy of synaptic contacts on dendrite spines of the cerebral cortex. *Nature* **183**, 1592–1593.

Gray, N.W., Fourgeaud, L., Huang, B., Chen, J., Cao, H., Oswald, B.J. et al. (2003). Dynamin 3 is a component of the postsynapse, where it interacts with mGluR5 and Homer. *Curr. Biol.* **13**, 510–515.

Gray, N.W., Kruchten, A.E., Chen, J., McNiven, M.A. (2005). A dynamin-3 spliced variant modulates the actin/cortactin-dependent morphogenesis of dendritic spines. *J. Cell Sci.* **118**, 1279–1290.

Greger, I.H., Khatri, L., Ziff, E.B. (2002). RNA editing at arg607 controls AMPA receptor exit from the endoplasmic reticulum. *Neuron* **34**, 759–772.

Groc, L., Heine, M., Cognet, L., Brickley, K., Stephenson, F.A., Lounis, B., Choquet, D. (2004). Differential activity-dependent regulation of the lateral mobilities of AMPA and NMDA receptors. *Nat. Neurosci.* **7**, 695–696.

Guillaud, L., Setou, M., Hirokawa, N. (2003). KIF17 dynamics and regulation of NR2B trafficking in hippocampal neurons. *J. Neurosci.* **23**, 131–140.

Haglund, K., Fiore, P.P.D., Dikic, I. (2003). Distinct monoubiquitin signals in receptor endocytosis. *Trends Biochem. Sci.* **28**, 598–603.

Haglund, K., Sigismund, S., Polo, S., Szymkiewicz, I., Fiore, P.P.D., Dikic, I. (2003). Multiple monoubiquitination of RTKs is sufficient for their endocytosis and degradation. *Nat. Cell Biol.* **5**, 461–466.

Hanus, C., Vannier, C., Triller, A. (2004). Intracellular association of glycine receptor with gephyrin increases its plasma membrane accumulation rate. *J. Neurosci.* **24**, 1119–1128.

Hayashi, Y., Shi, S.H., Esteban, J.A., Piccini, A., Poncer, J.C., Malinow, R. (2000). Driving AMPA receptors into synapses by LTP and CaMKII: Requirement for

GluR1 and PDZ domain interaction. *Science* **287**, 2262–2267.

Hegde, A.N. (2004). Ubiquitin-proteasome-mediated local protein degradation and synaptic plasticity. *Prog. Neurobiol.* **73**, 311–357.

Hering, H., Sheng, M. (2003). Activity-dependent redistribution and essential role of cortactin in dendritic spine morphogenesis. *J. Neurosci.* **23**, 11759–11769.

Hicke, L., Dunn, R. (2003). Regulation of membrane protein transport by ubiquitin and ubiquitin-binding proteins. *Annu. Rev. Cell Dev. Biol.* **19**, 141–172.

Hicke, L., Riezman, H. (1996). Ubiquitination of a yeast plasma membrane receptor signals its ligand-stimulated endocytosis. *Cell* **84**, 277–287.

Hoop, M.J.D., Huber, L.A., Stenmark, H., Williamson, E., Zerial, M., Parton, R.G., Dotti, C.G. (1994). The involvement of the small GTP-binding protein Rab5a in neuronal endocytosis. *Neuron* **13**, 11–22.

Horton, A.C., Ehlers, M.D. (2003). Dual modes of endoplasmic reticulum-to-Golgi transport in dendrites revealed by live-cell imaging. *J. Neurosci.* **23**, 6188–6199.

Horton, A.C., Ehlers, M.D. (2004). Secretory trafficking in neuronal dendrites. *Nat. Cell Biol.* **6**, 585–591.

Huang, J.D., Brady, S.T., Richards, B.W., Stenolen, D., Resau, J.H., Copeland, N.G., Jenkins, N.A. (1999). Direct interaction of microtubule- and actin-based transport motors. *Nature* **397**, 267–270.

Isaac, J.T., Nicoll, R.A., Malenka, R.C. (1995). Evidence for silent synapses: Implications for the expression of LTP. *Neuron* **15**, 427–434.

Jaskolski, F., Coussen, F., Nagarajan, N., Normand, E., Rosenmund, C., Mulle, C. (2004). Subunit composition and alternative splicing regulate membrane delivery of kainate receptors. *J. Neurosci.* **24**, 2506–2515.

Ju, W., Morishita, W., Tsui, J., Gaietta, G., Deerinck, T.J., Adams, S.R. et al. (2004). Activity-dependent regulation of dendritic synthesis and trafficking of AMPA receptors. *Nat. Neurosci.* **7**, 244–253.

Kabbani, N., Jeromin, A., Levenson, R. (2004). Dynamin-2 associates with the dopamine receptor signalplex and regulates internalization of activated D2 receptors. *Cell Signal* **16**, 497–503.

Kacharmina, J.E., Job, C., Crino, P., Eberwine, J. (2000). Stimulation of glutamate receptor protein synthesis and membrane insertion within isolated neuronal dendrites. *Proc. Natl. Acad. Sci. U S A* **97**, 11545–11550.

Kanaani, J., Diacovo, M.J., El-Husseini, A.E., Bredt, D.S., Baekkeskov, S. (2004). Palmitoylation controls trafficking of GAD65 from Golgi membranes to axon-specific endosomes and a Rab5a-dependent pathway to presynaptic clusters. *J. Cell Sci.* **117**, 2001–2013.

Kato, A., Rouach, N., Nicoll, R.A., Bredt, D.S. (2005). Activity-dependent NMDA receptor degradation mediated by retrotranslocation and ubiquitination. *Proc. Natl. Acad. Sci. U S A* **102**, 5600–5605.

Katzmann, D.J., Babst, M., Emr, S.D. (2001). Ubiquitin-dependent sorting into the multivesicular body pathway requires the function of a conserved endosomal protein sorting complex, ESCRT-I. *Cell* **106**, 145–155.

Katzmann, D.J., Odorizzi, G., Emr, S.D. (2002). Receptor downregulation and multivesicular-body sorting. *Nat. Rev. Mol. Cell Biol.* **3**, 893–905.

Kelleher, R.J., Govindarajan, A., Tonegawa, S. (2004). Translational regulatory mechanisms in persistent forms of synaptic plasticity. *Neuron* **44**, 59–73.

Keller, P., Simons, K. (1997). Post-Golgi biosynthetic trafficking. *J. Cell Sci.* **110 (Pt 24)**, 3001–3009.

Kessels, M., Engqvist-Goldstein, A., Drubin, D., Qualmann, B. (2001). Mammalian Abp1, a signal-responsive F-actin-binding protein, links the actin cytoskeleton to endocytosis via the GTPase dynamin. *J. Cell Biol.* **153**, 351–366.

Kirsch, J., Betz, H. (1998). Glycine-receptor activation is required for receptor clustering in spinal neurons. *Nature* **392**, 717–720.

Kittler, J.T., Delmas, P., Jovanovic, J.N., Brown, D.A., Smart, T.G., Moss, S.J. (2000). Constitutive endocytosis of GABA-A receptors by an association with the adaptin AP2 complex modulates inhibitory synaptic currents in hippocampal neurons. *J. Neurosci.* **20**, 7972–7977.

Kittler, J.T., Thomas, P., Tretter, V., Bogdanov, Y.D., Haucke, V., Smart, T.G., Moss, S.J. (2004). Huntingtin-associated protein 1 regulates inhibitory synaptic transmission by modulating gamma-aminobutyric acid type A receptor membrane trafficking. *Proc. Natl. Acad. Sci. U S A* **101**, 12736–12741.

Kneussel, M., Betz, H. (2000). Receptors, gephyrin and gephyrin-associated proteins: Novel insights into the assembly of inhibitory postsynaptic membrane specializations. *J. Physiol.* **525 (Pt 1)**, 1–9.

Kneussel, M., Haverkamp, S., Fuhrmann, J.C., Wang, H., Wässle, H., Olsen, R.W., Betz, H. (2000). The gamma-aminobutyric acid type A receptor (GABAAR)-associated protein GABARAP interacts with gephyrin but is not involved in receptor anchoring at the synapse. *Proc. Natl. Acad. Sci. U S A* **97**, 8594–8599.

Lanzetti, L., Palamidessi, A., Areces, L., Scita, G., Fiore, P.P.D. (2004). Rab5 is a signaling GTPase involved in actin remodelling by receptor tyrosine kinases. *Nature* **429**, 309–314.

Lavezzari, G., McCallum, J., Dewey, C.M., Roche, K.W. (2004). Subunit-specific regulation of NMDA receptor endocytosis. *J. Neurosci.* **24**, 6383–6391.

Lavezzari, G., McCallum, J., Lee, R., Roche, K.W. (2003). Differential binding of the AP-2 adaptor complex and PSD-95 to the C-terminus of the NMDA receptor subunit NR2B regulates surface expression. *Neuropharmacology* **45**, 729–737.

Lee, M.C.S., Miller, E.A., Goldberg, J., Orci, L., Schekman, R. (2004). Bi-directional protein transport between the ER and Golgi. *Annu. Rev. Cell Dev. Biol.* **20**, 87–123.

Lee, S.H., Liu, L., Wang, Y.T., Sheng, M. (2002). Clathrin adaptor AP2 and NSF interact with overlapping sites of GluR2 and play distinct roles in AMPA receptor trafficking and hippocampal LTD. *Neuron* **36**, 661–674.

Lee, S.H., Simonetta, A., Sheng, M. (2004). Subunit rules governing the sorting of internalized AMPA receptors in hippocampal neurons. *Neuron* **43**, 221–236.

Legendre-Guillemin, V., Metzler, M., Lemaire, J., Philie, J., Gan, L., Hayden, M.R., McPherson, P.S. (2005). Huntingtin interacting protein 1 (HIP1) regulates clathrin assembly through direct binding to the regulatory region of the clathrin light chain. *J. Biol. Chem.* **280**, 6101–6108.

Liao, D., Hessler, N.A., Malinow, R. (1995). Activation of postsynaptically silent synapses during pairing-induced LTP in CA1 region of hippocampal slice. *Nature* **375**, 400–404.

Lin, J.W., Ju, W., Foster, K., Lee, S.H., Ahmadian, G., Wyszynski, M. et al. (2000). Distinct molecular mechanisms and divergent endocytotic pathways of AMPA receptor internalization. *Nat. Neurosci.* **3**, 1282–1290.

Lippincott-Schwartz, J., Roberts, T.H., Hirschberg, K. (2000). Secretory protein trafficking and organelle dynamics in living cells. *Annu. Rev. Cell Dev. Biol.* **16**, 557–589.

Lissin, D.V., Carroll, R.C., Nicoll, R.A., Malenka, R.C., von Zastrow, M. (1999). Rapid, activation-induced redistribution of ionotropic glutamate receptors in cultured hippocampal neurons. *J. Neurosci.* **19**, 1263–1272.

Lüscher, C., Xia, H., Beattie, E.C., Carroll, R.C., von Zastrow, M., Malenka, R.C., Nicoll, R.A. (1999). Role of AMPA receptor cycling in synaptic transmission and plasticity. *Neuron* **24**, 649–658.

Malenka, R.C., Bear, M.F. (2004). LTP and LTD: An embarrassment of riches. *Neuron* **44**, 5–21.

Malinow, R., Malenka, R.C. (2002). AMPA receptor trafficking and synaptic plasticity. *Annu. Rev. Neurosci.* **25**, 103–126.

Man, H.Y., Lin, J.W., Ju, W.H., Ahmadian, G., Liu, L., Becker, L.E. et al. (2000). Regulation of AMPA receptor-mediated synaptic transmission by clathrin-dependent receptor internalization. *Neuron* **25**, 649–662.

Marsh, B.J., Volkmann, N., McIntosh, J.R., Howell, K.E. (2004). Direct continuities between cisternae at different levels of the Golgi complex in glucose-stimulated mouse islet beta cells. *Proc. Natl. Acad. Sci. U S A* **101**, 5565–5570.

Martin, S., Henley, J.M. (2004). Activity-dependent endocytic sorting of kainate receptors to recycling or degradation pathways. *EMBO J.* **23**, 4749–4759.

Matsuzaki, M., Ellis-Davies, G.C., Nemoto, T., Miyashita, Y., Iino, M., Kasai, H. (2001). Dendritic spine geometry is critical for AMPA receptor expression in hippocampal CA1 pyramidal neurons. *Nat. Neurosci.* **4**, 1086–1092.

Maxfield, F.R., McGraw, T.E. (2004). Endocytic recycling. *Nat. Rev. Mol. Cell Biol.* **5**, 121–132.

Meier, J., Vannier, C., Sergé, A., Triller, A., Choquet, D. (2001). Fast and reversible trapping of surface glycine receptors by gephyrin. *Nat. Neurosci.* **4**, 253–260.

Merrifield, C.J., Perrais, D., Zenisek, D. (2005). Coupling between clathrin-coated-pit invagination, cortactin recruitment, and membrane scission observed in live cells. *Cell* **121**, 593–606.

Meusser, B., Hirsch, C., Jarosch, E., Sommer, T. (2005). ERAD: The long road to destruction. *Nat. Cell Biol.* **7**, 766–772.

Miller, T.M., Heuser, J.E. (1984). Endocytosis of synaptic vesicle membrane at the frog neuromuscular junction. *J. Cell Biol.* **98**, 685–698.

Miyashiro, K., Dichter, M., Eberwine, J. (1994). On the nature and differential distribution of mRNAs in hippocampal neurites: implications for neuronal functioning. *Proc. Natl. Acad. Sci. U S A* **91**, 10800–10804.

Miyata, M., Finch, E.A., Khiroug, L., Hashimoto, K., Hayasaka, S., Oda, S.I. et al. (2000). Local calcium release in dendritic spines required for long-term synaptic depression. *Neuron* **28**, 233–244.

Mogelsvang, S., Marsh, B.J., Ladinsky, M.S., Howell, K.E. (2004). Predicting function from structure: 3D structure studies of the mammalian Golgi complex. *Traffic* **5**, 338–345.

Momiyama, A., Silver, R.A., Hausser, M., Notomi, T., Wu, Y., Shigemoto, R., Cull-Candy, S.G. (2003). The density of AMPA receptors activated by a transmitter quantum at the climbing fibre-Purkinje cell synapse in immature rats. *J. Physiol.* **549**, 75–92.

Mori, S., Heldin, C.H., Claesson-Welsh, L. (1992). Ligand-induced polyubiquitination of the platelet-derived growth factor beta-receptor. *J. Biol. Chem.* **267**, 6429–6434.

Mosesson, Y., Shtiegman, K., Katz, M., Zwang, Y., Vereb, G., Szollosi, J., Yarden, Y. (2003). Endocytosis of receptor tyrosine kinases is driven by monoubiquitylation, not polyubiquitylation. *J. Biol. Chem.* **278**, 21323–21326.

Mu, Y., Otsuka, T., Horton, A.C., Scott, D.B., Ehlers, M.D. (2003). Activity-dependent mRNA splicing controls ER export and synaptic delivery of NMDA receptors. *Neuron* **40**, 581–594.

Naisbitt, S., Valtschanoff, J., Allison, D.W., Sala, C., Kim, E., Craig, A.M. et al. (2000). Interaction of the postsynaptic density-95/guanylate kinase domain-associated protein complex with a light chain of myosin-V and dynein. *J. Neurosci.* **20**, 4524–4534.

Nimchinsky, E.A., Sabatini, B.L., Svoboda, K. (2002). Structure and function of dendritic spines. *Annu. Rev. Physiol.* **64**, 313–353.

Nimchinsky, E.A., Yasuda, R., Oertner, T.G., Svoboda, K. (2004). The number of glutamate receptors opened by synaptic stimulation in single hippocampal spines. *J. Neurosci.* **24**, 2054–2064.

O'Brien, R.J., Kamboj, S., Ehlers, M.D., Rosen, K.R., Fischbach, G.D., Huganir, R.L. (1998). Activity-dependent modulation of synaptic AMPA receptor accumulation. *Neuron* **21**, 1067–1078.

Okamoto, P., Gamby, C., Wells, D., Fallon, J., Vallee, R. (2001). Dynamin isoform-specific interaction with the shank/ProSAP scaffolding proteins of the postsynaptic density and actin cytoskeleton. *J. Biol. Chem.* **276**, 48458–48465.

Okano, A., Usuda, N., Furihata, K., Nakayama, K., Tian, Q.B., Okamoto, T., Suzuki, T. (2003). Huntingtin-interacting protein-1-related protein of rat (rHIP1R) is localized in the postsynaptic regions. *Brain Res.* **967**, 210–225.

Osterweil, E., Wells, D.G., Mooseker, M.S. (2005). A role for myosin VI in postsynaptic structure and glutamate receptor endocytosis. *J. Cell Biol.* **168**, 329–338.

Pak, D.T., Yang, S., Rudolph-Correia, S., Kim, E., Sheng, M. (2001). Regulation of dendritic spine morphology by SPAR, a PSD-95-associated RapGAP. *Neuron* **31**, 289–303.

Pak, D.T.S., Sheng, M. (2003). Targeted protein degradation and synapse remodeling by an inducible protein kinase. *Science* **302**, 1368–1373.

Park, M., Penick, E.C., Edwards, J.G., Kauer, J.A., Ehlers, M.D. (2004). Recycling endosomes supply AMPA receptors for LTP. *Science* **305**, 1972–1975.

Passafaro, M., Piëch, V., Sheng, M. (2001). Subunit-specific temporal and spatial patterns of AMPA receptor exocytosis in hippocampal neurons. *Nat. Neurosci.* **4**, 917–926.

Patrick, G.N., Bingol, B., Weld, H.A., Schuman, E.M. (2003). Ubiquitin-mediated proteasome activity is required for agonist-induced endocytosis of GluRs. *Curr. Biol.* **13**, 2073–2081.

Pearse, B., Smith, C., Owen, D. (2000). Clathrin coat construction in endocytosis. *Curr. Opin. Struct. Biol.* **10**, 220–228.

Pérez-Otaño, I., Ehlers, M.D. (2005). Homeostatic plasticity and NMDA receptor trafficking. *Trends Neurosci.* **28**, 229–238.

Petralia, R.S., Wang, Y., Wenthold, R.J. (2003). Internalization at glutamatergic synapses during development. *Eur. J. Neurosci.* **18**, 3207–3217.

Pierce, J., van Leyen, K., McCarthy, J. (2000). Translocation machinery for synthesis of integral membrane and secretory proteins in dendritic spines. *Nat. Neurosci.* **3**, 311–313.

Pierce, J., Mayer, T., McCarthy, J. (2001). Evidence for a satellite secretory pathway in neuronal dendritic spines. *Curr. Biol.* **11**, 351–355.

Polo, S., Sigismund, S., Faretta, M., Guidi, M., Capua, M.R., Bossi, G. et al. (2002). A single motif responsible for ubiquitin recognition and monoubiquitination in endocytic proteins. *Nature* **416**, 451–455.

Prekeris, R., Klumperman, J., Chen, Y.A., Scheller, R.H. (1998). Syntaxin 13 mediates cycling of plasma membrane proteins via tubulovesicular recycling endosomes. *J. Cell. Biol.* **143**, 957–971.

Prybylowski, K., Chang, K., Sans, N., Kan, L., Vicini, S., Wenthold, R.J. (2005). The synaptic localization of NR2B-containing NMDA receptors is controlled by interactions with PDZ proteins and AP-2. *Neuron* **47**, 845–857.

Puertollano, R., Bonifacino, J.S. (2004). Interactions of GGA3 with the ubiquitin sorting machinery. *Nat. Cell Biol.* **6**, 244–251.

Racca, C., Gardiol, A., Triller, A. (1997). Dendritic and postsynaptic localizations of glycine receptor alpha subunit mRNAs. *J. Neurosci.* **17**, 1691–1700.

Racca, C., Stephenson, F.A., Streit, P., Roberts, J.D., Somogyi, P. (2000). NMDA receptor content of synapses in stratum radiatum of the hippocampal CA1 area. *J. Neurosci.* **20**, 2512–2522.

Rácz, B., Blanpied, T.A., Ehlers, M.D., Weinberg, R.J. (2004). Lateral organization of endocytic machinery in dendritic spines. *Nat. Neurosci.* **7**, 917–918.

Rácz, B., Weinberg, R.J. (2004). The subcellular organization of cortactin in hippocampus. *J. Neurosci.* **24**, 10310–10317.

Ren, Z., Riley, N.J., Garcia, E.P., Sanders, J.M., Swanson, G.T., Marshall, J. (2003). Multiple trafficking signals regulate kainate receptor KA2 subunit surface expression. *J. Neurosci.* **23**, 6608–6616.

Ren, Z., Riley, N.J., Needleman, L.A., Sanders, J.M., Swanson, G.T., Marshall, J. (2003). Cell surface expression of GluR5 kainate receptors is regulated by an endoplasmic reticulum retention signal. *J. Biol. Chem.* **278**, 52700–52709.

Roche, K.W., Standley, S., McCallum, J., Ly, C.D., Ehlers, M.D., Wenthold, R.J. (2001). Molecular determinants of NMDA receptor internalization. *Nat. Neurosci.* **4**, 794–802.

Rosenberg, M., Meier, J., Triller, A., Vannier, C. (2001). Dynamics of glycine receptor insertion in the neuronal plasma membrane. *J. Neurosci.* **21**, 5036–5044.

Sala, C., Piëch, V., Wilson, N.R., Passafaro, M., Liu, G., Sheng, M. (2001). Regulation of dendritic spine morphology and synaptic function by Shank and Homer. *Neuron* **31**, 115–130.

Sampo, B., Kaech, S., Kunz, S., Banker, G. (2003). Two distinct mechanisms target membrane proteins to the axonal surface. *Neuron* **37**, 611–624.

Sanes, J., Lichtman, J. (1999). Can molecules explain long-term potentiation? *Nat. Neurosci.* **2**, 597–604.

Sans, N., Prybylowski, K., Petralia, R.S., Chang, K., Wang, Y., Racca, C. et al. (2003). NMDA receptor trafficking through an interaction between PDZ proteins and the exocyst complex. *Nat. Cell Biol.* **5**, 520–530.

Schafer, D.A., Weed, S.A., Binns, D., Karginov, A.V., Parsons, J.T., Cooper, J.A. (2002). Dynamin2 and cortactin regulate actin assembly and filament organization. *Curr. Biol.* **12**, 1852–1857.

Schröder, M., Kaufman, R.J. (2005). The Mammalian unfolded protein response. *Annu. Rev. Biochem.* **74**, 739–789.

Scott, D., Blanpied, T., Swanson, G., Zhang, C., Ehlers, M. (2001). An NMDA receptor ER retention signal regulated by phosphorylation and alternative splicing. *J. Neurosci.* **21**, 3063–3072.

Scott, D.B., Blanpied, T.A., Ehlers, M.D. (2003). Coordinated PKA and PKC phosphorylation suppresses RXR-mediated ER retention and regulates the surface delivery of NMDA receptors. *Neuropharmacology* **45**, 755–767.

Scott, D.B., Michailidis, I., Mu, Y., Logothetis, D., Ehlers, M.D. (2004). Endocytosis and degradative sorting of NMDA receptors by conserved membrane-proximal signals. *J. Neurosci.* **24**, 7096–7109.

Serge, A., Fourgeaud, L., Hemar, A., Choquet, D. (2003). Active surface transport of metabotropic glutamate receptors through binding to microtubules and actin flow. *J. Cell Sci.* **116**, 5015–5022.

Setou, M., Nakagawa, T., Seog, D.H., Hirokawa, N. (2000). Kinesin superfamily motor protein KIF17 and mLin-10 in NMDA receptor-containing vesicle transport. *Science* **288**, 1796–1802.

Shenoy, S.K., McDonald, P.H., Kohout, T.A., Lefkowitz, R.J. (2001). Regulation of receptor fate by ubiquitination of activated beta 2-adrenergic receptor and beta-arrestin. *Science* **294**, 1307–1313.

Shi, S.H., Hayashi, Y., Petralia, R.S., Zaman, S.H., Wenthold, R.J., Svoboda, K., Malinow, R. (1999). Rapid spine delivery and redistribution of AMPA receptors after synaptic NMDA receptor activation. *Science* **284**, 1811–1816.

Shim, J., Umemura, T., Nothstein, E., Rongo, C. (2004). The unfolded protein response regulates glutamate receptor export from the endoplasmic reticulum. *Mol. Biol. Cell* **15**, 4818–4828.

Slepnev, V.I., Camilli, P.D. (2000). Accessory factors in clathrin-dependent synaptic vesicle endocytosis. *Nat. Rev. Neurosci.* **1**, 161–172.

Snyder, E.M., Philpot, B.D., Huber, K.M., Dong, X., Fallon, J.R., Bear, M.F. (2001). Internalization of ionotropic glutamate receptors in response to mGluR activation. *Nat. Neurosci.* **4**, 1079–1085.

Soubeyran, P., Kowanetz, K., Szymkiewicz, I., Langdon, W.Y., Dikic, I. (2002). Cbl-CIN85-endophilin complex mediates ligand-induced downregulation of EGF receptors. *Nature* **416**, 183–187.

Spaargaren, M., Bos, J.L. (1999). Rab5 induces Rac-independent lamellipodia formation and cell migration. *Mol. Biol. Cell* **10**, 3239–3250.

Spacek, J., Harris, K.M. (1997). Three-dimensional organization of smooth endoplasmic reticulum in hippocampal CA1 dendrites and dendritic spines of the immature and mature rat. *J. Neurosci.* **17**, 190–203.

Standley, S., Roche, K.W., McCallum, J., Sans, N., Wenthold, R.J. (2000). PDZ domain suppression of an ER retention signal in NMDA receptor NR1 splice variants. *Neuron* **28**, 887–898.

Steward, O., Schuman, E.M. (2001). Protein synthesis at synaptic sites on dendrites. *Annu. Rev. Neurosci.* **24**, 299–325.

Sun, W., Ginovart, N., Ko, F., Seeman, P., Kapur, S. (2003). In vivo evidence for dopamine-mediated internalization of D2-receptors after amphetamine: Differential findings with [3H]raclopride versus [3H]spiperone. *Mol. Pharmacol.* **63**, 456–462.

Sur, C., McKernan, R., Triller, A. (1995). GABAA receptor-like immunoreactivity in the goldfish brainstem with emphasis on the Mauthner cell. *Neuroscience* **66**, 697–706.

Sutton, M.A., Schuman, E.M. (2005). Local translational control in dendrites and its role in long-term synaptic plasticity. *J. Neurobiol.* **64**, 116–131.

Svitkina, T.M., Bulanova, E.A., Chaga, O.Y., Vignjevic, D.M., Kojima, S., Vasiliev, J.M., Borisy, G.G. (2003). Mechanism of filopodia initiation by reorganization of a dendritic network. *J. Cell Biol.* **160**, 409–421.

Sytnyk, V., Leshchyns'ka, I., Dityatev, A., Schachner, M. (2004). Trans-Golgi network delivery of synaptic proteins in synaptogenesis. *J. Cell Sci.* **117**, 381–388.

Takumi, Y., Ramírez-León, V., Laake, P., Rinvik, E., Ottersen, O.P. (1999). Different modes of expression of AMPA and NMDA receptors in hippocampal synapses. *Nat. Neurosci.* **2**, 618–624.

Tanaka, J., Matsuzaki, M., Tarusawa, E., Momiyama, A., Molnar, E., Kasai, H., Shigemoto, R. (2005). Number and density of AMPA receptors in single synapses in immature cerebellum. *J. Neurosci.* **25**, 799–807.

Teng, H., Cole, J.C., Roberts, R.L., Wilkinson, R.S. (1999). Endocytic active zones: hot spots for endocytosis in vertebrate neuromuscular terminals. *J. Neurosci.* **19**, 4855–4866.

Terry-Lorenzo, R.T., Roadcap, D.W., Otsuka, T., Blanpied, T.A., Zamorano, P.L., Garner, C.C. et al. (2005). Neurabin/protein phosphatase-1 complex regulates dendritic spine morphogenesis and maturation. *Mol. Biol. Cell.* **16**, 2349–2362.

Thomas, P., Mortensen, M., Hosie, A.M., Smart, T.G. (2005). Dynamic mobility of functional GABA(A) receptors at inhibitory synapses. *Nat. Neurosci.* **8**, 889–897.

Toni, N., Buchs, P.A., Nikonenko, I., Povilaitite, P., Parisi, L., Muller, D. (2001). Remodeling of synaptic membranes after induction of long-term potentiation. *J. Neurosci.* **21**, 6245–6251.

Torre, E., Steward, O. (1996). Protein synthesis within dendrites: Glycosylation of newly synthesized proteins in dendrites of hippocampal neurons in culture. *J. Neurosci.* **16**, 5967–5978.

Tovar, K.R., Westbrook, G.L. (2002). Mobile NMDA receptors at hippocampal synapses. *Neuron* **34**, 255–264.

Triller, A., Choquet, D. (2005). Surface trafficking of receptors between synaptic and extrasynaptic membranes: and yet they do move! *Trends Neurosci.* **28**, 133–139.

Trucco, A., Polishchuk, R.S., Martella, O., Pentima, A.D., Fusella, A., Giandomenico, D.D. et al. (2004). Secretory traffic triggers the formation of tubular continuities across Golgi sub-compartments. *Nat. Cell Biol.* **6**, 1071–1081.

Turrigiano, G.G., Nelson, S.B. (2004). Homeostatic plasticity in the developing nervous system. *Nat. Rev. Neurosci.* **5**, 97–107.

van Rijnsoever, C., Täuber, M., Choulli, M.K., Keist, R., Rudolph, U., Mohler, H. (2004). Requirement of alpha5-GABAA receptors for the development of tolerance to the sedative action of diazepam in mice. *J. Neurosci.* **24**, 6785–6790.

van Vliet, C., Thomas, E.C., Merino-Trigo, A., Teasdale, R.D., Gleeson, P.A. (2003). Intracellular sorting and transport of proteins. *Prog. Biophys. Mol. Biol.* **83**, 1–45.

Vandenberghe, W., Nicoll, R.A., Bredt, D.S. (2005). Interaction with the unfolded protein response reveals a role for stargazin in biosynthetic AMPA receptor transport. *J. Neurosci.* **25**, 1095–1102.

Verkhratsky, A. (2005). Physiology and pathophysiology of the calcium store in the endoplasmic reticulum of neurons. *Physiol. Rev.* **85**, 201–279.

Wang, Y.T., Linden, D.J. (2000). Expression of cerebellar long-term depression requires postsynaptic clathrin-mediated endocytosis. *Neuron* **25**, 635–647.

Washbourne, P., Liu, X., Jones, E.G., McAllister, A.K. (2004). Cycling of NMDA receptors during trafficking in neurons before synapse formation. *J. Neurosci.* **24**, 8253–8264.

West, A., Neve, R., Buckley, K. (1997). Identification of a somatodendritic targeting signal in the cytoplasmic domain of the transferrin receptor. *J. Neurosci.* **17**, 6038–6047.

West, A., Neve, R., Buckley, K. (1997). Targeting of the synaptic vesicle protein synaptobrevin in the axon of cultured hippocampal neurons: Evidence for two distinct sorting steps. *J. Cell Biol.* **139**, 917–927.

White, S.H., von Heijne, G. (2005). Transmembrane helices before, during, and after insertion. *Curr. Opin. Struct. Biol.* **15**, 378–386.

Winckler, B., Forscher, P., Mellman, I. (1999). A diffusion barrier maintains distribution of membrane proteins in polarized neurons. *Nature* **397**, 698–701.

Wisco, D., Anderson, E.D., Chang, M.C., Norden, C., Boiko, T., Fölsch, H., Winckler, B. (2003). Uncovering multiple axonal targeting pathways in hippocampal neurons. *J. Cell Biol.* **162**, 1317–1328.

Yap, C.C., Murate, M., Kishigami, S., Muto, Y., Kishida, H., Hashikawa, T., Yano, R. (2003). Adaptor protein complex-4 (AP-4) is expressed in the central nervous system neurons and interacts with glutamate receptor delta2. *Mol. Cell Neurosci.* **24**, 283–295.

Yi, J.J., Ehlers, M.D. (2005). Ubiquitin and protein turnover in synapse function. *Neuron* **47**, 629–632.

Yoshida, Y., Chiba, T., Tokunaga, F., Kawasaki, H., Iwai, K., Suzuki, T. et al. (2002). E3 ubiquitin ligase that recognizes sugar chains. *Nature* **418**, 438–442.

Zerangue, N., Schwappach, B., Jan, Y.N., Jan, L.Y. (1999). A new ER trafficking signal regulates the subunit stoichiometry of plasma membrane K(ATP) channels. *Neuron* **22**, 537–548.

Zhu, J.J., Qin, Y., Zhao, M., Aelst, L.V., Malinow, R. (2002). Ras and Rap control AMPA receptor trafficking during synaptic plasticity. *Cell* **110**, 443–455.

Zito, K., Knott, G., Shepherd, G.M.G., Shenolikar, S., Svoboda, K. (2004). Induction of spine growth and synapse formation by regulation of the spine actin cytoskeleton. *Neuron* **44**, 321–334.

Synaptic Trafficking of AMPA Receptors

LU CHEN AND BITA MAGHSOODI

The roles of the α-amino-3-hydroxyl-5-methyl-4-isoxazolepropionic acid (AMPA)-type glutamate receptor in mediating basal synaptic transmission, synapse stabilization, and synaptic plasticity in the central nervous system have been established. As a dynamic component of the postsynaptic membrane, AMPA receptors are capable of rapid translocation in response to changes in synaptic activity. The molecular and cellular mechanisms that control the assembly and synaptic targeting of individual AMPA receptors are now beginning to be uncov-

ered. The goal of this chapter is to provide an overview of recent progress in understanding activity-dependent and activity-independent synaptic trafficking of AMPA receptors emphasizing the roles of specific AMPA receptor subunits and their interacting proteins.

I. INTRODUCTION

Glutamate is the main excitatory neurotransmitter in the mammalian central nervous system (CNS). Glutamate receptors are classified into two major categories: ligand-gated ion channels (ionotropic) and G-protein coupled receptors (metabotropic). Among different types of ionotropic glutamate receptors, the α-amino-3-hydroxyl-5-methyl-4-isoxazolepropionic acid (AMPA)-type and the kainate (KA)-type receptors mediate the majority of fast synaptic transmission in the mammalian CNS, and the N-methyl-D-aspartate (NMDA)-type receptor mediate a slower component of synaptic transmission, which is important for modulating synaptic function.

Like most proteins, AMPA receptors are synthesized in the endoplasmic reticulum (ER) and trafficked to the Golgi apparatus. Neurons are highly polarized cells in which proteins are sorted and transported to target locations after exiting from the Golgi. Multiple pathways and mechanisms are involved in AMPA receptor membrane targeting. This chapter provides an overview of mechanisms that regulate the synaptic trafficking of AMPA receptors.

A. AMPA Receptor Subunits, Topology, and Assembly

Four subunits, GluR1–4, contribute to the hetero-tetrameric assemblies of the AMPA receptor (Hollmann et al. 1994a; Keinanen et al. 1990; Mano and Teichberg 1998; Rosenmund et al. 1998). These four subunits are highly conserved with over 98 percent amino acid identity among mammals. Each subunit consists of about 900 amino acids with a calculated molecular weight of ~100 kD. The membrane topology of AMPA receptor subunits is depicted in Figure 9.1A (Bennett and Dingledine 1995; Wo and Oswald 1994). The N-terminal hydrophobic signal peptide is cleaved in the mature protein, rendering the N-terminus extracellular. There are four transmembrane domains (TM1–4) in each subunit. Three of them (TM1, TM3, and TM4) span the membrane, but TM2 forms a reentered loop structure that enters and exits the membrane on the intracellular side. This TM2 region is thought to line the pore in the assembled channel (Hollmann 1999).

The extracellular N-terminus can be divided into two domains: the first ~400 amino acids is called the N-terminal domain (NTD), and the ~150 amino acids preceding TM1 (S1), which, together with the extracellular loop between TM3 and TM4 (S2), forms the glutamate binding domain (Armstrong et al. 1998; Kuusinen et al. 1999; Stern-Bach et al. 1994). AMPA receptor assembly is initiated by dimerization of two subunits mediated by NTD regions (Kuusinen et al. 1999; Leuschner and Hoch 1999), and the full tetrameric assembly requires additional interactions among the membrane sectors and the C-terminal part of S2 (Ayalon and Stern-Bach 2001).

B. RNA Editing and Splicing of AMPA Receptor Subunits

Several processing events of the cognate gene transcripts increase the molecular complexity of the AMPA receptor. The S2 domain contains a region that modulates the desensitization of the AMPA receptor (see Figure 9.1A). Alternative splicing of the exons encoding this region yields two splice variants, termed flip and flop (Monyer et al. 1991; Sommer et al. 1990). Flip/flop splicing occurs in all AMPA receptor subunits. With no apparent differences in pharmacological selectivity, the flip forms of most subunits show slower and less desensitization than the flop forms (Lomeli et al. 1994; Mosbacher et al. 1994; Partin et al. 1996; Swanson et al. 1997). Flip/flop splicing is developmentally regulated with the flip variants being the predominant form during early development and the flop variants upregulated after birth. Both forms are present in adult animals (Monyer et al. 1991).

Immediately proceeding the flip/flop module in the S2 domain, all AMPA receptor subunit genes contain an arginine codon

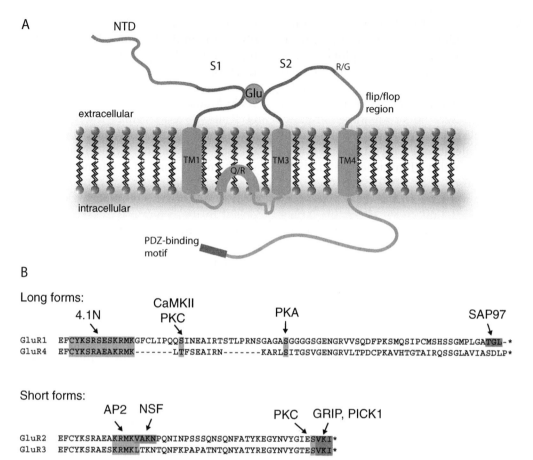

FIGURE 9.1. **AMPA receptor subunits. A.** Membrane topology of AMPA receptor subunits. All subunits are similar in structure with four membrane domains. TM1, TM3, and TM4 span the entire length of the membrane whereas TM2 forms a reentered loop and contributes to the channel pore. The N-terminal domain (NTD) is followed by S1, which forms the ligand-binding pocket together with S2. The Q/R editing, R/G editing, spliced flip/flop, and C-terminal PDZ binding sites are shown. **B.** Cytoplasmic tail regions of AMPA receptor subunits. Long- and short-tailed forms of AMPA receptors are aligned. Protein interaction sites and phosphorylation sites are indicated. See text for details.

(AGA) for the R/G editing site (see Figure 9.1A). Due to alternative splicing of the exons adjacent to this region, the first adenosine of the arginine codon can be altered by pre-mRNA editing to a GGA (Glycine) codon. R/G editing occurs only to GluR2, GluR3, and GluR4. The presence of R → G edited subunit in the AMPA receptor complex speeds up the channel recovery from desensitization and in most cases reduces the amount of desensitization (Lomeli et al. 1994). The extent of R/G editing increases developmentally and

reaches approximately 80 to 90% in the adult brain (Lomeli et al. 1994).

Residue 586 in the pore-lining TM2 domain of the GluR2 subunit is edited from CAG (glutamine/Q) to CGG (arginine/R) (Seeburg 1996; Sommer et al. 1991) (see Figure 9.1A). GluR1, GluR3, and GluR4 mRNAs are not edited at this position. The Q/R editing changes the rectification properties, single channel conductance, and Ca^{2+} permeability. Unedited GluR2 receptors (with R replaced by Q), as well as homomeric GluR1, GluR3, and GluR4 receptors, show inward rectification in their cation currents (Hollmann et al. 1991). In contrast, edited homomeric GluR2 receptors, as well as heteromeric receptors containing edited GluR2 subunit, have near linear I–V relationships (Hollmann et al. 1991; Verdoorn et al. 1991). Receptors containing GluR2(R) have smaller single-channel conductance at negative membrane potentials and lower Ca^{2+} permeability (Burnashev et al. 1992, 1996; Swanson et al. 1997).

Another region of the AMPA receptor subunits that undergoes alternative splicing is the cytoplasmic tail, resulting in C-terminal short and long forms (see Figure 9.1B). GluR1 and GluR4 subunits contain the long C-terminal tail (67 amino acids), and GluR2 and GluR3 subunits contain the short one (36 amino acids) (Gallo et al. 1992; Kohler et al. 1994). Various phosphorylation sites for protein kinase A (PKA), protein kinase C (PKC), and calcium/calmodulin-dependent kinase II (CaMKII) have been identified in the C-terminus of AMPA receptor subunits (Barria et al. 1997a; Blackstone et al. 1994; Greengard et al. 1991; Mammen et al. 1997; Roche et al. 1996; Tan et al. 1994; Wang et al. 1994). Phosphorylation of AMPA receptor subunits, GluR1 in particular, potentiates channel activation, and increases single channel conductance, channel open frequency, and mean open time (Banke et al. 2000; Blackstone et al. 1994; Derkach et al. 1999; Greengard et al. 1991; Oh and Derkach 2005). In addition, the C-terminal tails of the AMPA receptor subunits interact with a complex array of signaling and binding proteins (see Figure 9.1B). How these interactions influence AMPA receptor trafficking will be discussed next.

II. REGULATED RECEPTOR ASSEMBLY AND EXIT FROM THE ER

AMPA receptor channel properties are determined largely by their subunit com-

position. Tetrameric AMPA receptor assembly occurs in the ER immediately after synthesis. Interestingly, assembly of the AMPA receptor follows a specific stoichiometry and spatial configuration (Greger et al. 2003; Mansour et al. 2001). Editing at the Q/R site seems to regulate the AMPA receptor assembly at the tetramerization step. Edited R subunits (GluR2) are largely unassembled and ER-retained, whereas unedited Q subunits (GluR1, GluR3, and GluR4) are readily tetramerized and trafficked to the synapse (Greger et al. 2003). However, this Q/R-editing-regulated assembly does not result in high abundance of unedited homomeric AMPA receptors. In the adult hippocampus, GluR1-3 are found to be highly expressed (Wenthold et al. 1996). Immunoprecipitation experiments with subunit-specific antibodies suggest that there are two major populations of AMPA receptor complexes: those made up of GluR1 and GluR2 and those made up of GluR2 and GluR3. Very few receptors contain both GluR1 and GluR3, whereas approximately 8 percent of the total AMPA receptors are homomeric GluR1 (Wenthold et al. 1996). It has been suggested that a relatively high abundance of GluR2 in the ER facilitates the formation of GluR2-heterodimers, ultimately resulting in a preponderance of GluR2-containing heteromeric AMPA receptors (Dingledine et al. 1999; Ozawa et al. 1998; Seeburg 2002; Wenthold et al. 1996).

Exit of the AMPA receptor from the ER is dependent on the receptor composition. Glutamate receptors receive high-mannose glycosylation in the ER and are modified with more complex sugars in the Golgi (Hollmann et al. 1994b). Glycosylation patterns can be distinguished with endo-glycosidase-H (Endo-H), which digests immature high-mannose sugars. Most of GluR2-GluR3 receptors (60–80%) are EndoH-sensitive, but only a small fraction of GluR1-GluR2 receptors (~38%) show EndoH-sensitivity, indicating that GluR1-GluR2 receptors exit the ER much more effi-

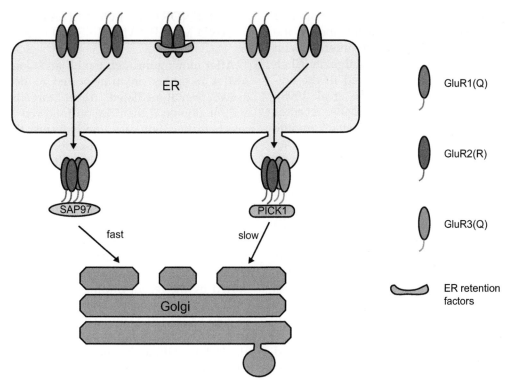

FIGURE 9.2.   **Q/R editing and early trafficking of AMPA receptors.** Retention of GluR2 homomeric receptors is mediated by Q → R editing, possibly involving interaction with an ER chaperon protein. GluR1-GluR2 heteromeric receptors exit more efficiently than GluR2-GluR3 heteromeric receptors, which may be assisted by GluR1-interacting protein SAP97.

ciently than GluR2-GluR3 receptors (Greger et al. 2002). ER retention of GluR2 containing receptors is controlled by Arg607 at the Q/R-editing site. Reversion of R to Q in GluR2 results in its rapid release from the ER and elevated surface expression (Greger et al. 2002). Coimmunoprecipitation of AMPA receptors and ER chaperons BiP and calnexin (Rubio and Wenthold 1999) suggests an involvement of chaperon proteins in AMPA receptor ER retention. Interaction of PDZ proteins with the NMDR receptor subunit NR-1 C-terminus has been shown to suppress ER retention of NMDA receptors (Scott et al. 2001; Standley et al. 2000) (see Chapter 10). Interestingly, the PDZ binding motifs in the AMPA receptor subunit C-terminal tails also are required for their ER exit (Greger et al. 2002). Differ-

ential interactions between PDZ proteins and GluR subunit C-terminal tails (e.g., short vs. long form) may allow receptors to exit the ER by overcoming retention, albeit with different kinetics. For example, SAP-97 (synapse-associated protein-97) and PICK1 (protein interacting with C kinase-1) interact with immature GluR1 and GluR2/3 subunits, respectively, early in the secretory pathway (Dev et al. 1999; Perez et al. 2001; Sans et al. 2001; Xia et al. 1999). These interactions may promote rapid (SAP-97 binding) or slow (PICK-1 binding) ER release, which results in preferential maturation of GluR1-GluR2 receptors compared to GluR2-GluR3 receptors (see Figure 9.2).

In addition to subunit-dependent ER exit mechanisms, a family of TARPs (transmembrane AMPA receptor regulatory proteins) also controls surface expression of AMPA receptors. The first member of the TARPs family identified to interact with AMPA receptor subunits is stargazin (Chen et al. 2000), which is mutated in the epileptic stargazer mouse (Letts et al. 1998). Stargazin is required for AMPA receptor surface expression in cerebellar granule cells (Chen et al. 1999; Hashimoto et al. 1999). The trafficking steps in which stargazin participates to achieve AMPA receptor surface expression remain unclear. Upregulation of ER chaperones as part of the protein unfolding response (UPR) seems to mimic the effect of stargazin in heterologous cells expressing GluR1, suggesting a role of stargazin in the early trafficking events of the biosynthetic pathway (Vandenberghe et al. 2005a).

It is not known whether there are trafficking checkpoints for AMPA receptors in the Golgi apparatus. In *C. elegans*, polarized sorting of glutamate receptor GLR1 is disrupted by mutations in LIN-10, a PDZ protein important for polarized trafficking in epithelia cells (Rongo et al. 1998). Enriched in *trans*-Golgi network, the mammalian homolog of LIN-10, mLIN-10, interacts with AMPA receptor subunits (Stricker and Huganir 2003). The role of mLIN-10/AMPA receptor interactions and AMPA receptor trafficking in mammalian cells is unknown. Another AMPA receptor trafficking-related protein nPIST (neuronal isoform of protein-interacting specifically with TC10) recently was identified by yeast two-hybrid screening using the C-terminus of stargazin as bait (Cuadra et al. 2004). When overexpressed, nPIST enhances synaptic clustering of AMPA receptors. nPIST is highly enriched in the Golgi apparatus, but also is found in dendrites where it partially localizes with synaptic markers such as PSD-95. Similar to other acidic cluster (AC)-domain containing proteins (Molloy et al. 1999), nPIST may mediate the trafficking of the stargazin-AMPA receptor complex between the Golgi and the plasma membrane through its AC domain.

III. DENDRITIC LOCALIZATION AND MEMBRANE INSERTION

After moving through the ER and Golgi, AMPA receptors are transported to dendrites. Signal-mediated direct targeting and/or universal distribution followed by selective retention are mechanisms that may mediate dendritic localization, and are used by other axon- or dendrite-specific neuronal proteins (Sampo et al. 2003; Stowell and Craig 1999).

Transportation of AMPA receptor-containing vesicles along dendritic shafts requires interactions with the microtubule cytoskeleton. The regulation of AMPA receptor trafficking by microtubules is supported by the recent finding that the light chain 2 (LC2) of microtubule-associated protein 1A (MAP1A) interacts directly with stargazin, a TARP family member (Ives et al. 2004). Although the LC2-stargazin-GluR2 complex can be immunoprecipitated from mouse cerebellar lysate, the mechanism by which this complex directs AMPA receptor trafficking remains to be determined.

Long-range bidirectional movement of cargo proteins along the microtubule tracks are mediated by motor proteins of the kinesin and dynein superfamily (Goldstein and Yang 2000) (see Chapters 2 and 3). In axons and distal dendrites, microtubules have their positive ends oriented toward the distal end of the process. However, in proximal dendrites, microtubule polarity is mixed (Baas et al. 1988). Recent findings indicate that kinesin sorting may provide a mechanism for dendrite-specific targeting of proteins. KIF21A and KIF21B are both kinesin-like proteins enriched in neurons. Although they share high amino acid sequence similarity, they have distinct localizations in neurons. KIF21A protein is localized throughout neurons, and KIF21B protein is highly abundant in dendrites (Marszalek et al. 1999).

Neuronal proteins that interact with motor proteins and direct their movements can mediate compartment-specific protein targeting. The Glutamate Receptor Interacting Protein 1 (GRIP1) binds directly to the GluR2 and GluR3 C-terminal PDZ-binding motif (Dong et al. 1997; Srivastava et al. 1998) and with members of KIF5 kinesin heavy chains (Setou et al. 2002). Interrupting the binding between GRIP1 and KIF5 leads to accumulation of kinesin in the somatodendritic area (Setou et al. 2002), suggesting that GRIP1 steers and drives KIF5 to dendrites. These "smart motors" and "smart motor complexes" may mediate AMPA receptors sorting to dendrites by recognizing dendrite-directed microtubules from axon-directed ones.

Over 90 percent excitatory synaptic inputs occur on dendritic spines, small protrusions emerging from dendrites (Gray 1959). Interaction with the actin-based cytoskeleton is required for delivering AMPA receptor-containing vesicles into spines because microtubules do not extend into spines. The juxtamembrane region in the C-terminal tails of GluR1 and GluR4 interact with 4.1N, a neuron-enriched homolog of the erythrocyte membrane cytoskeletal protein 4.1 (Coleman et al. 2003; Shen et al. 2000). The 4.1 family proteins are known to stabilize the spectrin/actin cytoskeleton by forming multimolecular complexes with transmembrane and membrane-associated proteins (Hoover and Bryant 2000). Surface expression of both GluR1 and GluR4 is reduced if their interaction with 4.1N is disrupted (Coleman et al. 2003; Shen et al. 2000). Interestingly, the region on GluR1 that mediates its binding with 4.1N has also been identified as a putative dendritic targeting signal for GluR1 (Ruberti and Dotti 2000).

Dendritically sorted AMPA receptor-containing vesicles fuse with the cytoplasmic membrane. The site of AMPA receptor exocytosis is unknown, but is speculated to be extrasynaptic because of the spatial obstruction presented by the dense protein network in the postsynaptic density (PSD). Several lines of evidence strongly support the notion that AMPA receptors are inserted at nonsynaptic sites and that these extrasynaptic receptors serve as a reservoir for synaptic receptors. First, smooth vesicles (likely to be exocytotic vesicles) have been observed in EM sections to be fusing with the plasma membrane next to the PSD and the spine head, indicating extrasynaptic insertion of cargo proteins (Spacek and Harris 1997). Second, synaptic NMDA receptor-mediated currents exhibit anomalous recovery following irreversible blockade by the activity-dependent blocker, MK801 (Tovar and Westbrook 2002). The recovery could not be attributed to MK-801 unbinding or insertion of newly synthesized receptors, suggesting that membrane receptors move laterally into synapses (see Chapter 1). Lateral diffusion of $GABA_A$ receptors recently has been observed with similar methods (Thomas et al. 2005). Third, in cultured neurons, a single receptor tracking method has been used to demonstrate the lateral movements of AMPA receptors between synapses and extrasynaptic membrane (Borgdorff and Choquet 2002; Tardin et al. 2003). Finally, a method to directly examine AMPA receptor insertion sites on the dendrites recently was developed by introducing an HA epitope tag, followed by a thrombin-specific protease cleavage site (HA/T-tag) into the N-terminus of GluR1 and GluR2. Because thrombin cleavage removes existing surface labeling (HA immunoreactivity) of HA/T-tagged AMPA receptors, sites where newly inserted AMPA receptor subunits accumulate are visualized (Passafaro et al. 2001). Surprisingly, the surface accumulation of GluR1, GluR2, and GluR3 subunits displayed strikingly different spatiotemporal patterns. The initial appearance of GluR1 and GluR3 in nonsynaptic sites preceded their accumulation at the synapse, demonstrating that GluR1- and GluR3-containing AMPA receptors are inserted at nonsynaptic sites before clustering at synaptic membrane. In contrast, surface accumulation of GluR2 was observed almost immediately (within 3 to 5 minutes) at synaptic sites, suggesting direct synaptic insertion or rapid synaptic translocation after exocytosis at nonsynaptic loci (Beretta et al. 2005; Passafaro et al. 2001). NSF (N-ethylmaleimide sensitive factor) is involved in synaptic cycling of AMPA receptors (see the following section). Deleting the NSF-binding site in the GluR2 C-terminus slowed the insertion rate of GluR2 and completely changed its surface delivery pattern, resulting in initial extrasynaptic appearance followed by gradual synaptic accumulation (Beretta et al. 2005). These data suggest that AMPA receptors initially are delivered to the extrasynaptic membrane and synaptic delivery is achieved through lateral diffusion.

IV. RETENTION AND CONSTITUTIVE CYCLING AT THE SYNAPSE

AMPA receptors are a highly dynamic component of the postsynaptic density. Imaging methods allowing visualization of receptor movements at the single-molecule level reveal a constant exchange of receptors between synaptic and extrasynaptic sites (Triller and Choquet 2005). Nonetheless, like most neurotransmitter receptors, AMPA receptors are highly concentrated at the synapse (Nusser 2000; Nusser et al. 1998). The seemingly contradicting observations between the continuous receptor movement and the apparent stability of synapses can be resolved by applying various types of restrictions to the movement at the synapse (Triller and Choquet 2005). The confinement of AMPA receptors at the synapse is the result of a number of factors such as anchorage by scaffolding proteins, and obstructed diffusion by other protein complexes and lipid rafts at the PSD. The contribution of proteins that interact with AMPA receptor subunits to maintain basal synaptic transmission may be to modulate activity-independent synap-

tic delivery and cycling of the receptors (see Fig. 9.3).

A. NSF Interaction with GluR2

The N-methylmaleimide sensitive factor (NSF) is a cytosolic ATPase implicated in the fusion of intracellular transport vesicles with their target membrane (Rothman 1994). NSF was first isolated because of its involvement in protein transport activity of Golgi membranes. A central role for NSF in the docking and/or fusion of synaptic vesicles was later discovered (Schiavo et al. 1995).

NSF is also a constituent of the PSD (Walsh and Kuruc 1992), and interacts with AMPA receptor subunit GluR2 (Henley et al. 1997; Nishimune et al. 1998; Osten et al. 1998). The NSF-binding site involves a 10-amino acid stretch in the proximal C-terminal domain of the GluR2 (Nishimune et al. 1998; Song et al. 1998). Loading neurons with peptides containing this binding site disrupts the interaction between NSF and GluR2, and causes a rapid and partial loss of synaptic AMPA receptors (Lee et al. 2002; Luscher et al. 1999; Luthi et al. 1999; Nishimune et al. 1998; Noel et al. 1999). Although the use of peptide-mediated disruption of the NSF-GluR2 interaction has revealed a potentially important role for NSF in AMPA receptor synaptic targeting, the interpretation of these results was based on the assumption that the peptides used in these studies specifically blocked the interaction of GluR2 with NSF. However, it was later found that AP2, a clathrin adaptor complex important for endocytosis, associates with a region of GluR2 that overlaps with the NSF binding site (Lee et al. 2002). Therefore, peptides used in most of the previous studies interfered with both NSF and AP2 binding with GluR2. Acutely applying a peptide that selectively dissociates the NSF-GluR2 interaction produced a rundown of AMPA receptor-mediated EPSCs, but does not affect extrasynaptic

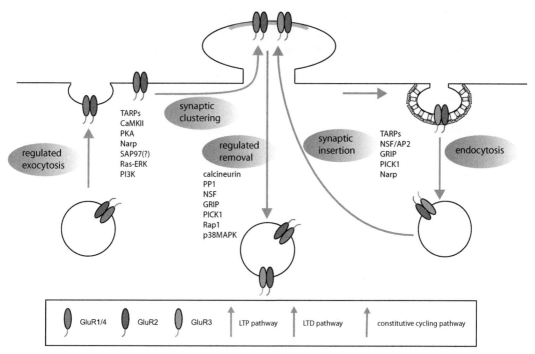

FIGURE 9.3. Constitutive and regulated synaptic trafficking of AMPA receptors. GluR1- or GluR4-containing AMPA receptors are inserted into synapses in an activity-dependent manner during LTP. GluR2-GluR3 oligomers constitutively cycle in and out of synapses. LTD involves activity-dependent removal of both types of AMPA receptors from synapses. Proteins thought to participate in these processes are listed. See text for details.

AMPA responses evoked by exogenous AMPA application (Lee et al. 2002). Furthermore, GluR2 mutants lacking NSF binding ability normally are expressed on the cell surface (Braithwaite et al. 2002; Lee et al. 2002; also see Shi et al. 2001), although their surface insertion rate is somewhat slower (Beretta et al. 2005), and internalization upon agonist stimulation is accelerated (Braithwaite et al. 2002; Lee et al. 2002). Thus, NSF plays an important role in maintaining synaptic strength during basal synaptic transmission by promoting delivery/recycling of GluR2/3-containing AMPA receptors to the synaptic membrane.

B. GRIP and PICK1 Interaction with GluR2 and GluR3

GRIP (glutamate receptor interacting protein) family proteins consist of AMPA receptor-binding protein (ABP) and GRIP1/2. GRIP2 is considered a splice variant of ABP with an extra PDZ domain (Bruckner et al. 1997). They were the first AMPA receptor-interacting proteins identified using a yeast two-hybrid screen (Dong et al. 1997). These proteins contain six (ABP) or seven (GRIP1/2) PDZ domains, of which domains 4 and 5 mediate binding to the extreme C-terminal PDZ binding motif (ESVKI) of GluR2 and GluR3. The multi-PDZ domain organization of GRIP suggests that it may serve as a scaffolding molecule (Pawson and Nash 2003), although the function of GRIP in GluR2-containing AMPA receptor trafficking is not fully understood. The subcellular distribution of GRIP originally was reported to be predominantly synaptic with additional immunoreactivity throughout the dendrites in cultured hippocampal neurons (Dong et al. 1997). Further study reveal that GRIP immunoreactivity is the strongest in inhibitory interneurons in hippocampus and neocortex, and is enriched at both excitatory and inhibitory synapses, suggesting additional functions of GRIP unrelated to its AMPA receptor binding (Burette et al. 2001; Wyszynski et al. 1998, 1999).

PICK1 (protein interacting with C kinase-1) (Staudinger et al. 1995) contains a single PDZ domain that binds to the same extreme C-terminal PDZ-binding domain of GluR2 and GluR3 as does GRIP (Dev et al. 1999; Xia et al. 1999). In addition, PICK1 may dimerize. When coexpressed with GluR2 in heterologous cells, PICK1 induces AMPA receptor clustering. A dimer of PICK1 recruits active PKCα close to AMPA receptors. Indeed, in cultured hippocampal neurons, PICK1 colocalizes with AMPA receptors and PKCα at excitatory synapses (Perez et al. 2001; Xia et al. 1999). GluR1, GluR2, and GluR3 all possess PKC phosphorylation sites in their C-terminal tails (Chung et al. 2000; Matsuda et al. 1999; Roche et al. 1996). Selective phosphorylation of AMPA receptors associated with the PICK1/PKCα complex provides an additional mechanism for regulating AMPA receptor function.

GluR2/3 binding to GRIP and PICK1 can be regulated independently. Phosphorylation of Ser 880 in the PDZ-binding motif of GluR2/3 (ESVKI) by PKC greatly reduces GRIP binding affinity for GluR2/3 subunits without affecting PICK1 binding, which leads to dissociation of GRIP from GluR2/3 in both heterologous cells and Purkinje neurons (Chung et al. 2000; Matsuda et al. 1999). Moreover, activation of PKC with phorbol ester causes a translocation of PICK1 and PKCα to the synapse and a reduction of synaptic GluR2 expression (Perez et al. 2001). Therefore, the phosphorylation state of GluR2/3 Serine 880 determines its preferential binding with GRIP or PICK1, which may be important for subcellular localization of AMPA receptors.

Peptides blocking GluR2/3 binding with GRIP/PICK1 and mutant GluR2/3 unable to bind GRIP/PICK1 have been used to address the roles of GRIP and PICK1 in AMPA receptor synaptic targeting. GRIP binding with AMPA receptors was proposed to be essential for maintaining synaptic accumulation of AMPA receptors (Osten et al. 2000). When a mutant GluR2 that does not bind GRIP was expressed in hippocampal primary cultures, it was targeted correctly to the synaptic membrane. However, its synaptic accumulation was greatly reduced compared to wild-type GluR2. Similar results were obtained in cultured hippocampal slices with a mutant GluR2 unable to bind both GRIP and PICK1 (Shi et al. 2001). The expression of this mutant GluR2 was not detected at the synaptic membrane, whereas exogenously expressed wild-type GluR2 was readily detected at synapses. These results suggest that GRIP binding to GluR2 is required for maintaining synaptic localization of AMPA receptors. However, another study found normal synaptic targeting and accumulation of GluR2 mutants lacking GRIP/PICK1 binding sites in cultured hippocampal neurons (Braithwaite et al. 2002). Moreover, GRIP is not only present at excitatory synapses, but also found in the cytosol and at inhibitory synapses, suggesting that it may serve multiple functions in AMPA receptor trafficking in addition to other synaptic events unrelated to AMPA receptors.

In addition to regulating AMPA receptor localization, GRIP can also act as a scaffold for other signaling molecules. Several GRIP-associated proteins (GRASPs) that interact with distinct PDZ domains of GRIP have been identified (Ye et al. 2000). GRASP-1 is a neuronal rasGEF associated with GRIP and AMPA receptors in vivo. Overexpression of GRASP-1 in cultured neurons specifically reduced the synaptic targeting of AMPA receptors. The subcellular distribution of AMPA receptors and GRASP-1 can be rapidly regulated by activation of NMDA receptors. GRIP1 also interacts with EphB receptors, a large family of receptor tyrosine kinase (RTKs) involved in morphogenesis, neural development, and plasticity (Klein 2004). A recent study revealed a role for GRIP1 in controlling dendrite morphogenesis by regulating EphB receptor trafficking (Hoogenraad et al. 2005). GluR2 and EphB receptors bind to different PDZ domains of GRIP, suggesting that they may simultaneously interact with signaling molecules, although no linkage between AMPA and EphB receptors has been demonstrated.

C. SAP97 Interaction with GluR1

SAP97 is a member of the synapse associated protein (SAP) family that interacts with AMPA receptor subunits. The interaction between SAP97 and AMPA receptors involves the PDZ domain of SAP97 and the extreme C-terminus of the GluR1 subunit (Leonard et al. 1998). Other members of the SAP family, such as SAP90 (also called PSD-95), chapsyn110 (PSD-93) and SAP102, interact with NMDA receptor subunits. SAP97 is localized at both pre- and postsynaptic sites (Muller et al. 1995; Valtschanoff et al. 2000). Though its function at the presynaptic site remains unknown, SAP97 is thought to play a role in GluR1-containing AMPA receptor trafficking via its direct binding to the AMPA receptor GluR1 subunit.

Using the N-terminal domain of SAP97 as bait in a yeast two-hybrid screen, myosin VI, a minus-end-directed actin-based motor, was identified as a SAP97 binding partner (Wu et al. 2002). In light membrane fractions prepared from rat brain, myosin VI and SAP97 form a trimeric complex with the AMPA receptor GluR1 subunit. These data suggest that SAP97 may serve as a molecular link between GluR1 and the actin-dependent motor protein myosin VI during the dynamic translocation of AMPA receptors to and from synapses. SAP97 associates with GluR1 early in the secretory pathway, and the receptors are present in the ER or cis-Golgi. In contrast, few synaptic AMPA receptors associate with SAP97, suggesting that SAP97 dissociates from the receptor complex at the plasma membrane (Sans et al. 2001). These data suggest that SAP97 is involved in early targeting of AMPA receptors, but not in their exo- or endocytosis, which is consistent with the notion that differential PDZ protein–AMPA receptor subunit interactions (i.e., SAP97-GluR1 vs. GRIP/PICK1-GluR2) may regulate AMPA receptor exit kinetics in the secretory pathway (Greger et al. 2002).

A number of studies performed in cultured neurons support a role for SAP97 in synaptic targeting of AMPA receptors (Hayashi et al. 2000; Nakagawa et al. 2004; Rumbaugh et al. 2003). When overexpressed, recombinant SAP97 is targeted to synapses and is enriched in spines in cultured hippocampal neurons (Nakagawa et al. 2004; Rumbaugh et al. 2003). The synaptic targeting of SAP97 resulted in spine head enlargement, an increase in the amount of surface GluR1-containing AMPA receptors, and an increase in the frequency of miniature excitatory synaptic currents (mEPSCs). Consistent with these results, expression of SAP97 in cultured hippocampal slice increased AMPA receptor-mediated EPSCs and drove GluR1-containing AMPA receptors into synapses (Nakagawa et al. 2004) (but see Schnell et al. 2002). Multimerization of SAP97 through its N-terminal L27 domain has been suggested to be required for potentiation of synaptic transmission. Conversely, RNAi knockdown of endogenous SAP97 reduced surface expression of both GluR1 and GluR2 and inhibited both AMPA and NMDA EPSCs (Nakagawa et al. 2004). The extreme C-terminal region of the GluR1 subunit that mediates GluR1 binding with SAP97 is required for synaptic delivery of GluR1-containing AMPA receptors (Hayashi et al. 2000) (but see Kim et al. 2005a).

In addition to binding GluR1 subunits, SAP97 also recruits PKA, through A kinase anchoring proteins (AKAPs) and CaMKII by direct binding (Colledge et al. 2000; Gardoni et al. 2003). It is likely that multimerization of SAP97 (Nakagawa et al. 2004) recruits glutamate receptors and protein kinases into the same protein complex. Phosphorylation of AMPA receptors is enhanced by the SAP97-AKAP79 complex, that directs PKA to GluR1 (Colledge et al. 2000). In addition, SAP97 can be phosphorylated by CaMKII, which drives the SAP97-GluR1 complex into synapses (Mauceri et al. 2004).

D. Stargazin and TARP Family Proteins

Stargazin and the TARP family proteins were first identified as membrane proteins essential for surface expression and synaptic targeting of AMPA receptors (Burgess et al. 1999; Chen et al. 2000; Klugbauer et al. 2000; Tomita et al. 2003).

Stargazin binds directly to AMPA receptors (Tomita et al. 2004). Synaptic targeting of the stargazin–AMPA receptor complex is mediated by synaptic membrane-associated guanylate kinases (MAGUKs) proteins, such as PSD-95 (Chen et al. 2000; Mi et al. 2004; Schnell et al. 2002). Overexpression of PSD-95 selectively increases AMPA receptor-mediated synaptic transmission in cultured hippocampal neurons (El-Husseini et al. 2000; Schnell et al. 2002). Acute removal of PSD-95 from the synapse with 2-bromopalmitate depletes synaptic AMPA receptors (El-Husseini Ael et al. 2002). This effect of PSD-95 is thought to be mediated through direct binding of the first two PDZ domains of synaptic PSD-95 to the extreme C-terminal PDZ-binding motif of stargazin. The surface and synaptic targeting of AMPA receptors can be dissociated into two distinct steps. Overexpression of a stargazin mutant lacking the PDZ-binding motif enhances surface AMPA receptor expression, but disrupts synaptic AMPA receptor localization (Chen et al. 2000). Increasing the level of synaptic PSD-95 recruits new AMPA receptors to synapses

without changing the number of surface AMPA receptors. Stargazin overexpression drastically increases the number of extra-synaptic AMPA receptors, but fails to alter synaptic currents if synaptic PSD-95 levels are kept constant (Schnell et al. 2002). These findings indicate that stargazin/TARPs control the number of surface AMPA receptors, whereas synaptic MAGUK proteins act as "slot" proteins that regulate the amount of AMPA receptors at the synapse.

In isolated AMPA receptor complexes from the brain, TARPs were found tightly associated with the receptor subunits (Fukata et al. 2005; Nakagawa et al. 2005; Vandenberghe et al. 2005b). Two populations of AMPA receptor were identified from cerebellar extracts: a functional form that contains the TARPs and an apo-form that lacks TARPs. Using the single-particle reconstruction, a three-dimensional structure of native AMPA receptors purified from rat brain was determined (Nakagawa et al. 2005). Members of the stargazin/TARPs family copurified with AMPA receptors and contributed to the density representing the transmembrane domain of the complex. The tight association of TARPS with AMPA receptors suggests that TARPs may function as more than just chaperon proteins for AMPA receptor trafficking. In this regard, stargazin not only mediates AMPA receptor surface expression, but also shapes synaptic responses by slowing channel deactivation and desensitization, as well as accelerating recovery from desensitization (Priel et al. 2005; Tomita et al. 2005a; Yamazaki et al. 2004). Separate domains of stargazin are involved in these two different roles. The C-terminal tail of stargazin regulates receptor trafficking, whereas the ectodomain modulates channel properties (Tomita et al. 2004; 2005a). Thus, stargazin/TARPs family proteins play important roles in controlling the efficacy of synaptic transmission in the brain by determining surface expression of AMPA receptors and shaping kinetics of synaptic responses.

E. Neuronal Activity-Regulated Pentraxin (Narp)

Narp is a neuronal immediate early gene (IEG) product that belongs to a secretory protein family called the pentraxins. It was cloned using differential cDNA technique due to its rapid induction in hippocampal and cortical neurons by physiological synaptic activity (Tsui et al. 1996). When expressed in HEK293 cells, Narp interacts with itself and forms large surface clusters that coaggregate with AMPA receptor subunits. Endogenous Narp is selectively enriched at the surface of excitatory synapses, and modulates synaptogenesis by clustering AMPA receptors (O'Brien et al. 1999; 2002). Synaptic Narp can be derived from both pre- and postsynaptic neurons, as suggested by ultrastructural studies showing Narp immunoreactivity in presynaptic boutons associated with synaptic vesicles, in the synaptic cleft, and in the postsynaptic density. Overexpression of recombinant Narp increases the number of excitatory but not inhibitory synapses in cultured spinal neurons. Interestingly, the secretion of Narp by hippocampal axons is restricted to contacts with interneurons. Exogenous application of Narp to cultured hippocampal neurons results in clusters of both NMDA- and AMPA-type glutamate receptors in hippocampal interneurons but not hippocampal pyramidal neurons (Mi et al. 2002). This suggested the intriguing possibility that excitatory synapses formed on dendritic shafts and spines are mediated by different molecules. Because Narp has no ability to directly aggregate NMDA receptors, it may aggregate NMDA receptors in hippocampal interneurons indirectly by cytoplasmic coupling of NMDA receptors to synaptic AMPA receptors.

V. REGULATED SYNAPTIC TARGETING

Much of the interest in regulated synaptic targeting of AMPA receptors stems from experimental evidence for a persistent enhancement in synaptic strength (referred to as long-term potentiation, or LTP) after repetitive activation of excitatory synapses in the dentate gyrus area of the hippocampus (Bliss and Lomo 1973). Activity-dependent long-lasting changes of synaptic strength are now found in many areas of the brain. Synaptic strength can also undergo long-term depression (LTD). Certain forms of LTP and LTD are expressed postsynaptically, which involves modulation of the amount of AMPA receptors at synapses (Malenka and Nicoll 1999; Nicoll 2003). Mechanisms that regulate activity-dependent insertion and removal of AMPA receptors at synapses are the focus of this section.

A. Regulated Removal of AMPA Receptors from the Synapse

Activity-Dependent Removal of Synaptic AMPA Receptors

Hebbian (associative) forms of plasticity have a strong tendency to destabilize network activity because of their positive feedback rules that tend to drive synaptic strength toward maximum or minimum values. Homeostatic regulation of neural activity has been considered a stabilizing mechanism operated at the neural circuit level to prevent Hebbian-type plasticity such as LTP and LTD from driving the neural activity toward runaway excitation or quiescence (Miller 1996; Turrigiano and Nelson 2004). As a consequence of LTP, increased synaptic drive leads to higher firing rates of the postsynaptic neuron. One mechanism to maintain neural firing rates after Hebbian-type plasticity is to globally regulate excitatory synaptic strengths (synaptic scaling).

Evidence supporting synaptic scaling is derived from experiments where synaptic activity was tested after chronic (2–3 days) manipulations of network activity with pharmacological reagents in primary neuronal cultures. Increasing network activity by blocking inhibitory synaptic transmission with GABA$_A$ receptor blocker causes a decrease in synaptic localization of AMPA receptors, as well as a decrease in the contribution of AMPA receptors to synaptic currents relative to NMDA receptors (Lissin et al. 1998; O'Brien et al. 1998; Turrigiano et al. 1998). Conversely, blocking neural activity with TTX or AMPA receptor antagonists leads to an increase in synaptic expression of AMPA receptors (O'Brien et al. 1998; Turrigiano et al. 1998). Translocation of AMPA receptors away from synapses occurs rapidly upon increase in synaptic activity. Minutes after the addition of glutamate to the culture media, a rapid redistribution of GluR1 subunits away from a subset of synaptic sites was observed (Lissin et al. 1999). Consistent with the idea that NMDA receptors are relatively stable at the synapse (Allison et al. 1998), such fast translocation was not observed for NMDA receptors. It now has been established that synaptic removal of AMPA receptors can be triggered by a number of stimuli that increase neural network activities, including chronic postsynaptic depolarization (Leslie et al. 2001) activation of receptors such as AMPA receptors (Carroll et al. 1999b; Lissin et al. 1999), NMDA receptors (Beattie et al. 2000; Carroll et al. 1999b), metabotropic glutamate receptors (Snyder et al. 2001), and insulin receptors (Man et al. 2000).

Removal of AMPA receptors from synapses is preceded by dissociation of AMPA receptors from their anchors on the postsynaptic membrane, which requires actin depolymerization (Zhou et al. 2001). Released AMPA receptors then presumably diffuse to a perisynaptic site where they may be internalized. Endocytosis of AMPA receptors is believed to be a dynamin-dependent process via clathrin-coated pits (Carroll et al. 1999a; Lee et al. 2002; Man et al. 2000; Wang and Linden 2000). This process can be blocked by biochemical inhibition of clathrin or by overexpression of a dominant-negative mutant form of

dynamin. In mature dendrites, clathrin coats are found in stable zones at sites of endocytosis, where clathrin undergoes continuous exchange with local cytosolic pools. In dendritic spines, endocytic zones are found lateral to the PSD where they develop and persist independent of synaptic activity (Blanpied et al. 2002). Immunogold labeling in the rat brain indicates that clathrin-coated pits are localized at the edge of the synaptic active zone and at further postsynaptic distances, including the side of the spines (Petralia et al. 2003; Racz et al. 2004). Perisynaptic localization of endocytic machinery suggests that receptor internalization occurs at extrasynaptic regions after lateral diffusion of the receptors. More direct evidence supporting extrasynaptic internalization comes from a recent study using ecliptic pHluorin-tagged GluR2 to visualize changes in AMPA receptor surface expression in real time (Ashby et al. 2004). Activation of NMDA receptors induced the rapid internalization of extrasynaptic AMPA receptors that recovered slowly upon NMDA washout. In contrast, pHluorin-GluR2 fluorescence from punctate, synaptic regions was stable during the period of brief NMDA application, but slowly declined over the course of several minutes. These data suggest that the removal of AMPA receptors from synapses is preceded by the internalization of AMPA receptors from the extrasynaptic sites.

Internalized AMPA receptors are differentially sorted in early endosomes between recycling and degradation pathways. Sorting is regulated by synaptic activity and differential activation of AMPA and NMDA receptors following subunit-dependent rules (Ehlers 2000; Lee et al. 2004). GluR1, GluR2, and GluR3 behave differently after endocytosis. GluR1 remains in the recycling pathway, and GluR3 is diverted to lysosomes. Internalized GluR2 enters the recycling pathway if induced by AMPA, but is targeted to late endosomes and lysosomes if induced by NMDA. It has been suggested that NSF binding to the GluR2 subunit prevents its targeting to lysosomes by promoting recycling of internalized GluR2 (Lee et al. 2004), although how NMDA receptor activation regulates NSF binding with GluR2 remains to be determined. Interestingly, changing basal activity levels of cultured neurons seems to reverse the sorting induced by AMPA and NMDA receptor activation. In cultured hippocampal neurons preincubated with tetradotoxin (TTX), NMDA receptor activation triggers Ca^{2+}-dependent AMPA receptor internalization followed by rapid membrane reinsertion. In the absence of NMDA activation, AMPA receptor activation caused targeting of AMPA receptors to late endosomes and lysosomes after a Ca^{2+}-independent AMPA receptor endocytosis (Ehlers 2000).

Synaptic Removal of AMPA Receptors and LTD

Two major mechanistically distinct forms of homosynaptic LTD are found in the brain. Induction of one requires NMDA receptor activation (Dudek and Bear 1992; Mulkey and Malenka 1992), whereas induction of the other requires activation of metabotropic glutamate receptors (mGluRs) (Bolshakov and Siegelbaum 1994; Oliet et al. 1997). Although there is experimental evidence supporting both pre- and postsynaptic expression of LTD, it is well accepted that removing AMPA receptors from synapses contributes to certain types of LTD, which have been described both in the hippocampus (Beattie et al. 2000; Carroll et al. 1999b; Luscher et al. 1999; Luthi et al. 1999; Man et al. 2000; Snyder et al. 2001) and the cerebellum (Matsuda et al. 2000; Wang and Linden 2000). Moreover, *in vivo* induction of LTD in the hippocampus resulted in a decrease in the number of AMPA receptors in synaptoneurosomes (Heynen et al. 2000).

A Ca^{2+}-dependent protein phosphatase cascade involving calcineurin (PP2B) and protein phosphatase 1 (PP1) is believed to be required for NMDA receptor-dependent LTD induction (Lisman 1989; Mulkey et al.

1993, 1994). NMDA- or AMPA-induced internalization of AMPA receptors can be blocked by removing extracellular Ca^{2+} or by calcineurin inhibitors (Beattie et al. 2000; Ehlers 2000; Lin et al. 2000). Calcineurin has been shown to facilitate endocytosis via its association with dynamin/amphiphysin and subsequent dephosphorylation of components of the endocytic machinery (Lai et al. 1999; Slepnev et al. 1998).

As discussed earlier, there seem to be subunit-specific rules for activity-dependent AMPA receptor internalization. It is unclear how these rules apply to LTD-related AMPA receptor removal. In the GluR2/3 knockout mouse, several forms of hippocampal synaptic plasticity, including LTD, are intact (Meng et al. 2003). This suggests that GluR1-containing AMPA receptors are capable of supporting long-lasting synaptic removal of AMPA receptors. Additionally, the state of GluR1 C-terminal phosphorylation may mark the history of activity at the synapse, and is associated with LTP and LTD (Lee et al. 2000). A naïve synapse contains mostly GluR1-containing AMPA receptors that are phosphorylated at the PKA site (Ser 845). LTD induction dephosphorylates Ser 845, a process that can be reversed by dedepression. LTP induction leads to increased phosphorylation of the CaMKII/PKC site (Ser 831), and the same site gets dephosphorylated during depotentiation. It is likely that SAP97 interaction with GluR1 helps to maintain the basal phosphorylation state of Ser 845 by recruiting PKA through AKAP (Colledge et al. 2000).

GluR2 interaction with PDZ proteins is required for the expression and maintenance of hippocampal and cerebellar LTD, although it is less clear which PDZ protein actually is involved. In hippocampal neurons, a peptide that blocks both GRIP and PICK1 binding with GluR2 impairs LTD and causes an increase in synaptic transmission in cells following the prior induction of NMDA receptor-dependent LTD, whereas a peptide that selectively blocks PICK1 binding with GluR2 does not have any effect on LTD or synaptic transmission (Daw et al. 2000). These findings suggest that GRIP binding with GluR2 plays an essential role in retaining AMPA receptors in an intracellular/extrasynaptic pool, which is important for expression and maintenance of LTD. Moreover, PKC inhibitors blocked the increase in synaptic transmission induced by the GRIP/PICK1 inhibitory peptide. These data suggest a role for GRIP in LTD; however, other studies in both hippocampal and cerebellar neurons demonstrated a role for PICK1 in LTD expression. Overexpression of PICK1 in cultured hippocampal neurons reduces surface expression of GluR2 (Perez et al. 2001). PICK1 interaction with GluR2 recruits PKC, and PKC phosphorylation of GluR2 Ser 880 disrupts its binding with GRIP but not PICK1 (Chung et al. 2000; Matsuda et al. 1999). Indeed, induction of LTD in hippocampal slices increases phosphorylation of Ser 880 within the GluR2 C-terminal PDZ ligand, suggesting that the modulation of GluR2 interaction with GRIP1 and PICK1 may regulate AMPA receptor internalization during LTD (Kim et al. 2001). Furthermore, postsynaptic intracellular perfusion of GluR2 C-terminal peptides that disrupt GluR2 interaction with PICK1 inhibits the expression of hippocampal LTD. In the hippocampus, the interaction of PICK1 with GluR2 seems to play a major role in cerebellar LTD. Acute application of a peptide that selectively blocks PICK1 binding with GluR2, as well as loading cells with antibodies directed against the PDZ domain of PICK1, impaired LTD induction in cultured Purkinje cells (Xia et al. 2000). PKC activation and phosphorylation of GluR2 Ser 880 are required for cerebellar LTD (Chung et al. 2003; Crepel and Krupa 1988; De Zeeuw et al. 1998; Linden and Connor 1991), supporting the hypothesis that PICK1 recruits PKC to GluR2 in cerebellar Purkinje cells. Phosphorylation of GluR2 by PKC may

prime AMPA receptors for internalization or stabilize the internalized receptors at an intracellular/extrasynaptic site.

The role of NSF in synaptic plasticity was discovered when it was found that peptide-mediated disruption of NSF-GluR2 binding reduces synaptic transmission and occludes hippocampal LTD (Luscher et al. 1999; Luthi et al. 1999). Saturation of LTD prevented further reduction of synaptic transmission by the peptide. Because the NSF interaction with GluR2 is also important for maintaining a mobile pool of recycling AMPA receptors (see earlier), it is possible that this mobile pool of AMPA receptors is specifically involved in LTD. Support for this hypothesis comes from a recent study using GluR2 deficient mice. In GluR2-/- Purkinje cells, basal synaptic transmission is reduced and LTD is absent (Chung et al. 2003). Both basal synaptic strength and LTD can be restored by transfecting Purkinje cells with the GluR2, but not the GluR3 subunit. Interestingly, when the NSF-binding sequence of GluR2 is transferred onto GluR3, the transfected receptor can enter synapses and rescue both basal synaptic transmission and LTD (Steinberg et al. 2004). These results demonstrate that NSF-GluR2 interaction incorporates AMPA receptors into synapses and renders them competent to undergo LTD. The role of NSF and PICK1 in AMPA receptor trafficking was clarified by the observation that the ATPase activity of NSF regulates PICK1-GluR2 interactions (Hanley et al. 2002).

B. Synaptic Insertion of AMPA Receptors

Subunit-Specific Trafficking Pathways and LTP

The synaptic delivery of AMPA receptors is involved in synapse formation, basal synaptic strength maintenance through constitutive cycling, and LTP expression. There appear to be two distinct subunit-specific trafficking pathways for AMPA receptors (Malinow et al. 2000). GluR2-GluR3 oligomers are constitutively delivered into synapses following an activity-independent pathway (Kakegawa et al. 2004; Passafaro et al. 2001; Shi et al. 2001). Thus, synaptic strength (total number of AMPA receptors at synapses) is maintained through continuous cycling of GluR2-GluR3 receptors (the constitutive pathway). In contrast, GluR1- and GluR4-containing receptors are inserted into synapses in an activity-dependent manner (the regulated pathway) (Hayashi et al. 2000; Kakegawa et al. 2004; Shi et al. 1999; Zhu et al. 2000). Upon plasticity induction, synaptic delivery of these two types of receptors leads to long-lasting enhancement of synaptic strength. The subunit-specific trafficking of AMPA receptors implies that the subcellular distribution of AMPA receptors may be different depending on their subunit composition. Furthermore, the rapid translocation of AMPA receptors through the regulated trafficking pathway suggests the existence of a pool of AMPA receptors that is extrasynaptic, a hypothesis supported by data derived from ultrastructural examination of the AMPA receptors (Baude et al. 1995; Kharazia et al. 1996; Martin et al. 1993; Molnar et al. 1993; Petralia and Wenthold 1992; Takumi et al. 1999). In GluR1-deficient mice, basal synaptic AMPA receptor-mediated response is normal, whereas extrasynaptic AMPA receptor-mediated response is greatly reduced and hippocampal CA1 LTP is absent (Zamanillo et al. 1999). These data strongly support the hypothesis that postsynaptic LTP is mediated through translocation of a pool of extrasynaptic GluR1-containing AMPA receptors into synapses. The differential distribution of AMPA receptor subunits has been observed in living tissues. GFP-tagged subunits were expressed in cultured hippocampal slices and their distribution was monitored with two-photon microscopy (Kakegawa

et al. 2004; Shi et al. 1999, 2001; Zhu et al. 2000). GFP-GluR1 is concentrated in the dendritic shaft region, in contrast to their synaptic/spine localization when expressed in dissociated cultures (Lissin et al. 1998; Shi et al. 1999). High-frequency LTP-generating stimulation induced translocation of GFP-GluR1 onto the plasma membrane surface and into dendritic spines. The function of these translocated receptors was confirmed by electrophysiological measurement of the synaptic currents and their rectification properties (Kakegawa et al. 2004; Shi et al. 1999). This activity-dependent delivery of the GluR1-containing AMPA receptor is thought to require its C-terminal PDZ-binding motif (Hayashi et al. 2000), although a knock-in mouse lacking the GluR1 PDZ-binding motif had normal synaptic localization of GluR1 and LTP (Kim et al. 2005a). In contrast, GFP-tagged GluR2 and GluR3 are found in synaptic membranes regardless of activity level (Kakegawa et al. 2004; Shi et al. 2001).

Signaling Cascades Underlying LTP

NMDA receptor-dependent LTP induction requires activation of NMDA receptors that leads to a rise in the intracellular calcium concentration (Malenka et al. 1989; Yang et al. 1999). This elevated calcium may result in activation of protein kinases, such as CaMKII and PKA that have been shown to play important roles in LTP (Blitzer et al. 1995; Lisman et al. 1997).

PKA activity is necessary but not sufficient for LTP expression. Phosphorylation of GluR1 Ser 845 by PKA is a prerequisite for AMPA receptor insertion (Ehlers 2000; Lee et al. 2000). Mutation of Ser 845 prevents CaMKII activity-induced synaptic delivery of AMPA receptors (Esteban et al. 2003). However, PKA activity alone is not sufficient for AMPA receptor insertion into synapses upon LTP induction. CaMKII activation also is required for LTP (Esteban et al. 2003). Therefore, GluR1 Ser 845 phosphorylation by PKA sets the state of naïve synapses, which can undergo either LTP (requiring CaMKII activity) or LTD (requiring dephosphorylation of Ser 845) (Hayashi et al. 2000; Lee et al. 2000).

CaMKII activity is required for LTP induction (Malinow et al. 1988; 1989; Silva et al. 1992). Activation of CaMKII drives extrasynaptic GluR1-containing AMPA receptors into synapses, leading to potentiated synaptic transmission (Hayashi et al. 2000). Several candidate proteins may serve as CaMKII substrates, whose phosphorylation leads to LTP. One of these candidate proteins is the GluR1 subunit. CaMKII phosphorylation of GluR1 Ser 831 increases the channel activity of receptors containing this subunit (Derkach et al. 1999). In addition, phosphorylation of GluR1 by CaMKII correlates with LTP induction (Barria et al. 1997b; Lee et al. 2000). LTP is impaired when GluR1 contains a mutation in Ser 831 that prevents its phosphorylation (Lee et al. 2003). However, mutations of GluR1 Ser 831 fail to prevent synaptic delivery of GluR1 by CaMKII activity (Hayashi et al. 2000). These contradicting results may be reconciled if there are two components to LTP, one that requires GluR1 phosphorylation and the other that does not. In GluR1 phosphorylation-independent LTP, phosphorylation of other target proteins by CaMKII must be involved. Both stargazin/TARPs family proteins and SynGAP, a synaptic rasGAP, are likely candidates for mediating CaMKII-dependent receptor delivery (Chen et al. 1998; Kim et al. 1998; Tomita et al. 2005b).

Stargazin is phosphorylated at its cytosolic C-terminus by CaMKII and PKC. Phosphorylation of stargazin promotes synaptic trafficking of AMPA receptors (Tomita et al. 2005b). Synaptic NMDA receptor activity can induce both stargazin phosphorylation via activation of CaMKII and PKC, and stargazin dephosphorylation by activation of PP1 downstream of calcineurin. In hippocampal neurons transfected with stargazin mutants that either lack phosphorylation sites or mimic phosphorylated forms, LTP and LTD are impaired. These results support a role of stargazin in synaptic plasticity, although the mechanism that regulates the synaptic insertion of AMPA receptors that is promoted by stargazin phosphorylation has not been determined.

Both LTP and LTD require NMDA receptor activation and postsynaptic calcium. Distinct signaling cascades that are associated with different NMDA receptor subunits may support these two forms of plasticity. Induction of LTP preferentially activates NR2A-containing NMDA receptors (Hrabetova et al. 2000; Liu et al. 2004; Yoshimura et al. 2003; also see Berberich et al. 2005). Subsequent activation of CaMKII results in phosphorylation and inhibition of SynGAP, a RasGAP (Chen et al. 1998; Kim et al. 1998) leading to Ras activation. The Ras-ERK or Ras-PI3K (phosphotidylinositol 3-kinase) pathway then drives synaptic insertion of GluR1 (Kim et al. 2005b; Krapivinsky et al. 2004; Man et al. 2003; Seger and Krebs 1995; Zhu et al. 2002). In contrast, LTD-inducing activity preferentially activates NR2B-containing NMDA receptors (Hrabetova et al. 2000; Liu et al. 2004; Yoshimura et al. 2003), which promotes the removal of synaptic AMPA receptors by both activating the Rap1-p38MAPK pathway and suppressing the Ras-ERK pathway through SynGAP activity (Bolshakov et al. 2000; Kawasaki et al. 1999; Kim et al. 2005b; Sawada et al. 2001; Zhu et al. 2002).

C. Activity-Dependent Switch of Receptor Subtype at the Synapse

Activity-dependent switching of receptor subtypes is a novel type of receptor modulation that adds complexity to the regulation of synaptic AMPA receptors. At cerebellar parallel fiber-stellate cell synapses, tetanic stimulation induces a replacement of calcium-permeable AMPA receptors with calcium-impermeable ones

(Liu and Cull-Candy 2000). Although this results in a slight reduction in AMPA receptor-mediated responses (largely due to the difference in single channel conductance of GluR2-lacking and GluR2-containing receptors) (Liu and Cull-Candy 2005), this type of plasticity mainly provides a protective mechanism against excitotoxicity by limiting synaptic calcium entry. Cytosolic proteins play an important role in the receptor subtype switch; both PICK1 and NSF interactions with AMPA receptors are required for the switch to occur (Gardner et al. 2005; Liu and Cull-Candy 2005; Terashima et al. 2004).

VI. CONCLUSIONS

Remarkable progress has been made toward understanding AMPA receptor trafficking in the last decade. Advanced imaging technologies have allowed direct visualization of receptor trafficking and their translocation related to synaptic plasticity. In parallel, multiple AMPA receptor binding proteins and their involvement in different steps of AMPA receptor trafficking gradually have been revealed. The biggest current challenge is to provide a coherent model that incorporates general trafficking rules and subunit-specific interactions into the regulation of AMPA receptor protein interactions. Perhaps with advanced technology that allows direct observation of genetically-tagged endogenous receptor trafficking *in vivo* (Rasse et al. 2005) we will be able to elucidate how AMPA receptors are targeted to synapses during synaptogenesis, and how AMPA receptors are dynamically regulated in synaptic plasticity.

Acknowledgements

L.C. is supported by the Arnold and Mabel Beckman Foundation, the David and Lucile Packard Foundation, and the National Institute of Health.

References

Allison, D.W., Gelfand, V.I., Spector, I., Craig, A.M. (1998). Role of actin in anchoring postsynaptic receptors in cultured hippocampal neurons: Differential attachment of NMDA versus AMPA receptors. *J. Neurosci.* **18**, 2423–2436.

Armstrong, N., Sun, Y., Chen, G.Q., Gouaux, E. (1998). Structure of a glutamate-receptor ligand-binding core in complex with kainate. *Nature* **395**, 913–917.

Ashby, M.C., De La Rue, S.A., Ralph, G.S., Uney, J., Collingridge, G.L., Henley, J.M. (2004). Removal of AMPA receptors (AMPARs) from synapses is preceded by transient endocytosis of extrasynaptic AMPARs. *J. Neurosci.* **24**, 5172–5176.

Ayalon, G., Stern-Bach, Y. (2001). Functional assembly of AMPA and kainate receptors is mediated by several discrete protein-protein interactions. *Neuron* **31**, 103–113.

Baas, P.W., Deitch, J.S., Black, M.M., Banker, G.A. (1988). Polarity orientation of microtubules in hippocampal neurons: uniformity in the axon and nonuniformity in the dendrite. *Proc. Natl. Acad. Sci. U S A* **85**, 8335–8339.

Banke, T.G., Bowie, D., Lee, H., Huganir, R.L., Schousboe, A., Traynelis, S.F. (2000). Control of GluR1 AMPA receptor function by cAMP-dependent protein kinase. *J. Neurosci.* **20**, 89–102.

Barria, A., Derkach, V., Soderling, T. (1997a). Identification of the Ca2+/calmodulin-dependent protein kinase II regulatory phosphorylation site in the alpha-amino-3-hydroxyl-5-methyl-4-isoxazole-propionate-type glutamate receptor. *J. Biol. Chem.* **272**, 32727–32730.

Barria, A., Muller, D., Derkach, V., Griffith, L.C., Soderling, T.R. (1997b). Regulatory phosphorylation of AMPA-type glutamate receptors by CaM-KII during long-term potentiation. *Science* **276**, 2042–2045.

Baude, A., Nusser, Z., Molnar, E., McIlhinney, R.A., Somogyi, P. (1995). High-resolution immunogold localization of AMPA type glutamate receptor subunits at synaptic and non-synaptic sites in rat hippocampus. *Neuroscience* **69**, 1031–1055.

Beattie, E.C., Carroll, R.C., Yu, X., Morishita, W., Yasuda, H., von Zastrow, M., Malenka, R.C. (2000). Regulation of AMPA receptor endocytosis by a signaling mechanism shared with LTD. *Nat. Neurosci.* **3**, 1291–1300.

Bennett, J.A., Dingledine, R. (1995). Topology profile for a glutamate receptor: Three transmembrane domains and a channel-lining reentrant membrane loop. *Neuron* **14**, 373–384.

Berberich, S., Punnakkal, P., Jensen, V., Pawlak, V., Seeburg, P.H., Hvalby, O., Kohr, G. (2005). Lack of NMDA receptor subtype selectivity for hippocampal long-term potentiation. *J. Neurosci.* **25**, 6907–6910.

Beretta, F., Sala, C., Saglietti, L., Hirling, H., Sheng, M., Passafaro, M. (2005). NSF interaction is important for direct insertion of GluR2 at synaptic sites. *Mol. Cell Neurosci.* **28**, 650–660.

Blackstone, C., Murphy, T.H., Moss, S.J., Baraban, J.M., Huganir, R.L. (1994). Cyclic AMP and synaptic activity-dependent phosphorylation of AMPA-preferring glutamate receptors. *J. Neurosci.* **14**, 7585–7593.

Blanpied, T.A., Scott, D.B., Ehlers, M.D. (2002). Dynamics and regulation of clathrin coats at specialized endocytic zones of dendrites and spines. *Neuron* **36**, 435–449.

Bliss, T.V., Lomo, T. (1973). Long-lasting potentiation of synaptic transmission in the dentate area of the anaesthetized rabbit following stimulation of the perforant path. *J. Physiol.* **232**, 331–356.

Blitzer, R.D., Wong, T., Nouranifar, R., Iyengar, R., Landau, E.M. (1995). Postsynaptic cAMP pathway gates early LTP in hippocampal CA1 region. *Neuron* **15**, 1403–1414.

Bolshakov, V.Y., Carboni, L., Cobb, M.H., Siegelbaum, S.A., Belardetti, F. (2000). Dual MAP kinase pathways mediate opposing forms of long-term plasticity at CA3-CA1 synapses. *Nat. Neurosci.* **3**, 1107–1112.

Bolshakov, V.Y., Siegelbaum, S.A. (1994). Postsynaptic induction and presynaptic expression of hippocampal long-term depression. *Science* **264**, 1148–1152.

Borgdorff, A.J., Choquet, D. (2002). Regulation of AMPA receptor lateral movements. *Nature* **417**, 649–653.

Braithwaite, S.P., Xia, H., Malenka, R.C. (2002). Differential roles for NSF and GRIP/ABP in AMPA receptor cycling. *Proc. Natl. Acad. Sci. U S A* **99**, 7096–7101.

Bruckner, M.K., Rossner, S., Arendt, T. (1997). Differential changes in the expression of AMPA receptors genes in rat brain after chronic exposure to ethanol: an in situ hybridization study. *J. Hirnforsch* **38**, 369–376.

Burette, A., Khatri, L., Wyszynski, M., Sheng, M., Ziff, E.B., Weinberg, R.J. (2001). Differential cellular and subcellular localization of ampa receptor-binding protein and glutamate receptor-interacting protein. *J. Neurosci.* **21**, 495–503.

Burgess, D.L., Davis, C.F., Gefrides, L.A., Noebels, J.L. (1999). Identification of three novel Ca(2+) channel gamma subunit genes reveals molecular diversification by tandem and chromosome duplication. *Genome Res.* **9**, 1204–1213.

Burnashev, N., Monyer, H., Seeburg, P.H., Sakmann, B. (1992). Divalent ion permeability of AMPA receptor channels is dominated by the edited form of a single subunit. *Neuron* **8**, 189–198.

Burnashev, N., Villarroel, A., Sakmann, B. (1996). Dimensions and ion selectivity of recombinant AMPA and kainate receptor channels and their dependence on Q/R site residues. *J. Physiol.* **496 (Pt 1)**, 165–173.

Carroll, R.C., Beattie, E.C., Xia, H., Luscher, C., Altschuler, Y., Nicoll, R.A. et al. (1999a). Dynamin-dependent endocytosis of ionotropic glutamate receptors. *Proc. Natl. Acad. Sci. U S A* **96**, 14112–14117.

Carroll, R.C., Lissin, D.V., von Zastrow, M., Nicoll, R.A., Malenka, R.C. (1999b). Rapid redistribution of glutamate receptors contributes to long-term depression in hippocampal cultures. *Nat. Neurosci.* **2**, 454–460.

Chen, H.J., Rojas-Soto, M., Oguni, A., Kennedy, M.B. (1998). A synaptic Ras-GTPase activating protein (p135 SynGAP) inhibited by CaM kinase II. *Neuron* **20**, 895–904.

Chen, L., Bao, S., Qiao, X., Thompson, R.F. (1999). Impaired cerebellar synapse maturation in waggler, a mutant mouse with a disrupted neuronal calcium channel gamma subunit. *Proc. Natl. Acad. Sci. U S A* **96**, 12132–12137.

Chen, L., Chetkovich, D.M., Petralia, R.S., Sweeney, N.T., Kawasaki, Y., Wenthold, R.J. et al. (2000). Stargazin regulates synaptic targeting of AMPA receptors by two distinct mechanisms. *Nature* **408**, 936–943.

Chung, H.J., Steinberg, J.P., Huganir, R.L., Linden, D.J. (2003). Requirement of AMPA receptor GluR2 phosphorylation for cerebellar long-term depression. *Science* **300**, 1751–1755.

Chung, H.J., Xia, J., Scannevin, R.H., Zhang, X., Huganir, R.L. (2000). Phosphorylation of the AMPA receptor subunit GluR2 differentially regulates its interaction with PDZ domain-containing proteins. *J. Neurosci.* **20**, 7258–7267.

Coleman, S.K., Cai, C., Mottershead, D.G., Haapalahti, J.P., Keinanen, K. (2003). Surface expression of GluR-D AMPA receptor is dependent on an interaction between its C-terminal domain and a 4.1 protein. *J. Neurosci.* **23**, 798–806.

Colledge, M., Dean, R.A., Scott, G.K., Langeberg, L.K., Huganir, R.L., Scott, J.D. (2000). Targeting of PKA to glutamate receptors through a MAGUK-AKAP complex. *Neuron* **27**, 107–119.

Crepel, F., Krupa, M. (1988). Activation of protein kinase C induces a long-term depression of glutamate sensitivity of cerebellar Purkinje cells. An in vitro study. *Brain Res.* **458**, 397–401.

Cuadra, A.E., Kuo, S.H., Kawasaki, Y., Bredt, D.S., Chetkovich, D.M. (2004). AMPA receptor synaptic targeting regulated by stargazin interactions with the Golgi-resident PDZ protein nPIST. *J. Neurosci.* **24**, 7491–7502.

Daw, M.I., Chittajallu, R., Bortolotto, Z.A., Dev, K.K., Duprat, F., Henley, J.M. et al. (2000). PDZ proteins interacting with C-terminal GluR2/3 are involved in a PKC-dependent regulation of AMPA receptors at hippocampal synapses. *Neuron* **28**, 873–886.

De Zeeuw, C.I., Hansel, C., Bian, F., Koekkoek, S.K., van Alphen, A.M., Linden, D.J., Oberdick, J. (1998). Expression of a protein kinase C inhibitor in Purkinje cells blocks cerebellar LTD and adaptation of the vestibulo-ocular reflex. *Neuron* **20**, 495–508.

Derkach, V., Barria, A., Soderling, T.R. (1999). Ca2+/calmodulin-kinase II enhances channel conductance of alpha-amino-3-hydroxy-5-methyl-4-isoxazolepropionate type glutamate receptors. *Proc. Natl. Acad. Sci. U S A* **96**, 3269–3274.

Dev, K.K., Nishimune, A., Henley, J.M., Nakanishi, S. (1999). The protein kinase C alpha binding protein PICK1 interacts with short but not long form alternative splice variants of AMPA receptor subunits. *Neuropharmacology* **38**, 635–644.

Dingledine, R., Borges, K., Bowie, D., Traynelis, S.F. (1999). The glutamate receptor ion channels. *Pharmacol. Rev.* **51**, 7–61.

Dong, H., O'Brien, R.J., Fung, E.T., Lanahan, A.A., Worley, P.F., Huganir, R.L. (1997). GRIP: A synaptic PDZ domain-containing protein that interacts with AMPA receptors. *Nature* **386**, 279–284.

Dudek, S.M., Bear, M.F. (1992). Homosynaptic long-term depression in area CA1 of hippocampus and effects of N-methyl-D-aspartate receptor blockade. *Proc. Natl. Acad. Sci. U S A* **89**, 4363–4367.

Ehlers, M.D. (2000). Reinsertion or degradation of AMPA receptors determined by activity-dependent endocytic sorting. *Neuron* **28**, 511–525.

El-Husseini, A.E., Schnell, E., Chetkovich, D.M., Nicoll, R.A., Bredt, D.S. (2000). PSD-95 involvement in maturation of excitatory synapses. *Science* **290**, 1364–1368.

El-Husseini A.E., Schnell, E., Dakoji, S., Sweeney, N., Zhou, Q., Prange, O. et al. (2002). Synaptic strength regulated by palmitate cycling on PSD-95. *Cell* **108**, 849–863.

Esteban, J.A., Shi, S.H., Wilson, C., Nuriya, M., Huganir, R.L., Malinow, R. (2003). PKA phosphorylation of AMPA receptor subunits controls synaptic trafficking underlying plasticity. *Nat. Neurosci.* **6**, 136–143.

Fukata, Y., Tzingounis, A.V., Trinidad, J.C., Fukata, M., Burlingame, A.L., Nicoll, R.A., Bredt, D.S. (2005). Molecular constituents of neuronal AMPA receptors. *J. Cell Biol.* **169**, 399–404.

Gallo, V., Upson, L.M., Hayes, W.P., Vyklicky, L., Jr., Winters, C.A., Buonanno, A. (1992). Molecular cloning and development analysis of a new glutamate receptor subunit isoform in cerebellum. *J. Neurosci.* **12**, 1010–1023.

Gardner, S.M., Takamiya, K., Xia, J., Suh, J.G., Johnson, R., Yu, S., Huganir, R.L. (2005). Calcium-permeable AMPA receptor plasticity is mediated by subunit-specific interactions with PICK1 and NSF. *Neuron* **45**, 903–915.

Gardoni, F., Mauceri, D., Fiorentini, C., Bellone, C., Missale, C., Cattabeni, F., Di Luca, M. (2003). CaMKII-dependent phosphorylation regulates SAP97/NR2A interaction. *J. Biol. Chem.* **278**, 44745–44752.

Goldstein, L.S., Yang, Z. (2000). Microtubule-based transport systems in neurons: The roles of kinesins and dyneins. *Annu. Rev. Neurosci.* **23**, 39–71.

Gray, E.G. (1959). Axo-somatic and axo-dendritic synapses of the cerebral cortex: An electron microscope study. *J. Anat.* **93**, 420–433.

Greengard, P., Jen, J., Nairn, A.C., Stevens, C.F. (1991). Enhancement of the glutamate response by cAMP-dependent protein kinase in hippocampal neurons. *Science* **253**, 1135–1138.

Greger, I.H., Khatri, L., Kong, X., Ziff, E.B. (2003). AMPA receptor tetramerization is mediated by Q/R editing. *Neuron* **40**, 763–774.

Greger, I.H., Khatri, L., Ziff, E.B. (2002). RNA editing at arg607 controls AMPA receptor exit from the endoplasmic reticulum. *Neuron* **34**, 759–772.

Hanley, J.G., Khatri, L., Hanson, P.I., Ziff, E.B. (2002). NSF ATPase and alpha-/beta-SNAPs disassemble the AMPA receptor-PICK1 complex. *Neuron* **34**, 53–67.

Hashimoto, K., Fukaya, M., Qiao, X., Sakimura, K., Watanabe, M., Kano, M. (1999). Impairment of AMPA receptor function in cerebellar granule cells of ataxic mutant mouse stargazer. *J. Neurosci.* **19**, 6027–6036.

Hayashi, Y., Shi, S.H., Esteban, J.A., Piccini, A., Poncer, J.C., Malinow, R. (2000). Driving AMPA receptors into synapses by LTP and CaMKII: requirement for GluR1 and PDZ domain interaction. *Science* **287**, 2262–2267.

Henley, J.M., Nishimune, A., Nash, S.R., Nakanishi, S. (1997). Use of the two-hybrid system to find novel proteins that interact with alpha-amino-3-hydroxy-5-methyl-4-isoxazole propionate (AMPA) receptor subunits. *Biochem. Soc. Trans.* **25**, 838–842.

Heynen, A.J., Quinlan, E.M., Bae, D.C., Bear, M.F. (2000). Bidirectional, activity-dependent regulation of glutamate receptors in the adult hippocampus in vivo. *Neuron* **28**, 527–536.

Hollmann, M. (1999). Structure of ionotropic glutamate receptors. *In* Ionotropic glutamate receptors in the CNS. Handbook of experimental pharmacology, P. Jonas, H. Monyer, eds., 3–98. Berlin, Springer Verlag.

Hollmann, M., Boulter, J., Maron, C., Heinemann, S. (1994a). Molecular biology of glutamate receptors. Potentiation of N-methyl-D-aspartate receptor splice variants by zinc. *Ren. Physiol. Biochem.* **17**, 182–183.

Hollmann, M., Hartley, M., Heinemann, S. (1991). Ca2+ permeability of KA-AMPA—Gated glutamate receptor channels depends on subunit composition. *Science* **252**, 851–853.

Hollmann, M., Maron, C., Heinemann, S. (1994b). N-glycosylation site tagging suggests a three transmembrane domain topology for the glutamate receptor GluR1. *Neuron* **13**, 1331–1343.

Hoogenraad, C.C., Milstein, A.D., Ethell, I.M., Henkemeyer, M., Sheng, M. (2005). GRIP1 controls dendrite morphogenesis by regulating EphB receptor trafficking. *Nat. Neurosci.* **8**, 906–915.

Hoover, K.B., Bryant, P.J. (2000). The genetics of the protein 4.1 family: Organizers of the membrane and cytoskeleton. *Curr. Opin. Cell Biol.* **12**, 229–234.

Hrabetova, S., Serrano, P., Blace, N., Tse, H.W., Skifter, D.A., Jane, D.E. et al. (2000). Distinct NMDA receptor subpopulations contribute to long-term potentiation and long-term depression induction. *J. Neurosci.* **20**, RC81.

Ives, J.H., Fung, S., Tiwari, P., Payne, H.L., Thompson, C.L. (2004). Microtubule-associated protein light chain 2 is a stargazin-AMPA receptor complex-interacting protein in vivo. *J. Biol. Chem.* **279**, 31002–31009.

Kakegawa, W., Tsuzuki, K., Yoshida, Y., Kameyama, K., Ozawa, S. (2004). Input- and subunit-specific AMPA receptor trafficking underlying long-term potentiation at hippocampal CA3 synapses. *Eur. J. Neurosci.* **20**, 101–110.

Kawasaki, H., Fujii, H., Gotoh, Y., Morooka, T., Shimohama, S., Nishida, E., Hirano, T. (1999). Requirement for mitogen-activated protein kinase in cerebellar long term depression. *J. Biol. Chem.* **274**, 13498–13502.

Keinanen, K., Wisden, W., Sommer, B., Werner, P., Herb, A., Verdoorn, T.A. et al. (1990). A family of AMPA-selective glutamate receptors. *Science* **249**, 556–560.

Kharazia, V.N., Wenthold, R.J., Weinberg, R.J. (1996). GluR1-immunopositive interneurons in rat neocortex. *J. Comp. Neurol.* **368**, 399–412.

Kim, C.H., Chung, H.J., Lee, H.K., Huganir, R.L. (2001). Interaction of the AMPA receptor subunit GluR2/3 with PDZ domains regulates hippocampal long-term depression. *Proc. Natl. Acad. Sci. U S A* **98**, 11725–11730.

Kim, C.H., Takamiya, K., Petralia, R.S., Sattler, R., Yu, S., Zhou, W. et al. (2005a). Persistent hippocampal CA1 LTP in mice lacking the C-terminal PDZ ligand of GluR1. *Nat. Neurosci.* **8**, 985–987.

Kim, J.H., Liao, D., Lau, L.F., Huganir, R.L. (1998). SynGAP: A synaptic RasGAP that associates with the PSD-95/SAP90 protein family. *Neuron* **20**, 683–691.

Kim, M.J., Dunah, A.W., Wang, Y.T., Sheng, M. (2005b). Differential roles of NR2A- and NR2B-containing NMDA receptors in Ras-ERK signaling and AMPA receptor trafficking. *Neuron* **46**, 745–760.

Klein, R. (2004). Eph/ephrin signaling in morphogenesis, neural development and plasticity. *Curr. Opin. Cell Biol.* **16**, 580–589.

Klugbauer, N., Dai, S., Specht, V., Lacinova, L., Marais, E., Bohn, G., Hofmann, F. (2000). A family of gamma-like calcium channel subunits. *FEBS Lett* **470**, 189–197.

Kohler, M., Kornau, H.C., Seeburg, P.H. (1994). The organization of the gene for the functionally dominant alpha-amino-3-hydroxy-5-methylisoxazole-4-propionic acid receptor subunit GluR-B. *J. Biol. Chem.* **269**, 17367–17370.

Krapivinsky, G., Medina, I., Krapivinsky, L., Gapon, S., Clapham, D.E. (2004). SynGAP-MUPP1-CaMKII synaptic complexes regulate p38 MAP kinase activity and NMDA receptor-dependent synaptic AMPA receptor potentiation. *Neuron* **43**, 563–574.

Kuusinen, A., Abele, R., Madden, D.R., Keinanen, K. (1999). Oligomerization and ligand-binding properties of the ectodomain of the alpha-amino-3-hydroxy-5-methyl-4-isoxazole propionic acid receptor subunit GluRD. *J. Biol. Chem.* **274**, 28937–28943.

Lai, M.M., Hong, J.J., Ruggiero, A.M., Burnett, P.E., Slepnev, V.I., De Camilli, P., Snyder, S.H. (1999). The calcineurin-dynamin 1 complex as a calcium sensor for synaptic vesicle endocytosis. *J. Biol. Chem.* **274**, 25963–25966.

Lee, H.K., Barbarosie, M., Kameyama, K., Bear, M.F., Huganir, R.L. (2000). Regulation of distinct AMPA receptor phosphorylation sites during bidirectional synaptic plasticity. *Nature* **405**, 955–959.

Lee, H.K., Takamiya, K., Han, J.S., Man, H., Kim, C.H., Rumbaugh, G. et al. (2003). Phosphorylation of the AMPA receptor GluR1 subunit is required for synaptic plasticity and retention of spatial memory. *Cell* **112**, 631–643.

Lee, S.H., Liu, L., Wang, Y.T., Sheng, M. (2002). Clathrin adaptor AP2 and NSF interact with overlapping sites of GluR2 and play distinct roles in AMPA receptor trafficking and hippocampal LTD. *Neuron* **36**, 661–674.

Lee, S.H., Simonetta, A., Sheng, M. (2004). Subunit rules governing the sorting of internalized AMPA receptors in hippocampal neurons. *Neuron* **43**, 221–236.

Leonard, A.S., Davare, M.A., Horne, M.C., Garner, C.C., Hell, J.W. (1998). SAP97 is associated with the alpha-amino-3-hydroxy-5-methylisoxazole-4-propionic acid receptor GluR1 subunit. *J. Biol. Chem.* **273**, 19518–19524.

Leslie, K.R., Nelson, S.B., Turrigiano, G.G. (2001). Postsynaptic depolarization scales quantal amplitude in cortical pyramidal neurons. *J. Neurosci.* **21**, RC170.

Letts, V.A., Felix, R., Biddlecome, G.H., Arikkath, J., Mahaffey, C.L., Valenzuela, A. et al. (1998). The mouse stargazer gene encodes a neuronal Ca2+-channel gamma subunit. *Nat. Genet.* **19**, 340–347.

Leuschner, W.D., Hoch, W. (1999). Subtype-specific assembly of alpha-amino-3-hydroxy-5-methyl-4-isoxazole propionic acid receptor subunits is mediated by their n-terminal domains. *J. Biol. Chem.* **274**, 16907–16916.

Lin, J.W., Ju, W., Foster, K., Lee, S.H., Ahmadian, G., Wyszynski, M. et al. (2000). Distinct molecular mechanisms and divergent endocytotic pathways of AMPA receptor internalization. *Nat. Neurosci.* **3**, 1282–1290.

Linden, D.J., Connor, J.A. (1991). Participation of postsynaptic PKC in cerebellar long-term depression in culture. *Science* **254**, 1656–1659.

Lisman, J. (1989). A mechanism for the Hebb and the anti-Hebb processes underlying learning and memory. *Proc. Natl. Acad. Sci. U S A* **86**, 9574–9578.

Lisman, J., Malenka, R.C., Nicoll, R.A., Malinow, R. (1997). Learning mechanisms: The case for CaM-KII. *Science* **276**, 2001–2002.

Lissin, D.V., Carroll, R.C., Nicoll, R.A., Malenka, R.C., von Zastrow, M. (1999). Rapid, activation-induced redistribution of ionotropic glutamate receptors in cultured hippocampal neurons. *J. Neurosci.* **19**, 1263–1272.

Lissin, D.V., Gomperts, S.N., Carroll, R.C., Christine, C.W., Kalman, D., Kitamura, M. et al. (1998). Activity differentially regulates the surface expression of synaptic AMPA and NMDA glutamate receptors. *Proc. Natl. Acad. Sci. U S A* **95**, 7097–7102.

Liu, L., Wong, T.P., Pozza, M.F., Lingenhoehl, K., Wang, Y., Sheng, M. et al. (2004). Role of NMDA receptor subtypes in governing the direction of hippocampal synaptic plasticity. *Science* **304**, 1021–1024.

Liu, S.J., Cull-Candy, S.G. (2005). Subunit interaction with PICK and GRIP controls Ca2+ permeability of AMPARs at cerebellar synapses. *Nat. Neurosci.* **8**, 768–775.

Liu, S.Q., Cull-Candy, S.G. (2000). Synaptic activity at calcium-permeable AMPA receptors induces a switch in receptor subtype. *Nature* **405**, 454–458.

Lomeli, H., Mosbacher, J., Melcher, T., Hoger, T., Geiger, J.R., Kuner, T. et al. (1994). Control of kinetic properties of AMPA receptor channels by nuclear RNA editing. *Science* **266**, 1709–1713.

Luscher, C., Xia, H., Beattie, E.C., Carroll, R.C., von Zastrow, M., Malenka, R.C., Nicoll, R.A. (1999). Role of AMPA receptor cycling in synaptic transmission and plasticity. *Neuron* **24**, 649–658.

Luthi, A., Chittajallu, R., Duprat, F., Palmer, M.J., Benke, T.A., Kidd, F.L. et al. (1999). Hippocampal LTD expression involves a pool of AMPARs regulated by the NSF-GluR2 interaction. *Neuron* **24**, 389–399.

Malenka, R.C., Kauer, J.A., Perkel, D.J., Nicoll, R.A. (1989). The impact of postsynaptic calcium on synaptic transmission—Its role in long-term potentiation. *Trends Neurosci.* **12**, 444–450.

Malenka, R.C., Nicoll, R.A. (1999). Long-term potentiation—A decade of progress? *Science* **285**, 1870–1874.

Malinow, R., Madison, D.V., Tsien, R.W. (1988). Persistent protein kinase activity underlying long-term potentiation. *Nature* **335**, 820–824.

Malinow, R., Mainen, Z.F., Hayashi, Y. (2000). LTP mechanisms: from silence to four-lane traffic. *Curr. Opin. Neurobiol.* **10**, 352–357.

Malinow, R., Schulman, H., Tsien, R.W. (1989). Inhibition of postsynaptic PKC or CaMKII blocks induction but not expression of LTP. *Science* **245**, 862–866.

Mammen, A.L., Kameyama, K., Roche, K.W., Huganir, R.L. (1997). Phosphorylation of the alpha-amino-3-hydroxy-5-methylisoxazole4-propionic acid receptor GluR1 subunit by calcium/cal-modulin-dependent kinase II. *J. Biol. Chem.* **272**, 32528–32533.

Man, H.Y., Lin, J.W., Ju, W.H., Ahmadian, G., Liu, L., Becker, L.E. et al. (2000). Regulation of AMPA receptor-mediated synaptic transmission by clathrin-dependent receptor internalization. *Neuron* **25**, 649–662.

Man, H.Y., Wang, Q., Lu, W.Y., Ju, W., Ahmadian, G., Liu, L. et al. (2003). Activation of PI3-kinase is required for AMPA receptor insertion during LTP of mEPSCs in cultured hippocampal neurons. *Neuron* **38**, 611–624.

Mano, I., Teichberg, V.I. (1998). A tetrameric subunit stoichiometry for a glutamate receptor-channel complex. *Neuroreport* **9**, 327–331.

Mansour, M., Nagarajan, N., Nehring, R.B., Clements, J.D., Rosenmund, C. (2001). Heteromeric AMPA receptors assemble with a preferred subunit stoichiometry and spatial arrangement. *Neuron* **32**, 841–853.

Marszalek, J.R., Weiner, J.A., Farlow, S.J., Chun, J., Goldstein, L.S. (1999). Novel dendritic kinesin sorting identified by different process targeting of two related kinesins: KIF21A and KIF21B. *J. Cell Biol.* **145**, 469–479.

Martin, L.J., Blackstone, C.D., Levey, A.I., Huganir, R.L., Price, D.L. (1993). AMPA glutamate receptor subunits are differentially distributed in rat brain. *Neuroscience* **53**, 327–358.

Matsuda, S., Launey, T., Mikawa, S., Hirai, H. (2000). Disruption of AMPA receptor GluR2 clusters following long-term depression induction in cerebellar Purkinje neurons. *EMBO. J.* **19**, 2765–2774.

Matsuda, S., Mikawa, S., Hirai, H. (1999). Phosphorylation of serine-880 in GluR2 by protein kinase C prevents its C terminus from binding with glutamate receptor-interacting protein. *J. Neurochem.* **73**, 1765–1768.

Mauceri, D., Cattabeni, F., Di Luca, M., Gardoni, F. (2004). Calcium/calmodulin-dependent protein kinase II phosphorylation drives synapse-associated protein 97 into spines. *J. Biol. Chem.* **279**, 23813–23821.

Meng, Y., Zhang, Y., Jia, Z. (2003). Synaptic transmission and plasticity in the absence of AMPA glutamate receptor GluR2 and GluR3. *Neuron* **39**, 163–176.

Mi, R., Sia, G.M., Rosen, K., Tang, X., Moghekar, A., Black, J.L. et al. (2004). AMPA receptor-dependent clustering of synaptic NMDA receptors is mediated by Stargazin and NR2A/B in spinal neurons and hippocampal interneurons. *Neuron* **44**, 335–349.

Mi, R., Tang, X., Sutter, R., Xu, D., Worley, P., O'Brien, R.J. (2002). Differing mechanisms for glutamate receptor aggregation on dendritic spines and shafts in cultured hippocampal neurons. *J. Neurosci.* **22**, 7606–7616.

Miller, K.D. (1996). Synaptic economics: competition and cooperation in synaptic plasticity. *Neuron* **17**, 371–374.

Molloy, S.S., Anderson, E.D., Jean, F., Thomas, G. (1999). Bi-cycling the furin pathway: From TGN localization to pathogen activation and embryogenesis. *Trends Cell Biol.* **9**, 28–35.

Molnar, E., Baude, A., Richmond, S.A., Patel, P.B., Somogyi, P., McIlhinney, R.A. (1993). Biochemical and immunocytochemical characterization of antipeptide antibodies to a cloned GluR1 glutamate receptor subunit: Cellular and subcellular distribution in the rat forebrain. *Neuroscience* **53**, 307–326.

Monyer, H., Seeburg, P.H., Wisden, W. (1991). Glutamate-operated channels: developmentally early and mature forms arise by alternative splicing. *Neuron* **6**, 799–810.

Mosbacher, J., Schoepfer, R., Monyer, H., Burnashev, N., Seeburg, P.H., Ruppersberg, J.P. (1994). A molecular determinant for submillisecond desensitization in glutamate receptors. *Science* **266**, 1059–1062.

Mulkey, R.M., Endo, S., Shenolikar, S., Malenka, R.C. (1994). Involvement of a calcineurin/inhibitor-1 phosphatase cascade in hippocampal long-term depression. *Nature* **369**, 486–488.

Mulkey, R.M., Herron, C.E., Malenka, R.C. (1993). An essential role for protein phosphatases in hippocampal long-term depression. *Science* **261**, 1051–1055.

Mulkey, R.M., Malenka, R.C. (1992). Mechanisms underlying induction of homosynaptic long-term depression in area CA1 of the hippocampus. *Neuron* **9**, 967–975.

Muller, B.M., Kistner, U., Veh, R.W., Cases-Langhoff, C., Becker, B., Gundelfinger, E.D., Garner, C.C. (1995). Molecular characterization and spatial distribution of SAP97, a novel presynaptic protein homologous to SAP90 and the Drosophila discs-large tumor suppressor protein. *J. Neurosci.* **15**, 2354–2366.

Nakagawa, T., Cheng, Y., Ramm, E., Sheng, M., Walz, T. (2005). Structure and different conformational states of native AMPA receptor complexes. *Nature* **433**, 545–549.

Nakagawa, T., Futai, K., Lashuel, H.A., Lo, I., Okamoto, K., Walz, T. et al. (2004). Quaternary structure, protein dynamics, and synaptic function of SAP97 controlled by L27 domain interactions. *Neuron* **44**, 453–467.

Nicoll, R.A. (2003). Expression mechanisms underlying long-term potentiation: A postsynaptic view. *Philos. Trans. R. Soc. Lond. B. Biol. Sci.* **358**, 721–726.

Nishimune, A., Isaac, J.T., Molnar, E., Noel, J., Nash, S.R., Tagaya, M. et al. (1998). NSF binding to GluR2 regulates synaptic transmission. *Neuron* **21**, 87–97.

Noel, J., Ralph, G.S., Pickard, L., Williams, J., Molnar, E., Uney, J.B. et al. (1999). Surface expression of AMPA receptors in hippocampal neurons is regulated by an NSF-dependent mechanism. *Neuron* **23**, 365–376.

Nusser, Z. (2000). AMPA and NMDA receptors: Similarities and differences in their synaptic distribution. *Curr. Opin. Neurobiol.* **10**, 337–341.

Nusser, Z., Lujan, R., Laube, G., Roberts, J.D., Molnar, E., Somogyi, P. (1998). Cell type and pathway dependence of synaptic AMPA receptor number and variability in the hippocampus. *Neuron* **21**, 545–559.

O'Brien, R., Xu, D., Mi, R., Tang, X., Hopf, C., Worley, P. (2002). Synaptically targeted narp plays an essential role in the aggregation of AMPA receptors at excitatory synapses in cultured spinal neurons. *J. Neurosci.* **22**, 4487–4498.

O'Brien, R.J., Kamboj, S., Ehlers, M.D., Rosen, K.R., Fischbach, G.D., Huganir, R.L. (1998). Activity-dependent modulation of synaptic AMPA receptor accumulation. *Neuron* **21**, 1067–1078.

O'Brien, R.J., Xu, D., Petralia, R.S., Steward, O., Huganir, R.L., Worley, P. (1999). Synaptic clustering of AMPA receptors by the extracellular immediate-early gene product Narp. *Neuron* **23**, 309–323.

Oh, M.C., Derkach, V.A. (2005). Dominant role of the GluR2 subunit in regulation of AMPA receptors by CaMKII. *Nat Neurosci.* **8**, 853–854.

Oliet, S.H., Malenka, R.C., Nicoll, R.A. (1997). Two distinct forms of long-term depression coexist in CA1 hippocampal pyramidal cells. *Neuron* **18**, 969–982.

Osten, P., Khatri, L., Perez, J.L., Kohr, G., Giese, G., Daly, C. et al. (2000). Mutagenesis reveals a role for ABP/GRIP binding to GluR2 in synaptic surface accumulation of the AMPA receptor. *Neuron* **27**, 313–325.

Osten, P., Srivastava, S., Inman, G.J., Vilim, F.S., Khatri, L., Lee, L.M. et al. (1998). The AMPA receptor GluR2 C terminus can mediate a reversible, ATP-dependent interaction with NSF and alpha- and beta-SNAPs. *Neuron* **21**, 99–110.

Ozawa, S., Kamiya, H., Tsuzuki, K. (1998). Glutamate receptors in the mammalian central nervous system. *Prog. Neurobiol.* **54**, 581–618.

Partin, K.M., Fleck, M.W., Mayer, M.L. (1996). AMPA receptor flip/flop mutants affecting deactivation, desensitization, and modulation by cyclothiazide, aniracetam, and thiocyanate. *J. Neurosci.* **16**, 6634–6647.

Passafaro, M., Piech, V., Sheng, M. (2001). Subunit-specific temporal and spatial patterns of AMPA receptor exocytosis in hippocampal neurons. *Nat. Neurosci.* **4**, 917–926.

Pawson, T., Nash, P. (2003). Assembly of cell regulatory systems through protein interaction domains. *Science* **300**, 445–452.

Perez, J.L., Khatri, L., Chang, C., Srivastava, S., Osten, P., Ziff, E.B. (2001). PICK1 targets activated protein kinase C alpha to AMPA receptor clusters in spines of hippocampal neurons and reduces surface levels of the AMPA-type glutamate receptor subunit 2. *J. Neurosci.* **21**, 5417–5428.

Petralia, R.S., Wang, Y.X., Wenthold, R.J. (2003). Internalization at glutamatergic synapses during development. *Eur. J. Neurosci.* **18**, 3207–3217.

Petralia, R.S., Wenthold, R.J. (1992). Light and electron immunocytochemical localization of AMPA-selective glutamate receptors in the rat brain. *J. Comp. Neurol.* **318**, 329–354.

Priel, A., Kolleker, A., Ayalon, G., Gillor, M., Osten, P., Stern-Bach, Y. (2005). Stargazin reduces desensitization and slows deactivation of the AMPA-type glutamate receptors. *J. Neurosci.* **25**, 2682–2686.

Racz, B., Blanpied, T.A., Ehlers, M.D., Weinberg, R.J. (2004). Lateral organization of endocytic machinery in dendritic spines. *Nat. Neurosci.* **7**, 917–918.

Rasse, T.M., Fouquet, W., Schmid, A., Kittel, R.J., Mertel, S., Sigrist, C.B. et al. (2005). Glutamate receptor dynamics organizing synapse formation in vivo. *Nature Neurosci.* **8**, 898–905.

Roche, K.W., O'Brien, R.J., Mammen, A.L., Bernhardt, J., Huganir, R.L. (1996). Characterization of multiple phosphorylation sites on the AMPA receptor GluR1 subunit. *Neuron* **16**, 1179–1188.

Rongo, C., Whitfield, C.W., Rodal, A., Kim, S.K., Kaplan, J.M. (1998). LIN-10 is a shared component of the polarized protein localization pathways in neurons and epithelia. *Cell* **94**, 751–759.

Rosenmund, C., Stern-Bach, Y., Stevens, C.F. (1998). The tetrameric structure of a glutamate receptor channel. *Science* **280**, 1596–1599.

Rothman, J.E. (1994). Mechanisms of intracellular protein transport. *Nature* **372**, 55–63.

Ruberti, F., Dotti, C.G. (2000). Involvement of the proximal C terminus of the AMPA receptor subunit GluR1 in dendritic sorting. *J. Neurosci.* **20**, RC78.

Rubio, M.E., Wenthold, R.J. (1999). Calnexin and the immunoglobulin binding protein (BiP) coimmunoprecipitate with AMPA receptors. *J. Neurochem.* **73**, 942–948.

Rumbaugh, G., Sia, G.M., Garner, C.C., Huganir, R.L. (2003). Synapse-associated protein-97 isoform-specific regulation of surface AMPA receptors and synaptic function in cultured neurons. *J. Neurosci.* **23**, 4567–4576.

Sampo, B., Kaech, S., Kunz, S., Banker, G. (2003). Two distinct mechanisms target membrane proteins to the axonal surface. *Neuron* **37**, 611–624.

Sans, N., Racca, C., Petralia, R.S., Wang, Y.X., McCallum, J., Wenthold, R.J. (2001). Synapse-associated protein 97 selectively associates with a subset of AMPA receptors early in their biosynthetic pathway. *J. Neurosci.* **21**, 7506–7516.

Sawada, Y., Nakamura, K., Doi, K., Takeda, K., Tobiume, K., Saitoh, M. et al. (2001). Rap1 is involved in cell stretching modulation of p38 but not ERK or JNK MAP kinase. *J. Cell Sci.* **114**, 1221–1227.

Schiavo, G., Gmachl, M.J., Stenbeck, G., Sollner, T.H., Rothman, J.E. (1995). A possible docking and fusion particle for synaptic transmission. *Nature* **378**, 733–736.

Schnell, E., Sizemore, M., Karimzadegan, S., Chen, L., Bredt, D.S., Nicoll, R.A. (2002). Direct interactions between PSD-95 and stargazin control synaptic AMPA receptor number. *Proc. Natl. Acad. Sci. U S A* **99**, 13902–13907.

Scott, D.B., Blanpied, T.A., Swanson, G.T., Zhang, C., Ehlers, M.D. (2001). An NMDA receptor ER retention signal regulated by phosphorylation and alternative splicing. *J. Neurosci.* **21**, 3063–3072.

Seeburg, P.H. (1996). The role of RNA editing in controlling glutamate receptor channel properties. *J. Neurochem.* **66**, 1–5.

Seeburg, P.H. (2002). A-to-I editing: new and old sites, functions and speculations. *Neuron* **35**, 17–20.

Seger, R., Krebs, E.G. (1995). The MAPK signaling cascade. *FASEB. J.* **9**, 726–735.

Setou, M., Seog, D.H., Tanaka, Y., Kanai, Y., Takei, Y., Kawagishi, M., Hirokawa, N. (2002). Glutamate-receptor-interacting protein GRIP1 directly steers kinesin to dendrites. *Nature* **417**, 83–87.

Shen, L., Liang, F., Walensky, L.D., Huganir, R.L. (2000). Regulation of AMPA receptor GluR1 subunit surface expression by a 4. 1N-linked actin cytoskeletal association. *J. Neurosci.* **20**, 7932–7940.

Shi, S., Hayashi, Y., Esteban, J.A., Malinow, R. (2001). Subunit-specific rules governing AMPA receptor trafficking to synapses in hippocampal pyramidal neurons. *Cell* **105**, 331–343.

Shi, S.H., Hayashi, Y., Petralia, R.S., Zaman, S.H., Wenthold, R.J., Svoboda, K., Malinow, R. (1999). Rapid spine delivery and redistribution of AMPA receptors after synaptic NMDA receptor activation. *Science* **284**, 1811–1816.

Silva, A.J., Stevens, C.F., Tonegawa, S., Wang, Y. (1992). Deficient hippocampal long-term potentiation in alpha-calcium-calmodulin kinase II mutant mice. *Science* **257**, 201–206.

Slepnev, V.I., Ochoa, G.C., Butler, M.H., Grabs, D., De Camilli, P. (1998). Role of phosphorylation in regulation of the assembly of endocytic coat complexes. *Science* **281**, 821–824.

Snyder, E.M., Philpot, B.D., Huber, K.M., Dong, X., Fallon, J.R., Bear, M.F. (2001). Internalization of ionotropic glutamate receptors in response to mGluR activation. *Nat. Neurosci.* **4**, 1079–1085.

Sommer, B., Keinanen, K., Verdoorn, T.A., Wisden, W., Burnashev, N., Herb, A. et al. (1990). Flip and flop: A cell-specific functional switch in glutamate-operated channels of the CNS. *Science* **249**, 1580–1585.

Sommer, B., Kohler, M., Sprengel, R., Seeburg, P.H. (1991). RNA editing in brain controls a determinant of ion flow in glutamate-gated channels. *Cell* **67**, 11–19.

Song, I., Kamboj, S., Xia, J., Dong, H., Liao, D., Huganir, R.L. (1998). Interaction of the N-ethylmaleimide-sensitive factor with AMPA receptors. *Neuron* **21**, 393–400.

Spacek, J., Harris, K.M. (1997). Three-dimensional organization of smooth endoplasmic reticulum in hippocampal CA1 dendrites and dendritic spines of the immature and mature rat. *J. Neurosci.* **17**, 190–203.

Srivastava, S., Osten, P., Vilim, F.S., Khatri, L., Inman, G., States, B. et al. (1998). Novel anchorage of

GluR2/3 to the postsynaptic density by the AMPA receptor-binding protein ABP. *Neuron* 21, 581–591.

Standley, S., Roche, K.W., McCallum, J., Sans, N., Wenthold, R.J. (2000). PDZ domain suppression of an ER retention signal in NMDA receptor NR1 splice variants. *Neuron* 28, 887–898.

Staudinger, J., Zhou, J., Burgess, R., Elledge, S.J., Olson, E.N. (1995). PICK1: A perinuclear binding protein and substrate for protein kinase C isolated by the yeast two-hybrid system. *J. Cell Biol.* 128, 263–271.

Steinberg, J.P., Huganir, R.L., Linden, D.J. (2004). N-ethylmaleimide-sensitive factor is required for the synaptic incorporation and removal of AMPA receptors during cerebellar long-term depression. *Proc. Natl. Acad. Sci. U S A* 101, 18212–18216.

Stern-Bach, Y., Bettler, B., Hartley, M., Sheppard, P.O., O'Hara, P.J., Heinemann, S.F. (1994). Agonist selectivity of glutamate receptors is specified by two domains structurally related to bacterial amino acid-binding proteins. *Neuron* 13, 1345–1357.

Stowell, J.N., Craig, A.M. (1999). Axon/dendrite targeting of metabotropic glutamate receptors by their cytoplasmic carboxy-terminal domains. *Neuron* 22, 525–536.

Stricker, N.L., Huganir, R.L. (2003). The PDZ domains of mLin-10 regulate its trans-Golgi network targeting and the surface expression of AMPA receptors. *Neuropharmacology* 45, 837–848.

Swanson, G.T., Kamboj, S.K., Cull-Candy, S.G. (1997). Single-channel properties of recombinant AMPA receptors depend on RNA editing, splice variation, and subunit composition. *J. Neurosci.* 17, 58–69.

Takumi, Y., Ramirez-Leon, V., Laake, P., Rinvik, E., Ottersen, O.P. (1999). Different modes of expression of AMPA and NMDA receptors in hippocampal synapses. *Nat. Neurosci.* 2, 618–624.

Tan, S.E., Wenthold, R.J., Soderling, T.R. (1994). Phosphorylation of AMPA-type glutamate receptors by calcium/calmodulin-dependent protein kinase II and protein kinase C in cultured hippocampal neurons. *J. Neurosci.* 14, 1123–1129.

Tardin, C., Cognet, L., Bats, C., Lounis, B., Choquet, D. (2003). Direct imaging of lateral movements of AMPA receptors inside synapses. *EMBO. J.* 22, 4656–4665.

Terashima, A., Cotton, L., Dev, K.K., Meyer, G., Zaman, S., Duprat, F. et al. (2004). Regulation of synaptic strength and AMPA receptor subunit composition by PICK1. *J. Neurosci.* 24, 5381–5390.

Thomas, P., Mortensen, M., Hosie, A.M., Smart, T.G. (2005). Dynamic mobility of functional GABA(A) receptors at inhibitory synapses. *Nat. Neurosci.* 8, 889–897.

Tomita, S., Adesnik, H., Sekiguchi, M., Zhang, W., Wada, K., Howe, J.R. et al. (2005a). Stargazin modulates AMPA receptor gating and trafficking by distinct domains. *Nature* 435, 1052–1058.

Tomita, S., Chen, L., Kawasaki, Y., Petralia, R.S., Wenthold, R.J., Nicoll, R.A., Bredt, D.S. (2003). Functional studies and distribution define a family of transmembrane AMPA receptor regulatory proteins. *J. Cell Biol.* 161, 805–816.

Tomita, S., Fukata, M., Nicoll, R.A., Bredt, D.S. (2004). Dynamic interaction of stargazin-like TARPs with cycling AMPA receptors at synapses. *Science* 303, 1508–1511.

Tomita, S., Stein, V., Stocker, T.J., Nicoll, R.A., Bredt, D.S. (2005b). Bidirectional synaptic plasticity regulated by phosphorylation of stargazin-like TARPs. *Neuron* 45, 269–277.

Tovar, K.R., Westbrook, G.L. (2002). Mobile NMDA receptors at hippocampal synapses. *Neuron* 34, 255–264.

Triller, A., Choquet, D. (2005). Surface trafficking of receptors between synaptic and extrasynaptic membranes: and yet they do move! *Trends Neurosci.* 28, 133–139.

Tsui, C.C., Copeland, N.G., Gilbert, D.J., Jenkins, N.A., Barnes, C., Worley, P.F. (1996). Narp, a novel member of the pentraxin family, promotes neurite outgrowth and is dynamically regulated by neuronal activity. *J. Neurosci.* 16, 2463–2478.

Turrigiano, G.G., Leslie, K.R., Desai, N.S., Rutherford, L.C., Nelson, S.B. (1998). Activity-dependent scaling of quantal amplitude in neocortical neurons. *Nature* 391, 892–896.

Turrigiano, G.G., Nelson, S.B. (2004). Homeostatic plasticity in the developing nervous system. *Nat. Rev. Neurosci.* 5, 97–107.

Valtschanoff, J.G., Burette, A., Davare, M.A., Leonard, A.S., Hell, J.W., Weinberg, R.J. (2000). SAP97 concentrates at the postsynaptic density in cerebral cortex. *Eur. J. Neurosci.* 12, 3605–3614.

Vandenberghe, W., Nicoll, R.A., Bredt, D.S. (2005a). Interaction with the unfolded protein response reveals a role for stargazin in biosynthetic AMPA receptor transport. *J. Neurosci.* 25, 1095–1102.

Vandenberghe, W., Nicoll, R.A., Bredt, D.S. (2005b). Stargazin is an AMPA receptor auxiliary subunit. *Proc. Natl. Acad. Sci. U S A* 102, 485–490.

Verdoorn, T.A., Burnashev, N., Monyer, H., Seeburg, P.H., Sakmann, B. (1991). Structural determinants of ion flow through recombinant glutamate receptor channels. *Science* 252, 1715–1718.

Walsh, M.J., Kuruc, N. (1992). The postsynaptic density: constituent and associated proteins characterized by electrophoresis, immunoblotting, and peptide sequencing. *J. Neurochem.* 59, 667–678.

Wang, L.Y., Dudek, E.M., Browning, M.D., MacDonald, J.F. (1994). Modulation of AMPA/kainate receptors in cultured murine hippocampal neurones by protein kinase C. *J. Physiol.* 475, 431–437.

Wang, Y.T., Linden, D.J. (2000). Expression of cerebellar long-term depression requires postsynaptic clathrin-mediated endocytosis. *Neuron* 25, 635–647.

Wenthold, R.J., Petralia, R.S., Blahos, J., II, Niedzielski, A.S. (1996). Evidence for multiple AMPA receptor complexes in hippocampal CA1/CA2 neurons. *J. Neurosci.* **16**, 1982–1989.

Wo, Z.G., Oswald, R.E. (1994). Transmembrane topology of two kainate receptor subunits revealed by N-glycosylation. *Proc. Natl. Acad. Sci. U S A* **91**, 7154–7158.

Wu, H., Nash, J.E., Zamorano, P., Garner, C.C. (2002). Interaction of SAP97 with minus-end-directed actin motor myosin VI. Implications for AMPA receptor trafficking. *J. Biol. Chem.* **277**, 30928–30934.

Wyszynski, M., Kim, E., Yang, F.C., Sheng, M. (1998). Biochemical and immunocytochemical characterization of GRIP, a putative AMPA receptor anchoring protein, in rat brain. *Neuropharmacology* **37**, 1335–1344.

Wyszynski, M., Valtschanoff, J.G., Naisbitt, S., Dunah, A.W., Kim, E., Standaert, D.G. et al. (1999). Association of AMPA receptors with a subset of glutamate receptor-interacting protein in vivo. *J. Neurosci.* **19**, 6528–6537.

Xia, J., Chung, H.J., Wihler, C., Huganir, R.L., Linden, D.J. (2000). Cerebellar long-term depression requires PKC-regulated interactions between GluR2/3 and PDZ domain-containing proteins. *Neuron* **28**, 499–510.

Xia, J., Zhang, X., Staudinger, J., Huganir, R.L. (1999). Clustering of AMPA receptors by the synaptic PDZ domain-containing protein PICK1. *Neuron* **22**, 179–187.

Yamazaki, M., Ohno-Shosaku, T., Fukaya, M., Kano, M., Watanabe, M., Sakimura, K. (2004). A novel action of stargazin as an enhancer of AMPA receptor activity. *Neurosci. Res.* **50**, 369–374.

Yang, S.N., Tang, Y.G., Zucker, R.S. (1999). Selective induction of LTP and LTD by postsynaptic [Ca2+]i elevation. *J. Neurophysiol.* **81**, 781–787.

Ye, B., Liao, D., Zhang, X., Zhang, P., Dong, H., Huganir, R.L. (2000). GRASP-1: A neuronal RasGEF associated with the AMPA receptor/GRIP complex. *Neuron* **26**, 603–617.

Yoshimura, Y., Ohmura, T., Komatsu, Y. (2003). Two forms of synaptic plasticity with distinct dependence on age, experience, and NMDA receptor subtype in rat visual cortex. *J. Neurosci.* **23**, 6557–6566.

Zamanillo, D., Sprengel, R., Hvalby, O., Jensen, V., Burnashev, N., Rozov, A. et al. (1999). Importance of AMPA receptors for hippocampal synaptic plasticity but not for spatial learning. *Science* **284**, 1805–1811.

Zhou, Q., Xiao, M., Nicoll, R.A. (2001). Contribution of cytoskeleton to the internalization of AMPA receptors. *Proc. Natl. Acad. Sci. U S A* **98**, 1261–1266.

Zhu, J.J., Esteban, J.A., Hayashi, Y., Malinow, R. (2000). Postnatal synaptic potentiation: delivery of GluR4-containing AMPA receptors by spontaneous activity. *Nat. Neurosci.* **3**, 1098–1106.

Zhu, J.J., Qin, Y., Zhao, M., Van Aelst, L., Malinow, R. (2002). Ras and Rap control AMPA receptor trafficking during synaptic plasticity. *Cell* **110**, 443–455.

10

Subunit-Specific NMDA Receptor Trafficking to Synapses

ANDRES BARRIA

I. INTRODUCTION

The N-methyl-D-aspartate receptor (NMDA-R) is a member of the glutamate-activated ion channel family. The receptor forms a cation-selective channel with high calcium permeability that is tightly regulated by oxidizing agents, protons, zinc, polyamines, protein kinases, calmodulin, and most notably magnesium (for a review see Dingledine et al. 1999; Hollmann and Heinemann 1994). At resting membrane potentials, the channel is blocked by physiological concentration of extracellular magnesium in manner a strongly dependent on voltage. Partial depolarization of the plasma membrane relieves the magnesium block and allows the flux of ions through the channel (Mayer et al. 1984; Wollmuth et al. 1998). It is this property that allows the receptor to be a coincidence detector of pre- and postsynaptic activity required in Hebbian models of plasticity (Bliss et al. 2003). Varied levels of synaptic NMDA-R activation with corresponding degrees of calcium influx can result in multiple effects: low levels of NMDA-R activation may produce depression of synaptic transmission, higher levels of activation with a larger influx of calcium may potentiate synaptic transmission (Cummings et al. 1996; Zucker 1999), and very high levels of calcium influx can result in cell death (Choi 1995). Thus, the number and properties of NMDA-R at a particular synapse must be well controlled in order to regulate the amount of calcium entry.

It is now clear that the NMDA receptor is a tetrameric complex (Dingledine et al. 1999; Rosenmund et al. 1998; Wollmuth and Sobolevsky 2004) and an obligate heteromultimer composed of NR1 subunits and one or more NR2 subunits (Ishii et al. 1993; Meguro et al. 1992; Monyer et al. 1992). NR1 subunits contain a binding site for glycine, and NR2 subunits are responsible for glutamate binding. The NR1 subunit is ubiquitous and encoded by a single gene that gives rise to eight different splice variants (reviewed in Zukin and Bennett 1995). Some properties, such as modulation by zinc, polyamines, and protein kinase C (PKC), are modified by incorporation of different splice variants into the NMDA-R complex (Dingledine et al. 1999; Zukin and Bennett 1995). Four different genes encode the NR2 subunits (NR2A-D) and their expression is developmentally and regionally regulated (Dingledine et al. 1999; Monyer et al. 1994; Sheng et al. 1994). NR2 subunits modify the biophysical properties of the channel such as conductance, mean open time of the channel, and sensitivity to external magnesium (Dingledine et al. 1999; Ishii et al. 1993; Monyer et al. 1992). NR3 subunits have been shown to coassemble with NR1 and NR2 subunits and modulate some channel properties (Ciabarra et al. 1995; Das et al. 1998). Thus, a receptor complex assembled with different combinations of NR1, NR2, and NR3 subunits allows for considerable regulation of channel properties.

The postsynaptic density is a proteinaceous specialization underneath the postsynaptic membrane of excitatory synapses that is composed of different ionotropic and metabotropic receptors, scaffolding proteins, and signaling complexes (Kennedy 1997; Walikonis et al. 2000). The NMDA-R is an integral part of the postsynaptic density and it is essential for synapse formation and maturation of neuronal connectivity. The targeting of the NMDA-R to synapses and its stabilization there depends on a series of interactions with other proteins. The cytoplasmic carboxy termini of synaptic receptors like GABA receptors and AMPA-type glutamate receptors play critical roles in directing their trafficking to, and stabilization at, synaptic sites (Calver et al. 2001; Osten et al. 2000; Pagano et al. 2001; Shi et al. 2001). Similarly, considerable evidence, including studies with transgenic animals (Mori et al. 1998; Sprengel et al. 1998; Steigerwald et al. 2000), indicates that the carboxy termini domain of NMDA-R subunits control their trafficking and stabilization at synapses. The carboxy terminal domain of NMDA-R subunits are intracellular domains capable of interacting with different scaffolding and signaling proteins (Dingledine et al. 1999; Sheng and Sala 2001). In particular, the carboxy termini of NR2 subunits influence trafficking, localization, and internalization of the receptor. Despite the fact that the global homology among NR2 subunits is very high, at the carboxy terminal region it is only approximately 20 to 30 percent (Ishii et al. 1993; Meguro et al. 1992; Monyer et al. 1992), suggesting the potential for many interactions with regulatory proteins in a subunit-specific manner. Thus, different subunits not only regulate the properties of the receptor, but also regulate trafficking, surface expression, and insertion at synapses, as well as intracellular signaling by recruiting enzymes to the postsynaptic density.

II. ASSEMBLY OF NMDA-Rs

Assembly of NMDA-Rs occurs early in the secretory pathway as individual subunits are synthesized and folded in the endoplasmic reticulum (ER). Only correctly folded and assembled receptors are exported from the ER, whereas partially or incorrectly assembled complexes are retained in the ER and degraded (Ellgaard and Helenius 2003). In order to form a functional receptor complex, the NR1 subunit must assemble with one of the four NR2 subunits (Dingledine et al. 1999; Ishii et al. 1993; Meguro et al. 1992; Monyer et al. 1992). NR3 subunits may also be incorporated, resulting in a receptor with diminished activity (Das et al. 1998). Since only NR2 subunits contain a glutamate binding site, receptors composed of NR1 and NR3 alone are considered functional glycine receptors (Chatterton et al. 2002). Studies of recombinant or native receptors using pharmacological, electrophysiological, or biochemical techniques have shown the existence of both heterodimeric (NR1/NR2) and heterotrimeric (NR1/NR2B/NR2D; NR1/NR2A/NR2B; NR1/NR2A/NR3A) assemblies (Brickley et al. 2003; Kew et al. 1998; Luo et al. 1997; Monyer et al. 1994; Perez-Otano et al. 2001; Sheng et al. 1994; Tovar and Westbrook 1999; Vicini et al. 1998), although the far more abundant NMDA-Rs are heterodimeric NR1/NR2 complexes.

Neurons have an excess of intracellular NR1 subunits relative to NR2 subunits. Roughly 50 percent of the NR1 present in neurons exist unassembled as a monomer in intracellular compartments, not accessible to surface labels (Hall and Soderling 1997). This intracellular pool is degraded rapidly if it is not assembled into complexes (Chazot and Stephenson 1997; Huh and Wenthold 1999). In contrast, NR2 subunits are always present as heteromers with NR1, and over 90 percent are present at the surface of the cell (Hall and Soderling 1997). The most widely distributed NR2 subunits are NR2A and NR2B. The NR2A and NR2B subunits have long intracellular tails that enable many protein-protein interactions. NR2A and NR2B subunits affect the biophysical properties of the NMDA-R complex with NR2A conferring faster kinetics of

decay than NR2B (Dingledine et al. 1999). The NR2B subunit is expressed prenatally and is required for normal neuronal pattern formation and viability of the animal (Kutsuwada et al. 1996), and the NR2A subunit progressively increases its expression (Monyer et al. 1994; Sheng et al. 1994) and synaptic incorporation (Flint et al. 1997; Stocca and Vicini 1998; Tovar and Westbrook 1999) during development. NMDA-R mediated excitatory postsynaptic currents (EPSCs) acquire a faster decay time with age. These faster kinetics result in decreased calcium influx, resulting in the hypothesis that the change in incorporation of NR2B to NR2A subunits in synapses may be responsible for the lack of plasticity observed in older animals (Carmignoto and Vicini 1992; Crair and Malenka 1995; Philpot et al. 2001a; Quinlan et al. 2004). NMDA-Rs containing the NR2C subunit are characterized by a low sensitivity to blockade by magnesium and largely are restricted to the cerebellum (Dingledine et al. 1999; Ishii et al. 1993; Monyer et al. 1992). The NR2D subunit is most highly expressed early in development and confers a prolonged deactivation and a low magnesium sensitivity and conductance to the receptor (Dingledine et al. 1999).

Once synthesized and assembled, NMDA-Rs are transported in vesicles along dendrites using microtubules and adaptor proteins such as mLin10 that couple the vesicle to kinesin motors (Setou et al. 2000) (see Chapters 2, 4, and 5). Shortly after an axon contacts a dendrite, mobile transport packages containing NMDA-Rs are recruited to the nascent synapse (Washbourne et al. 2002). These new synapses do not always have a postsynaptic density opposed to the release site, suggesting a role for the NMDA-R in recruiting proteins and serving as a focal point for the assembly of the postsynaptic specialization. In addition, NMDA-Rs can be removed or inserted into existing synapses, modifying the type of receptor present and affecting the biophysical parameters of NMDA-R responses. Because of the large intracellular carboxy termini of NR2 subunits, they likely have a role in regulating the trafficking of NMDA-R complexes.

III. EXITING THE ENDOPLASMIC RETICULUM—THE ROLE OF NR1 SUBUNIT

NMDA-Rs undergo folding, maturation, and oligomerization in the endoplasmic reticulum (ER). Molecular chaperones retain unfolded, misfolded, and unassembled proteins in the ER (Ellgaard and Helenius 2003). Some multimeric proteins require only chaperone-based quality control for proper oligomerization. However, many transmembrane proteins use an ER retention signal consisting of a di-lysine motif (KKXX-COOH) or a di-arginine motif (NH2-XXRR) (Teasdale and Jackson 1996). A steric masking mechanism allows only the fully assembled protein to be released from the retention signal, exit the ER, and reach the plasma membrane. Thus, correct assembly of receptor subunits is a mechanism that masks the retention signal and allows the complex to exit from the ER. Another ER retention signal, RXR, has been defined in multitransmembrane domain proteins such as the ATP-sensitive potassium channel (Zerangue et al. 1999).

The NR1 subunit also has an RXR retention signal in its carboxy terminal domain (intralumenal). Thus, when expressed alone in heterologous cells, the NR1 subunit does not form functional receptors and does not reach the cell surface (McIlhinney et al. 1998) or the synapse (Barria and Malinow 2002). However, certain variants of NR1 have an alternatively spliced form that overrides the retention signal, allowing the receptor to reach the surface (Ehlers et al. 1995; Okabe et al. 1999). The NR1 gene has a total of 22 exons (Hollmann et al. 1993), three of which (exons 5, 21, 22) undergo alternative splicing to generate the eight

NR1 splice variants (reviewed in Zukin and Bennett 1995). Exon 5 encodes a splice cassette of 21 amino acids, termed N1, that is inserted in the amino terminus domain. Exons 21 and 22 encode two independent splice cassettes of 37 amino acids (C1) and 38 amino acids (C2), respectively, that make the last stretch of the carboxy terminus domain. Splice variants in the carboxy terminus are generated by deletion of exon 21 and/or make use of an alternative splice site within exon 22. Thus, the Cl and/or C2 cassette can be removed. Removal of the exon segment that encodes the C2 insert results in a new open reading frame that encodes an unrelated sequence of 22 amino acids (C2') before a second stop codon is reached. Importantly, the last six amino acids of this new sequence is a conserved PDZ binding domain. Thus, the NR1 splice variants differ in two important regions: the distal end of the amino terminus domain with two configurations, and the carboxy terminus with four configurations (C1-C2; C1-C2'; C2; C2').

The most abundant variant of NR1 contains cassettes C1 and C2. This variant, named NR1-1, does not reach the cell surface when expressed without NR2 in heterologous cells and is retained in the ER (McIlhinney et al. 1996). When expressed alone, NR1-1 is not inserted at synapses, but is instead retained in the dendritic shaft (Barria and Malinow 2002), most likely because the ER retention signal of NR1 is found on the C1 cassette (Scott et al. 2001; Standley et al. 2000; Xia et al. 2001), and homodimers of NR1-1 (Meddows et al. 2001) therefore cannot mask the ER retention signal. The other three C-tail splice variants of NR1 reach the cell surface. Thus, the ER retention signal present on C1 explains the retention of NR1-1 (C1-C2), and its absence in C2 explains the cell surface expression of NR1-2 (C2) and NR1-4 (C2'). However, NR1-3 (C1-C2') is expressed on the cell surface suggesting that another signal, present in C2', can override the RXR retention signal in C1. This overriding signal is a PDZ-binding domain at the distal end of C2′ (STVV) (Standley et al. 2000). Two mechanisms may explain the override of the retention signal. One possibility is that the PDZ proteins interacting with C2′ cassette mask the RXR motif. Another possibility is that the PDZ protein may be acting as an export signal that simply overrides the ER retention signal. The last six amino acids of C2′ are sufficient to suppress ER retention. Thus, soluble fusion proteins containing the PDZ interacting domain of C2′ effectively blocks the surface expression of NR1-4, suggesting that saturation of the trafficking pathway leads to an intracellular buildup of NR1-4 (Holmes et al. 2002). Interestingly, interaction of NR1-4 with NR1-1 results in limited surface expression of NR1-1, suggesting that the subunits interact and disrupt the ER retention of NR1-1 (Okabe et al. 1999). The interaction of a PDZ protein early in the secretory pathway provides an interesting mechanism by which the receptor overcomes ER retention (see Fig. 10.1).

IV. RELEASING THE RETENTION FROM THE ER—THE ROLE OF NR2 SUBUNITS

NR2 subunits do not form functional receptors when expressed alone in heterologous cells (Monyer et al. 1992), and those that have been studied (NR2A and NR2B) do not reach the cell surface (McIlhinney et al. 1998). Similarly, NR2 subunits do not appear to function independently of NR1, since no NR2 homodimers are found on neurons (Hall and Soderling 1997). The NR2 subunit is retained in the ER when expressed in heterologous cells, and like NR1, its retention depends on a signal in its carboxy terminus, although no peptide sequence has been identified. Recent studies using a chimera of the NR2B

carboxy terminal domain and the α chain of the interleukin-2 receptor (Tac) suggest that either multiple sites are required or that a more complex structure is responsible for the retention (Hawkins et al. 2004). Expression of tagged NR2 subunits can reach the surface of hippocampal cells and be inserted at synapses likely due to assembly of NR2 subunits (NR2A or NR2B) with endogenous NR1 subunits (Barria and Malinow, personal observation). Thus, NR1 retention may serve as a checkpoint such that NR1 subunits may remain in the ER and become associated with newly synthesized NR2 subunits to allow ER exit. Thus, coexpression of NR2 subunits (NR2A or NR2B) along with NR1 subunits allows the receptor to be assembled and reach the cell surface (McIlhinney et al. 1996, 1998), or to reach dendritic spines and be inserted at synapses (Barria and Malinow 2002), indicating that assembly of the complex suppresses the retention signal from NR1. The mechanism by which assembly overcomes ER retention of both subunits is not clear, but the experimental evidence does not support a simple masking of the ER retention signal via complementary cytoplasmic tails (Hawkins et al. 2004; Krupp et al. 1999; Sprengel et al. 1998; Vissel et al. 2001; Zheng et al. 1999). An HLFY motif that follows immediately after transmembrane domain 4 of the NR2 subunits functions as a necessary signal for ER release of the assembled functional NMDA-R (Hawkins et al. 2004). In addition, NR2 subunits also have a PDZ binding motif at the distal end of the carboxy terminal tail, suggesting a mechanism similar to the one used by the NR1 subunit containing an alternatively spliced C2′ cassette. NR1 subunits also can assemble with NR3 subunits, which is sufficient to relieve the ER retention of NR1 and to allow the trafficking of the receptor to the cytoplasmic membrane. However, this assembly produces functional glycine receptors (Perez-Otano et al. 2001).

V. TRAFFICKING FROM THE ER TO THE PLASMA MEMBRANE

After they are released from the ER, NMDA-Rs are further processed in the Golgi apparatus and then distributed to the *trans*-Golgi network (TGN) and endosomes (see Fig. 10.1). Dendrites, and in some cases even dendritic spines, contain functional ER and Golgi complexes. Thus some receptor processing may occur near the synapse, providing a mechanism for a rapid and local response (see Chapter 8).

The kinesin, KIF17, interacts directly with NMDA-Rs (Setou et al. 2000). A complex of mLin-7, mLin-2, and mLin-10 associates with a cargo vesicle containing NMDA-Rs and with KIF17, which transports it along microtubules to the synapse. Little is known about the nature or formation of vesicles that contain newly processed receptors. However, because it would require creating a large number of discrete vesicles, it is unlikely that each type of receptor is parceled into its own vesicle population.

PDZ proteins interact with the NMDA-R before it reaches the synapse, and this interaction is involved in the synaptic delivery of NMDA-Rs (Setou et al. 2000; Standley et al. 2000). A yeast two-hybrid analysis showed that one of the PDZ partners of the NR2 subunit, SAP102, interacts through its PDZ domains with sec8, a protein of the exocyst or sec6/8 complex (Sans et al. 2003). The exocyst is a complex of eight proteins that was shown to be involved in targeting of secretory vesicles to the plasma membrane (Hsu et al. 1999) (see Chapter 13). The fact that SAP102 interacts with sec8 suggests that the NMDA-R indirectly associates with the exocyst complex through SAP102. Several studies, including functional analyses, confirm that the NMDA-R, SAP102, and sec8 form a complex in the brain and suggest that formation of this complex is involved in the synaptic delivery of NMDA-Rs. Interestingly, SAP102 is developmentally regulated, being highly expressed early in development at the same time as NR2B (Petralia et al. 2005; Sans et al. 2000). With age, Sap102 expression decreases as the expression of PSD-95 and NR2A increases. Coimmunoprecipitation studies in the adult hippocampus have shown a preferential interaction between NR2B and Sap102 and PSD95 with NR2A. This suggests that a mechanism used by NR2 subunits to control the trafficking of NMDA-R complexes involves association with different PDZ proteins that are developmentally regulated. However, receptors without the PDZ interacting domain are able to bypass the exocyst interaction and reach the cell surface (Sans et al. 2000). Experiments using a tagged NMDA-R also have shown that complexes composed of NR1 and NR2 subunits (NR2A or NR2B) lacking the PDZ binding domain can be inserted at synapses with a 50 percent efficiency (Barria and Malinow 2002). Together, the evidence indicates that there may be two mechanisms for delivery of NMDA-Rs to the cell surface, one of which involves the PDZ interacting domain and one that does not.

VI. TARGETING NMDA RECEPTORS TO THE SYNAPSE

NMDA-Rs play many important roles at synapses and their number and composition are tightly regulated. Insertion at synapses happens after NMDA-Rs are assembled and released from the ER, however the exact pathway of insertion is not known. Receptors may be inserted at the cell surface in the dendritic shaft where they wait for the proper signal to be incorporated into synapses in a lateral movement within the plane of the lipid bilayer (see Chapter 1). Alternatively, they may be inserted directly into the synapse from intracellular compartments (see Fig. 10.1). It is important to point out that both pathways are not necessarily exclusive, and both may be regulated depending on

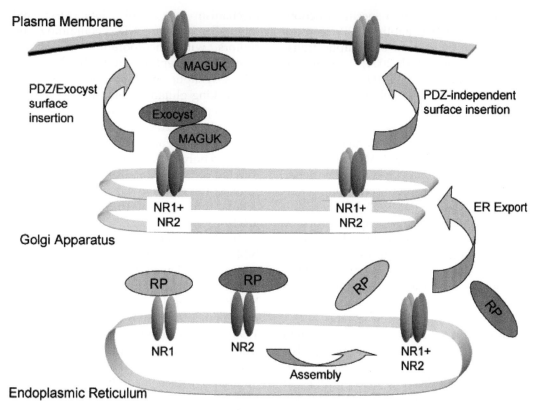

FIGURE 10.1. **Controlled exit from the endoplasmic reticulum and surface insertion.** Unassembled NR1 and NR2 subunits are retained in the endoplasmic reticulum (ER) through C-tail retention signals and unknown retention proteins (RP). Upon assembly, the heterologous NMDA-R is released and exits the ER. Interaction of NR2 subunits with the Exocyst complex via MAKUS like SAP-102 and PSD-95 allows further trafficking and insertion into the plasma membrane. A pathway that does not require PDZ-interacting domains has also been described.

subunit composition, cell type, or developmental expression of other interacting proteins.

At early stages of development, extrasynaptic receptors composed of NR1 and NR2B greatly outnumber synaptic receptors (Tovar and Westbrook 1999). As synapses develop, the number of extrasynaptic receptors decreases and NMDA-Rs are found mainly at synapses. The increase in synaptic NMDA-Rs is accompanied by a decline in NR2B expression and an increase in receptors containing NR2A (Liao et al. 1999; Rosenmund et al. 1995; Tovar and Westbrook 1999). However, functional and immunocytochemical studies have shown that NMDA-Rs are still present at extrasy-

naptic sites. These extrasynaptic receptors contain NR2B subunits that have the ability to diffuse laterally in and out of synapses, whereas NR2A has closer association with the synapse (Rumbaugh and Vicini 1999; Tovar and Westbrook 1999). Receptors containing NR2D subunits are kept at extrasynaptic sites (Vicini and Rumbaugh 2000). Perhaps the extrasynaptic population simply represents receptors that have been delivered to the plasma membrane and are awaiting incorporation into the synapse. NMDA-Rs can rapidly move between synaptic and extrasynaptic sites, likely by lateral diffusion of the receptors (Tovar and Westbrook 2002). However, it is possible that there is a distinct population of extrasynaptic receptors with a specific function. For example, extrasynaptic NR2B receptors may induce depression of synaptic transmission (Massey et al. 2004). Evidence of selective coupling of NMDA channels at extrasynaptic sites to inhibitory currents may limit excitation during periods of intense activity (Isaacson and Murphy 2001). NR2A and NR2B confer different kinetic properties to the NMDA-R. NR2A-containing receptors have a faster decay time than those receptors containing NR2B subunits (Meguro et al. 1992; Monyer et al. 1992). This electrophysiological property has been used to distinguish the type of receptor present at synapses or at extrasynaptic sites. The decay of NMDA-R EPSCs is faster during development in the central nervous system and correlates with an increase in expression of the NR2A subunit (Flint et al. 1997) and a decrease in sensitivity to NR2B-selective antagonists (Rumbaugh and Vicini 1999; Stocca and Vicini 1998), indicating incorporation of NR2A into the synapse. However, extrasynaptic receptors are still blocked by these agents (Momiyama 2000; Rumbaugh and Vicini 1999; Tovar and Westbrook 1999). These results are consistent with synaptic and extrasynaptic receptor pools having different subunit compositions and forming distinct populations of receptors.

NR2A- and NR2B-containing receptors are added to the synapse using different mechanisms. NR2B-containing receptors are delivered into synapses in a constitutive form that is not limited by synaptic activity and does not affect the kinetics of NMDA-Rs EPSCs, indicating that NR2B receptors are replacing only preexisting NR2B receptors. In contrast, incorporation of NR2A-containing receptors into synapses requires ligand binding to preexisting synaptic NMDA-Rs. The incorporation of NR2A subunits decreases synaptic transmission and the decay time of NMDA-R EPSCs, indicating that NR2B receptors are being replaced (Barria and Malinow 2002). This observation suggests the existence of a

mechanism that couples ligand-triggered internalization of NR2B-containing receptors with incorporation of NR2A-containing receptors. Interestingly, when NR2A synaptic incorporation is inhibited by blocking glutamate binding to preexisting receptors, NR2A-containing receptors accumulate intracellularly at dendritic spines. Thus, ligand binding appears to be required only for incorporation of receptors into the synapse. Moreover, NR2A is incorporated directly into synapses from intracellular compartments (Barria and Malinow 2002).

Transgenic mice have been used to study NMDA-R trafficking and targeting to synapses (reviewed in Sprengel and Single 1999). Knockout mice lacking NR1 and NR2B subunits die shortly after birth due to developmental failure (Kutsuwada et al. 1996). It is possible to prepare cell cultures from these mice by hand-feeding pups for a few days. Microisland cultures that form autaptic synapses made from mice lacking NR2B subunits revealed the existence of NMDA-R responses after five to six days in culture. These responses are significantly faster than normal, indicating that they are composed of NR2A-containing receptors. Thus, in this special preparation, functional NR2A subunits can be delivered to the synapse in the absence of the NR2B subunit (Tovar et al. 2000). Mice lacking NR2A subunits are viable, have NMDA-Rs with slow kinetics, and are characterized by reduced long-term potentiation (LTP) (Kadotani et al. 1996; Sakimura et al. 1995). Mice lacking the NR2C subunit (that is confined to the cerebellum) have increased NMDA-R-EPSCs with faster kinetics (Ebralidze et al. 1996). Uncoordinated motor behavior is observed with loss of both NR2A and NR2C, and deletion of either individual subunit results in normal motor behavior (Kadotani et al. 1996). NR2D knockout mice develop normally, but have deficits in certain locomotor activities and monoamine metabolism (Ikeda et al. 1995; Miyamoto et al. 2002). The ability of mice that lack different NR2 subunits to form functional synapses implies that subunit trafficking to the synapse can occur effectively even in the absence of the normal complement of NR2 subunits. However, because compensatory mechanisms may be involved in these genetically modified animals, it is difficult to address exactly the role of different subunits under normal conditions. Transgenic mice expressing NR2A with a deletion of its carboxy terminus are characterized by an absence of NMDA-Rs at the synapse, and the receptors that are expressed are limited to extrasynaptic sites (Steigerwald et al. 2000). Mice expressing truncated NR2B subunits have a lethal phenotype, but cultures made from these animals show a decrease in localization of NR2B containing receptors at synapses (Mori et al. 1998; Sprengel et al. 1998). These transgenic mice lost a large region of the NR2 carboxy terminus, therefore the region responsible for synaptic delivery cannot be determined from these studies.

Receptor clustering at the synapse involves interactions of the receptor with proteins that are part of the PSD. Identification of the interacting proteins, and how these interactions are regulated, is central to understanding the trafficking of NMDA-Rs at the synapse. Major components of the PSD are proteins of the MAGUK family (PSD-95, PSD-93, SAP97, and SAP102) (Sheng and Sala 2001) that bind directly to NMDA-Rs through their PDZ domains (Kornau et al. 1997). This basic scaffold stabilizes the receptor at the synapse and allows other proteins to be recruited, forming the post-synaptic specialization. The PDZ-binding domain on the NR2 subunits is a likely candidate for controlling synaptic delivery because it mediates interaction with several MAGUK proteins that are localized to the synapse (Sheng and Sala 2001). However, as discussed previously, expression of truncated proteins in slices (Barria and Malinow 2002), or transgenic animals with deletions of the carboxy ter-

minus (Mori et al. 1998; Sprengel et al. 1998; Steigerwald et al. 2000) have not conclusively established their role in the delivery or stabilization of NMDA-Rs at the synapses. Moreover, studies in which some of the MAGUK proteins have been deleted or overexpressed also have proved inconclusive with regard to synaptic localization of NMDARs (El-Husseini et al. 2000; McGee et al. 2001; Migaud et al. 1998). Compensatory mechanisms for regulating MAGUKs may be confounding interpretation of these experiments (Misawa et al. 2001). The effect of NMDA-R subunit overexpression on NMDA-R properties in cultured cerebellar granule cells recently has been investigated. Overexpression of the NR1 subunit does not alter the total number of functional channels in neurons, but overexpression of either NR2A or NR2B subunits produces an increase in receptor number, implying that synthesis of NR2 subunits controls the number of functional channels expressed in these neurons (Prybylowski et al. 2002). However, the amplitude of NMDA-R-EPSCs is not increased by overexpression of NR2 subunits, indicating that subunit availability is not the major factor in determining the number of synaptic NMDA-Rs. Additionally, expression of NR1 and NR2B subunits in hippocampal slices does not increase the size of NMDA-R EPSCs, although expression of NR1/NR2A in organotypic slices decreases synaptic EPSCs. Variations in responses may be traced to differences in the two systems used (dissociated cells vs. cultured slices). In cerebellar granule cells, transfected NR2 subunits are targeted to the synapse because the deactivation kinetics of NMDA-EPSCs is controlled by the NR2 subunit overexpressed. In this system, targeting to the synapse is dependent on the PDZ binding domain (Prybylowski et al. 2002). This differs from expressing NR2 subunits in hippocampal slices, where lack of the PDZ binding domain reduces but does not eliminate incorporation of the receptor into synapses (Barria and Malinow

2002). Thus, the role of PDZ binding and MAGUK proteins in targeting NMDA-R to synapses remains unclear.

Other proteins that may link the NMDA-R to other glutamate receptors and to ion channels in the postsynaptic membrane have been described. For example, GKAP binds to the GK-like domain found in MAGUKs (Hirao et al. 1998) and to Shank. Shank binds to dimers of Homer that interact with type 1 metabotropic glutamate receptors in the postsynaptic membrane. Shank also binds to Homer dimers that interact with inositol 1,4,5-triphosphate (IP3) receptors in extensions of the reticulum that lie adjacent to the PSD (Petralia et al. 2001; Sheng 2001). NMDA-Rs may interact with AMPA receptors via a combination of calcium/calmodulin-dependent protein kinase II (CaMKII) and an assembly of proteins, consisting of actinin, actin, 4.1N protein, and SAP97/GluR1 (Lisman et al. 2002). NMDA and AMPA receptors may also be linked through PSD-95 and other MAGUKs via Stargazin, a protein that binds to the PDZ domains of MAGUKs and mediates synaptic targeting of AMPA receptors (Chen et al. 2000). The postsynaptic complex can be linked in several ways to actin filaments that control the overall structure of the postsynaptic spine and form pathways for transport of proteins to and from the postsynaptic membrane (Ehlers 2002). The NMDA-R directly interacts with the actin cytoskeleton through a complex of NMDA-R-actinin-actin. The interactions of NMDA-Rs with scaffolding proteins and with the actin cytoskeleton stabilize the receptor complex. It is predicted that diffusion is facilitated when the actin cytoskeleton is transiently dissociated. Whereas actin-protein associations play significant roles in synaptic structure and function, anchoring of NMDA-R/PSD-95 complexes at synapses are independent of actin associations (Allison et al. 2000). Moreover, studies of NR2B lateral movement revealed that this phenomenon is not affected by actin-depolymerizing

agents in immature cultured neurons (Tovar and Westbrook 2002).

VII. ACTIVITY-DEPENDENT CHANGES IN NMDA RECEPTOR SUBUNIT LOCALIZATION

Synaptic activity plays a major role in NMDA-R subunit exchange from receptors containing NR2B to those containing NR2A. This change in subunit composition not only occurs following spontaneous activity in organotypic slices (Barria and Malinow 2002), but also after sensory experience (Philpot et al. 2001a; Quinlan et al. 1999a,b), learning (Quinlan et al. 2004) and during development (Carmignoto and Vicini 1992; Monyer et al. 1994). Sensory experience can strongly regulate NMDA-R subunit expression and incorporation at the synapse (see Fig. 10.2). For example, ocular dominance (Bear and Rittenhouse 1999) modifies NR2 subunit composition at synapses. Rats reared in the dark have significantly lower expression of the NR2A subunit (Quinlan et al. 1999a), and visual experience decreases NR2B-antagonist sensitivity of EPSCs and shortens EPSC duration, suggesting NR2A incorporation into synapses (Philpot et al. 2001a). This change in subunit composition may explain the lack of plasticity in older animals.

Blockade of NMDA-R activity can cause a dramatic increase in expression of NR2 subunits and an increase in surface expression of the NR1 subunit (Crump et al. 2001; Follesa and Ticku 1996; Rao and Craig 1997). Blockade of synaptic activity or treatment with NMDA-R antagonists also has been shown to increase NMDA-R colocalization with PSD-95 (a marker of excitatory synapses) without an apparent increase in the number of total synapses (Rao and Craig 1997). Because significant synaptic rearrangement occurs in the absence of protein synthesis, changes in subunit trafficking may be involved (Crump et al. 2001). PKA may control this activity-

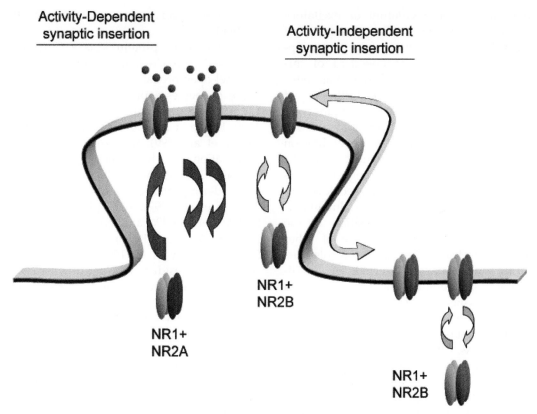

Activity-Dependent synaptic insertion

Activity-Independent synaptic insertion

NR1+ NR2A

NR1+ NR2B

NR1+ NR2B

FIGURE 10.2. **Model describing NR2-controlled synaptic incorporation of NMDA-Rs.** NR2B-containing receptors are incorporated into synapses in a manner that does not require synaptic activity and replaces only NR2B-containing receptors. The receptors could be incorporated directly into synapses or after extrasynaptic incorporation into the plasma membrane and lateral diffusion. As NR2A expression increases, NR2A-containing receptors are incorporated into synapses in a manner that requires ligand binding to synaptic receptors. NR2A-containing receptors replace synaptic NR2B-containing receptors and reduce synaptic NMDA-R responses.

dependent change in receptor localization downstream of NMDA-R activation. Long-term blockade of NMDA-Rs during development in culture leads to a substantial increase in synapse number and results in a more complex dendritic arborization of CA1 pyramidal cells (Luthi et al. 2001). The increase in synapse number is observed physiologically as a large increase in the frequency of NMDA-R-mediated miniature excitatory postsynaptic currents (NMDA-R mEPSCs) with no change in amplitude. In contrast, blockade of synaptic transmission with tetrodotoxin (TTX) leads to an increase in the amplitude of NMDA-R mEPSCs with no change in frequency (Luthi et al. 2001).

Changes in synaptic activity during the culture period induces parallel changes in AMPA and NMDA-R currents, implying that there is activity-dependent scaling of the number of AMPA-Rs and NMDA-Rs at the synapse to maintain a constant ratio between these two receptor types (Watt et al. 2000). Accordingly, induction of AMPA-R potentiation is followed with a delay by a proportional increase in NMDA-Rs (Watt et al. 2004).

Excitatory synaptic innervation is critical for clustering of AMPA receptors, although PSD-95 and NMDA-Rs can cluster in the absence of this input (Rao et al. 2000). Both synaptic activity and NMDA-R activation are critical regulators of excitatory synapses, but the initial clustering of NMDA-Rs and formation of the postsynaptic density may be independent of this activity. A similar conclusion has been reached when studying the trafficking of NMDA-Rs into synapses in organotypic slices. Insertion of NMDA-Rs into synaptic sites follows different rules depending on receptor subunit composition. Synaptic insertion of NR2B-containing receptors does not increase with increased levels of expression, nor does it require glutamatergic synaptic activity. Apposition of the presynaptic terminal or release of other substances (Dalva et al. 2000) controls clustering and synaptic insertion of NR2B-containing receptors. The data suggest a model in which newly formed synapses initially acquire NR2B-containing receptors in a manner not requiring glutamatergic transmission. As NR2A expression increases, either through a developmental program or increased neural activity (Quinlan et al. 1999b), NR2B-containing receptors are replaced by NR2A-containing receptors. This replacement is use-dependent, consistent with *in vivo* (Carmignoto and Vicini 1992; Quinlan et al. 1999a,b) and *in vitro* (Chavis and Westbrook 2001) studies. Interestingly, replacement of NR2B with NR2A reduces synaptic NMDA-R currents to single stimuli (Barria and Malinow 2002) as well as bursts (Philpot et al. 2001b) of afferent activity. The reduction in responses to single stimuli suggests that with activity, more NR2B-containing receptors are removed than NR2A-containing receptors are inserted, although these differences may be due to differences in NR2A/NR2B single-channel properties.

VIII. TRAFFICKING REGULATION BY PHOSPHORYLATION

Phosphorylation of NMDA-Rs by different kinases can modulate their biophysical properties (for a review see Dingledine et al. 1999), and may regulate NMDA-R trafficking.

Activation of protein kinase C (PKC) increases NMDA-elicited currents and channel-open probability in neurons and in *Xenopus* oocytes expressing NMDA receptors (Chen and Huang 1992; Gerber et al. 1989; Roche et al. 2001; Xiong et al. 1998; Zheng et al. 1999). Kinase-induced phosphorylation sites have been identified on the NR1, NR2A, and NR2B subunits (Hall and Soderling 1997; Leonard and Hell 1997; Tingley et al. 1997). PKC phosphorylates Ser890 and Ser896 and cAMP-dependent protein kinase (PKA) phosphorylates Ser897 (Tingley et al. 1997). Interestingly, these serine residues are just downstream from (and adjacent to) the ER-retention signal in the NR1 C-terminus. Phosphorylation of Ser890 within C1 by PKC disperses surface-associated clusters of NR1 subunits (Tingley et al. 1997), at apparent odds with observed PKC-induced potentiation of NMDA-receptor function. Thus, PKC modulates channel activity not only by changing intrinsic channel properties, but also by regulating receptor and channel trafficking. Two distinct mechanisms promote PKC-induced NMDA-R exocytosis. First, PKC-mediated phosphorylation of Ser896 and Ser897 increases levels of NMDA receptors at the surface of COS7 cells with a relatively slow time course of two to three hours (Scott et al. 2001). Mutation of the serine residues to alanine abolishes PKC-induced surface expression; mutation to negatively charged glutamate residues (which mimics phosphorylation) increases cell surface levels of NR1 (Scott et al. 2001). These data suggest that phosphorylation may disrupt the interaction of NR1 with the retention protein(s). PKC can also rapidly deliver new NMDA receptors to the cell surface of dendritic shafts and spines of hippocampal neurons in a rapid manner (within minutes) (Lan et al. 2001). This PKC-regulated receptor trafficking occurs via SNARE (synaptosome-associated-protein receptor)-dependent exocytosis. Studies involving site-directed mutagenesis and deletion indicate that PKC-mediated potentiation does not occur by direct phosphorylation of receptor subunits, but instead has been hypothesized to act by phosphorylating the trafficking protein SNAP-25.

PKA increases the activity of NMDA-Rs and phosphorylates NR1, NR2A, and NR2B (Leonard and Hell 1997) in addition to inducing synaptic targeting of NMDA-Rs (Crump et al. 2001). Protein phosphatases generally down-regulate the function of NMDA-Rs (Westphal et al. 1999). However, anchoring PKA to NMDA-Rs via Yotiao and the C1 exon cassette of NR1 allows it to overcome the constitutive activity of protein phosphatase type 1 (PP1), resulting in rapid enhancement of NMDA-R currents (Westphal et al. 1999).

Biochemical studies have demonstrated a high affinity binding between the catalytic domain of calcium-dependent kinase II (CaMKII) and the carboxy terminal tail of the NR2B subunit of the NMDA-R. CaMKII phosphorylates NR2B *in vitro* and *in vivo* (Omkumar et al. 1996) and both proteins interact (Bayer et al. 2001; Kim et al. 2005; Leonard et al. 1999; Strack et al. 2000) in a Ca^{2+}/CaM-dependent manner (Leonard et al. 1999; Strack and Colbran 1998; Strack et al. 2000). Interestingly, the interaction of CaMKII and NR2B initially requires Ca^{2+}/CaM to expose the CaMKII catalytic site, and then renders the enzyme constitutively active independent of its phosphorylation state, suppresses inhibitory autophosphorylation, and increases the affinity of calmodulin for the enzyme (Bayer et al. 2001). Given that persistently active CaMKII can mimic LTP, the association between NMDA-Rs and active CaMKII could be a key component of the persistently enhanced transmission (Lisman et al. 2002). Activated CaMKII that is anchored to NR2B subunits may phosphorylate synaptic AMPA-Rs (Barria et al. 1997; Lee et al. 2000), increasing their single-channel conductance (Derkach et al. 1999) and/or assemble membrane sites for newly deliv-

ered AMPA-Rs (Lisman et al. 2002). The role of CaMKII in trafficking of NMDA-Rs, or their stabilization at synapses, it is not clear.

IX. INTERNALIZATION OF NMDA RECEPTORS

Glutamate receptors at synapses seem to be in constant movement. However, NMDA-Rs are relatively more stable components of the postsynaptic membrane compared to AMPA-Rs. For example, the basal rate of endocytosis of surface AMPA receptors in cultured cortical neurons is nearly threefold that of NMDA-Rs (Ehlers 2000; Huh and Wenthold 1999). As previously mentioned, NR2B receptors can be inserted into synapses by replacing NR2B receptors in a constitutive manner. In contrast, NR2A-containing receptors replace NR2B in a process that requires ligand-binding to preexisting receptors (Barria and Malinow 2002). Thus, two different mechanisms are responsible for removal of receptors. The first does not require activity and exchanges NR2B with NR2B. The second engages an internalization process that is coupled to the insertion of NR2A. Studies in heterologous cells suggest a mechanism for internalization of NMDA-Rs that involves a ring of tyrosines on the C-termini of NMDA-R subunits, just distal to the last transmembrane domain (Vissel et al. 2001). Dephosphorylation of these tyrosines allows AP-2 binding, leading to clathrin-mediated endocytosis of the NMDA-R. Interestingly, agonist binding to the NMDA-R mediates this dephosphorylation independent of ion flux, suggesting that NMDA-Rs could trigger intracellular signal cascades and would behave, in this way, like a metabotropic glutamate receptor (mGluR). In fact, group 1 mGluRs also induce NMDA-R internalization, perhaps by indirectly affecting the binding of NMDA-Rs to PSD-95 (Snyder et al. 2001).

NMDA-R internalization is regulated by its association with PSD-95 and other PDZ proteins. NMDA-Rs that are bound to these proteins are stabilized and are less likely to be internalized, whereas NMDA-Rs that are not bound to these proteins are internalized readily. This is a useful mechanism for removing surface NMDA-Rs that are not tethered at the synapse. A chimera of the integral membrane protein Tac and the distal carboxy terminus of NR2B revealed that the last seven amino acids (the PDZ binding domain) are important in stabilizing the protein by interacting with MAGUK proteins such as PSD-95 (Roche et al. 2001). Control of internalization of NMDA-R internalization via binding to PSD-95 is likely a dynamic event, which may occur under conditions leading to the dissociation of PSD-95 from other components of the postsynaptic complex. Binding of NMDA-Rs to PSD-95 is affected by phosphorylation of the receptor. Interestingly, the main site of tyrosine phosphorylation of NR2B is Y1472, which is very close to the PDZ-binding domain on the C-terminal (Nakazawa et al. 2001). Increased phosphorylation of this site occurs following tetanic stimulation of Schaffer collaterals in the CA1 region of the hippocampus, and might interfere with PSD-95 binding, allowing the receptor to be more available for internalization. Indeed, Y1472 is part of a motif, YEKL, that may mediate clathrin-mediated endocytosis by binding to AP-2 adaptor complexes (Roche et al. 2001).

X. CONCLUSION

The NMDA-R is critically involved in synaptic transmission; therefore, it is not surprising that its trafficking is a tightly regulated process. NMDA-R subunit composition controls aspects of the biophysical properties of the ligand-activated ionotropic receptor, as well as its trafficking. The retention of NR1 in the ER is an important checkpoint, ensuring that only assembled heterocomplexes are processed. Distinct NR2 subunits can direct the receptor into synapses or to extrasynaptic sites. Despite the knowledge that NMDA-Rs interact with proteins that may aid in their trafficking, the precise role of these proteins in receptor delivery is unclear. The expression of NMDA-R interacting proteins is regionally and developmentally regulated, suggesting that elucidating the dynamics of these interactions will aid in our understanding of NMDA-R trafficking.

References

Allison, D.W., Chervin, A.S., Gelfand, V.I., Craig, A.M. (2000). Postsynaptic scaffolds of excitatory and inhibitory synapses in hippocampal neurons: Maintenance of core components independent of actin filaments and microtubules. *J. Neurosci.* **20**, 4545–4554.

Barria, A., Derkach, V., Soderling, T. (1997). Identification of the Ca2+/calmodulin-dependent protein kinase II regulatory phosphorylation site in the alpha-amino-3-hydroxyl-5-methyl-4-isoxazole-propionate-type glutamate receptor. *J. Biol. Chem.* **272**, 32727–32730.

Barria, A., Malinow, R. (2002). Subunit-specific NMDA receptor trafficking to synapses. *Neuron* **35**, 345–353.

Bayer, K.U., De Koninck, P., Leonard, A.S., Hell, J.W., Schulman, H. (2001). Interaction with the NMDA receptor locks CaMKII in an active conformation. *Nature* **411**, 801–805.

Bear, M.F., Rittenhouse, C.D. (1999). Molecular basis for induction of ocular dominance plasticity. *J. Neurobiol.* **41**, 83–91.

Benke, D., Honer, M., Michel, C., Bettler, B., Mohler, H. (1999). Gamma-aminobutyric acid type B receptor splice variant proteins GBR1a and GBR1b are both associated with GBR2 in situ and display differential regional and subcellular distribution. *J. Biol. Chem.* **274**, 27323–27330.

Bliss, T.V., Collingridge, G.L., Morris, R.G. (2003). Introduction. Long-term potentiation and structure of the issue. *Philos. Trans. R. Soc. Lond. B. Biol. Sci.* **358**, 607–611.

Brickley, S.G., Misra, C., Mok, M.H., Mishina, M., Cull-Candy, S.G. (2003). NR2B and NR2D subunits coassemble in cerebellar Golgi cells to form a distinct NMDA receptor subtype restricted to extrasynaptic sites. *J. Neurosci.* **23**, 4958–4966.

Calver, A.R., Robbins, M.J., Cosio, C., Rice, S.Q., Babbs, A.J., Hirst, W.D. et al. (2001). The C-terminal domains of the GABA(b) receptor subunits mediate intracellular trafficking but are not required for receptor signaling. *J. Neurosci.* **21**, 1203–1210.

Carmignoto, G., Vicini, S. (1992). Activity-dependent decrease in NMDA receptor responses during development of the visual cortex. *Science* **258**, 1007–1011.

Chatterton, J.E., Awobuluyi, M., Premkumar, L.S., Takahashi, H., Talantova, M., Shin, Y. et al. (2002). Excitatory glycine receptors containing the NR3 family of NMDA receptor subunits. *Nature* **415**, 793–798.

Chavis, P., Westbrook, G. (2001). Integrins mediate functional pre- and postsynaptic maturation at a hippocampal synapse. *Nature* **411**, 317–321.

Chazot, P.L., Stephenson, F.A. (1997). Molecular dissection of native mammalian forebrain NMDA receptors containing the NR1 C2 exon: Direct demonstration of NMDA receptors comprising NR1, NR2A, and NR2B subunits within the same complex. *J. Neurochem.* **69**, 2138–2144.

Chen, L., Chetkovich, D.M., Petralia, R.S., Sweeney, N.T., Kawasaki, Y., Wenthold, R.J. et al. (2000). Stargazin regulates synaptic targeting of AMPA receptors by two distinct mechanisms. *Nature* **408**, 936–943.

Chen, L., Huang, L.Y. (1992). Protein kinase C reduces Mg2+ block of NMDA-receptor channels as a mechanism of modulation. *Nature* **356**, 521–523.

Choi, D.W. (1995). Calcium: still center-stage in hypoxic-ischemic neuronal death. *Trends Neurosci.* **18**, 58–60.

Ciabarra, A.M., Sullivan, J.M., Gahn, L.G., Pecht, G., Heinemann, S., Sevarino, K.A. (1995). Cloning and characterization of chi-1: A developmentally regulated member of a novel class of the ionotropic glutamate receptor family. *J. Neurosci.* **15**, 6498–6508.

Crair, M.C., Malenka, R.C. (1995). A critical period for long-term potentiation at thalamocortical synapses. *Nature* **375**, 325–328.

Crump, F.T., Dillman, K.S., Craig, A.M. (2001). cAMP-dependent protein kinase mediates activity-regulated synaptic targeting of NMDA receptors. *J. Neurosci.* **21**, 5079–5088.

Cummings, J.A., Mulkey, R.M., Nicoll, R.A., Malenka, R.C. (1996). Ca2+ signaling requirements for long-term depression in the hippocampus. *Neuron* **16**, 825–833.

Dalva, M.B., Takasu, M.A., Lin, M.Z., Shamah, S.M., Hu, L., Gale, N.W., Greenberg, M.E. (2000). EphB receptors interact with NMDA receptors and regulate excitatory synapse formation. *Cell* **103**, 945–956.

Das, S., Sasaki, Y.F., Rothe, T., Premkumar, L.S., Takasu, M., Crandall, J.E. et al. (1998). Increased NMDA current and spine density in mice lacking the NMDA receptor subunit NR3A. *Nature* **393**, 377–381.

Derkach, V., Barria, A., Soderling, T.R. (1999). Ca2+/calmodulin-kinase II enhances channel conductance of alpha-amino-3-hydroxy-5-methyl-4-isoxazolepropionate type glutamate receptors. *Proc. Natl. Acad. Sci. U S A* **96**, 3269–3274.

Dingledine, R., Borges, K., Bowie, D., Traynelis, S.F. (1999). The glutamate receptor ion channels. *Pharmacol. Rev.* **51**, 7–61.

Ebralidze, A.K., Rossi, D.J., Tonegawa, S., Slater, N.T. (1996). Modification of NMDA receptor channels and synaptic transmission by targeted disruption of the NR2C gene. *J. Neurosci.* **16**, 5014–5025.

Ehlers, M.D. (2000). Reinsertion or degradation of AMPA receptors determined by activity-dependent endocytic sorting. *Neuron* **28**, 511–525.

Ehlers, M.D. (2002). Molecular morphogens for dendritic spines. *Trends Neurosci.* **25**, 64–67.

Ehlers, M.D., Tingley, W.G., Huganir, R.L. (1995). Regulated subcellular distribution of the NR1 subunit of the NMDA receptor. *Science* **269**, 1734–1737.

El-Husseini, A.E., Schnell, E., Chetkovich, D.M., Nicoll, R.A., Bredt, D.S. (2000). PSD-95 involvement in maturation of excitatory synapses. *Science* **290**, 1364–1368.

Ellgaard, L., Helenius, A. (2003). Quality control in the endoplasmic reticulum. *Nat. Rev. Mol. Cell Biol.* **4**, 181–191.

Flint, A.C., Maisch, U.S., Weishaupt, J.H., Kriegstein, A.R., Monyer, H. (1997). NR2A subunit expression shortens NMDA receptor synaptic currents in developing neocortex. *J. Neurosci.* **17**, 2469–2476.

Follesa, P., Ticku, M.K. (1996). NMDA receptor upregulation: Molecular studies in cultured mouse cortical neurons after chronic antagonist exposure. *J. Neurosci.* **16**, 2172–2178.

Gardoni, F., Schrama, L.H., Kamal, A., Gispen, W.H., Cattabeni, F., Di Luca, M. (2001). Hippocampal synaptic plasticity involves competition between Ca2+/calmodulin-dependent protein kinase II and postsynaptic density 95 for binding to the NR2A subunit of the NMDA receptor. *J. Neurosci.* **21**, 1501–1509.

Gerber, G., Kangrga, I., Ryu, P.D., Larew, J.S., Randic, M. (1989). Multiple effects of phorbol esters in the rat spinal dorsal horn. *J. Neurosci.* **9**, 3606–3617.

Hall, R.A., Soderling, T.R. (1997). Differential surface expression and phosphorylation of the N-methyl-D-aspartate receptor subunits NR1 and NR2 in cultured hippocampal neurons. *J. Biol. Chem.* **272**, 4135–4140.

Hawkins, L.M., Prybylowski, K., Chang, K., Moussan, C., Stephenson, F.A., Wenthold, R.J. (2004). Export from the endoplasmic reticulum of assembled N-methyl-d-aspartic acid receptors is controlled by a motif in the C-terminus of the NR2 subunit. *J. Biol. Chem.* **279**, 28903–28910.

Hirao, K., Hata, Y., Ide, N., Takeuchi, M., Irie, M., Yao, I. et al. (1998). A novel multiple PDZ domain-containing molecule interacting with N-methyl-D-aspartate receptors and neuronal cell adhesion proteins. *J. Biol. Chem.* **273**, 21105–21110.

Hollmann, M., Boulter, J., Maron, C., Beasley, L., Sullivan, J., Pecht, G., Heinemann, S. (1993). Zinc potentiates agonist-induced currents at certain splice variants of the NMDA receptor. *Neuron* **10**, 943–954.

Hollmann, M., Heinemann, S. (1994). Cloned glutamate receptors. *Annu. Rev. Neurosci.* **17**, 31–108.

Holmes, K.D., Mattar, P.A., Marsh, D.R., Weaver, L.C., Dekaban, G.A. (2002). The N-methyl-D-aspartate receptor splice variant NR1-4 C-terminal domain. Deletion analysis and role in subcellular distribution. *J. Biol. Chem.* **277**, 1457–1468.

Hsu, S.C., Hazuka, C.D., Foletti, D.L., Scheller, R.H. (1999). Targeting vesicles to specific sites on the plasma membrane: the role of the sec6/8 complex. *Trends Cell Biol.* **9**, 150–153.

Huh, K.H., Wenthold, R.J. (1999). Turnover analysis of glutamate receptors identifies a rapidly degraded pool of the N-methyl-D-aspartate receptor subunit, NR1, in cultured cerebellar granule cells. *J. Biol. Chem.* **274**, 151–157.

Ikeda, K., Araki, K., Takayama, C., Inoue, Y., Yagi, T., Aizawa, S., Mishina, M. (1995). Reduced spontaneous activity of mice defective in the epsilon 4 subunit of the NMDA receptor channel. *Brain Res. Mol. Brain Res.* **33**, 61–71.

Isaacson, J.S., Murphy, G.J. (2001). Glutamate-mediated extrasynaptic inhibition: direct coupling of NMDA receptors to Ca(2+)-activated K+ channels. *Neuron* **31**, 1027–1034.

Ishii, T., Moriyoshi, K., Sugihara, H., Sakurada, K., Kadotani, H., Yokoi, M. et al. (1993). Molecular characterization of the family of the N-methyl-D-aspartate receptor subunits. *J. Biol. Chem.* **268**, 2836–2843.

Kadotani, H., Hirano, T., Masugi, M., Nakamura, K., Nakao, K., Katsuki, M., Nakanishi, S. (1996). Motor discoordination results from combined gene disruption of the NMDA receptor NR2A and NR2C subunits, but not from single disruption of the NR2A or NR2C subunit. *J. Neurosci.* **16**, 7859–7867.

Kennedy, M.B. (1997). The postsynaptic density at glutamatergic synapses. *Trends Neurosci.* **20**, 264–268.

Kew, J.N., Richards, J.G., Mutel, V., Kemp, J.A. (1998). Developmental changes in NMDA receptor glycine affinity and ifenprodil sensitivity reveal three distinct populations of NMDA receptors in individual rat cortical neurons. *J. Neurosci.* **18**, 1935–1943.

Kim, M.J., Dunah, A.W., Wang, Y.T., Sheng, M. (2005). Differential roles of NR2A- and NR2B-containing NMDA receptors in Ras-ERK signaling and AMPA receptor trafficking. *Neuron* **46**, 745–760.

Kornau, H.C., Seeburg, P.H., Kennedy, M.B. (1997). Interaction of ion channels and receptors with PDZ domain proteins. *Curr. Opin. Neurobiol.* **7**, 368–373.

Krupp, J.J., Vissel, B., Thomas, C.G., Heinemann, S.F., Westbrook, G.L. (1999). Interactions of calmodulin and alpha-actinin with the NR1 subunit modulate Ca2+-dependent inactivation of NMDA receptors. *J. Neurosci.* **19**, 1165–1178.

Kutsuwada, T., Sakimura, K., Manabe, T., Takayama, C., Katakura, N., Kushiya, E. et al. (1996). Impair-

ment of suckling response, trigeminal neuronal pattern formation, and hippocampal LTD in NMDA receptor epsilon 2 subunit mutant mice. *Neuron* **16**, 333–344.

Lan, J.Y., Skeberdis, V.A., Jover, T., Grooms, S.Y., Lin, Y., Araneda, R.C. et al. (2001). Protein kinase C modulates NMDA receptor trafficking and gating. *Nat. Neurosci.* **4**, 382–390.

Laurie, D.J., Seeburg, P.H. (1994). Regional and developmental heterogeneity in splicing of the rat brain NMDAR1 mRNA. *J. Neurosci.* **14**, 3180–3194.

Lee, H.K., Barbarosie, M., Kameyama, K., Bear, M.F., Huganir, R.L. (2000). Regulation of distinct AMPA receptor phosphorylation sites during bidirectional synaptic plasticity. *Nature* **405**, 955–959.

Leonard, A.S., Hell, J.W. (1997). Cyclic AMP-dependent protein kinase and protein kinase C phosphorylate N-methyl-D-aspartate receptors at different sites. *J. Biol. Chem.* **272**, 12107–12115.

Leonard, A.S., Lim, I.A., Hemsworth, D.E., Horne, M.C., Hell, J.W. (1999). Calcium/calmodulin-dependent protein kinase II is associated with the N-methyl-D-aspartate receptor. *Proc. Natl. Acad. Sci. U S A* **96**, 3239–3244.

Liao, D., Zhang, X., O'Brien, R., Ehlers, M.D., Huganir, R.L. (1999). Regulation of morphological postsynaptic silent synapses in developing hippocampal neurons. *Nat. Neurosci.* **2**, 37–43.

Lisman, J., Schulman, H., Cline, H. (2002). The molecular basis of CaMKII function in synaptic and behavioural memory. *Nat. Rev. Neurosci.* **3**, 175–190.

Luo, J., Wang, Y., Yasuda, R.P., Dunah, A.W., Wolfe, B.B. (1997). The majority of N-methyl-D-aspartate receptor complexes in adult rat cerebral cortex contain at least three different subunits (NR1/NR2A/NR2B). *Mol. Pharmacol.* **51**, 79–86.

Luthi, A., Schwyzer, L., Mateos, J.M., Gahwiler, B.H., McKinney, R.A. (2001). NMDA receptor activation limits the number of synaptic connections during hippocampal development. *Nat. Neurosci.* **4**, 1102–1107.

Margeta-Mitrovic, M., Jan, Y.N., Jan, L.Y. (2000). A trafficking checkpoint controls GABA(B) receptor heterodimerization. *Neuron* **27**, 97–106.

Massey, P.V., Johnson, B.E., Moult, P.R., Auberson, Y.P., Brown, M.W., Molnar, E. et al. (2004). Differential roles of NR2A- and NR2B-containing NMDA receptors in cortical long-term potentiation and long-term depression. *J. Neurosci.* **24**, 7821–7828.

Mayer, M.L., Westbrook, G.L., Guthrie, P.B. (1984). Voltage-dependent block by Mg2+ of NMDA responses in spinal cord neurones. *Nature* **309**, 261–263.

McGee, A.W., Topinka, J.R., Hashimoto, K., Petralia, R.S., Kakizawa, S., Kauer, F.W. et al. (2001). PSD-93 knock-out mice reveal that neuronal MAGUKs are not required for development or function of parallel fiber synapses in cerebellum. *J. Neurosci.* **21**, 3085–3091.

McIlhinney, R.A., Le Bourdelles, B., Molnar, E., Tricaud, N., Streit, P., Whiting, P.J. (1998). Assembly intracellular targeting and cell surface expression of the human N-methyl-D-aspartate receptor subunits NR1A and NR2A in transfected cells. *Neuropharmacology* **37**, 1355–1367.

McIlhinney, R.A., Molnar, E., Atack, J.R., Whiting, P.J. (1996). Cell surface expression of the human N-methyl-D-aspartate receptor subunit 1a requires the co-expression of the NR2A subunit in transfected cells. *Neuroscience* **70**, 989–997.

Meddows, E., Le Bourdelles, B., Grimwood, S., Wafford, K., Sandhu, S., Whiting, P., McIlhinney, R.A. (2001). Identification of molecular determinants that are important in the assembly of N-methyl-D-aspartate receptors. *J. Biol. Chem.* **276**, 18795–18803.

Meguro, H., Mori, H., Araki, K., Kushiya, E., Kutsuwada, T., Yamazaki, M. et al. (1992). Functional characterization of a heteromeric NMDA receptor channel expressed from cloned cDNAs. *Nature* **357**, 70–74.

Migaud, M., Charlesworth, P., Dempster, M., Webster, L.C., Watabe, A.M., Makhinson, M. et al. (1998). Enhanced long-term potentiation and impaired learning in mice with mutant postsynaptic density-95 protein. *Nature* **396**, 433–439.

Misawa, H., Kawasaki, Y., Mellor, J., Sweeney, N., Jo, K., Nicoll, R.A., Bredt, D.S. (2001). Contrasting localizations of MALS/LIN-7 PDZ proteins in brain and molecular compensation in knockout mice. *J. Biol. Chem.* **276**, 9264–9272.

Miyamoto, Y., Yamada, K., Noda, Y., Mori, H., Mishina, M., Nabeshima, T. (2002). Lower sensitivity to stress and altered monoaminergic neuronal function in mice lacking the NMDA receptor epsilon 4 subunit. *J. Neurosci.* **22**, 2335–2342.

Momiyama, A. (2000). Distinct synaptic and extra-synaptic NMDA receptors identified in dorsal horn neurones of the adult rat spinal cord. *J. Physiol.* **523 (Pt 3)**, 621–628.

Monyer, H., Burnashev, N., Laurie, D.J., Sakmann, B., Seeburg, P.H. (1994). Developmental and regional expression in the rat brain and functional properties of four NMDA receptors. *Neuron* **12**, 529–540.

Monyer, H., Sprengel, R., Schoepfer, R., Herb, A., Higuchi, M., Lomeli, H. et al. (1992). Heteromeric NMDA receptors: Molecular and functional distinction of subtypes. *Science* **256**, 1217–1221.

Mori, H., Manabe, T., Watanabe, M., Satoh, Y., Suzuki, N., Toki, S. et al. (1998). Role of the carboxy-terminal region of the GluR epsilon2 subunit in synaptic localization of the NMDA receptor channel. *Neuron* **21**, 571–580.

Nakazawa, T., Komai, S., Tezuka, T., Hisatsune, C., Umemori, H., Semba, K. et al. (2001). Characterization of Fyn-mediated tyrosine phosphorylation sites on GluR epsilon 2 (NR2B) subunit of the N-methyl-D-aspartate receptor. *J. Biol. Chem.* **276**, 693–699.

Okabe, S., Miwa, A., Okado, H. (1999). Alternative splicing of the C-terminal domain regulates cell surface expression of the NMDA receptor NR1 subunit. *J. Neurosci.* **19**, 7781–7792.

Omkumar, R.V., Kiely, M.J., Rosenstein, A.J., Min, K.T., Kennedy, M.B. (1996). Identification of a phosphorylation site for calcium/calmodulin-dependent protein kinase II in the NR2B subunit of the N-methyl-D-aspartate receptor. *J. Biol. Chem.* **271**, 31670–31678.

Osten, P., Khatri, L., Perez, J.L., Kohr, G., Giese, G., Daly, C. et al. (2000). Mutagenesis reveals a role for ABP/GRIP binding to GluR2 in synaptic surface accumulation of the AMPA receptor. *Neuron* **27**, 313–325.

Pagano, A., Rovelli, G., Mosbacher, J., Lohmann, T., Duthey, B., Stauffer, D. et al. (2001). C-terminal interaction is essential for surface trafficking but not for heteromeric assembly of GABA(b) receptors. *J. Neurosci.* **21**, 1189–1202.

Paupard, M.C., Friedman, L.K., Zukin, R.S. (1997). Developmental regulation and cell-specific expression of N-methyl-D-aspartate receptor splice variants in rat hippocampus. *Neuroscience* **79**, 399–409.

Perez-Otano, I., Schulteis, C.T., Contractor, A., Lipton, S.A., Trimmer, J.S., Sucher, N.J., Heinemann, S.F. (2001). Assembly with the NR1 subunit is required for surface expression of NR3A-containing NMDA receptors. *J. Neurosci.* **21**, 1228–1237.

Petralia, R.S., Sans, N., Wang, Y.X., Wenthold, R.J. (2005). Ontogeny of postsynaptic density proteins at glutamatergic synapses. *Mol. Cell Neurosci.* **29**, 436–452.

Petralia, R.S., Wang, Y.X., Sans, N., Worley, P.F., Hammer, J.A., 3rd, Wenthold, R.J. (2001). Glutamate receptor targeting in the postsynaptic spine involves mechanisms that are independent of myosin Va. *Eur. J. Neurosci.* **13**, 1722–1732.

Philpot, B.D., Sekhar, A.K., Shouval, H.Z., Bear, M.F. (2001a). Visual experience and deprivation bidirectionally modify the composition and function of NMDA receptors in visual cortex. *Neuron* **29**, 157–169.

Philpot, B.D., Weisberg, M.P., Ramos, M.S., Sawtell, N.B., Tang, Y.P., Tsien, J.Z., Bear, M.F. (2001b). Effect of transgenic overexpression of NR2B on NMDA receptor function and synaptic plasticity in visual cortex. *Neuropharmacology* **41**, 762–770.

Prybylowski, K., Fu, Z., Losi, G., Hawkins, L.M., Luo, J., Chang, K. et al. (2002). Relationship between availability of NMDA receptor subunits and their expression at the synapse. *J. Neurosci.* **22**, 8902–8910.

Quinlan, E.M., Lebel, D., Brosh, I., Barkai, E. (2004). A molecular mechanism for stabilization of learning-induced synaptic modifications. *Neuron* **41**, 185–192.

Quinlan, E.M., Olstein, D.H., Bear, M.F. (1999a). Bidirectional, experience-dependent regulation of N-methyl-D-aspartate receptor subunit composition in the rat visual cortex during postnatal development. *Proc. Natl. Acad. Sci. U S A* **96**, 12876–12880.

Quinlan, E.M., Philpot, B.D., Huganir, R.L., Bear, M.F. (1999b). Rapid, experience-dependent expression of synaptic NMDA receptors in visual cortex in vivo. *Nat. Neurosci.* **2**, 352–357.

Rao, A., Cha, E.M., Craig, A.M. (2000). Mismatched appositions of presynaptic and postsynaptic components in isolated hippocampal neurons. *J. Neurosci.* **20**, 8344–8353.

Rao, A., Craig, A.M. (1997). Activity regulates the synaptic localization of the NMDA receptor in hippocampal neurons. *Neuron* **19**, 801–812.

Roche, K.W., Standley, S., McCallum, J., Dune Ly, C., Ehlers, M.D., Wenthold, R.J. (2001). Molecular determinants of NMDA receptor internalization. *Nat. Neurosci.* **4**, 794–802.

Rosenmund, C., Feltz, A., Westbrook, G.L. (1995). Synaptic NMDA receptor channels have a low open probability. *J. Neurosci.* **15**, 2788–2795.

Rosenmund, C., Stern-Bach, Y., Stevens, C.F. (1998). The tetrameric structure of a glutamate receptor channel. *Science* **280**, 1596–1599.

Rumbaugh, G., Vicini, S. (1999). Distinct synaptic and extrasynaptic NMDA receptors in developing cerebellar granule neurons. *J. Neurosci.* **19**, 10603–10610.

Sakimura, K., Kutsuwada, T., Ito, I., Manabe, T., Takayama, C., Kushiya, E. et al. (1995). Reduced hippocampal LTP and spatial learning in mice lacking NMDA receptor epsilon 1 subunit. *Nature* **373**, 151–155.

Sans, N., Petralia, R.S., Wang, Y.X., Blahos, J., 2nd, Hell, J.W., Wenthold, R.J. (2000). A developmental change in NMDA receptor-associated proteins at hippocampal synapses. *J. Neurosci.* **20**, 1260–1271.

Sans, N., Prybylowski, K., Petralia, R.S., Chang, K., Wang, Y.X., Racca, C. et al. (2003). NMDA receptor trafficking through an interaction between PDZ proteins and the exocyst complex. *Nat. Cell Biol.* **5**, 520–530.

Scott, D.B., Blanpied, T.A., Swanson, G.T., Zhang, C., Ehlers, M.D. (2001). An NMDA receptor ER retention signal regulated by phosphorylation and alternative splicing. *J. Neurosci.* **21**, 3063–3072.

Setou, M., Nakagawa, T., Seog, D.H., Hirokawa, N. (2000). Kinesin superfamily motor protein KIF17 and mLin-10 in NMDA receptor-containing vesicle transport. *Science* **288**, 1796–1802.

Sheng, M. (2001). The postsynaptic NMDA-receptor–PSD-95 signaling complex in excitatory synapses of the brain. *J. Cell Sci.* **114**, 1251.

Sheng, M., Cummings, J., Roldan, L.A., Jan, Y.N., Jan, L.Y. (1994). Changing subunit composition of heteromeric NMDA receptors during development of rat cortex. *Nature* **368**, 144–147.

Sheng, M., Sala, C. (2001). PDZ domains and the organization of supramolecular complexes. *Annu. Rev. Neurosci.* **24**, 1–29.

Shi, S., Hayashi, Y., Esteban, J.A., Malinow, R. (2001). Subunit-specific rules governing AMPA receptor trafficking to synapses in hippocampal pyramidal neurons. *Cell* **105**, 331–343.

Snyder, E.M., Philpot, B.D., Huber, K.M., Dong, X., Fallon, J.R., Bear, M.F. (2001). Internalization of ionotropic glutamate receptors in response to mGluR activation. *Nat. Neurosci.* **4**, 1079–1085.

Sprengel, R., Single, F.N. (1999). Mice with genetically modified NMDA and AMPA receptors. *Ann. N. Y. Acad. Sci.* **868**, 494–501.

Sprengel, R., Suchanek, B., Amico, C., Brusa, R., Burnashev, N., Rozov, A. et al. (1998). Importance of the intracellular domain of NR2 subunits for NMDA receptor function in vivo. *Cell* **92**, 279–289.

Standley, S., Roche, K.W., McCallum, J., Sans, N., Wenthold, R.J. (2000). PDZ domain suppression of an ER retention signal in NMDA receptor NR1 splice variants. *Neuron* **28**, 887–898.

Steigerwald, F., Schulz, T.W., Schenker, L.T., Kennedy, M.B., Seeburg, P.H., Kohr, G. (2000). C-Terminal truncation of NR2A subunits impairs synaptic but not extrasynaptic localization of NMDA receptors. *J. Neurosci.* **20**, 4573–4581.

Stocca, G., Vicini, S. (1998). Increased contribution of NR2A subunit to synaptic NMDA receptors in developing rat cortical neurons. *J. Physiol.* **507 (Pt 1)**, 13–24.

Strack, S., Colbran, R.J. (1998). Autophosphorylation-dependent targeting of calcium/calmodulin-dependent protein kinase II by the NR2B subunit of the N-methyl-D-aspartate receptor. *J. Biol. Chem.* **273**, 20689–20692.

Strack, S., McNeill, R.B., Colbran, R.J. (2000). Mechanism and regulation of calcium/calmodulin-dependent protein kinase II targeting to the NR2B subunit of the N-methyl-D-aspartate receptor. *J. Biol. Chem.* **275**, 23798–23806.

Teasdale, R.D., Jackson, M.R. (1996). Signal-mediated sorting of membrane proteins between the endoplasmic reticulum and the Golgi apparatus. *Annu. Rev. Cell Dev. Biol.* **12**, 27–54.

Tingley, W.G., Ehlers, M.D., Kameyama, K., Doherty, C., Ptak, J.B., Riley, C.T., Huganir, R.L. (1997). Characterization of protein kinase A and protein kinase C phosphorylation of the N-methyl-D-aspartate receptor NR1 subunit using phosphorylation site-specific antibodies. *J. Biol. Chem.* **272**, 5157–5166.

Tovar, K.R., Sprouffske, K., Westbrook, G.L. (2000). Fast NMDA receptor-mediated synaptic currents in neurons from mice lacking the epsilon2 (NR2B) subunit. *J. Neurophysiol.* **83**, 616–620.

Tovar, K.R., Westbrook, G.L. (1999). The incorporation of NMDA receptors with a distinct subunit composition at nascent hippocampal synapses in vitro. *J. Neurosci.* **19**, 4180–4188.

Tovar, K.R., Westbrook, G.L. (2002). Mobile NMDA receptors at hippocampal synapses. *Neuron* **34**, 255–264.

Vicini, S., Rumbaugh, G. (2000). A slow NMDA channel: In search of a role. *J. Physiol.* **525 (Pt 2)**, 283.

Vicini, S., Wang, J.F., Li, J.H., Zhu, W.J., Wang, Y.H., Luo, J.H. et al. (1998). Functional and pharmacological differences between recombinant N-methyl-D-aspartate receptors. *J. Neurophysiol.* **79**, 555–566.

Vissel, B., Krupp, J.J., Heinemann, S.F., Westbrook, G.L. (2001). A use-dependent tyrosine dephosphorylation of NMDA receptors is independent of ion flux. *Nat. Neurosci.* **4**, 587–596.

Walikonis, R.S., Jensen, O.N., Mann, M., Provance, D.W., Jr., Mercer, J.A., Kennedy, M.B. (2000). Identification of proteins in the postsynaptic density fraction by mass spectrometry. *J. Neurosci.* **20**, 4069–4080.

Washbourne, P., Bennett, J.E., McAllister, A.K. (2002). Rapid recruitment of NMDA receptor transport packets to nascent synapses. *Nat. Neurosci.* **5**, 751–759.

Watt, A.J., Sjostrom, P.J., Hausser, M., Nelson, S.B., Turrigiano, G.G. (2004). A proportional but slower NMDA potentiation follows AMPA potentiation in LTP. *Nat. Neurosci.* **7**, 518–524.

Watt, A.J., van Rossum, M.C., MacLeod, K.M., Nelson, S.B., Turrigiano, G.G. (2000). Activity coregulates quantal AMPA and NMDA currents at neocortical synapses. *Neuron* **26**, 659–670.

Westphal, R.S., Tavalin, S.J., Lin, J.W., Alto, N.M., Fraser, I.D., Langeberg, L.K. et al. (1999). Regulation of NMDA receptors by an associated phosphatase-kinase signaling complex. *Science* **285**, 93–96.

Wollmuth, L.P., Kuner, T., Sakmann, B. (1998). Adjacent asparagines in the NR2-subunit of the NMDA receptor channel control the voltage-dependent block by extracellular Mg2+. *J. Physiol.* **506 (Pt 1)**, 13–32.

Wollmuth, L.P., Sobolevsky, A.I. (2004). Structure and gating of the glutamate receptor ion channel. *Trends Neurosci.* **27**, 321–328.

Xia, H., Hornby, Z.D., Malenka, R.C. (2001). An ER retention signal explains differences in surface expression of NMDA and AMPA receptor subunits. *Neuropharmacology* **41**, 714–723.

Xiong, Z.G., Raouf, R., Lu, W.Y., Wang, L.Y., Orser, B.A., Dudek, E.M. et al. (1998). Regulation of N-methyl-D-aspartate receptor function by constitutively active protein kinase C. *Mol. Pharmacol.* **54**, 1055–1063.

Zerangue, N., Schwappach, B., Jan, Y.N., Jan, L.Y. (1999). A new ER trafficking signal regulates the subunit stoichiometry of plasma membrane K(ATP) channels. *Neuron* **22**, 537–548.

Zheng, X., Zhang, L., Wang, A.P., Bennett, M.V., Zukin, R.S. (1999). Protein kinase C potentiation of N-methyl-D-aspartate receptor activity is not mediated by phosphorylation of N-methyl-D-aspartate receptor subunits. *Proc. Natl. Acad. Sci. U S A* **96**, 15262–15267.

Zucker, R.S. (1999). Calcium- and activity-dependent synaptic plasticity. *Curr. Opin. Neurobiol.* **9**, 305–313.

Zukin, R.S., Bennett, M.V. (1995). Alternatively spliced isoforms of the NMDARI receptor subunit. *Trends Neurosci.* **18**, 306–313.

TRAFFICKING OF ION CHANNELS AND TRANSPORTERS

Trafficking of membrane proteins can profoundly influence neuronal function. In this section the trafficking of two examples of nonreceptor membrane proteins, neurotransmitter transporters, and potassium channels are discussed.

Neurotransmitter transporters are found on both plasma membrane, where they act to remove secreted neurotransmitters from the extracellular space, and on intracellular membranes, like synaptic vesicles, where they are critical for neurotransmitter filling. Changes in membrane trafficking influence the activity of neurotransmitter transporters. For example, the localization of plasma membrane transporters in different cellular compartments influences the time course of synaptic transmission, the recycling of transmitter to replenish depleted stores, and can affect synaptic plasticity (Fremeau and Edwards).

Neurons express a wide variety of voltage-dependent ion channels on their surface membranes that together determine intrinsic membrane excitability. The expression of these channels at discrete functional sites in the neuronal membrane is actively regulated. The dynamic control of intracellular trafficking, polarized expression, and local clustering of potassium channels may couple various stimuli to membrane excitability (Vacher et al.).

TRAFFICKING OF ION CHANNELS AND TRANSPORTERS

11

Membrane Trafficking of Vesicular Neurotransmitter Transporters

ROBERT T. FREMEAU, JR. AND ROBERT H. EDWARDS

I. INTRODUCTION

Much of the current interest in neurotransmitter transporters derives from their original identification as the sites of action for multiple psychoactive drugs. Cocaine acts by inhibiting the reuptake of monoamine transmitters across the plasma membrane (Axelrod et al. 1961). Inhibition of the dopamine transporter (DAT) may be most important, but recent work also has suggested that cocaine can act by inhibiting the serotonin transporter (SERT) and possibly the norepinephrine transporter (Giros et al. 1996; Rocha et al. 1998; Sora et al. 2001; Moron et al. 2002). Tricyclic antidepressants inhibit norepinephrine reuptake and the more recent serotonin-selective reuptake inhibitors (SSRIs) specifically target SERT

(Amara and Kuhar 1993). Methamphetamine acts by promoting flux reversal mediated by DAT, and MDMA ("Ecstasy") by promoting efflux through SERT (Rudnick and Wall 1992a; Sulzer et al. 1995). However, the ability of amphetamines to release monoamines through this nonvesicular mechanism depends in part on the storage of monoamine in secretory vesicles. Amphetamines act as weak bases to dissipate the proton electrochemical gradient that drives transport across the vesicle membrane, allowing the efflux of stored monoamine into the cytoplasm before reverse transport across the plasma membrane (Sulzer and Rayport 1990; Sulzer et al. 1995; Fon et al. 1997). Thus, amphetamines also act indirectly to interfere with vesicular monoamine transport.

The pronounced psychoactive effects of cocaine, antidepressants, and amphetamines indicate the potential for regulation of neurotransmitter transport to influence normal behavior and contribute to neuropsychiatric illness. Many groups have studied the regulation of neurotransmitter transport. A number of neurotransmitter transporters undergo phosphorylation, but the physiological consequences of this posttranslational modification remains uncertain. In the case of the dopamine transporter, phosphorylation of the N-terminus was identified several years ago (Vaughan et al. 1997), but only recently has been shown to affect transport activity. N-terminal phosphorylation increases the rate of efflux mediated by DAT—uptake is not affected—and this contributes to the efflux of dopamine produced by amphetamines, through a channel-like mechanism (Gnegy et al. 2004; Khoshbouei et al. 2004; Kahlig et al. 2005). A number of other neurotransmitter transporters also undergo phosphorylation. Phosphorylation of the GABA transporter GAT1 by protein kinase C (PKC) influences its ability to be modulated by the t-SNARE syntaxin 1A (Beckman et al. 1998). In many cases, however, phosphorylation affects trafficking rather than transport activity. Indeed, changes in membrane trafficking remain among the best-understood mechanisms that regulate neurotransmitter transport.

II. PLASMA MEMBRANE NEUROTRANSMITTER TRANSPORTERS

Several members of the Na^+/Cl^--dependent neurotransmitter transporter family reside primarily on intracellular membranes. Since these transporters depend on Na^+ and Cl^- gradients present only at the plasma membrane, their activity is inherently regulated by membrane trafficking. The proline transporter (PROT) was one of the first such neurotransmitter transporters to be recognized. PROT is localized selectively to a subset of presynaptic terminals that form asymmetric, excitatory-type synapses typical of glutamatergic nerve terminals (Renick et al. 1999). Surprisingly, within these terminals, PROT localizes preferentially to small clear synaptic vesicles rather than the plasma membrane. The morphological and biochemical properties of PROT-containing vesicles appear similar to synaptic vesicles (Renick et al. 1999). Nonetheless, its activity can easily be measured at the cell surface in transfected cells, presumably due in part to its overexpression (Fremeau et al. 1992). The synaptic vesicle location of PROT indicates the potential for regulation by neural activity, and further suggests that constitutive cell surface expression may have deleterious effects. Since the biological role of PROT is unknown, however, the rationale for regulation by activity is also unclear.

Recent work on the hemicholinium-sensitive choline transporter (CHT) provides another striking example of a plasma membrane transporter localized to intracellular membranes. Originally identified in C. elegans, CHT takes up the choline produced by acetylcholinesterase (Apparsundaram et al. 2000; Okuda et al. 2000). Although CHT does not belong to the same Na^+/Cl^--coupled neurotransmitter transporter family as DAT, NET, and SERT, it has a similar role in neurotransmitter recycling. In its absence, neuromuscular transmission fails shortly after birth (Ferguson et al. 2004). CHT also localizes to the nerve terminal and in particular to a subset of synaptic vesicles (Ferguson et al. 2003). CHT contains a dileucine endocytosis motif that is presumably required for targeting to synaptic vesicles, and localization to these membranes confers a coupling between high rates of transmitter release and choline reuptake (Ferguson et al. 2003). Interestingly, the localization of CHT to a subset of cholinergic synaptic vesicles begins to suggest a heterogeneity of function, and protein composition for what otherwise appear to be a homogeneous population of synaptic vesicles.

In addition to the baseline expression of PROT and CHT on synaptic vesicles, several other members of the Na$^+$/Cl$^-$-coupled plasma membrane transporter undergo regulation by changes in trafficking. DAT interacts with the PDZ-containing protein PICK1, which has been proposed to regulate the internalization of DAT (Torres et al. 2001). However, it also appears that the interaction may serve primarily to localize PICK1 to the plasma membrane (Bjerggaard et al. 2004). In addition, activation of protein kinase C induces the redistribution of DAT away from the cell surface, although apparently not through direct phosphorylation of the transporter (Vaughan et al. 1997). Further, drugs of abuse influence the trafficking of DAT. Amphetamines reduce the activity of DAT in transfected cells and dopamine neurons by redistributing the protein away from the plasma membrane (Saunders et al. 2000; Kahlig et al. 2004). Recent work suggests an opposing role for insulin acting through the Akt kinase (Garcia et al. 2005). Cocaine also increases cell surface expression of DAT (Daws et al. 2002). Despite the abundant evidence for regulation of DAT by changes in membrane trafficking, the underlying cellular mechanisms remain unclear. In many cases, the kinases involved do not appear to phosphorylate DAT directly. Indeed, the sequences responsible for endocytosis of DAT have been identified only very recently, revealing a noncanonical internalization motif—FREKLAYAIA—that also appears in the other members of the Na$^+$/Cl$^-$-coupled neurotransmitter transporter family (Holton et al. 2005). It may now be possible to work backward from these sequences to understand the molecular mechanisms involved in the regulation of DAT trafficking.

Similar to DAT, the closely related SERT and NET are internalized after stimulation of PKC (Ramamoorthy et al. 1998; Jayanthi et al. 2004). In the case of SERT, substrates such as amphetamines as well as monoamines can reduce the internalization produced by activation of PKC (Ramamoorthy and Blakely 1999), suggesting a mechanism to adjust SERT expression to the extracellular concentration of its substrate. Nontransported SERT antagonists do not have this effect. Conversely, adenosine receptor stimulation and activation of protein kinase G by nitric oxide increase SERT function at least in part by increasing its localization to the plasma membrane (Miller and Hoffman 1994; Zhu et al. 2004). Dephosphorylation also has an important role in the regulation of cell surface SERT expression, and the transporter associates with the catalytic subunit of protein phosphatase-2A (Bauman et al. 2000). Recent work has suggested that NET internalization produced by activation of PKC does not involve typical clathrin- and dynamin-dependent endocytosis, but rather a pathway involving lipid rafts (Jayanthi et al. 2004).

Changes in the trafficking of NET and SERT also appear to affect behavior. Familial orthostatic intolerance involves a mutation in NET that reduces cell surface expression of the transporter (Hahn et al. 2003). This defect causes an increase in serum norepinephrine levels and heart rate that apparently underlies the clinical condition. In addition, a mutation in SERT has been identified in families with obsessive-compulsive disorder and other psychiatric conditions (Kilic et al. 2003). The mutation causes a constitutive up-regulation of transport activity that occludes the activation by protein kinase G (Kilic et al. 2003; Prasad et al. 2005). However, others affect localization, supporting a role for the regulation of SERT trafficking in normal behavior (Prasad et al. 2005).

In addition to an interaction with the t-SNARE syntaxin 1A, the GABA transporter GAT1 undergoes regulation at the level of membrane trafficking. Indeed, the association with syntaxin increases cell surface GAT1, as well as reduces its intrinsic transport activity (Horton and Quick 2001). In neurons, GAT1 normally resides at both the plasma membrane (60%) and on small vesicles that undergo activity- and Ca^{2+}-dependent regulated exocytosis (40%) (Bernstein and Quick 1999; Deken et al. 2003). These vesicles contain several proteins usually associated with synaptic vesicles, but not others such as synaptophysin and the vesicular GABA transporter, indicating that they are distinct from the synaptic vesicles that release GABA. Like SERT, GAT1 undergoes internalization in response to activation of PKC (Wang and Quick 2005), and substrates increase the proportion on the plasma membrane (Bernstein and Quick 1999). Although it remains unclear whether PKC phosphorylates GAT1 directly, serine phosphorylation of the transporter increases after activation of PKC, and tyrosine phosphorylation of the transporter, which increases cell surface expression and exhibits an inverse correlation with serine phosphorylation (Quick et al. 2004). The role of this regulation remains uncertain, but it presumably contributes to the inhibitory tone in various brain regions.

The excitatory amino acid transporters (EAATs) that remove glutamate from the extracellular space also undergo regulation by changes in membrane trafficking. This family of transporters depends on Na^+, but not Cl^-, and shows no sequence similarity to the Na^+/Cl^--coupled neurotransmitter transporters (Amara and Fontana 2002; Yernool et al. 2004; Yamashita et al. 2005). In contrast to DAT, NET, SERT, and GAT1, the neuronal isoform EAAT3 increases in function and cell surface expression after activation of PKC, in the C6 glioma cell line and in synaptosomes (Davis et al. 1998). Redistribution to the plasma membrane may not involve direct phosphorylation of EAAT3, but it does depend on phosphatidylinositol-3-kinase (PI3K) (Davis et al. 1998). Tyrosine kinase signaling appears to activate a similar pathway (Sims et al. 2000). Although PKC may not phosphorylate

EAAT3 directly, the PKCα isoform appears to associate with the transporter after stimulation with phorbol esters (Gonzalez et al. 2003), and may inhibit endocytosis, as well as promote exocytosis (Fournier et al. 2004).

What function does the regulation of EAAT3 trafficking serve? In addition to a potential role in synaptic homeostasis, EAAT3 may serve as a guardian for synaptic spillover. Glutamate escaping from glutamate transporters activates metabotropic receptors at parallel and climbing fiber synapses of the cerebellum and in the hippocampus (Brasnjo and Otis 2001; Dzubay and Otis 2002; Selkirk et al. 2003). These receptors in turn may activate PKC and so increase EAAT3 activity, thus reducing the opportunity for spillover. In this case, EAAT3 trafficking provides a mechanism to control the amount of synaptic spillover. In contrast, the glial isoforms, such as EAAT2 that account for most of the glutamate uptake observed in brain extracts, may not be regulated by PKC (Tan et al. 1999).

III. VESICULAR MONOAMINE AND ACETYLCHOLINE TRANSPORTERS

Changes in trafficking regulate the activity of plasma membrane transporters, but vesicular neurotransmitter transporters undergo exocytosis and endocytosis as part of their basic function, independent of their potential for regulation. Since these transporters fill secretory vesicles with transmitters, they appear on the cell surface with exocytosis, then recycle to form new vesicles. They must therefore contain signals targeting them to secretory vesicles, and recycling them from the cell surface. Surprisingly, it has been very difficult to identify these signals in other synaptic vesicle proteins. Synaptobrevin contains a nonconsensus, amphipathic signal that promotes targeting to synaptic vesicles (Grote et al. 1995). Synaptotagmin also appears to contain both tyrosine- and dileucine-based signals for endocytosis, and indeed recruits the adaptor protein AP-2, presumably to promote the endocytosis of other vesicle proteins (Zhang et al. 1994; Jorgensen et al. 1995; Blagoveshchenskaya et al. 1999; Haucke and De Camilli 1999). However, even when expressed alone in nonneural cells, most synaptic vesicle proteins localize to endocytic membranes (Cameron et al. 1991). Thus, synaptic vesicle proteins each appear to contain their own signals for internalization from the plasma membrane, presumably because synaptic vesicles form locally at the nerve terminal by endocytosis. In addition, it may be difficult to identify the trafficking signals in many synaptic vesicle proteins like synaptobrevin and synaptotagmin as a result of their dual role in exocytosis and endocytosis (Jarousse and Kelly 2001; Jarousse et al. 2003). In particular, distinguishing the effects of mutations on exocytosis and endocytosis seems to require more precise approaches that separate these two processes temporally (Poskanzer et al. 2003).

In contrast to many other synaptic vesicle proteins, certain vesicular neurotransmitter transporters contain consensus endocytosis motifs that have been relatively easy to identify. In particular, the vesicular monoamine transporters (VMATs) and the vesicular acetylcholine transporter (VAChT) that are closely related in sequence, both contain a C-terminal, cytoplasmic dileucine-like motif that promotes constitutive internalization from the plasma membrane in a range of species from *Drosophila* to mammals (Tan et al. 1998; Greer et al. 2005). This sequence is both necessary and sufficient to target the transporters to endocytic vesicles. Interestingly, the sequences upstream of the dileucine motif itself (bold, next) are highly conserved as well, and slightly divergent between the VMATs and VAChT:

rat VMAT2   K E E K M A **I L**
rat VAChT   R S E R D V **L L**

In a number of other membrane proteins, acidic residues 4 and 5 positions upstream of the dileucine (underlined) apparently contribute to endocytosis (Pond et al. 1995; Dietrich et al. 1997). Despite their conservation, however, these upstream acidic residues do not influence the internalization of VMAT2 or VAChT from the plasma membrane (Tan et al. 1998), they affect other trafficking events.

Rat pheochromocytoma PC12 cells contain both large dense core vesicles (LDCVs) and synaptic-like microvesicles (SLMVs). Prior to the molecular cloning of VMATs and VAChT, the LDCVs of PC12 cells were shown to contain dopamine, and the SLMVs acetylcholine (Bauerfeind et al. 1993). Consistent with these previous observations, the nonneuronal isoform VMAT1 and the neuronal isoform VMAT2 both reside almost exclusively on LDCVs when expressed in PC12 cells (Liu et al. 1994; Weihe et al. 1994). VAChT also resides predominantly on SLMVs in these cells (Liu and Edwards 1997b). The VMATs and VAChT are very closely related in sequence, which suggests that a divergent sequence might be responsible for the differences in localization. Indeed, the C-terminus of VAChT can redirect VMAT2 to SLMVs in PC12 cells, although it is less clear what happens when the C-terminus of VAChT is replaced by that of VMAT2 (Varoqui and Erickson 1998). Further, VAChT contains a serine five residues upstream of its dileucine motif, in contrast to a glutamate in VMAT, raising the possibility that an acidic residue directs the protein to LDCVs and a neutral residue to SLMVs. Supporting this hypothesis, replacement of both upstream glutamates in VMAT2 by alanine redistributes the protein away from LDCVs in PC12 cells (Krantz et al. 2000). Interestingly, the residue five positions upstream of the dileucine motif in VAChT (Ser-480) undergoes phosphorylation that is stimulated by activation of PKC and inhibited by activation of PKA. Replacement of Ser-480 by aspartate to mimic phosphorylation and the glutamate present at the equivalent position in VMAT shifts VAChT toward LDCVs (Krantz et al. 2000). Conversely, replacement of Ser-480 by alanine to prevent phosphorylation reduces the small amount of VAChT present on LDCVs. Thus, residues upstream of the dileucine motif influence trafficking to LDCVs. These residues do not seem important for targeting to SLMVs in PC12 cells; rather, neutral replacements seem to shift the transporters from LDCVs toward SLMVs (Krantz et al. 2000). Synaptic vesicles release transmitter at the active zone, whereas LDCVs release transmitter at any site along the plasma membrane, including dendrites, as well as the axon. Moreover, synaptic vesicles and LDCVs differ in their mechanisms of release (Martin 1994) (see Chapter 14). The differences in trafficking of VMAT2 and VAChT have important implications for the site and mode of transmitter release. In addition, the identification of these trafficking signals has provided insight into the biogenesis of these specialized neurosecretory vesicles.

In contrast to synaptic vesicles that recycle locally at the nerve terminal, LDCVs bud directly from the *trans*-Golgi network (TGN) (Orci et al. 1987). An *in vitro* assay recapitulates features of this sorting event, in particular the selective sorting of secretogranin (also known as chromogranin) to LDCVs in the regulated secretory pathway, and heparin sulfate proteoglycan to constitutive secretory vesicles (Tooze and Huttner 1990). Recent work has further suggested that chromogranin A is sufficient to direct the formation of LDCVs, even in nonneural cells (Kim et al. 2001). The lumenal carboxypeptidase E also has been suggested to recruit particular peptide cargo (Zhang et al. 2003; Lou et al. 2005). Since chromogranins and carboxypeptidase E are soluble rather than integral membrane proteins, these observations have given rise to the hypothesis that LDCVs form from within, by wrapping membrane around a dense core. In contrast, the identification of a cyto-

plasmic trafficking signal in VMAT2 suggests that LDCVs form in much the same way as other vesicles, by recruiting cytosolic sorting machinery. Thus, it is very important to determine whether this signal acts in the TGN to promote sorting to LDCVs. Metabolic labeling of the transporter with ^{35}S-amino acids during biosynthesis in the endoplasmic reticulum, combined with surface detection, indeed has indicated that the motif targets VMAT2 directly to LDCVs, without first passing through the plasma membrane (Li et al. 2005) (see Figure 11.1). However, it is also possible that other proteins missorted to LDCVs are removed after their formation in the TGN. Consistent with this possibility, it has been proposed that LDCVs form by selective retention rather than by selective sorting (Arvan and Castle 1998).

LDCVs undergo a process of maturation in which proteins destined for other membranes are removed. This process has been well characterized for the protease furin that normally resides in the TGN (Molloy et al. 1999). In the case of furin, removal from maturing LDCVs requires a highly acidic sequence that also contains serines phosphorylated by casein kinase (Jones et al. 1995). This acidic cluster then recruits the cytosolic adaptor PACS, AP-1, and clathrin to promote the retrieval of furin back to the TGN (Dittie et al. 1996, 1997; Wan et al. 1998). VMAT2 contains a remarkably similar sequence that is also phosphorylated by casein kinase (Krantz et al. 1997). Replacement of these serines by aspartate to mimic phosphorylation promotes the retrieval of VMAT2, very similar to furin (Waites et al. 2001). In contrast to furin, however, wild-type VMAT2 remains on LDCVs through maturation. We thus infer that VMAT2 remains dephosphorylated in maturing LDCVs, whereas furin undergoes phosphorylation and these differences in phosphorylation state regulate retrieval. Deletion of the entire acidic cluster from VMAT2 also promotes retrieval, suggesting that the sequence behaves more like a

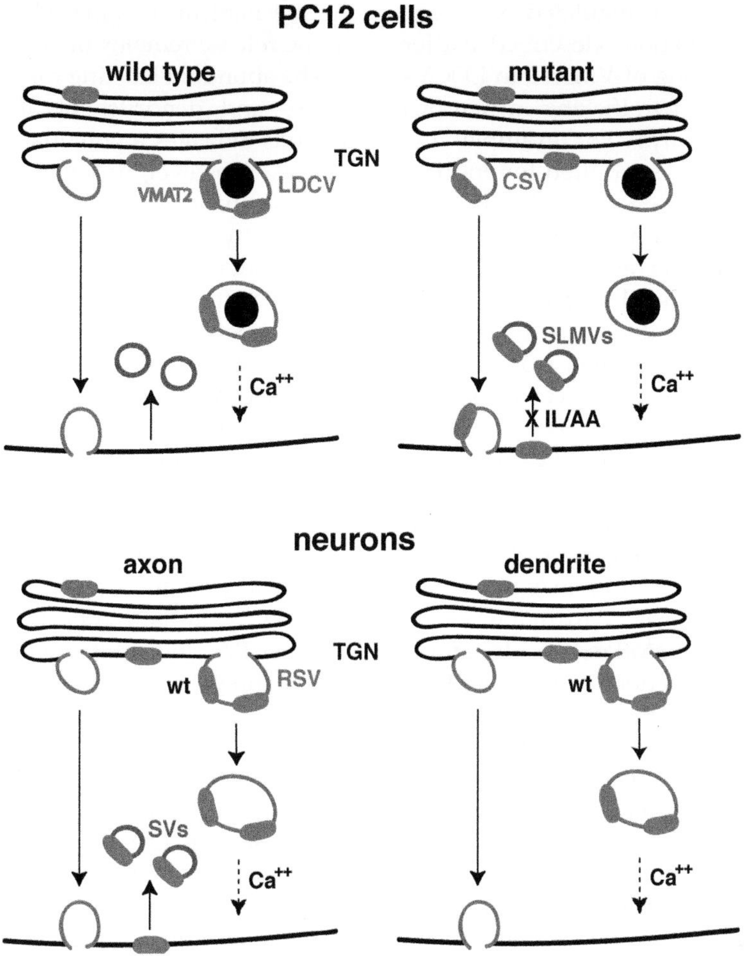

FIGURE 11.1.   **Trafficking of VMAT2 in PC12 cells and in neurons.** Wild-type VMAT2 sorts directly and specifically to LDCVs in the biosynthetic pathway of PC12 cells (top left panels). Wild-type VMAT2 does not have the chance to reach synaptic-like microvesicles (SLMVs), which form at the plasma membrane, since it does not reach the plasma membrane in the absence of stimulation. Both IL/AA and EE/AA mutations (top right panel) disrupt sorting to LDCVs, diverting VMAT2 to constitutive secretory vesicles (CSVs). The IL/AA mutant remains trapped on the plasma membrane because it is defective in endocytosis, but the EE/AA mutant can enter SLMVs. In neurons (bottom panels), VMAT2 traffics to regulated secretory vesicles (RSVs) very similar in function to the LDCVs of PC12 cells but without morphologically identifiable dense cores. In addition, VMAT2 sorts to synaptic vesicles (SVs), possibly through the regulated exocytosis of VMAT2+ RSVs. Axons (left) exhibit regulated exocytosis of both SVs and RSVs, but dendrites (right) contain only RSVs. Interference with sorting to RSVs (due to the EE/AA mutation) thus eliminates regulated exocytosis of VMAT2 in dendrites, but impairs it only slightly in axons. Reprinted from Li et al. 2005, with permission of Elsevier.

retention signal than a sorting motif, and phosphorylation inactivates the retention (Waites et al. 2001).

It is important to determine the significance of the sorting events identified and characterized in PC12 cells for neurons. The nerve terminal contains abundant synaptic vesicles. In contrast to PC12 cells, however, neurons contain relatively few if any morphologically identifiable LDCVs. Nonetheless, neurons as well as endocrine cells release a wide variety of peptides, and these require sorting to a regulated secretory pathway. The mutations described earlier that influence sorting of VMAT2 to LDCVs in PC12 cells affect trafficking in neurons in very similar ways. For example, the dileucine-like motif is required for endocytosis in neurons as well as PC12 cells (Li et al. 2005). In addition, the upstream acidic residues influence the rate of VMAT2 delivery to the plasma membrane without substantially affecting endocytosis, just as if the mutations divert the protein from LDCVs in neurons as well as PC12 cells (Li et al. 2005).

The trafficking of VMAT2 also contributes to the release of monoamines from dendrites. Considerable work has demonstrated the dendritic release of dopamine, norepinephrine, and serotonin (Geffen et al. 1976; Adell and Artigas 1998). In the ventral tegmental area, somatodendritic dopamine release and the activation of presynaptic D1 dopamine receptors is required to induce behavioral sensitization to amphetamines, a form of plasticity considered a model for drug addiction (Vezina 2004). Dopamine release in the midbrain activates D2 autoreceptors (Cragg and Greenfield 1997; Bergquist et al. 2003; Beckstead et al. 2004). However, the mechanism of somatodendritic monoamine release has remained unclear. Somatodendritic monoamine release may involve either regulated exocytosis of vesicles filled with transmitter, or efflux through a nonvesicular mechanism. Supporting exocytosis, somatodendritic release by electrical stimulation may require Ca^{2+} and vesicular storage (Cheramy et al. 1981; Heeringa and Abercrombie 1995; Rice et al. 1997; Jaffe et al. 1998; Beckstead et al. 2004). However, other experiments have suggested only partial Ca^{2+}-dependence, and insensitivity to clostridial neurotoxins that block vesicle fusion (Chen and Rice 2001; Bergquist et al. 2002). Furthermore, a recent study has shown that synaptically driven somatodendritic release is sensitive to inhibitors of DAT (Falkenburger et al. 2001). Therefore,

the mechanism of somatodendritic monoamine release remains unclear.

The abundant labeling for VMAT2 in cell bodies and dendrites of all monoamine neurons indicates a potential for exocytotic release. However, ultrastructural studies have localized VMAT2 in dendrites predominantly to tubulovesicular structures that do not resemble any well-characterized neurosecretory vesicle (Nirenberg et al. 1996). Whether these vesicles and VMAT2 undergo regulated exocytosis in dendrites is unclear. Interestingly, endogenous VAChT exhibits similar, strong somatodendritic labeling of cholinergic neurons (Gilmor et al. 1996; Weihe et al. 1996), raising the possibility that acetylcholine may also undergo release from cell body and dendrites. In contrast, most synaptic vesicle proteins and in particular the vesicular GABA and glutamate transporters, which show no sequence similarity to the VMATs and VAChT, exhibit an almost exclusively axonal distribution. VMAT2 undergoes regulated exocytosis in the dendrites of cultured hippocampal neurons, and these cells can release dopamine in a quantal manner after loading with L-dopa (Li et al. 2005) (see Figure 11.1). Further, the upstream acidic residues of the dileucine motif are required for regulated dendritic exocytosis—in their absence, dendritic exocytosis is constitutive but axonal release remains regulated, albeit slightly reduced (Li et al. 2005) (see Figure 11.1). Regulated dendritic exocytosis thus appears very similar to the regulated secretory pathway in PC12 cells. Although hippocampal neurons do not express typical LDCV markers, such as the chromogranins (Fischer-Colbrie et al. 1993; Calegari et al. 1999), transfection of the chromogranins yields a distribution essentially identical to that of VMAT2 (Li et al. 2005). This compartment also contains neurotrophins, suggesting that VMAT2 defines a retrograde signaling pathway that may contribute to synapse development, function, and plasticity.

At the nerve terminal, VMAT2 appears to undergo several types of regulation by changes in membrane trafficking. Cocaine and other DAT inhibitors increase the activity of VMAT2 *in vivo*, by redistributing the protein into a population of synaptic vesicles that can be isolated by differential centrifugation (Brown et al. 2001b; Sandoval et al. 2002). This redistribution depends on the activation of D2 dopamine receptors, suggesting that physiological stimulation may have a similar effect (Brown et al. 2001a; Truong et al. 2004). Conversely, amphetamines reduce the activity of VMAT2 and redistribute the protein away from this same population of synaptic vesicles (Brown et al. 2000; Riddle et al. 2002). Surprisingly, D2 dopamine receptors also appear responsible for this change in localization (Ugarte et al. 2003). Since VMAT protects against damage from the parkinsonian neurotoxin MPTP (Liu et al. 1992; Gainetdinov et al. 1998) and methamphetamine causes considerable toxicity to dopamine neurons (Seiden et al. 1993), the loss of VMAT activity may contribute to methamphetamine toxicity. Further, methamphetamine neurotoxicity involves the release of endogenous dopamine and production of oxidative stress (Cubells et al. 1994), which may contribute to Parkinson's disease. The effects of D2 dopamine receptor stimulation on VMAT distribution and function therefore may have an important role in neurodegeneration, as well as in physiological signaling.

IV. VESICULAR GLUTAMATE TRANSPORTERS

Glutamate is a ubiquitous amino acid that is required by all cells for its role in protein synthesis and intermediary metabolism. Neurons and other cells that release glutamate as an extracellular signal have evolved specialized mechanisms for its regulated secretion. Synaptic vesicles in excitatory nerve terminals actively accumulate glutamate, enabling its quantal release by exocytosis. A distinct family of proteins has been identified that mediates vesicular glutamate transport (VGLUTs) (reviewed in Fremeau et al. 2004b). Originally isolated as Na^+-dependent plasma membrane transporters for inorganic phosphate, VGLUT1 and VGLUT2 confer glutamate transport into neurosecretory vesicles with all the same properties as native synaptic vesicles (Bellocchio et al. 2000; Takamori et al. 2000; Bai et al. 2001; Fremeau et al. 2001; Herzog et al. 2001; Takamori et al. 2001; Varoqui et al. 2002). These include dependence on a H^+ electrochemical gradient as the driving force for vesicular uptake, selectivity for glutamate but not aspartate, and a biphasic dependence on chloride. Furthermore, the ectopic expression of VGLUT1 or VGLUT2 in inhibitory neurons is sufficient to confer exocytotic glutamate release (Takamori et al. 2000, 2001). The identification of VGLUT1 and VGLUT2 thus has provided the first molecular markers for glutamatergic neurons.

The transport properties of VGLUT1 and VGLUT2 are very similar and together account for the exocytotic release of glutamate by most established glutamatergic neurons. However, they exhibit a striking, complementary pattern of expression in the adult vertebrate brain. In particular, glutamate neurons in the cerebral cortex, hippocampus, and cerebellar cortex express abundant VGLUT1, whereas neurons in the thalamus, deep cerebellar nuclei, and brainstem primarily express VGLUT2 (Fremeau et al. 2001). Interestingly, even within the same adult brain region, the VGLUT1 and VGLUT2 proteins are localized to distinct synapses with different physiological properties (Fremeau et al. 2001; Hartig et al. 2003). These findings suggest that VGLUT1 and VGLUT2 may contribute to differences in transmitter release which extend beyond their identified role in glutamate uptake.

The loss of VGLUT1 selectively impairs transmission at a subset of excitatory synapses in the hippocampus and cerebellum (Fremeau et al. 2004a). However, residual VGLUT1-independent activity was detected in both CA1 stratum radiatum of the hippocampus and in parallel fibers of the cerebellum. The existence of VGLUT1-independent glutamate release in these pathways is unexpected because Schaffer collaterals are thought to form a uniform population of excitatory nerve terminals in stratum radiatum. Similarly, parallel fiber synapses onto Purkinje cells in the cerebellum are widely believed to represent a homogeneous population of excitatory synapses. The residual glutamate release in the absence of VGLUT1 is due to the transient expression of VGLUT2 at distinct release sites in the same cells that normally express VGLUT1 early in postnatal development (Fremeau et al. 2004a). Consistent with these results, during early postnatal development of the cortex, hippocampus, and cerebellar granule cells, VGLUT2 expression precedes the expression of VGLUT1 and then disappears (Miyazaki et al. 2003).

The residual excitatory transmission observed in the hippocampus of VGLUT1 knockout mice provided an opportunity to explore the functional differences between VGLUT1 and VGLUT2. Although evoked release measured by field potential was severely impaired in the absence of VGLUT1, and the frequency of spontaneous events was reduced due to the silencing of many synapses, the residual responses observed in knockout mice resembled those of wild-type in terms of quantal size. The probability of evoked release was also normal, suggesting that VGLUT1 and VGLUT2 localize to distinct synaptic release sites. However, the VGLUT1-independent (VGLUT2) synapses in stratum radiatum depress more rapidly in response to high frequency stimulation than wild-type (almost entirely VGLUT1) synapses formed by the same neuron (Fremeau et al. 2004a). These results suggest that the two VGLUT isoforms differ in their recycling at the nerve terminal,

either as a result of their own intrinsic signals, or due to their association with other proteins that confer differences in the rate of recycling. Because a single packet of transmitter does not saturate postsynaptic glutamate receptors, the speed of synaptic vesicle recycling and/or filling may be an important determinant of synaptic efficacy at many glutamatergic synapses (Ishikawa et al. 2002).

Synapses lacking VGLUT1 exhibit a reduction of synaptic vesicle number, particularly affecting those vesicles that are further from the active zone (Fremeau et al. 2004a). This reduction does not contribute to the increased synaptic depression observed in knockouts because it presumably affects only the synapses that would otherwise express VGLUT1, which are physiologically silent in the knockout. Nerve terminals in the hippocampus and cerebellum of VGLUT1 knockout animals have irregular, tubulovesicular membranes that are not seen in wild-type mice. However, wild-type and VGLUT1 knockout mice do not differ in the number of docked vesicles, the number of asymmetric synapses, or the size of the postsynaptic density.

Neurons typically synthesize, store, and release multiple neuroactive substances (reviewed in Hokfelt et al. 1984). In most cases, neurons release a small, classical transmitter stored in synaptic vesicles and one or more peptides stored in large dense core vesicles. For instance, hypothalamic neurons that release corticotropin-releasing factor (CRF), orexin, oxytocin, and vasopressin also express VGLUT2, and therefore have the capacity for exocytotic glutamate release (Ziegler et al. 2002; Rosin et al. 2003; Hrabovszky et al. 2005). However, the functional significance of glutamate corelease from hypothalamic neuroendocrine neurons is not understood. A single neuron can corelease two classical transmitters with distinct postsynaptic actions. For example, spinal motor neurons express VGLUT2 and paired recordings from motor neurons and

Renshaw cells, or other motor neurons revealed postsynaptic currents that were mediated by the activation of glutamate, as well as acetylcholine receptors (Nishimaru et al. 2005). However, glutamate release was not detected at the peripheral neuromuscular junction. Motor neurons express VGLUT2 at central terminals onto Renshaw cells and motor neurons, but not at peripheral terminals in the muscle. Interestingly, VGLUT2 does not colocalize with VAChT in the central collaterals, suggesting segregation of the release sites for glutamate and acetylcholine.

Several catecholamine cell groups in the brainstem, including C1-3 and A2, which regulate cardiovascular function, coexpress VGLUT2 and tyrosine hydroxylase (Stornetta et al. 2002a, b). Bulbospinal C1 neurons of the rostral ventral lateral medulla (RVLM) that control blood pressure by regulating sympathetic tone appear to release glutamate (Morrison et al. 1991; Huangfu et al. 1994; Deuchars et al. 1995). Consistent with their release of glutamate as their primary transmitter, the bulbospinal C1 RVLM cells differ from other catecholamine neurons in that they do not express an identified plasma membrane catecholamine transporter (Lorang et al. 1994). VGLUT2 mRNA has not been detected in noradrenergic neurons of the locus coeruleus, indicating that only a subset of noradrenergic neurons may release glutamate (Stornetta et al. 2002a).

The third VGLUT isoform, VGLUT3, shows a surprising pattern of expression in the vertebrate brain (Fremeau et al. 2002; Gras et al. 2002; Schafer et al. 2002; Takamori et al. 2002). VGLUT3 is expressed by many cells thought to release a classical transmitter other than the excitatory amino acid glutamate. In particular, monoamine neurons, GABAergic interneurons in the hippocampus and cortex (particularly layers 2 and 6), a subset of amacrine cells in the retina, cholinergic interneurons in the striatum, and a subset of astrocytes all express VGLUT3 mRNA (Fremeau et al.

2002; Gras et al. 2002; Schafer et al. 2002). VGLUT3 also is expressed outside the nervous system, in the liver, and perhaps the kidney.

Within the brain, VGLUT3 mRNA is expressed in the median and dorsal raphe nuclei, suggesting expression by serotonergic neurons (Fremeau et al. 2002; Gras et al. 2002; Schafer et al. 2002). However, the cell bodies of these neurons do not contain detectable VGLUT3 protein. In single cell cultures from the raphe nucleus, at least 70 percent of the serotonin neurons examined *in vitro* express VGLUT3 (Fremeau et al. 2002), and individual axonal processes of isolated raphe neurons contain either VLUT3 or serotonin, suggesting segregation of release sites. In brain sections, neuronal processes that label with VMAT2 and VGLUT3 (but negative for tyrosine hydroxylase) are present in several projection areas, supporting the expression of VGLUT3 by serotonin neurons *in vivo* (Schafer et al. 2002; Somogyi et al. 2004).

Although compelling evidence indicates that dopamine neurons also form glutamatergic synapses *in vitro* (Sulzer et al. 1998) and mediate a fast excitatory synaptic signal *in vivo* (Chuhma et al. 2004; Lavin et al. 2005), it has been difficult to identify the VGLUT subtype that they express, as expression of VGLUT2 (Dal Bo et al. 2004) and VGLUT3 (Fremeau et al. 2002) have been reported. Considering the prominent role of dopamine in neuropsychiatric illness and drug addiction, it is important to elucidate the physiological role of glutamate release from dopamine neurons and to determine whether release of both transmitters occurs from the same synaptic sites or vesicles.

Scattered neurons in stratum radiatum of the hippocampus and layers 2 and 3 of the cortex express VGLUT3, suggesting that it is expressed by interneurons (Fremeau et al. 2002). Although interneurons can be excitatory, VGLUT3 was found to colocalize with glutamic acid decarboxylase (GAD), the biosynthetic enzyme for the inhibitory neurotransmitter GABA, in both cell bodies and processes of a subset of hippocampal interneurons (Fremeau et al. 2002). Subsequently, the VGLUT3-expressing GABAergic interneurons were shown to belong to a subset that express cholecystokinin, preprotachykinin B, and the CB1 cannabinoid receptor, but not parvalbumin, calretinin, vasoactive intestinal peptide, or somatostatin (Hioki et al. 2004; Somogyi et al. 2004). Ultrastructural studies showed that the VGLUT3 positive processes form symmetric synapses in the pyramidal cell layer of the hippocampus and within layers 2 and 3 of the cortex (Fremeau et al. 2002; Hioki et al. 2004; Somogyi et al. 2004). Asynchronous release at these synapses may reflect the corelease of glutamate (Hefft and Jonas 2005), but the relationship between the release of GABA and glutamate by the same neuron remains unknown.

In the retina, a subset of amacrine cells expresses VGLUT3 (Fremeau et al. 2002; Fyk-Kolodziej et al. 2004; Johnson et al. 2004). Generally considered inhibitory, amacrine cells that express VGLUT3 do not colocalize with GAD. Rather, VGLUT3 colocalizes with glycine immunoreactivity in the cell bodies, but not the processes of selected amacrine cells (Johnson et al. 2004). These results suggest that VGLUT3 may define a previously unidentified subset of amacrine cells capable of exocytotic glutamate release.

In the striatum, VGLUT3 colocalizes almost completely with the biosynthetic enzyme for acetylcholine, choline acetyltransferase (ChAT), and the vesicular acetylcholine transporter (VAChT) (Fremeau et al. 2002; Gras et al. 2002), indicating that VGLUT3 is expressed by tonically active cholinergic interneurons. Previous work has focused on the neuromodulatory actions of acetycholine released from these cells (previously reviewed in Kawaguchi et al. 1995), but their abundant expression of VGLUT3 indicates the potential for additional synaptic signaling mediated by glutamate. VGLUT3

colocalizes with VAChT in both cell bodies and processes (Fremeau et al. 2002; Gras et al. 2002), but it remains to be determined whether cholinergic interneurons segregate the proteins to different release sites and/or different synaptic vesicles.

Why do monoamine neurons, GABAergic interneurons, and cholinergic interneurons possess the capability for exocytotic glutamate release? The corelease of glutamate may serve a distinct role in these circuits. In particular, the glutamatergic component may confer the potential for activity-dependent plasticity.

In contrast to the exclusively axonal localization of VGLUT1 and VGLUT2, VGLUT3 also localizes to subsynaptic vesicles in dendrites that may mediate retrograde synaptic signaling (Fremeau et al. 2002). Previous studies have demonstrated glutamate release from the dendrites of pyramidal neurons in the cortex (Zilberter 2000). More recently, VGLUT3 was found in the dendrites of layer 2/3 pyramidal neurons that form reciprocal synapses with fast-spiking interneurons (Harkany et al. 2004). The release of glutamate from pyramidal cell dendrites provides a retrograde signal that regulates the efficacy of inhibitory transmission. In particular, glutamate released from the pyramidal cell dendrite activates metabotropic receptors on the axon terminals of fast-spiking interneurons, inhibiting Ca^{2+} influx, and decreasing the probability of GABA release. Thus, VGLUT3-mediated dendritic release of glutamate regulates retrograde synaptic signaling in the cortex.

V. CONCLUSIONS

Changes in membrane trafficking dramatically influence the activity of neurotransmitter transporters. The localization of plasma membrane transporters, in different cellular compartments, influences the time course of synaptic transmission, the recycling of transmitter to replenish depleted stores, and, by regulating spillover and the activation of metabotropic receptors, the potential for synaptic plasticity. Vesicular transporters undergo exocytosis as an intrinsic component of transmitter release, but also define the vesicles capable of release and hence the biological role of transmitters released by different cells and from different sites made by the same cell.

References

Adell, A., Artigas, F. (1998). A microdialysis study of the in vivo release of 5-Ht in the median raphe nucleus of the rat. *Br. J. Pharmacol.* **125**, 1361–1367.

Amara, S.G., Fontana, A.C. (2002). Excitatory amino acid transporters: Keeping up with glutamate. *Neurochem. Int.* **41**, 313–318.

Amara, S.G., Kuhar, M.J. (1993). Neurotransmitter transporters: Recent progress. *Ann. Rev. Neurosci.* **16**, 73–93.

Apparsundaram, S., Ferguson, S.M., George, A.L., Jr., Blakely, R.D. (2000). Molecular cloning of a human, hemicholinium-3-sensitive choline transporter. *Biochem. Biophys. Res. Commun.* **276**, 862–867.

Arvan, P., Castle, D. (1998). Sorting and storage during secretory granule biogenesis: Looking backward and looking forward. *Biochem. J.* **332**, 593–610.

Axelrod, J., Whitby, L., Hertting, G. (1961). Effect of psychotropic drugs on the uptake of 3h-norepinephrine by tissues. *Science* **133**, 383–384.

Bai, L., Xu, H., Collins, J.F., Ghishan, F.K. (2001). Molecular and functional analysis of a novel neuronal vesicular glutamate transporter. *J. Biol. Chem.* **276**, 36764–36769.

Bauerfeind, R., Regnier-Vigouroux, A., Flatmark, T., Huttner, W.B. (1993). Selective storage of acetylcholine, but not catecholamines, in neuroendocrine synaptic-like microvesicles of early endosomal origin. *Neuron* **11**, 105–121.

Bauman, A.L., Apparsundaram, S., Ramamoorthy, S., Wadzinski, B.E., Vaughan, R.A., Blakely, R.D. (2000). Cocaine and antidepressant-sensitive biogenic amine transporters exist in regulated complexes with protein phosphatase 2a. *J. Neurosci.* **20**, 7571–7578.

Beckman, M.L., Bernstein, E.M., Quick, M.W. (1998). Protein kinase C regulates the interaction between a GABA transporter and syntaxin 1a. *J. Neurosci.* **18**, 6103–6112.

Beckstead, M.J., Grandy, D.K., Wickman, K., Williams, J.T. (2004). Vesicular dopamine release elicits an inhibitory postsynaptic current in midbrain dopamine neurons. *Neuron* **42**, 939–946.

Bellocchio, E.E., Reimer, R.J., Fremeau, R.T.J., Edwards, R.H. (2000). Uptake of glutamate into synaptic vesi-cles by an inorganic phosphate transporter. *Science* **289**, 957–960.

Bergquist, F., Niazi, H.S., Nissbrandt, H. (2002). Evidence for different exocytosis pathways in dendritic and terminal dopamine release in vivo. *Brain Res.* **950**, 245–253.

Bergquist, F., Shahabi, H.N., Nissbrandt, H. (2003). Somatodendritic dopamine release in rat substantia nigra influences motor performance on the accelerating rod. *Brain Res.* **973**, 81–91.

Bernstein, E.M., Quick, M.W. (1999). Regulation of gamma-aminobutyric acid (GABA) transporters by extracellular GABA. *J. Biol. Chem.* **274**, 889–895.

Bjerggaard, C., Fog, J.U., Hastrup, H., Madsen, K., Loland, C.J., Javitch, J.A., Gether, U. (2004). Surface targeting of the dopamine transporter involves discrete epitopes in the distal C terminus but does not require canonical Pdz domain interactions. *J. Neurosci.* **24**, 7024–7036.

Blagoveshchenskaya, A.D., Hewitt, E.W., Cutler, D.F. (1999). Di-Leucine signals mediate targeting of tyrosinase and synaptotagmin to synaptic-like microvesicles within Pc12 cells. *Mol. Biol. Cell* **10**, 3979–3990.

Brasnjo, G., Otis, T.S. (2001). Neuronal glutamate transporters control activation of postsynaptic metabotropic glutamate receptors and influence cerebellar long-term depression. *Neuron* **31**, 607–616.

Brown, J.M., Hanson, G.R., Fleckenstein, A.E. (2000). Methamphetamine rapidly decreases vesicular dopamine uptake. *J. Neurochem.* **74**, 2221–2223.

Brown, J.M., Hanson, G.R., Fleckenstein, A.E. (2001a). Cocaine-induced increases in vesicular dopamine uptake: role of dopamine receptors. *J. Pharmacol. Exp. Ther.* **298**, 1150–1153.

Brown, J.M., Hanson, G.R., Fleckenstein, A.E. (2001b). Regulation of the vesicular monoamine transporter-2: A novel mechanism for cocaine and other psychostimulants. *J. Pharmacol. Exp. Ther.* **296**, 762–767.

Calegari, F., Coco, S., Taverna, E., Bassetti, M., Verderio, C., Corradi, N. et al. (1999). A regulated secretory pathway in cultured hippocampal astrocytes. *J. Biol. Chem.* **274**, 22539–22547.

Cameron, P.L., Sudhof, T.C., Jahn, R., De Camilli, P. (1991). Colocalization of synaptophysin with transferrin receptors: Implications for synaptic vesicle biogenesis. *J. Cell Biol.* **115**, 151–164.

Chen, B.T., Rice, M.E. (2001). Novel Ca++ dependence and time course of somatodendritic dopamine release: Substantia nigra versus striatum. *J. Neurosci.* **21**, 7841–7847.

Cheramy, A., Leviel, V., Glowinski, J. (1981). Dendritic release of dopamine in the substantia nigra. *Nature* **289**, 537–542.

Chuhma, N., Zhang, H., Masson, J., Zhuang, X., Sulzer, D., Hen, R., Rayport, S. (2004). Dopamine neurons mediate a fast excitatory signal via their glutamatergic synapses. *J. Neurosci.* **24**, 972–981.

Cragg, S.J., Greenfield, S.A. (1997). Differential autoreceptor control of somatodendritic and axon terminal dopamine release in substantia nigra, ventral tegmental area, and striatum. *J. Neurosci.* **17**, 5738–5746.

Cubells, J.F., Rayport, S., Rajendran, G., Sulzer, D. (1994). Methamphetamine neurotoxicity involves vacuolation of endocytic organelles and dopamine-dependent intracellular oxidative stress. *J. Neurosci.* **14**, 2260–2271.

Dal Bo, G., St-Gelais, F., Danik, M., Williams, S., Cotton, M., Trudeau, L.E. (2004). Dopamine neurons in culture express vglut2 explaining their capacity to release glutamate at synapses in addition to dopamine. *J. Neurochem.* **88**, 1398–1405.

Davis, K.E., Straff, D.J., Weinstein, E.A., Bannerman, P.G., Correale, D.M., Rothstein, J.D., Robinson, M.B. (1998). Multiple signaling pathways regulate cell surface expression and activity of the excitatory amino acid carrier 1 subtype of Glu transporter in C6 Glioma. *J. Neurosci.* **18**, 2475–2485.

Daws, L.C., Callaghan, P.D., Moron, J.A., Kahlig, K.M., Shippenberg, T.S., Javitch, J.A., Galli, A. (2002). Cocaine increases dopamine uptake and cell surface expression of dopamine transporters. *Biochem. Biophys. Res. Commun.* **290**, 1545–1550.

Deken, S.L., Wang, D., Quick, M.W. (2003). Plasma membrane GABA transporters reside on distinct vesicles and undergo rapid regulated recycling. *J. Neurosci.* **23**, 1563–1568.

Deuchars, S.A., Morrison, S.F., Gilbey, M.P. (1995). Medullary-evoked epsps in neonatal rat sympathetic preganglionic neurones in vitro. *J. Physiol.* **487 (Pt 2)**, 453–463.

Dietrich, J., Kastrup, J., Nielsen, B.L., Odum, N., Geisler, C. (1997). Regulation and function of the Cd3gamma Dxxxll motif: A binding site for adaptor protein-1 and adaptor protein-2 in vitro. *J. Cell Biol.* **138**, 271–281.

Dittie, A.S., Hajibagheri, N., Tooze, S.A. (1996). The Ap-1 adaptor complex binds to immature secretory granules from Pc12 cells and is regulated by Adp-ribosylation factor. *J. Cell Biol.* **132**, 523–536.

Dittie, A.S., Thomas, L., Thomas, G., Tooze, S.A. (1997). Interaction of furin in immature secretory granules from neuroendocrine cells with the Ap-1 adaptor complex is modulated by casein kinase II phosphorylation. *EMBO J.* **16**, 4859–4870.

Dzubay, J.A., Otis, T.S. (2002). Climbing fiber activation of metabotropic glutamate receptors on cerebellar Purkinje neurons. *Neuron* **36**, 1159–1167.

Falkenburger, B.H., Barstow, K.L., Mintz, I.M. (2001). Dendrodendritic inhibition through reversal of dopamine transport. *Science* **293**, 2465–2470.

Ferguson, S.M., Bazalakova, M., Savchenko, V., Tapia, J.C., Wright, J., Blakely, R.D. (2004). Lethal impairment of cholinergic neurotransmission in hemicholinium-3-sensitive choline transporter knockout mice. *Proc. Natl. Acad. Sci. U S A* **101**, 8762–8767.

Ferguson, S.M., Savchenko, V., Apparsundaram, S., Zwick, M., Wright, J., Heilman, C.J. et al. (2003). Vesicular localization and activity-dependent trafficking of presynaptic choline transporters. *J. Neurosci.* **23**, 9697–9709.

Fischer-Colbrie, R., Kirchmair, R., Schobert, A., Olenik, C., Meyer, D.K., Winkler, H. (1993). Secretogranin II is synthesized and secreted in astrocyte cultures. *J. Neurochem.* **60**, 2312–2314.

Fon, E.A., Pothos, E.N., Sun, B.-C., Killeen, N., Sulzer, D., Edwards, R.H. (1997). Vesicular transport regulates monoamine storage and release but is not essential for amphetamine action. *Neuron* **19**, 1271–1283.

Fournier, K.M., Gonzalez, M.I., Robinson, M.B. (2004). Rapid trafficking of the neuronal glutamate transporter, Eaac1: Evidence for distinct trafficking pathways differentially regulated by protein kinase C and platelet-derived growth factor. *J. Biol. Chem.* **279**, 34505–34513.

Fremeau, R.T., Jr., Burman, J., Qureshi, T., Tran, C.H., Proctor, J., Johnson, J. et al. (2002). The identification of vesicular glutamate transporter 3 suggests novel modes of signaling by glutamate. *Proc. Natl. Acad. Sci. U S A* **99**, 14488–14493.

Fremeau, R.T., Jr., Kam, K., Qureshi, T., Johnson, J., Copenhagen, D.R., Storm-Mathisen, J. et al. (2004a). Vesicular glutamate transporters 1 and 2 target to functionally distinct synaptic release sites. *Science* **304**, 1815–1819.

Fremeau, R.T., Jr., Troyer, M.D., Pahner, I., Nygaard, G.O., Tran, C.H., Reimer, R.J. et al. (2001). The expression of vesicular glutamate transporters defines two classes of excitatory synapse. *Neuron* **31**, 247–260.

Fremeau, R.T., Jr., Voglmaier, S., Seal, R.P., Edwards, R.H. (2004b). VGLUTs define subsets of excitatory neurons and suggest novel roles for glutamate. *Trends Neurosci.* **27**, 98–103.

Fremeau, R.T.J., Caron, M.G., Blakely, R.D. (1992). Molecular cloning and expression of a high affinity L-proline transporter expressed in putative glutamatergic pathways of rat brain. *Neuron* **8**, 915–926.

Fyk-Kolodziej, B., Dzhagaryan, A., Qin, P., Pourcho, R.G. (2004). Immunocytochemical localization of three vesicular glutamate transporters in the cat retina. *J. Comp. Neurol.* **475**, 518–530.

Gainetdinov, R.R., Fumagalli, F., Wang, Y.M., Jones, S.R., Levey, A.I., Miller, G.W., Caron, M.G. (1998). Increased MPTP neurotoxicity in vesicular monoamine transporter 2 heterozygote knockout mice. *J. Neurochem.* **70**, 1973–1978.

Garcia, B.G., Wei, Y., Moron, J.A., Lin, R.Z., Javitch, J.A., Galli, A. (2005). AKT is essential for insulin modulation of amphetamine-induced human dopamine transporter cell-surface redistribution. *Mol. Pharmacol.* **68**, 102–109.

Geffen, L.B., Jessell, T.M., Cuello, A.C., Iversen, L.L. (1976). Release of dopamine from dendrites in rat substantia nigra. *Nature* **260**, 258–260.

Gilmor, M.L., Nash, N.R., Roghani, A., Edwards, R.H., Yi, H., Hersch, S.M., Levey, A.I. (1996). Expression of the putative vesicular acetylcholine transporter in rat brain and localization in cholinergic synaptic vesicles. *J. Neurosci.* **16**, 2179–2190.

Giros, B., Jaber, M., Jones, S.R., Wightman, R.M., Caron, M.G. (1996). Hyperlocomotion and indifference to cocaine and amphetamine in mice lacking the dopamine transporter. *Nature* **379**, 606–612.

Gnegy, M.E., Khoshbouei, H., Berg, K.A., Javitch, J.A., Clarke, W.P., Zhang, M., Galli, A. (2004). Intracellular Ca2+ regulates amphetamine-induced dopamine efflux and currents mediated by the human dopamine transporter. *Mol. Pharmacol.* **66**, 137–143.

Gonzalez, M.I., Bannerman, P.G., Robinson, M.B. (2003). Phorbol myristate acetate-dependent interaction of protein kinase C alpha and the neuronal glutamate transporter Eaac1. *J. Neurosci.* **23**, 5589–5593.

Gras, C., Herzog, E., Bellenchi, G.C., Bernard, V., Ravassard, P., Pohl, M. et al. (2002). A third vesicular glutamate transporter expressed by cholinergic and serotoninergic neurons. *J. Neurosci.* **22**, 5442–5451.

Greer, C.L., Grygoruk, A., Patton, D.E., Ley, B., Romero-Calderon, R., Chang, H.Y. et al. (2005). A splice variant of the drosophila vesicular monoamine transporter contains a conserved trafficking domain and functions in the storage of dopamine, serotonin, and octopamine. *J. Neurobiol.* **64**, 239–258.

Grote, E., Hao, J.C., Bennett, M.K., Kelly, R.B. (1995). A targeting signal in vamp regulating transport to synaptic vesicles. *Cell* **81**, 581–589.

Hahn, M.K., Robertson, D., Blakely, R.D. (2003). A mutation in the human norepinephrine transporter gene (Slc6a2) associated with orthostatic intolerance disrupts surface expression of mutant and wild-type transporters. *J. Neurosci.* **23**, 4470–4478.

Harkany, T., Holmgren, C., Hartig, W., Qureshi, T., Chaudhry, F.A., Storm-Mathisen, J. et al. (2004). Endocannabinoid-independent retrograde signaling at inhibitory synapses in layer 2/3 of neocortex: involvement of vesicular glutamate transporter 3. *J. Neurosci.* **24**, 4978–4988.

Hartig, W., Riedel, A., Grosche, J., Edwards, R.H., Fremeau, R.T., Jr., Harkany, T. et al. (2003). Complementary distribution of vesicular glutamate transporters 1 and 2 in the nucleus accumbens of rat: Relationship to calretinin-containing extrinsic innervation and calbindin-immunoreactive neurons. *J. Comp. Neurol.* **465**, 1–10.

Haucke, V., De Camilli, P. (1999). Ap-2 recruitment to synaptotagmin stimulated by tyrosine-based endocytic motifs. *Science* **285**, 1268–1271.

Heeringa, M.J., Abercrombie, E.D. (1995). Biochemistry of somatodendritic dopamine release in substantia nigra: An in vivo comparison with striatal dopamine release. *J. Neurochem.* **65**, 192–200.

Hefft, S., Jonas, P. (2005). Asynchronous GABA release generates long-lasting inhibition at a hippocampal interneuron-principal neuron synapse. *Nat. Neurosci.* **8**, 1319–1328.

Herzog, E., Bellenchi, G.C., Gras, C., Bernard, V., Ravassard, P., Bedet, C. et al. (2001). The existence of a second vesicular glutamate transporter specifies subpopulations of glutamatergic neurons. *J. Neurosci.* **21**, RC181.

Hioki, H., Fujiyama, F., Nakamura, K., Wu, S.X., Matsuda, W., Kaneko, T. (2004). Chemically specific circuit composed of vesicular glutamate transporter 3- and preprotachykinin B-producing interneurons in the rat neocortex. *Cereb. Cortex* **14**, 1266–1275.

Hokfelt, T., Johansson, O., Goldstein, M. (1984). Chemical anatomy of the brain. *Science* **225**, 1326–1334.

Holton, K.L., Loder, M.K., Melikian, H.E. (2005). Nonclassical, distinct endocytic signals dictate constitutive and PKC-regulated neurotransmitter transporter internalization. *Nat. Neurosci.* **8**, 881–888.

Horton, N., Quick, M.W. (2001). Syntaxin 1a upregulates GABA transporter expression by subcellular redistribution. *Mol. Membr. Biol.* **18**, 39–44.

Hrabovszky, E., Wittmann, G., Turi, G.F., Liposits, Z., Fekete, C. (2005). Hypophysiotropic thyrotropin-releasing hormone and corticotropin-releasing hormone neurons of the rat contain vesicular glutamate transporter-2. *Endocrinology* **146**, 341–347.

Huangfu, D., Hwang, L.J., Riley, T.A., Guyenet, P.G. (1994). Role of serotonin and catecholamines in sympathetic responses evoked by stimulation of rostral medulla. *Am. J. Physiol.* **266**, R338–352.

Ishikawa, T., Sahara, Y., Takahashi, T. (2002). A single packet of transmitter does not saturate postsynaptic glutamate receptors. *Neuron* **34**, 613–621.

Jaffe, E.H., Marty, A., Schulte, A., Chow, R.H. (1998). Extrasynaptic vesicular transmitter release from the somata of substantia nigra neurons in rat midbrain slices. *J. Neurosci.* **18**, 3548–3553.

Jarousse, N., Kelly, R.B. (2001). The Ap2 binding site of synaptotagmin 1 is not an internalization signal but a regulator of endocytosis. *J. Cell Biol.* **154**, 857–866.

Jarousse, N., Wilson, J.D., Arac, D., Rizo, J., Kelly, R.B. (2003). Endocytosis of synaptotagmin 1 is mediated by a novel, tryptophan-containing motif. *Traffic* **4**, 468–478.

Jayanthi, L.D., Samuvel, D.J., Ramamoorthy, S. (2004). Regulated internalization and phosphorylation of the native norepinephrine transporter in response to phorbol esters. Evidence for localization in lipid rafts and lipid raft-mediated internalization. *J. Biol. Chem.* **279**, 19315–19326.

Johnson, J., Sherry, D.M., Liu, X., Fremeau, R.T., Jr., Seal, R.P., Edwards, R.H., Copenhagen, D.R. (2004). Vesicular glutamate transporter 3 expression identifies glutamatergic amacrine cells in the rodent retina. *J. Comp. Neurol.* **477**, 386–398.

Johnson, M.D. (1994). Synaptic glutamate release by postnatal rat serotonergic neurons in microculture. *Neuron* **12**, 433–442.

Jones, B.G., Thomas, L., Molloy, S.S., Thulin, C.D., Fry, M.D., Walsh, K.A., Thomas, G. (1995). Intracellular trafficking of furin is modulated by the phosphorylation state of a casein kinase II site in its cytoplasmic tail. *EMBO J.* **14**, 5869–5883.

Jorgensen, E.M., Hartwieg, E., Schuske, K., Nonet, M.L., Jin, Y., Horvitz, H.R. (1995). Defective recycling of synaptic vesicles in synaptotagmin mutants of caenorhabditis elegans. *Nature* **378**, 196–199.

Kahlig, K.M., Binda, F., Khoshbouei, H., Blakely, R.D., McMahon, D.G., Javitch, J.A., Galli, A. (2005). Amphetamine induces dopamine efflux through a dopamine transporter channel. *Proc. Natl. Acad. Sci. U S A* **102**, 3495–3500.

Kahlig, K.M., Javitch, J.A., Galli, A. (2004). Amphetamine regulation of dopamine transport. Combined measurements of transporter currents and transporter imaging support the endocytosis of an active carrier. *J. Biol. Chem.* **279**, 8966–8975.

Kawaguchi, Y., Wilson, C.J., Augood, S.J., Emson, P.C. (1995). Striatal interneurones: Chemical, physiological and morphological characterization. *TINS.* **18**, 527–535.

Khoshbouei, H., Sen, N., Guptaroy, B., Johnson, L., Lund, D., Gnegy, M.E. et al. (2004). N-terminal phosphorylation of the dopamine transporter is required for amphetamine-induced efflux. *PLOS. Biol.* **2**, E78.

Kilic, F., Murphy, D.L., Rudnick, G. (2003). A human serotonin transporter mutation causes constitutive activation of transport activity. *Mol. Pharmacol.* **64**, 440–446.

Kim, T., Tao-Cheng, J.H., Eiden, L.E., Loh, Y.P. (2001). Chromogranin A, an "on/off" switch controlling dense-core granule biogenesis. *Cell* **106**, 499–509.

Krantz, D.E., Peter, D., Liu, Y., Edwards, R.H. (1997). Phosphorylation of a vesicular monoamine transporter by casein kinase II. *J. Biol. Chem.* **272**, 6752–6759.

Krantz, D.E., Waites, C., Oorschot, V., Liu, Y., Wilson, R.I., Tan, P.K. et al. (2000). A phosphorylation site in the vesicular acetylcholine transporter regulates sorting to secretory vesicles. *J. Cell Biol.* **149**, 379–395.

Lavin, A., Nogueira, L., Lapish, C.C., Wightman, R.M., Phillips, P.E., Seamans, J.K. (2005). Mesocortical dopamine neurons operate in distinct temporal domains using multimodal signaling. *J. Neurosci.* **25**, 5013–5023.

Li, H., Waites, C.L., Staal, R.G., Dobryy, Y., Park, J., Sulzer, D.L., Edwards, R.H. (2005). Sorting of vesicular monoamine transporter 2 to the regulated secretory pathway confers the somatodendritic exocytosis of monoamines. *Neuron* **48**, 619–633.

Liu, Q.R., Lopez-Corcuera, B., Nelson, H., Mandiyan, S., Nelson, N. (1992). Cloning and expression of a CDNA encoding the transporter of taurine and beta-alanine in mouse brain. *Proc. Natl. Acad. Sci. U S A* **89**, 12145–12149.

Liu, Y., Edwards, R.H. (1997b). Differential localization of vesicular acetylcholine and monoamine transporters in PC12 cells but not CHO cells. *J. Cell Biol.* **139**, 907–916.

Liu, Y., Schweitzer, E.S., Nirenberg, M.J., Pickel, V.M., Evans, C.J., Edwards, R.H. (1994). Preferential localization of a vesicular monoamine transporter to dense core vesicles in PC12 cells. *J. Cell Biol.* **127**, 1419–1433.

Lorang, D., Amara, S.G., Simerly, R.B. (1994). Cell-type-specific expression of catecholamine transporters in the rat brain. *J. Neurosci.* **14**, 4903–4914.

Lou, H., Kim, S.K., Zaitsev, E., Snell, C.R., Lu, B., Loh, Y.P. (2005). Sorting and activity-dependent secretion of BDNF require interaction of a specific motif with the sorting receptor carboxypeptidase E. *Neuron* **45**, 245–255.

Martin, T.F.J. (1994). The molecular machinery for fast and slow neurosecretion. *Curr. Opin. Neurobiol.* **4**, 626–632.

Miller, K.J., Hoffman, B.J. (1994). Adenosine A3 receptors regulate serotonin transport via nitric oxide and CGMP. *J. Biol. Chem.* **269**, 27351–27356.

Miyazaki, T., Fukaya, M., Shimizu, H., Watanabe, M. (2003). Subtype switching of vesicular glutamate transporters at parallel fibre-Purkinje cell synapses in developing mouse cerebellum. *Eur. J. Neurosci.* **17**, 2563–2572.

Molloy, S.S., Anderson, E.D., Jean, F., Thomas, G. (1999). Bi-cycling the furin pathway: From TGN localization to pathogen activation and embryogenesis. *Trends Cell Biol.* **9**, 28–35.

Moron, J.A., Brockington, A., Wise, R.A., Rocha, B.A., Hope, B.T. (2002). Dopamine uptake through the norepinephrine transporter in brain regions with low levels of the dopamine transporter: Evidence from knock-out mouse lines. *J. Neurosci.* **22**, 389–395.

Morrison, S.F., Callaway, J., Milner, T.A., Reis, D.J. (1991). Rostral ventrolateral medulla: A source of the glutamatergic innervation of the sympathetic intermediolateral nucleus. *Brain Res.* **562**, 126–135.

Nirenberg, M.J., Chan, J., Liu, Y., Edwards, R.H., Pickel, V.M. (1996). Ultrastructural localization of the vesicular monoamine transporter-2 in midbrain dopaminergic neurons: Potential sites for somatodendritic storage and release of dopamine. *J. Neurosci.* **16**, 4135–4145.

Nishimaru, H., Restrepo, C.E., Ryge, J., Yanagawa, Y., Kiehn, O. (2005). Mammalian motor neurons co-release glutamate and acetylcholine at central synapses. *Proc. Natl. Acad. Sci. U S A* **102**, 5245–5249.

Okuda, T., Haga, T., Kanai, Y., Endou, H., Ishihara, T., Katsura, I. (2000). Identification and characterization of the high-affinity choline transporter. *Nat. Neurosci.* **3**, 120–125.

Orci, L., Ravazzola, M., Amherdt, M., Perrelet, A., Powell, S.K., Quinn, D.L., Moore, H.P. (1987). The trans-most cisternae of the golgi complex: A compartment for sorting of secretory and plasma membrane proteins. *Cell* 51, 1039–1051.

Pond, L., Kuhn, L.A., Teyton, L., Schutze, M.P., Tainer, J.A., Jackson, M.R., Peterson, P.A. (1995). A role for acidic residues in di-leucine motif-based targeting to the endocytic pathway. *J. Biol. Chem.* 270, 19989–19997.

Poskanzer, K.E., Marek, K.W., Sweeney, S.T., Davis, G.W. (2003). Synaptotagmin I is necessary for compensatory synaptic vesicle endocytosis in vivo. *Nature* 426, 559–563.

Prasad, H.C., Zhu, C.B., McCauley, J.L., Samuvel, D.J., Ramamoorthy, S., Shelton, R.C. et al. (2005). Human serotonin transporter variants display altered sensitivity to protein kinase G and P38 mitogen-activated protein kinase. *Proc. Natl. Acad. Sci. U S A* 102, 11545–11550.

Quick, M.W., Hu, J., Wang, D., Zhang, H.Y. (2004). Regulation of a gamma-aminobutyric acid transporter by reciprocal tyrosine and serine phosphorylation. *J. Biol. Chem.* 279, 15961–15967.

Ramamoorthy, S., Blakely, R.D. (1999). Phosphorylation and sequestration of serotonin transporters differentially modulated by psychostimulants. *Science* 285, 763–766.

Ramamoorthy, S., Giovanetti, E., Qian, Y., Blakely, R.D. (1998). Phosphorylation and regulation of antidepressant-sensitive serotonin transporters. *J. Biol. Chem.* 273, 2458–2466.

Renick, S.E., Kleven, D.T., Chan, J., Stenius, K., Milner, T.A., Pickel, V.M., Fremeau, R.T.J. (1999). The mammalian brain high-affinity L-proline transporter is enriched preferentially in synaptic vesicles in a subpopulation of excitatory nerve terminals in rat forebrain. *J. Neurosci.* 19, 21–33.

Rice, M.E., Cragg, S.J., Greenfield, S.A. (1997). Characteristics of electrically evoked somatodendritic dopamine release in substantia nigra and ventral tegmental area in vitro. *J. Neurophysiol.* 77, 853–862.

Riddle, E.L., Topham, M.K., Haycock, J.W., Hanson, G.R., Fleckenstein, A.E. (2002). Differential trafficking of the vesicular monoamine transporter-2 by methamphetamine and cocaine. *Eur. J. Pharmacol.* 449, 71–74.

Rocha, B.A., Fumagalli, F., Gainetdinov, R.R., Jones, S.R., Ator, R., Giros, B. et al. (1998). Cocaine self-administration in dopamine-transporter knockout mice. *Nature Neurosci.* 1, 132–137.

Rosin, D.L., Weston, M.C., Sevigny, C.P., Stornetta, R.L., Guyenet, P.G. (2003). Hypothalamic orexin (hypocretin) neurons express vesicular glutamate transporters VGLUT1 or VGLUT2. *J. Comp. Neurol.* 465, 593–603.

Rudnick, G., Wall, S.C. (1992a). The molecular mechanism of "Ecstasy" [3,4-methylenedioxymethamphetamin (MDMA)]: Serotonin transporters are targets for mdma-induced serotonin release. *Proc. Natl. Acad. Sci. U S A* 89, 1817–1821.

Sandoval, V., Riddle, E.L., Hanson, G.R., Fleckenstein, A.E. (2002). Methylphenidate redistributes vesicular monoamine transporter-2: Role of dopamine receptors. *J. Neurosci.* 22, 8705–8710.

Saunders, C., Ferrer, J.V., Shi, L., Chen, J., Merrill, G., Lamb, M.E. et al. (2000). Amphetamine-induced loss of human dopamine transporter activity: An internalization-dependent and cocaine-sensitive mechanism. *Proc. Natl. Acad. Sci. U S A* 97, 6850–6855.

Schafer, M.K., Varoqui, H., Defamie, N., Weihe, E., Erickson, J.D. (2002). Molecular cloning and functional identification of mouse vesicular glutamate transporter 3 and its expression in subsets of novel excitatory neurons. *J. Biol. Chem.* 277, 50734–50748.

Seiden, L.S., Sabol, K.E., Ricaurte, G.A. (1993). Amphetamine: Effects on catecholamine systems and behavior. *Ann. Rev. Pharmacol. Toxicol.* 32, 639–677.

Selkirk, J.V., Naeve, G.S., Foster, A.C. (2003). Blockade of excitatory amino acid transporters in the rat hippocampus results in enhanced activation of group I and group III metabotropic glutamate receptors. *Neuropharmacology* 45, 885–894.

Sims, K.D., Straff, D.J., Robinson, M.B. (2000). Platelet-derived growth factor rapidly increases activity and cell surface expression of the EAAC1 subtype of glutamate transporter through activation of phosphatidylinositol 3-kinase. *J. Biol. Chem.* 275, 5228–5237.

Somogyi, J., Baude, A., Omori, Y., Shimizu, H., Mestikawy, S.E., Fukaya, M. et al. (2004). GABAergic basket cells expressing cholecystokinin contain vesicular glutamate transporter type 3 (VGLUT3) in their synaptic terminals in hippocampus and iso-cortex of the rat. *Eur. J. Neurosci.* 19, 552–569.

Sora, I., Hall, F.S., Andrews, A.M., Itokawa, M., Li, X.F., Wei, H.B. et al. (2001). Molecular mechanisms of cocaine reward: Combined dopamine and serotonin transporter knockouts eliminate cocaine place preference. *Proc. Natl. Acad. Sci. U S A* 98, 5300–5305.

Stornetta, R.L., Sevigny, C.P., Guyenet, P.G. (2002a). Vesicular glutamate transporter DNPI/VGLUT2 MRNA is present in C1 and several other groups of brainstem catecholaminergic neurons. *J. Comp. Neurol.* 444, 191–206.

Stornetta, R.L., Sevigny, C.P., Schreihofer, A.M., Rosin, D.L., Guyenet, P.G. (2002b). Vesicular glutamate transporter DNPI/VGLUT2 is expressed by both C1 adrenergic and nonaminergic presympathetic vasomotor neurons of the rat medulla. *J. Comp. Neurol.* 444, 207–220.

Sulzer, D., Chen, T.-K., Lau, Y.Y., Kristensen, H., Rayport, S., Ewing, A. (1995). Amphetamine redistributes dopamine from synaptic vesicles to the cytosol and promotes reverse transport. *J. Neurosci.* 15, 4102–4108.

Sulzer, D., Joyce, M.P., Lin, L., Geldwert, D., Haber, S.N., Hattori, T., Rayport, S. (1998). Dopamine neurons make glutamatergic synapses in vitro. *J. Neurosci.* **18**, 4588–4602.

Sulzer, D., Rayport, S. (1990). Amphetamine and other psychostimulants reduce ph gradients in midbrain dopaminergic neurons and chromaffin granules: A mechanism of action. *Neuron* **5**, 797–808.

Takamori, S., Malherbe, P., Broger, C., Jahn, R. (2002). Molecular cloning and functional characterization of human vesicular glutamate transporter 3. *Embo Rep.* **3**, 798–803.

Takamori, S., Rhee, J.S., Rosenmund, C., Jahn, R. (2000). Identification of a vesicular glutamate transporter that defines a glutamatergic phenotype in neurons. *Nature* **407**, 189–194.

Takamori, S., Rhee, J.S., Rosenmund, C., Jahn, R. (2001). Identification of differentiation-associated brain-specific phosphate transporter as a second vesicular glutamate transporter. *J. Neurosci.* **21**, Rc182: 1–6.

Tan, J., Zelenaia, O., Correale, D., Rothstein, J.D., Robinson, M.B. (1999). Expression of the GLT-1 subtype of Na+-dependent glutamate transporter: Pharmacological characterization and lack of regulation by protein kinase C. *J. Pharmacol. Exp. Ther.* **289**, 1600–1610.

Tan, P.K., Waites, C., Liu, Y., Krantz, D.E., Edwards, R.H. (1998). A leucine-based motif mediates the endocytosis of vesicular monoamine and acetylcholine transporters. *J. Biol. Chem.* **273**, 17351–17360.

Tooze, S.A., Huttner, W.B. (1990). Cell-free protein sorting to the regulated and constitutive secretory pathways. *Cell* **60**, 837–847.

Torres, G.E., Yao, W.D., Mohn, A.R., Quan, H., Kim, K.M., Levey, A.I. et al. (2001). Functional interaction between monoamine plasma membrane transporters and the synaptic PDZ domain-containing protein PICK1. *Neuron* **30**, 121–134.

Truong, J.G., Newman, A.H., Hanson, G.R., Fleckenstein, A.E. (2004). Dopamine D2 receptor activation increases vesicular dopamine uptake and redistributes vesicular monoamine transporter-2 protein. *Eur. J. Pharmacol.* **504**, 27–32.

Ugarte, Y.V., Rau, K.S., Riddle, E.L., Hanson, G.R., Fleckenstein, A.E. (2003). Methamphetamine rapidly decreases mouse vesicular dopamine uptake: Role of hyperthermia and dopamine D2 receptors. *Eur. J. Pharmacol.* **472**, 165–171.

Varoqui, H., Erickson, J.D. (1998). The cytoplasmic tail of the vesicular acetylcholine transporter contains a synaptic vesicle targeting signal. *J. Biol. Chem.* **273**, 9094–9098.

Varoqui, H., Schafer, M.K.-H., Zhu, H., Weihe, E., Erickson, J.D. (2002). Identification of the differentiation-associated Na+/Pi transporter as a novel vesicular glutamate transporter expressed in a distinct set of glutamatergic synapses. *J. Neurosci.* **22**, 142–155.

Vaughan, R.A., Huff, R.A., Uhl, G.R., Kuhar, M.J. (1997). Protein kinase C-mediated phosphorylation and functional regulation of dopamine transporters in striatal synaptosomes. *J. Biol. Chem.* **272**, 15541–15546.

Vezina, P. (2004). Sensitization of midbrain dopamine neuron reactivity and the self-administration of psychomotor stimulant drugs. *Neurosci. Biobehav. Rev.* **27**, 827–839.

Waites, C.L., Mehta, A., Tan, P.K., Thomas, G., Edwards, R.H., Krantz, D.E. (2001). An acidic motif retains vesicular monoamine transporter 2 on large dense core vesicles. *J. Cell Biol.* **152**, 1159–1168.

Wan, L., Molloy, S.S., Thomas, L., Liu, G., Xiang, Y., Rybak, S.L., Thomas, G. (1998). PACS-1 defines a novel gene family of cytosolic sorting proteins required for trans-Golgi network localization. *Cell* **94**, 205–216.

Wang, D., Quick, M.W. (2005). Trafficking of the plasma membrane gamma-aminobutyric acid transporter GAT1. Size and rates of an acutely recycling pool. *J. Biol. Chem.* **280**, 18703–18709.

Weihe, E., Schafer, M.K., Erickson, J.D., Eiden, L.E. (1994). Localization of vesicular monoamine transporter isoforms (VMAT1 and VMAT2) to endocrine cells and neurons in rat. *J. Mol. Neurosci.* **5**, 149–164.

Weihe, E., Tao-Cheng, J.-H., Schafer, M.K.-H., Erickson, J.D., Eiden, L.E. (1996). Visualization of the vesicular acetylcholine transporter in cholinergic nerve terminals and its targeting to a specific population of small synaptic vesicles. *Proc. Natl. Acad. Sci. U S A* **93**, 3547–3552.

Yamashita, A., Singh, S.K., Kawate, T., Jin, Y., Gouaux, E. (2005). Crystal structure of a bacterial homologue of Na(+)/Cl(–)-dependent neurotransmitter transporters. *Nature* **437**, 215–223.

Yernool, D., Boudker, O., Jin, Y., Gouaux, E. (2004). Structure of a glutamate transporter homologue from pyrococcus horikoshii. *Nature* **431**, 811–818.

Zhang, C.F., Dhanvantari, S., Lou, H., Loh, Y.P. (2003). Sorting of carboxypeptidase E to the regulated secretory pathway requires interaction of its transmembrane domain with lipid rafts. *Biochem. J.* **369**, 453–460.

Zhang, J.Z., Davletov, B.A., Sudhof, T.C., Anderson, R.G. (1994). Synaptotagmin I is a high affinity receptor for clathrin AP-2: Implications for membrane recycling. *Cell* **78**, 751–760.

Zhu, C.B., Hewlett, W.A., Feoktistov, I., Biaggioni, I., Blakely, R.D. (2004). Adenosine receptor, protein kinase G, and P38 mitogen-activated protein kinase-dependent up-regulation of serotonin transporters involves both transporter trafficking and activation. *Mol. Pharmacol.* **65**, 1462–1474.

Ziegler, D.R., Cullinan, W.E., Herman, J.P. (2002). Distribution of vesicular glutamate transporter MRNA in rat hypothalamus. *J. Comp. Neurol.* **448**, 217–229.

Zilberter, Y. (2000). Dendritic release of glutamate suppresses synaptic inhibition of pyramidal neurons in rat neocortex. *J. Physiol.* **528**, 489–496.

12

Determinants of Voltage-Gated Potassium Channel Distribution in Neurons

HELENE VACHER, HIROAKI MISONOU,
AND JAMES S. TRIMMER

Neurons express a wide variety of voltage-dependent ion channels in their surface membranes that together determine intrinsic membrane excitability. The expression of these channels at discrete functional sites in the neuronal membrane is dynamically regulated at a number of cellular levels. Understanding how neurons regulate the expression and localization of ion channels is critical to elucidating the diversity of normal neuronal function, as well as the pathophysiological dysfunction arising from defects in these processes. Voltage-dependent K+ (Kv) channels act as potent modulators of excitatory events, such as action potentials, excitatory synaptic potentials, and Ca^{2+} influx. Mammalian central neurons express a large number of different Kv channel complexes, with diverse patterns of cellular expression and subcellular localization that impact their contribution to regulating cell excitability. Here we review the mechanisms regulating intracellular trafficking and localization of Kv channels in mammalian neurons and discuss how dynamic regulation of these events impacts neuronal signaling.

I. INTRODUCTION

Electrical excitability is a fundamental property of neurons. Diversity in intrinsic neuronal excitability and function is generated by variable expression, subcellular localization, and activity of a complex repertoire of neuronal ion channels. Dynamic regulation of intrinsic excitability can further alter the behavior of neurons and confer plasticity to neuronal signaling. Moreover, aberrant expression, localization, and function of ion channels can result in channel-based pathophysiologies or channelopathies. Thus, an understanding of how neurons regulate the expression and localization of ion channels is critical to understanding the diversity of normal neuronal function, its dynamic modulation to achieve plasticity, and how defects in these processes lead to neuronal dysfunction.

Among the large complement of ion channels expressed in mammalian neurons are an especially wide variety of voltage-dependent potassium (Kv) channels. Kv channels can regulate diverse electrical properties in neurons, including action potential amplitude and duration, the fre-

quency of cell firing, the kinetics and amount of neurotransmitter release, and the cells resting membrane potential (Hille 2001). Kv channels function as supramolecular protein complexes composed of four pore-forming and voltage-sensing principal (or α) subunits, plus up to four associated auxiliary (or β) subunits. In heterologous cells, these Kv α and β subunits can heteromultimerize to yield biophysically and pharmacologically distinct $\alpha_4\beta_4$ channel complexes. Kv channels can also associate with various scaffolding proteins and enzymes, which can impact channel localization, turnover, and function. As such, the functional characteristics, abundance, and subcellular localization of Kv channel complexes are determined by diverse protein-protein interactions, both between constituent α and β subunits, and between these subunits and a wide variety of interacting proteins (Song 2002).

Expression of Kv channel genes is highly regulated, with specific promoter elements acting in concert with transcriptional machinery to achieve precise temporal and spatial cellular patterns of expression (Mandel et al. 1993). Although evidence for dynamic regulation of ion channel translation has not been observed, it is clear that multiple posttranslational processes, including the protein-protein interactions mediating subunit assembly, folding, export from the endoplasmic reticulum (ER), intracellular trafficking, localization in the plasma membrane, and removal via endocytosis, regulate channel function. In addition, interactions with specific intracellular enzymes lead to covalent modification of Kv channel α and β subunit polypeptides, which can also impact diverse characteristics of ion channel proteins in neuronal membranes.

Many Kv channel α subunits carry specific intracellular trafficking signals that regulate their exit from the rough ER, as well as their stepwise transit through the biosynthetic pathway. Polarized sorting of channel proteins to axon- or dendrite-directed cargo vesicles in the *trans*-Golgi network, targeted insertion at discrete sites within axonal or dendritic plasma membrane subdomains, and active retention at these sites to maintain discrete patterns of channel localization in the face of membrane fluidity are also mediated by discrete signals on Kv channel α subunits. Kv channel β subunits can also profoundly influence these characteristics.

Mechanisms that regulate channel assembly, folding, and ER export of Kv channels are beginning to be understood in much greater detail, in contrast to the mechanisms used to establish the correct subcellular localization of Kv channels. ER export competence seems to be determined by two major mechanisms, the chaperone-based quality control machinery that senses and acts on the folding state of proteins while in the ER, and machinery that recognizes specific trafficking determinants. Whether these systems operate independently to control exit of proteins from the ER, or are simply different aspects of the same system is not clear. Specific intracellular mechanisms also exist to sort newly synthesized Kv channels into axon- or dendrite-destined transport vesicles. It seems likely that, given the highly restricted localization of many Kv channels, mechanisms exist for insertion of channel transporting vesicles at specific sites (e.g., perisynaptic, juxtaparanodal, axon initial segment, distal dendritic). Once present in the plasma membrane, such restricted localization, as well as inclusion of Kv channels in high-density clusters, must be maintained in the face of the dynamic lateral mobility of the lipid bilayer (see Chapter 1).

This chapter focuses on recent insights into the mechanisms mediating intracellular trafficking of Kv channels, as well as those involved in generating and maintaining the distinct subcellular localization of Kv channels in neurons. This review focuses on the principal subunits of the classical *Shaker*-related mammalian Kv family (i.e., Kv1–Kv4).

II. MOLECULAR PROPERTIES OF Kv CHANNELS

A. Nomenclature of Kv Channel Genes

The nomenclature system for Kv channel α subunits, originally proposed by Chandy and colleagues and now widely accepted (Chandy 1991; Gutman et al. 2003), is based primarily on the relatedness of amino acid sequences between the different Kv α subunits. The principal physiologically permeant ion of these channels is denoted by the chemical symbol (K for potassium), followed by the abbreviation of the ligand, which in this case is voltage (v). The remainder of the nomenclature relates to the gene families within these ion channel groups. For example, the prototypical Kv channels have been divided into four families (Kv1–Kv4) based on their relative similarity of amino acid sequence, and on relatedness to their single gene orthologues in *Drosophila*: Kv1 (*Shaker*), Kv2 (*Shab*), Kv3 (*Shaw*), and Kv4 (*Shal*). A parallel nomenclature for Kv channel α subunit genes has also been developed, with genes named KCN*, the four gene families assigned the letters A through D (i.e., Kv1–Kv4 = KCNA–KCND) and the specific gene numbers following the Kv nomenclature (e.g., Kv1.1 = KCNA1, Kv1.4 = KCNA4, Kv2.1 = KCNB1, Kv4.2 = KCND2, etc.).

B. Kv Subunit Gene Expression: Diversity, Classification, and Splice Variants

The human genome contains a total of 16 KCNA–D or Kv α subunit genes. Some of these genes generate messages that are subject to alternative splicing. In mammalian brain, the expression of many of these Kv channel subunits is restricted to neurons, although glial cells may also express a subset of the neuronal repertoire. In general, Kv channels exhibit subfamily-specific patterns of subcellular localization. Kv1 channels are found predominantly on axons and nerve terminals; Kv2 channels on the soma and dendrites; Kv3 channels can be found in dendritic or axonal domains, depending on the subunit and cell type; and Kv4 channels are concentrated in somatodendritic membranes. However, the cellular locations and extent of coassembly of α subunits to yield heteromeric channels are highly variable within each subfamily. For example, Kv1 family members exhibit extensive coassembly, but Kv2 family members do not. Alternative splicing can contribute to this complexity by generating a number of functionally distinct α subunit isoforms (Pongs et al. 1999). In addition, assembly with auxiliary subunits can dramatically impact expression, localization, and function of the resultant channel complexes. For example, inclusion of the Kvβ1.1 subunit in Kv1 channel complexes containing Kv1.1 or Kv1.2 dramatically alters channel gating properties, converting the channels from sustained, or delayed-rectifier type, to rapidly inactivating, or A-type. The specific subunit composition of native complexes can dramatically impact both the expression level, localization, and function of Kv1 channels in mammalian neurons (Trimmer 1998b).

The three most abundant Kv1 α subunits expressed in mammalian brain, Kv1.1, Kv1.2, and Kv1.4, are predominantly localized to axons and nerve terminals (Trimmer et al. 2004) (see Figure 12.3). These α subunits form heterotetrameric channel complexes, as Kv1.1, Kv1.2, and Kv1.4 (together with Kvβ subunits) and exhibit very similar patterns of localization in many brain regions, and biochemical association (Rhodes et al. 1997). Immuno-electron microscopy has revealed that Kv1.1, Kv1.2, and Kv1.4 are concentrated along axons and in the axonal membrane immediately preceding and/or within axon terminals (Sheng et al. 1992; Wang et al. 1994; Cooper et al. 1998). The localization of these channels on axons and nerve terminals in each of the hippocampal subfields has been confirmed in excitotoxic lesion studies

(Monaghan et al. 2001). Activity of Kv1 channels at these sites can play a critical role in regulating the extent of nerve terminal depolarization and neurotransmitter release.

Kv2 family members form delayed rectifier Kv channels that are expressed prominently in mammalian brain, where they are localized in the somatodendritic domain of neurons. Kv2.1 was the first member of this family identified in molecular cloning studies and is unique in that it was identified and isolated by expression cloning (Frech et al. 1989). Kv2.2 is expressed in many of the same cells that express Kv2.1. However, unlike other Kv channels (Kv1, Kv3, and Kv4 family members), the two members of the mammalian Kv2 family apparently do not readily form heteromultimers in native neurons (although, see Blaine et al. 1998), such that the subcellular localizations of Kv2.1 and Kv2.2 expressed in the same cells are distinct (Hwang et al. 1993a,b; Lim et al. 2000). Kv2 family α subunits are prominently expressed in mammalian brain. Within individual neurons, the localization of Kv2.1 is highly restricted and is present on only the somatic and proximal dendritic membrane, but it is absent from the distal dendrites, axons, and nerve terminals. These light microscope level analyses have been confirmed by excitotoxic lesions (Monaghan et al. 2001) and immuno-electron microscopy (Du et al. 1998). Intriguingly, within these membrane domains, Kv2.1 is present in large clusters (Trimmer 1991; Hwang et al. 1993b; Maletic-Savatic et al. 1995; Scannevin et al. 1996; Lim et al. 2000), which are present on the cell surface membrane immediately facing astrocytic processes. KV2.1 is also present in clusters over subsurface cisterns (Du et al. 1998), which are intracellular membrane compartments rich in inositol triphosphate, ryanodine receptors, and intracellular Ca^{2+} stores (Antonucci et al. 2001).

Kv3 currents can have either sustained (Kv3.1, Kv3.2) or transient (Kv3.3, Kv3.4) characteristics, and can form heterooligomeric channels with intermediate gating characteristics (Rudy et al. 1999). Kv3 mRNAs are somewhat unusual among Kv α subunit mRNAs in that they are subjected to extensive alternative splicing to generate subunits that differ only at their cytoplasmic carboxyl termini (Luneau et al. 1991). For example, studies on the exogenous expression of three different Kv3.2 splice variants (Kv3.2a, b, and c) in polarized epithelial cells revealed that alternative splicing led to differences in subcellular localization. The Kv3.2a variant was localized to the basolateral membrane, whereas the Kv3.2b and Kv3.2c isoforms were found apically (Ponce et al. 1997). The epithelial cell : neuron analogous membrane hypothesis (Dotti et al. 1990) (also see Chapter 13) predicts that as such, in neurons, Kv3.2a would be localized to the somatodendritic domain, and Kv3.2b and Kv3.2c would be localized to the axon. Alternative splicing of mammalian brain Kv3.1 leads to differences in the polarized expression of Kv3.1 variants in mammalian central neurons (Ozaita et al. 2002).

The Kv4 α subunits Kv4.1, Kv4.2, and Kv4.3 form transient or A-type Kv channels. Experimental knockdown of Kv4 α subunit expression in mammalian neurons results in suppression of A-type Kv channels (Malin et al. 2000; 2001). Kv4.1 is expressed at very low levels in mammalian brain (Serodio et al. 1998), and mRNA expression that can be detected in neurons does not correlate with A-type current density (Hattori et al. 2003). In contrast, Kv4.2 and Kv4.3 are expressed at relatively high levels (Serodio et al. 1998), and their expression patterns correlate well with neuronal A-type current density in a number of neuronal types (Shibata et al. 1999; Tkatch et al. 2000; Liss et al. 2001; Song 2002; Hattori et al. 2003). Kv4.2 is expressed at high levels in many principal cells, whereas Kv4.3 is found in a subset of principal cells and in many interneurons (Trimmer et al. 2004). *In situ* hybridization analysis reveals that the expression of Kv4.2 and Kv4.3 is widespread throughout the brain and that in many brain regions the cellular expression of these two Kv4 genes is reciprocal or complementary. However, there are numerous cells in which Kv4.2 and Kv4.3 are coexpressed (Serodio et al. 1998). Although a detailed report of the expression and localization of the Kv4 auxiliary subunit DPPX has not yet been published, localization of KChIPs has been studied extensively. These studies have revealed that KChIPs are concentrated in somatodendritic membranes where their distribution corresponds closely with that described for Kv4.2 and/or Kv4.3. The distribution of KChIP1 closely matches the distribution of Kv4.3, particularly in neocortical, hippocampal, and striatal interneurons and in the dendrites of cerebellar granule cells (Rhodes et al. 2004; Strassle et al. 2005).

C. Structure of Kv Subunits

Kv channels exhibit extensive of amino acid sequence identity within the transmembrane and pore-forming domains. The *Shaker* Kv channel from *Drosophila melanogaster* was the first Kv channel to be characterized (Tempel et al. 1987), and remains one of the best studied Kv channels. *Shaker* channels are tetramers of α subunits (MacKinnon 1991), and each α subunit consists of six transmembrane segments, S1–S6 (see Figure 12.1). The fourth transmembrane segment, or S4, of each subunit acts as the main component of the *voltage sensor* that detects the electrical potential across the membrane and controls the voltage-dependent gating of the channel (Papazian et al. 1991; Yellen 1998). The pore, which is responsible for rapid and selective potassium ion conduction, is formed by the close association of the last two transmembrane segments (S5 and S6) from each of the four subunits around a central water-filled cavity (Zhou et al. 2001; Nelson 2003; Noskov et al. 2004). The molecular conformations of the voltage sensor in resting and

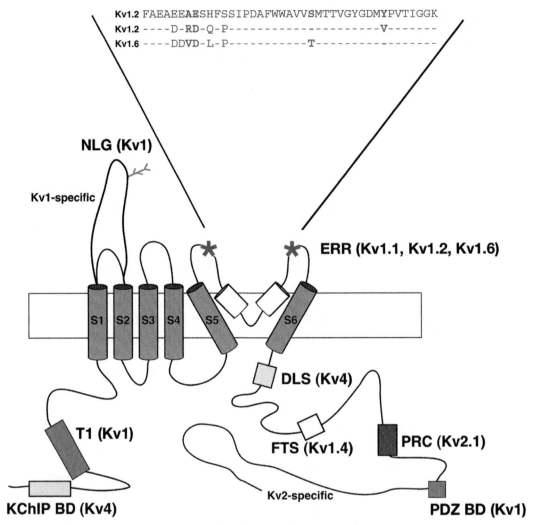

Kv1.2 FAEAEEAESHFSSIPDAFWWAVVSMTTVGYGDMYPVTIGGK
Kv1.2 ----D-RD-Q-P--------------------V-------
Kv1.6 ----DDVD-L-P-----------T----------------

FIGURE 12.1. **Cartoon summary of predicted topology and trafficking signals on Kv α subunits.** The predicted transmembrane topology of Kv α subunits and sites defined as important determinants of intracellular trafficking and polarized expression are shown. KChIP BD, binding domain for KChIP auxiliary subunits that are critical for regulating intracellular trafficking of Kv4 α subunits; T1, tetramerization domain important for Kvβ–mediated intracellular and polarized trafficking of Kv1 α subunits; NLG, N-linked glycosylation site of Kv1 α subunits; ERR, ER retention motif found in Kv1.1, Kv1.2, and Kv1.6 α subunits. Residues within the 41 amino acid P-domain of Kv1.1, Kv1.2, and Kv1.6 are aligned. The critical residues of the ERR motif are colored in red. DLS, dendritic localization signal in Kv4 α subunits; FTS, forward-trafficking signal in Kv1.4; PRC, proximal dendritic clustering signal in Kv2.1; PDZ BD, PDZ-binding motif in Kv1 α subunits.

active state, and how they are allosterically coupled to gating of the central pore domain are not well understood. One model of voltage-dependent gating derived from the structure of the bacterial KvAP channel suggests that the S4 segment is located at the periphery of the protein, where positively charged S4 residues that sense changes in membrane potential are in contact with the hydrocarbon lipids, and S2 lies close to the pore. Other models assume that S4 is a transmembrane segment surrounded by other parts of the protein (Cohen et al. 2003). The mechanisms for coupling of voltage sensor movement to channel opening are not yet clear.

III. DETERMINANTS OF INTRACELLULAR TRAFFICKING OF Kv CHANNELS

Neurons use diverse mechanisms to control their Kv channel cell surface expression levels, which can profoundly affect neuronal excitability and signaling (Hille 2001). Expression of Kv channel α subunit genes at varying levels results in varying stoichiometries of α subunits incorporated into tetrameric channels, and can influence channel trafficking, localization, and function. Kv1 homomeric channels formed by different α subunits exhibit distinct protein stability and trafficking. These patterns are altered upon coassembly into heteromeric complexes, such that the precise α subunit composition affects both the overall level of the protein and its trafficking characteristics. Subunit composition can affect Kv channel cell surface expression levels and function. The overall steady-state cell surface expression levels and subunit composition appear to be regulated by a hierarchical system of regulatory steps, many of which operate at the level of the ER. The ER is a highly versatile organelle equipped with chaperones and folding enzymes that are essential for protein folding and export, and also for retention and degradation of misfolded or aberrantly assembled protein complexes.

A. Trafficking Signals Responsible for ER Retention

Each of the Kv1 α subunits exhibit high amino acid sequence identity (Stuhmer et al. 1989), but show differences in trafficking and function (Papazian 1999; Manganas et al. 2000; Trimmer et al. 2004). The primary determinant for regulating trafficking of Kv1 α subunits appears to be a potent ER retention (ERR) signal consisting of residues in the turret region (see Figure 12.1) at the external face of the channel pore domain (Manganas et al. 2001b; Zhu et al. 2001). Cell surface expression of Kv1 sub-

units can also be influenced by a cytoplasmic C-terminal VXXSL motif (Li et al. 2000), and by auxiliary Kvβ subunits (Shi et al. 1996) (see Figure 12.1). In heterologous cells, these factors lead to a steady-state distribution whereby homotetrameric Kv1.1, Kv1.6 channels are localized in the ER, Kv1.2 channels in both the ER and at the cell surface, and Kv1.4 channels mainly on the cell surface (Manganas et al. 2000; Zhu et al. 2003a) (see Figure 12.2). Kv1.1 homomeric channels in the ER appear to be properly folded and assembled as tetramers with no evidence of aggregation or gross misfolding (Manganas et al. 2001a; Zhu et al. 2003a). The generation of a chimera between Kv1.1 and Kv1.4 revealed that any Kv1.4 α subunit containing the Kv1.1 pore region (P-loop), including the turret domain, is ER-retained (Manganas et al. 2001b) (see Figure 12.2). Conversely, any Kv1.1 α subunit containing the Kv1.4 P-loop is exported efficiently from the ER. An alignment of the Kv1.1 and Kv1.4 P-loop sequences revealed three key positions, all in the turret domain, which differed between Kv1.1 and Kv1.4. Mutation of those residues in Kv1.1 to those in Kv1.4 yields functional Kv1 channels that have Kv1.4-like trafficking patterns, and *vice versa* (Manganas et al. 2001b). Studies of these determinants for Kv1 channel surface expression have been conducted primarily in heterologous expression systems, although critical experiments have been reproduced in cultured neurons. This Kv1.1 ERR signal has also been shown to function in cultured hippocampal neurons (Manganas et al. 2001b). Analysis of a large number of Kv1.1/Kv1.4 chimeras and truncation, deletion, and point mutants suggests that this luminal ERR motif is dominant over the cytoplasmic C-terminal forward trafficking signal, and also dominant over the effects of Kvβ subunit coexpression on Kv1 α subunit trafficking (Manganas et al. 2001b). Although small but measurable increases of Kv1.1 surface expression have been observed in cells coexpressing Kvβ2 subunits (Shi et al. 1996),

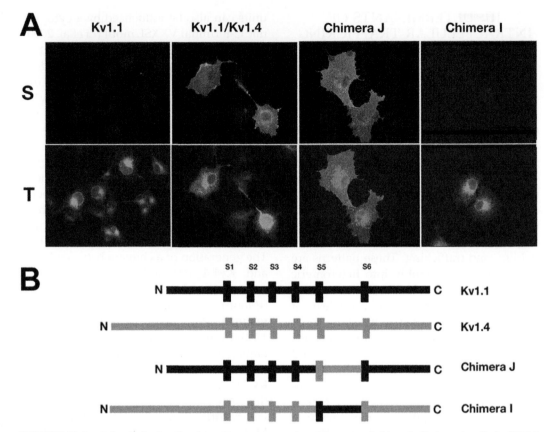

FIGURE 12.2.   **Subcellular localization of homotetrameric wild-type and chimeric Kv1 a subunits in COS-1 cells. A.** Kv1.1 and Chimera I accumulate in the ER, whereas Kv1.4 and Chimera J are present on the cell surface. S, surface staining with ectodomain-directed antibody in the absence of detergent permeabilization; T, total staining with cytoplasmic domain-directed antibody after detergent permeabilization. **B.** Cartoon of the segments used to generate these Kv1 chimeras, Kv1.1 segments are in black, Kv1.4 segments in grey: Chimera J, Kv1.1 N(1–321)-Kv1.4 (475–542)- Kv1.1 C(390–495); Chimera I Kv1.4 N(1–474)- Kv1.1 (322–389)- Kv1.4 C(543–654).

the effects are small compared to auxiliary subunit effects on other Kv channels. Manganas and collaborators showed that three amino acids residues of the Kv1.1 highly conserved pore region are critical in the ERR: A352, E353, and Y379 (see Figure 12.3) (Manganas et al. 2001b; Zhu et al. 2003a, 2005). Subsequent studies yielded similar results and also revealed an important role of S369 in the Kv1.1 ERR (Zhu et al. 2003a, 2005). The Kv1.1 ERR appears to be dictated only by these four amino acid residues of the pore region. Point mutations of other Kv1 subunits at equivalent position of Kv1.1 S369 and Y379 affect cell surface trafficking efficiency (Zhu et al. 2005). This

study proposed that the Kv1s could be controlled by a positive trafficking code located only at equivalent position of Kv1.1 S369 and Y379 (see Figures 12.1 and 12.3). The amino acid replacement of S369 and Y379 in different constructs suggests that the positive trafficking code for the relative surface expression levels may be T369, K379 > T369, V379 > T369, Y379 > T369, R379 ~ T369, H379 > S369, V379 > S369, Y379 > S369, K379 (Zhu et al. 2005). This trafficking code is not transferable to Kv2, Kv3, and Kv4 subfamily members; it seems to be unique to the Kv1 subfamily (Zhu et al. 2005). However, this study did not include the potential role of the two other critical amino acid residues

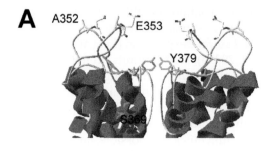

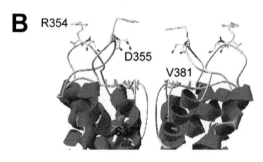

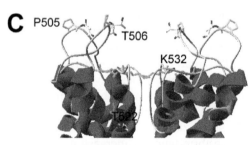

FIGURE 12.3. **Comparison of the X-ray crystallographic structure of the pore region of wild-type Kv1.2, and that predicted for Kv1.2 point mutants in critical residues of the ERR motif.** The Kv1.2 crystal structure was generated with Swiss PDB viewer using the coordinates (Protein Data Bank accession number 2A79) of Long et al. (2005). Views of pore helices (S5 and S6 in red) of two subunits and the side chains of the four critical residues of ERR motif of each subunit. **A.** Kv1.2 with the four critical ERR residues of Kv1.1 at equivalent positions; **B.** wild-type Kv1.2; **C.** Kv1.2 with four ERR critical residues of Kv1.4 at equivalent positions.

A352, E353 in the Kv1 trafficking code (see Figure 12.3). Currently, this ERR signal is the first example of an ERR signal localized to an ionic channel pore, and is unusual among membrane proteins in its luminal localization.

One intriguing and surprising aspect of the identity of the three of the four P-loop residues that dictate Kv1 family trafficking is that they are the same as those that determine high-affinity binding to the mamba snake neurotoxin dendrotoxin or DTX (in Kv1.1 A352, E353, and Y379) (Hurst et al. 1991; Tytgat et al. 1995; Imredy et al. 2000). Moreover, each of the Kv1 family members that bind DTX (Kv1.1, Kv1.2, and Kv1.6) exhibit a strong degree of ERR relative to the Kv1 family members (Kv1.3, Kv1.4, and Kv1.5) that lack the critical binding residues and as such do not bind a DTX (Dolly et al. 1996; Manganas et al. 2000, 2001b; Tiffany et al. 2000). This allows for speculation that the ERR of Kv1.1, and of Kv1.2 and Kv1.6, may be mediated by a resident ERR receptor protein that binds to the turret domain of Kv1.1 in a fashion similar to the binding of a DTX. This mechanism is attractive in that should the resident ERR receptor resemble DTX, it may block the pore of ER-retained Kv1 channels, preventing any deleterious effects of ER-localized Kv1 channels on the membrane integrity of the ER. This hypothetical ERR receptor for Kv1.1 and other DTX-sensitive Kv1 α subunits remains to be identified.

A specific perinuclear localization (i.e., ER and *cis*-Golgi complex) also has been described for Kv4.2 homotetramers in heterologous cells (An et al. 2000; Bahring et al. 2001; O'Callaghan et al. 2003; Shibata et al. 2003). Kv4.2 tends to misfold and/or aggregate, as suggested by perinuclear localization, lack of cell surface expression, and insolubility in nonionic detergents (Trimmer 1998a; Shibata et al. 2003). Moreover, Kv4.2 exhibits a markedly shorter half-life in the absence of the auxiliary subunit KChIPs coexpression (see later), consistent with its misfolded and ER-retained phenotype (Shibata et al. 2003). One important signal for the trafficking of membrane channel proteins is the RXR ERR motif. This motif has been shown to be necessary for ERR in both KATP channels (Zerangue et al. 1999) and the NR1 subunit of N-methyl-D-aspartate (NMDA) receptors (Scott et al. 2001). Although an RXR

ERR signal has not been identified specifically in Kv4.2, there exists an RKR in its N-terminus, an RYR in an intracellular loop between the second and third transmembrane domains, and an RIR in its carboxyl terminus, any of which would serve as a good candidate for ER retention. Mutation of the N-terminal RKR potential ERR signal did not alter the ER localization of Kv4.2 in COS cells (Shibata et al. 2003), suggesting that this particular site is not involved in ER retention; however, deletion of the Kv4.2 N-terminus produced a redistribution of Kv4 immunoreactivity to the cell membrane (Bahring et al. 2001). These observations strengthen the possibility that the Kv4 N-terminus may contain an ERR signal. Deletion of the Kv4 N-terminus might attenuate the ERR signal and redistribute Kv4 channels to the cell surface (Bahring et al. 2001).

B. Trafficking Signals Responsible for ER Export

Intrinsic sequences and/or interaction with auxiliary subunits can regulate ER export. Kv1.4 homotetrameric channels expressed in a wide variety of heterologous expression systems (COS-1, HEK293, CHO, MDCK cell lines), and in cultured hippocampal neurons, are efficiently expressed on the cell surface (Bekele-Arcuri et al. 1996; Shi et al. 1999; Li et al. 2000; Manganas et al. 2000). A large proportion of the steady-state Kv1.4 cellular pool (between 80 and 90%) is present on the cell surface. This is due to the lack of the critical amino acids in the turret region that mediate ER retention in Kv1.1 (Manganas et al. 2001b), combined with the presence of a unique ER export or forward trafficking signal (FTS) in the cytoplasmic carboxyl-terminal region (Li et al. 2000). This cytoplasmic FTS appears to be recessive to the turret domain ERR signal (which would be luminal in the ER), since Kv1.4 chimeras with active ERR signals from Kv1.1 but that are otherwise composed of Kv1.4 (including an intact FTS) are efficiently retained in the ER. However,

simple deletion of this VKESL ER export/FTS signal decreases the level of Kv1.4 surface expression (Li et al. 2000; Zhu et al. 2003b). Kvβ subunits do not dramatically affect the trafficking of wild-type Kv1.4 α subunits. This may be due to the fact that the inherent trafficking properties of these proteins appear to be near the maximum efficiency possible, and Kvβ subunit effects on trafficking of FTS-deficient Kv1.4 have not been investigated. Kvβ1 and Kvβ2 can promote the stability, N-linked glycosylation, and surface expression of homomeric Kv1.2 channels (Shi et al. 1996; Campomanes et al. 2002). The effect of Kvβ subunits such as Kvβ2 are to increase the overall levels of those subunit combinations that already exhibit some intrinsic trafficking efficiency. Thus, Kvβ2 effects are obvious for Kv1.2 and for heteromeric channels containing Kv1.1, Kv1.2, and Kv1.4, but are not detectable for Kv1.1 homotetramers. One possible explanation is that the main consequence of Kvβ2 association is to stabilize these Kv1 channel complexes, a phenomenon shown previously for Kv1.2 homotetramers (Shi et al. 1996). This stabilization could indirectly impact surface abundance by increasing the chance that a Kv1 channel complex would have to achieve ER export competence. This may be accomplished through chaperone-assisted folding and/or saturation of retention receptor sites by increasing the lifetime of the channel complex in the ER. Kvβ subunits such as Kvβ2 are able to promote plasma membrane abundance of most Kv1 channels regardless of subunit composition, suggesting that these cytoplasmic subunits are not directly involved in masking retention signals, but mediate their effects indirectly via other mechanisms, such as stabilization or folding.

C. Posttranslational Events Affecting Kv Channels

An enormous variety of posttranslational modifications of mammalian membrane proteins have been described, including glycosylation, phosphorylation, methylation, and ubiquitination. Posttranslational modification of proteins can play an important role in the proper folding, assembly, and trafficking, as well as in dynamic regulation of function. In this section, we will discuss the role of glycosylation and phosphorylation of Kv channels in channel trafficking.

Asparagine (N)-linked glycosylation (NLG) of membrane and secreted proteins has been shown (Noskov et al. 2004) to promote proper protein folding in the ER, as well as to alter protein transport and targeting (Helenius et al. 2001). In general, glycosylation of Kv channels is not required for efficient trafficking to the cell surface, but does appear to increase surface expression of the channel by decreasing channel turnover and increasing channel stability. For example, blocking NLG of Shaker-type Kv channels dramatically decreases channel stability and steady-state cell surface expression levels, but does not effect the folding (Khanna et al. 2001) and assembly of functional channels, or their transport to the cell surface (Santacruz-Toloza et al. 1994). Kv1.1–Kv1.5, but not Kv1.6, α subunits contain a single consensus site for NLG, which is located in the extracellular linker segment between transmembrane segments S1 and S2 (see Figure 12.1). In native mammalian brain Kv1.1, Kv1.2, and Kv1.4 α subunits, this NLG site carries a sialic acid-bearing oligosaccharide chain, conferring a strong negative charge to the extracellular glycan chain (Sheng et al. 1993; Thornhill et al. 1996; Shi et al. 1999; Manganas et al. 2000). Moreover, the NLG of these Kv1 α subunits, and the spatial segregation of oligosaccharide-processing enzymes within the secretory pathway, allows for tracking of the biosynthetic trafficking of Kv1 α subunits by straightforward biochemical analyses of the glycan chain (Trimmer 1998a). Although Kv2.1 α subunits also carry a single consensus NLG site, on the S3-S4 linker segment, native

brain Kv2.1 channels and recombinant Kv2.1 channels expressed in heterologous systems are not N-glycosylated (Shi et al. 1999). This is consistent with systematic analyses of other polytopic membrane proteins that show a marked preference for glycosylation site usage in the first extracellular loop (as in the Kv1 α subunits) over sites in other extracellular loops (Landolt-Marticorena et al. 1994). Kv3 α subunits contain two N-linked glycosylation sites in the S1-S2 linker. A comprehensive biochemical analysis of the oligosaccharide chains on these sites has not been performed. The full nature and extent of the association and colocalization of Kv3 channels in mammalian neurons is not yet known, in part due to the additional complexity conferred to such analyses by extensive alternative splicing. None of the Kv4-family channels contains a consensus sequence for NLG on their extracellular surface. There are no known strong consensus sequences for O-linked glycosylation, and O-glycosylation of Kv channels has not been reported.

Kv channel α subunits are extensively modified by covalent phosphorylation. This posttranslational modification, governed by a diverse array of protein kinases and phosphatases, can modulate the expression and the function of supramolecular complexes formed from α subunits, auxiliary subunits, and interacting proteins (as discussed later). One important question relevant for the present discussion is whether phosphorylation can dynamically regulate Kv trafficking. Steady-state Kv1.1 current density increases in transfected cells upon stimulation of specific protein kinases (Winklhofer et al. 2003). However, whether this is mediated through effects on biosynthetic trafficking, endocytosis/turnover, or through changes in biophysical properties of active surface channels has not been directly assessed. Recently, Varga and colleagues (2004) identified Kv4.2 as a novel target for calcium–calmodulin-dependent kinase II (CaMKII) regulation. The specific phosphorylation of the identified sites (Ser438 and Ser459) on the C-terminus of Kv4.2 by CaMKII resulted in an increase in channel surface expression and peak current amplitudes, suggesting an effect on trafficking and/or turnover.

D. Auxiliary Subunits of Kv Channels

The best characterized of auxiliary subunits are the cytoplasmic Kvβ subunits associated with Kv1 family members (Pongs et al. 1999). The bulk of Kv1 channel complexes in mammalian brain have associated Kvβ subunits (Rhodes et al. 1996). Kvβ subunits lack putative transmembrane domains and potential glycosylation sites, suggesting that they are cytoplasmic proteins (Scott et al. 1994). Moreover, cryoelectron microscopy studies place their localization on the cytoplasmic face of the channel complex (Sokolova et al. 2003). Three Kvβ genes have been identified: Kvβ1, Kvβ2, and Kvβ3 (Heinemann et al. 1995; Leicher et al. 1998). Kvβ1.1–1.3 proteins arise by alternative splicing from the same gene (England et al. 1995), whereas Kvβ2.1 and Kvβ3.1 are derived from distinct genes. The Kvβ subunits are each approximately 300 amino acids in length and share a common conserved core (over 85% amino acid identity), with the highest degree of variability in the amino termini. The Kvβ subunits are a tetramer of oxidoreductase-like proteins arranged with four-fold rotational symmetry like the integral membrane α subunits (Gulbis et al. 1999). Each oxidoreductase-like Kvβ subunit contains an active site with critical catalytic residues, and a bound NADPH (the reduced form of nicotinamide adenine dinucleotide phosphate) cofactor, but the specific substrate for any enzymatic activity is unknown, and the biological function of the oxidoreductase-like structure of Kvβ subunits remains a mystery (Gulbis et al. 1999). Studies of Kv channel biosynthesis have shown that α and β subunits coassemble in the ER and remain together as a permanent complex (Shi et al. 1996; Nagaya et al. 1997). Kvβ subunits attach to a Kv channel through an interaction with the N-terminal cytoplasmic T1 domain (see Figure 12.1) (Gulbis et al. 2000). Differences in the functional effects of Kvβ subunits often can be ascribed to the variations in the amino terminal variable domain. Arguably the most dramatic functional effect conferred by the association of Kvβ subunits with Kv channels is an increase in the kinetics of channel inactivation gating (Heinemann et al. 1995). In a striking example, Kvβ1 subunits can convert a normally noninactivating delayed rectifier Kv channel to a rapidly inactivating channel (Rettig et al. 1994; Leicher et al. 1998). Kvβ1 and Kvβ2 also have been shown to modulate voltage dependence of Kv channels in heterologous expression systems (England et al. 1995). A number of additional roles have been proposed for the function of Kvβ/α subunit interactions. All the Kvβ subunits promote the cell surface expression of coexpressed α subunits (Levin et al. 1996; Nakahira et al. 1996; Shi et al. 1996; Nagaya et al. 1997; Trimmer 1998b), including Kv4.3 channels (Yang et al. 2001). In these cases, Kvβ subunits may aid in proper protein folding and/or α subunit assembly and thus enhance transport through the secretory pathway from ER-Golgi-plasma membrane.

Calnexin, the resident ER trafficking protein, also regulates Kv1 channel trafficking (Manganas et al. 2004). Calnexin is a type I integral membrane protein found in the ER that functions as a molecular chaperone to facilitate the folding and assembly of newly synthesized membrane proteins (Bergeron et al. 1994; Hammond et al. 1994). Calnexin has been defined as a lectin chaperone that binds to membrane glycoproteins. The interaction occurs between the N-terminal ER luminal domain of the calnexin and the luminal N-linked oligosaccharide chains of glycoproteins (Ou et al. 1993). Homotetrameric Kv1.2 channels, possessing a weak pore-localized ER retention (Manganas et al. 2001b) and weak VXXSN (Li et al. 2000), respond to calnexin coexpression with increased intracellular trafficking and cell surface expression (Manganas et al. 2004). This suggests that calnexin may be acting in its classical role to assist in efficient folding and/or increased stability of Kv1.2 α subunit glycoprotein in the ER. The interaction of the Kv1.2 N-terminus with Kvβ2 subunits increases the efficiency of addition of the N-linked oligosaccharide chain to Kv1.2 α subunits (Shi et al. 1999), presumably through a conformational change at or near the site of addition of the N-linked oligosaccharide chain. This may lead to a higher proportion of Kv1.2 α subunits that are competent for interaction with calnexin, and a subsequent increase in the number of channels that would be correctly folded and exported from the ER. Kvβ2 subunits and calnexin coexpression was epistatic, suggesting that they share a common pathway for promoting Kv1.2 channel surface expression (Manganas et al. 2004). In contrast, the trafficking of homotetrameric Kv1.1 channels that contain strong ER retention motifs (and no VXXSL export motifs) is not influenced by calnexin coexpression.

Auxiliary subunits play important modulatory roles on surface expression and gating of Kv4 channels. Two distinct sets of auxiliary subunits for Kv4 channels have been identified. One set is a family of cytoplasmic calcium-binding proteins, called KChIPs, that are members of the neuronal calcium sensor gene family (An et al. 2000). At least four KChIP genes exist in mammals (An et al. 2000; Holmqvist et al. 2002), and multiple alternatively spliced isoforms of each KChIP gene product have been reported. In heterologous expression systems, all KChIP isoforms increase the surface density (Shibata et al. 2003), slow the inactivation gating, and speed the kinetics of recovery from inactivation (An et al. 2000) of Kv4 channels, with the exception of the KChIP4a splice variant. The molecular

mechanism by which KChIP1, KChIP2, and KChIP3 modify Kv4 expression and gating is beginning to be understood. When heterologously expressed alone in COS-1 or CHO cells, Kv4.2 channels are found mostly retained in the perinuclear ER and/or Golgi apparatus with minimal trafficking to cell surface (An et al. 2000; Bahring et al. 2001; Shibata et al. 2003). Deletion of residues 2–40 in the N-terminal domain of Kv4.2 yields increased Kv4.2 currents in HEK-293 cells, and a positive correlation exists between the size of the N-terminal deletion and current enhancement (Bahring et al. 2001). This suggests that Kv4 channel trafficking is determined primarily by a highly conserved hydrophobic N-terminal sequence. Based on mutagenesis studies, and on the three-dimensional X-ray crystallographic structure of KChIPs, the binding of KChIP masks the ER-retaining N-terminus and allows efficient targeting of Kv4 channels to the cell surface (Kim et al. 2004; Zhou et al. 2004). Possible scenarios for the Kv4–KChIP complex include a dimer of dimers binding to the tetrameric Kv channel, or four KChIP molecules individually associated with each Kv4 subunit in fourfold symmetric fashion. The coexpression of KChIP1-3 with Kv4.2 releases Kv4.2 from its intrinsic ER retention and results in robust trafficking to the Golgi and plasma membrane (An et al. 2000; Bahring et al. 2001; Shibata et al. 2003). A unique splice variant of the KChIP4 gene, KChIP4a, fails to increase surface expression of Kv4 (Holmqvist et al. 2002; Shibata et al. 2003). KChIPs1–3 exhibit diffuse cytoplasmic distribution in COS-1 cells, but KChIP4a immunostaining is perinuclear, suggesting an association with the ER. The N-terminal domain of KChIP4a possesses a long stretch of hydrophobic residues named the Kv channel inactivation suppressor (KIS). It appears that the KIS domain may also direct cytoplasmic KChIP4a to the ER where it interferes with surface trafficking of Kv4 channels. Interestingly, NCS-1 (also known as frequenin), a member of the EF-hand family of Ca^{2+} sensing proteins that includes KChIPs, is also expressed in mouse brain and coimmunoprecipitates with Kv4.2 and Kv4.3 (Nakamura et al. 2001; Guo et al. 2002a). In heterologous systems, NCS-1 coexpression with Kv4 α subunits increases the current density and slows the rate of inactivation of the Kv4.x current. In contrast to KChIPs, however, NCS-1 does not affect the voltage dependence of inactivation or rate of recovery from inactivation of the channel. NCS-1 also can alter the cellular distribution of Kv4 channel (Nakamura et al. 2001; Guo et al. 2002b), suggesting that in addition to regulating the rate of current inactivation for Kv4.2 and Kv4.3, NCS-1 can regulate A-type potassium current by enhancing the surface expression of these channels.

Recently, a new Kv4 auxiliary subunit, a dipeptidyl-peptidase-like protein (DPPX), was identified by immunoprecipitation of Kv4.2 from rat cerebellar membranes (Nadal et al. 2003). DPPX is a type II transmembrane protein belonging to the prolyl oligopeptidase family of serine proteases (Kin et al. 2001; Nadal et al. 2003). This family includes proteins such as DPP-IV (CD26) (Gorrell 2003), fibroblast activation protein (FAP) (Scanlan et al. 1994), prolyl oligopeptidase (POP) (Fulop et al. 1998), DPP8 (Abbott et al. 2000), DPP9, and DPP10 (DPPY) (Qi et al. 2003). Both DPPX and DPPY proteins have a substitution at the catalytic serine (aspartic acid in DPPX and glycine in DPPY) that appears to render them inactive as proteases (Yokotani et al. 1993; Qi et al. 2003). In the absence of a known proteolytic role, DPPX has been identified as an associated component of A-type Kv channels. DPPX coexpression alters the cellular localization and function of Kv4 channels (Nadal et al. 2003; Jerng et al. 2004). Coexpression of DPPX and Kv4.2 increased the surface expression levels 20-fold, increased speed of recovery from inactivation, and shifted the inactivation voltage dependence (Nadal et al. 2003; Jerng et al. 2004). The binding site of DPPX

on the Kv4 channel is unknown at present. DPPX is a single pass, type I membrane protein with a very short intracellular C-terminal tail (31–95 residues depending on the isoform), one putative transmembrane helix, and a large 749 amino acid extracellular domain. The interaction between DPPX and Kv4 is most likely mediated through the transmembrane helix and the extracellular domain. Strop and collaborators (2004) determined the structure of the extracellular domain of DPPX at 3.0 Å resolution. Based on their structure, as well as structural and biochemical data for other Kv channel interacting proteins (fourfold symmetry of the channel as seen for KChIPs and Kvβ), they proposed a different stoichiometry of Kv4 and DPPX: either one dimer (2:4 stoichiometry), or a dimer of dimers (4:4 stoichiometry) of DPPX forms the DPPX:Kv4 complex.

Additional interacting proteins can also affect Kv4 channel trafficking to the plasma membrane or their turnover rate. In general, these accessory factors do not exhibit the ability to alter channel properties, and they have not been as well studied as the modulatory auxiliary KChIP and DPPX subunits. In the brain, Kv4 accessory factors including Kvβs also interact with Kv1 α subunits, PSD-95, and filamin. Confounding reports have suggested that, although they have no effects on Kv4.3 channel gating, Kvβ1 and Kvβ2 interact with Kv4.3 channels in rat brain and mildly enhance their surface expression (by less than twofold) in heterologous systems by interacting with the cytoplasmic C-terminus (later reported to be the N-terminus) (Yang et al. 2001; Wang et al. 2003). PSD-95 proteins contain PDZ domains that bind to the canonical C-terminal PDZ binding motifs present on Kv1.1, Kv1.2, and Kv1.4 channels (Kim et al. 1995), and coexpression of these Kv1 channels with PSD-95 in heterologous cells results in high density plasma membrane channel clusters (Kim et al. 1996; Tiffany et al. 2000). Coimmunoprecipitation of Kv4.2 and PSD-95, that were coexpressed in a mammalian cell line suggests that their association occurs through a noncanonical C-terminal valine–serine–alanine–leucine (VSAL) motif (Wong et al. 2002). PSD-95 coexpression increased the surface expression of Kv4.2 channels by approximately twofold and induced channel clustering. Filamin, another cytoskeletal structural protein, directly binds to actin and Kv4.2 channels at a proline-rich region (PTPP) in the cytoplasmic C-terminal region and increases current density approximately 2.7-fold (Petrecca et al. 2000). Therefore, filamin may act as an intermediary between Kv4.2 and cytoskeletal actin at or near the postsynaptic density.

In addition to these auxiliary subunits, a number of "electrically silent" α subunit-like polypeptides exist in the genome (Drewe et al. 1992). In heterologous expression systems, these polypeptides can coassemble with and functionally modify bona fide Kv α subunits (Patel et al. 1997). However, though there is some information available about expression patterns of these genes in mammalian tissues (Drewe et al. 1992; Salinas et al. 1997), very little is known of their expression at the protein level or their contribution to native mammalian brain Kv channels.

IV. DETERMINANTS OF POLARIZED DISTRIBUTION OF Kv CHANNELS

Polarized localization of ion channels is essential for neurons to generate and maintain local and specialized signaling domains, such as synapses and the nodes of Ranvier. In fact, different Kv channels show distinct patterns of subcellular localization in neurons, as well as distinct patterns of regional distribution. In this section, we discuss possible mechanisms underlying the polarized localization of Kv channels that is observed in neurons, and its dynamic regulation. Studying the intracellular trafficking of Kv channels is difficult because:

(1) Kv channels form complex hetero-multimeric channels in neurons.
(2) Both the N- and C-termini are in the cytoplasm, and extracellular domains are limited to relatively short segments between transmembrane segments, which complicates analyses of surface pool of the proteins by using epitope tags, biotinylation, or ectodomain-directed antibodies.
(3) Expression levels of Kv channels are relatively low.

Despite these complications, there are several compelling studies describing trafficking determinants of Kv channels that provide insights into molecular mechanisms that determine the polarized localization of Kv channels in neurons.

A. Determinants of Somatodendritic Localization of Kv Channels

A number of studies have focused on defining the determinants of the polarized dendritic localization of Kv2.1 and Kv4.2, two major somatodendritic Kv channels. As shown in Figure 12.4, in CA1 pyramidal neurons, Kv2.1 and Kv4.2 show a striking difference in their subcellular localization; Kv2.1 is expressed only in the soma and proximal dendrites, whereas the majority of Kv4.2 is localized in the distal dendrites.

Initial studies of the polarized and clustered localization of Kv2.1 were performed in polarized epithelial MDCK cells. In these cells, Kv2.1 localizes in the basolateral membrane (Scannevin et al. 1996). These data are consistent with the analogous membrane hypothesis, whereby the apical and basolateral membranes of epithelial cells correspond to the axonal and somatodendritic membranes of neurons (Dotti et al. 1990). Interestingly, Kv2.1 is also present in large clusters in the basolateral domains of MDCK cells (Scannevin et al. 1996). Expression of truncation mutants that lack relatively large portions of the cytoplasmic Kv2.1 carboxy terminus revealed that a 130

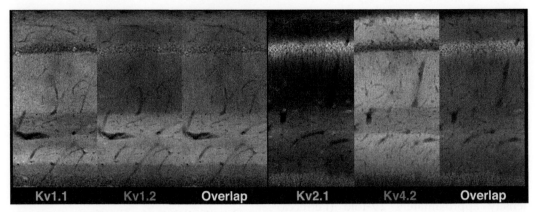

FIGURE 12.4. **Distinct localization of Kv channel subunits in brain.** Saggittal sections from adult rat brain were double stained with antibodies against Kv1.1 (green in the overlap image) and Kv1.2 (red), or Kv2.1 (green) and Kv4.2 (red). Images are from the CA1 region to dentate granule layer of the hippocampus, from the top to the bottom of the image.

amino acid region that lies approximately midway in the 440 amino acid carboxyl terminus of Kv2.1 is necessary for both polarized expression and clustering in MDCK cells (Scannevin et al. 1996). Studies in primary cultures of hippocampal neurons (Lim et al. 2000) revealed that expression of epitope-tagged wild-type Kv2.1 and the two aforementioned mutants in hippocampal neurons yielded results consistent with those in MDCK cells. Thus, wild-type Kv2.1 was found in large clusters on the soma and proximal dendrites of cultured neurons, and the truncation mutants delineated the same 130 amino acid segment as being critical to this localization. Wild-type Kv2.2 exhibited a uniform localization on axons and dendrites, and Kv2.1/Kv2.2 chimeras revealed that the disparate carboxyl-terminal regions of the channels controlled these subtype-specific localizations (Lim et al. 2000).

Additional deletion analysis of Kv2.1 revealed a 25 amino acid segment that is necessary for polarized and clustered localization. Generation of chimeric Kv1.5 α subunits, having this segment appended to the carboxy terminus, is sufficient for dendritic localization and clustering, unlike uniformly expressed wild-type Kv1.5 that is localized to Kv2.1-like clusters on the soma

and proximal dendrites. Interestingly, this targeting signal does not function autonomously when appended to single pass type I transmembrane proteins, suggesting that the structural background of Kv channels, perhaps the tetrameric quaternary structure, is required for proper targeting signal function (Lim et al. 2000). The amino acid sequence of this targeting signal is well conserved in Kv2.2, suggesting that determinants in addition to the primary structure of the targeting signal contribute to its function. An alanine scan through this segment revealed that three of the four critical amino acids are serine residues, suggesting that changes in phosphorylation state may dynamically regulate Kv2.1 localization (see later). Ultrastructural analysis has shown that plasma membrane Kv2.1 clusters on the somata and proximal dendrites of pyramidal neurons lie over subsurface cisternae (Du et al. 1998). These sites, where intracellular Ca^{2+} stores come into close apposition with the plasma membrane, represent a specialized neuronal signaling domain that also may contain elevated levels of voltage-dependent Ca^{2+} channels (Westenbroek et al. 1990). A functional relationship between Kv2.1 and dendritic $[Ca^{2+}]_i$ transients was revealed by antisense depletion of Kv2.1 in rat hippocampal neurons (Du et al. 1998). Moreover, ryanodine receptor intracellular Ca^{2+} release channels colocalize with Kv2.1 clusters in cultured hippocampal neurons (Antonucci et al. 2001), suggesting that the clustered localization of Kv2.1 could affect dendritic Ca^{2+} signaling at these sites.

As described earlier, Kv4 channels exhibit strict polarized somatodendritic localization in neurons (Trimmer et al. 2004). The determinants of dendritic targeting of Kv4 α subunits have been investigated in organotypic cortical slice cultures using an epitope tag incorporated in the extracellular loop, and several different chimeras between Kv4.2 and axonally localized Kv1.3 (Rivera et al. 2003). Analyses of these chimeras revealed a critical 16-amino acid dileucine-containing motif in the cytoplasmic carboxyl-terminal region of Kv4.2 that is conserved in all Kv4 family members, from nematodes to mammals (Rivera et al. 2003). Unlike the determinant for polarized localization and clustering of dendritic Kv2.1 described previously, the Kv4 targeting signal also targets a type 1 membrane protein (CD8) to dendrites (Rivera et al. 2003). It should be noted that these results were obtained in a neuronal background that presumably contains endogenous KChIPs and/or possibly DPPX; therefore, the contribution of these interacting proteins to the effects of this targeting signal cannot be determined. Thus, it will be important to determine whether this trafficking signal directs polarized Kv4 targeting in an expression background lacking these proteins (e.g., MDCK cells), or whether the Kv4.2 mutants that lack the ER extension motif, and the KChIP binding site, localize to the dendrites (Shibata et al. 2003). A direct determination of the role of KChIPs and DPPX in contributing to the polarized expression of Kv4 channels is also needed. Other proteins that may interact with Kv4 α subunits, such as PSD-95 and filamin, that enhance the steady-state cell surface expression level and clustering of Kv4.2 also may play a role in polarized

localization (Wong et al. 2002). Interestingly, in cultured hippocampal neurons, the obvious clusters of Kv4.2 and PSD-95 (as well as filamin) do not colocalize, but rather appear interdigitated along the dendrites (Shibata et al. 2003). The roles of these proteins in the polarized expression of Kv4 channels remain to be elucidated.

B. Determinants of Axonal Localization of Kv Channels

The precise determinants for the polarized expression and local clustering of Kv1 channels are not well characterized. Difficulty recapitulating the heteromeric channels found in native neurons in model cell systems (e.g., polarized epithelial cells), compounded by the inefficient intracellular trafficking inherent to many of the Kv1 α subunits have limited progress on this question. Moreover, for reasons that are not well understood, efficiently expressed Kv1 α subunits, such as Kv1.4, when expressed in such model systems do not "behave"; for example, axonal Kv1.4 is found in the basolateral membrane of polarized MDCK epithelial cells (Le Maout et al. 1996). However, recent studies have provided compelling data suggesting that cytoplasmic Kvβ subunits are involved in polarized trafficking of Kv1 α subunits. Kv1.2 α subunits overexpressed in cultured hippocampal neurons exhibit somatodendritic localization, however, cotransfection of the Kvβ2 subunit yields a pronounced axonal localization of Kv1.2 (Campomanes et al. 2002). Consistent with these studies, analysis of Kv1.2 deletion mutants and chimeric channels revealed that the amino terminal T1 domain that comprises the Kvβ subunit binding site (Sewing et al. 1996; Yu et al. 1996; Gulbis et al. 2000) is essential for axonal expression of Kv1.2 channels (Gu et al. 2003). Specific mutations that disrupt Kvβ2 binding also disrupt axonal expression of Kv1 α subunits in transfected neurons, further suggesting that endogenous Kvβ subunits contribute to the axonal

Kv1 localization (Gu et al. 2003). Remarkably, the wild-type Kv1.2 T1 domain, but not mutants with altered Kvβ binding, is able to direct the axonal localization of single-pass transmembrane reporter proteins (Gu et al. 2003). Thus, the cytoplasmic Kvβ subunits may affect not only early biosynthetic processing events and ER export (Shi et al. 1996), but also axonal localization (Campomanes et al. 2002; Gu et al. 2003) of Kv1 channels. The importance of Kvβ2 is underscored by the finding that deletion of the human Kvβ2 gene in chromosome 1p36 deletion syndrome is closely linked to epilepsy that is present in a subset of these patients (Heilstedt et al. 2001). Additionally, Kvβ2 null mice exhibit occasional seizures, and cold swim-induced tremors similar to that observed in Kv1.1-null mice (McCormack et al. 2002). However, Kv1.2 localization is only minimally affected in cerebellar basket cell synaptic terminals in Kvβ2 null mice, although Kvβ1 can substitute for Kvβ2 in effects on Kv1 trafficking (Shi et al. 1996; Campomanes et al. 2002).

MAGUKs such as PSD-95 and SAP97 function as scaffolding molecules to promote coclustering of membrane receptors and ion channels (Sala et al. 2001). Mutational and structural analyses showed that PSD-95 and other MAGUKs bind to the PDZ-binding motif (S/TxV) in the carboxy-termini of all Kv1 α subunits (Doyle et al. 1996; Kim et al. 1996). In most mammalian central neurons, MAGUKs are found at synapses, generally in the postsynaptic density. Kv1 channels are localized to axons, suggesting that Kv1 : MAGUK interaction does not occur frequently in native cells. There are a few examples where Kv1 channels and PSD-95 are colocalized in neurons, such as in cerebellar basket cell terminals (Kim et al. 1996; Laube et al. 1996) and at the juxtaparanodal regions of nodes of Ranvier (Rasband et al. 2001). Mutation of the *Drosophila* discs-large (*dlg*) gene, a homologue of PSD-95, results in failure to localize Shaker potassium channels to the

neuromuscular junction of flies, suggesting dependence on *dlg* for potassium channel localization (Rasband et al. 2002). However, the localization of Kv1.1 and Kv1.2 in basket cell terminals and juxtaparanodes is normal in transgenic mice expressing a truncated form of PSD-95 that is no longer expressed at these sites. These data suggest that MAGUKs are not required for Kv1 channel localization in neurons and the interacting proteins that presumably cluster Kv1 channels in axons and near nerve terminals remain to be identified.

V. PATHOLOGICAL ALTERATIONS IN Kv CHANNEL TRAFFICKING AND DISTRIBUTION

Kv channels control the membrane excitability of neurons. Failure to regulate Kv channels may result in severe neurologic disorders with hyperexcitability, such as ataxia or epilepsy. The sorting mechanisms responsible for directing Kv channels to different compartments within the neuronal membrane can impact local channel availability and the resultant aberrations in neuronal excitability can lead to neuronal dysfunction.

A. Episodic Ataxia

Episodic ataxia type-1 (EA-1) is a neurological syndrome characterized by continuous myokymia and cerebellar ataxia (Kullmann, 2002). EA-1 is caused by recessively inherited mutations in the Kv1.1 subunit. Genetic linkage studies in EA-1 patients have identified at least 14 mutations in the Kv1.1 gene. Most of the EA-1 mutations lead to Kv1.1 channels with altered biophysical characteristics. However, the single EA-1 nonsense mutation that results in the truncation of the last 79 amino acid residues in the C-terminus causes loss of expression, misfolding, and resultant intracellular aggregation of Kv1 channels (Manganas et al. 2001a). Interestingly, expression of this mutant also results in intracellular aggregation and retention of coassembled wild-type Kv1 α and Kvβ subunits. These dominant negative properties of the EA-1 mutant Kv1.1 resemble those of the mutant CFTR protein encoded by the predominant mutation found in cystic fibrosis patients. Furthermore, in cardiac hypertrophy, a reduction in outward transient potassium currents (I_{to}) primarily carried by Kv4.2 channels occurs along with the reduction of KChIP2 expression level (Kuo et al. 2001), which may decrease the trafficking of Kv4.2 to the cell surface (Shibata et al. 2003). Diverse channelopathies therefore result from defects in protein trafficking of Kv channels.

B. Epilepsy

Epilepsy is a common neurological disorder, and in some cases can be linked to channel dysfunction. Nine ion channel genes have been implicated in pathogenesis of human idiopathic epilepsy (Mulley et al. 2003), of which two encode Kv channels (KCNQ2 and KCNQ3). Deletion of Kv1.1 channels in mice results in early postnatal epilepsy (Smart et al. 1998), suggesting a significant role for Kv channels in the development of epilepsy. Deletion of Kv3 channels affects excitability and leads to ataxia and other behavioral changes but does not result in seizures (Matsukawa et al. 2003). There is currently no genetic evidence for the involvement of Kv2 or Kv4 family members in seizures and epilepsy. However, recent studies have shown that both Kv2.1 and Kv4.2 are dynamically modulated under hyperexcitable conditions.

Native Kv2.1 is extensively modified by phosphorylation (Murakoshi et al. 1997; Tiran et al. 2003). There are 75 serine, 36 threonine, and 13 tyrosine residues within the 620 amino acids in the large cytoplasmic N- and C-termini of rat Kv2.1. In animal models of *status epilepticus*, this constitutively high level of Kv2.1 phosphorylation is dynamically reduced (Misonou et al. 2004). In cultured hippocampal neurons, relatively short (10 min) glutamate treatment causes a similar reduction of the phosphorylated form of Kv2.1 by dephosphorylation events mediated through calcineurin. Intriguingly, this dephosphorylation dramatically changes the localization of Kv2.1 from clusters to a uniform distribution on the cell surface. Kv2.1 dephosphorylation also results in an overall increase of Kv2.1 channel availability, which leads to suppression of neuronal excitability. As mentioned earlier, Kv2.1 clusters are localized in close proximity to intracellular Ca^{2+} stores where calcineurin is also enriched (Cameron et al. 1995). This microdomain formation may enable rapid and efficient modulation of Kv2.1 upon signaling, and dispersion of the clusters enables efficient recovery of Kv2.1 phosphorylation away from the local "hot spot" of Ca^{2+} and calcineurin. Given the restricted localization of Kv2.1, these findings indicate that Kv2.1 is a potential "on/off switch" in the proximal dendrites and soma for excitatory electric signals that arise from more distal portions of the dendrites, since all electrical signals must pass through these regions to the site of somal integration and axon potential initiation in the axon hillock.

Distinct from Kv2.1 channels, Kv4.2 channels are expressed in the distal parts of dendrites (Trimmer et al. 2004). The excitability of the distal dendrites of CA1 pyramidal neurons is increased in an animal model of epilepsy, presumably by suppression of Kv4.2 activity through enhanced phosphorylation of the α subunits by extracellular signal-regulated kinases (ERK) (Bernard et al. 2004). Dynamic phosphorylation of Kv4.2 also is implicated in trafficking of this Kv channel (Shibata et al. 2003). Immunolocalization studies of various phosphorylated forms of Kv4.2 reveal the distinct subcellular localization of different phosphorylated forms in hippocampal pyramidal neurons (Varga et al. 2000), indicating that differential phosphorylation may regulate localization of Kv4.2. Epileptic activity therefore may change the phosphorylation state of Kv4.2 and its trafficking pattern, and thus rapidly change the local availability of Kv4.2 in the dendrites. Interestingly, interaction with the Ca^{2+}-binding KChIP auxiliary subunit regulates Kv4.2 phosphorylation, suggesting a complex interplay between synaptic activity, kinase activity, intracellular Ca^{2+}, and dendritic excitability. Such activity-dependent suppression of Kv4.2 by enhanced phosphorylation may underlie the pathology of epilepsy, whereas activity-dependent modulation of Kv2.1 by calcineurin-mediated dephosphorylation may protect against hyperexcitability.

VI. FUTURE DIRECTIONS

A great deal is known of the molecular mechanisms that regulate the biophysical properties of Kv channels. However, much less is known of the mechanisms underlying the trafficking and localization of these channels. Some determinants that control intracellular trafficking, polarized expression, and local clustering of Kv channels have been identified. Recent studies have suggested that trafficking and localization of Kv channels are dynamically modulated in response to both physiological and pathological stimuli that affect membrane excitability. Future studies will focus on identifying cellular components and molecular mechanisms that mediate changes in electrical excitability that occur during neuronal development, in response to neuronal plasticity, in aging and under pathophysiological conditions.

A. Biochemical Analysis of Kv Channels

Phosphorylation is the major mechanism controlling trafficking and localization of neuronal proteins. For example, the phosphorylation state of ionotropic glutamate receptors dramatically changes their intracellular trafficking, thus modulating the availability of these receptors at synapses (Scott et al. 2001; Malinow 2003). In contrast, little is known about the dynamic phosphorylation events of Kv channels in neurons. However, recent progress in identification and quantitation of phosphorylation at specific sites in channel proteins will provide better insights into these posttranslational modifications. Moreover, generation of reliable phosphospecific antibodies will enable a better understanding of dynamic regulation of Kv channel phosphorylation in native cells.

B. Proteomic Identification of Interacting Proteins

Knowing the individual components of Kv channel-containing multiprotein complexes is a first step toward understanding the role of the individual proteins in the protein trafficking, clustering, and regulation of Kv channels. Proteomic approaches can help to identify interacting partners for proteins of interest. The discovery of DPPX as a component of Kv4 channel complexes biochemically purified from mammalian brain suggests that this approach may be applied successfully to other Kv channels (Nadal et al. 2003).

C. Transgenic Animals

Although the functional roles of the diverse family of Kv channels may be partially redundant, and the role of individual subunits difficult to assess in the context of complicated heteromeric channel complexes, transgenic or knock-in mice expressing mutant Kv channels represent an attractive way to study trafficking of Kv channels in neurons. Ectopic expression of Kv channels in certain types of neurons by using cell type specific promoters may help to elucidate the regulatory mechanism responsible for Kv channel targeting *in vivo*

(Mitsui et al. 2005). Knockout animals may also be useful as a source of cultured neurons lacking channels into which various mutants can be expressed to analyze the function of particular domains.

D. Analysis of Kv Channels in Living Neurons

Imaging of fluorescently tagged α subunit polypeptides in live cells will help to analyze trafficking of Kv channels in detail. Green fluorescent proteins (GFPs) and their variants have been used extensively to track the dynamics of glutamate receptors and their interacting proteins. Intrinsically fluorescent proteins, especially those proteins with photo-convertible properties, seem to be very useful for live or time-lapse imaging. Kaede (Ando et al. 2002) and photoactivated-GFP (PA-GFP) (Patterson et al. 2002) are convertible, either from green to red or faint to bright fluorescence upon the photo-conversion, which enable pulse-chase imaging studies. Although there are other variants of GFP, such as a photoswitchable cyan fluorescent protein (Chudakov et al. 2004), these remain to be tested in neuronal cells. Studies using these fluorescent tags likely will yield important information about the molecular mechanisms of Kv channel trafficking in neurons.

References

Abbott, C.A., Yu, D.M., Woollatt, E., Sutherland, G.R., McCaughan, G.W., Gorrell, M.D. (2000). Cloning, expression and chromosomal localization of a novel human dipeptidyl peptidase (DPP) IV homolog, DPP8. *Eur. J. Biochem.* **267**, 6140–6150.

An, W.F., Bowlby, M.R., Betty, M., Cao, J., Ling, H.P., Mendoza, G. et al. (2000). Modulation of A-type potassium channels by a family of calcium sensors. *Nature* **403**, 553–556.

Ando, R., Hama, H., Yamamoto-Hino, M., Mizuno, H., Miyawaki, A. (2002). An optical marker based on the UV-induced green-to-red photoconversion of a fluorescent protein. *Proc. Natl. Acad. Sci. U. S. A.* **99**, 12651–12656.

Antonucci, D.E., Lim, S.T., Vassanelli, S., Trimmer, J.S. (2001). Dynamic localization and clustering of dendritic Kv2.1 voltage-dependent potassium channels in developing hippocampal neurons. *Neuroscience* **108**, 69–81.

Bahring, R., Dannenberg, J., Peters, H.C., Leicher, T., Pongs, O., Isbrandt, D. (2001). Conserved Kv4 N-terminal domain critical for effects of Kv channel-interacting protein 2.2 on channel expression and gating. *J. Biol. Chem.* **276**, 23888–23894.

Bekele-Arcuri, Z., Matos, M.F., Manganas, L., Strassle, B.W., Monaghan, M.M., Rhodes, K.J., Trimmer, J.S. (1996). Generation and characterization of subtype-specific monoclonal antibodies to K+ channel alpha- and beta-subunit polypeptides. *Neuropharmacology* **35**, 851–865.

Bergeron, J.J., Brenner, M.B., Thomas, D.Y., Williams, D.B. (1994). Calnexin: a membrane-bound chaperone of the endoplasmic reticulum. *Trends Biochem. Sci.* **19**, 124–128.

Bernard, C., Anderson, A., Becker, A., Poolos, N.P., Beck, H., Johnston, D. (2004). Acquired dendritic channelopathy in temporal lobe epilepsy. *Science* **305**, 532–535.

Blaine, J.T., Ribera, A.B. (1998). Heteromultimeric potassium channels formed by members of the Kv2 subfamily. *J. Neurosci.* **18**, 9585–9593.

Cameron, A.M., Steiner, J.P., Roskams, A.J., Ali, S.M., Ronnett, G.V., Snyder, S.H. (1995). Calcineurin associated with the inositol 1,4,5-trisphosphate receptor-FKBP12 complex modulates Ca2+ flux. *Cell* **83**, 463–472.

Campomanes, C.R., Carroll, K.I., Manganas, L.N., Hershberger, M.E., Gong, B., Antonucci, D.E. et al. (2002). Kv beta subunit oxidoreductase activity and Kv1 potassium channel trafficking. *J. Biol. Chem.* **277**, 8298–8305.

Chandy, K.G. (1991). Simplified gene nomenclature. *Nature* **352**, 26.

Chudakov, D.M., Verkhusha, V.V., Staroverov, D.B., Souslova, E.A., Lukyanov, S., Lukyanov, K.A. (2004). Photoswitchable cyan fluorescent protein for protein tracking. *Nat. Biotechnol.* **22**, 1435–1439.

Cohen, B.E., Grabe, M., Jan, L.Y. (2003). Answers and questions from the KvAP structures. *Neuron* **39**, 395–400.

Cooper, E.C., Milroy, A., Jan, Y.N., Jan, L.Y., Lowenstein, D.H. (1998). Presynaptic localization of Kv1.4-containing A-type potassium channels near excitatory synapses in the hippocampus. *J. Neurosci.* **18**, 965–974.

Dolly, J.O., Parcej, D.N. (1996). Molecular properties of voltage-gated K+ channels. *J. Bioenerg. Biomembr.* **28**, 231–253.

Dotti, C.G., Simons, K. (1990). Polarized sorting of viral glycoproteins to the axon and dendrites of hippocampal neurons in culture. *Cell* **62**, 63–72.

Doyle, D.A., Lee, A., Lewis, J., Kim, E., Sheng, M., MacKinnon, R. (1996). Crystal structures of a complexed and peptide-free membrane protein-binding domain: molecular basis of peptide recognition by PDZ. *Cell* **85**, 1067–1076.

Drewe, J.A., Verma, S., Frech, G., Joho, R.H. (1992). Distinct spatial and temporal expression patterns of K+ channel mRNAs from different subfamilies. *J. Neurosci.* **12**, 538–548.

Du, J., Tao-Cheng, J.H., Zerfas, P., McBain, C.J. (1998). The K+ channel, Kv2.1, is apposed to astrocytic processes and is associated with inhibitory postsynaptic membranes in hippocampal and cortical principal neurons and inhibitory interneurons. *Neuroscience* **84**, 37–48.

England, S.K., Uebele, V.N., Shear, H., Kodali, J., Bennett, P.B., Tamkun, M.M. (1995). Characterization of a voltage-gated K+ channel beta subunit expressed in human heart. *Proc. Natl. Acad. Sci. U. S. A.* **92**, 6309–6313.

Frech, G.C., VanDongen, A.M., Schuster, G., Brown, A.M., Joho, R.H. (1989). A novel potassium channel with delayed rectifier properties isolated from rat brain by expression cloning. *Nature* **340**, 642–645.

Fulop, V., Bocskei, Z., Polgar, L. (1998). Prolyl oligopeptidase: an unusual beta-propeller domain regulates proteolysis. *Cell* **94**, 161–170.

Gorrell, M.D. (2003). First bite. *Nat. Struct. Biol.* **10**, 3–5.

Gu, C., Jan, Y.N., Jan, L.Y. (2003). A conserved domain in axonal targeting of Kv1 (Shaker) voltage-gated potassium channels. *Science* **301**, 646–649.

Gulbis, J.M., Mann, S., MacKinnon, R. (1999). Structure of a voltage-dependent K+ channel beta subunit. *Cell* **97**, 943–952.

Gulbis, J.M., Zhou, M., Mann, S., MacKinnon, R. (2000). Structure of the cytoplasmic beta subunit-T1 assembly of voltage- dependent K+ channels. *Science* **289**, 123–127.

Guo, W., Li, H., Aimond, F., Johns, D.C., Rhodes, K.J., Trimmer, J.S., Nerbonne, J.M. (2002a). Role of heteromultimers in the generation of myocardial transient outward K+ currents. *Circ. Res.* **90**, 586–593.

Guo, W., Malin, S.A., Johns, D.C., Jeromin, A., Nerbonne, J.M. (2002b). Modulation of Kv4-encoded K(+) currents in the mammalian myocardium by neuronal calcium sensor-1. *J. Biol. Chem.* **277**, 26436–26443.

Gutman, G.A., Chandy, K.G., Adelman, J.P., Aiyar, J., Bayliss, D.A., Clapham, D.E. et al. (2003). International union of pharmacology. XLI. Compendium of voltage-gated ion channels: Potassium channels. *Pharmacol. Rev.* **55**, 583–586.

Hammond, C., Braakman, I., Helenius, A. (1994). Role of N-linked oligosaccharide recognition, glucose trimming, and calnexin in glycoprotein folding and quality control. *Proc. Natl. Acad. Sci. U. S. A.* **91**, 913–917.

Hattori, S., Murakami, F., Song, W.J. (2003). Quantitative relationship between kv4.2 mRNA and A-type K+ current in rat striatal cholinergic interneurons during development. *J. Neurophysiol* **90**, 175–183.

Heilstedt, H.A., Burgess, D.L., Anderson, A.E., Chedrawi, A., Tharp, B., Lee, O. et al. (2001). Loss of the potassium channel beta-subunit gene, KCNAB2, is associated with epilepsy in patients with 1p36 deletion syndrome. *Epilepsia* **42**, 1103–1111.

Heinemann, S.H., Rettig, J., Wunder, F., Pongs, O. (1995). Molecular and functional characterization of a rat brain Kv beta 3 potassium channel subunit. *FEBS Lett.* **377**, 383–389.

Helenius, A., Aebi, M. (2001). Intracellular functions of N-linked glycans. *Science* **291**, 2364–2369.

Hille, B. (2001). Ionic channels of excitable membranes. Sinauer, Sunderland, MA.

Holmqvist, M.H., Cao, J., Hernandez-Pineda, R., Jacobson, M.D., Carroll, K.I., Sung, M.A. et al. (2002). Elimination of fast inactivation in Kv4 A-type potassium channels by an auxiliary subunit domain. *Proc. Natl. Acad. Sci. U. S. A.* **99**, 1035–1040.

Hurst, R.S., Busch, A.E., Kavanaugh, M.P., Osborne, P.B., North, R.A., Adelman, J.P. (1991). Identification of amino acid residues involved in dendrotoxin block of rat voltage-dependent potassium channels. *Mol. Pharmacol.* **40**, 572–576.

Hwang, P.M., Cunningham, A.M., Peng, Y.W., Snyder, S.H. (1993a). CDRK and DRK1 K+ channels have contrasting localizations in sensory systems. *Neuroscience* **55**, 613–620.

Hwang, P.M., Fotuhi, M., Bredt, D.S., Cunningham, A.M., Snyder, S.H. (1993b). Contrasting immunohistochemical localizations in rat brain of two novel K+ channels of the *Shab* subfamily. *J. Neurosci.* **13**, 1569–1576.

Imredy, J.P., MacKinnon, R. (2000). Energetic and structural interactions between delta-dendrotoxin and a voltage-gated potassium channel. *J. Mol. Biol.* **296**, 1283–1294.

Jerng, H.H., Qian, Y., Pfaffinger, P.J. (2004). Modulation of Kv4.2 channel expression and gating by dipeptidyl peptidase 10 (DPP10). *Biophys. J.* **87**, 2380–2396.

Khanna, R., Myers, M.P., Laine, M., Papazian, D.M. (2001). Glycosylation increases potassium channel stability and surface expression in mammalian cells. *J. Biol. Chem.* **276**, 34028–34034.

Kim, E., Niethammer, M., Rothschild, A., Jan, Y.N., Sheng, M. (1995). Clustering of Shaker-type K+ channels by interaction with a family of membrane-associated guanylate kinases. *Nature* **378**, 85–88.

Kim, E., Sheng, M. (1996). Differential K+ channel clustering activity of PSD-95 and SAP97, two related membrane-associated putative guanylate kinases. *Neuropharmacology* **35**, 993–1000.

Kim, L.A., Furst, J., Gutierrez, D., Butler, M.H., Xu, S., Goldstein, S.A., Grigorieff, N. (2004). Three-dimensional structure of I(to); Kv4.2-KChIP2 ion channels by electron microscopy at 21 Angstrom resolution. *Neuron* **41**, 513–519.

Kin, Y., Misumi, Y., Ikehara, Y. (2001). Biosynthesis and characterization of the brain-specific membrane protein DPPX, a dipeptidyl peptidase IV-related protein. *J. Biochem. (Tokyo)* **129**, 289–295.

Kullmann, D.M. (2002). The neuronal channelopathies. *Brain* **125**, 1177–1195.

Kuo, H.C., Cheng, C.F., Clark, R.B., Lin, J.J., Lin, J.L., Hoshijima, M. et al. (2001). A defect in the Kv channel-interacting protein 2 (KChIP2) gene leads to a complete loss of I(to) and confers susceptibility to ventricular tachycardia. *Cell* **107**, 801–813.

Landolt-Marticorena, C., Reithmeier, R.A. (1994). Asparagine-linked oligosaccharides are localized to single extracytosolic segments in multi-span membrane glycoproteins. *Biochem. J.* **302 (Pt 1)**, 253–260.

Laube, G., Roper, J., Pitt, J.C., Sewing, S., Kistner, U., Garner, C.C. et al. (1996). Ultrastructural localization of Shaker-related potassium channel subunits and synapse-associated protein 90 to septate-like junctions in rat cerebellar Pinceaux. *Brain Res. Mol. Brain Res.* **42**, 51–61.

Le Maout, S., Sewing, S., Coudrier, E., Elalouf, J.M., Pongs, O., Merot, J. (1996). Polarized targeting of a shaker-like (A-type) K(+)-channel in the polarized epithelial cell line MDCK. *Molecular Membrane Biology* **13**, 143–147.

Leicher, T., Bahring, R., Isbrandt, D., Pongs, O. (1998). Coexpression of the KCNA3B gene product with Kv1.5 leads to a novel A-type potassium channel. *J. Biol. Chem.* **273**, 35095–35101.

Levin, G., Chikvashvili, D., Singer-Lahat, D., Peretz, T., Thornhill, W.B., Lotan, I. (1996). Phosphorylation of a K+ channel alpha subunit modulates the inactivation conferred by a beta subunit. Involvement of cytoskeleton. *J. Biol. Chem.* **271**, 29321–29328.

Li, D., Takimoto, K., Levitan, E.S. (2000). Surface expression of Kv1 channels is governed by a C-terminal motif. *J. Biol. Chem.* **275**, 11597–11602.

Lim, S.T., Antonucci, D.E., Scannevin, R.H., Trimmer, J.S. (2000). A novel targeting signal for proximal clustering of the Kv2.1 K+ channel in hippocampal neurons. *Neuron* **25**, 385–397.

Liss, B., Franz, O., Sewing, S., Bruns, R., Neuhoff, H., Roeper, J. (2001). Tuning pacemaker frequency of individual dopaminergic neurons by Kv4.3L and KChip3.1 transcription. *EMBO J.* **20**, 5715–5724.

Long, S.B., Campbell, E.B., Mackinnon, R. (2005). Crystal structure of a mammalian voltage-dependent shaker family K+ channel. *Science* (in press).

Luneau, C.J., Williams, J.B., Marshall, J., Levitan, E.S., Oliva, C., Smith, J.S. et al. (1991). Alternative splicing contributes to K+ channel diversity in the mammalian central nervous system. *Proc. Natl. Acad. Sci. U. S. A.* **88**, 3932–3936.

MacKinnon, R. (1991). Determination of the subunit stoichiometry of a voltage-activated potassium channel. *Nature* **350**, 232–235.

Maletic-Savatic, M., Lenn, N.J., Trimmer, J.S. (1995). Differential spatiotemporal expression of K+ channel polypeptides in rat hippocampal neurons developing in situ and in vitro. *J. Neurosci.* **15**, 3840–3851.

Malin, S.A., Nerbonne, J.M. (2000). Elimination of the fast transient in superior cervical ganglion neurons with expression of KV4.2W362F: molecular dissection of IA. *J. Neurosci.* **20**, 5191–5199.

Malin, S.A., Nerbonne, J.M. (2001). Molecular heterogeneity of the voltage-gated fast transient outward K+ current, I(Af), in mammalian neurons. *J. Neurosci.* **21**, 8004–8014.

Malinow, R. (2003). AMPA receptor trafficking and long-term potentiation. *Philos. Trans. R. Soc. Lond. B. Biol. Sci.* **358**, 707–714.

Mandel, G., McKinnon, D. (1993). Molecular basis of neural-specific gene expression. *Annu. Rev. Neurosci.* **16**, 323–345.

Manganas, L.N., Trimmer, J.S. (2000). Subunit composition determines Kv1 potassium channel surface expression. *J. Biol. Chem.* **275**, 29685–29693.

Manganas, L.N., Trimmer, J.S. (2004). Calnexin regulates mammalian Kv1 channel trafficking. *Biochem. Biophys. Res. Commun.* **322**, 577–584.

Manganas, L.N., Akhtar, S., Antonucci, D.E., Campomanes, C.R., Dolly, J.O., Trimmer, J.S. (2001a). Episodic ataxia type-1 mutations in the Kv1.1 potassium channel display distinct folding and intracellular trafficking properties. *J. Biol. Chem.* **276**, 49427–49434.

Manganas, L.N., Wang, Q., Scannevin, R.H., Antonucci, D.E., Rhodes, K.J., Trimmer, J.S. (2001b). Identification of a trafficking determinant localized to the Kv1 potassium channel pore. *Proc. Natl. Acad. Sci. U. S. A.* **98**, 14055–14059.

Matsukawa, H., Wolf, A.M., Matsushita, S., Joho, R.H., Knopfel, T. (2003). Motor dysfunction and altered synaptic transmission at the parallel fiber-Purkinje cell synapse in mice lacking potassium channels Kv3.1 and Kv3.3. *J. Neurosci.* **23**, 7677–7684.

McCormack, K., Connor, J.X., Zhou, L., Ho, L.L., Ganetzky, B., Chiu, S.Y., Messing, A. (2002). Genetic analysis of the mammalian K+ channel beta subunit Kvbeta 2 (Kcnab2). *J. Biol. Chem.* **277**, 13219–13228.

Misonou, H., Mohapatra, D.P., Park, E.W., Leung, V., Zhen, D., Misonou, K. et al. (2004). Regulation of ion channel localization and phosphorylation by neuronal activity. *Nat. Neurosci.* **7**, 711–718.

Mitsui, S., Saito, M., Hayashi, K., Mori, K., Yoshihara, Y. (2005). A novel phenylalanine-based targeting signal directs telencephalin to neuronal dendrites. *J. Neurosci.* **25**, 1122–1131.

Monaghan, M.M., Trimmer, J.S., Rhodes, K.J. (2001). Experimental localization of Kv1 family voltage-gated K+ channel alpha and beta subunits in rat hippocampal formation. *J. Neurosci.* **21**, 5973–5983.

Mulley, J.C., Scheffer, I.E., Petrou, S., Berkovic, S.F. (2003). Channelopathies as a genetic cause of epilepsy. *Curr. Opin. Neurol.* **16**, 171–176.

Murakoshi, H., Shi, G., Scannevin, R.H., Trimmer, J.S. (1997). Phosphorylation of the Kv2.1 K+ channel alters voltage-dependent activation. *Mol. Pharmacol.* **52**, 821–828.

Nadal, M.S., Ozaita, A., Amarillo, Y., de Miera, E.V., Ma, Y., Mo et al. (2003). The CD26-related dipeptidyl aminopeptidase-like protein DPPX is a critical component of neuronal A-type K+ channels. *Neuron* **37**, 449–461.

Nagaya, N., Papazian, D.M. (1997). Potassium channel alpha and beta subunits assemble in the endoplasmic reticulum. *J. Biol. Chem.* **272**, 3022–3027.

Nakahira, K., Shi, G., Rhodes, K.J., Trimmer, J.S. (1996). Selective interaction of voltage-gated K+ channel beta-subunits with alpha-subunits. *J. Biol. Chem.* **271**, 7084–7089.

Nakamura, T.Y., Pountney, D.J., Ozaita, A., Nandi, S., Ueda, S., Rudy, B., Coetzee, W.A. (2001). A role for frequenin, a Ca^{2+}-binding protein, as a regulator of Kv4 K+ -currents. *Proc. Natl. Acad. Sci. U. S. A.* **98**, 12808–12813.

Nelson, P.H. (2003). Modeling the concentration-dependent permeation modes of the KcsA potassium ion channel. *Phys. Rev. E. Stat. Nonlin. Soft. Matter. Phys.* **68**, 061908.

Noskov, S.Y., Berneche, S., Roux, B. (2004). Control of ion selectivity in potassium channels by electrostatic and dynamic properties of carbonyl ligands. *Nature* **431**, 830–834.

O'Callaghan, D.W., Hasdemir, B., Leighton, M., Burgoyne, R.D. (2003). Residues within the myristoylation motif determine intracellular targeting of the neuronal Ca2+ sensor protein KChIP1 to post-ER transport vesicles and traffic of Kv4 K+ channels. *J. Cell Sci.* **116**, 4833–4845.

Ou, W.J., Cameron, P.H., Thomas, D.Y., Bergeron, J.J. (1993). Association of folding intermediates of glycoproteins with calnexin during protein maturation. *Nature* **364**, 771–776.

Ozaita, A., Martone, M.E., Ellisman, M.H., Rudy, B. (2002). Differential subcellular localization of the two alternatively spliced isoforms of the Kv3.1 potassium channel subunit in brain. *J. Neurophysiol.* **88**, 394–408.

Papazian, D.M. (1999). Potassium channels: Some assembly required. *Neuron* **23**, 7–10.

Papazian, D.M., Timpe, L.C., Jan, Y.N., Jan, L.Y. (1991). Alteration of voltage-dependence of Shaker potassium channel by mutations in the S4 sequence. *Nature* **349**, 305–310.

Patel, A.J., Lazdunski, M., Honore, E. (1997). Kv2.1/Kv9.3, a novel ATP-dependent delayed-rectifier K+ channel in oxygen-sensitive pulmonary artery myocytes. *EMBO J.* **16**, 6615–6625.

Patterson, G.H., Lippincott-Schwartz, J. (2002). A photoactivatable GFP for selective photolabeling of proteins and cells. *Science* **297**, 1873–1877.

Petrecca, K., Miller, D.M., Shrier, A. (2000). Localization and enhanced current density of the Kv4.2 potassium channel by interaction with the actin-binding protein filamin. *J. Neurosci.* **20**, 8736–8744.

Ponce, A., Vega-Saenz de Miera, E., Kentros, C., Moreno, H., Thornhill, B., Rudy, B. (1997). K+ channel subunit isoforms with divergent carboxy-terminal sequences carry distinct membrane targeting signals. *J. Membr. Biol.* **159**, 149–159.

Pongs, O., Leicher, T., Berger, M., Roeper, J., Bahring, R., Wray, D. et al. (1999) Functional and molecular aspects of voltage-gated K+ channel beta subunits. *Ann. N. Y. Acad. Sci.* **868**, 344–355.

Qi, S.Y., Riviere, P.J., Trojnar, J., Junien, J.L., Akinsanya, K.O. (2003). Cloning and characterization of dipeptidyl peptidase 10, a new member of an emerging subgroup of serine proteases. *Biochem. J.* **373**, 179–189.

Rasband, M.N., Park, E.W., Zhen, D., Arbuckle, M.I., Poliak, S., Peles, E. et al. (2002) Clustering of neuronal potassium channels is independent of their interaction with PSD-95. *J. Cell Biol.* **159**, 663–672.

Rasband, M.N., Trimmer, J.S. (2001). Developmental clustering of ion channels at and near the node of Ranvier. *Dev. Biol.* **236**, 5–16.

Rettig, J., Heinemann, S.H., Wunder, F., Lorra, C., Parcej, D.N., Dolly, J.O., Pongs, O. (1994). Inactivation properties of voltage-gated K+ channels altered by presence of beta-subunit. *Nature* **369**, 289–294.

Rhodes, K.J., Carroll, K.I., Sung, M.A., Doliveira, L.C., Monaghan, M.M., Burke, S.L. et al. (2004). KChIPs and Kv4 alpha subunits as integral components of A-type potassium channels in mammalian brain. *J. Neurosci.* **24**, 7903–7915.

Rhodes, K.J., Monaghan, M.M., Barrezueta, N.X., Nawoschik, S., Bekele-Arcuri, Z., Matos, M.F. et al. (1996). Voltage-gated K+ channel beta subunits: Expression and distribution of Kv beta 1 and Kv beta 2 in adult rat brain. *J. Neurosci.* **16**, 4846–4860.

Rhodes, K.J., Strassle, B.W., Monaghan, M.M., Bekele-Arcuri, Z., Matos, M.F., Trimmer, J.S. (1997). Association and colocalization of the Kv beta1 and Kv beta2 beta-subunits with Kv1 alpha-subunits in mammalian brain K+ channel complexes. *J. Neurosci.* **17**, 8246–8258.

Rivera, J.F., Ahmad, S., Quick, M.W., Liman, E.R., Arnold, D.B. (2003). An evolutionarily conserved dileucine motif in Shal K+ channels mediates dendritic targeting. *Nat. Neurosci.* **6**, 243–250.

Rudy, B., Chow, A., Lau, D., Amarillo, Y., Ozaita, A., Saganich, M. et al. (1999). Contributions of Kv3 channels to neuronal excitability. *Ann. N. Y. Acad. Sci.* **868**, 304–343.

Sala, C., Piech, V., Wilson, N.R., Passafaro, M., Liu, G., Sheng, M. (2001). Regulation of dendritic spine morphology and synaptic function by Shank and Homer. *Neuron* **31**, 115–130.

Salinas, M., Duprat, F., Heurteaux, C., Hugnot, J.P., Lazdunski, M. (1997). New modulatory alpha subunits for mammalian Shab K+ channels. *J. Biol. Chem.* **272**, 24371–24379.

Santacruz-Toloza, L., Huang, Y., John, S.A., Papazian, D.M. (1994). Glycosylation of shaker potassium channel protein in insect cell culture and in Xenopus oocytes. *Biochemistry* 33, 5607–5613.

Scanlan, M.J., Raj, B.K., Calvo, B., Garin-Chesa, P., Sanz-Moncasi, M.P., Healey, J.H. et al. (1994). Molecular cloning of fibroblast activation protein alpha, a member of the serine protease family selectively expressed in stromal fibroblasts of epithelial cancers. *Proc. Natl. Acad. Sci. U. S. A.* 91, 5657–5661.

Scannevin, R.H., Murakoshi, H., Rhodes, K.J., Trimmer, J.S. (1996). Identification of a cytoplasmic domain important in the polarized expression and clustering of the Kv2.1 K+ channel. *J. Cell Biol.* 135, 1619–1632.

Scott, D.B., Blanpied, T.A., Swanson, G.T., Zhang, C., Ehlers, M.D. (2001). An NMDA receptor ER retention signal regulated by phosphorylation and alternative splicing. *J. Neurosci.* 21, 3063–3072.

Scott, V.E., Rettig, J., Parcej, D.N., Keen, J.N., Findlay, J.B., Pongs, O., Dolly, J.O. (1994). Primary structure of a beta subunit of alpha-dendrotoxin-sensitive K+ channels from bovine brain. *Proc. Natl. Acad. Sci. U. S. A.* 91, 1637–1641.

Serodio, P., Rudy, B. (1998). Differential expression of Kv4 K+ channel subunits mediating subthreshold transient K+ (A-type) currents in rat brain. *J. Neurophysiol.* 79, 1081–1091.

Sewing, S., Roeper, J., Pongs, O. (1996). Kv beta 1 subunit binding specific for shaker-related potassium channel alpha subunits. *Neuron* 16, 455–463.

Sheng, M., Liao, Y.J., Jan, Y.N., Jan, L.Y. (1993). Presynaptic A-current based on heteromultimeric K+ channels detected in vivo. *Nature* 365, 72–75.

Sheng, M., Tsaur, M.L., Jan, Y.N., Jan, L.Y. (1992). Subcellular segregation of two A-type K+ channel proteins in rat central neurons. *Neuron* 9, 271–284.

Shi, G., Nakahira, K., Hammond, S., Rhodes, K.J., Schechter, L.E., Trimmer, J.S. (1996). Beta subunits promote K+ channel surface expression through effects early in biosynthesis. *Neuron* 16, 843–852.

Shi, G., Trimmer, J.S. (1999). Differential asparagine-linked glycosylation of voltage-gated K+ channels in mammalian brain and in transfected cells. *J. Membr. Biol.* 168, 265–273.

Shibata, R., Misonou, H., Campomanes, C.R., Anderson, A.E., Schrader, L.A., Doliveira, L.C. et al. (2003). A fundamental role for KChIPs in determining the molecular properties and trafficking of Kv4.2 potassium channels. *J. Biol. Chem.* 278, 3645–3654.

Shibata, R., Wakazono, Y., Nakahira, K., Trimmer, J.S., Ikenaka, K. (1999). Expression of Kv3.1 and Kv4.2 genes in developing cerebellar granule cells. *Dev. Neurosci.* 21, 87–93.

Smart, S.L., Lopantsev, V., Zhang, C.L., Robbins, C.A., Wang, H., Chiu, S.Y. et al. (1998). Deletion of the Kv1.1 potassium channel causes epilepsy in mice. *Neuron* 20, 809–819.

Sokolova, O., Accardi, A., Gutierrez, D., Lau, A., Rigney, M., Grigorieff, N. (2003). Conformational changes in the C terminus of Shaker K+ channel

bound to the rat Kv beta2-subunit. *Proc. Natl. Acad. Sci. U. S. A.* 100, 12607–12612.

Song, W.J. (2002). Genes responsible for native depolarization-activated K+ currents in neurons. *Neurosci. Res.* 42, 7–14.

Strassle, B.W., Menegola, M., Rhodes, K.J., Trimmer, J.S. (2005). Light and electron microscopic analysis of KChIP and Kv4 localization in rat cerebellar granule cells. *J. Comp. Neurol.* 484, 144–155.

Strop, P., Bankovich, A.J., Hansen, K.C., Garcia, K.C., Brunger, A.T. (2004). Structure of a human A-type potassium channel interacting protein DPPX, a member of the dipeptidyl aminopeptidase family. *J. Mol. Biol.* 343, 1055–1065.

Stuhmer, W., Ruppersberg, J.P., Schroter, K.H., Sakmann, B., Stocker, M., Giese, K.P. et al. (1989). Molecular basis of functional diversity of voltage-gated potassium channels in mammalian brain. *EMBO J.* 8, 3235–3244.

Tempel, B.L., Papazian, D.M., Schwarz, T.L., Jan, Y.N., Jan, L.Y. (1987). Sequence of a probable potassium channel component encoded at Shaker locus of Drosophila. *Science* 237, 770–775.

Thornhill, W.B., Wu, M.B., Jiang, X., Wu, X., Morgan, P.T., Margiotta, J.F. (1996). Expression of Kv1.1 delayed rectifier potassium channels in Lec mutant Chinese hamster ovary cell lines reveals a role for sialidation in channel function. *J. Biol. Chem.* 271, 19093–19098.

Tiffany, A.M., Manganas, L.N., Kim, E., Hsueh, Y.P., Sheng, M., Trimmer, J.S. (2000). PSD-95 and SAP97 exhibit distinct mechanisms for regulating K(+) channel surface expression and clustering. *J. Cell Biol.* 148, 147–158.

Tiran, Z., Peretz, A., Attali, B., Elson, A. (2003). Phosphorylation-dependent regulation of Kv2.1 channel activity at tyrosine 124 by Src and by protein-tyrosine phosphatase epsilon. *J. Biol. Chem.* 278, 17509–17514.

Tkatch, T., Baranauskas, G., Surmeier, D.J. (2000). Kv4.2 mRNA abundance and A-type K(+) current amplitude are linearly related in basal ganglia and basal forebrain neurons. *J. Neurosci.* 20, 579–588.

Trimmer, J.S. (1991). Immunological identification and characterization of a delayed rectifier K+ channel polypeptide in rat brain. *Proc. Natl. Acad. Sci. U. S. A.* 88, 10764–10768.

Trimmer, J.S. (1998a). Analysis of K+ channel biosynthesis and assembly in transfected mammalian cells. *Methods Enzymol.* 293, 32–49.

Trimmer, J.S. (1998b). Regulation of ion channel expression by cytoplasmic subunits. *Curr. Opin. Neurobiol.* 8, 370–374.

Trimmer, J.S., Rhodes, K.J. (2004). Localization of voltage-gated ion channels in mammalian brain. *Annu. Rev. Physiol.* 66, 477–519.

Tytgat, J., Debont, T., Carmeliet, E., Daenens, P. (1995). The alpha-dendrotoxin footprint on a mammalian potassium channel. *J. Biol. Chem.* 270, 24776–24781.

Varga, A.W., Anderson, A.E., Adams, J.P., Vogel, H., Sweatt, J.D. (2000). Input-specific immunolocalization of differentially phosphorylated Kv4.2 in the mouse brain. *Learn. Mem.* **7**, 321–332.

Varga, A.W., Yuan, L.L., Anderson, A.E., Schrader, L.A., Wu, G.Y., Gatchel, J.R. et al. (2004). Calcium-calmodulin-dependent kinase II modulates Kv4.2 channel expression and upregulates neuronal A-type potassium currents. *J. Neurosci.* **24**, 3643–3654.

Wang, H., Kunkel, D.D., Schwartzkroin, P.A., Tempel, B.L. (1994). Localization of Kv1.1 and Kv1.2, two K channel proteins, to synaptic terminals, somata, and dendrites in the mouse brain. *J. Neurosci.* **14**, 4588–4599.

Wang, L., Takimoto, K., Levitan, E.S. (2003). Differential association of the auxiliary subunit Kvbeta2 with Kv1.4 and Kv4.3 K+ channels. *FEBS Lett.* **547**, 162–164.

Westenbroek, R.E., Ahlijanian, M.K., Catterall, W.A. (1990). Clustering of L-type Ca2+ channels at the base of major dendrites in hippocampal pyramidal neurons. *Nature* **347**, 281–284.

Winklhofer, M., Matthias, K., Seifert, G., Stocker, M., Sewing, S., Herget, T. et al. (2003). Analysis of phosphorylation-dependent modulation of Kv1.1 potassium channels. *Neuropharmacology* **44**, 829–842.

Wong, W., Newell, E.W., Jugloff, D.G., Jones, O.T., Schlichter, L.C. (2002). Cell surface targeting and clustering interactions between heterologously expressed PSD-95 and the Shal voltage-gated potassium channel, Kv4.2. *J. Biol. Chem.* **277**, 20423–20430.

Yang, E.K., Alvira, M.R., Levitan, E.S., Takimoto, K. (2001). Kvbeta subunits increase expression of Kv4.3 channels by interacting with their C termini. *J. Biol. Chem.* **276**, 4839–4844.

Yellen, G. (1998). The moving parts of voltage-gated ion channels. *Q. Rev. Biophys.* **31**, 239–295.

Yokotani, N., Doi, K., Wenthold, R.J., Wada, K. (1993). Non-conservation of a catalytic residue in a dipeptidyl aminopeptidase IV-related protein encoded by a gene on human chromosome 7. *Hum. Mol. Genet.* **2**, 1037–1039.

Yu, W., Xu, J., Li, M. (1996). NAB domain is essential for the subunit assembly of both alpha-alpha and alpha-beta complexes of shaker-like potassium channels. *Neuron* **16**, 441–453.

Zerangue, N., Schwappach, B., Jan, Y.N., Jan, L.Y. (1999). A new ER trafficking signal regulates the subunit stoichiometry of plasma membrane K(ATP) channels. *Neuron* **22**, 537–548.

Zhou, W., Qian, Y., Kunjilwar, K., Pfaffinger, P.J., Choe, S. (2004). Structural insights into the functional interaction of KChIP1 with Shal-type K(+) channels. *Neuron* **41**, 573–586.

Zhou, Y., Morais-Cabral, J.H., Kaufman, A., MacKinnon, R. (2001). Chemistry of ion coordination and hydration revealed by a K+ channel-Fab complex at 2.0 A resolution. *Nature* **414**, 43–48.

Zhu, J., Gomez, B., Watanabe, I., Thornhill, W.B. (2005). Amino acids in the pore region of Kv1 potassium channels dictate cell surface protein levels: a possible trafficking code in the Kv1 subfamily. *Biochem. J.* **388**, 355–362.

Zhu, J., Watanabe, I., Gomez, B., Thornhill, W.B. (2001). Determinants involved in Kv1 potassium channel folding in the endoplasmic reticulum, glycosylation in the Golgi, and cell surface expression. *J. Biol. Chem.* **276**, 39419–39427.

Zhu, J., Watanabe, I., Gomez, B., Thornhill, W.B. (2003a). Heteromeric Kv1 potassium channel expression: Amino acid determinants involved in processing and trafficking to the cell surface. *J. Biol. Chem.* **278**, 25558–25567.

Zhu, J., Watanabe, I., Gomez, B., Thornhill, W.B. (2003b). Trafficking of Kv1.4 potassium channels: Interdependence of a pore region determinant and a cytoplasmic C-terminal VXXSL determinant in regulating cell surface trafficking. *Biochem. J.* **375**, 761–768.

TRAFFICKING IN OTHER CELL TYPES

Comparison of protein trafficking mechanisms in different cell types has enhanced our understanding of neuronal trafficking. Studies in polarized epithelial cells have revealed mechanisms that ensure exocytosis at apical and basolateral plasma membrane domains that are responsible for maintaining the functional polarity of these cells (Yeaman). Examination of calcium-dependent exocytosis of dense-core vesicles in neuroendocrine cells has revealed biochemical stages required for secretion and led to an enhanced understanding of the role of lipids in exocytosis (Martin). Secretion from nonelectrically excitable glial cells recently has been appreciated and the implications of bidirectional signaling between neurons and glia for neuronal function are significant (Lee and Parpura).

TRAFFICKING IN OTHER CELL TYPES

13

Protein Trafficking in the Exocytic Pathway of Polarized Epithelial Cells

CHARLES YEAMAN

Polarized trafficking mechanisms that ensure faithful and efficient exocytosis at apical and basolateral plasma membrane domains of epithelial cells are responsible for maintaining the functional polarity of these cells. Studies performed over the past quarter century have identified the sorting organelles and transport itineraries pursued by many proteins; defined several categories of sorting signals and mechanisms responsible for segregating proteins; and highlighted functions for cytoskeletal components and identified numerous proteins required for tethering, docking, and fusion of transport intermediates at select sites on the plasma membrane. Much of what has been learned through analysis of polarized trafficking in epithelial cells has provided insight into mechanisms operating during sorting and trafficking to axonal and somatodendritic membrane domains in neurons.

I. INTRODUCTION

Like neurons, epithelial cells are highly polarized. A conspicuous feature of both cell types is the division of their surfaces into structurally and functionally distinct membrane domains. Epithelial cells constitute a protective barrier against the harsh external environment, defending the organism against assault by microbes, high osmotic pressure, and digestive enzymes. They also serve as exchange interfaces with that environment, regulating the homeostatic transport of solutes, electrolytes, and water. To perform these different functions, epithelial cells have evolved two distinct membrane domains, a lumen-facing apical domain and a basolateral domain, which contacts neighboring cells and an underlying basal lamina. Separating apical and basolateral surfaces are tight junctions that help to prevent passive mixing of proteins and lipids between the two domains, and form a tight seal that regulates transport of materials between cells in the epithelium.

Maintenance of polarized membrane domains requires highly efficient sorting mechanisms that function in both the biosynthetic and recycling pathways, as well as selective retention and removal mechanisms operating at each plasma membrane domain. Together, these mechanisms ensure that the correct proteins and lipids are delivered to and accumulate within the appropriate membrane domain. Epithelial cells provide a paradigm of cell polarization, and advances in our understanding of the fundamental mechanisms that underlie polarized trafficking in epithelial cells provide important clues about how these processes might function in neurons. This chapter will focus on how epithelial cells sort and deliver proteins to apical and basolateral surfaces, and high-

light similarities to mechanisms operating in neurons.

II. POLARIZED TRAFFICKING ROUTES IN EPITHELIAL CELLS

Proteins and lipids destined for apical and basolateral plasma membrane domains are sorted in two distinct sites in epithelial cells, the *trans*-Golgi network (TGN) and the endosomal system. Sorting in each compartment is likely to involve similar, or overlapping sorting determinants on cargo proteins that are recognized and decoded by sorting machinery inherent to each location. Although direct biosynthetic sorting pathways leading from the TGN to apical and basolateral plasma membrane domains were the first to be identified (Matlin and Simons 1984; Misek et al. 1984; Rindler et al. 1984), it is now appreciated that the endosomal system of epithelial cells also has an indispensable polarized sorting function. This fact is not surprising, given that nearly 50 percent of a typical membrane protein undergoes endocytosis every hour (Mellman 1996). If specific endosomal sorting mechanisms were not in place to insure that internalized proteins were recycled back to the appropriate membrane domain, randomization of surface content would occur rapidly, leading to a breakdown of epithelial polarity. In fact, certain epithelial cells, such as hepatocytes, rely almost exclusively on postendocytic sorting pathways to segregate apical and basolateral proteins (Tuma and Hubbard 2003). In these cells, most apical proteins are targeted indirectly by transcytosis following initial insertion into the basolateral membrane. In this section, the overall characteristics of each organelle responsible for polarized sorting in epithelial cells are described, and evidence for the existence of multiple pathways by which proteins are delivered to different plasma membrane domains is discussed. The next section will address the nature of sorting signals and mechanisms responsible for targeted delivery to those domains.

A. Biosynthetic Sorting Pathways

Early studies on Madin-Darby canine kidney (MDCK) cells infected with enveloped viruses such as VSV and influenza established that these viruses budded asymmetrically from the basolateral and apical surfaces, respectively, because the steady state distributions of viral envelope glycoproteins were polarized (Rodriguez-Boulan and Sabatini 1978; Rodriguez-Boulan and Pendergast 1980). Envelope glycoproteins were inserted into basolateral or apical plasma membrane domains vectorially, without prior passage through the contralateral surface (Matlin and Simons 1984; Misek et al. 1984; Rindler et al. 1984; Rodriguez-Boulan et al. 1984; Pfeiffer et al. 1985). These pioneering studies established that biosynthetic sorting in the exocytic pathway occurs, and that apical and basolateral proteins are segregated from one another in an intracellular compartment. Further studies on a host of endogenous and exogenous proteins were largely confirmatory, and extended the repertoire of proteins known to undergo polarized sorting in the secretory pathway (Caplan et al. 1986; Wessels et al. 1989, 1990; Casanova et al. 1991).

Identification of the TGN as the major sorting organelle for newly synthesized apical and basolateral proteins is derived from several types of experimental evidence accumulated over more than two decades. Early ultrastructural studies of doubly infected MDCK cells showed that the major viral glycoproteins of VSV (VSVG protein) and influenza (hemagglutin, HA) were intermixed from their site of synthesis in the endoplasmic reticulum through the Golgi apparatus, but segregated into distinct post-Golgi transport intermediates prior to arrival at the plasma membrane (Rindler et al. 1984). Furthermore, a clever biochemical assay for colocalization of VSVG and influenza neuraminidase showed that these proteins colocalized within the TGN (Fuller et al. 1985). These studies revealed that apical and basolateral proteins traveled together through much of the secretory pathway, and that polarized sorting did not occur prior to arrival in the TGN.

What is the evidence that apical and basolateral proteins exit the TGN in distinct carriers? Incubation of cells at a reduced temperature of 20°C prevents fission and release of post-Golgi transport intermediates and leads to accumulation of apical and basolateral proteins in the TGN (Matlin and Simons 1983; Griffiths et al. 1985). Assays have been established that reconstitute the production of post-Golgi carriers in semi-intact cells or cell-free systems (Bennett et al. 1988; Wandinger-Ness et al. 1990; Pimplikar and Simons 1993; Pimplikar et al. 1994; Musch et al. 1996). This has facilitated the characterization of distinct types of TGN-derived vesicles produced *in vitro* (Bennett et al. 1988; Wandinger-Ness et al. 1990), and the demonstration that apical and basolateral vesicle budding from the TGN is sensitive to different peptide inhibitors and pharmacological manipulations (Pimplikar and Simons 1993; Pimplikar et al. 1994; Musch et al. 1996). More recently, expression of kinase-dead mutants of protein kinase D, which associate with the TGN but not endosomes (Liljedahl et al. 2001; Maeda et al. 2001; Baron and Malhotra 2002), were shown to selectively inhibit post-Golgi trafficking of basolateral, but not apical proteins in MDCK cells (Yeaman et al. 2004). Collectively, these studies established that apical and basolateral proteins exit the TGN by distinct mechanisms in physically separable carriers.

Time-lapse fluorescence imaging studies support the conclusion that sorting of apical and basolateral cargo occurs in the TGN. In both nonpolarized and polarized MDCK cells, apical and basolateral proteins have been observed to segregate from each other within TGN domains, and finally exit in separate tubulovesicular carriers that most often proceeded directly to the plasma membrane without passage through intermediate compartments (Keller et al. 2001; Kreitzer et al. 2003). Structural evidence for the existence of distinctly coated sorting domains within the TGN is provided by HVEM tomography. This powerful technique has revealed the presence of both clathrin-coated domains and distinct regions enriched in a novel and yet undefined "lace-like" coat (Ladinsky et al. 1994; 1999; 2002). It has been suggested that sorting of apical and basolateral proteins in the TGN may proceed via a process similar to that of cisternal maturation within the Golgi apparatus (Mironov et al. 1998). This model postulates that basolateral proteins are removed from tubulo-saccular domains of the TGN through the budding of clathrin-coated vesicles containing specific adaptors for basolateral proteins (Rodriguez-Boulan and Musch 2005). This would leave behind a "mature" tubule enriched in apical proteins that would be transported to the apical plasma membrane.

Although there is substantial experimental evidence supporting a direct pathway by which newly synthesized apical and basolateral proteins, sorted in the TGN, are vectorially delivered to the correct surface domain, there is also evidence that some proteins travel through recycling endosomes en route to the plasma membrane. This was suggested originally by results of pulse-chase/cell fractionation experiments that showed certain newly synthesized basolateral proteins, such as receptors for transferrin (Futter et al. 1995), asialoglycoproteins (Leitinger et al. 1995), and IgA (Orzech et al. 2000), are accessible to internalized ligands before they are delivered to the plasma membrane. More recently, time-lapse imaging studies have shown that newly synthesized VSVG codistributes with recycling endosome markers during its transport to the plasma membrane

(Folsch et al. 2003; Ang et al. 2004). In addition, E-cadherin and an apically sorted mutant lacking a basolateral sorting signal were observed to pass through recycling endosomes before delivery to the plasma membrane in nonpolarized MDCK cells (Lock and Stow 2005). Endosome ablation experiments have supported the notion that transit through recycling endosomes represents an obligatory step in the biosynthetic delivery of VSVG to the basolateral membrane (Ang et al. 2004). Though provocative, these results are controversial because imaging studies by other groups support a direct delivery route for VSVG and other basolateral proteins (Keller et al. 2001; Kreitzer et al. 2003; Polishchuk et al. 2004).

B. Postendocytic Sorting Pathways

Although the accumulated evidence is strong for a model in which newly synthesized proteins are sorted in the TGN or recycling endosomes and directly targeted to apical and basolateral surfaces, the strict reliance on these biosynthetic sorting pathways varies substantially between different types of epithelia. This phenomenon has been termed the *flexible epithelial phenotype* (Mostov et al. 2003; Rodriguez-Boulan et al. 2005). It has been known for many years that some epithelial cells, most notably hepatocytes, transport most of their proteins to the basolateral surface initially, then retrieve a subset of these for delivery to the apical surface via the process of transcytosis. A recent study has shown that even MDCK cells, long believed to employ the transcytotic pathway only minimally, in fact may deliver many proteins to the apical surface after first depositing them in the basolateral plasma membrane (Polishchuk et al. 2004). Apical proteins with a propensity to associate with lipid rafts have been found to exit the TGN in the same long tubular carriers that mediate basolateral protein exit, although apical and basolateral cargoes were segregated within the tubules. Using a mild tannic acid treatment to selectively fix and inhibit apical or basolateral membrane proteins, including t-SNAREs, raft-associated apical markers accumulated within the cytoplasm of cells, whether tannic acid was applied to the apical or the basolateral surface. When tannic acid was applied basolaterally apical and basolateral markers coaccumulated near the intercellular junctions. However, when applied apically, tannic acid did not block the exocytosis of basolateral cargo; instead, apical cargo accumulated in distinct cytoplasmic organelles in a region immediately subjacent to the apical plasma membrane. These data support a model in which raft-associated apical proteins exit the TGN with basolateral cargo, and following initial delivery to the basolateral surface, are internalized and rerouted to the apical surface.

The various post-endocytic trafficking pathways and organelles that mediate them are diagrammed in Figure 13.1. Omitted for simplicity are late endosomes and lysosomes. Late endosomes receive specific biosynthetic cargo from the TGN (e.g., mannose-6-phosphate receptors and associated lysosomal hydrolases), as well as soluble proteins that are sorted away from membrane receptors in distinct apical and basolateral early (sorting) endosomes (AEE and BEE) (Dunn et al. 1989). These substances are ultimately transferred to lysosomes for degradation. Because these do not function in the trafficking of proteins to apical and basolateral surfaces, they are not considered further in this chapter. However, numerous excellent reviews on the biogenesis and function of late endosomes and lysosomes have appeared recently, and the reader is referred to these (Stahl and Barbieri 2002; Bonifacino and Traub 2003; Luzio et al. 2003).

Studies in model epithelial cell lines, such as MDCK (kidney), CaCo-2 (intestinal), and WIF-B (hepatocyte) have revealed the existence of distinct sets of early endosomes for apical and basolateral membrane proteins, as well as a common set of recycling endosomes, in which apical

and basolateral membrane proteins are segregated from one another. Careful kinetic, morphological, and functional analyses of intracellular trafficking itineraries pursued by recycling cargo (such as transferrin and its receptor, TfnR) and transcytosing markers (such as IgA and its receptor, pIgR) have revealed the existence of at least three, and more likely four distinct endosome populations (Rojas and Apodaca 2002). Material that is internalized from opposite poles of epithelial cells through either clathrin-dependent or clathrin-independent pathways first accumulates within spatially segregated populations of early endosomes located near the basolateral and apical membranes (Bomsel et al. 1989; Parton et al. 1989). Both early endosome populations have properties that distinguish them from other endosomes, most notably the accumulation of fluid-phase markers. The small GTPases Rab4 and Rab5, as well as the Rab5 effector EEA1, are associated with these compartments (Bucci et al. 1994; Simonsen et al. 1998; Sheff et al. 1999; Leung et al. 2000). Each is capable of rapidly recycling membrane proteins back to the surface of origin, without further trafficking to the common recycling endosome (CRE), or sending membrane proteins to the CRE for further sorting (Sheff et al. 1999). Although they share many features, AEE and BEE are biochemically and physiologically distinct organelles exhibiting unique recycling rates, fusigenic properties, and overall protein compositions (Fialka et al. 1999; Sheff et al. 1999).

CREs are the major sorting station for membrane proteins internalized from apical and basolateral surfaces, and as mentioned in the foregoing section they also may participate in polarized sorting in the biosynthetic pathway. These are tubular compartments located primarily within an apico-nuclear region of polarized cells (Ihrke et al. 1993; Hemery et al. 1996; Odorizzi et al. 1996; Sheff et al. 1999; Brown et al. 2000; Leung et al. 2000; Wang et al. 2000). These organelles are clearly distinct

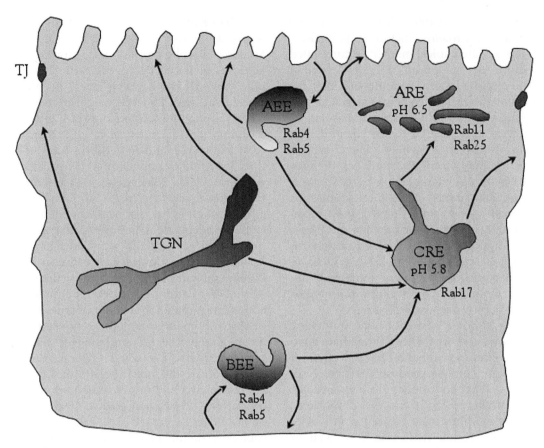

FIGURE 13.1. **Sorting compartments and trafficking itineraries in epithelial cells.** In the biosynthetic pathway, apical and basolateral proteins are segregated from one another in the *trans*-Golgi network (TGN) or the common recycling endosome (CRE), and directly delivered to appropriate plasma membrane domains. Post-endocytic pathways, comprising both recycling and transcytosis, involve passage through distinct sets of apical and basolateral early endosomes (AEE and BEE, respectively), as well as the CRE. The CRE receives endocytosed membrane proteins from both the AEE and BEE, and newly synthesized proteins from the TGN, and is the major polarized sorting hub in the cell. Cargo leaving the CRE may be recycled back to apical or basolateral surfaces, although transcytosis likely involves transit through a distinct compartment, the apical recycling endosome (ARE). Each of these endosomes have distinct morphologies, localizations, and associated Rab GTPases. Transport through the cytoplasm is facilitated by distinct cytoskeletal components for apical and basolateral routes. Tethering, docking, and fusion of transport intermediates with each membrane domain is specified by unique protein complexes, such as the Exocyst, and domain-specific fusion machines, composed minimally of t-SNAREs and associated proteins (not shown here). Fusion of transport intermediates in the basolateral routes is enriched in the top third of the lateral plasma membrane, below the tight junction (TJ).

from early endosomes, because they do not accumulate fluid-phase markers, they are associated with a distinct subset of Rab GTPases (Rab17 and perhaps Rab11), and because it is possible to resolve these organelles physically, biochemically, and pharmacologically (Hunziker and Peters 1998; Zacchi et al. 1998; Sheff et al. 1999; Leung et al. 2000). The CRE receives membrane-bound cargo from both apical and basolateral early endosomes via an actin- and microtubule-dependent transport pathway, and may receive newly synthesized apical and basolateral proteins from the TGN as well. Whether the CRE represents a major sorting station in the biosynthetic pathway, this organelle is almost certainly the primary site at which apically destined transcytotic cargo, such as pIgR-dIgA complexes, is segregated from receptors, like TfnR and LDLR, that will recycle back to the basolateral plasma membrane. Thus, the CRE appears to be a major hub of polarized sorting activity in epithelial cells.

Mathematic modeling of the kinetic behavior of pIgR-dIgA complexes is consistent with the hypothesis that transcytosing cargo is delivered directly from the CRE to the apical plasma membrane (Sheff et al. 1999; 2002). However, several lines of evidence support an alternate pathway in which pIgR-dIgA passes through an intermediate compartment, variously termed the apical recycling endosome (ARE), or subapical compartment. The ARE is composed of long tubulo-vesicular elements located high in the apical cytoplasm, just beneath the plasma membrane (Apodaca et al. 1994; Gibson et al. 1998). These organelles accumulate internalized dIgA but not transferrin (Barroso and Sztul 1994; Brown et al. 2000; Leung et al. 2000; Wang et al. 2000), and have a different lumenal pH (6.5) than that of the CRE (5.8) (Wang et al. 2000). Endogenous Rab11 and the closely related GTPase Rab25 are found primarily on ARE in polarized epithelial cells, and expression of either dominant-negative or constitutively active mutants of these disrupt pIgR-dIgA transcytosis but have little or no effect on recycling of TfnR to the basolateral surface (Casanova et al. 1999; Brown et al. 2000; Leung et al. 2000; Wang et al. 2000).

A central question that remains to be answered is how various sorting signals and adaptors, described in the following section, function in polarized sorting within this complex system of TGN and endosomal membranes. There is evidence that related, but distinct mechanisms function in different sorting organelles. However, the significance of multiple sorting steps with partially overlapping mechanisms is not altogether clear, unless these pathways evolved to provide multiple layers of quality control or flexibility to the sorting process. Nevertheless, identification of various endocytic compartments with sorting functions in epithelial cells provides a framework to begin dissecting the increasingly complex problem of polarized trafficking in these cells.

III. CARGO SORTING MECHANISMS IN THE BIOSYNTHETIC ROUTES

Segregation of apical and basolateral proteins into different transport intermediates in the TGN and endosomal system involves two elements: cis-acting signals within the structure of the sorted protein that contain the necessary information for proper localization, and trans-acting factors that mediate the concentration of cargo into nascent carriers. Although it was once thought that transport to the basolateral surface occurs by default (Simons and Wandinger-Ness 1990), it is now clear that transport to both basolateral and apical surfaces occur by signal-mediated processes. In general, basolateral sorting is thought to occur by a process conceptually similar to receptor-ligand interactions, in which proteinaceous motifs on basolateral cargo are recognized and bound by other proteins, or sorting "receptors," and this promotes incorporation into a basolateral transport vesicle. In contrast, much of the available data supports a very different mechanism for apical sorting. The lipid-raft hypothesis states that many proteins are sorted apically if they have an intrinsic affinity for membrane microdomains rich in glycosphingolipids and cholesterol (van Meer and Simons 1988; Simons and Ikonen 1997; Schuck and Simons 2004).

A. Basolateral Sorting Mechanisms

Basolateral sorting signals are comprised of short peptide sequences present within cytoplasmic domains of transported

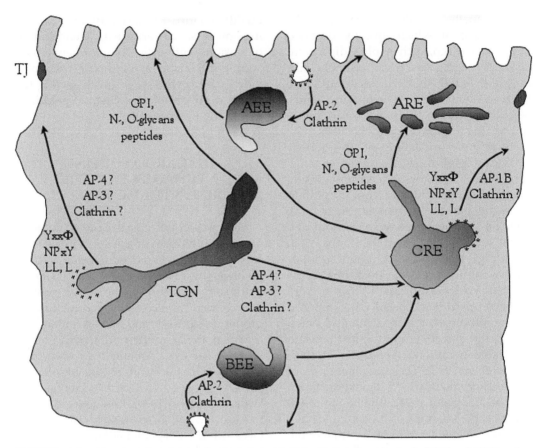

FIGURE 13.2. **Signals and sorting mechanisms in epithelial cell.** Apical sorting in the TGN and CRE involves mechanisms that recognize glycosylphosphatidylinositol (GPI) anchors, N- and O-glycans, or peptide moieties on the sorted cargo. For many apical membrane proteins, sorting into this pathway is facilitated by inclusion into small lipid rafts and clustering of small rafts into larger functional lipid rafts. Raft clustering could be mediated by caveolin oligomers, VIP17/MAL or lumenal lectins (see text). Basolateral sorting likely involves both clathrin (AP-1B, AP-3) or nonclathrin (AP-4) adaptors functioning at the TGN and/or CRE, which bind either tyrosine-dependent (YxxΦ, NPxY) or hydrophobic (LL, L) signals and mediate their inclusion into basolateral transport intermediates. A role for clathrin in these processes is speculative.

proteins. Several types of signals have been identified through mutagenesis of different basolateral proteins and analysis of the polarized trafficking phenotype that results (see Figure 13.2). By definition, basolateral signals are sequences that, when mutated, disrupt basolateral localization. In addition to showing that a sequence is necessary for basolateral targeting, the best-established signals are those also shown to be sufficient for sorting. These sequences direct baso-lateral sorting when transplanted onto

reporter proteins that would otherwise be delivered in a nonpolarized fashion or sorted apically. The first such signal to be identified was that responsible for basolateral sorting of pIgR. Deletion of the cytoplasmic domain caused truncated pIgR to be delivered apically (Mostov et al. 1986), and transfer of a 14-amino-acid peptide from this domain to placental alkaline phosphatase, an apical protein, caused the chimera to be redirected to the basolateral surface (Casanova et al. 1991).

Two broad categories of autonomous basolateral sorting signals have been defined from studies of a large number of unrelated proteins expressed in MDCK cells (Campo et al. 2005). The first group includes signals that are related to those responsible for clathrin-mediated endocytosis and protein trafficking to the endolysosomal compartment (Matter and Mellman 1994). The surprising finding that basolateral sorting signals are related to endocytic motifs was made concurrently by three groups studying three different proteins (Brewer and Roth 1991; Hunziker et al. 1991; Le Bivic et al. 1991). Further analysis of one of these proteins, the LDL receptor, revealed the presence of two independently acting signals, both of which required critical tyrosine residues (Matter et al. 1992). It was later shown that one form of hypercholesterolemia results from a mutation within the LDLR cytoplasmic domain that inactivates one of these sorting signals, causing the receptor to be mistargeted to the apical surface (Koivisto et al. 2001). Extensive analysis of several basolateral proteins has firmly established that two types of tyrosine-based motifs exist, NPxY and YxxΦ (where x is any residue and Φ is any bulky hydrophobic amino acid) (Bonifacino and Traub 2003). Whereas the NPxY motif functions in basolateral sorting and endocytosis of LDL receptors, the YxxΦ motif mediates basolateral targeting of VSVG (Thomas et al. 1993; Thomas and Roth 1994) and the asialoglycoprotein receptor (Geffen et al. 1993), and basolateral recycling of TfnR (Odorizzi and Trowbridge 1997). This signal also is found in the cytoplasmic domain of lysosomal membrane glycoproteins (LAMPs), which initially are targeted to the basolateral plasma membrane before being internalized and delivered to lysosomes (Nabi et al. 1991).

A second type of signal that participates in basolateral sorting, endocytosis and lysosomal targeting are dileucine-based signals, which fit [D/E]xxxL[L/I] or DxxLL consen-

sus motifs (see Figure 13.2) (Bonifacino and Traub 2003). The function of dileucine signals in basolateral targeting is incompletely understood. The best characterized examples are those present on mannose-6-phosphate receptors (Johnson and Kornfeld 1992). These mediate transport from the TGN to late endosomes and bind the VHS domain of the Golgi-localized, gamma ear containing, ADP ribosylation factor binding protein (GGA) adaptors (Puertollano et al. 2001; Takatsu et al. 2001; Shiba et al. 2002). A dileucine signal lacking a requirement for acidic residues is responsible for basolateral sorting of Fc receptors in MDCK cells (Hunziker and Fumey 1994; Matter et al. 1994). Similar signals also have been characterized on several basolateral membrane proteins, including E-cadherin (Miranda et al. 2001), CD44 (Sheikh and Isacke 1996), MHC class II and Ii molecules (Odorizzi and Trowbridge 1997; Simonsen et al. 1998, 1999), and ectopically expressed norepinephrine receptors (Gu et al. 2001). Amazingly, basolateral sorting of stem cell factor receptor and of Emmprin/CD136 has been shown to depend on a single leucine motif (Wehrle-Haller and Imhof 2001; Deora et al. 2004). Whether the same sorting machinery is responsible for sorting each of these proteins to the basolateral surface is not known.

In addition to signals resembling endocytic or lysosomal targeting motifs, there is a growing list of basolateral sorting signals that are unrelated to these known signals (Muth and Caplan 2003). This heterogeneous group includes sequences that interact with PSD-95/Discs Large/ZO-1 (PDZ) domains. PDZ domains commonly are found in large, multivalent cytoplasmic proteins and have well-documented activities in establishing cell polarity and organizing functional membrane domains (Fanning and Anderson 1996; Altschuler et al. 2003; Roh and Margolis 2003). In addition, one PDZ-containing protein, Lin-7, recently was shown to regulate basolateral sorting and retention of the ErbB receptor tyrosine kinase through interactions with the carboxyl-terminus of the receptor (Straight et al. 2001; Shelly et al. 2003). Other unique sorting signals have been defined on pIgG (Casanova et al. 1991; Aroeti et al. 1993), TfnR (Odorizzi and Trowbridge 1997), and the neural cell adhesion molecule N-CAM (Le Gall et al. 1997). However, machinery that binds and sorts proteins with these signals remains unknown.

The finding that many basolateral sorting motifs are related to clathrin-dependent endocytic signals suggested that both might be recognized by similar proteins (Matter and Mellman 1994). Indeed, early experiments demonstrated that certain basolateral sorting signals operated in both the biosynthetic and recycling pathways (Matter et al. 1993; Aroeti and Mostov 1994). It is now understood that basolateral sorting involves signal recognition by adaptor protein (AP) complexes and clathrin (Folsch 2005; Rodriguez-Boulan et al. 2005). Four distinct classes of organelle-specific heterotetrameric AP complexes are encoded by the genome (AP-1 thru AP-4) (Robinson and Bonifacino 2001), and one variant form expressed exclusively in polarized epithelial cells (AP-1B) expands the sorting repertoire of these cells (Ohno et al. 1999). Each of these complexes consists of two large subunits (α, β1-4, δ, ϵ or γ), one medium subunit (μ1-4), and one small subunit (σ1-4).

Cargo selection is based on interactions between sorting signals and specific domains on AP complexes. Medium subunits (μ1-4) of AP complexes bind to the YxxΦ sorting signal (Ohno et al. 1995; 1996). Binding involves two hydrophobic pockets on the μ chains, which accommodate the tyrosine and the bulky hydrophobic residues at positions 1 and 4 (Owen and Evans 1998; Owen et al. 2001, 2004). Structurally distinct subsets of signals bind different μ chains with overlapping but differing preferences. Dileucine signals of the [D/E]xxxL[L/I] type, in contrast, inter-

act with either the β subunits (Rapoport et al. 1998), or with AP-1 or AP-3 hemicomplexes comprised of either γ/σ1 or δ/σ3 heterodimes (Janvier et al. 2003). NPxY motifs are capable of binding AP-2 complexes and clathrin directly (Bonifacino and Traub 2003), but signals of this type interact more robustly with the phosphotyrosine-binding (PTB) domains of accessory proteins, such as Dab2 and ARH (Mishra et al. 2002; Mishra et al. 2002).

AP-2 is found exclusively at the plasma membrane and mediates clathrin-dependent endocytosis, but each of the other AP complexes is present at the TGN or CRE, and potentially could function in basolateral sorting (Robinson 2004). Available evidence implicates just three AP complexes, AP-1B, AP-3, and AP-4, in sorting basolateral membrane proteins, although data for AP-3 has been provided only in nonpolarized cells (Folsch et al. 1999; Nishimura et al. 2002; Simmen et al. 2002). Since AP-1 and AP-3 have clathrin-binding sites, it is likely that clathrin also is required for basolateral protein sorting. However, a direct demonstration that clathrin functions in basolateral sorting has not been provided. AP-4 does not have a clathrin binding site and therefore could mediate basolateral sorting of LDL and Tfn receptors by promoting the formation of long tubules that direct these proteins to the basolateral membrane (Simmen et al. 2002).

The AP-1B complex shares β1, γ, and σ1 subunits with the AP-1A complex, which functions in trafficking between the TGN and late endosomes, but has an epithelial-specific medium subunit (μ1B) (Ohno et al. 1999). Evidence that AP-1B is required for efficient basolateral sorting originally was provided by genetic complementation experiments (Folsch et al. 1999). Porcine kidney LLC-PK1 cells lacking μ1B expression missort a number of basolateral proteins, including receptors for LDL and Tfn (Folsch et al. 1999), asialoglycoproteins (Sugimoto et al. 2002), and VSVG (Folsch et al. 2003) to the apical surface. Exogenous μ1B assembles into a complex with other AP-1 components when stably expressed in LLC-PK1 cells, and the resultant AP-1B complexes are biochemically and spatially distinct from endogenous AP-1A complexes (Folsch et al. 2003). Basolateral cargo colocalizes with AP-1B adaptors within clathrin-coated buds and vesicles in the perinuclear vicinity, and AP-1B complexes coimmunoprecipitate with LDL receptors (Folsch et al. 2001). Importantly, ectopic expression of μ1B restores correct polarity of missorted basolateral proteins (Folsch et al. 1999, 2003; Sugimoto et al. 2002). These results provide convincing evidence for a basolateral sorting process dependent on the AP-1B complex. The location of this sorting activity appears to be the CRE, rather than the TGN, based on the finding that newly synthesized LDL receptors initially are targeted to the basolateral surface of LLC-PK1 cells, but require μ1B function to maintain basolateral polarity during postendocytic recycling (Gan et al. 2002).

Several open questions remain about the function of AP-1B in basolateral sorting. First, if AP-1B functions in the CRE, then what is the nature of the sorting machinery that segregates basolateral from apical proteins in the TGN? It is unclear at present whether another AP complex, such as AP-3 or AP-4, has this function. Novel proteins recently have been described that could function to sort proteins in the TGN, including Four-phosphate-adaptor-protein (FAPP)1 and FAPP2. These proteins are recruited to the TGN through interactions with phosphatidylinositol 4-phosphate (PI-4P) and Arf, and participate in VSVG exit, at least in nonpolarized cells (De Matteis and Godi 2004; Godi et al. 2004). Second, how are AP-1A and AP-1B targeted to different organelles? The sole difference between these adaptors, the unique μ1A and μ1B subunits, are greater than 80 percent identical (Ohno et al. 1999). These small differences could account for their unique localizations if different μ chains are shown to bind TGN- or CRE-specific

phosphatidylinositol phosphate lipids. Interestingly, localization of AP-1A to the TGN is dependent on μ1A binding to PI-4P (Wang et al. 2003). Third, how does μ1B function to sort basolateral proteins whose sorting signals do not fit the consensus μ1B-binding site? For example, LDL receptors do not have a YxxΦ signal, yet basolateral polarity of this protein is absolutely dependent on μ1B expression. It is possible that μ1B binds non-YxxΦ signals at a distinct site, and indeed a mutant form of μ1B in which the YxxΦ binding site was disrupted retained its ability to support basolateral polarity of LDL receptor, but not of VSVG (Sugimoto et al. 2002). It also has been proposed that μ1B functions to maintain basolateral polarity of non-YxxΦ proteins through an indirect mechanism involving recruitment to the CRE of other factors involved in basolateral trafficking, such as Cdc42, Rab8, and the Exocyst (Ang et al. 2003; Folsch et al. 2003). Alternatively, AP-1B could recruit accessory proteins to the CRE that have a high affinity for non-YxxΦ signals, such as EpsinR/CLINT/enthoprotin (Kalthoff et al. 2002; Wasiak et al. 2002; Hirst et al. 2003; Mills et al. 2003), and these accessory proteins might collaborate with AP-1B to facilitate basolateral sorting of LDL receptor and other proteins. A simpler model is also possible. Perhaps μ1B expression promotes elevated clathrin assembly on the CRE, relative to cells that do not express μ1B. Because NPxY signals can bind directly to clathrin, this would be expected to enhance the sorting capacity of the CRE for proteins having an NPxY motif, even if the YxxΦ binding site on μ1B was inactivated.

B. Apical Sorting Mechanisms

In contrast to basolateral sorting signals, which consist of short, linear sequences of amino acid residues in the cytoplasmic domain of transported proteins, apical sorting signals comprise a diverse collection of moieties, including glycosylphosphatidylinositol (GPI) anchors, N- and O-linked sugars, and peptides, which may be present within the luminal, membrane-spanning, or cytoplasmic domains of proteins (Rodriguez-Boulan et al. 2005). This complexity may help to explain why apical sorting signals, though hypothesized earlier than basolateral signals, proved more difficult to define. Complicating matters further, signals responsible for apical sorting can be present on proteins that also have conventional basolateral signals (Matter and Mellman 1994; Fiedler and Simons 1995). Apical signals are, in general, hierarchically weaker than basolateral signals, and often are uncovered only upon removal of dominant basolateral signals. Finally, multiple apical sorting signals often are required to act cooperatively to mediate sorting (Sarnataro et al. 2002). Due to the overall pleiomorphic nature of apical sorting signals, a comprehensive understanding of the mechanisms underlying apical sorting processes has evolved more slowly than basolateral sorting. Nevertheless, a model is emerging to explain apical sorting of a major class of proteins, those whose sorting depends on an association with cholesterol/glycosphingolipid microdomains (lipid rafts) (Schuck and Simons 2004; Rodriguez-Boulan et al. 2005). For a smaller group of proteins, represented by the example of ectopically expressed rhodopsin, apical sorting is achieved by direct interaction between the transported cargo and a microtubule-associated motor protein (discussed later) (Tai et al. 1999).

For the majority of proteins analyzed, apical sorting involves multiple signals that act cooperatively to promote their incorporation into functional lipid rafts. The first defined apical sorting determinant was the GPI anchor. Early studies demonstrated that many endogenous GPI-anchored proteins were localized at the apical surface of MDCK cells and other epithelial cell lines (Lisanti et al. 1988). Attachment of GPI anchors to nonpolarized secretory proteins

was sufficient to target the membrane-bound chimeric proteins to the apical surface (Brown et al. 1989; Lisanti et al. 1989).

A second class of apical signals is represented by N- and O-linked sugars (Fiedler and Simons 1995; Rodriguez-Boulan and Gonzalez 1999). The widespread and generic nature of these signals helps to explain why many basolateral proteins become apically targeted following removal or inactivation of their basolateral sorting signals. A function for N- and O-glycans in apical sorting originally was suggested by the observation that deglycosylated secretory proteins are frequently secreted in a nonpolarized fashion (Rindler and Traber 1988; Fiedler and Simons 1995; Scheiffele et al. 1995). N-glycosylation of a nonglycosylated, nonpolarized secretory protein (growth hormone) was sufficient to promote its apical sorting (Scheiffele et al. 1995). An important function of N-glycans in facilitating apical protein exit from the TGN was revealed by the discovery that deglycosylated membrane proteins became trapped within this organelle (Gut et al. 1998). Interestingly, correct sorting of one protein, endolyn, is dependent on specific types of N-glycans (Potter et al. 2004). Although N-glycans function anywhere in the extracellular domain of proteins, O-glycans typically are clustered in a juxtamembrane segment (Yeaman et al. 1997; Jacob et al. 2000; Breuza et al. 2002). This positioning has prompted the suggestion that O-glycans contribute to apical sorting by helping to orient the protein in a plane perpendicular to the lipid bilayer (Rodriguez-Boulan and Gonzalez 1999).

The third group of apical sorting signals consists of proteinaceous motifs. These may be present in either the lumenal, transmembrane, or cytoplasmic domains of the protein. An early clue to the existence of such signals was the realization that certain transmembrane and secretory proteins were sorted apically, even in the complete absence of glycans (Marzolo et al. 1997;

Rodriguez-Boulan and Gonzalez 1999; Bravo-Zehnder et al. 2000). Transmembrane domains of influenza HA and certain other proteins function as sorting signals by contributing to their association with lipid rafts (Scheiffele et al. 1997; Jacob et al. 2000); however, other transmembrane domains appear to function by a mechanism that is independent of raft-association (Dunbar et al. 2000). Recently, several apical membrane proteins have been shown to have cytoplasmic sorting signals that resemble basolateral sorting signals (Rodriguez-Boulan et al. 2004). For example, apical sorting of the rat ileal bile acid transporter is dependent on a peptide signal that adopts a beta turn structure, similar to that responsible for pIgR sorting (Aroeti et al. 1993; Sun et al. 2003). Furthermore, apical sorting of megalin and LRP require tyrosine-based signals of the NPxY type (Marzolo et al. 2003; Nagai et al. 2003; Takeda et al. 2003). Therefore, long-standing distinctions between signals responsible for basolateral and apical sorting are becoming blurred, and it is possible that certain apical proteins might be sorted by means of a clathrin-dependent processes. Further research is necessary to characterize these new apical sorting signals.

It is conceivable that certain types of apical sorting signals, such as N- and O-glycans, function as conventional sorting signals that are recognized by lectin-like receptors for packaging into nascent apical carrier vesicles. VIP36 was isolated as a component of detergent-insoluble lipid rafts in MDCK cells and proposed to serve as a TGN-based sorting lectin with specificity for either GalNAc or high mannose sugars (Fiedler et al. 1994; Fiedler and Simons 1994; Yamashita et al. 1999). When overexpressed, VIP36 partially localizes to the TGN and endosomes (Fiedler and Simons 1996), but endogenous VIP36 is expressed primarily in the *cis*-Golgi in MDCK cells (Fullekrug et al. 1999), where it is proposed to function in quality control of N-glycan processing (Hauri et al. 2000).

Therefore, the collective data indicates that VIP36 probably functions at an early station in the secretory pathway and is likely not to be directly involved in apical sorting. Nevertheless, it is still possible that apical secretory proteins are sorted through association with an as-yet unidentified receptor. It is noteworthy that polarized secretion of soluble apical proteins is saturable at high expression levels (Marmorstein et al. 2000).

In contrast to the sorting receptor model for apical protein trafficking, the lipid raft hypothesis states that apical proteins are sorted by affinity for microdomains enriched in glycosphingolipids and cholesterol (lipid rafts) (van Meer and Simons 1988; Simons and Ikonen 1997; Schuck and Simons 2004). According to the raft hypothesis, lipid rafts and associated proteins form a sorting platform in the TGN and CRE that include apical proteins but exclude basolateral proteins. Ultimately, rafts enriched in apically targeted proteins detach from the sorting organelle, forming transport intermediates that are delivered to the apical plasma membrane. Several lines of evidence, accumulated over the past 20 years, support the conclusion that raft-association is necessary, but not sufficient, for apical sorting of many proteins (Rodriguez-Boulan et al. 2005). First, there is a very strong correlation between apical sorting fidelity and raft-association, as assessed by detergent insolubility (Skibbens et al. 1989; Brown and Rose 1992) or fluorescence recovery after photobleaching (FRAP) analysis (Shvartsman et al. 2003). Second, in some epithelial cell lines GPI-anchored proteins fail to incorporate into lipid rafts, and in these cells GPI-anchored proteins are not sorted apically (Zurzolo et al. 1994). Third, agents that disrupt lipid rafts, either by causing cholesterol depletion or preventing glycosphingolipid synthesis, cause GPI-anchored proteins and HA to be missorted (Mays et al. 1995; Keller and Simons 1998).

However, numerous observations indicate that raft-association is not sufficient for apical sorting (Benting et al. 1999;

Rodriguez-Boulan et al. 2005), and that a second event involving clustering or oligomerization of raft-associated proteins, also is required (Hannan et al. 1993; Paladino et al. 2004). Recent studies indicate that lipid rafts are very small (50 nm), highly dynamic structures capable of holding only three to five GPI-anchored proteins (Sharma et al. 2004). The mechanisms responsible for clustering rafts into larger transport intermediates are not fully understood, but several models have been proposed. First, it is possible that VIP21/caveolin-1, a cholesterol-binding component of lipid rafts that forms large homo-oligomers in the apical transport pathway in MDCK cells, mediates raft clustering (Scheiffele et al. 1998). Second, VIP17/MAL and MAL2 are raft-associated proteins present in apical transport intermediates and they have a propensity to form clusters (Zacchetti et al. 1995; de Marco et al. 2002). Antisense-mediated depletion of VIP17/MAL in MDCK and Fisher rat thyroid (FRT) cells disrupted raft-association and apical sorting of several proteins (Cheong et al. 1999; Puertollano and Alonso 1999; Martin-Belmonte et al. 2000). However, it remains to be determined whether VIP17/MAL functions in biosynthetic or postendocytic sorting to the apical surface (Tall et al. 2003). Finally, it is possible that N-glycans promote clustering of proteins through interactions with lectins in the TGN and CRE. Interestingly, a concanavalin A-resistant mutant MDCK cell line that does not glycosylate proteins appropriately also fails to cluster GPI-anchored proteins and missorts them to the basolateral surface (Zurzolo et al. 1994; Hannan and Edidin 1996).

IV. ROLE OF THE CYTOSKELETON IN POLARIZED TRAFFICKING IN EPITHELIAL CELLS

For efficient trafficking to occur, the packaging and production of transport intermediates in the TGN and endosomes must be coordinated with their movement through the cytoplasm and delivery to fusion sites at the plasma membrane. Although it is convenient to consider these processes separately, the sorting of proteins, and the fission, release, trafficking, docking, and fusion of transport intermediates with different plasma membrane domains comprise seamless trafficking itineraries. Coordination of these processes is achieved by several components, including organelle-specific Rab GTPases and their effectors, the cytoskeleton, a vesicle tethering complex known as the Exocyst, and distinct sets of cognate SNAREs associated with apical and basolateral membranes. Together, these factors ensure that polarized trafficking is both accurate and efficient. Actin and microtubule cytoskeletons are required to coordinate exocytic transport pathways in epithelial cells. Furthermore, cytoskeletal elements appear to be required for both exit of cargo-laden transport intermediates from sorting organelles as well as their transport across the cytosol (Allan et al. 2002). Finally, accumulating evidence indicates that microtubules also are involved in the *de novo* establishment of a polarized trafficking route to the apical surface (Rodriguez-Boulan et al. 2005).

A. Microtubules and Associated Motors

Although they lack the long cytoplasmic processes characteristic of most neurons, epithelial cells rely on microtubules and associated motors to deliver post-Golgi exocytic vesicles to both apical and basolateral plasma membrane domains (Musch 2004). The unique asymmetrical distribution of microtubules in epithelial cells provides an infrastructure upon which polarized trafficking itineraries could, in principle, be organized. In contrast to many fibroblastic cells, which nucleate microtubules at a juxtanuclear centrosome, most microtubules in polarized epithelial cells are noncentrosomal and are aligned along the apico-basal axis of polarity (Bacallao et al. 1989; Bre et al. 1990). These are uniformly oriented with their minus ends facing the apical membrane and their plus ends facing the basal surface. Thus, the selective association of minus-end-directed motors with transport intermediates carrying apical cargo would result in apical delivery of these proteins, whereas plus-end-directed motors would be expected to drive intermediates carrying basolateral cargo toward the basal surface, assuming that appropriate vesicle docking and fusion machineries were in place to accommodate the vesicles once they arrived at their destinations. In addition to the longitudinal bundles of polarized microtubules, apical and basal cytoplasmic spaces are densely packed with webs of relatively short, randomly oriented microtubules. What function, if any, do these different microtubule populations actually serve in polarized trafficking? At what stage in the trafficking process are they involved?

Evidence that microtubule-based transport facilitates protein trafficking to different surfaces in epithelial cells has been accumulating for more than two decades (Musch 2004). There is consensus in the literature that microtubule disruption following treatment with nocodazole or colchicine inhibits delivery of newly synthesized proteins to the apical plasma membrane (Rindler et al. 1987; Achler et al. 1989; Eilers et al. 1989; Matter et al. 1990; van Zeijl and Matlin 1990; Gilbert et al. 1991; Ojakian and Schwimmer 1992; Lafont et al. 1994; Grindstaff et al. 1998). In some of these studies, protein trafficking to the basolateral surface was similarly impaired (Lafont et al. 1994; Grindstaff et al. 1998), although most of these reports concluded that basolateral exocytosis was independent of microtubule-based transport (Salas et al. 1986; Rindler et al. 1987; Achler et al. 1989; Eilers et al. 1989; Matter et al. 1990; van Zeijl and Matlin 1990). However, TGN-derived vesicles carrying either an apical marker (influenza hemagglutinin), or a basolateral

marker (VSVG) were shown to specifically bind microtubules *in vitro* (van der Sluijs et al. 1990). Furthermore, immunodepletion of cytosolic kinesin, a plus-end-directed motor, from streptolysin O-permeabilized MDCK cells impaired basolateral transport of VSVG (Lafont et al. 1994). Live cell imaging experiments in both nonpolarized and polarized MDCK cells have shown that GFP-tagged apical and basolateral markers exit the TGN in long tubules and small vesicles (Hirschberg et al. 1998; Toomre et al. 1999; Kreitzer et al. 2000, 2003; Keller et al. 2001; Polishchuk et al. 2004). These carriers move away from the TGN toward the cell periphery along microtubule tracks with a speed of ~1 μm/s, characteristic of microtubule-based transport. Therefore, it seems likely that microtubules facilitate exocytic protein transport to both apical and basolateral surfaces, and that discrepancies between early reports likely resulted from variability in the degree to which the Golgi apparatus was fragmented and scattered following microtubule disruption (Musch 2004).

As noted earlier, there is evidence that a plus-end-directed motor, kinesin, is required for efficient basolateral transport of VSVG (Lafont et al. 1994). A number of studies indicate that specific plus-end and minus-end-directed motors are also required for efficient transport to the apical membrane. First, immunodepletion of either kinesin or of dynein inhibits transport of influenza HA to the apical membrane of permeabilized MDCK cells (Lafont et al. 1994). Second, careful time-lapse imaging studies have provided evidence that an unconventional kinesin may have multiple functions in apical trafficking (Kreitzer et al. 2000). Antibody inhibition experiments showed that this motor is required for production of long TGN-derived tubules containing an apical marker (p75^{NTR}-GFP), and also for the transport of small vesicles containing the same marker from the TGN to the cell periphery. Note that kinesin function was not required for the budding of the small vesicles from the TGN. It is likely that these were produced by an independent mechanism involving cytoplasmic dynein, a motor that had been shown previously to bind Golgi stacks *in vitro* and to promote the budding of small vesicles upon addition of ATP (Fath et al. 1997). Third, minus-end-directed motors including dynein and the unconventional kinesin KIFC3 are associated with post-Golgi transport vesicles containing apical cargo (Tai et al. 1999; Noda et al. 2001). As discussed in the previous section, the light chain of cytoplasmic dynein, TCTEX, interacts directly with a carboxyl-terminal domain of rhodopsin and this interaction is essential for apical trafficking of exogenously expressed rhodopsin in MDCK cells (Tai et al. 1999). Moreover, expression of a dominant-negative KIFC3 mutant impairs apical trafficking of two raft-associated proteins, influenza HA and of annexin-XIIIb (Noda et al. 2001).

These studies indicate that microtubule-based transport is important for trafficking to both basolateral and apical surfaces, and that several microtubule-based motors are likely to have multiple functions in generating transport carriers in the TGN (and recycling endosomes; Gibson et al. 1998), and in moving these carriers along microtubule tracks to the cell periphery. It is also clear that different types of post-Golgi transport intermediates associate with different motor proteins. It remains to be conclusively determined whether both plus-end- and minus-end-directed kinesins operate in apical transport pathways. A challenging problem for future research in this area will be to determine whether specific microtubules within the mixed polarity subapical array serve as tracks for different transport carriers bearing plus-end- or minus-end-directed motors. Attention should be focused on this population of microtubules, because evidence has been presented that longitudinal bundles of microtubules with uniform polarity are not essential for polarized trafficking in the

exocytic pathway (Grindstaff et al. 1998). Using a calcium-switch protocol to initiate polarity development in MDCK cells, it was shown that longitudinal microtubule bundles develop only after extended time (>72 hr) in culture. Using classic pulse-chase, plasma membrane domain-selective biotinylation methods to assess targeted delivery of several endogenous apical and basolateral proteins during development of cell polarity, it was demonstrated that efficient and faithful delivery of proteins to the correct surface domain occurred days before an epithelial-like microtubule organization was established. Furthermore, polarized trafficking proceeded just as efficiently in a strain of MDCK cells that never acquires an epithelial-like organization. Therefore, although it is clear that intact microtubules are required for efficient polarized trafficking in epithelial cells, the specific function of longitudinal microtubule bundles in this process is uncertain.

Recent studies also highlight a role for the microtubule cytoskeleton in establishing the apical trafficking route *de novo* during establishment of epithelial cell polarity. In single, contact-naïve MDCK cells, marker proteins of the apical and basolateral membrane are randomly distributed over the entire cell surface (Rodriguez-Boulan and Nelson 1989). Live imaging experiments using total internal reflection fluorescence microscopy have revealed active fusion of post-Golgi transport intermediates containing apical (p75[NTR]-GFP) or basolateral (LRL receptor-GFP) proteins with the basal plasma membrane of nonpolarized cells (Kreitzer et al. 2003). Once cells have developed polarity, however, all membrane fusion activity ceased at the basal surface. Fusion of apical transport intermediates became restricted to the noncontacting (apical) surface, and basolateral transport intermediates fused exclusively with the top third of the lateral membrane, in the vicinity of the apical junctional complex. Reorganization of trafficking routes coincides with establishment of

E-cadherin-mediated cell-cell adhesion, and accompanying changes in plasma membrane distributions of two t-SNAREs, syntaxin-3 and syntaxin-4 (Kreitzer et al. 2003). In nonpolarized cells, these SNAREs are localized uniformly on the plasma membrane. Fusion sites of apical and basolateral carriers also are distributed randomly. Restriction of fusion sites to apical and lateral membrane domains in polarized cells mirrors the redistribution of syntaxin-3 and syntaxin-4, respectively, to these sites. Importantly, when microtubules were disrupted, syntaxin-3 became randomly localized on both apical and basolateral surfaces and apical carriers again underwent fusion at the basal plasma membrane, as in nonpolarized cells. Neither syntaxin-4 localization nor basolateral fusion sites were affected by microtubule disruption. Therefore, it seems likely that an intact microtubule cytoskeleton is required in some capacity to establish or maintain a polarized trafficking pathway to the apical membrane.

B. Actin and Associated Motors

In contrast to the well-studied and generally accepted functions of microtubules in polarized membrane transport, much less is known about the functions of the actin cytoskeleton in exocytic trafficking in epithelial cells. This may be due, in part, to inherent differences in the two filament systems. Like microtubules, actin filaments are polarized, with fast-growing plus (barbed) ends and slower-growing minus (pointed) ends. However, unlike microtubules, actin filaments are not organized into parallel bundles that align with the major polarity axis of the cell. Instead, they assemble dense networks of filaments that are typically much shorter than microtubules, or loose nonparallel bundles (stress fibers). The motors that move cargo on actin (myosins) are relatively slow compared to microtubule-based motors, and all except one (myosin VI) move cargo toward the barbed end. Therefore, a role for actin-based motility in facilitating vesicle transport across the relatively long distance between the sorting compartment and the plasma membrane has been elusive.

Actin filaments may function as tracks for the myosin-driven transport of vesicles across shorter distances, akin to the function they have in smaller eukaryotes such as budding yeast. For example, in intestinal epithelial cells, myosin 1A has been implicated in trafficking certain apical vesicles (i.e., those carrying raft-associated cargo such as sucrase-isomaltase (SI)), but not of others (i.e., those carrying nonraft-associated cargo such as lactase-phlorizin hydrolase (LPH)) (Jacob et al. 2003). Both types of transport intermediates required microtubule-based motility to reach the cell periphery, but only SI trafficking also relied on actin. Furthermore, myosin 1A and SI were shown to interact within lipid rafts and expression of a dominant-negative myosin 1A mutant impaired SI trafficking to the apical surface of Caco-2 intestinal epithelial cells (Tyska and Mooseker 2004). Phosphorylation of myosin 1A by a raft-associated protein kinase, alpha-kinase 1 (ALPK1), is required for polarized trafficking of SI to the apical surface (Heine et al. 2005). These findings are consistent with a function for myosin 1A in vesicle trafficking through the dense cortical actin network in the terminal web at the apex of intestinal epithelial cells.

Recycling of membrane receptors to the apical plasma membrane also appears to rely on actin-based motility. Myosin Vb recently has been shown to facilitate receptor trafficking from the apical recycling endosome to the apical plasma membrane through associations with Rab11a and its binding partner Rab11-FIP2 (Lapierre et al. 2001; Hales et al. 2002). The C-terminal tail of myosin Vb interacts with both Rab11a and Rab11-FIP2, and overexpression of the myosin Vb tail impaired both apical recycling and basolateral-to-apical transcytosis of pIgR-IgA complexes (Lapierre et al.

2001). A recent study has also highlighted a possible role for actin-based motility in basolateral receptor recycling (Sheff et al. 2002).

In addition to functions associated with short-distance transport near the cell cortex, there is mounting evidence that actin and associated proteins are actively involved in assembling transport vesicles at the TGN and recycling endosomes. First, actin and numerous actin-binding proteins, including myosins I, II, and VI, spectrin, ankyrin, dynamin2, and cortactin are found on the Golgi or TGN (Fath et al. 1997; Ikonen et al. 1997; Musch et al. 1997; Buss et al. 1998, 2002; Allan et al. 2002; Stamnes 2002; Cao et al. 2005). Additionally, an *in vitro* vesicle budding assay has revealed a potential function for a myosin II activity in the production of basolateral, but not of apical transport vesicles (Musch et al. 1997). However the mechanism by which myosin II promotes basolateral vesicle production is not completely understood.

Cdc42, a Rho-family GTPase that has essential and highly conserved functions in regulating actin dynamics, also is recruited to Golgi membranes, and in its active GTP-bound state recruits effectors to the Golgi, such as IQGAPs, N-WASP, and CIP-4 (Erickson et al. 1996; McCallum et al. 1998; Fucini et al. 2002; Luna et al. 2002; Larocca et al. 2004). Active Cdc42 promotes actin assembly on the Golgi, and this is associated with an accelerated rate of exit of apical cargo from the TGN (Musch et al. 2001). In contrast, perturbing Cdc42 function with either constitutively active or inactive mutants slowed the exit of basolateral protein from the TGN (Musch et al. 2001), and disrupted the polarized targeting of basolateral vesicles from the TGN and endosomes (Kroschewski et al. 1999; Cohen et al. 2001). Mechanisms by which Cdc42 and actin regulate basolateral membrane trafficking are unknown. However, it has been suggested that Cdc42 promotes fission and release of apical transport intermediates from the TGN via recruitment of

dynamins to these sites. Dynamins are large GTPases that are involved in deforming membranes to form tubules and vesicles (Schafer 2004). A complex consisting of dynamin2 and cortactin is recruited to the Golgi in response to Arf1 activity (Cao et al. 2005). In support of a role in polarized trafficking, dynamin function is required for the release of apical carriers, as a dominant-negative dynamin mutant impaired the release of p75[NTR]-GFP from the TGN (Kreitzer et al. 2000).

V. TETHERING, DOCKING, AND FUSION OF TRANSPORT INTERMEDIATES

Polarized trafficking to apical and basolateral membrane domains of epithelial cells involves multiple levels of regulation that ensure that the overall process is both efficient and accurate. Although cargo segregation in the TGN and CRE, and association of transport intermediates with specific cytoskeletal components represent two important layers of quality control in this process, specification of membrane fusion between transport vesicles and target membranes provides a final critical step in a hierarchy of regulatory mechanisms that ensure that vesicles fuse with the correct target membrane and are not incorrectly delivered to the wrong acceptor membrane. Over the past decade, much experimental evidence has accumulated to support a model in which specific SNAREs and associated regulatory factors function with or without the hetero-octameric Exocyst tethering complex to specify fusion of transport intermediates with correct plasma membrane domains.

A. SNAREs and Associated Proteins

The core machinery for vesicle fusion throughout the secretory and endocytic pathway is comprised of SNARE proteins, which are described in depth elsewhere in this book (see Chapters 6, 7, and 14). Early studies showed that various treatments that functionally inhibit SNARE machinery have differential inhibitory effects on apical and basolateral trafficking in polarized MDCK cells (Ikonen et al. 1995). The logical implication of these data was that apical membrane trafficking occurred by a novel mechanism that was independent of known SNARE machinery. However, further studies established that SNAREs are required for all trafficking routes to the apical membrane, including the biosynthetic (Lafont et al. 1999), recycling (Low et al. 1998), and transcytotic (Apodaca et al. 1996; Low et al. 1998) pathways in MDCK and Caco-2 cells (Breuza et al. 2000). Regulated apical trafficking of proton pumps in inner medullary collecting duct cells in the kidney (Banerjee et al. 1999, 2001; Frank et al. 2002), as well as vasopressin-stimulated trafficking of vesicles carrying AQP2 water channels to the apical surface of renal collecting duct cells (Jo et al. 1995; Nielsen et al. 1995; Mandon et al. 1996; Inoue et al. 1998; Gouraud et al. 2002) also were shown to require components of the SNARE machinery. Therefore, fusion machinery governing exocytosis at the apical plasma membrane, like that at the basolateral plasma membrane, is similar to that responsible for most other membrane fusion events in cells.

Specificity in the fusigenic properties of apical and basolateral membrane domains is likely to be established, in part, by the polarized distribution of individual t-SNAREs. For example, endogenous or exogenously expressed Syntaxins 2A and 3 are enriched in the apical plasma membrane of MDCK cells, whereas Syntaxin 4 accumulates in the basolateral membrane (Low et al. 1996; Quinones et al. 1999). Assembly and function of Syntaxin 3-containing SNARE complexes at the apical membrane are regulated by Munc18-2, a member of the Sec1/Munc18 family of SNARE-associated proteins (Riento et al. 1996, 1998, 2000). In contrast, function of

Syntaxin 4 at the basolateral surface may be regulated by mLgl, the mammalian homologue of Lethal giant larvae, a *D. melanogaster* protein involved in epithelial polarity development (Musch et al. 2002). Other t-SNARE components, including Syntaxin 2B, Syntaxin 11, and SNAP-23 are expressed in a nonpolarized fashion on both apical and basolateral surfaces in these cells (Low et al. 1996, 2000; Quinones et al. 1999). These observations have led to the testable hypothesis that differential localization of t-SNAREs is responsible for polarized vesicle fusion processes in epithelial cells. Consistent with this model, overexpression of Syntaxin 3, but not Syntaxins 2 or 4, reduced the efficiency of trafficking to the apical surface of MDCK cells, but had no effect on basolateral exocytosis (Low et al. 1998). A similar finding was made when a Syntaxin 3-TfnR chimera was overexpressed in Caco-2 cells (Breuza et al. 2000) that, like MDCK cells, express Syntaxin 3 on the apical plasma membrane (Delgrossi et al. 1997). Furthermore, addition of function-blocking antibodies to Syntaxin 3 or Syntaxin 4 to permeabilized MDCK cells reportedly inhibited exocytosis to the apical and basolateral surface, respectively (Lafont et al. 1999).

Note that epithelial cells isolated from various sources also display polarized distributions of individual t-SNAREs, but no single pattern describes all epithelia. For example, pancreatic acinar cells express Syntaxin 2 and 4 on the apical and basolateral surface, respectively, but express Syntaxin 3 on zymogen granules in the cytoplasm (Gaisano et al. 1996). Retinal pigment epithelial cells do not express Syntaxin 3 at all, but instead place Syntaxins 1A and 1B on the apical surface and localize Syntaxins 2 and 4 within a narrow zone near the tight junctions on the lateral plasma membrane (Low et al. 2002). In native epithelial cells of the kidney, localization of Syntaxins 3 and 4 are controversial. Several papers document a basolateral distribution of Syntaxin 3 and an apical distribution of Syntaxin 4 (Mandon et al. 1996, 1997; Breton et al. 2002). However, another study shows that Syntaxins 3 and 4 are expressed exclusively on the apical and basolateral surfaces, respectively, of every cell in the nephron (Li et al. 2002). Given that multiple alternatively spliced forms of these proteins have been described (Ibaraki et al. 1995), and alternatively spliced Syntaxins 2A and 2B show different localization patterns in MDCK cells (Quinones et al. 1999), it is possible that differences in t-SNARE localization observed in native kidney epithelial cells results from isoform-specific localization patterns.

Progress in identifying SNARE machinery responsible for polarized trafficking in epithelial cells has been restricted largely to defining localizations of t-SNAREs, described earlier. In contrast, relatively little progress has been made in identifying specific v-SNAREs responsible for fusion of different transport intermediates trafficking between the TGN or CRE and apical or basolateral plasma membranes. Data has been presented for three v-SNAREs that could function in different apical pathways, but more work is needed to define the exact trafficking step each one mediates. Apical exocytosis in the biosynthetic pathway could be mediated by a tetanus-toxin-insensitive v-SNARE, ti-VAMP (VAMP 7) (Galli et al. 1998). This protein localizes, in part, to vesicles in the apical cytoplasm and the apical plasma membrane, where it forms complexes with SNAP-23 and Syntaxin 3 (Galli et al. 1998; Lafont et al. 1999). In addition, function-blocking antibodies to VAMP 7 partially inhibited apical trafficking of HA (Lafont et al. 1999). A distinct v-SNARE, VAMP 2, is implicated in vasopressin-regulated AQP-2 trafficking to the apical surface of renal cortical collecting duct cells and proton pump trafficking in intercalated cells (Nielsen et al. 1995; Inoue et al. 1998; Banerjee et al. 2001). Finally, a third v-SNARE, VAMP 8/endobrevin localizes to the apical endosomal system in nearly all segments of the nephron, and

cycles between there and the apical plasma membrane of MDCK cells (Steegmaier et al. 2000). In contrast to the apical pathways, essentially nothing is known regarding the identity of v-SNAREs functioning in biosynthetic or postendocytic trafficking to the basolateral surface.

B. The Exocyst

Screens for secretory mutants in the budding yeast *S. cerevisiae* uncovered a group of eight genes encoding a hetero-octameric protein complex termed the Exocyst, which is required for polarized exocytosis within the growing daughter cell (TerBush and Novick 1995; TerBush et al. 1996; Guo et al. 1999). Epithelial cells express a related complex (also known as the Sec6/8 complex), and this has been proposed to serve as a spatial determinant for fusion sites during polarized membrane growth (Grindstaff et al. 1998; Hsu et al. 1999; Yeaman et al. 1999). Consistent with this overall concept, MDCK cells undergoing lateral membrane expansion recruit the Exocyst from the cytosol to the lateral, but not basal or apical membranes, and function-blocking antibodies to the Sec8 subunit inhibit delivery of LDL receptors to the lateral membrane, but not of p75[NTR] to the apical membrane (Grindstaff et al. 1998). Furthermore, fusion of exocytic transport vesicles with the basolateral plasma membrane is limited to the upper third of the lateral membrane (Kreitzer et al. 2003), in close proximity to the Exocyst that accumulates at the apical junctional complex of these cells (Grindstaff et al. 1998; Yeaman et al. 2004). The Exocyst similarly localizes to the junctional vicinity of actively growing epithelial tubules in a three-dimensional culture model of kidney tubulogenesis (Lipschutz et al. 2000). Exocyst-mediated membrane trafficking to these sites is likely to be important for tissue morphogenesis and polarized epithelial functions. In cells derived from kidney tubules of patients with autosomal-dominant polycystic kidney disease, the Exocyst is present in the cytosol rather than at the plasma membrane, and basolateral trafficking is severely impaired (Charron et al. 2000).

Exocyst function in epithelial cells is regulated by RalA (Moskalenko et al. 2001; Shipitsin and Feig 2004), a small GTPase that interacts with both Sec5 and Exo84, two subunits of the Exocyst (Sugihara et al. 2002; Fukai et al. 2003; Moskalenko et al. 2003; Mott et al. 2003; Jin et al. 2005). Induction of constitutively active RalA72L expression in a conditional cell line enhances basolateral exocytosis of E-cadherin, but has no effect on apical trafficking of endogenous proteins (Shipitsin and Feig 2004). Conversely, interfering with RalA-Sec5 interactions by a variety of techniques causes basolateral proteins to become aberrantly expressed on the apical plasma membrane (Moskalenko et al. 2001). The mechanisms by which RalA regulates Exocyst activity are not fully understood. One hypothesis that has some experimental support is that RalA promotes vesicle tethering through dual subunit interactions involving an Exo84-containing hemicomplex on transport vesicles with a Sec5-containing hemicomplex at the plasma membrane (Moskalenko et al. 2003). Structural analysis of RalA in complex with Sec5 or Exo84 has shown that RalA cannot simultaneously bind both subunits (Fukai et al. 2003; Mott et al. 2003; Jin et al. 2005), and gel filtration analyses using recombinant proteins indicates that Sec5 and Exo84 compete for RalA binding (Jin et al. 2005). Therefore, it is possible that RalA forms distinct complexes with both Sec5 and Exo84 and these regulate Exocyst functions by unique mechanisms, or that RalA binds Sec5 and Exo84 in a sequential fashion, and this somehow regulates spatiotemporal aspects of vesicle docking and fusion. In addition to the well-characterized interactions between RalA and Sec5/Exo84, at least three other interactions involving small GTPases and Exocyst subunits have been described in mammalian cells, including Rab11-Sec15 (Zhang et al. 2004), Arf6-Sec10 (Prigent et al. 2003), and TC10-Exo70 (Inoue et al. 2003). How each of these interactions regulates vesicle docking and fusion events involving the Exocyst represents a major unsolved question in the field.

A thorough understanding of Exocyst functions in epithelial membrane trafficking has been made still more complicated by numerous recent studies documenting the association of one or more Exocyst subunits with several organelles other than the lateral plasma membrane. These include both the TGN (Yeaman et al. 2001, 2004) and CRE (Folsch et al. 2003; Shipitsin and Feig 2004), as well as the endoplasmic reticulum (Lipschutz et al. 2003), primary cilium (Rogers et al. 2004), and a novel ring-like structure at the midbody of cells undergoing cytokinesis (Gromley et al. 2005). Because the Exocyst associates with both of the sorting organelles involved in polarized exocytosis, it is possible that it has a unique function in formation of transport vesicles. Disruption of exocyst function at these sites impairs trafficking of basolateral proteins in nonpolarized cells (Yeaman et al. 2001). Expression of constitutively active RalA (Shipitsin and Feig 2004), or overexpression of AP-1B clathrin adaptors (Folsch et al. 2003) promote heightened levels of Exocyst proteins on the CRE, suggesting that these factors might regulate vesicle budding, in part, by recruiting the tether. In contrast, expression of dominant inhibitory PKD1 causes Sec6 to accumulate on the TGN (Yeaman et al. 2004), similar to its distribution in nonpolarized fibroblasts (Yeaman et al. 2004). Therefore, interaction of the Exocyst with organelles associated with polarized sorting is likely to be subject to multiple levels of regulation.

Trafficking to the apical plasma membrane of MDCK cells likely requires a tethering factor also, but the identity of this factor is not yet known. Exocyst function was examined only for a single, nonraft-associated protein trafficking in the biosynthetic pathway (Grindstaff et al. 1998), so it is possible that other apical routes rely on the Exocyst. It is also important to mention that, although the Exocyst is enriched on the lateral plasma membrane of MDCK cells, it is found associated with specialized regions of the apical plasma membrane in other cell types, and these are sites of regulated exocytosis and polarized membrane growth. For example, the Exocyst is present at the apical pole of pancreatic acinar cells, where zymogen granules will ultimately fuse (Shin et al. 2000). In *Drosophila* photoreceptor cells, Exocyst subunits are enriched at the rhabdomere, the light sensing subdomain of the apical membrane, and function to target secretory vesicles to this region only, and not to the apical stalk membrane or the basolateral membrane (Beronja et al. 2005). Rhodopsin trafficking to the rhabdomere also was shown to require the function of Rab11 (Satoh et al. 2005), a small GTPase recently shown to interact with Exocyst subunit Sec15 (Zhang et al. 2004). Thus, as in budding yeast, Exocyst activity in epithelial tissues is associated with membrane domains engaged in active remodeling processes, where a large number of exocytic transport vesicles must be consumed quickly and efficiently. Whether distinct tethering complexes function in constitutive trafficking to other surface domains remains to be determined.

VI. CONCLUDING REMARKS: NEURONAL POLARITY AND THE EPITHELIAL CELL PARADIGM

Studies conducted over the past quarter century have revealed a number of important details about the processes involved in protein sorting and trafficking to apical and basolateral domains of polarized epithelial cells. They also have served to highlight just how much we still do not understand about mechanisms controlling these processes. Nevertheless, research in epithelial cells has served as a useful paradigm for investigat-

ing mechanisms by which polarized exocytic pathways function in other cell types, including neurons. Studies of epithelial trafficking mechanisms also have benefited from insights provided by studies in neurons. This is especially true of research into the functions of molecular motors in polarized trafficking, where investigations in epithelial cells have profited immensely from examples first defined in neuronal transport pathways (Hirokawa and Takemura 2005). Moreover, an appreciation of the functional role of the Exocyst in polarized membrane growth was gained by direct interactions between laboratories using epithelial and neuronal cell models (Grindstaff et al. 1998; Matern et al. 2001).

Early studies showing that epithelial cells and hippocampal neurons sort viral glycoproteins and GPI-anchored proteins similarly prompted analogies between the apical and axonal transport routes and the basolateral and dendritic routes (Dotti and Simons 1990; Dotti et al. 1991). This model was put to vigorous tests, and notable exceptions were reported (Craig and Banker 1994; Higgins et al. 1997). However, there are numerous cases in which proteins displaying polarized expression in one epithelial cell type exhibit a reversed polarity in another epithelial cell type (Zurzolo et al. 1993; Roush et al. 1998). Understanding the mechanisms underlying those polarity differences, such as changes in expression levels of clathrin adaptors and accessory proteins, or differences in mechanisms leading to lipid raft clustering in the sorting organelle, will provide important clues to how sorting takes place in both epithelial cells and neurons.

References

Achler, C., Filmer, D., Merte, C., Drenckhahn, D. (1989). Role of microtubules in polarized delivery of apical membrane proteins to the brush border of the intestinal epithelium. *J. Cell Biol.* **109**, 179–189.

Allan, V.J., Thompson, H.M., McNiven, M.A. (2002). Motoring around the Golgi. *Nat. Cell Biol.* **4**, E236–242.

Altschuler, Y., Hodson, C., Milgram, S.L. (2003). The apical compartment: Trafficking pathways, regulators and scaffolding proteins. *Curr. Opin. Cell Biol.* **15**, 423–429.

Ang, A.L., Folsch, H., Koivisto, U.M., Pypaert, M., Mellman, I. (2003). The Rab8 GTPase selectively regulates AP-1B-dependent basolateral transport in polarized Madin-Darby canine kidney cells. *J. Cell Biol.* **163**, 339–350.

Ang, A.L., Taguchi, T., Francis, S., Folsch, H., Murrells, L.J., Pypaert, M. et al. (2004). Recycling endosomes can serve as intermediates during transport from the Golgi to the plasma membrane of MDCK cells. *J. Cell Biol.* **167**, 531–543.

Apodaca, G., Cardone, M.H., Whiteheart, S.W., DasGupta, B.R., Mostov, K.E. (1996). Reconstitution of transcytosis in SLO-permeabilized MDCK cells: Existence of an NSF-dependent fusion mechanism with the apical surface of MDCK cells. *EMBO J.* **15**, 1471–1481.

Apodaca, G., Katz, L.A., Mostov, K.E. (1994). Receptor-mediated transcytosis of IgA in MDCK cells is via apical recycling endosomes. *J. Cell Biol.* **125**, 67–86.

Aroeti, B., Kosen, P.A., Kuntz, I.D., Cohen, F.E., Mostov, K.E. (1993). Mutational and secondary structural analysis of the basolateral sorting signal of the polymeric immunoglobulin receptor. *J. Cell Biol.* **123**, 1149–1160.

Aroeti, B., Mostov, K.E. (1994). Polarized sorting of the polymeric immunoglobulin receptor in the exocytotic and endocytotic pathways is controlled by the same amino acids. *EMBO J.* **13**, 2297–2304.

Bacallao, R., Antony, C., Dotti, C., Karsenti, E., Stelzer, E.H., Simons, K. (1989). The subcellular organization of Madin-Darby canine kidney cells during the formation of a polarized epithelium. *J. Cell Biol.* **109**, 2817–2832.

Banerjee, A., Li, G., Alexander, E.A., Schwartz, J.H. (2001). Role of SNAP-23 in trafficking of H+-ATPase in cultured inner medullary collecting duct cells. *Am. J. Physiol. Cell Physiol.* **280**, C775–781.

Banerjee, A., Shih, T., Alexander, E.A., Schwartz, J.H. (1999). SNARE proteins regulate H(+)-ATPase redistribution to the apical membrane in rat renal inner medullary collecting duct cells. *J. Biol. Chem.* **274**, 26518–26522.

Baron, C.L., Malhotra, V. (2002). Role of diacylglycerol in PKD recruitment to the TGN and protein transport to the plasma membrane. *Science* **295**, 325–328.

Barroso, M., Sztul, E.S. (1994). Basolateral to apical transcytosis in polarized cells is indirect and involves BFA and trimeric G protein sensitive passage through the apical endosome. *J. Cell Biol.* **124**, 83–100.

Bennett, M.K., Wandinger-Ness, A., Simons, K. (1988). Release of putative exocytic transport vesicles from perforated MDCK cells. *EMBO J.* **7**, 4075–4085.

Benting, J.H., Rietveld, A.G., Simons, K. (1999). N-Glycans mediate the apical sorting of a GPI-anchored, raft-associated protein in Madin-Darby canine kidney cells. *J. Cell Biol.* **146**, 313–320.

Beronja, S., Laprise, P., Papoulas, O., Pellikka, M., Sisson, J., Tepass, U. (2005). Essential function of Drosophila Sec6 in apical exocytosis of epithelial photoreceptor cells. *J. Cell Biol.* **169**, 635–646.

Bomsel, M., Prydz, K., Parton, R.G., Gruenberg, J., Simons, K. (1989). Endocytosis in filter-grown Madin-Darby canine kidney cells. *J. Cell Biol.* 3243–3258.

Bonifacino, J.S., Traub, L.M. (2003). Signals for sorting of transmembrane proteins to endosomes and lysosomes. *Annu. Rev. Biochem.* **72**, 395–447.

Bravo-Zehnder, M., Orio, P., Norambuena, A., Wallner, M., Meera, P., Toro, L. et al. (2000). Apical sorting of a voltage- and Ca2+-activated K+ channel alpha-subunit in Madin-Darby canine kidney cells is independent of N-glycosylation. *Proc. Natl. Acad. Sci. U S A* **97**, 13114–13119.

Bre, M.H., Pepperkok, R., Hill, A.M., Levilliers, N., Ansorge, W., Stelzer, E.H., Karsenti, E. (1990). Regulation of microtubule dynamics and nucleation during polarization in MDCK II cells. *J. Cell Biol.* 3013–3021.

Breton, S., Inoue, T., Knepper, M.A., Brown, D. (2002). Antigen retrieval reveals widespread basolateral expression of syntaxin 3 in renal epithelia. *Am. J. Physiol. Renal Physiol.* **282**, F523–529.

Breuza, L., Fransen, J., Le Bivic, A. (2000). Transport and function of syntaxin 3 in human epithelial intestinal cells. *Am. J. Physiol. Cell Physiol.* **279**, C1239–1248.

Breuza, L., Garcia, M., Delgrossi, M.H., Le Bivic, A. (2002). Role of the membrane-proximal O-glycosylation site in sorting of the human receptor for neurotrophins to the apical membrane of MDCK cells. *Exp. Cell Res.* **273**, 178–186.

Brewer, C.B., Roth, M.G. (1991). A single amino acid change in the cytoplasmic domain alters the polarized delivery of influenza virus hemagglutinin. *J. Cell Biol.* **114**, 413–421.

Brown, D.A., Crise, B., Rose, J.K. (1989). Mechanism of membrane anchoring affects polarized expression of two proteins in MDCK cells. *Science* **245**, 1499–1501.

Brown, D.A., Rose, J.K. (1992). Sorting of GPI-anchored proteins to glycolipid-enriched membrane subdomains during transport to the apical cell surface. *Cell* **68**, 533–544.

Brown, P.S., Wang, E., Aroeti, B., Chapin, S.J., Mostov, K.E., Dunn, K.W. (2000). Definition of distinct compartments in polarized Madin-Darby canine kidney (MDCK) cells for membrane-volume sorting, polarized sorting and apical recycling. *Traffic* **1**, 124–140.

Bucci, C., Wandinger-Ness, A., Lutcke, A., Chiariello, M., Bruni, C.B., Zerial, M. (1994). Rab5a is a common component of the apical and basolateral endocytic machinery in polarized epithelial cells. *Proc. Natl. Acad. Sci. U S A* **91**, 5061–5065.

Buss, F., Kendrick-Jones, J., Lionne, C., Knight, A.E., Cote, G.P., Paul Luzio, J. (1998). The localization of myosin VI at the golgi complex and leading edge of fibroblasts and its phosphorylation and recruitment into membrane ruffles of A431 cells after growth factor stimulation. *J. Cell Biol.* **143**, 1535–1545.

Buss, F., Luzio, J.P., Kendrick-Jones, J. (2002). Myosin VI, an actin motor for membrane traffic and cell migration. *Traffic* **3**, 851–858.

Campo, C., Mason, A., Maouyo, D., Olsen, O., Yoo, D., Welling, P.A. (2005). Molecular mechanisms of membrane polarity in renal epithelial cells. *Rev. Physiol. Biochem. Pharmacol.* **153**, 47–99.

Cao, H., Weller, S., Orth, J.D., Chen, J., Huang, B., Chen, J.L. et al. (2005). Actin and Arf1-dependent recruitment of a cortactin-dynamin complex to the Golgi regulates post-Golgi transport. *Nat. Cell Biol.* **7**, 483–492.

Caplan, M.J., Anderson, H.C., Palade, G.E., Jamieson, J.D. (1986). Intracellular sorting and polarized cell surface delivery of (Na+,K+)ATPase, an endogenous component of MDCK cell basolateral plasma membranes. *Cell* **46**, 623–631.

Casanova, J.E., Apodaca, G., Mostov, K.E. (1991). An autonomous signal for basolateral sorting in the cytoplasmic domain of the polymeric immunoglobulin receptor. *Cell* **66**, 65–75.

Casanova, J.E., Mishumi, Y., Ikehara, Y., Hubbard, A.L., Mostov, K.E. (1991). Direct apical sorting of rat liver dipeptidylpeptidase IV expressed in Madin-Darby canine kidney cells. *J. Biol. Chem.* **266**, 24428–24432.

Casanova, J.E., Wang, X., Kumar, R., Bhartur, S.G., Navarre, J., Woodrum, J.E. et al. (1999). Association of Rab25 and Rab11a with the apical recycling system of polarized Madin-Darby canine kidney cells. *Mol. Biol. Cell* **10**, 47–61.

Charron, A.J., Nakamura, S., Bacallao, R., Wandinger-Ness, A. (2000). Compromised cytoarchitecture and polarized trafficking in autosomal dominant polycystic kidney disease cells. *J. Cell Biol.* **149**, 111–124.

Cheong, K.H., Zacchetti, D., Schneeberger, E.E., Simons, K. (1999). VIP17/MAL, a lipid raft-associated protein, is involved in apical transport in MDCK cells. *Proc. Natl. Acad. Sci. U S A* **96**, 6241–6248.

Cohen, D., Musch, A., Rodriguez-Boulan, E. (2001). Selective control of basolateral membrane protein polarity by cdc42. *Traffic* **2**, 556–564.

Craig, A.M., Banker, G. (1994). Neuronal polarity. *Annu. Rev. Neurosci.* **17**, 267–310.

de Marco, M.C., Martin-Belmonte, F., Kremer, L., Albar, J.P., Correas, I., Vaerman, J.P. et al. (2002). MAL2, a novel raft protein of the MAL family, is an essential component of the machinery for transcytosis in hepatoma HepG2 cells. *J. Cell Biol.* **159**, 37–44.

De Matteis, M.A., Godi, A. (2004). PI-loting membrane traffic. *Nat. Cell Biol.* **6**, 487–492.

Delgrossi, M.H., Breuza, L., Mirre, C., Chavrier, P., Le Bivic, A. (1997). Human syntaxin 3 is localized apically in human intestinal cells. *J. Cell Sci.* **110 (Pt 18)**, 2207–2214.

Deora, A.A., Gravotta, D., Kreitzer, G., Hu, J., Bok, D., Rodriguez-Boulan, E. (2004). The basolateral targeting signal of CD147 (EMMPRIN) consists of a single leucine and is not recognized by retinal pigment epithelium. *Mol. Biol. Cell* **15**, 4148–4165.

Dotti, C.G., Parton, R.G., Simons, K. (1991). Polarized sorting of glypiated proteins in hippocampal neurons. *Nature* **349**, 158–161.

Dotti, C.G., Simons, K. (1990). Polarized sorting of viral glycoproteins to the axon and dendrites of hippocampal neurons in culture. *Cell* **62**, 63–72.

Dunbar, L.A., Aronson, P., Caplan, M.J. (2000). A transmembrane segment determines the steady-state localization of an ion-transporting adenosine triphosphatase. *J. Cell Biol.* **148**, 769–778.

Dunn, K.W., McGraw, T.E., Maxfield, F.R. (1989). Iterative fractionation of recycling receptors from lysosomally destined ligands in an early sorting endosome. *J. Cell Biol.* 3303–3314.

Eilers, U., Klumperman, J., Hauri, H.P. (1989). Nocodazole, a microtubule-active drug, interferes with apical protein delivery in cultured intestinal epithelial cells (Caco-2). *J. Cell Biol.* **108**, 13–22.

Erickson, J.W., Zhang, C., Kahn, R.A., Evans, T., Cerione, R.A. (1996). Mammalian Cdc42 is a brefeldin A-sensitive component of the Golgi apparatus. *J. Biol. Chem.* **271**, 26850–26854.

Fanning, A.S., Anderson, J.M. (1996). Protein-protein interactions: PDZ domain networks. *Curr. Biol.* **6**, 1385–1388.

Fath, K.R., Trimbur, G.M., Burgess, D.R. (1997). Molecular motors and a spectrin matrix associate with Golgi membranes in vitro. *J. Cell Biol.* **139**, 1169–1181.

Fialka, I., Steinlein, P., Ahorn, H., Bock, G., Burbelo, P.D., Haberfellner, M. et al. (1999). Identification of syntenin as a protein of the apical early endocytic compartment in Madin-Darby canine kidney cells. *J. Biol. Chem.* **274**, 26233–26239.

Fiedler, K., Parton, R.G., Kellner, R., Etzold, T., Simons, K. (1994). VIP36, a novel component of glycolipid rafts and exocytic carrier vesicles in epithelial cells. *EMBO J.* **13**, 1729–1740.

Fiedler, K., Simons, K. (1994). A putative novel class of animal lectins in the secretory pathway homologous to leguminous lectins. *Cell* **77**, 625–626.

Fiedler, K., Simons, K. (1995). The role of N-glycans in the secretory pathway. *Cell* **81**, 309–312.

Fiedler, K., Simons, K. (1996). Characterization of VIP36, an animal lectin homologous to leguminous lectins. *J. Cell Sci.* **109**, 271–276.

Folsch, H. (2005). The building blocks for basolateral vesicles in polarized epithelial cells. *Trends Cell Biol.* **15**, 222–228.

Folsch, H., Ohno, H., Bonifacino, J.S., Mellman, I. (1999). A novel clathrin adaptor complex mediates basolateral targeting in polarized epithelial cells. *Cell* **99**, 189–198.

Folsch, H., Pypaert, M., Maday, S., Pelletier, L., Mellman, I. (2003). The AP-1A and AP-1B clathrin adaptor complexes define biochemically and functionally distinct membrane domains. *J. Cell Biol.* **163**, 351–362.

Folsch, H., Pypaert, M., Schu, P., Mellman, I. (2001). Distribution and function of AP-1 clathrin adaptor complexes in polarized epithelial cells. *J. Cell Biol.* **152**, 595–606.

Frank, A.E., Wingo, C.S., Andrews, P.M., Ageloff, S., Knepper, M.A., Weiner, I.D. (2002). Mechanisms through which ammonia regulates cortical collecting duct net proton secretion. *Am. J. Physiol. Renal Physiol.* **282**, F1120–1128.

Fucini, R.V., Chen, J.L., Sharma, C., Kessels, M.M., Stamnes, M. (2002). Golgi vesicle proteins are linked to the assembly of an actin complex defined by mAbp1. *Mol. Biol. Cell* **13**, 621–631.

Fukai, S., Matern, H.T., Jagath, J.R., Scheller, R.H., Brunger, A.T. (2003). Structural basis of the interaction between RalA and Sec5, a subunit of the sec6/8 complex. *EMBO J.* **22**, 3267–3278.

Fullekrug, J., Scheiffele, P., Simons, K. (1999). VIP36 localisation to the early secretory pathway. *J. Cell Sci.* **112** (Pt 17), 2813–2821.

Fuller, S.D., Bravo, R., Simons, K. (1985). An enzymatic assay reveals that proteins destined for the apical or basolateral domains of an epithelial cell line share the same late Golgi compartments. *EMBO J.* **4**, 297–307.

Futter, C.E., Connolly, C.N., Cutler, D.F., Hopkins, C.R. (1995). Newly synthesized transferrin receptors can be detected in the endosome before they appear on the cell surface. *J. Biol. Chem.* **270**, 10999–11003.

Gaisano, H.Y., Ghai, M., Malkus, P.N., Sheu, L., Bouquillon, A., Bennett, M.K., Trimble, W.S. (1996). Distinct cellular locations of the syntaxin family of proteins in rat pancreatic acinar cells. *Mol. Biol. Cell* **7**, 2019–2027.

Galli, T., Zahraoui, A., Vaidyanathan, V.V., Raposo, G., Tian, J.M., Karin, M., Niemann, H., Louvard, D. (1998). A novel tetanus neurotoxin-insensitive vesicle-associated membrane protein in SNARE complexes of the apical plasma membrane of epithelial cells. *Mol. Biol. Cell* **9**, 1437–1448.

Gan, Y., McGraw, T.E., Rodriguez-Boulan, E. (2002). The epithelial-specific adaptor AP1B mediates postendocytic recycling to the basolateral membrane. *Nat. Cell Biol.* **4**, 605–609.

Geffen, I., Fuhrer, C., Leitinger, B., Weiss, M., Huggel, K., Griffiths, G., Spiess, M. (1993). Related signals for endocytosis and basolateral sorting of the asialoglycoprotein receptor. *J. Biol. Chem.* **268**, 20772–20777.

Gibson, A., Futter, C.E., Maxwell, S., Allchin, E.H., Shipman, M., Kraehenbuhl, J.P. et al. (1998). Sorting mechanisms regulating membrane protein traffic in the apical transcytotic pathway of polarized MDCK cells. *J. Cell Biol.* **143**, 81–94.

Gilbert, T., Le Bivic, A., Quaroni, A., Rodriguez-Boulan, E. (1991). Microtubular organization and its involvement in the biogenetic pathways of plasma membrane proteins in Caco-2 intestinal epithelial cells. *J. Cell Biol.* **113**, 275–288.

Godi, A., Di Campli, A., Konstantakopoulos, A., Di Tullio, G., Alessi, D.R., Kular, G.S. et al. (2004). FAPPs control Golgi-to-cell-surface membrane traffic by binding to ARF and PtdIns(4)P. *Nat. Cell Biol.* **6**, 393–404.

Gouraud, S., Laera, A., Calamita, G., Carmosino, M., Procino, G., Rossetto, O. et al. (2002). Functional involvement of VAMP/synaptobrevin-2 in cAMP-stimulated aquaporin 2 translocation in renal collecting duct cells. *J. Cell Sci.* **115**, 3667–3674.

Griffiths, G., Pfeiffer, S., Simons, K., Matlin, K. (1985). Exit of newly synthesized membrane proteins from the trans cisterna of the Golgi complex to the plasma membrane. *J. Cell Biol.* **101**, 949–964.

Grindstaff, K.K., Bacallao, R.L., Nelson, W.J. (1998). Apiconuclear organization of microtubules does not specify protein delivery from the trans-Golgi network to different membrane domains in polarized epithelial cells. *Mol. Biol. Cell* **9**, 685–699.

Grindstaff, K.K., Yeaman, C., Anandasabapathy, N., Hsu, S.C., Rodriguez-Boulan, E., Scheller, R.H., Nelson, W.J. (1998). Sec6/8 complex is recruited to cell-cell contacts and specifies transport vesicle delivery to the basal-lateral membrane in epithelial cells. *Cell* **93**, 731–740.

Gromley, A., Yeaman, C., Rosa, J., Redick, S., Chen, C.-T., Mirabelle, S. et al. (2005). Centriolin-anchoring of exocyst and SNARE complexes at the midbody is required for localized secretion and abscission during cytokinesis. *Cell*. In press.

Gu, H.H., Wu, X., Giros, B., Caron, M.G., Caplan, M.J., Rudnick, G. (2001). The NH(2)-terminus of norepinephrine transporter contains a basolateral localization signal for epithelial cells. *Mol. Biol. Cell* **12**, 3797–3807.

Guo, W., Roth, D., Walch-Solimena, C., Novick, P. (1999). The exocyst is an effector for Sec4p, targeting secretory vesicles to sites of exocytosis. *EMBO J.* **18**, 1071–1080.

Gut, A., Kappeler, F., Hyka, N., Balda, M.S., Hauri, H.-P., Matter, K. (1998). Carbohydrate-mediated Golgi to cell surface transport and apical targeting of membrane proteins. *EMBO J.* **17**, 1919–1929.

Hales, C.M., Vaerman, J.P., Goldenring, J.R. (2002). Rab11 family interacting protein 2 associates with Myosin Vb and regulates plasma membrane recycling. *J. Biol. Chem.* **277**, 50415–50421.

Hannan, L.A., Edidin, M. (1996). Traffic, polarity, and detergent solubility of a glycosylphosphatidylinositol-anchored protein after LDL-deprivation of MDCK cells. *J. Cell Biol.* **133**, 1265–1276.

Hannan, L.A., Lisanti, M.P., Rodriguez-Boulan, E., Edidin, M. (1993). Correctly sorted molecules of a GPI-anchored protein are clustered and immobile when they arrive at the apical surface of MDCK cells. *J. Cell Biol.* **120**, 353–358.

Hauri, H., Appenzeller, C., Kuhn, F., Nufer, O. (2000). Lectins and traffic in the secretory pathway. *FEBS Lett.* **476**, 32–37.

Heine, M., Cramm-Behrens, C.I., Ansari, A., Chu, H.P., Ryazanov, A.G., Naim, H.Y., Jacob, R. (2005). Alpha-kinase 1, a new component in apical protein transport. *J. Biol. Chem.*

Hemery, I., Durand, S.A., Feldmann, G., Vaerman, J.P., Maurice, M. (1996). The transcytotic pathway of an apical plasma membrane protein (B10) in hepatocytes is similar to that of IgA and occurs via a tubular pericentriolar compartment. *J. Cell Sci.* 1215–1227.

Higgins, D., Burack, M., Lein, P., Banker, G. (1997). Mechanisms of neuronal polarity. *Curr. Opin. Neurobiol.* **7**, 599–604.

Hirokawa, N., Takemura, R. (2005). Molecular motors and mechanisms of directional transport in neurons. *Nat. Rev. Neurosci.* **6**, 201–214.

Hirschberg, K., Miller, C.M., Ellenberg, J., Presley, J.F., Siggia, E.D., Phair, R.D., Lippincott-Schwartz, J. (1998). Kinetic analysis of secretory protein traffic and characterization of golgi to plasma membrane transport intermediates in living cells. *J. Cell Biol.* **143**, 1485–1503.

Hirst, J., Motley, A., Harasaki, K., Peak Chew, S.Y., Robinson, M.S. (2003). EpsinR: An ENTH domain-containing protein that interacts with AP-1. *Mol. Biol. Cell* **14**, 625–641.

Hsu, S.C., Hazuka, C.D., Foletti, D.L., Scheller, R.H. (1999). Targeting vesicles to specific sites on the plasma membrane: The role of the sec6/8 complex. *Trends. Cell Biol.* **9**, 150–153.

Hunziker, W., Fumey, C. (1994). A di-leucine motif mediates endocytosis and basolateral sorting of macrophage IgG Fc receptors in MDCK cells. *EMBO J.* **13**, 2963–2967.

Hunziker, W., Harter, C., Matter, K., Mellman, I. (1991). Basolateral sorting in MDCK cells requires a distinct cytoplasmic domain determinant. *Cell* **66**, 907–920.

Hunziker, W., Peters, P.J. (1998). Rab17 localizes to recycling endosomes and regulates receptor-mediated transcytosis in epithelial cells. *J. Biol. Chem.* **273**, 15734–15741.

Ibaraki, K., Horikawa, H.P., Morita, T., Mori, H., Sakimura, K., Mishina, M., Saisu, H., Abe, T. (1995). Identification of four different forms of syntaxin 3. *Biochem. Biophys. Res. Commun.* **211**, 997–1005.

Ihrke, G., Neufeld, E.B., Meads, T., Shanks, M.R., Cassio, D., Laurent, M. et al. (1993). WIF-B cells: An in vitro model for studies of hepatocyte polarity. *J. Cell Biol.* 1761–1775.

Ikonen, E., de Almeid, J.B., Fath, K.F., Burgess, D.R., Ashman, K., Simons, K., Stow, J.L. (1997). Myosin II is associated with Golgi membranes: Identification of p200 as nonmuscle myosin II on Golgi-derived vesicles. *J. Cell Sci.* **110**, 2155–2164.

Ikonen, E., Tagaya, M., Ullrich, O., Montecucco, C., Simons, K. (1995). Different requirements for NSF, SNAP, and Rab proteins in apical and basolateral transport in MDCK cells. *Cell* **81**, 571–580.

Inoue, M., Chang, L., Hwang, J., Chiang, S.H., Saltiel, A.R. (2003). The exocyst complex is required for targeting of Glut4 to the plasma membrane by insulin. *Nature* **422**, 629–633.

Inoue, T., Nielsen, S., Mandon, B., Terris, J., Kishore, B.K., Knepper, M.A. (1998). SNAP-23 in rat kidney: Colocalization with aquaporin-2 in collecting duct vesicles. *Am. J. Physiol.* **275**, F752–760.

Jacob, R., Alfalah, M., Grunberg, J., Obendorf, M., Naim, H.Y. (2000). Structural determinants required for apical sorting of an intestinal brush-border membrane protein. *J. Biol. Chem.* **275**, 6566–6572.

Jacob, R., Heine, M., Alfalah, M., Naim, H.Y. (2003). Distinct cytoskeletal tracks direct individual vesicle populations to the apical membrane of epithelial cells. *Curr. Biol.* **13**, 607–612.

Janvier, K., Kato, Y., Boehm, M., Rose, J.R., Martina, J.A., Kim, B.Y. et al. (2003). Recognition of dileucine-based sorting signals from HIV-1 Nef and LIMP-II by the AP-1 gamma-sigma1 and AP-3 delta-sigma3 hemicomplexes. *J. Cell Biol.* **163**, 1281–1290.

Jin, R., Junutula, J.R., Matern, H.T., Ervin, K.E., Scheller, R.H., Brunger, A.T. (2005). Exo84 and Sec5 are competitive regulatory Sec6/8 effectors to the RalA GTPase. *EMBO J.* **24**, 2064–2074.

Jo, I., Harris, H.W., Amendt-Raduege, A.M., Majewski, R.R., Hammond, T.G. (1995). Rat kidney papilla contains abundant synaptobrevin protein that participates in the fusion of antidiuretic hormone-regulated water channel-containing endosomes in vitro. *Proc. Natl. Acad. Sci. U S A* **92**, 1876–1880.

Johnson, K.F., Kornfeld, S. (1992). The cytoplasmic tail of the mannose 6-phosphate/insulin-like growth factor-II receptor has two signals for lysosomal enzyme sorting in the Golgi. *J. Cell Biol.* **119**, 249–257.

Kalthoff, C., Groos, S., Kohl, R., Mahrhold, S., Ungewickell, E.J. (2002). Clint: A novel clathrin-binding ENTH-domain protein at the Golgi. *Mol. Biol. Cell* **13**, 4060–4073.

Keller, P., Simons, K. (1998). Cholesterol is required for surface transport of influenza virus hemagglutinin. *J. Cell Biol.* **140**, 1357–1367.

Keller, P., Toomre, D., Diaz, E., White, J., Simons, K. (2001). Multicolour imaging of post-Golgi sorting and trafficking in live cells. *Nat. Cell Biol.* **3**, 140–149.

Koivisto, U.M., Hubbard, A.L., Mellman, I. (2001). A novel cellular phenotype for familial hypercholesterolemia due to a defect in polarized targeting of LDL receptor. *Cell* **105**, 575–585.

Kreitzer, G., Marmorstein, A., Okamoto, P., Vallee, R., Rodriguez-Boulan, E. (2000). Kinesin and dynamin are required for post-Golgi transport of a plasma-membrane protein. *Nat. Cell Biol.* **2**, 125–127.

Kreitzer, G., Schmoranzer, J., Low, S.H., Li, X., Gan, Y., Weimbs, T. et al. (2003). Three-dimensional analysis of post-Golgi carrier exocytosis in epithelial cells. *Nat. Cell Biol.* **5**, 126–136.

Kroschewski, R., Hall, A., Mellman, I. (1999). Cdc42 controls secretory and endocytic transport to the basolateral plasma membrane of MDCK cells. *Nat. Cell Biol.* **1**, 8–13.

Ladinsky, M.S., Kremer, J.R., Furcinitti, P.S., McIntosh, J.R., Howell, K.E. (1994). HVEM tomography of the trans-Golgi network: Structural insights and identification of a lace-like vesicle coat. *J. Cell Biol.* **127**, 29–38.

Ladinsky, M.S., Mastronarde, D.N., McIntosh, J.R., Howell, K.E., Staehelin, L.A. (1999). Golgi structure in three dimensions: Functional insights from the normal rat kidney cell. *J. Cell Biol.* **144**, 1135–1149.

Ladinsky, M.S., Wu, C.C., McIntosh, S., McIntosh, J.R., Howell, K.E. (2002). Structure of the Golgi and distribution of reporter molecules at 20 degrees C reveals the complexity of the exit compartments. *Mol. Biol. Cell* **13**, 2810–2825.

Lafont, F., Burkhardt, J.K., Simons, K. (1994). Involvement of microtubule motors in basolateral and apical transport in kidney cells. *Nature* **372**, 801–803.

Lafont, F., Verkade, P., Galli, T., Wimmer, C., Louvard, D., Simons, K. (1999). Raft association of SNAP receptors acting in apical trafficking in Madin-Darby canine kidney cells. *Proc. Natl. Acad. Sci. U S A* **96**, 3734–3738.

Lapierre, L.A., Kumar, R., Hales, C.M., Navarre, J., Bhartur, S.G., Burnette, J.O. et al. (2001). Myosin vb is associated with plasma membrane recycling systems. *Mol. Biol. Cell* **12**, 1843–1857.

Larocca, M.C., Shanks, R.A., Tian, L., Nelson, D.L., Stewart, D.M., Goldenring, J.R. (2004). AKAP350 interaction with cdc42 interacting protein 4 at the Golgi apparatus. *Mol. Biol. Cell* **15**, 2771–2781.

Le Bivic, A., Sambuy, Y., Patzak, A., Patil, N., Chao, M., Rodriguez-Boulan, E. (1991). An internal deletion in the cytoplasmic tail reverses the apical localization of human NGF receptor in transfected MDCK cells. *J. Cell Biol.* **115**, 607–618.

Le Gall, A.H., Powell, S.K., Yeaman, C.A., Rodriguez-Boulan, E. (1997). The neural cell adhesion molecule expresses a tyrosine-independent basolateral sorting signal. *J. Biol. Chem.* **272**, 4559–4567.

Leitinger, B., Hille-Rehfeld, A., Spiess, M. (1995). Biosynthetic transport of the asialoglycoprotein receptor H1 to the cell surface occurs via endosomes. *Proc. Natl. Acad. Sci. U S A* **92**, 10109–10113.

Leung, S.M., Ruiz, W.G., Apodaca, G. (2000). Sorting of membrane and fluid at the apical pole of polarized Madin-Darby canine kidney cells. *Mol. Biol. Cell* **11**, 2131–2150.

Li, X., Low, S.H., Miura, M., Weimbs, T. (2002). SNARE expression and localization in renal epithelial cells suggest mechanism for variability of trafficking phenotypes. *Am. J. Physiol. Renal. Physiol.* **283**, F1111–1122.

Liljedahl, M., Maeda, Y., Colanzi, A., Ayala, I., Van Lint, J., Malhotra, V. (2001). Protein kinase D regulates the fission of cell surface destined transport carriers from the trans-Golgi network. *Cell* **104**, 409–420.

Lipschutz, J.H., Guo, W., O'Brien, L.E., Nguyen, Y.H., Novick, P., Mostov, K.E. (2000). Exocyst is involved in cystogenesis and tubulogenesis and acts by modulating synthesis and delivery of basolateral plasma membrane and secretory proteins. *Mol. Biol. Cell* **11**, 4259–4275.

Lipschutz, J.H., Lingappa, V.R., Mostov, K.E. (2003). The exocyst affects protein synthesis by acting on the translocation machinery of the endoplasmic reticulum. *J. Biol. Chem.* **278**, 20954–20960.

Lisanti, M.P., Caras, I.W., Davitz, M.A., Rodriguez-Boulan, E. (1989). A glycophospholipid membrane anchor acts as an apical targeting signal in polarized epithelial cells. *J. Cell Biol.* **109**, 2145–2156.

Lisanti, M.P., Sargiacomo, M., Graeve, L., Saltiel, A.R., Rodriguez-Boulan, E. (1988). Polarized apical distribution of glycosyl-phosphatidylinositol-anchored proteins in a renal epithelial cell line. *Proc. Natl. Acad. Sci. U S A* **85**, 9557–9561.

Lock, J.G., Stow, J.L. (2005). Rab11 in recycling endosomes regulates the sorting and basolateral transport of E-cadherin. *Mol. Biol. Cell* **16**, 1744–1755.

Low, S.H., Chapin, S.J., Weimbs, T., Komuves, L.G., Bennett, M.K., Mostov, K.E. (1996). Differential localization of syntaxin isoforms in polarized Madin-Darby canine kidney cells. *Mol. Biol. Cell* **7**, 2007–2018.

Low, S.H., Chapin, S.J., Wimmer, C., Whiteheart, S.W., Komuves, L.G., Mostov, K.E., Weimbs, T. (1998). The SNARE machinery is involved in apical plasma membrane trafficking in MDCK cells. *J. Cell Biol.* **141**, 1503–1513.

Low, S.H., Marmorstein, L.Y., Miura, M., Li, X., Kudo, N., Marmorstein, A.D., Weimbs, T. (2002). Retinal pigment epithelial cells exhibit unique expression and localization of plasma membrane syntaxins which may contribute to their trafficking phenotype. *J. Cell Sci.* **115**, 4545–4553.

Low, S.H., Miura, M., Roche, P.A., Valdez, A.C., Mostov, K.E., Weimbs, T. (2000). Intracellular redirection of plasma membrane trafficking after loss of epithelial cell polarity. *Mol. Biol. Cell* **11**, 3045–3060.

Luna, A., Matas, O.B., Martinez-Menarguez, J.A., Mato, E., Duran, J.M., Ballesta, J. et al. (2002). Regulation of protein transport from the Golgi complex to the endoplasmic reticulum by CDC42 and N-WASP. *Mol. Biol. Cell* **13**, 866–879.

Luzio, J.P., Poupon, V., Lindsay, M.R., Mullock, B.M., Piper, R.C., Pryor, P.R. (2003). Membrane dynamics and the biogenesis of lysosomes. *Mol. Membr. Biol.* **20**, 141–154.

Maeda, Y., Beznoussenko, G.V., Van Lint, J., Mironov, A.A., Malhotra, V. (2001). Recruitment of protein kinase D to the trans-Golgi network via the first cysteine-rich domain. *EMBO J.* **20**, 5982–5990.

Mandon, B., Chou, C.L., Nielsen, S., Knepper, M.A. (1996). Syntaxin-4 is localized to the apical plasma membrane of rat renal collecting duct cells: possible role in aquaporin-2 trafficking. *J. Clin. Invest.* **98**, 906–913.

Mandon, B., Nielsen, S., Kishore, B.K., Knepper, M.A. (1997). Expression of syntaxins in rat kidney. *Am. J. Physiol.* **273**, F718–730.

Marmorstein, A.D., Csaky, K.G., Baffi, J., Lam, L., Rahaal, F., Rodriguez-Boulan, E. (2000). Saturation of, and competition for entry into, the apical secretory pathway. *Proc. Natl. Acad. Sci. U S A* **97**, 3248–3253.

Martin-Belmonte, F., Puertollano, R., Millan, J., Alonso, M.A. (2000). The MAL proteolipid is necessary for the overall apical delivery of membrane proteins in the polarized epithelial Madin-Darby canine kidney and Fischer rat thyroid cell lines. *Mol. Biol. Cell* **11**, 2033–2045.

Marzolo, M.P., Bull, P., Gonzalez, A. (1997). Apical sorting of hepatitis B surface antigen (HBsAg) is independent of N-glycosylation and glycosylphosphatidylinositol-anchored protein segregation. *Proc. Natl. Acad. Sci. U S A* **94**, 1834–1839.

Marzolo, M.P., Yuseff, M.I., Retamal, C., Donoso, M., Ezquer, F., Farfan, P. et al. (2003). Differential distribution of low-density lipoprotein-receptor-related protein (LRP) and megalin in polarized epithelial cells is determined by their cytoplasmic domains. *Traffic* **4**, 273–288.

Matern, H.T., Yeaman, C., Nelson, W.J., Scheller, R.H. (2001). The Sec6/8 complex in mammalian cells: characterization of mammalian Sec3, subunit interactions, and expression of subunits in polarized cells. *Proc. Natl. Acad. Sci. U S A* **98**, 9648–9653.

Matlin, K.S., Simons, K. (1983). Reduced temperature prevents transfer of a membrane glycoprotein to the cell surface but does not prevent terminal glycosylation. *Cell* **34**, 233–243.

Matlin, K.S., Simons, K. (1984). Sorting of an apical plasma membrane glycoprotein occurs before it reaches the cell surface in cultured epithelial cells. *J. Cell Biol.* **99**, 2131–2139.

Matter, K., Bucher, K., Hauri, H.P. (1990). Microtubule perturbation retards both the direct and the indirect apical pathway but does not affect sorting of plasma membrane proteins in intestinal epithelial cells (Caco-2). *Embo. J.* **9**, 3163–3170.

Matter, K., Hunziker, W., Mellman, I. (1992). Basolateral sorting of LDL receptor in MDCK cells: The cytoplasmic domain contains two tyrosine-dependent targeting determinants. *Cell* **71**, 741–753.

Matter, K., Mellman, I. (1994). Mechanisms of cell polarity: Sorting and transport in epithelial cells. *Curr. Opin. Cell Biol.* **6**, 545–554.

Matter, K., Whitney, J.A., Yamamoto, E.M., Mellman, I. (1993). Common signals control low density lipoprotein receptor sorting in endosomes and the Golgi complex of MDCK cells. *Cell* **74**, 1053–1064.

Matter, K., Yamamoto, E.M., Mellman, I. (1994). Structural requirements and sequence motifs for polarized sorting and endocytosis of LDL and Fc receptors in MDCK cells. *J. Cell Biol.* **126**, 991–1004.

Mays, R.W., Siemers, K.A., Fritz, B.A., Lowe, A.W., van Meer, G., Nelson, W.J. (1995). Hierarchy of mechanisms involved in generating Na/K-ATPase polarity in MDCK epithelial cells. *J. Cell Biol.* **130**, 1105–1115.

McCallum, S.J., Erickson, J.W., Cerione, R.A. (1998). Characterization of the association of the actin-binding protein, IQGAP, and activated Cdc42 with Golgi membranes. *J. Biol. Chem.* **273**, 22537–22544.

Mellman, I. (1996). Endocytosis and molecular sorting. *Ann. Rev. Cell Dev. Biol.* **12**, 575–625.

Mills, I.G., Praefcke, G.J., Vallis, Y., Peter, B.J., Olesen, L.E., Gallop, J.L. et al. (2003). EpsinR: an AP1/clathrin interacting protein involved in vesicle trafficking. *J. Cell Biol.* **160**, 213–222.

Miranda, K.C., Khromykh, T., Christy, P., Le, T.L., Gottardi, C.J., Yap, A.S. et al. (2001). A dileucine motif targets E-cadherin to the basolateral cell surface in Madin-Darby canine kidney and LLC-PK1 epithelial cells. *J. Biol. Chem.* **276**, 22565–22572.

Mironov, A., Jr., Luini, A., Mironov, A. (1998). A synthetic model of intra-Golgi traffic. *FASEB J.* **12**, 249–252.

Misek, D.E., Bard, E., Rodriguez-Boulan, E. (1984). Biogenesis of epithelial cell polarity: Intracellular sorting and vectorial exocytosis of an apical plasma membrane glycoprotein. *Cell* 537–546.

Mishra, S.K., Keyel, P.A., Hawryluk, M.J., Agostinelli, N.R., Watkins, S.C., Traub, L.M. (2002). Disabled-2 exhibits the properties of a cargo-selective endocytic clathrin adaptor. *EMBO J.* **21**, 4915–4926.

Mishra, S.K., Watkins, S.C., Traub, L.M. (2002). The autosomal recessive hypercholesterolemia (ARH) protein interfaces directly with the clathrin-coat machinery. *Proc. Natl. Acad. Sci. U S A* **99**, 16099–16104.

Moskalenko, S., Henry, D.O., Rosse, C., Mirey, G., Camonis, J.H., White, M.A. (2002). The exocyst is a Ral effector complex. *Nat. Cell Biol.* **4**, 66–72.

Moskalenko, S., Tong, C., Rosse, C., Mirey, G., Formstecher, E., Daviet, L. et al. (2003). Ral GTPases regulate exocyst assembly through dual subunit interactions. *J. Biol. Chem.* **278**, 51743–51748.

Mostov, K.E., Su, T., ter Beest, M. (2003). Polarized epithelial membrane traffic: Conservation and plasticity. *Nat. Cell Biol.* **5**, 287–293.

Mostov, K.E., de Bruyn Kops, A., Deitcher, D.L. (1986). Deletion of the cytoplasmic domain of the polymeric immunoglobulin receptor prevents basolateral localization and endocytosis. *Cell* **47**, 359–364.

Mott, H.R., Nietlispach, D., Hopkins, L.J., Mirey, G., Camonis, J.H., Owen, D. (2003). Structure of the GTPase-binding domain of Sec5 and elucidation of its Ral binding site. *J. Biol. Chem.* **278**, 17053–17059.

Musch, A. (2004). Microtubule organization and function in epithelial cells. *Traffic* **5**, 1–9.

Musch, A., Cohen, D., Kreitzer, G., Rodriguez-Boulan, E. (2001). cdc42 regulates the exit of apical and basolateral proteins from the trans-Golgi network. *EMBO J.* **20**, 2171–2179.

Musch, A., Cohen, D., Rodriguez-Boulan, E. (1997). Myosin II is involved in the production of constitutive transport vesicles from the TGN. *J. Cell Biol.* **138**, 291–306.

Musch, A., Cohen, D., Yeaman, C., Nelson, W.J., Rodriguez-Boulan, E., Brennwald, P.J. (2002). Mammalian homolog of Drosophila tumor suppressor lethal (2) giant larvae interacts with basolateral exocytic machinery in Madin-Darby canine kidney cells. *Mol. Biol. Cell* **13**, 158–168.

Musch, A., Xu, H., Shields, D., Rodriguez-Boulan, E. (1996). Transport of vesicular stomatis virus G protein to the cell surface is signal mediated in polarized and nonpolarized cells. *J. Cell Biol.* **133**, 543–558.

Muth, T.R., Caplan, M.J. (2003). Transport protein trafficking in polarized cells. *Annu. Rev. Cell Dev. Biol.* **19**, 333–366.

Nabi, I.R., Le Bivic, A., Fambrough, D., Rodriguez-Boulan, E. (1991). An endogenous MDCK lysosomal membrane glycoprotein is targeted basolaterally before delivery to lysosomes. *J. Cell Biol.* **115**, 1573–1584.

Nagai, M., Meerloo, T., Takeda, T., Farquhar, M.G. (2003). The adaptor protein ARH escorts megalin to and through endosomes. *Mol. Biol. Cell* **14**, 4984–4996.

Nielsen, S., Marples, D., Birn, H., Mohtashami, M., Dalby, N.O., Trimble, M., Knepper, M. (1995). Expression of VAMP-2-like protein in kidney collecting duct intracellular vesicles. Colocalization with Aquaporin-2 water channels. *J. Clin. Invest.* **96**, 1834–1844.

Nishimura, N., Plutner, H., Hahn, K., Balch, W.E. (2002). The delta subunit of AP-3 is required for efficient transport of VSV-G from the trans-Golgi network to the cell surface. *Proc. Natl. Acad. Sci. U S A* **99**, 6755–6760.

Noda, Y., Okada, Y., Saito, N., Setou, M., Xu, Y., Zhang, Z., Hirokawa, N. (2001). KIFC3, a microtubule minus end-directed motor for the apical transport of annexin XIIIb-associated Triton-insoluble membranes. *J. Cell Biol.* **155**, 77–88.

Odorizzi, G., Pearse, A., Domingo, D., Trowbridge, I.S., Hopkins, C.R. (1996). Apical and basolateral endosomes of MDCK cells are interconnected and contain a polarized sorting mechanism. *J. Cell Biol.* **135**, 139–152.

Odorizzi, G., Trowbridge, I.S. (1997). Structural requirements for basolateral sorting of the human transferrin receptor in the biosynthetic and endocytic pathways of Madin-Darby canine kidney cells. *J. Cell Biol.* **137**, 1255–1264.

Odorizzi, G., Trowbridge, I.S. (1997). Structural requirements for major histocompatibility complex class II invariant chain trafficking in polarized Madin-Darby canine kidney cells. *J. Biol. Chem.* **272**, 11757–11762.

Ohno, H., Fournier, M.C., Poy, G., Bonifacino, J.S. (1996). Structural determinants of interaction of tyrosine-based sorting signals with the adaptor medium chains. *J. Biol. Chem.* **271**, 29009–29015.

Ohno, H., Stewart, J., Fournier, M.C., Bosshart, H., Rhee, I., Miyatake, S. et al. (1995). Interaction of tyrosine-based sorting signals with clathrin-associated proteins. *Science* **269**, 1872–1875.

Ohno, H., Tomemori, T., Nakatsu, F., Okazaki, Y., Aguilar, R.C., Foelsch, H. et al. (1999). Mu1B, a novel adaptor medium chain expressed in polarized epithelial cells. *FEBS Lett.* **449**, 215–220.

Ojakian, G.K., Schwimmer, R. (1992). Antimicrotubule drugs inhibit the polarized insertion of an intracellular glycoprotein pool into the apical membrane of Madin-Darby canine kidney (MDCK) cells. *J. Cell Sci.* **103**, 677–687.

Orzech, E., Cohen, S., Weiss, A., Aroeti, B. (2000). Interactions between the exocytic and endocytic pathways in polarized Madin-Darby canine kidney cells. *J. Biol. Chem.* **275**, 15207–15219.

Owen, D.J., Collins, B.M., Evans, P.R. (2004). Adaptors for clathrin coats: Structure and function. *Annu. Rev. Cell Dev. Biol.* **20**, 153–191.

Owen, D.J., Evans, P.R. (1998). A structural explanation for the recognition of tyrosine-based endocytotic signals. *Science* **282**, 1327–1332.

Owen, D.J., Setiadi, H., Evans, P.R., McEver, R.P., Green, S.A. (2001). A third specificity-determining site in mu 2 adaptin for sequences upstream of Yxx phi sorting motifs. *Traffic* **2**, 105–110.

Paladino, S., Sarnataro, D., Pillich, R., Tivodar, S., Nitsch, L., Zurzolo, C. (2004). Protein oligomerization modulates raft partitioning and apical sorting of GPI-anchored proteins. *J. Cell Biol.* **167**, 699–709.

Parton, R.G., Prydz, K., Bomsel, M., Simons, K., Griffiths, G. (1989). Meeting of the apical and basolateral endocytic pathways of the Madin-Darby canine kidney cell in late endosomes. *J. Cell Biol.* 3259–3272.

Pfeiffer, S., Fuller, S.D., Simons, K. (1985). Intracellular sorting and basolateral appearance of the G protein of vesicular stomatitis virus in Madin-Darby canine kidney cells. *J. Cell Biol.* **101**, 470–476.

Pimplikar, S.W., Ikonen, E., Simons, K. (1994). Basolateral protein transport in streptolysin O-permeabilized MDCK cells. *J. Cell Biol.* **125**, 1025–1035.

Pimplikar, S.W., Simons, K. (1993). Role of heterotrimeric G proteins in polarized membrane transport. *J. Cell Sci. Suppl.* **17**, 27–32.

Polishchuk, R., Di Pentima, A., Lippincott-Schwartz, J. (2004). Delivery of raft-associated, GPI-anchored proteins to the apical surface of polarized MDCK cells by a transcytotic pathway. *Nat. Cell Biol.* **6**, 297–307.

Potter, B.A., Ihrke, G., Bruns, J.R., Weixel, K.M., Weisz, O.A. (2004). Specific N-glycans direct apical delivery of transmembrane, but not soluble or glycosylphosphatidylinositol-anchored forms of endolyn in Madin-Darby canine kidney cells. *Mol. Biol. Cell* **15**, 1407–1416.

Prigent, M., Dubois, T., Raposo, G., Derrien, V., Tenza, D., Rosse, C. et al. (2003). ARF6 controls postendocytic recycling through its downstream exocyst complex effector. *J. Cell Biol.* **163**, 1111–1121.

Puertollano, R., Aguilar, R.C., Gorshkova, I., Crouch, R.J., Bonifacino, J.S. (2001). Sorting of mannose 6-phosphate receptors mediated by the GGAs. *Science* **292**, 1712–1716.

Puertollano, R., Alonso, M.A. (1999). MAL, an integral element of the apical sorting machinery, is an itinerant protein that cycles between the trans-Golgi network and the plasma membrane. *Mol. Biol. Cell* **10**, 3435–3447.

Quinones, B., Riento, K., Olkkonen, V.M., Hardy, S., Bennett, M.K. (1999). Syntaxin 2 splice variants exhibit differential expression patterns, biochemical properties and subcellular localizations. *J. Cell Sci.* **112 (Pt 23)**, 4291–4304.

Rapoport, I., Chen, Y.C., Cupers, P., Shoelson, S.E., Kirchhausen, T. (1998). Dileucine-based sorting signals bind to the b chain of AP-1 at a site distinct and regulated differently from the tyrosine-based motif-binding site. *EMBO J.* **17**, 2148–2155.

Riento, K., Galli, T., Jansson, S., Ehnholm, C., Lehtonen, E., Olkkonen, V.M. (1998). Interaction of Munc-18-2 with syntaxin 3 controls the association of apical SNAREs in epithelial cells. *J. Cell Sci.* **111 (Pt 17)**, 2681–2688.

Riento, K., Jantti, J., Jansson, S., Hielm, S., Lehtonen, E., Ehnholm, C. et al. (1996). A sec1-related vesicle-transport protein that is expressed predominantly in epithelial cells. *Eur. J. Biochem.* **239**, 638–646.

Riento, K., Kauppi, M., Keranen, S., Olkkonen, V.M. (2000). Munc18-2, a functional partner of syntaxin 3, controls apical membrane trafficking in epithelial cells. *J. Biol. Chem.* **275**, 13476–13483.

Rindler, M.J., Ivanov, I.E., Plesken, H., Rodriguez-Boulan, E., Sabatini, D.D. (1984). Viral glycoproteins destined for apical or basolateral plasma membrane domains traverse the same Golgi apparatus during their intracellular transport in doubly infected Madin-Darby canine kidney cells. *J. Cell Biol.* **98**, 1304–1319.

Rindler, M.J., Ivanov, I.E., Sabatini, D.D. (1987). Microtubule-acting drugs lead to the nonpolarized delivery of the influenza hemagglutinin to the cell surface of polarized Madin-Darby canine kidney cells. *J. Cell Biol.* **104**, 231–241.

Rindler, M.J., Traber, M.G. (1988). A specific sorting signal is not required for the polarized secretion of newly synthesized proteins from cultured intestinal epithelial cells. *J. Cell Biol.* **107**, 471–479.

Robinson, M.S. (2004). Adaptable adaptors for coated vesicles. *Trends Cell Biol.* **14**, 167–174.

Robinson, M.S., Bonifacino, J.S. (2001). Adaptor-related proteins. *Curr. Opin. Cell Biol.* **13**, 444–453.

Rodriguez-Boulan, E., Gonzalez, A. (1999). Glycans in post-Golgi apical targeting: Sorting signals or structural props? *Trends Cell Biol.* **9**, 291–294.

Rodriguez-Boulan, E., Kreitze, G., Musch, A. (2005). Organization of vesicular trafficking in epithelia. *Nat. Rev. Mol. Cell Biol.* **6**, 233–247.

Rodriguez-Boulan, E., Musch, A. (2005). Protein sorting in the Golgi complex: Shifting paradigms. *Biochim. Biophys. Acta.*

Rodriguez-Boulan, E., Musch, A., Le Bivic, A. (2004). Epithelial trafficking: New routes to familiar places. *Curr. Opin. Cell Biol.* **16**, 436–442.

Rodriguez-Boulan, E., Nelson, W.J. (1989). Morphogenesis of the polarized epithelial cell phenotype. *Science* **245**, 718–725.

Rodriguez-Boulan, E., Paskiet, K.T., Salas, P.J., Bard, E. (1984). Intracellular transport of influenza virus hemagglutinin to the apical surface of Madin-Darby canine kidney cells. *J. Cell Biol.* **98**, 308–319.

Rodriguez-Boulan, E., Pendergast, M. (1980). Polarized distribution of viral envelope proteins in the plasma membrane of infected epithelial cells. *Cell* **20**, 45–54.

Rodriguez-Boulan, E., Sabatini, D.D. (1978). Asymmetric budding of viruses in epithelial monlayers: A model system for study of epithelial polarity. *Proc. Natl. Acad. Sci. U S A* **75**, 5071–5075.

Rogers, K.K., Wilson, P.D., Snyder, R.W., Zhang, X., Guo, W., Burrow, C.R., Lipschutz, J.H. (2004). The exocyst localizes to the primary cilium in MDCK cells. *Biochem. Biophys. Res. Commun.* **319**, 138–143.

Roh, M.H., Margolis, B. (2003). Composition and function of PDZ protein complexes during cell polarization. *Am. J. Physiol. Renal. Physiol.* **285**, F377–387.

Rojas, R., Apodaca, G. (2002). Immunoglobulin transport across polarized epithelial cells. *Nat. Rev. Mol. Cell Biol.* **3**, 944–955.

Roush, D.L., Gottardi, C.J., Naim, H.Y., Roth, M.G., Caplan, M.J. (1998). Tyrosine-based membrane protein sorting signals are differentially interpreted by polarized Madin-Darby canine kidney and LLC-PK1 epithelial cells. *J. Biol. Chem.* **273**, 26862–26869.

Salas, P.J., Misek, D.E., Vega-Salas, D.E., Gundersen, D., Cereijido, M., Rodriguez-Boulan, E. (1986). Microtubules and actin filaments are not critically involved in the biogenesis of epithelial cell surface polarity. *J. Cell Biol.* **102**, 1853–1867.

Sarnataro, D., Paladino, S., Campana, V., Grassi, J., Nitsch, L., Zurzolo, C. (2002). PrPC is sorted to the basolateral membrane of epithelial cells independently of its association with rafts. *Traffic* **3**, 810–821.

Satoh, A.K., O'Tousa, J.E., Ozaki, K., Ready, D.F. (2005). Rab11 mediates post-Golgi trafficking of rhodopsin to the photosensitive apical membrane of Drosophila photoreceptors. *Development* **132**, 1487–1497.

Schafer, D.A. (2004). Regulating actin dynamics at membranes: a focus on dynamin. *Traffic* **5**, 463–469.

Scheiffele, P., Peranen, J., Simons, K. (1995). N-glycans as apical sorting signals in epithelial cells. *Nature* **378**, 96–98.

Scheiffele, P., Roth, M.G., Simons, K. (1997). Interaction of influenza virus haemagglutinin with sphingolipid-cholesterol membrane domains via its transmembrane domain. *EMBO J.* **16**, 5501–5508.

Scheiffele, P., Verkade, P., Fra, A.M., Virta, H., Simons, K., Ikonen, E. (1998). Caveolin-1 and -2 in the exocytic pathway of MDCK cells. *J. Cell Biol.* **140**, 795–806.

Schuck, S., Simons, K. (2004). Polarized sorting in epithelial cells: raft clustering and the biogenesis of the apical membrane. *J. Cell Sci.* **117**, 5955–5964.

Sharma, P., Varma, R., Sarasij, R.C., Ira, Gousset, K., Krishnamoorthy, G. et al. (2004). Nanoscale organization of multiple GPI-anchored proteins in living cell membranes. *Cell* **116**, 577–589.

Sheff, D.R., Daro, E.A., Hull, M., Mellman, I. (1999). The receptor recycling pathway contains two distinct populations of early endosomes with different sorting functions. *J. Cell Biol.* **145**, 123–139.

Sheff, D.R., Kroschewski, R., Mellman, I. (2002). Actin dependence of polarized receptor recycling in Madin-Darby canine kidney cell endosomes. *Mol. Biol. Cell.* **13**, 262–275.

Sheikh, H., Isacke, C.M. (1996). A di-hydrophobic Leu-Val motif regulates the basolateral localization of CD44 in polarized Madin-Darby canine kidney epithelial cells. *J. Biol. Chem.* **271**, 12185–12190.

Shelly, M., Mosesson, Y., Citri, A., Lavi, S., Zwang, Y., Melamed-Book, N. et al. (2003). Polar expression of ErbB-2/HER2 in epithelia. Bimodal regulation by Lin-7. *Dev. Cell* **5**, 475–486.

Shiba, T., Takatsu, H., Nogi, T., Matsugaki, N., Kawasaki, M., Igarashi, N. et al. (2002). Structural basis for recognition of acidic-cluster dileucine sequence by GGA1. *Nature* **415**, 937–941.

Shin, D.M., Zhao, X.S., Zeng, W., Mozhayeva, M., Muallem, S. (2000). The mammalian Sec6/8 complex interacts with Ca(2+) signaling complexes and regulates their activity. *J. Cell Biol.* **150**, 1101–1112.

Shipitsin, M., Feig, L.A. (2004). RalA but not RalB enhances polarized delivery of membrane proteins to the basolateral surface of epithelial cells. *Mol. Cell Biol.* **24**, 5746–5756.

Shvartsman, D.E., Kotler, M., Tall, R.D., Roth, M.G., Henis, Y.I. (2003). Differently anchored influenza hemagglutinin mutants display distinct interaction dynamics with mutual rafts. *J. Cell Biol.* **163**, 879–888.

Simmen, T., Honing, S., Icking, A., Tikkanen, R., Hunziker, W. (2002). AP-4 binds basolateral signals and participates in basolateral sorting in epithelial MDCK cells. *Nat. Cell Biol.* **4**, 154–159.

Simons, K., Ikonen, E. (1997). Functional rafts in cell membranes. *Nature* **387**, 569–572.

Simons, K., Wandinger-Ness, A. (1990). Polarized sorting in epithelia. *Cell* **62**, 207–210.

Simonsen, A., Bremnes, B., Nordeng, T.W., Bakke, O. (1998). The leucine-based motif DDQxxLI is recognized both for internalization and basolateral sorting of invariant chain in MDCK cells. *Eur. J. Cell Biol.* **76**, 25–32.

Simonsen, A., Lippe, R., Christoforidis, S., Gaullier, J.M., Brech, A., Callaghan, J. et al. (1998). EEA1 links PI(3)K function to Rab5 regulation of endosome fusion. *Nature* **394**, 494–498.

Simonsen, A., Pedersen, K.W., Nordeng, T.W., von der Lippe, A., Stang, E., Long, E.O., Bakke, O. (1999). Polarized transport of MHC class II molecules in Madin-Darby canine kidney cells is directed by a leucine-based signal in the cytoplasmic tail of the beta-chain. *J. Immunol.* **163**, 2540–2548.

Skibbens, J.E., Roth, M.G., Matlin, K.S. (1989). Differential extractability of influenza virus hemagglutinin during intracellular transport in polarized epithelial cells and nonpolar fibroblasts. *J. Cell Biol.* **108**, 821–832.

Stahl, P.D., Barbieri, M.A. (2002). Multivesicular bodies and multivesicular endosomes: the "ins and outs" of endosomal traffic. *Sci. STKE.* **2002**, PE32.

Stamnes, M. (2002). Regulating the actin cytoskeleton during vesicular transport. *Curr. Opin. Cell Biol.* **14**, 428–433.

Steegmaier, M., Lee, K.C., Prekeris, R., Scheller, R.H. (2000). SNARE protein trafficking in polarized MDCK cells. *Traffic* **1**, 553–560.

Straight, S.W., Chen, L., Karnak, D., Margolis, B. (2001). Interaction with mLin-7 alters the targeting of endocytosed transmembrane proteins in mammalian epithelial cells. *Mol. Biol. Cell* **12**, 1329–1340.

Sugihara, K., Asano, S., Tanaka, K., Iwamatsu, A., Okawa, K., Ohta, Y. (2002). The exocyst complex binds the small GTPase RalA to mediate filopodia formation. *Nat. Cell Biol.* **4**, 73–78.

Sugimoto, H., Sugahara, M., Folsch, H., Koide, Y., Nakatsu, F., Tanaka, N. et al. (2002). Differential recognition of tyrosine-based basolateral signals by AP-1B subunit mu1B in polarized epithelial cells. *Mol. Biol. Cell* **13**, 2374–2382.

Sun, A.Q., Salkar, R., Sachchidanand, Xu, S., Zeng, L., Zhou, M.M., Suchy, F.J. (2003). A 14-amino acid sequence with a beta-turn structure is required for apical membrane sorting of the rat ileal bile acid transporter. *J. Biol. Chem.* **278**, 4000–4009.

Tai, A.W., Chuang, J.Z., Bode, C., Wolfrum, U., Sung, C.H. (1999). Rhodopsin's carboxy-terminal cytoplasmic tail acts as a membrane receptor for cytoplasmic dynein by binding to the dynein light chain Tctex-1. *Cell* **97**, 877–887.

Takatsu, H., Katoh, Y., Shiba, Y., Nakayama, K. (2001). Golgi-localizing, gamma-adaptin ear homology domain, ADP-ribosylation factor-binding (GGA) proteins interact with acidic dileucine sequences within the cytoplasmic domains of sorting receptors through their Vps27p/Hrs/STAM (VHS) domains. *J. Biol. Chem.* **276**, 28541–28545.

Takeda, T., Yamazaki, H., Farquhar, M.G. (2003). Identification of an apical sorting determinant in the cytoplasmic tail of megalin. *Am. J. Physiol. Cell Physiol.* **284**, C1105–1113.

Tall, R.D., Alonso, M.A., Roth, M.G. (2003). Features of influenza HA required for apical sorting differ from those required for association with DRMs or MAL. *Traffic* **4**, 838–849.

TerBush, D.R., Maurice, T., Roth, D., Novick, P. (1996). The Exocyst is a multiprotein complex required for exocytosis in Saccharomyces cerevisiae. *EMBO J.* **15**, 6483–6494.

TerBush, D.R., Novick, P. (1995). Sec6, Sec8, and Sec15 are components of a multisubunit complex which localizes to small bud tips in Saccharomyces cerevisiae. *J. Cell Biol.* **130**, 299–312.

Thomas, D.C., Brewer, C.B., Roth, M.G. (1993). Vesicular stomatitis virus glycoprotein contains a dominant cytoplasmic basolateral sorting signal critically dependent upon a tyrosine. *J. Biol. Chem.* **268**, 3313–3320.

Thomas, D.C., Roth, M.G. (1994). The basolateral targeting signal in the cytoplasmic domain of glycoprotein G from vesicular stomatitis virus resembles a variety of intracellular targeting motifs related by primary sequence but having diverse targeting activities. *J. Biol. Chem.* **269**, 15732–15739.

Toomre, D., Keller, P., White, J., Olivo, J.C., Simons, K. (1999). Dual-color visualization of trans-Golgi network to plasma membrane traffic along microtubules in living cells. *J. Cell Sci.* **112 (Pt 1)**, 21–33.

Tuma, P.L., Hubbard, A.L. (2003). Transcytosis: crossing cellular barriers. *Physiol. Rev.* **83**, 871–932.

Tyska, M.J., Mooseker, M.S. (2004). A role for myosin-1A in the localization of a brush border disaccharidase. *J. Cell Biol.* **165**, 395–405.

van der Sluijs, P., Bennett, M.K., Antony, C., Simons, K., Kreis, T.E. (1990). Binding of exocytic vesicles from MDCK cells to microtubules in vitro. *J. Cell Sci.* **95**, 545–553.

van Meer, G., Simons, K. (1988). Lipid polarity and sorting in epithelial cells. *J. Cell Biochem.* **36**, 51–58.

van Zeijl, M.J., Matlin, K.S. (1990). Microtubule perturbation inhibits intracellular transport of an apical membrane glycoprotein in a substrate-dependent

manner in polarized Madin-Darby canine kidney epithelial cells. *Cell Regul.* **1**, 921–936.

Wandinger-Ness, A., Bennett, M.K., Antony, C., Simons, K. (1990). Distinct transport vesicles mediate the delivery of plasma membrane proteins to the apical and basolateral domains of MDCK cells. *J. Cell Biol.* **111**, 987–1000.

Wang, E., Brown, P.S., Aroeti, B., Chapin, S.J., Mostov, K.E., Dunn, K.W. (2000). Apical and basolateral endocytic pathways of MDCK cells meet in acidic common endosomes distinct from a nearly-neutral apical recycling endosome. *Traffic* **1**, 480–493.

Wang, Y.J., Wang, J., Sun, H.Q., Martinez, M., Sun, Y.X., Macia, E. et al. (2003). Phosphatidylinositol 4 phosphate regulates targeting of clathrin adaptor AP-1 complexes to the Golgi. *Cell* **114**, 299–310.

Wasiak, S., Legendre-Guillemin, V., Puertollano, R., Blondeau, F., Girard, M., de Heuvel, E. et al. (2002). Enthoprotin: A novel clathrin-associated protein identified through subcellular proteomics. *J. Cell Biol.* **158**, 855–862.

Wehrle-Haller, B., Imhof, B.A. (2001). Stem cell factor presentation to c-Kit. Identification of a basolateral targeting domain. *J. Biol. Chem.* **276**, 12667–12674.

Wessels, H.P., Geffen, I., Spiess, M. (1989). A hepatocyte-specific basolateral membrane protein is targeted to the same domain when expressed in Madin-Darby canine kidney cells. *J. Biol. Chem.* **264**, 17–20.

Wessels, H.P., Hansen, G.H., Fuhrer, C., Look, A.T., Sjostrom, H., Noren, O., Spiess, M. (1990). Aminopeptidase N is directly sorted to the apical domain in MDCK cells. *J. Cell Biol.* 2923–2930.

Yamashita, K., Hara-Kuge, S., Ohkura, T. (1999). Intracellular lectins associated with N-linked glycoprotein traffic. *Biochim. Biophys. Acta* **1473**, 147–160.

Yeaman, C., Ayala, M.I., Wright, J.R., Bard, F., Bossard, C., Ang, A. et al. (2004). Protein kinase D regulates basolateral membrane protein exit from trans-Golgi network. *Nat. Cell Biol.* **6**, 106–112.

Yeaman, C., Grindstaff, K.K., Nelson, W.J. (1999). New perspectives on mechanisms involved in generating epithelial cell polarity. *Physiol. Rev.* **79**, 73–98.

Yeaman, C., Grindstaff, K.K., Nelson, W.J. (2004). Mechanism of recruiting Sec6/8 (exocyst) complex to the apical junctional complex during polarization of epithelial cells. *J. Cell Sci.* **117**, 559–570.

Yeaman, C., Grindstaff, K.K., Wright, J.R., Nelson, W.J. (2001). Sec6/8 complexes on trans-Golgi network and plasma membrane regulate late stages of exocytosis in mammalian cells. *J. Cell Biol.* **155**, 593–604.

Yeaman, C., Le Gall, A.H., Baldwin, A.N., Monlauzeur, L., Le Bivic, A., Rodriguez-Boulan, E. (1997). The O-glycosylated stalk domain is required for apical sorting of neurotrophin receptors in polarized MDCK cells. *J. Cell Biol.* **139**, 929–940.

Zacchetti, D., Peranen, J., Murata, M., Fiedler, K., Simons, K. (1995). VIP17/MAL, a proteolipid in apical transport vesicles. *FEBS Lett.* **377**, 465–469.

Zacchi, P., Stenmark, H., Parton, R.G., Orioli, D., Lim, F., Giner, A. et al. (1998). Rab17 regulates membrane trafficking through apical recycling endosomes in polarized epithelial cells. *J. Cell Biol.* **140**, 1039–1053.

Zhang, X.M., Ellis, S., Sriratana, A., Mitchell, C.A., Rowe, T. (2004). Sec15 is an effector for the Rab11 GTPase in mammalian cells. *J. Biol. Chem.* **279**, 43027–43034.

Zurzolo, C., Lisanti, M.P., Caras, I.W., Nitsch, L., Rodriguez-Boulan, E. (1993). Glycosylphosphatidylinositol-anchored proteins are preferentially targeted to the basolateral surface in Fischer rat thyroid epithelial cells. *J. Cell Biol.* **121**, 1031–1039.

Zurzolo, C., van't Hof, W., van Meer, G., Rodriguez-Boulan, E. (1994). Glycosphingolipid clusters and the sorting of GPI-anchored proteins in epithelial cells. *Braz. J. Med. Biol. Res.* **27**, 317–322.

Zurzolo, C., van't Hof, W., van Meer, G., Rodriguez-Boulan, E. (1994). VIP21/caveolin, glycosphingolipid clusters and the sorting of glycosylphosphatidylinositol-anchored proteins in epithelial cells. *EMBO J.* **13**, 42–53.

Trafficking in Neuroendocrine Cells

T.F.J. MARTIN

The major signaling role of neuroendocrine cells is mediated by the secretion of peptide and biogenic amine hormones. Peptides, as well as amine transporters, are delivered in the Golgi to dense-core secretory vesicles (DCVs), which are conveyed to the plasma membrane where they undergo exocytic fusion triggered by Ca^{2+} elevations. This chapter will review recent developments in understanding the mechanisms underlying the trafficking and targeting of DCVs to, and their fusion with, the plasma membrane.

I. INTRODUCTION

Ca^{2+}-dependent vesicle exocytosis is a major membrane trafficking event that is essential for intercellular communication in the nervous and endocrine systems. The two major classes of regulated exocytic vesicles are synaptic vesicles (SVs) and dense-core vesicles (DCVs). Although both vesicle types are widely utilized throughout the nervous system, synaptic communication relies primarily on small (~35 nm) SVs that deliver classical neurotransmitters, such as acetylcholine, glutamate, and GABA into a synaptic cleft or neuromuscular junction. Neuroendocrine and endocrine cells rely primarily on larger (~100–300 nm) DCVs that mediate the release of a broad array of contents, such as biogenic amines, neuropeptides, neurotrophins, and peptide hormones into the extracellular space (Sudhof 2004; An and Zenisek 2004).

The anterograde exocytic pathway of vesicles consists of their biogenesis, trafficking to the plasma membrane, tethering or docking at the plasma membrane, priming in preparation for fusion, and Ca^{2+}-triggered fusion. Following exocytosis, vesicles are retrieved by endocytosis and either recycled locally for Ca^{2+}-dependent exocytosis or reclaimed by retrograde endosomal pathways. SVs and DCVs are of distinct biogenic origin, with SVs derived from endosomes and DCVs generated from the *trans*-Golgi cisternae. In spite of these differences, the molecular components for exocytic vesicle fusion and its Ca^{2+} regulation are the same. SVs and DCVs utilize a common fusion machinery consisting of the neuronal SNARE (soluble N-ethylmaleimide sensitive factor attachment protein receptor) proteins (synaptobrevin-2/VAMP-2, SNAP-25 and syntaxin-1) regulated by the Ca^{2+}-dependent synaptotagmins (Chen and Scheller 2001).

There are a number of differences between Ca^{2+}-triggered SV and DCV exocytosis, which include sites of fusion and fusion speeds. SV exocytosis occurs at highly specialized synaptic active zones, whereas DCV exocytosis in neurons occurs outside of active zones, as well as in cell bodies and dendrites (Ludwig and Pittman 2003). Neuroendocrine cells lack anatomically distinct preferential release sites for DCVs. Ca^{2+}-activated SV exocytosis exhibits shorter latencies to fusion and higher fusion rates than DCV exocytosis (Martin 2003).

Initial fusion pore diameters for SVs may also be smaller than those for DCVs (Klyachko and Jackson 2002). It is likely that there are molecular components that are distinct for either SV or DCV exocytosis, but our current knowledge of either process is too limited to be able to attribute specific differences in the molecular components that underlie vesicle docking and tethering, priming, and fusion pore regulation.

Studies of the DCV exocytic pathway have been greatly facilitated by the availability of immortalized cell lines of adrenal and pancreatic β cell origin, the utility of cell lines for reverse genetic strategies, the development of permeable cell secretion systems, the exploitation of capacitance to detect DCV-plasma membrane merger, the use of amperometry to detect oxidizable content release, and the large size of DCVs for microscope imaging studies. In this chapter, recent developments toward understanding the transport, tethering, docking, and fusion mechanisms for DCVs will be reviewed.

II. BIOGENESIS OF DCVs IN THE GOLGI AND THEIR MATURATION

DCVs are generated *de novo* in the trans-compartment of the Golgi. Many studies of this process have employed adrenal chromaffin cell-derived PC12 cells. The budding process that generates immature DCVs is thought to involve $PI(4,5)P_2$ synthesis and ARF1, but the details of this process are incompletely understood (Tooze et al. 2001). DCV formation appears to be a "coatless" budding event that does not employ clathrin or clathrin adapters. It has been suggested that DCV formation in the Golgi may be cargo-driven (Kim et al. 2001; see also Day and Gorr 2003). Cargo (e.g., chromogranins, peptide hormones, processing enzymes) are prone to aggregation at the pH and Ca^{2+} concentrations within the *trans-* Golgi lumen (Dannies 1999). It is striking that overexpression of cargo, such as chromogranin proteins, can induce the generation of DCV-like organelles in non-neuroendocrine cell types (Beuret et al. 2004; Huh et al. 2003; Kim et al. 2001; see also Day and Gorr 2003). In addition to specific cargo, mature DCVs possess a characteristic array of membrane proteins. There appear to be both positive as well as negative sorting events that establish the membrane protein composition of mature DCVs. VMAT2, the vesicular monoamine transporter that is required for catecholamine uptake by mature adrenal chromaffin DCVs, is retained on DCVs during maturation based on a C-terminal cytosolic acidic cluster motif (Waites et al. 2001). However, phosphorylation of this motif by casein kinase II enables interactions with PACS-1 (phosphofurin acidic cluster sorting protein), which promotes AP-1 recruitment and the formation of clathrin-coated vesicles that sort VMAT2 back to the Golgi (Waites et al. 2001). The Golgi enzyme furin and the SNARE VAMP-4 are similarly retrieved from immature DCVs and recycled back to the Golgi via a casein kinase II-phosphorylated acidic cluster motif that recruits PACS-1 (Thomas 2002; Hinners et al. 2003).

Another Golgi SNARE protein, syntaxin-6, is present on immature DCVs and has been proposed to mediate homotypic fusion between immature DCVs (Wendler et al. 2001). Syntaxin-6, VAMP-4, and synaptotagmin-4 are present on immature DCVs, but are removed by the clathrin-dependent sorting that converts immature to mature DCVs (Eaton et al. 2000; Tooze et al. 2001; Moore et al. 2002). How VAMP-2 and synaptotagmins-1/9, which are essential for the fusion competence of mature DCVs (see later), are retained in the flux of these other sorting events is unknown. Overexpression of synaptotagmins in PC12 cells does appear to overwhelm the sorting machinery such that overexpressed isoforms are retained on mature DCVs.

Several studies indicate that immature DCVs rapidly transit to the cell periphery and complete their conversion to mature DCVs at that location (Rudolf et al. 2001; Duncan et al. 2003). Studies with a neuropeptide-timer-dsRED-E5 fusion protein found that DCVs undergo a time-dependent recirculation in the cell. Newly synthesized DCVs preferentially populate the plasma membrane, whereas older DCVs are found deeper in the cytoplasm. The basis for this age-dependent segregation is unclear. Nonetheless, stimulation by Ca^{2+}-influx leads to the preferential release of the newer docked pool of immobile DCVs (Duncan et al. 2003).

III. CYTOPLASMIC TRANSPORT OF DCVs IS MEDIATED BY KINESIN AND MYOSIN

Following their biogenesis in the *trans*-Golgi, DCVs are transported rapidly (up to 2 μm/sec) to the cell periphery and cell surface. The initial stages of transport from the Golgi involve microtubule-based transport that utilizes kinesin-1 (Wacker et al. 1997; Rudolf et al. 2001; Varadi et al. 2003). The nature of the receptor on DCVs that mediates kinesin-1 interactions is presently unknown.

In the cell periphery, DCVs transfer to the actin cytoskeleton to continue their progress to the plasma membrane. Undocked DCVs in the cell periphery are embedded in a cortical actin cytoskeleton that impedes their free diffusion (Lang et al. 2000; Oheim and Stuhmer 2000; Steyer and Almers 1999; Ng et al. 2002). DCV movement within the actin cytoskeletal network depends upon the dynamic disassembly/reassembly of F actin "cages," and DCV movement is suppressed by stabilizing F actin with jasplakinolide treatment. However, DCV motility also requires F actin and is mediated by myosin motors. A subset of DCVs in the periphery exhibit directed movement on actin filaments

(Lang et al. 2000). Overall, peripheral cytoskeletal-associated DCVs constitute a pool that can undergo recruitment to the plasma membrane. Ca^{2+} influx increases the number of peripheral DCVs that exhibit directed motility (Oheim and Stuhmer 2000; Ng et al. 2002; Li et al. 2004), and these represent a recruitment pool to replenish the plasma membrane-docked DCVs that undergo exocytosis.

Myosin Va is the principal motor protein responsible for DCV movement in the actin cortex (Rudolf et al. 2003; Rose et al. 2003; Ivarsson et al. 2005; Varadi et al. 2005). Interference with myosin Va function by overexpression of myosin Va tail fragments or by siRNA-mediated downregulation of myosin Va was found to abort the trafficking of DCVs in the periphery and to reduce the number of stimulus-induced DCV exocytic events (Rudolf et al. 2003; Rose et al. 2003; Ivarsson et al. 2005; Varadi et al. 2005). Myosin II also has been implicated in DCV motility and may be involved in regulating fusion pore opening (Neco et al. 2004).

Interactions between DCVs and myosin Va appear to be mediated by Rab27a on the vesicles. Studies of melanosome trafficking in skin melanocytes revealed a requirement for Rab27a on melanosomes, which links to myosin Va via melanophilin/Slac-2a, a member of the exophilin/Slp-Slac family of proteins (Fukuda et al. 2002c; Strom et al. 2002). Rab27a on melanosomes in retinal pigment epithelim alternatively links to myosin VIIa by MyRIP/Slac-2c, another member of this protein family (El Amraoui et al. 2002). The exophilin/Slp-Slac family of proteins contain an N-terminal Rab27-binding domain. The Slp (synaptotagmin-like proteins) contain C-terminal tandem C2 domains, whereas the Slac (synaptotagmin-like with absent C2 domains) proteins contain the N-terminal Rab27-binding domain and diverse C-terminal domains.

Recent studies in chromaffin and PC12 cells (which lack myosin VIIa) suggest that Rab27a interactions with myosin Va utilizes MyRIP for trafficking on the actin cytoskeleton (Fukuda and Kuroda 2002; Desnos et al. 2003). Overexpression of a constitutively active Rab27a or MyRIP decreased the diffusion coefficient for DCVs and inhibited Ca^{2+}-activated secretion (Desnos et al. 2003). Conversely, the down-regulation of MyRIP in pancreatic β cells reduced stimulus-evoked DCV exocytosis (Waselle et al. 2003; Ivarsson et al. 2005). These results indicate that DCV recruitment to the plasma membrane for exocytosis in neuroendocrine cells employs actin-based transport via myosin Va molecules that are linked to Rab27a on DCVs by MyRIP/Slac-2c.

IV. DCVs ARE TETHERED AT THE PLASMA MEMBRANE PRIOR TO FUSION

DCVs that are within a vesicle diameter of the plasma membrane are morphologically docked, as determined by electron microscopy (Banerjee et al. 1996a; Plattner et al. 1997; Steyer et al. 1997). Live-cell imaging studies utilizing TIRF microscopy indicate that DCVs undergo a dramatic reduction in mobility upon docking and become immobilized (Steyer et al. 1997; Oheim and Stuhmer 2000; Oheim et al. 1998, 1999; Johns et al. 2001). The restricted mobility observed in the x, y plane of the plasma membrane, as well as in the z axis, suggests that DCVs are captured by a tether of approximately 25 nm of unknown identity (Oheim and Stuhmer 2000; Johns et al. 2001). Clostridial neurotoxin proteolysis of SNARE proteins had little impact on the immobility of docked DCVs, indicating that the tether employed to dock DCVs does not consist of SNARE complexes (Tsuboi et al. 2000; Johns et al. 2001).

Even though docked DCVs are the first to undergo exocytosis upon Ca^{2+} elevation (Oheim et al. 1998, 1999; Oheim and Stuhmer 2000), they do so surprisingly slowly compared to SVs docked at active zones. In PC12 cells, where a major fraction of the DCVs are docked at the plasma membrane (Martin and Kowalchyk 1997), DCV exocytosis exhibits a time constant of 10 to 30s, whereas SVs in the active zone exhibit submillisecond time constants (Martin 2003). It is likely that docking of SVs in the active zone helps to confer submillisecond fusion reactions by placing SVs precisely in the right place near sites of Ca^{2+} entry. Though tethering/docking of DCVs does appear to confer a kinetic advantage for fusion near sites of Ca^{2+} entry (Becherer et al. 2003), there may also be other functions for DCV docking. These could include maintaining DCVs for reuse in fusion (see later), or fused DCVs may be used as recipients for cytoplasmic DCVs that engage in compound exocytosis (Kishimoto et al. 2005). Undocked DCVs that are newly recruited to the plasma membrane are also capable of undergoing exocytosis in stimulated cells, but they are immobile and have a delay (~1sec) between arrival and fusion indicating that tethering/docking or priming events may be a prerequisite for successful fusion (Oheim and Stuhmer 2000; Allersma et al. 2004).

V. WHAT MEDIATES DCV TETHERING TO THE PLASMA MEMBRANE?

The molecular basis of DCV tethering/docking is unknown, although fine strands connecting DCVs to the plasma membrane in chromaffin cells have been detected in quick-freeze deep-etch EM analysis (Nakata et al. 1990). Recent work has characterized several proteins and protein complexes that may be part of a tethering machinery that contributes to DCV docking.

At many trafficking stations in the secretory pathway, transport vesicles become tethered to acceptor membranes as an essential prelude to membrane fusion (Sztul and Lupashin 2006). Membrane tethering complexes consist of proteins containing coiled-coil domains or large multisubunit oligomers that are recruited to donor vesicle and acceptor membrane compartments. Such tethering complexes are distinct and highly characteristic for each trafficking station (Sztul and Lupashin 2006). It is thought that tethering complexes are important for conferring the specificity of donor and acceptor membrane interactions. Typically Rab proteins and compartment-specific phosphoinositide lipids play key roles in promoting the recruitment and assembly of tethering complexes (Zerial and McBride 2001). Proteins in tethering complexes characteristically exhibit interactions with members of the Sec1 family of proteins, which in turn interact with syntaxin SNAREs. Thus, tethering complexes represent an important bridge between vesicle arrival at a compartment and the set-up for SNARE protein-mediated fusion reactions (Pfeffer 1999; Sztul and Lupashin 2006). Rab-promoted tethering complexes have been characterized for ER-to-Golgi and intra-Golgi trafficking, endosome, and yeast vacuolar fusion, and for Golgi-derived transport vesicles with the plasma membrane (Gillingham and Munro 2003; Sztul and Lupashin 2006). Based on the highly distinct composition of tethering complexes at various trafficking stations, it is likely that tethering complexes are highly individualized for SV and DCV exocytosis and possibly individualized for DCVs in different cell types.

In regulated DCV exocytosis, Munc18-1 is the Sec1 family protein of interest. Although its role has been studied for many years, exactly where and how Munc18-1 participates in vesicle transport, priming, or fusion remains enigmatic (Rizo and Sudhof 2002). *In vitro* binding studies and structural studies characterized the interaction of Munc18-1 with syntaxin-1. Munc18-1 binds and stabilizes one of the conformers of syntaxin-1 in which intramolecular interactions between an N-terminal domain and the membrane-proximal SNARE motif of syntaxin-1 forms a closed conformation.

The closed conformation of syntaxin-1 does not interact with SNAP-25 or VAMP-2 to form SNARE complexes. A paradigm emerged that Munc18-bound syntaxin-1 in a closed form would need to transition to an open form to allow syntaxin-1 to enter into fusogenic SNARE complexes (Rizo and Sudhof 2002). The mechanism by which the closed form of syntaxin-1 is primed to the open form, presumed to involve dissociation of Munc18-1, remains to be determined. Recent studies on DCV exocytosis have identified some syntaxin-1-independent roles for Munc18-1, in addition to its syntaxin-1-dependent roles (Ciufo et al. 2005; Schutz et al. 2005).

A role for Munc18-1 as essential for the docking of DCVs to the plasma membrane was recently characterized (Voets et al. 2001b). Chromaffin cells from Munc18-1 knockout mice exhibited marked decreases in Ca^{2+}-triggered DCV exocytosis that corresponded to large decreases in the number of docked DCVs at the plasma membrane (Voets et al. 2001b). This work suggests that a Munc18-1/syntaxin-1 complex represents a docking site for DCVs. Similarly, in the pituitary gland of Munc18-1 knockout mice, decreased peptide hormone secretion is accompanied by decreases in plasma membrane-proximal DCVs (Korteweg et al. 2005). In contrast to the strong but incomplete loss of DCV exocytosis in chromaffin cells from the Munc18-1 knockout mouse, SV exocytosis in neurons from the knockout was reported to be completely abolished, but without any loss of SV docking (Verhage et al. 2000). If confirmed, these results would indicate an important difference in the role of Munc18-1 in DCV and SV exocytosis. The studies on DCVs in chromaffin cells indicate that Munc18-1, like other Sec-1 family proteins at other trafficking stations, play a positive (rather than inhibitory) role in fusion by mediating DCV tethering or docking (Voets et al. 2001b). In recent work, components that interact with Munc18-1 also have been implicated as part of a presumptive DCV tethering complex.

Rab3 and Rab27 isoforms reside on DCVs (Fukuda 2005; Izumi et al. 2003; Cheviet et al. 2004) and a diverse family of Rab3/Rab27 effector proteins that may function in some manner to tether DCVs prior to fusion have recently been described. In pancreatic β cells, granuphilin/Slp4 was identified as a Rab27a-binding protein that resides on DCVs and interacts with syntaxin-1 (Wang et al. 1999; Yi et al. 2002; Torii et al. 2002; Izumi et al. 2003). The overexpression of granuphilin in β cells increased the number of docked DCVs but surprisingly inhibited Ca^{2+}-triggered DCV exocytosis (Torii et al. 2004). In contrast, studies on β cells from a granuphilin knockout mouse confirmed that the absence of granuphilin decreased the number of docked DCVs, but enhanced their fusion probability (Gomi et al. 2005). Because granuphilin interacts only with the closed form of syntaxin-1 (Torii et al. 2002), it may simultaneously mediate tethering between Rab27a and plasma membrane syntaxin-1, while stabilizing syntaxin-1 in a closed form. Granuphilin/Slp4 was also found to bind Munc18-1 directly (Fukuda 2003b), suggesting that granuphilin/Munc18-1 may be a tethering complex. This complex, via its syntaxin-1 interactions, could promote proximity of newly arrived DCVs with SNAREs in the plasma membrane, however, it is unclear what interactions mediate transitions of the DCV to enable formation of fusogenic SNARE complexes. Granuphilin is expressed in pancreatic β (but not α) cells and in pituitary cells (and possibly in adrenal cells; see Wang et al. 1999 but also Fukuda et al. 2002b) but not in neurons, indicating that a granuphilin-mediated tethering mechanism is quite specialized for a subset of DCVs and would not function in SV tethering.

Another Rab27a-interacting component has been implicated in DCV tethering in adrenal PC12 cells. Rabphilin, initially discovered as a putative Rab3 effector, but more recently characterized as a likely Rab27a effector (Fukuda 2003a), was reported to interact with Rab27a via an N-terminal domain and with SNAP-25 via the second of its C-terminal tandem C2 domains (C2B) (Tsuboi and Fukuda 2005). Overexpression of rabphilin in PC12 cells was found to increase the number of docked DCVs and to increase their fusion probability upon Ca^{2+} influx (Tsuboi and Fukuda 2005). This provides a model for DCV tethering in which rabphilin bridges Rab27a on DCVs to plasma membrane SNAP-25. Neurons from rabphilin knockout mice were reported to exhibit no deficits in SV exocytosis (Schluter et al. 1999). In PC12 cells, Tsuboi and Fukuda (2005) identified a long-sought role for rabphilin in DCV exocytosis that is either redundant or not present in SV exocytosis. Whereas pancreatic β cells apparently lack rabphilin, granuphilin/Slp4 is expressed in PC12 cells and has also been reported to function in DCV exocytosis (Fukuda et al. 2002b; Fukuda 2003b). This could indicate that the granuphilin- and rabphilin-mediated tethering systems function in parallel in PC12 cells but not in other neuroendocrine cells. Rab27-mediated mechanisms for vesicle transport and docking/tethering may be distinct for DCVs outside the nervous system. Rab27a is expressed in peptide-secreting endocrine cells and the adrenal medulla, but little is expressed in brain. Low levels of Rab27b are present in brain (Tolmachova et al. 2004).

Recent work has suggested other possible DCV tethering mechanisms for pancreatic β cells, as well as PC12 cells. These involve the mammalian exocyst complex (termed Sec6-Sec8 complex). Extensive studies in *S. cerevisae* on the polarized constitutive fusion of Golgi-derived vesicles with the plasma membrane revealed a tethering complex of eight proteins, termed the Exocyst, that functions under the direction of the vesicular Rab Sec4 (Guo et al. 2000).

In yeast, Sec4 recruits Sec15p, which in turn recruits other exocyst proteins to the vesicle, which assemble with a Sec3/Exo70 complex at the plasma membrane (Boyd et al. 2004). The recruitment of Sec3 and Exo70 to sites on the plasma membrane is controlled by other polarity-establishing GTPases (Lipschutz and Mostov 2002). This allows targeting to specific sites on the plasma membrane, as well as the tethering of vesicles prior to SNARE-mediated fusion. It has been proposed that the mammalian counterpart of this complex may function similarly in polarized or regulated secretory pathways (see Lipschutz and Mostov 2002; also see Chapter 13). A recent study indicated that mammalian Sec6 localizes to insulin granules in pancreatic β cells, that Sec8 localizes to the plasma membrane, and that Sec5 is largely cytosolic (Tsuboi et al. 2005). Overexpression of truncated forms of Sec6 or Sec8 reduced the number of docked insulin granules and inhibited stimulated insulin secretion (Tsuboi et al. 2005). In other studies, the GTPase Ral A, which interacts directly with the Sec5 protein component of the Exocyst, was found to affect regulated DCV exocytosis in PC12 cells (Moskalenko et al. 2001; Wang et al. 2004). Although the detailed paradigm involving the Exocyst complex in yeast constitutive exocytosis provides an attractive model for docking/tethering in the DCV pathway, much additional work is needed to assess this possibility.

VI. DOCKED DCVs REQUIRE ATP-DEPENDENT PRIMING PRIOR TO FUSION

Direct imaging studies have shown that not all DCVs that are docked at the plasma membrane are equally capable of undergoing Ca^{2+}-triggered fusion (Oheim et al. 1999; Becherer et al. 2003; Barg et al. 2002). Prior capacitance studies of DCV exocytosis had inferred that only a small subset of docked

DCVs were in a ready release state termed the ready releasable pool (RRP) (Rettig and Neher 2002). It generally is considered that biochemical reactions are needed to render docked DCVs competent for Ca^{2+}-triggered exocytosis, a process referred to as *priming*. In adrenal chromaffin (Xu et al. 1998), PC12 (Hay and Martin 1992), and pancreatic β cells (Olsen et al. 2003), DCV priming requires ATP and is reversible.

DCV priming reactions that are ATP-dependent have been characterized in permeable neuroendocrine cells (Holz et al. 1989; Hay and Martin 1992). In permeable cells, Ca^{2+}-triggered DCV exocytosis is found to consist of early ATP-independent and later ATP-dependent phases. In the absence of ATP, Ca^{2+}-triggered DCV exocytosis undergoes a time-dependent decrease that represents the progressive reversal of priming and the loss of an RRP. Inclusion of ATP in the permeable cell incubations allows the characterization of ATP-dependent priming processes, many of which may already be completed in energy-repleted intact cells. Several ATP-dependent processes have been characterized. In pancreatic β cells, DCV acidification driven by a V-type ATPase was reported to be essential based on the ability of ATPase and chloride channel inhibitors to block early evoked capacitance increases measuring the RRP (Barg et al. 2001). Studies in yeast had implicated a subunit of the V-ATPase Vo complex VPH1 as a potential protein fusion pore element in vacuolar fusion that functioned following SNARE complex formation (Bayer et al. 2003). In addition, mutations in the *Drosophila* homolog vha100-1 were reported to reduce evoked SV exocytosis at the neuromuscular junction (Hiesinger et al. 2005). Whether the Vo ATPase functions directly or indirectly (via acidification) in DCV fusion is unclear at present.

In PC12 cells, reactions dependent upon αSNAP and NSF that mediated the disassembly of *cis*-SNARE complexes were identified as essential for priming (Banerjee et al. 1996a), and in chromaffin cells αSNAP expression was observed to enhance the RRP (Xu et al. 1999a). More recently, Lang et al. (2002) found that SNAP-25 and syntaxin-1 were largely unassociated in membrane lawns from PC12 cells, but associated into *cis*-SNARE complexes in incubations without ATP. These studies indicate that an important role for ATP in DCV exocytosis is to maintain SNARE proteins in a disassembled state in preparation for *trans*-SNARE complex formation in fusion.

VII. PI(4,5)P₂ SYNTHESIS IS AN ESSENTIAL COMPONENT OF DCV PRIMING

There is now considerable evidence that lipid phosphorylation is an important reaction underlying the ATP-dependent priming of DCVs. The finding that PI (phosphatidylinositol) was required for ATP-dependent DCV exocytosis in chromaffin cells (Eberhard et al. 1990) was clarified when a phosphatidylinositol transfer protein and a type I phosphatidylinositol(4)monophosphate 5-kinase (PI(4)P 5-kinase I), components needed for PI(4,5)P₂ synthesis, were identified as required cytosolic DCV priming factors (Hay and Martin 1993; Hay et al. 1995). Of three isoforms of PI(4)P 5-kinase type I, the type Iα and Iγ isoforms exhibit the greatest priming activity in DCV exocytosis (Hay et al. 1995; Aikawa and Martin 2003; Wang et al. 2005b). PI(4,5)P₂ itself, rather than its hydrolysis products diacylglycerol or inositol(1,4,5)P₃ is essential for DCV exocytosis (Eberhard et al. 1990; Hay et al. 1995; Holz et al. 2000). The ATP-primed state is fully reversible (Hay and Martin 1992) and the reversal was accelerated by a type II phosphoinositide 5-phosphatase that converts PI(4,5)P₂ to PI(4)P (Martin et al. 1997).

PI(4,5)P₂ conversion to PI(3,4,5)P₃ is unlikely to play an essential role in DCV exocytosis because the PI 3-kinase inhibitors wortmannin and LY294002 did not inhibit ATP-dependent priming of DCVs in PC12 cells (Martin et al. 1997). Recent work has indicated that very high concentrations of these inhibitors do alter secretion in chromaffin and PC12 cells (Chasserot-Golaz et al. 1998; Meunier et al. 2005). However, these compounds inhibit many enzymes and LY294002 treatment may decrease PI(4,5)P₂ levels in chromaffin cells (Milosevic et al. 2005). More direct studies have shown that a type II PI 3-kinase, a relatively drug-insensitive kinase, may play a role in modulating secretion in chromaffin cells (Meunier et al. 2005). Future studies need to clarify the basis of this modulatory effect.

The essential role of PI(4,5)P₂ in the priming of DCVs has been extensively confirmed in intact cell studies in which it was shown that PI(4,5)P₂ formation is required for refilling of the RRP. PIP₂ antibodies prevented refilling of the RRP in pancreatic β cells, and the requirement for ATP in RRP refilling was bypassed by direct injection of PI(4)P or PI(4,5)P₂ (Olsen et al. 2003). Similarly, in chromaffin cells, microinjection of PI(4,5)P₂ or PIP 5-kinase Iγ overexpression increased the RRP, whereas the 5-phosphatase-mediated hydrolysis of PI(4,5)P₂ reduced the RRP (Milosevic et al. 2005). In PC12 cells, depletion of plasma membrane PI(4,5)P₂ or its increased synthesis by expression of a constitutively active PI(4)P 5-kinase Iγ correspondingly depressed or enhanced rates of DCV exocytosis (Aikawa and Martin 2003). The RRP and its refilling rates were reduced in chromaffin cells from a PI(4)P 5-kinase Iγ knockout mouse (Gong et al. 2005). Loss of PI(4)P 5-kinase Iγ did not alter the number of docked DCVs at the plasma membrane, but reduced the RRP, reduced the frequency of stimulated exocytic events, and prolonged fusion pore opening times (Gong et al. 2005). These studies indicate that PI(4,5)P₂ plays an essential role in exocytosis at a point beyond DCV docking in rendering the fusion apparatus competent for Ca²⁺ triggered exocytosis.

VIII. WHY IS PI(4,5)P₂ REQUIRED FOR DCV EXOCYTOSIS?

Key questions about the involvement of PI(4,5)P₂ in regulated DCV exocytosis concern the location of this phosphoinositide with respect to the fusion machinery, and the identity of relevant interacting components. DCVs contain a type II PI-4 kinase α (Barylko et al. 2001) that catalyzes conversion of PI to PI(4)P. PI(4)P can undergo conversion to PI(4,5)P₂ if DCVs are provided with ATP and a soluble type I PI(4)P 5-kinase. However, this conversion apparently does not occur in DCV-containing cells. Rather, the PI(4,5)P₂ synthesized during ATP-dependent priming resides on the plasma membrane (Holz et al. 2000; Aikawa and Martin 2003). Plasma membrane PI(4,5)P₂ localizes to ~300 nm microdomain clusters in neuroendocrine cells (Aikawa and Martin 2003; Aoyagi et al. 2005; Milosevic et al. 2005). Recent work in PC12 cells identified subsets of docked DCVs that colocalize with plasma membrane microdomains containing PI(4,5)P₂, syntaxin-1, or both. The DCVs that colocalized with both PI(4,5)P₂ and syntaxin-1 undergo preferential depletion upon Ca²⁺-influx and these DCVs presumably represent the RRP (Aoyagi et al. 2005). These data suggest that priming consists of the process of localizing DCVs to specific sites on the plasma membrane where PI(4,5)P₂ is concentrated and syntaxin-1 is clustered. Thus, PI(4,5)P₂ microdomains are important determinants for localizing DCV exocytosis to specific sites on the plasma membrane. Because of the high rates of lipid diffusion, maintaining PI(4,5)P₂ in microdomains in the plasma membrane requires its sequestration by binding proteins. Whether SNARE proteins (such as syntaxin-1 with its membrane-proximal polybasic domains) or other constituents serve this function will be important to determine.

Although a central essential role for PI(4,5)P₂ in DCV exocytosis has been established, the precise molecular mechanisms mediating its role remain to be determined. In other membrane trafficking events and in cytoskeletal regulation, specific phosphoinositides function as landmarks on the membrane that recruit proteins with phosphoinositide-binding domains (Martin 1998). Examples include EEA1 that contains a FYVE domain and localizes to endosomal membrane PI(3)P, or FAPP that contains a PH domain that mediates recruitment to Golgi membrane PI(4)P (De Matteis and Godi 2004). In the DCV exocytic pathway, several proteins that bind PI(4,5)P$_2$ have been characterized. Synaptotagmin-1, a calcium sensor for DCV exocytosis, interacts with PI(4,5)P$_2$ via its C2B domain and has been proposed to "steer" DCVs to specific locations in the plasma membrane containing PI(4,5)P$_2$ (Bai et al. 2004a). Rabphilin, a Rab27a- and Rab3-binding protein, contains tandem C2 domains that exhibit Ca^{2+}-dependent interactions with PI(4,5)P$_2$ (Chung et al. 1998). CAPS functions in DCV priming through interactions with both DCVs and plasma membrane (Grishanin et al. 2002; 2004). Plasma membrane interactions are mediated through a PH domain in CAPS that binds PI(4,5)P$_2$ (Grishanin et al. 2002). CAPS has been proposed to function as a low affinity PI(4,5)P$_2$ sensor for priming DCV exocytosis (Grishanin et al. 2004). Studies in pancreatic β cells showed that the direct enhancing effects of microinjected PI(4,5)P$_2$ on priming DCVs were blocked by CAPS antibodies (Olsen et al. 2003). Studies critically assessing the role of PI(4,5)P$_2$ binding in the function of synaptotagmin-1, rabphilin, and CAPS in the DCV pathway are needed.

Because of the central role of PI(4,5)P$_2$ in priming DCVs, the regulation of PI(4)P 5-kinase activity may be an important determinant of rates of DCV-mediated secretion. During stimulation by Ca^{2+}-influx, the RRP pool of DCVs is depleted rapidly during the first second of stimulation. After this, rates of secretion become dependent upon rates of RRP refilling. Because PI(4,5)P$_2$ levels are determinants for RRP refilling rates, regulation of the synthesis of PI(4,5)P$_2$ or its hydrolysis may have a strong influence on rates of secretion beyond the first second of stimulation. The GTPase ARF6 is a major factor that regulates plasma membrane PI(4,5)P$_2$ synthesis. PI(4)P 5-kinase isoforms directly interact with ARF6 and recruit these enzymes to the plasma membrane for PI(4,5)P$_2$ synthesis (Honda et al. 1999; Aikawa and Martin 2003). Ca^{2+} influx, which enhances PI(4,5)P$_2$ synthesis (Eberhard and Holz 1991), increases the association of ARF6 with PI(4)P 5-kinase (Aikawa and Martin 2003) and with SCAMP2, a proposed scaffolding protein, that localizes to plasma membrane sites containing syntaxin-1 near docked DCVs in PC12 cells (Liu et al. 2005). SCAMP2 also regulates PLD activity, which generates PA, an activator of PI(4)P 5-kinase. An ARF6-dependent cascade may play an important role in establishing high concentration domains of PI(4,5)P$_2$ in the plasma membrane that become sites for DCV recruitment and exocytosis.

IX. SNARES ARE AT THE CORE OF DCV FUSION

As is true for Ca^{2+}-dependent SV exocytosis, the Ca^{2+}-regulated fusion of DCVs with the plasma membrane in neuroendocrine cells requires the neural SNARE proteins synaptobrevin-2/VAMP-2 on the vesicle, and SNAP-25 plus syntaxin-1 on the plasma membrane. (See Chapter 6 and recent reviews Chen and Scheller 2001; Rettig and Neher 2002; Jahn et al. 2003). The essential role of SNAREs for DCV fusion initially was established from studies in which the clostridial neurotoxin proteases selective for VAMP-2, SNAP-25 or syntaxin-1 were introduced into neuroendocrine cells to inhibit Ca^{2+}-triggered DCV exocytosis (see Xu et al. 1998b for references). In addition, genetic studies showed that Ca^{2+}-

triggered DCV exocytosis in adrenal chromaffin cells from SNAP-25 knockout mice was nearly abolished (Sorensen et al. 2003a). Ca^{2+}-triggered DCV exocytosis in VAMP-2/synaptobrevin-2 knockout mice was more complex due to compensation by cellubrevin, but it was nonetheless concluded that VAMP-2/synaptobrevin-2 played a dominant role in DCV exocytosis (Borisovska et al. 2005). The finding that neuronal SNAREs reconstituted into proteoliposomes are capable of mediating lipid bilayer mixing (Weber et al. 1998) suggest a model in which SNARE protein complexes, formed in *trans* across DCV and plasma membranes, directly catalyze membrane fusion. The number of SNARE complexes minimally required for mediating DCV exocytosis is not precisely known; however, it has been estimated at ≥ 3 in inhibition studies (Hua et al. 2001) and at 5–8 in amperometry studies in PC12 cells (Han et al. 2004). A current model envisions a rosette assembly of SNARE complexes that compose part of the lining of the fusion pore.

The transmembrane segments of VAMP-2/synaptobrevin 2 and syntaxin-1 may provide part of the lining for the fusion pore that develops at the time of fusion. Systematic replacement of residues in the transmembrane segment of syntaxin-1 with a bulky tryptophan residue revealed positions along one face of an α-helix that reduced the amplitude of the prespike foot in amperometric recordings of DCV exocytic events in PC12 cells (Han et al. 2004). The prespike foot is thought to represent the diffusion of catecholamines through the initial fusion pore. Presumably, the transmembrane domain of VAMP-2/synaptobrevin-2 would constitute the remaining half of the fusion pore, an assumption that requires direct experimental testing.

It remains unclear precisely how and when *trans*-SNARE complexes form during DCV exocytosis. PC12 cells exhibit a slow mode of Ca^{2+}-dependent exocytosis ($\tau \sim$ 10–30 s) that appears to be rate-limited by

DCV priming reactions that are Ca^{2+}-dependent and do not exhibit cooperativity for Ca^{2+} (Grishanin et al. 2004). Plasma membrane syntaxin-1 and SNAP-25 in PC12 cells are uncomplexed and distributed in partially overlapping microdomain clusters of ~200 nm where they are susceptible to cleavage by clostridial neurotoxins (that only operate on the free proteins) (Banerjee et al. 1996b; Lang et al. 2001, 2002). A majority of the docked DCVs on the plasma membrane colocalize with either syntaxin-1 or SNAP25 clusters (Lang et al. 2001; Aoyagi et al. 2005). In permeable PC12 cells, fusogenic SNARE complexes do not appear to form until DCV exocytosis is triggered by Ca^{2+} (Banerjee et al. 1996b; Chen et al. 1999, 2001). These results suggest that SNARE complex formation for fusion occurs during a prefusion Ca^{2+}-dependent priming step (Grishanin et al. 2004). Consistent with this idea, Ca^{2+} influx was reported to promote the formation of a possible SNARE complex precursor (SNAP-25/syntaxin-1) detected by FRET of a SNAP-25 construct (An and Almers 2004). Overall, the idea that a Ca^{2+}-dependent priming step promotes SNARE complex formation would account for the long latencies (1–3 sec) between Ca^{2+} rises and DCV fusion observed in PC12 cells (Grishanin et al. 2004; Ninomiya et al. 1997; Wang et al. 2001).

In adrenal chromaffin cells, a faster burst component of Ca^{2+}-dependent exocytosis is observed. A subset of docked DCVs undergo Ca^{2+}-dependent priming into slower-release (SRP) and ready-release (RRP) pools prior to fusion (Rettig and Neher 2002). The SRP/RRP is dynamic but stable in chromaffin cells unlike the situation in PC12 cells where an SRP/RRP is either extremely small or unstable. Ca^{2+}-triggered exocytosis of DCVs from the SRP and RRP is rapid ($\tau \sim 0.3$ and 0.03 s, respectively) and exhibits cooperativity for Ca^{2+}. Based upon antibody neutralization and clostridial neurotoxin inhibition studies, it has been proposed that DCVs in the SRP and RRP state contain loose and tight *trans*-SNARE complexes (Xu et al. 1998; 1999b). In this case, Ca^{2+} rises may be sensed by synaptotagmin-1 (see later), which promotes the final stages of SNARE complex tightening, leading to DCV fusion with short (10 msec) latencies (Rettig and Neher 2002).

X. SYNAPTOTAGMINS SENSE CALCIUM AND TRIGGER SNARE-DEPENDENT DCV FUSION

The exocytic fusion of DCVs, like that for SVs, is completely dependent upon elevations in cytoplasmic Ca^{2+}. Imaging studies in chromaffin cells, revealing Ca^{2+} microdomains at sites of channel-mediated entry, found that DCVs within 300 nm of Ca^{2+} microdomains exhibited finite probabilities for Ca^{2+}-dependent fusion (Becherer et al. 2003). What confers the Ca^{2+} regulation on SNARE-dependent DCV exocytosis? Increasing evidence for both SV and DCV exocytosis indicates that synaptotagmins are the Ca^{2+} sensors for fusion. Synaptotagmins are vesicle-associated, membrane-spanning proteins with an N-terminal lumenal domain and cytoplasmic tandem C2 domains that bind Ca^{2+}. Their properties and the evidence that they represent Ca^{2+}-sensors for exocytosis have been reviewed (Sudhof 2004; Chapman 2002). A recent issue has been determining which of 16 synaptotagmins (Craxton 2004) plays a role in mediating Ca^{2+}-triggered DCV exocytosis in neuroendocrine cells.

A substantial body of work has demonstrated an essential role for synaptotagmin-1 in mediating SV exocytosis in mice, flies, and worms (Yoshihara et al. 2003). Synaptotagmin-1 plays a role in fast Ca^{2+}-triggered fusion of SVs at a stage beyond vesicle docking or priming that corresponds to the triggering of SNARE-dependent fusion (Geppert et al. 1994). However, additional Ca^{2+} sensors, likely other synaptotagmin isoforms, must be present on SVs because Ca^{2+}-triggered neurotransmitter release persists, or is even enhanced, in the form of delayed asynchronous release in synaptotagmin-1 nulls (Yoshihara and Littleton 2002; Nishiki and Augustine 2004; Maximov and Sudhof 2005).

A large number of different synaptotagmin isoforms are expressed in neuroendocrine cells with some localized to DCVs (Tucker and Chapman 2002; Marqueze et al. 2000). For example, PC12 cells express synaptotagmins 1, 4, 7, and 9 at different levels and locations (Zhang et al. 2002; Tucker et al. 2003). Early studies indicated that loss of synaptotagmin-1, which localizes to DCVs, did not affect Ca^{2+}-triggered norepinephrine secretion (Shoji-Kasai et al. 1992). However, PC12 cells lack fast components of Ca^{2+}-triggered exocytosis (Grishanin et al. 2004). In adrenal chromaffin cells from synaptotagmin-1 knockout mice, the fastest exocytic burst component of the Ca^{2+}-triggered capacitance increase was reduced, whereas the slower burst and sustained components were unaltered (Voets et al. 2001a; Rettig and Neher 2002). These studies are consistent with the idea that synaptotagmin-1 is required for the fastest mode of DCV exocytosis but that other Ca^{2+} sensors that operate with slower kinetics are present and capable of mediating Ca^{2+}-triggered exocytosis.

In PC12 cells, a second major synaptotagmin isoform, synaptotagmin-9, was localized to DCVs and found to overlap with the distribution of synaptotagmin-1 (Fukuda et al. 2002a; Zhang et al. 2002; Tucker et al. 2003). Antibody neutralization studies in permeable PC12 cells indicated that both synaptotagmins 1 and 9 function in Ca^{2+}-triggered DCV exocytosis (Fukuda et al. 2002a). Moreover, siRNA studies have shown that synaptotagmins 1 and 9 function redundantly for DCV exocytosis in PC12 cells (K. Lynch and T. Martin). Although either synaptotagmin-1 or -9 alone mediated Ca^{2+}-triggered DCV exocytosis, each isoform independently conferred distinct kinetics to the cumulative probabilities for DCV fusion. It may generally be the case that DCVs contain multiple synaptotagmin isoforms that collectively contribute distinct properties to fusion events.

Although synaptotagmin-7 is a minor isoform in PC12 cells (Tucker et al. 2003), its overexpression as a higher affinity Ca^{2+}-binding isoform resulted in enhanced secretory responses at lower Ca^{2+} levels (Fukuda et al. 2004; Wang et al. 2005c). In pancreatic β cells and cell lines, synaptotagmins 3, 5, 7, 8, and 9 have been proposed to mediate Ca^{2+}-dependent insulin secretion (Mizuta et al. 1997; Brown et al. 2000; Gut et al. 2001; Iezzi et al. 2004). Recent siRNA studies showed that downregulation of either synaptotagmin-5 or -9 reduced Ca^{2+}-dependent insulin secretion from INS-1E cells (Iezzi et al. 2004). The overall characteristics of secretion mediated by DCVs (kinetics, Ca^{2+} sensitivity) may be conferred in part by the precise composition of synaptotagmin isoforms on DCVs.

The mechanisms by which synaptotagmins mediate Ca^{2+} triggering of DCV exocytosis are being defined. The major biochemical properties of synaptotagmins thought to be important for Ca^{2+}-dependent exocytosis include C2 domain-mediated Ca^{2+}-dependent binding to acidic phospholipids, such as PS and PI(4,5)P_2, Ca^{2+}-dependent and -independent interactions with the t-SNARE proteins SNAP-25 and syntaxin-1, and oligomerization (Sudhof 2004; Chapman 2002). Current evidence favors a mechanism in which synaptotagmin interactions with phospholipids and t-SNAREs play an essential role in mediating synaptotagmin's triggering roles.

A transgenic mouse expressing a R233Q synaptotagmin-1 mutant was found to exhibit a twofold decrease in the Ca^{2+}-dependent probability of SV release in correlation with a twofold increase in the K_D for Ca^{2+}-dependent PS binding by the R233Q mutant (Fernandez-Chacon et al. 2001). Studies in chromaffin cells from the mutant mouse revealed a twofold higher Ca^{2+} threshold for triggering DCV exocytosis (Sorensen et al. 2003b). These studies provide strong evidence that synaptotagmin-1 is a Ca^{2+} sensor for rapid DCV exocytosis. However, the finding that the R233Q mutant synaptotagmin-1 also exhibits reduced Ca^{2+}-dependent interactions with SNAP-25 (Wang et al. 2003b) weakens the conclusion that PS is the major effector for synaptotagmin-1.

Evidence that SNAP-25 is a major Ca^{2+}-dependent effector of synaptotagmins was provided by studies in PC12 cells expressing mutant SNAP-25 proteins (Zhang et al. 2002). C-terminal acidic residues in SNAP-25 are essential for Ca^{2+}-dependent synaptotagmin binding (Gerona et al. 2000). Replacement of wild-type SNAP-25 with binding-deficient mutants in PC12 cells resulted in a strong loss of Ca^{2+}-triggered DCV exocytosis in correlation with the loss of binding (Zhang et al. 2002). Evidence that t-SNAREs are important effectors was also provided by the overexpression of synaptotagmin-1 linker mutants in PC12 cells (Bai et al. 2004b). Synaptotagmin-1 mutants with lengthened inter-C2 domain linkers exhibit impaired interactions with syntaxin/SNAP-25 heterodimers but not with phospholipids. Overexpression of these mutants in PC12 cells resulted in strong reductions in Ca^{2+}-dependent catecholamine secretion accompanied by an apparent destabilization of fusion pores (Bai et al. 2004b). Future studies utilizing synaptotagmin mutants with well-characterized biochemical deficits to rescue function in null backgrounds should provide more direct information to assess the importance of phospholipids and SNAREs as effectors in Ca^{2+}-triggered DCV exocytosis.

XI. DCV FUSION PORES DILATE OR CLOSE

At least two distinct pathways for exo-/endocytosis have been proposed for neuronal SVs (see Chapters 6 and 7). The classical pathway of full fusion followed by clathrin-mediated endocytosis appears to predominate with intense stimulation. At lower stimulation levels, a process known as kiss-and-run has been described (see Rizzoli and Betz 2005). The kiss-and-run mode for exo-/endocytosis involves rapid fusion pore closure and SV retrieval. In its purest form, kiss-and-run is envisioned to capture the SV with its membrane lipid and protein content intact. This would facilitate reuse of the SV, after transmitter reloading and repriming, without the need for intervening protein sorting events. Studies on this mode of SV exocytosis and controversies surrounding them recently have been reviewed (Ryan 2003; An and Zenisek 2004; Rizzoli and Betz 2005).

Early studies of DCV exocytosis also had characterized a classical pathway involving full fusion of the DCV into the plasma membrane with discharge of contents. The DCV membrane and constituents were reclaimed by clathrin-mediated endocytosis followed by endosomal trafficking of DCV membrane constituents and slow recycling through the Golgi (Bauer et al. 2004). However, recent studies of DCV exocytosis in neuroendocrine cells have characterized exo-/endocytic modes that are similar to kiss-and-run. These have been termed *cavi-capture* because they involve a range of fusion pore closure times that allow escape of DCV membrane constituents into the plasma membrane (Taraska et al. 2003). This mode of exocytosis differs from the classical pathway of full fusion because the DCV may remain at its exocytic site at the plasma membrane after fusion pore closure. In addition, the DCVs can variably retain some dense-core constituents and can recycle through another round of exocytosis upon restimulation. This contrasts with the classical pathway in which endocytosed DCV membrane recycles through the endocytic pathway back to the Golgi.

Although the cavicapture mode of release may correspond to rapid endocytosis characterized in studies of DCVs

in patch-clamp capacitance recordings (Palfrey and Artalejo 2003), this mode of DCV exocytosis recently has been imaged directly in cells expressing various GFP variant fusion proteins that are targeted to the lumen or the membrane of the DCVs. In PC12 and chromaffin cells, a content marker NPY (~4 kDa) generally was released during an exocytic event, whereas a larger tPA (~59 kDa) content marker frequently was not released. Instead the tPA reported proton fluxes (via pH effects on fluorescence) through the open fusion pore until pore closure occurred (Taraska et al. 2003; Perrais et al. 2004; Taraska and Almers 2004). Phogrin, a DCV membrane protein, was largely retained in post-fusion DCVs, whereas a membrane-bound dye (FM4-64) was rapidly lost even from DCVs that experienced rapid fusion pore closure (Taraska and Almers 2004). In chromaffin cells, the majority of DCV exocytic events were accompanied by the rapid diffusion of synaptobrevin-2/VAMP-2 into the plasma membrane (Allersma et al. 2004). By expressing a cytosolic CFP protein, post-fusion DCVs were detected as intact organelles (shadows) linked to the plasma membrane at fusion sites (Taraska et al. 2003). Two-photon microscopy with polar fluophores in PC12 cells also detected long-lived (>30 s) open fusion pores on DCVs stably maintained at the plasma membrane (Kishimoto et al. 2005).

Similar cavicapture modes for pancreatic β cell DCVs have been reported for a wide range of markers (Tsuboi and Rutter 2003; Tsuboi et al. 2004). Whereas NPY release was observed in the majority of exocytic events, phogrin exodus to the plasma membrane occurred only slowly in events where fusion pore closure was delayed. In contrast, synaptobrevin/VAMP and synaptotagmin I, two distinct DCV proteins, were observed to spread into the plasma membrane for most of the DCV exocytic events (Tsuboi and Rutter 2003; Tsuboi et al. 2004).

A wide variety of distinct behaviors for individual DCV exocytic events was observed in the preceding studies. Full release of content markers accompanied by extensive diffusion of DCV membrane constituents into the plasma membrane probably represent full fusion events. In contrast, the incomplete release of content with fusion pore reclosure and the retention of the DCV at exocytic sites indicates a distinct mode of exo-/endocytosis in the form of cavicapture. Unlike pure kiss-and-run, cavicapture events were accompanied by diffusion of DCV membrane constituents into the plasma membrane. A lipid marker diffused rapidly whereas membrane protein constituents diffused more slowly. These studies indicate that the fusion pore apparatus does not constitute a barrier to diffusion from the merged DCV to plasma membrane. The studies also imply that DCVs retrieved following fusion pore closure would contain membranes with altered protein constituents.

Amperometry studies of catecholamine release from adrenal chromaffin cells had indicated that high extracellular Ca^{2+} may shift DCV exocytosis from full fusion to kiss-and-run (here termed cavicapture) (Ales et al. 1999). This effect was potentially attributable to elevated intracellular Ca^{2+}, but it may instead be due to divalent cation effects on the DCV matrix to impede release of DCV constituents (Perrais et al. 2004). Recent studies have indicated that high frequency stimulation in chromaffin cells is associated with a transition from cavicapture to full fusion (Fulop et al. 2005). The mechanisms underlying cavicapture as well as its regulation remain to be fully elucidated. However, recent studies have implicated dynamin as the basis for postfusion fission reactions in cavicapture.

XII. DYNAMIN MEDIATES PRECOCIOUS DCV FUSION PORE CLOSURE

On plasma membrane lawns derived from PC12 cells, Ca^{2+} was found to trigger DCV exocytosis with incomplete release of an NPY-GFP fusion protein (Holroyd et al. 2002). Incomplete release of NPY-GFP was associated with the uptake of sulforhodamine and HRP into DCVs that had apparently resealed in a mode of cavicapture. Dynamin-1 was found to be associated with some of the resealed DCVs, and a nonhydrolyzable GTP analog was shown to inhibit the sulforhodamine uptake by DCVs fusing with the plasma membrane (Holroyd et al. 2002). The conclusion from these studies, that dynamin mediated the precocious resealing of DCVs that have fused with the plasma membrane, is consistent with other studies reporting that injected dynamin PH domain, GTPγS, or dynamin-binding amphiphysin SH3 domain protein each inhibited rapid DCV endocytosis detected by capacitance (Artalejo et al. 2002), or fusion pore dilation detected by amperometry (Graham et al. 2002). Recent studies in β cells (Tsuboi et al. 2004) firmly implicated dynamin in cavicapture by demonstrating that overexpression of a GTPase dynamin mutant K44E or of a dynamin PH domain mutant K535A slowed or eliminated, respectively, fusion pore closure during cavicapture. Dynamin-1 appeared to be recruited to DCV exocytic sites whereas clathrin accessory factors (epsin, amphipysin) and clathrin were not. These data indicate that cavicapture represents a rapid, clathrin-independent, dynamin-dependent mode of DCV recapture (also see Artalejo et al. 2002). Dynamin may "sweep" across the plasma membrane in response to Ca^{2+} elevations and pinch off invaginated membranes (Tsuboi et al. 2002). Perhaps dynamin, through its PH domain, moves into regions of high $PI(4,5)P_2$ concentrations searching for membrane substrates of high curvature. Thus, DCVs that are fusing with the plasma membrane are resealed by a dynamin-based mechanism that does not represent the reversal of SNARE-mediated fusion (Holroyd et al. 2002). Future work will need to define the factors that dictate whether fusion pore dilation or dynamin-dependent resealing predominate in this apparent competition for a single DCV exocytic event.

Studies in mast cells had shown that cytoplasmic Ca^{2+} enhances fusion pore dilation (Fernandez-Chacon and Alvarez de Toledo 1995). Recent studies in PC12 cells indicate that synaptotagmins play key roles as Ca^{2+} sensors that regulate fusion pore dilation, in addition to their roles in regulating Ca^{2+}-dependent fusion probabilities. Overexpression of synaptotagmin-1 or synaptotagmin-4 in PC12 cells results in a prolongation or shortening, respectively, of the time from fusion pore opening to dilation as measured by "foot" currents in amperometric determinations of catecholamine secretion from individual DCVs (Wang et al. 2001). Subsequently, synaptotagmin-4 overexpression was found to induce an increased number of "stand-alone feet" at the expense of complete spike events (Wang et al. 2003a; see also Wang et al. 2005a). Consistent with this, in pancreatic β cells, synaptotagmin-4 overexpression slowed the release kinetics of NPY and the diffusion of synaptobrevin/VAMP into the plasma membrane (Tsuboi and Rutter 2003). Because synaptotagmin-4 contains a C2A domain with reduced Ca^{2+}-dependent phospholipid binding, these results suggest that phospholipid binding is needed for the transition from fusion pore opening to dilation. Overexpression of synaptotagmin mutants with Ca^{2+}-binding defects in C2B arrest fusion pore dilation (Wang et al. 2005a). Overall, these studies have characterized fusion pore opening and fusion pore dilation as distinct Ca^{2+}-dependent events that are differentially influenced by mutations in synaptotagmin C2A and C2B domains. These data indicate that distinct synaptotagmin interactions with its several effectors (PS, PIP_2, SNAREs) drive fusion pore opening as well as subsequent dilation.

XIII. CAVICAPTURE ALLOWS SELECTIVE RELEASE OF DCV CONSTITUENTS

DCVs contain a broad mixture of signaling molecules exhibiting a large size range from small molecules (ATP), to intermediate (peptides), to macromolecules (protein hormones, enzymes). The precise pathway for fusion pore progression (stabilization at fixed size, dilation, reclosure) varies for single DCV exocytic events and is subject to regulation. Postfusion events of fusion pore progression impart at least two additional characteristics to the secretion of molecules by DCV exocytosis. The first is to segregate the kinetics of secretion of small molecules into a faster time domain than the secretion of large molecules. For full fusion events in pancreatic β cells, the time constants for fusion (capacitance changes, ~0.02 sec), proton flux and ATP release (~0.4–0.9 sec), and peptide (IAPP-EGFP, ~1–10 sec) release differed by three orders of magnitude (Barg et al. 2002; Obermuller et al. 2005). Electrophysiological measurement of the DCV fusion pore indicate an initial size of ~2 nm (Lindau and Almers 1995). Progression of fusion pore dilation to ≥4 nm would be required for the release of all but the smallest peptide or protein constituents (Barg et al. 2002), which would impart a significant kinetic delay to the process of peptide/protein secretion. Two-photon microscopy in PC12 cells detected stable fusion pores of ~6 nm on exocytosing DCVs that persisted for >30 s (Kishimoto et al. 2005). An additional requirement of solubilizing DCV constituents from an intragranular matrix would impose added delays for constituent release (Angleson et al. 1999; Brigadski et al. 2005).

The precocious closure of a fusion pore through cavicapture would allow small molecule release but preclude peptide/protein release. The second important characteristic for DCV exocytosis is the potential for regulating small molecule and macromolecule release differentially. Stimulus strength shifts the dynamics of DCV exocytosis between cavicapture and full fusion events (Fulop et al. 2005). At low frequency stimulation of adrenal chromaffin cells, catecholamine release was subquantal and little chromogranin B was released. In contrast, at high frequency stimulation, catecholamines and chromogranin B were coreleased. Moreover, low frequency stimulation is associated with DCVs that are retrieved and capable of recycling into new exocytic events upon restimulation (Fulop et al. 2005). Overall, the ability of DCVs to undergo regulated transitions between cavicapture and full fusion based on stimulation strength provides neuroendocrine/endocrine tissues the capacity to inform hormonal target tissues about their activation status by encoding this information into differing mixtures of signaling molecules.

References

Aikawa, Y., Martin, T.F.J. (2003). ARF6 regulates a plasma membrane pool of phosphatidylinositol(4,5)bisphosphate required for regulated exocytosis. *J. Cell Biol.* **162**, 647–659.

Ales, E., Tabares, L., Poyato, J.M., Valero, V., Lindau, M., Alvarez de Toledo, G. (1999). High calcium concentrations shift the mode of exocytosis to the kiss-and-run mechanism. *Nature Cell Biol.* **1**, 40–44.

Allersma, M.W., Wang, Li, Axelrod, D., Holz, R.W. (2004). Visualization of regulated exocytosis with a granule-membrane probe using total internal reflection microscopy. *Mol. Biol. Cell* **15**, 4658–4668.

An, S.J., Almers, W. (2004). Tracking SNARE complex formation in live endocrine cells. *Science* **306**, 1042–1046.

An, S., Zenisek, D. (2004). Regulation of exocytosis in neurons and neuroendocrine cells. *Curr. Opin. Neurobiol.* **14**, 522–530.

Angleson, J.K., Cochilla, A.J., Kilic, G., Nussinovitch, I., Betz, W.J. (1999). Regulation of dense core release from neuroendocrine cells revealed by imaging single exocytic events. *Nature Neurosci.* **2**, 440–446.

Aoyagi, K., Sugaya, T., Umeda, M., Yamamoto, S., Terakawa, S., Takahashi, M. (2005). The activation of exocytotic sites by the formation of phosphatidylinositol 4,5-bisphosphate microdomains at syntaxin clusters. *J. Biol. Chem.* **280**, 17346–17352.

Artalejo, C.R., Elhamdani, A., Palfrey, H.C. (2002). Sustained stimulation shifts the mechanism of endocytosis from dynamin-1-dependent rapid endocytosis to clathrin- and dynamin-2-mediated slow endocytosis in chromaffin cells. *Proc. Natl. Acad. Sci. U S A* **99**, 6358–6363.

Bai, J., Tucker, W.C., Chapman, E.R. (2004a). PIP$_2$ increases the speed of response of synaptotagmin and steers its membrane-penetration activity toward the plasma membrane. *Nature Struct. Mol. Biol.* **11**, 36–44.

Bai, J., Wang, C.-T., Richards, D.A., Jackson, M.B., Chapman, E.R. (2004b). Fusion pore dynamics are regulated by synaptotagmin-t-SNARE interactions. *Neuron* **41**, 929–942.

Banerjee, A., Barry, V.A., DasGupta, B.R., Martin, T.F.J. (1996a). N-ethylmaleimide-sensitive factor acts as a perfusion ATP-dependent step in Ca^{2+}-activated exocytosis. *J. Biol. Chem.* **271**, 20223–20226.

Banerjee, A., Kowalchyk, J.A., DasGupta, B.R., Martin, T.F.J. (1996b). SNAP-25 is required for a late post-docking step in Ca^{2+}-dependent exocytosis. *J. Biol. Chem.* **271**, 20227–20230.

Barg, S., Huang, P., Eliasson, L., Nelson, D.J., Obermuller, S., Rorsman, P. et al. (2001). Priming of insulin granules for exocytosis by granular Cl$^-$ uptake and acidification. *J. Cell Sci.* **114**, 2145–2154.

Barg, S., Olofsson, C.S., Schriever-Abeln, Wendt, A., Gebre-Medhin, S., Renstrom, E., Rorsman, P. (2002). Delay between fusion pore opening and peptide release from large dense-core vesicles in neuroendocrine cells. *Neuron* **33**, 287–299.

Barylko, B., Gerber, S.H., Binns, D.D., Grichine, N., Khvotchev, M., Sudhof, T.C., Albanesi, J.P. (2001). A novel family of phosphatidylinositol 4-kinases conserved from yeast to humans. *J. Biol. Chem.* **276**, 7705–7708.

Bauer, R.A., Overlease, R.L., Lieber, J.L., Angleson, J.K. (2004). Retention and stimulus-dependent recycling of dense core vesicle content in neuroendocrine cells. *J. Cell Sci.* **117**, 2193–2202.

Bayer, M.J., Reese, C., Buhler, S., Peters, C., Mayer, A. (2003). Vacuole membrane fusion: Vo functions after trans-SNARE pairing and is coupled to the Ca^{2+}-releasing channel. *J. Cell Biol.* **162**, 211–222.

Becherer, U., Moser, T., Stuhmer, W., Oheim, M. (2003). Calcium regulates exocytosis at the level of single vesicles. *Nature Neurosci.* **6**, 846–853.

Beuret, N., Stettler, H., Renold, A., Rutishauser, J., Spiess, M. (2004). Expression of regulated secretory proteins is sufficient to generate granule-like structures in constitutively secreting cells. *J. Biol. Chem.* **279**, 20242–20249.

Borisovska, M., Zhao, Y., Tsytsyura, Y., Glyvuk, N., Takamori, S., Matti, U. et al. (2005). v-SNAREs control exocytosis of vesicles from priming to fusion. *EMBO J.* **24**, 2114–2126.

Boyd, C., Hughes, T., Pypaert, M., Novick, P. (2004). Vesicles carry most exocyst subunits to exocytic sites marked by the remaining two subunits, Sec3p and Exo70p. *J. Cell Biol.* **167**, 889–901.

Brigadski, T., Hartmann, M., Lessmann, V. (2005). Differential vesicular targeting and time course of synaptic secretion of the mammalian neurotrophins. *J. Neurosci.* **25**, 7601–7614.

Brose, N., Rosenmund, C., Rettig, J. (2000). Regulation of transmitter release by Unc-13 and its homologues. *Curr. Opin. Neurobiol.* **10**, 303–311.

Brown, H., Meister, B., Deeney, J., Corkey, B.E., Yang, S.-N., Larsson, O. et al. (2000). Synaptotagmin III isoform is compartmentalized in pancreatic β cells and has a functional role in exocytosis. *Diabetes* **49**, 383–391.

Chapman, E.R. (2002). Synaptotagmin: A Ca^{2+} sensor that triggers exocytosis? *Nature Rev. Mol. Cell Biol.* **3**, 1–11.

Chen, Y.A., Scales, S.J., Patel, S.M., Doung, Y.-C., Scheller, R.H. (1999). SNARE complex formation is triggered by Ca^{2+} and drives membrane fusion. *Cell* **97**, 165–174.

Chen, Y.A., Scales, S.J., Scheller, R.H. (2001). Sequential SNARE assembly underlies priming and triggering of exocytosis. *Neuron* **30**, 161–170.

Chen, Y.A., Scheller, R.H. (2001). SNARE-mediated membrane fusion. *Nature Rev. Mol. Cell Biol.* **2**, 98–106.

Cheviet, S., Waselle, L., Regazzi, R. (2004). Noc-king out exocrine and endocrine secretion. *Trends Cell Biol.* **14**, 525–528.

Chung, S.H., Song, W.J., Kim, K., Bednarski, J.J., Chen, J., Prestwich, G.D., Holz, R.W. (1998). The C2 domains of rabphilin 3A specifically bind phosphatidylinositol 4,5-bisphosphate containing vesicles in a Ca2+-dependent manner. *J. Biol. Chem.* **273**, 10240–10248.

Ciufo, L.F., Barclay, J.W., Burgoyne, R.D., Morgan, A. (2005). Munc18-1 regulates early and late stages of exocytosis via syntaxin-independent protein interactions. *Mol. Biol. Cell* **16**, 470–482.

Craxton, M. (2004). Synaptotagmin gene content of the sequenced genomes. *BMC Genomics* **5**, 43–56.

Dannies, P.S. (1999). Protein hormone storage in secretory granules: Mechanisms for concentration and sorting. *Endocr. Rev.* **20**, 3–21.

Day, R., Gorr, S.-U. (2003). Secretory granule biogenesis and chromogranin A: Master gene, on/off switch or assembly factor? *Trends Endocrinol. Metab.* **14**, 10–13.

De Matteis, M.A., Godi, A. (2004). PI-loting membrane traffic. *Nature Cell Biol.* **6**, 487–492.

Desnos, C., Schonn, J.-S., Huet, S., Tran, V.S., El-Amraoui, A., Raposo, G. et al. (2003). Rab27A and its effector MyRIP link secretory granules to F-actin and control their motion toward release sites. *J. Cell Biol.* **163**, 559–570.

Duncan, R.R., Greaves, J., Wiegand, U.K., Matskevich, I., Bodammer, G., Apps, D.K. et al. (2003). Functional and spatial segregation of secretory vesicle pools according to vesicle age. *Nature* **422**, 176–180.

Eaton, B.A., Haugwitz, M., Lau, D., Moore, H.-P. (2000). Biogenesis of regulated exocytic carriers in neuroendocrine cells. *J. Neurosci.* **20**, 7334–7344.

Eberhard, D.A., Cooper, C.L., Low, M.G., Holz, R.W. (1990). Evidence that the inositol phospholipids are necessary for exocytosis. *Biochem. J.* **268**, 15–25.

Eberhard, D.A., Holz, R.W. (1991). Calcium promotes the accumulation of polyphosphoinositides in intact and permeabilized bovine adrenal chromaffin cells. *Cell. Mol. Neurobiol.* **11**, 357–370.

El-Amraoui, A., Schonn, J.S., Kussel-Andermann, P., Blanchard, S., Desnos, C., Henry, J.P. et al. (2002). MyRIP, a novel Rab effector, enables myosin VIIa recruitment to retinal melanosomes. *EMBO Rep.* **3**, 463–470.

Fernandez-Chacon, R., Alvarez de Toledo, G. (1995). Cytosolic calcium facilitates release of secretory products after exocytotic vesicle fusion. *FEBS Lett.* **363**, 221–225.

Fernandez-Chacon, R., Konigstorfer, A., Gerber, S.H., Garcia, J., Matos, M.F., Stevens, C.F. et al. (2001). Synaptotagmin I functions as a calcium regulator of release probability. *Nature* **410**, 41–49.

Fukuda, M., Kuroda, T.S. (2002). Slac2-c (synaptotagmin-like protein homologue lacking C2 domains-c), a novel linker protein that interacts with Rab27, myosin Va/VIIa, and actin. *J. Biol. Chem.* **277**, 39673–39678.

Fukuda, M., Kowalchyk, J.A., Zhang, X., Martin, T.F.J., Mikoshiba, K. (2002a). Synaptotagmin IX regulates Ca^{2+}-dependent secretion in PC12 cells. *J. Biol. Chem.* **277**, 4601–4604.

Fukuda, M., Kanno, E., Saegusa, C., Ogata, Y., Kuroda, T.S. (2002b). Slp4-a/granuphilin regulates dense-core vesicle exocytosis in PC12 cells. *J. Biol. Chem.* **277**, 39673–39678.

Fukuda, M., Kuroda, T.S., Mikoshiba, K. (2002c). Slac2a/melanophilin, the missing link between Rab27 and myosin Va: Implications of a tripartite protein complex for melanosome transport. *J. Biol. Chem.* **277**, 12432–12436.

Fukuda, M. (2003a). Distinct Rab binding specificity of Rim1, Rim2, rabphilin, and Noc2. Identification of a critical determinant of Rab3A/Rab27A recognition by Rim2. *J. Biol. Chem.* **278**, 15373–15380.

Fukuda, M. (2003b). Slp4-a/granuphilin-a inhibits dense-core vesicle exocytosis through interaction with the GDP-bound form of Rab27A in PC12 cells. *J. Biol. Chem.* **278**, 15390–15396.

Fukuda, M., Kanno, E., Satoh, M. Saegusa, C., Yamamoto, A. (2004). Synaptotagmin VII is targeted to dense-core vesicles and regulates their Ca^{2+}-dependent exocytosis in PC12 cells. *J. Biol. Chem.* **279**, 52677–52684.

Fukuda, M. (2005). Versatile role of Rab27 in membrane trafficking: Focus on the Rab27 effector families. *J. Biochem.* **137**, 9–16.

Fulop, T., Radabaugh, S., Smith, C. (2005). Activity-dependent differential transmitter release in mouse adrenal chromaffin cells. *J. Neurosci.* **25**, 7324–7332.

Geppert, M., Goda, Y., Hammer, R.E., Li, C., Rosahl, T.W., Stevens, C.F., Sudhof, T.C. (1994). Synaptotagmin I: A major Ca^{2+} sensor for transmitter release at a central synapse. *Cell* **79**, 717–727.

Gerona, R.R.L., Larsen, E.C., Kowalchyk, J.A., Martin, T.F.J. (2000). The C terminus of SNAP25 is essential for Ca^{2+}-dependent binding of synaptotagmin to SNARE complexes. *J. Biol. Chem.* **275**, 6328–6336.

Gillingham, A.K., Munro, S. (2003). Long coiled-coil proteins and membrane traffic. *Biochim. Biophys. Acta* **1641**, 71–85.

Gomi, H., Mizutani, S., Kasai, K., Itohara, S., Izumi, T. (2005). Granuphilin molecularly docks insulin granules to the fusion machinery. *J. Cell Biol.* **171**, 99–109.

Gong, L.W., DiPaolo, G., Diaz, E., Cestra, G., Diaz, M.E., Lindau, M. et al. (2005). Phosphatidylinositol phosphate kinase type I gamma regulates dynamics of large dense-core vesicle fusion. *Proc. Natl. Acad. Sci. USA* **102**, 5204–5209.

Graham, M.E., O'Callaghan, D.W., McMahon, H.T., Burgoyne, R.D. (2002). Dynamin-dependent and dynamin-independent processes contribute to the regulation of single vesicle release kinetics and quantal size. *Proc. Natl. Acad. Sci. U S A* **99**, 7124–7129.

Grishanin, R.N., Klenchin, V.A., Loyet, K.M., Kowalchyk, J.A., Ann, K.-S., Martin, T.F.J. (2002). Membrane association domains in CAPS mediate plasma membrane and dense-core vesicle binding required for Ca^{2+}-dependent exocytosis. *J. Biol. Chem.* **277**, 22025–22034.

Grishanin, R.N., Kowalchyk, J.A., Klenchin, V.A., Ann, K.-S., Earles, C.A., Chapman, E.R. et al. (2004). CAPS acts at a perfusion step in dense-core vesicle exocytosis as PIP_2 binding protein. *Neuron* **43**, 551–562.

Guo, W., Sacher, M., Barrowman, J., Ferro-Novick, S., Novick, P. (2000). Protein complexes in transport vesicle targeting. *Trends Cell Biol.* **10**, 251–255.

Gut, A., Kiraly, C.E., Fukuda, M., Mikoshiba, K., Wollheim, C.B., Lang, J. (2001). Expression and localization of synaptotagmin isoforms in endocrine β cells: their function in insulin exocytosis. *J. Cell Sci.* **114**, 1709–1716.

Han, X., Wang, C.-T., Bai, J., Chapman, E.R., Jackson, M.B. (2004). Transmembrane segments of syntaxin line the fusion pore of Ca^{2+}-triggered exocytosis. *Science* **304**, 289–292.

Hay, J.C., Martin, T.F.J. (1992). Resolution of regulated secretion into sequential MgATP-dependent and calcium-dependent stages mediated by distinct cytosolic proteins. *J. Cell Biol.* **119**, 139–151.

Hay, J.C., Martin, T.F.J. (1993). Phosphatidylinositol transfer protein required for ATP-dependent priming of Ca^{2+}-activated secretion. *Nature* **366**, 572–575.

Hay, J.C., Fisette, P.L., Jenkins, G.H., Fukami, K., Takenawa, T., Anderson, R.A., Martin, T.F.J. (1995). ATP-dependent inositide phosphorylation required for Ca^{2+}-activated secretion. *Nature* **374**, 173–177.

Hiesinger, P.R., Fayyazuddin, A., Mehta, S.Q., Rosenmund, T., Schulze, K.L., Zhai, R.G. et al. (2005). The v-ATPase Vo subunit a1 is required for a late step in synaptic vesicle exocytosis in *Drosophila*. *Cell* **121**, 607–620.

Hinners, I., Wendler, F., Fei, H., Thomas, L., Thomas, G., Tooze, S.A. (2003). AP-1 recruitment to VAMP4 is modulated by phosphorylation-dependent binding of PACS-1. *EMBO Rep.* **4**, 1182–1189.

Holz, R.W., Bittner, M.A., Peppers, S.C., Senter, R.A., Eberhard, D.A. (1989). MgATP-independent and MgATP-dependent exocytosis. *J. Biol. Chem.* **264**, 5412–5419.

Holz, R.W., Hlubek, M.D., Sorensen, S.D., Fisher, S.K., Balla, T., Ozaki, S. et al. (2000). A pleckstrin homology domain specific for $PtdIns-4,5-P_2$ and fused to green fluorescent protein identifies plasma membrane $PtdIns-4,5-P_2$ as being important in exocytosis. *J. Biol. Chem.* **275**, 17878–17885.

Holroyd, P., Lang, T., Wenzel, D., De Camilli. P., Jahn. R. (2002). Imaging direct, dynamin-dependent recapture of fusing secretory granules on plasma membrane lawns from PC12 cells. *Proc. Natl. Acad. Sci. U S A* **99**, 16806–16811.

Honda, A., Nogami, M., Yokozeki, T., Yamazaki, M., Nakamura, H., Watanabe, H. et al. (1999). Phosphatidylinositol 4-phosphate 5-kinase alpha is a downstream effector of the small G protein ARF6 in membrane ruffle formation. *Cell* **99**, 521–532.

Hua, Y., Scheller, R.H. (2001). Three SNARE complexes cooperate to mediate membrane fusion. *Proc. Natl. Acad. Sci. USA* **98**, 8065–8070.

Huh, Y.H., Jeon, S.H., Yoo, S.H. (2003). Chromogranin B-induced secretory granule biogenesis: comparison with the similar role of chromogranin A. *J. Biol. Chem.* **278**, 40581–40589.

Iezzi, M., Kouri, G., Fukuda, M., Wollheim, C.B. (2004) Synaptotagmin V and IX isoforms control Ca^{2+}-dependent insulin exocytosis. *J. Cell Sci.* **117**, 3119–3127.

Ivarsson, R., Jing, X., Waselle, L., Regazzi, R., Renstrom, R. (2005). Myosin 5a controls insulin granule recruitment during late-phase secretion. *Traffic* **6**, 1027–1035.

Izumi, T., Gomi, H., Kasai, K., Mizutani, S., Torii, S. (2003). The roles of Rab27 and its effectors in the regulated secretory pathways. *Cell Struct. Funct.* **28**, 465–474.

Jahn, R., Lang, T., Sudhof, T.C. (2003). Membrane fusion. *Cell* **112**, 519–533.

Johns, L.M., Levitan, E.S., Shelden, E.A., Holz, R.W., Axelrod, D. (2001). Restriction of secretory granule motion near the plasma membrane of chromaffin cells. *J. Cell Biol.* **153**, 177–190.

Kim, T., Tao-Cheng, J.-H., Eiden, L.E., Loh, Y.P. (2001). Chromogranin A, an "on/off" switch controlling dense-core secretory granule biogenesis. *Cell* **106**, 499–509.

Kishimoto, T., Liu, T.-T., Hatakeyama, H., Nemoto, T., Takahashi, N., Kasai, H. (2005). Sequential compound exocytosis of large dense-core vesicles in PC12 cells studied with TEPIQ (two-photon extracellular polar-tracer imaging-based quantification) analysis. *J. Physiol.* **568**, 905–915.

Klyachko, V.A., Jackson, M.B. (2002). Capacitance steps and fusion pores of small and large-dense-core vesicles in nerve terminals. *Nature* **418**, 89–92.

Korteweg, N., Maia, A.S., Thompson, B., Roubos, E.W., Burbach, J.P., Verhage, M. (2005). The role of Munc18-1 in docking and exocytosis of peptide hormone vesicles in the anterior pituitary. *Biol. Cell.* **97**, 445–455.

Lang, T., Wacker, I., Wunderlich, I., Rohrbach, A., Giese, G., Soldati, T., Almers, W. (2000). Role of actin cortex in the subplasmalemmal transport of secretory granules in PC12 cells. *Biophys. J.* **78**, 2863–2877.

Lang, T., Margittai, M., Holzler, H., Jahn, R. (2002). SNAREs in native plasma membranes are active and readily form core complexes with endogenous and exogenous SNAREs. *J. Cell Biol.* **158**, 751–760.

Li, D., Xiong, J., Qu, A., Xu, T. (2004). Three-dimensional tracking of single secretory granules in live PC12 cells. *Biophys. J.* **87**, 1991–2001.

Lindau, M., Almers, W. (1995). Structure and function of fusion pores in exocytosis and ectoplasmic membrane fusion. *Curr. Opin. Cell Biol.* **7**, 509–517.

Lipschutz, J.H., Mostov, K.E. (2002). Exocyst: The many masters of the exocyst. *Curr. Biol.* **12**, R212–R214.

Liu, L., Liao, H., Castle, A., Zhang, J., Casanova, J., Szabo, G., Castle, D. (2005). SCAMP2 interacts with ARF6 and phospholipase D1 and links their function to exocytic fusion pore formation in PC12 cells. *Mol. Biol. Cell* **16**, 4463–4472.

Ludwig, M., Pittman, Q.J. (2003). Talking back: Dendritic neurotransmitter release. *Trends Neurosci.* **26**, 255–261.

Martin, T.F.J. (1998). Phosphoinositide lipids as signaling molecules: Common themes for signal transduction, cytoskeletal regulation, and membrane trafficking. *Annu. Rev. Cell Dev. Biol.* **14**, 231–264.

Martin, T.F.J. (2003). Tuning exocytosis for speed: Fast and slow modes. *Biochim. Biophys. Acta* **1641**, 157–165.

Martin, T.F.J., Kowalchyk, J.A. (1997). Docked secretory vesicles undergo Ca^{2+}-activated exocytosis in a cell-free system. *J. Biol. Chem.* **272**, 14447–14453.

Martin, T.F.J., Loyet, K.M., Barry, V.A., Kowalchyk, J.A. (1997). The role of PtdIns(4,5)P2 in exocytotic membrane fusion. *Biochem. Soc. Trans.* **25**, 1137–1141.

Marqueze, B., Berton, F., Seagar, M. (2000). Synaptotagmins in membrane traffic: Which vesicles do the tagmins tag? *Biochimie* **82**, 409–420.

Maximov, A., Sudhof, T.C. (2005). Autonomous function of synaptotagmin I in triggering synchronous release independent of asynchronous release. *Neuron* **48**, 547–554.

Meunier, F.A., Osborne, S.L., Hammond, G.R.V., Cooke, F.T., Parker, P.J., Domin, J., Schiavo, G. (2005). Phosphatidylinositol 3-kinase C2α is essential for ATP-dependent priming of neurosecretory granule exocytosis. *Mol. Biol. Cell* **16**, 4841–4851.

Milosevic, I., Sorensen, J.B., Lang, T., Krauss, M., Nagy, G., Haucke, V. et al. (2005). Plasmalemmal phosphatidylinositol-4,5-bisphosphate level regulates the releasable vesicle pool in chromaffin cells. *J. Neurosci.* **25**, 2557–2565.

Mizuta, M., Kurose, T., Miki, T., Shoji-Kasai, Y., Takahashi, M., Seino, S., Matsukura, S. (1997). Localization and functional role of synaptotagmin III in insulin secretory vesicles in pancreatic β cells. *Diabetes* **46**, 2002–2006.

Moore, H.-P., Andresen, J.M., Eaton, B.A., Grabe, M., Haugwitz, M., Wu, M.M., Machen, T.E. (2002). Biosynthesis and secretion of pituitary hormones: dynamics and regulation. *Archiv. Physiol. Biochem.* **110**, 16–25.

Moskalenko, S., Henry, D.O., Rosse, C., Mirey, G., Camonis, J.H., White, M.A. (2001). The exocyst is a Ral effector complex. *Nature Cell Biol.* **4**, 66–72.

Nakata, T., Sobue, K., Hirokawa, N. (1990). Conformational change and localization of calpactin I complex involved in exocytosis as revealed by quick-freeze, deep-etch electron microscopy and immunocytochemistry. *J. Cell Biol.* **110**, 13–25.

Neco, P., Giner, D., Viniegra, S., Borges, R., Villarroel, A., Gutierrez, L.M. (2004). New roles of myosin II during vesicle transport and fusion in chromaffin cells. *J. Biol. Chem.* **279**, 27450–27457.

Ng, Y.-K., Lu, X., Levitan, E.S. (2002). Physical mobilization of secretory vesicles facilitates neuropeptide release by nerve growth factor-differentiated PC12 cells. *J. Physiol.* **542**, 395–402.

Ninomiya, Y., Kishimoto, T., Yamazawa, T., Ikeda, H., Miyashita, Y., Kasai, H. (1997). Kinetic diversity in the fusion of exocytic vesicles. *EMBO J.* **16**, 929–934.

Nishiki, T.-I., Augustine, G.J. (2004). Synaptotagmin I synchronizes transmitter release in mouse hippocampal neurons. *J. Neurosci.* **24**, 6127–6132.

Obermuller, S., Lindqvist, A., Karanauskaite, J., Galvanovskis, J., Rorsman, P., Barg, S. (2005). Selective nucleotide-release from dense-core granules in insulin-secreting cells. *J. Cell Sci.* **118**, 4271–4282.

Oheim, M., Loerke, D., Stuhmer, W., Chow, R.H. (1998). The last few milliseconds in the life of a secretory granule. *Eur. Biophys. J.* **27**, 83–98.

Oheim, M., Loerke, D., Stuhmer, W., Chow, R.H. (1999). Multiple stimulation-dependent processes regulate the size of the releasable pool of vesicles. *Eur. Biophys. J.* **28**, 91–101.

Oheim, M., Stuhmer, W. (2000). Tracking chromaffin granules on their way through the actin cortex. *Eur. Biophys. J.* **29**, 67–89.

Olsen, H.L., Hoy, M., Zhang, W., Bertorello, A.M., Bokvist, K., Capito, K. et al. (2003). Phosphatidylinositol 4-kinase serves as a metabolic sensor and regulates priming of secretory granules in pancreatic β cells. *Proc. Natl. Acad. Sci. USA* **100**, 5187–5192.

Palfrey, H.C., Artalejo, C.R. (2003). Secretion: kiss and run caught on film. *Curr. Biol.* **13**, R397–R399.

Perrais, D., Kleppe, I.C., Taraska, J.W., Almers, W. (2004). Recapture after exocytosis causes differential retention of protein in granules of bovine chromaffin cells. *J. Physiol.* **560**, 413–428.

Pfeffer, S.R. (1999). Transport vesicle targeting: Tethers before SNAREs. *Nature Cell Biol.* **1**, E17–E22.

Plattner, H., Artalejo, A.R., Neher, E. (1997). Ultrastructural organization of bovine chromaffin cell cortex-analysis by cryofixation and morphometry of aspects pertinent to exocytosis. *J. Cell Biol.* **139**, 1709–1717.

Rettig, J., Neher, E. (2002). Emerging roles of presynaptic proteins in Ca^{2+}-triggered exocytosis. *Science* **298**, 781–785.

Rizo, J., Sudhof, T.C. (2002). SNAREs and Munc18 in synaptic vesicle fusion. *Nature Rev. Neurosci.* **3**, 641–653.

Rizzoli, S.O., Betz, W.J. (2005). Synaptic vesicle pools. *Nature Rev. Neurosci.* **6**, 57–69.

Rose, S.D., Lejen, T., Casaletti, L., Larson, R.E., Pene, T.D., and Trifaro, J.M. (2003). Myosins II and V in chromaffin cells: Myosin V is a chromaffin vesicle molecular motor involved in secretion. *J. Neurochem.* **85**, 287–298.

Rudolf, R., Salm, T., Rustom, A., Gerdes, H.-H. (2001). Dynamics of immature secretory granules: Role of cytoskeletal elements during transport, cortical restriction, and F-actin-dependent tethering. *Mol. Biol. Cell* **12**, 1353–1365.

Rudolf, R., Kogel, T., Kuznetsov, S.A., Salm, T., Schlicker, O., Hellwig, A. et al. (2003). Myosin Va facilitates the distribution of secretory granules in the F-actin rich cortex of PC12 cells. *J. Cell Sci.* **116**, 1339–1348.

Ryan, T.A. (2003). Kiss-and-run, fuse-pinch-and-linger, fuse-and-collapse: The life and times of a neurosecretory granule. *Proc. Natl. Acad. Sci. USA* **100**, 2171–2173.

Schluter, O.M., Schnell, E., Verhage, M., Tzonopoulos, T., Nicoll, R.A., Janz, R. et al. (1999). Rabphilin knock-out mice reveal that rabphilin is not required for Rab3 function in regulating neurotransmitter release. *J. Neurosci.* **19**, 5834–5846.

Schutz, D., Zilly, F., Lang, T., Jahn, R., Bruns, D. (2005). A dual function for Munc-18 in exocytosis of PC12 cells. *Eur. J. Neurosci.* **21**, 2419–2432.

Shoji-Kasai, Y., Yoshida, A., Sato, K., Hoshino, T., Ogura, A., Kondo, S. et al. (1992). Neurotransmitter release from synaptotagmin-deficient clonal variants of PC12 cells. *Science* **256**, 1821–1823.

Sorensen, J.B., Nagy, G., Varoqueaux, F., Nehring, R.B., Brose, N., Wilson, M.C., Neher, E. (2003a). Differential control of the releasable vesicle pools by SNAP-25 splice variants and SNAP-23. *Cell* **114**, 75–86.

Sorensen, J.B., Fernandez-Chacon, R., Sudhof, T.C., Neher, E. (2003b). Examining synaptotagmin 1 function in dense core vesicle exocytosis under direct control of Ca^{2+}. *J. Gen. Physiol.* **122**, 265–276.

Steyer, J.A., Horstmann, H., Almers, W. (1997). Transport, docking and exocytosis of single secretory granules in live chromaffin cells. *Nature* **388**, 474–478.

Steyer, J.A., Almers, W. (1999). Tracking single secretory granules in live chromaffin cells by evanescent-field fluorescence microscopy. *Biophys. J.* **76**, 2262–2271.

Strom, M., Hume, A.N., Tarafder, A.K., Barkagianni, E., Seabra, M.C. (2002). A family of Rab27-binding proteins: Melanophilin links Rab27a and myosin Va function in melanosome transport. *J. Biol. Chem.* **277**, 25423–25430.

Sudhof, T.C. (2004). The synaptic vesicle cycle. *Annu. Rev. Neurosci.* **27**, 509–547.

Sztul, E., Lupashin, V. (2006). Role of tethering factors in secretory membrane traffic. *Am. J. Physiol. Cell Physiol.* **290**, C11–C26.

Taraska, J.W., Perrais, D., Ohara-Imaizumi, M., Nagamatsu, S., Almers, W. (2003). Secretory granules are recaptured largely intact after stimulated exocytosis in cultured endocrine cells. *Proc. Natl. Acad. Sci. USA* **100**, 2070–2075.

Taraska, J.W., Almers, W. (2004). Bilayers merge even when exocytosis is transient. *Proc. Natl. Acad. Sci. USA* **101**, 8780–8785.

Thomas, G. (2002). Furin at the cutting edge: From protein traffic to embryogenesis and disease. *Nature Rev. Mol. Cell Biol.* **3**, 753–766.

Tolmachova, T., Anders, R., Stinchcombe, J., Bossi, G., Griffiths, G.M., Huxley, C., Seabra, M.C. (2004). A general role for Rab27a in secretory cells. *Mol. Biol. Cell* **15**, 332–344.

Tooze, S.A., Martens, G.J.M., Huttner, W.B. (2001). Secretory granule biogenesis: Rafting to the SNARE. *Trends Cell Biol.* **11**, 116–122.

Torii, S., Zhao, S., Yi, Z., Takeuchi, T., Izumi, T. (2002). Granuphilin modulates the exocytosis of secretory granules through interaction with syntaxin 1a. *Mol. Cell. Biol.* **22**, 5518–5526.

Torii, S., Takeuchi, T., Nagamatsu, S., Izumi, T. (2004). Rab27 effector granuphilin promotes the plasma membrane targeting of insulin granules via interaction with syntaxin 1a. *J. Biol. Chem.* **279**, 22532–22538.

Tsuboi, T., Zhao, C., Terakawa, S., Rutter, G.A. (2000). Simultaneous evanescent wave imaging of insulin vesicle membrane and cargo during a single exocytic event. *Curr. Biol.* **10**, 1307–1310.

Tsuboi, T., Terakawa, S., Scalettar, B.A., Fantus, C., Roder, J., Jeromin, A. (2002). Sweeping model of dynamin activity. Visualization of coupling between exocytosis and endocytosis under an evanescent wave microscope with green fluorescent proteins. *J. Biol. Chem.* **277**, 15957–15961.

Tsuboi, T., Rutter, G.A. (2003). Multiple forms of kiss-and-run exocytosis revealed by evanescent wave microscopy. *Curr. Biol.* **13**, 563–567.

Tsuboi, T., McMahon, H.T., Rutter, G.A. (2004). Mechanisms of dense core vesicle recapture following kiss and run (cavicapture) exocytosis in insulin-secreting cells. *J. Biol. Chem.* **279**, 47115–47124.

Tsuboi, T., Fukuda, M. (2005). The C2B domain of rabphilin directly interacts with SNAP-25 and regulates the docking step of dense core vesicle exocytosis in PC12 cells. *J. Biol. Chem.* **280**, 39253–39259.

Tsuboi, T., Ravier, M.A., Xie, H., Ewart, M.-A., Gould, G.W., Baldwin, S.A., Rutter, G.A. (2005). Mammalian exocyst complex is required for the docking step of insulin vesicle exocytosis. *J. Biol. Chem.* **280**, 25565–25570.

Tucker, W.C., Chapman, E.R. (2002). Role of synaptotagmin in Ca^{2+}-triggered exocytosis. *Biochem. J.* **366**, 1–13.

Tucker, W.C., Edwardson, J.M., Bai, J., Kim, H.-J., Martin, T.F.J., Chapman, E.R. (2003). Identification of synaptotagmin effectors via acute inhibition of secretion from cracked PC12 cells. *J. Cell Biol.* **162**, 199–209.

Varadi, A., Tsuboi, T., Johnson-Cadwell, L.I., Allan, V.J., Rutter, G.A. (2003). Kinesin 1 and cytoplasmic dynein orchestrate glucose-stimulated insulin-containing vesicle movements in clonal MIN6 β cells. *Biochem. Biophys. Res. Commun.* **311**, 272–282.

Varadi, A., Tsuboi, T., Rutter, G.A. (2005). Myosin Va transports dense core secretory vesicles in pancreatic MIN6 β cells. *Mol. Biol. Cell* **16**, 2670–2680.

Verhage, M., Maia, A.S., Plomp, J.J., Brussard, A.B., Heeroma, J.H., Vermeer, H. et al. (2000). Synaptic assembly of the brain in the absence of neurotransmitter secretion. *Science* **287**, 864–869.

Voets, T., Moser, T., Lund, P.-E., Chow, R.H., Geppert, M., Sudhof, T.C., Neher, E. (2001a). Intracellular calcium dependence of large dense-core vesicle exocytosis in the absence of synaptotagmin I. *Proc. Natl. Acad. Sci. USA* **98**, 11680–11685.

Voets, T., Toonen, R.F., Brian, E.C., de Wit, H., Moser, T., Rettig, J. et al. (2001b). Munc18-1 promotes large dense-core vesicle docking. *Neuron* **31**, 581–591.

Wacker, I., Kaether, C., Kromer, A., Migala, A., Almers, W., Gerdes, H.-H. (1997). Microtubule-dependent transport of secretory vesicles visualized in real time

with a GFP-tagged secretory protein. *J. Cell Sci.* **110**, 1453–1463.

Waites, C.L., Mehta, A., Tan, P.K., Thomas, G., Edwards, R.H., Krantz, D.E. (2001). An acidic motif retains vesicular monoamine transporter 2 on large dense core vesicles. *J. Cell Biol.* **152**, 1159–1168.

Wang, C.-T., Grishanin, R., Earles, C.A., Chang, P.Y., Martin, T.F.J., Chapman, E.R., Jackson, M.B. (2001). Synaptotagmin modulation of fusion pore kinetics in regulated exocytosis of dense-core vesicles. *Science* **294**, 1111–1115.

Wang, C.T., Lu, J.C., Bai, J., Chang, P.Y., Martin, T.F., Chapman, E.R., Jackson, M.B. (2003a). Different domains of synaptotagmin control the choice between kiss-and-run and full fusion. *Nature* **424**, 943–947.

Wang, C.T., Bai, J., Chang, P.Y., Chapman, E.R., Jackson, M.B. (2005a). Synaptotagmin Ca^{2+} triggers two sequential steps in regulated exocytosis in rat PC12 cells: Fusion pore opening and fusion pore dilation. *J Physiol.* Nov 17 (epub ahead of print).

Wang, J., Takeuchi, T., Yokota, H., Izumi, T. (1999). Novel rabphilin-3-like protein associates with insulin-containing granules in pancreatic β cells. *J. Biol. Chem.* **274**, 28542–28548.

Wang, L., Li, G., Sugita, S. (2004). RalA-Exocyst interaction mediates GTP-dependent exocytosis. *J. Biol. Chem.* **279**, 19875–19881.

Wang, L., Li, G., Sugita, S. (2005b). A central kinase domain of Type I phosphatidylinositol phosphate kinases is sufficient to prime exocytosis. *J. Biol. Chem.* **280**, 16522–16527.

Wang, P., Wang, C.-T., Bai, J., Jackson, M.B., Chapman, E.R. (2003b). Mutations in the effector binding loops in the C2A and C2B domains of synaptotagmin I disrupt exocytosis in a nonadditive manner. *J. Biol. Chem.* **278**, 47030–47037.

Wang, P., Chicka, M.C., Bhalla, A., Richards, D.A., Chapman, E.R. (2005c). Synaptotagmin VII is targeted to secretory organelles in PC12 cells, where it functions as a high-affinity calcium sensor. *Mol. Cell. Biol.* **25**, 8693–8702.

Waselle, L., Coppola, T., Fukuda, M., Iezzi, M., El-Amraoui, A., Petit, C., Regazzi, R. (2003). Involvement of the Rab27 binding protein Slac2c/MyRIP in insulin exocytosis. *Mol. Biol. Cell* **14**, 4103–4113.

Wendler, F., Page, L., Urbe, S., Tooze, S.A. (2001). Homotypic fusion of immature secretory granules during maturation requires syntaxin 6. *Mol. Biol. Cell* **12**, 1699–1709.

Xu, T., Binz, T., Niemann, H., Neher, E. (1998). Multiple kinetic components of exocytosis distinguished by neurotoxin sensitivity. *Nature Neurosci.* **1**, 192–200.

Xu, T., Ashery, U., Burgoyne, R.D., Neher, E. (1999a). Early requirement for a-SNAP and NSF in the secretory cascade in chromaffin cells. *EMBO J.* **18**, 3293–3304.

Xu, T., Rammner, B., Margittai, M., Artalejo, A.R., Neher, E., Jahn, R. (1999b). Inhibition of SNARE complex assembly differentially affects kinetic components of exocytosis. *Cell* **99**, 713–722.

Yi, Z., Yokota, H., Torii, S., Aoki, T., Hosaka, M., Zhao, S. et al. (2002). The Rab27a/granuphilin complex regulates the exocytosis of insulin-containing dense-core granules. *Mol. Cell. Biol.* **22**, 1858–1867.

Yoshihara, M., Littleton, J.T. (2002). Synaptotagmin I functions as a calcium sensor to synchronize neurotransmitter release. *Neuron* **36**, 897–908.

Yoshihara, M., Adolfsen, B., Littleton, J.T. (2003). Is synaptotagmin the calcium sensor? *Curr. Opin. Neurobiol.* **13**, 315–323.

Zerial, M., McBride, H. (2001). Rab proteins as membrane organizers. *Nature Rev. Mol. Cell. Biol.* **2**, 107–117.

Zhang, X., Kim-Miller, M.J., Fukuda, M., Kowalchyk, J.A., Martin, T.F.J. (2002). Ca^{2+}-dependent synaptotagmin binding to SNAP-25 is essential for Ca^{2+}-triggered exocytosis. *Neuron* **34**, 599–611.

15

Exocytic Release of Glutamate from Astrocytes: Comparison to Neurons

WILLIAM LEE AND VLADIMIR PARPURA

Astrocytes, a subtype of glial cell, have long been neglected as active participants in intercellular communication and information processing in the central nervous system, in part due to their lack of electrical excitability. However, astrocytes exhibit a form of excitability based on intracellular Ca^{2+} elevations, which can stimulate glutamate release from astrocytes. The mechanism underlying this release is exocytosis. Indeed, astrocytes express protein components of exocytic secretory machinery, including the core fusion complex as well as transporters and pumps necessary for filling astrocytic vesicles with glutamate.

The characteristics of exocytosis in astrocytes are distinct from those observed in neurons due to differences in expression of subtypes of exocytic protein subtypes. The morphological arrangements of exocytic secretory machinery and functional neurotransmitter receptors in astrocytic processes enable them to receive signals, focally, from adjacent synaptic terminals and respond back to terminals/dendrites via exocytic glutamate release. This bidirectional astrocyte-neuron signaling functionally and morphologically forms a tripartite synapse that includes the neuronal pre- and postsynaptic elements as well as surrounding astrocytic elements. Thus, astrocytes have the potential to play an active role in the computational power of the brain.

I. INTRODUCTION

Glial cells were first described in 1856 by the German pathologist Rudolf Virchow, who termed them neuroglia (see p. 890 of Virchow 1856). Virchow proposed that *glia* (Greek for glue or slime) serve as cement (in German *Kitt* means putty) that binds the nervous elements together. This original concept has changed radically, although the name has survived. Mammalian glial cells clearly exceed neurons, not only in their preponderance but also in the variety of their types. In this chapter we focus on astrocytes, which can actively participate in synaptic transmission. This function is endowed to astrocytes by their intimate structural relationship with neurons and by the ability of astrocytes to release glutamate and other compounds via exocytosis. Unlike neurons, astrocytes appear to be electrically silent and display their excitability through variations in intracellular Ca^{2+} concentration ($[Ca^{2+}]_i$). We briefly outline bidirectional glutamate-mediated signaling between neurons and astrocytes, followed by a description of the characteristics of Ca^{2+}-dependent exocytic release of glutamate from astrocytes, with a comparison to exocytosis in neurons and some neuroendocrine cells.

II. EXCITABILITY AND INTERCELLULAR COMMUNICATION

Neurons convey their excitability electrically with transient changes in membrane potential mediated by the opening and closing of voltage-gated ion channels located on their plasma membranes. This excitability can be transmitted from neuron to neuron via electrical or chemical synapses. In electrical synapses, the electrical impulse can spread rapidly via gap junction channels; these synapses account for a very small subset of all synapses in the mammalian brain. On the other hand, chemical synapses, which are far more prevalent in the mammalian brain, are slower, but provide a variety of possibilities for pre- and postsynaptic modulation. Synaptic transmission at chemical synapses begins with the initiation of action potentials at the spike-initiation zone (commonly at the axonal hillock), which are then conducted along the axon. There is a limit to the rate at which a neuron can discharge action potentials, defined as the absolute refractory period. Once an action potential is generated it is not possible to experimentally initiate another discharge for about 1 msec, thus, the maximum frequency of action potential discharge is about 1000 Hz, although more commonly and physiologically this firing rate only reaches 100 Hz. Although the conduction velocity of an action potential can vary, depending on several parameters such as the diameter of the axon or the extent of myelination, to name a few, 10 m/sec is considered a typical rate. After the arrival of action potentials at the presynaptic terminals, voltage-gated Ca^{2+} channels (VGCC) open to allow an influx of Ca^{2+} from the extracellular space. The resulting elevation of $[Ca^{2+}]_i$ triggers the fusion of neurotransmitter-filled synaptic vesicles with the plasma membrane, releasing neurotransmitter into the synaptic cleft where it can diffuse and bind to postsynaptic receptors to exert its effect (see Chapter 6). This process of exocytosis can occur rapidly, within 0.2 msec of Ca^{2+} influx at an active zone decorated with synaptic vesicles ready to release their content (a so-called, readily releasable vesicle pool). In this local area, $[Ca^{2+}]_i$ can reach high levels, exceeding 100 μm; the transfer of information between neurons at synapses is dependent on the level and duration of elevated $[Ca^{2+}]_i$ in synaptic terminals. This $[Ca^{2+}]_i$ elevation can be shaped via electrical

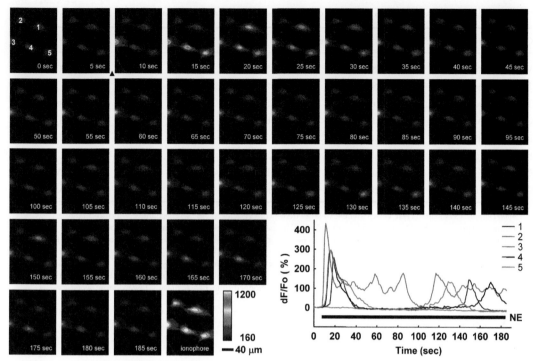

FIGURE 15.1. **Astrocytes display [Ca^{2+}]$_i$ oscillations when exposed to extracellular application of norepinephrine (NE).** Purified astrocytes from the visual cortex were loaded with calcium indicator, fluo-3. NE (25 μm) was applied through a glass pipette using a pressure injection system after the acquisition of the image at 5 sec (arrowhead). This stimulation increases the [Ca^{2+}]$_i$ (images 10–185 sec) from its resting level (images 0–5 sec) (raw data). All cells, with the exception of the cell 2, responded to NE application with an increase of fluo-3 fluorescence. This is not due to nonresponsiveness of fluo-3, since addition of a Ca^{2+} ionophore, 4-bromo-A23197 (10 μM, 2 min) caused an increase in fluo-3 fluorescence; it could be due to a suboptimal concentration of NE and/or density of NE receptors. The pseudocolor scale is a linear representation of the fluorescence intensities ranging from 160 to 1200. This scale does not cover the entire dynamic range (0–4095) of the camera; none of responses reached this saturation level. The numbers on the first image (0 sec) mark astrocytes, whose corresponding time courses of fluo-3 fluorescence, reporting on [Ca^{2+}]$_i$ are shown in the graph. Horizontal bar indicates addition of NE. Changes in fluo-3 fluorescence are expressed as dF/F$_o$ (%) after background subtraction. Scale bar, 40 μm.

excitability encoded in action potential frequencies, where higher frequencies correlate to a higher level of [Ca^{2+}]$_i$ and subsequent recruitment of additional vesicles (reserve pool) to fuse with the plasma membrane. Hence, the frequency and pattern of action potentials represent the code for the transfer of information among neurons.

In contrast to neurons, astrocytes display very stable membrane potential and are not capable of generating action potentials despite the presence of voltage-gated ion channels (reviewed in Verkhratsky and Steinhauser 2000). Thus, astrocytes classically have been regarded as nonexcitable

cells. However, several studies in the early 1990s have demonstrated that astrocytes display a different form of excitability based on [Ca^{2+}]$_i$ elevations (Cornell-Bell et al. 1990; Jensen and Chiu 1990; Charles et al. 1991). Astrocytes possess a plethora of receptors for a wide range of transmitters (including glutamate, norepinephrine, γ-aminobutyric acid [GABA], histamine, adenosine 5'-triphosphate [ATP], acetylcholine, etc.); the activation of these receptors can cause [Ca^{2+}]$_i$ elevations (reviewed in Verkhratsky and Kettenmann 1996; Verkhratsky et al. 1998). An example of such receptor-mediated excitability is shown in Figure 15.1, where

TABLE 15.1. Comparison of amperometric measurements obtained from neurons and astrocytes.

Cell type	Neuron*		Astrocyte**	
Vesicle diameter (nm)	38	55–150	310***	310***
Type of vesicle	SSV	LDCV	?	?
Type of fusion	full	full	full	kiss-and-run
Quantal charge (pC)	0.003	0.047	4.3	0.4
Rise time (ms)	0.09	0.475	7.7	1.9
Half-width (ms)	0.595	3.66	54	17
Original referral	small events	large events	l-AS_M	s-AS_M
Corresponding to	spike	spike	spike	SAF

Abbreviations: (SSV) small clear vesicles; (LDCV) large dense core vesicles; (l-AS_M) mechanically-induced large amperometric spikes; (s-AS_M) mechanically-induced small amperometric spikes; (SAF) stand-alone foot.

*Measurements obtained from isolated Retzius cells (serotonin-releasing neurons) of the leech *Hirudo medicinalis*; for details see Bruns and Jahn (1995).

**Measurements obtained from cultured hippocampal astrocytes isolated from Sprague-Dawley rats and pre-loaded with dopamine (70 mM, 45 minutes); for details see Chen et al. (2005). Since dopamine causes an increase in astrocytic $[Ca^{2+}]_i$, this preloading technique stimulated membrane recycling for a prolonged time, hence, biasing toward large secretory vesicles/granules.

***Note that various sizes of astrocytic vesicle diameters have been reported elsewhere (Maienschein et al. 1999; Bezzi et al. 2004; Crippa et al. 2005), as discussed in the text.

extracellular application of norepinephrine onto cultured astrocytes caused oscillatory $[Ca^{2+}]_i$ elevations. Most neuroligands known to stimulate inositol-1,4,5-triphosphate (IP_3) production via activation of G-protein-coupled receptors (GPCRs), such as norepinephrine and glutamate, can induce $[Ca^{2+}]_i$ oscillations in astrocytes with reported frequencies of 0.01–0.1 Hz (summarized in Table 15.1 of Parpura 2004).

Cornell-Bell et al. (1990) made the seminal observation that extracellular glutamate can cause an increase of $[Ca^{2+}]_i$ in cultured hippocampal astrocytes. These elevations exhibited several distinct patterns:

• Sustained oscillations with constant or decreasing frequencies
• Damped-oscillations
• Step-like sustained elevated $[Ca^{2+}]_i$

In addition to increases of $[Ca^{2+}]_i$, they also observed a spread of $[Ca^{2+}]_i$ in the form of a wave, which was dependent on the extracellular glutamate concentration

$([Glut]_o)$. When astrocytes were exposed to low $[Glut]_o$ (<1 µM) only asynchronous flickers of $[Ca^{2+}]_i$ were observed, with intracellular waves typically propagating through portions of an astrocyte. At higher $[Glut]_o$ (1–10 µM), the intracellular Ca^{2+} waves propagated throughout the entire astrocyte and started spreading to adjacent astrocytes in the form of an intercellular Ca^{2+} wave. At even higher $[Glut]_o$ (10–100 µM), intercellular Ca^{2+} waves could spread several hundred µm from the origin of the wave. However, at $[Glut]_o$ of 1 mM there was no propagation of intercellular Ca^{2+} waves and astrocytes mainly showed step-like increases in $[Ca^{2+}]_i$, albeit with a few oscillations. These findings opened the possibility that astrocytes could receive signals from neurons mediated by glutamate, where the amount of glutamate could control the pattern of $[Ca^{2+}]_i$ elevations in astrocytes.

Evidence for glutamate-mediated neuron-astrocyte signaling was shown in organotypic slice cultures of rat hippocampus (Dani et al. 1992). Electrical stimuli were applied to mossy fibers originating from the dentate gyrus while observing changes in $[Ca^{2+}]_i$ of cells from the CA3 region that were loaded with the Ca^{2+} indicator, fluo-3, and visualized using confocal microscopy. Following electrical stimulation of fibers, astrocytes displayed increases in $[Ca^{2+}]_i$ and intercellular Ca^{2+} waves. Some astrocytes exhibited $[Ca^{2+}]_i$ oscillations with a periodicity and frequency similar to those seen in cultured astrocytes exposed to extracellular glutamate. Astrocytic responses after stimulation of mossy fibers were blocked by kynurenic acid, a glutamate receptor (GluR) antagonist, supporting the notion that a spillover of synaptically released glutamate most likely mediated this neuron-astrocyte signaling.

Porter and McCarthy (1996) investigated neuron-astrocyte signaling in acute hippocampal slices. They loaded slices with the Ca^{2+} indicator, Calcium Green-1, and monitored $[Ca^{2+}]_i$ by confocal microscopy. Electrical stimulation of the Schaeffer collaterals caused an increase in $[Ca^{2+}]_i$ of astrocytes located in the stratum radiatum of the CA1 region. The use of tetrodotoxin (TTX), an antagonist of voltage-gated Na^+ channels, and ω-conotoxin MVIIC, a VGCC blocker, inhibited the astrocytic responses to electrical stimulation of the Schaeffer collaterals, associating these astrocytic responses with neuronal synaptic activity. To test whether this neuron-astrocyte signaling was mediated by glutamate released from neurons, α-methyl-4-carboxyphenylglycine (MCPG), a selective metabotropic GluR (mGluR) receptor antagonist, was used during electrical stimulation of the Schaeffer collaterals; MCPG blocked astrocytic responses. Additional experiments showed that at high levels of neuronal activity both metabotropic and ionotropic GluRs on astrocytes were involved in the $[Ca^{2+}]_i$ increase in astrocytes. These results indicate that hippocampal astrocytes can respond to glutamate released from synaptic terminals in acute slices.

The studies of Cornel-Bell et al. and Dani et al. also revealed the presence of a long-range (several hundred μm) signaling pathway between astrocytes based on intercellular Ca^{2+} waves (reviewed in Arcuino et al. 2004; Giaume et al. 2004; Kettenmann and Schipke 2004), which propagate among astrocytes at average speed of ~15 μm/sec. This is about three orders of magnitude slower than neuronal signal propagation. Two different mechanisms have been hypothesized to underlie this homotypic (astrocyte-astrocyte) signaling: (1) diffusion of second messengers via gap junctions combined with mobilization of Ca^{2+} from intracellular stores (Finkbeiner 1992; Venance et al. 1997) and (2) the release of ATP from astrocytes into the extracellular space with subsequent purinergic receptor activation in neighboring astrocytes (Guthrie et al. 1999; Arcuino et al. 2002; Zhang et al. 2003). Interestingly, the propagation of Ca^{2+} waves among astrocytes correlate with the progression of an extracellular wave of glutamate (Innocenti et al. 2000).

The responsiveness of astrocytes to synaptically released glutamate exhibits activity-dependent changes. Pasti et al. (1997), using acute slices from both visual cortex and hippocampus, have demonstrated that astrocytes can respond to neuronal activity with $[Ca^{2+}]_i$ oscillations mediated through mGluRs. The frequency of these astrocytic $[Ca^{2+}]_i$ oscillations can be correlated to the level of synaptic activity since increased frequency or intensity of stimuli applied to presynaptic afferents produced an increase in the frequency of astrocytic $[Ca^{2+}]_i$ oscillations. Additionally, astrocytes exhibited plastic changes in $[Ca^{2+}]_i$ oscillations, consistent with the studies on cultured cells (Pasti et al. 1995), where a long-lasting increase in the frequency of $[Ca^{2+}]_i$ oscillations was recorded when astrocytes were exposed to multiple applications of an mGluR agonist (1S, 3R)-1-aminocyclopentane-1, 3-dicarboxylic acid (t-ACPD). Astrocytes in acute slices adjacent to repeatedly stimulated neurons exhibited increases in $[Ca^{2+}]_i$ oscillation frequency, which developed over a period of minutes and lasted for at least three hours. Thus, intense synaptic activity can induce long-lasting changes (plasticity) in both neurons and astrocytes. In neurons this plasticity is observed as an increase in synaptic strength, referred to as long-term potentiation (LTP), and is exhibited as an increase in frequency of $[Ca^{2+}]_i$ oscillations in astrocytes. Both of these phenomena have been postulated to represent a form of cellular learning. Although we have a substantial understanding of the molecular and cellular mechanisms of LTP in neurons, in contrast, we lack understanding of the underlying mechanism for the changes in the frequency of $[Ca^{2+}]_i$ oscillations in astrocytes. Even though these phenomena are likely to be mechanistically different, it is tempting to compare them. Besides the obvious fact that both can be mediated by synaptically released glutamate, they also exhibit persistency, lasting for hours after their induction. However, the induction of LTP occurs within several milliseconds, whereas the increased frequency of $[Ca^{2+}]_i$ oscillations in astrocytes requires minutes to develop.

There is growing evidence that astrocytic increases in $[Ca^{2+}]_i$ are an important part of astrocyte-neuron signaling (Nedergaard 1994; Parpura et al. 1994), which can result in the modulation of synaptic transmission. Parpura et al. (1994) used the neuroligand, bradykinin, to stimulate cultured astrocytes from the visual cortex, causing an increase in astrocytic $[Ca^{2+}]_i$ and subsequent glutamate-dependent elevation of neuronal $[Ca^{2+}]_i$. A broad spectrum GluR antagonist, D-glutamylglycine, prevented bradykinin-induced Ca^{2+} accumulations in neurons, without altering astrocytic Ca^{2+} responses to bradykinin. Additional study identified

that activation of N-methyl-D-aspartate receptors (NMDARs) mediated this effect. These results showed that $[Ca^{2+}]_i$ increases in astrocytes lead to glutamate release that in turn signaled to adjacent neurons. The existence of this pathway was confirmed in hippocampal astrocyte-neuron cocultures (Hassinger et al. 1995; Araque et al. 1998a, b), and in acute hippocampal (Pasti et al. 1997; Bezzi et al. 1998) and ventrobasal thalamic slices (Parri et al. 2001). These studies suggest that astrocytes can receive neuronal signals that can stimulate astrocytic receptors to cause an increase in $[Ca^{2+}]_i$. Additionally, $[Ca^{2+}]_i$ increases in astrocytes lead to release of glutamate from these cells, which can serve to signal adjacent neurons. This implies that astrocytes could be integral elements in the computational power of the brain. To have a full partnership in this possible bidirectional communication between neurons and astrocytes, astrocytes should be able to initiate astrocyte-neuron signaling based on their intrinsic activity.

Parri et al. (2001) demonstrated that the astrocytes in acute slices of the rat ventrobasal thalamus displayed spontaneous $[Ca^{2+}]_i$ oscillations; these oscillations were not mediated by discharge of action potentials, because the use of TTX did not abolish them. Interestingly, these spontaneous $[Ca^{2+}]_i$ oscillations in astrocytes could initiate Ca^{2+} waves that propagated to neighboring astrocytes and also could cause inward currents in adjacent neurons, mediated by activation of NMDARs. Most of the spontaneous oscillations observed in astrocytes have irregular frequencies and amplitudes. However, a subset of astrocytes in the ventrobasal thalamus displayed rhythmic pacemaker $[Ca^{2+}]_i$ oscillations with an average frequency of 0.019 Hz (10–90% range, 0.012–0.02 Hz; Parri and Crunelli 2001). Astrocytes in acute hippocampal slices also exhibit spontaneous $[Ca^{2+}]_i$ oscillations (Aguado et al. 2002; Nett et al. 2002). These spontaneous oscillations are mediated by the activation of astrocytic IP_3

receptors located on internal Ca^{2+} stores (Nett et al. 2002; Parri and Crunelli 2003).

Glutamate-mediated astrocyte-neuron signaling can result in the modulation of synaptic transmission (Araque et al. 1998a, b; Kang et al. 1998; Newman and Zahs 1998; Fiacco and McCarthy 2004; Liu et al. 2004). This may not be surprising, since the ultrastructure of the CNS shows that astrocytes enwrap nerve terminals (Peters et al. 1991; Grosche et al. 1999; Ventura and Harris 1999), making them perfectly positioned to exchange information with synapses. As discussed earlier, it has been demonstrated that astrocytes can respond to glutamatergic synaptic activation (Dani et al. 1992; Porter and McCarthy 1996). Additionally, astrocytes, by releasing glutamate, have been shown to act on synapses and modulate spontaneous (Araque et al. 1998b) and action-potential evoked synaptic transmission (Araque et al. 1998a). These data led to the concept that astrocytes should be viewed as an integral element of a tripartite synapse (reviewed in Araque et al. 1999), along with neuronal pre- and postsynaptic elements (see Figure 15.2).

The initial investigations into astrocytic modulation of spontaneous synaptic transmission used mixed cultures of hippocampal neurons and astrocytes to reveal that glutamate released from astrocytes can increase the frequency of spontaneous postsynaptic currents (sPSCs) in adjacent cultured neurons (Araque et al. 1998b). Since these experiments were performed in the presence of TTX to prevent action potentials, these spontaneous events were limited to miniature PSCs (mPSCs); the effect on the frequency of mPSCs was mediated by extrasynaptic NMDARs. In addition to modulating spontaneous synaptic events, glutamate release from astrocytes was also implicated in affecting action potential-evoked synaptic transmission (Araque et al. 1998a). Thus, after obtaining the basal measurement of synaptic transmission between a pair of neurons, surrounding

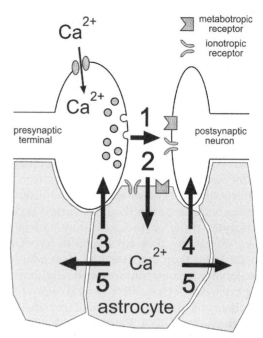

FIGURE 15.2. **Scheme of glutamate-mediated bidirectional astrocyte-neuron signaling.** After arrival of action potentials at a presynaptic terminal and opening of voltage gated Ca^{2+} channels, glutamate containing synaptic vesicles (filled circles) fuse to the plasma membrane to release glutamate into the synaptic cleft. The released glutamate signals to the postsynaptic neuron by acting on receptors (**1**). A portion of the released glutamate can reach astrocytes surrounding the synapse, and can stimulate their receptors (**2**) leading to an increase in astrocytic intracellular Ca^{2+} levels. Additionally, glutamate can be taken up by astrocytes via plasma membrane glutamate transporters. Through Ca^{2+}-dependent glutamate release, astrocytes in turn can signal to neuronal pre- (**3**) and postsynaptic (**4**) sites. The increase of $[Ca^{2+}]_i$ in a single astrocyte can spread to adjacent astrocytes in the form of a wave of elevated $[Ca^{2+}]_i$, a signaling predominately mediated by ATP (**5**). Adapted from Parpura (2004).

astrocytes were stimulated to cause an increase in their $[Ca^{2+}]_i$. This stimulus caused a transient reduction in the amplitude of action potential-evoked PSCs that lasted tens of seconds. The observation that astrocytes in culture are able to influence spontaneous synaptic transmission has been corroborated by data obtained in acute hippocampal slice preparations (Kang et al.

1998; Fiacco and McCarthy 2004; Liu et al. 2004).

It is now well established that astrocytes cannot only sense neuronal activity, but also are capable of signaling back to neurons to engage in glutamate-mediated bidirectional communication (reviewed in Araque et al. 1999; Carmignoto 2000; Haydon 2001; Newman 2003; Parpura 2004; Volterra and Meldolesi 2005). This astrocyte-neuron signaling stems from Ca^{2+}-mediated glutamate release from astrocytes. Thus, even though neurons and astrocytes differ in the way they convey their excitability, they are similar in respect to the way they can use Ca^{2+}-dependent excitability to communicate with neighboring cells. Astrocytes utilize several different mechanisms to release glutamate (reviewed in Evanko et al. 2004; Parpura et al. 2004). We will focus our discussion on the Ca^{2+}-dependent exocytic release pathway.

III. Ca^{2+}-DEPENDENT EXOCYTIC RELEASE OF GLUTAMATE

A. Ca^{2+} Dependency and Ca^{2+} Sources

Neurons release glutamate via Ca^{2+}-dependent exocytosis. Experimental manipulations that block $[Ca^{2+}]_i$ elevations in neurons or that elevate $[Ca^{2+}]_i$ directly reveal that $[Ca^{2+}]_i$ elevation is necessary and sufficient to cause glutamate release. Similar experimental evidence has been collected for astrocytes, supporting the notion that astrocytes can release glutamate in a Ca^{2+}-dependent manner, by a mechanism that has characteristics of exocytosis.

The first evidence for Ca^{2+}-dependent glutamate release from astrocytes was derived from studies in purified cortical astrocytic cultures (Parpura et al. 1994). Astrocytes were equilibrated for prolonged periods of time (40–60 minutes) either in a solution containing normal external free Ca^{2+} (2.4 mM), or in a solution with low external free Ca^{2+} (24 nM); the latter solution caused depletion of internal Ca^{2+} stores and prevented Ca^{2+} entry from the extracellular space. Supernatants collected from the cultured astrocytes were subjected to high-performance liquid chromatography. Under resting conditions there was a basal (constitutive) release of glutamate from astrocytes into extracellular saline. Addition of the Ca^{2+} ionophore, ionomycin, in the presence of external Ca^{2+}, but not in its absence (low Ca^{2+} solution), produced an increase in the release of glutamate from astrocytes, indicating that elevated $[Ca^{2+}]_i$ is necessary and sufficient to stimulate glutamate release. Indeed, other stimuli that directly increased astrocytic $[Ca^{2+}]_i$, such as mechanical stimulation (Parpura et al. 1994; Araque et al. 1998a,b; Hua et al. 2004; Montana et al. 2004), photostimulation (Parpura et al. 1994), and photolysis of Ca^{2+} cages (Araque et al. 1998b; Parpura and Haydon 2000) all caused release of glutamate. Additionally, increasing the astrocytic buffering capacity for cytosolic Ca^{2+} using the Ca^{2+} chelator, BAPTA (Araque et al. 1998a; Bezzi et al. 1998; Innocenti et al. 2000; Hua et al. 2004; Montana et al. 2004), reduced the evoked glutamate release from astrocytes (see Figure 15.3).

The process of exocytosis and its Ca^{2+} dependency have been extensively studied in neurons (for review see Sudhof 2004). Pioneering work by Dodge and Rahamimoff (1967) has established the quantitative relationship between external Ca^{2+} levels and the amplitude of end-plate potentials. During nerve terminal stimulation the amplitude of end-plate potential can be used to assess the Ca^{2+} level at a synaptic terminal. Since the level of external Ca^{2+} controls the driving force for Ca^{2+} influx, it is reflected in the amount of neurotransmitter release. Although the established relationship was highly nonlinear, when plotted on a logarithm-logarithm plot, the initial part was represented by a straight line showing a slope of ~4. This slope, commonly referred to as the Hill coefficient, suggested a cooperativity of at least 4 Ca^{2+} in the process of transmitter release. Since then, this coefficient in neuronal terminals has been reported to be between 2 and 4 (Augustine and Eckert 1984; Smith et al. 1985; Stanley 1986; Zhong and Wu 1991; Trudeau et al. 1998).

The quantitative analysis of the relationship between $[Ca^{2+}]_i$ and release of trans-

mitter(s) also has been examined in astrocytes (Parpura and Haydon 2000; Kreft et al. 2004). Parpura and Haydon (2000) examined the Ca^{2+} dependency of glutamate release from astrocytes by using flash photolysis of caged Ca^{2+} to control $[Ca^{2+}]_i$ in astrocytes, while electrophysiologically measuring currents in adjacent neurons, which were utilized as extracellular glutamate sensors. The reported Hill coefficient for glutamate release of 2.1 to 2.7 was in good agreement with values observed in neurons. The delay between the delivery of the flash and the glutamatergic response in neurons was under 10 msec, the period imposed by the sampling rate in these experiments. In a different study, Kreft et al. (2004) examined the Ca^{2+} dependency of exocytosis in astrocytes by using capacitance measurements to monitor changes in the membrane area of solitary astrocytes, presumably due to vesicular fusions (full fusion of any vesicles, not only glutamate-containing vesicles), as a consequence of photolysis of caged Ca^{2+} utilized to raise $[Ca^{2+}]_i$. Here, the measured Hill coefficient was 5.3; the apparent difference in the

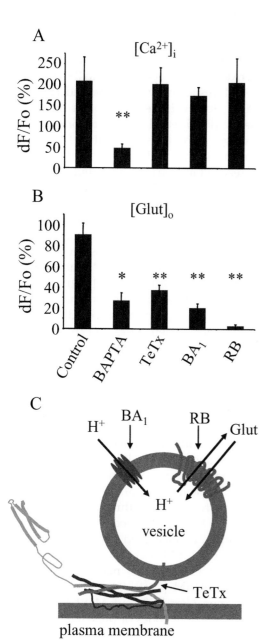

FIGURE 15.3. **Ca^{2+}-dependent glutamate release from astrocytes utilizes a vesicular exocytic pathway.** Mechanical stimulation caused (**A**) the increase of $[Ca^{2+}]_i$ and (**B**) the release of glutamate into extracellular space ($[Glut]_o$). This release of glutamate is Ca^{2+}-dependent, because incubation of astrocytes with BAPTA-AM, a membrane permeable Ca^{2+} chelator, reduced the mechanically-induced $[Ca^{2+}]_i$ and $[Glut]_o$ increases, implicating exocytosis as the underlying mechanism for this release. Pretreatment of astrocytes with a holoprotein of tetanus toxin (TeTx), that cleaves synaptobrevin 2 (**C**; red) reduced the release of glutamate. Target-SNARE proteins shown are syntaxin 1 (**C**; green) and SNAP-23 (**C**; blue). Similar reductions in glutamate release can be obtained when treating astrocytes with bafilomycin A_1 (BA_1) a specific inhibitor of V-ATPase (**C**; brown) and Rose Bengal (RB), an allosteric site modulator of VGLUTs (**C**; pink). (**C**) A drawing of the exocytic process in astrocytes. Arrow indicates sites of actions for pharmacological agents. * and ** Significant at $p < 0.05$ and $p < 0.01$, respectively. **A** and **B** are modified from Montana et al. (2004).

measurements of the Hill coefficients between the two studies may be due to the fact that they measured different phenomena in response to the rise of $[Ca^{2+}]_i$ in astrocytes (discussed in Kreft et al. 2004). Additionally, the capacitance measurements revealed a delay of 50 to 100 msec between the flash delivery and the maximal rate of membrane capacitance (C_m) increase, reported at 3000 fF/s. Therefore, the exocytic kinetics in astrocytes were about two orders of magnitude slower than the kinetics of exocytosis at synaptic terminals (maximal rate of up to 100,000–150,000 fF/s; Heidelberger et al. 1994) and chromaffin cells (maximal rates up to 100,000 fF/s; Chow et al. 1994; Heinemann et al. 1994), but similar to those measured in pituitary melanotrophs (maximal rates ~2000 fF/s; Thomas et al. 1993a, 1993b; Rupnik et al. 2000). The delay between the $[Ca^{2+}]_i$ increase and C_m increase in neurons was in the submillisecond to millisecond time range (Heidelberger et al. 1994), consistent with short synaptic delays reported elsewhere (Katz and Miledi 1965; Llinas et al. 1981; Delaney and Zucker 1990; reviewed in Sabatini and Regehr 1999).

Interestingly, the threshold for Ca^{2+}-dependent release of glutamate in astrocytes was reported at submicromolar $[Ca^{2+}]_i$ (Parpura and Haydon 2000; Pasti et al. 2001). Using photolysis of caged Ca^{2+} to cause step-like increases in astrocytic $[Ca^{2+}]_i$ (Parpura and Haydon 2000), it was found that a modest increase of $[Ca^{2+}]_i$, from a resting level of ~80 nM to 140 nM, was sufficient to trigger release of glutamate from hippocampal astrocytes. Similarly, relatively low amplitudes, ~550 nM, of $[Ca^{2+}]_i$ oscillations in cortical astrocytes, evoked by t-ACPD, caused exocytic release of glutamate from astrocytes. The amplitude of these astrocytic $[Ca^{2+}]_i$ oscillations controlled the efficacy of glutamate release, and the frequency of $[Ca^{2+}]_i$ oscillations controlled the frequency of glutamate release (Pasti et al. 2001). However, the capacitance measurements revealed that the half-

maximal response of the exocytic rate in cortical astrocytes occurs at ~20 μM $[Ca^{2+}]_i$. Such levels of $[Ca^{2+}]_i$ should be reachable during physiologically relevant $[Ca^{2+}]_i$ dynamics in astrocytes. Note that the average cytosolic $[Ca^{2+}]$ increases in astrocytes due to physiological stimulation of their neuroligand receptors are commonly reported only in the low- to mid-micromolar range. For example, application of the three major endogenous agonists within the hippocampus—glutamate, norepinephrine, and dopamine—caused, on average, increases of astrocytic $[Ca^{2+}]_i$ to 2.0, 1.8, and 3.5 μM, respectively (Parpura and Haydon 2000). However, these and other measurements (Pasti et al. 2001) report on the average cytosolic values; therefore, it is very likely that higher levels of $[Ca^{2+}]$ may be reached in restricted subcellular regions, such as close to IP_3 receptors on the endoplasmic reticulum, which mediate Ca^{2+} efflux from the internal Ca^{2+} stores to the cytosol. Indeed, Ca^{2+} released from the IP_3-sensitive intracellular stores in mucosal mast cells caused local $[Ca^{2+}]_i$ increases of over 16 μm (Csordas et al. 1999). Thus, it seems reasonable to suggest that similar localized $[Ca^{2+}]_i$ increases would occur in astrocytes. Nonetheless, the micromolar range of $[Ca^{2+}]_i$ changes, up to ~20 μM, would result only in a slow rate of exocytosis in astrocytes. Similarly, minimal $[Ca^{2+}]_i$, within the submicromolar range of 0.5–1 μM, can cause detectable secretion in chromaffin cells and pituitary melanotrophs, with half-maximal rates of 7–21 μM and 27 μM $[Ca^{2+}]_i$, respectively (Thomas et al. 1993a,b; Chow et al. 1994; Heinemann et al. 1994). In contrast with astrocytes, the threshold for $[Ca^{2+}]_i$ required to activate exocytosis at synaptic terminals, as measured by the increase in C_m, is over 10 μM, and typically greater than 20 μM, with the half maximal response being ~194 μM $[Ca^{2+}]_i$ (Heidelberger et al. 1994; von Gersdorff and Matthews 1994). Since, at active zones of presynaptic terminals, $[Ca^{2+}]_i$ is thought to reach several hundred micromolar during exocytosis (Roberts et al. 1990; Adler et al. 1991; Llinas et al. 1992), this would guarantee the full exocytic rate. Thus, neuronal exocytic release occurs at levels of $[Ca^{2+}]_i$ over 100-fold higher than its resting levels (~100 nM). Such a difference in the rate of exocytosis and the required $[Ca^{2+}]$ dynamics between neurons and astrocytes suggests that the fast synchronized release of glutamate from neurons is designed for fast synaptic transmission, whereas the slow release of glutamate from astrocytes is geared toward modulation of synaptic transmission (Haydon 2001; Zhang and Haydon 2005).

The source of cytosolic Ca^{2+} necessary for exocytic release differs in neurons and in astrocytes. In neurons, the majority of the $[Ca^{2+}]_i$ elevation necessary for synchronized release of transmitters comes from the influx of Ca^{2+} through opened VGCCs, although Ca^{2+}-induced Ca^{2+} release (ryanodine/caffeine sensitive) and IP_3-induced Ca^{2+} release from internal stores play a role as well (reviewed in Verkhratsky 2002). In astrocytes, the cytosolic $[Ca^{2+}]$ elevation necessary for exocytic release of glutamate is supported primarily by Ca^{2+} efflux from internal Ca^{2+} stores to the cytoplasm, which can be activated via GPCR-mediated pathways. Indeed, preincubation of astrocytes with thapsigargin, an inhibitor of Ca^{2+}-ATPase specific for internal stores, greatly reduced evoked glutamate release from astrocytes (Araque et al. 1998a; Innocenti et al. 2000; Hua et al. 2004), indicating that the internal stores represent the predominant source of Ca^{2+} necessary for glutamate release from astrocytes. However, astrocytes can express various channels and/or receptors permeable to Ca^{2+}, such as VGCCs (MacVicar 1984; Latour et al. 2003; but see Carmignoto et al. 1998), transient receptor potential genes (TRP) (Pizzo et al. 2001; Golovina 2005; Malarkey and Parpura 2005), and ionotropic GluRs (iGluRs) (reviewed in Verkhratsky et al. 1998) that allow Ca^{2+} entry across the plasma mem-brane. Indeed, dual sources exist that produce the Ca^{2+} required for exocytic glutamate release from astrocytes; the predominant source is the release of Ca^{2+} from internal stores, but entry of Ca^{2+} from the extracellular space also plays a role (Hua et al. 2004). These sources were revealed by the reduction in mechanically induced glutamate release in the presence of thapsigargin and Cd^{2+}, a blocker of Ca^{2+} entry from the extracellular space. Both IP_3- and caffeine/ryanodine-sensitive stores have been implicated in supplying Ca^{2+} for glutamate release. Thus, the application of diphenylboric acid 2-aminoethyl ester (2-APB) solution, a cell-permeant IP_3 receptor antagonist, greatly reduced mechanically induced glutamate release. Additionally, incubation of astrocytes with ryanodine (10 µM), which can block the release of Ca^{2+} from the ryanodine/caffeine-sensitive stores, attenuates mechanically induced glutamate release. Furthermore, the sustained presence of caffeine (10 mM) that depletes ryanodine/caffeine stores also reduced mechanically induced glutamate release. However, when astrocytes were pretreated using a combination of 2-APB with ryanodine or caffeine, there was no additive effect on reduction of mechanically induced glutamate release when compared to treatments with one pharmacological agent only. These data suggest that mechanically induced Ca^{2+}-dependent exocytic glutamate release from astrocytes is mediated through both IP_3- and ryanodine/caffeine-sensitive internal Ca^{2+} stores; the most likely interpretation is that glutamate release requires the coactivation of both stores and that these stores are operating jointly.

B. Protein Components of Exocytic Secretory Machinery

The exocytic release of transmitter from nerve terminals critically depends on the presence of the protein components of the soluble N-ethyl maleimide-sensitive fusion protein (NSF) attachment protein receptor (SNARE) complex, including synapto-brevin 2, syntaxin and synaptosome-associated protein of 25 kDa (SNAP-25), or in some cases, SNAP-23 (the latter mediating hippocampal GABA-ergic synaptic transmission in culture; see Verderio et al. 2004) (also see Chapter 6). Astrocytes possess secretory apparatus similar to that found in nerve terminals (reviewed in Malarkey and Parpura 2006).

Astrocytes express synaptobrevin 2 (also known as vesicle-associated membrane protein 2 (VAMP 2)), its homologue, cellubrevin (VAMP 3), syntaxin 1, and SNAP-23 (Parpura et al. 1995a; Jeftinija et al. 1997; Hepp et al. 1999; Maienschein et al. 1999; Araque et al. 2000; Pasti et al. 2001; Bezzi et al. 2004; Montana et al. 2004; Zhang et al. 2004b). Synaptobrevin 2 was found in processes of freshly isolated astrocytes (Zhang et al. 2004b), whereas processes of astrocytes *in situ* contained cellubrevin (Bezzi et al. 2004). There have been discrepancies with regard to some of the proteins detected in astrocytes: SNAP-25, synaptophysin, and synaptotagmin (Syt) 1, which often are used as neuron-specific markers, were found by some groups in cultured astrocytes (Jeftinija et al. 1997; Maienschein et al. 1999; Wilhelm et al. 2004; Anlauf and Derouiche 2005), whereas others were unable to detect these molecules (Parpura et al. 1994, 1995a; Latour et al. 2003; Montana et al. 2004; Zhang et al. 2004a,b; Crippa et al. 2005). These discrepancies may result from culturing induced expression of proteins that would not be produced by astrocytes *in vivo* (Maienschein et al. 1999; Wilhelm et al. 2004).

Clostridial toxins, metalloproteases that cleave SNARE proteins, reduce the level of Ca^{2+}-dependent glutamate release in astrocytes (Jeftinija et al. 1997; Bezzi et al. 1998, 2001, 2004; Araque et al. 2000; Pasti et al. 2001; Hua et al. 2004; Montana et al. 2004; also see Figure 15.3). The expression of the cytoplasmic tail of synaptobrevin 2

(containing the SNARE domain, but lacking the ability to anchor to the vesicular membrane) inhibits evoked glutamate release from astrocytes (Zhang et al. 2004b), consistent with the prevention of SNARE complex assembly and vesicular fusion. Additionally, tetanus toxin reduces the C_m increase (Kreft et al. 2004) and reduces the number of amperometric spikes (Chen et al. 2005) from astrocytes. Furthermore, α-Latrotoxin, which induces the release of neurotransmitter by directly stimulating the secretory machinery (reviewed in Sudhof 2001), causes glutamate release from astrocytes (Parpura et al. 1995b; Jeftinija et al. 1996).

The Ca^{2+} sensor for exocytosis in astrocytes seems to be synaptotagmin 4 (Syt 4), rather than Syt 1 as it is in neurons. Syt 4 has been detected in processes of freshly isolated astrocytes and astrocytes *in situ* (Zhang et al. 2004a), as well as in isolated synaptobrevin 2-containing vesicles from these cells (Crippa et al. 2005). Reduction of Syt 4 expression in cultured astrocytes using RNA interference decreases Ca^{2+}-dependent glutamate release. Similarly, expression of a Syt 4 mutant, manipulated within its putative Ca^{2+}-binding C2B domain (Thomas et al. 1999; but see Dai et al. 2004), acts in a dominant-negative manner affecting Ca^{2+}-dependent glutamate release from astrocytes (Zhang et al. 2004a). Thus, Syt 4 seems to function in astrocytes as a Ca^{2+} sensor for vesicular fusion with the plasma membrane, similar to the function of Syt 1 at neuronal presynaptic terminals. Interestingly, exogenously expressed Syts 1 and 4 are interchangeable in supporting synaptic transmission in *Drosophila* (Robinson et al. 2002) and exocytic release of glutamate from murine astrocytes (Zhang et al. 2004a). However, in pheochromocytoma PC12 cells, Syt 4 has been implicated in mediating kiss-and-run events, via the formation of a transient fusion pore, whereas Syt 1 supports full fusion events, where the vesicle collapses into the plasma membrane (Wang et al.

2003). Both forms of the exocytic events have been described in neurons (Klyachko and Jackson 2002; Aravanis et al. 2003a,b; Gandhi and Stevens 2003) and astrocytes (Bezzi et al. 2004; Kreft et al. 2004; Zhang et al. 2004b; Chen et al. 2005; Crippa et al. 2005). Consequently, differential expression of Syts within a particular cell type can lead to distinctive modes of exocytosis: (1) vesicles that fully fuse with the plasma membrane and discharge the entire content of transmitter stored within their vesicular lumen, followed by their retrieval from the plasma membrane and refilling before they fuse again, which is an inherently slower process than a kiss-and-run event; (2) vesicles undergoing kiss-and-run events that only partially release their transmitter content, so that refilling of these transmitter-depleted vesicles would be relatively quick, providing a mechanism for very rapid recycling of vesicles (Ales et al. 1999; Aravanis et al. 2003a). Astrocytes, by expressing Syt 4, therefore, could support a rapidly recycling pool of vesicles, requiring fewer vesicles (see later) for Ca²⁺-dependent exocytosis than the Syt 1-dependent transmission executed by glutamatergic neurons.

A recent study that examined activity-dependent synaptic plasticity at *Drosophila* glutamatergic neuromuscular junctions (NMJs) described a retrograde signaling pathway mediated by Syt 4 (Yoshihara et al. 2005). In this system, Ca²⁺ entry into the postsynaptic site triggered fusion of Syt 4-containing vesicles with the postsynaptic membrane. This resulted in the activation of a cyclic adenosine monophosphate-dependent protein kinase (PKA) pathway in the presynaptic terminal. Active PKA produced cytoskeletal changes resulting in the induction of presynaptic cell growth and differentiation. Additionally, PKA enhanced secretion, consistent with previous reports that presynaptic activation of PKA can enhance the amount of neurotransmitter release (Capogna et al. 1995; Trudeau et al. 1996). Since astrocytes

possess Syt 4-containing vesicles (Zhang et al. 2004a; Crippa et al. 2005), it is tempting to speculate that a similar retrograde signaling mechanism might be present in astrocytes, which could play a role in the formation and plasticity of the tripartite synapse.

In addition to the differential expression of Syts 1 and 4, it is possible that the differential expression of SNAP-25 and SNAP-23 in glutamatergic neurons and astrocytes, respectively, also dictates the characteristics of exocytosis in these cells. SNAP-25 and SNAP-23 have been reported to play distinct roles in exocytic release, where SNAP-25B (like SNAP-25A, though with two to three times higher efficiency), but not SNAP-23, can support an exocytic burst. SNAP-25 appears to control the size of the primed (readily releasable) vesicle pool to allow for synchronous fusion, or exocytic burst, upon a rise in [Ca²⁺]ᵢ (Sorensen et al. 2003). In chromaffin cells of *Snap25 null* and wild-type mice, Sorensen et al. (2003) showed that the absence of SNAP-25 did not disrupt the docking of large dense core vesicles, as examined by electron microscopy. In chromaffin cells from *Snap25 null* mice expressing SNAP-25B or SNAP-23, exocytic events were triggered by photolysis of caged-Ca²⁺, and exocytosis was monitored simultaneously using capacitance and amperometry. The absence of SNAP-25 in mutant cells almost completely abolished Ca²⁺-dependent exocytosis, which was rescued in full by expressing SNAP-25B. However, when SNAP-23 was expressed in mutant cells, there was a small but significant recovery of overall exocytosis, but the burst was absent. Thus, SNAP-25B, but not SNAP-23 stabilized vesicles in the primed, readily releasable state and allowed for fast synchronized release. Therefore, one would expect that cells expressing the SNAP-25 isoform (e.g., glutamatergic neurons), but not SNAP-23 expressing cells (e.g., astrocytes), would have a readily releasable vesicle pool with the capacity for fast Ca²⁺-sensitive synchronous release, in conjunction with the Ca^{2+} sensor Syt 1 (Voets et al. 2001). Interestingly, Bezzi et al. (2004) using total internal reflection fluorescence microcopy (TIRFM) recorded exocytic bursts in cultured astrocytes. Within 100 to 150 msec after stimulation of these cells with an mGluR agonist to cause an increase in astrocytic $[Ca^{2+}]_i$, there was an increase in the fusion rate, which peaked within 50 msec and then declined to basal rates. These bursts lasted 500 to 600 msec, although most of the fusion events occurred within the first 200 msec. This is an exciting finding, because it indicates that astrocytes might be capable of establishing a readily releasable pool of vesicles to support a slow exocytic burst (consult, e.g., Sorensen et al. 2003 for the time constants of slow and fast exocytic bursts in chromaffin cells) in the absence of SNAP-25/Syt 1 using alternative molecules. However, since this work did not assess the composition of the core SNARE complex (i.e., the presence of SNAP-25/SNAP-23) and the identity of the Ca^{2+} sensor (Syt 1/Sty 4) involved in slow exocytic burst, this finding possibly could reflect a culturing effect with the expression of the neuron-specific proteins SNAP-25 and Syt 1 in astrocytic cultures. Clearly, future experiments will have to rigorously address the issue of the existence and characteristics of astrocytic readily releasable vesicle pools and exocytic bursts. Nevertheless, the molecular composition of astrocytic secretory machinery and the display of slow exocytic bursts and slow rates of C_m increase (see earlier) in astrocytes favor a relatively slow exocytic release in these cells, rather than the fast synchronous release seen in neurons.

By interacting with different Syt isoforms, SNAP-25 and SNAP-23 were shown to mediate docking of secretory granules to the plasma membrane *in vitro* (Chieregatti et al. 2004). The association of SNAP-25 with Syt 1 (as well as with Syts 3 and 7) requires micromolar (1–10 μM) $[Ca^{2+}]$ for granule docking, whereas the interaction of SNAP-23 with Syt 3 and 7 allows granule docking at ~100 nM $[Ca^{2+}]$, which corresponds to resting $[Ca^{2+}]_i$ in mammalian cells. Indeed, when SNAP-23 is exogenously expressed in intact endocrine cells, it produces high-level secretion under basal conditions. These data suggest that SNAP-23/SNAP-25 expression might control the mode of release (basal/constitutive versus regulated) by forming docking/fusion complexes at different $[Ca^{2+}]_i$ thresholds. Interestingly, astrocytes exhibit a substantial level of glutamate secretion under basal conditions, compared with the amount of glutamate release during regulated (evoked) secretion (basal $[Glu]_o$ ~50 nM; peak $[Glu]_o$ ~140 nM following evoked release; for details see Parpura et al. 1994; 1995b). Additionally, astrocytes have submicromolar $[Ca^{2+}]_i$ thresholds for glutamate release (Parpura and Haydon 2000; Pasti et al. 2001). Both of these properties of astrocytic glutamate release seem to be in good agreement with SNAP-23 mediating exocytosis. However, Syts 4, 5, and 11 are expressed by astrocytes (Zhang et al. 2004a) and do not interact with SNAP-23 *in vitro* (with perhaps the exception of Syt 5, which shows weak interaction with SNAP-23; Chieregatti et al. 2004); Syt 5 is expressed throughout the astrocytic body, but not in processes (Zhang et al. 2004a). Thus, it remains to be elucidated how differences in SNAP-23/SNAP-25 interactions with other molecules of the core SNARE complex, Ca^{2+} sensors, and various ancillary molecules to the SNARE complex regulate properties of astrocytic exocytosis.

C. Filling of Vesicles with Glutamate

Exocytic release of glutamate at presynaptic terminals requires a mechanism by which glutamate can be concentrated inside synaptic vesicles. There are two essential protein components necessary for this mechanism: the vacuolar-type H^+-ATPase (V-ATPase) and the vesicular glutamate transporters (VGLUTs) (see Chapter 11). V-ATPase creates a proton gradient by pumping H$^+$ into the vesicular lumen leading to its acidification (pH of ~5.5 at rest compared to a pH of 7.2 in the cytoplasm). The proton gradient created between the vesicular lumen and cytoplasm allows VGLUTs to drive glutamate into vesicles while transporting protons to the cytoplasm. This mechanism for filling vesicles with glutamate also exists in astrocytes, since these cells express functional V-ATPase and VGLUTs.

Using transgenic mice in which the enhanced green fluorescent protein (EGFP) was placed under the control of the glial fibrillary acidic protein (GFAP; an astrocytic marker) promoter (Nolte et al. 2001), Wilhelm et al. (2004) found punctate expression of V-ATPase in the cell periphery and processes of astrocytes in tissue sections, indicating the presence of acidic organelles, including endosomes, lysosomes, and secretory vesicles (Finbow and Harrison 1997). Similar immunopositive labeling for V-ATPase was reported in cultured astrocytes (Parpura 2004). Montana et al. (2004) addressed the functional contribution of V-ATPase to the acidification of the astrocytic vesicular pool by transfecting cultured cortical astrocytes using a plasmid encoding for synapto-pHluorin (Miesenbock et al. 1998; Sankaranarayanan et al. 2000), a chimeric protein that contains a pH-sensitive fluorescent protein, pHluorin, attached to the C-terminus of synaptobrevin 2 (see Figure 15.4). Since pHluorin is attached to the lumenal (intravesicular) portion of synaptobrevin 2, it senses intravesicular pH with its fluorescence mostly quenched at resting intravesicular pH. However, blockade of V-ATPase using bafilomycin A$_1$ can collapse the proton gradient leading to the alkalinization of the intravesicular lumen and an increase in pHluorin's fluorescence (Sankaranarayanan and Ryan 2001) (see Chapter 6). Incubation of astrocytes with bafilomycin A$_1$ caused a time-dependent increase in synapto-pHluorin fluorescence (see Figure 15.4), consistent with the increase of intravesicular pH. Furthermore, synapto-pHluorin fluorescence was present in a punctate pattern consistent with its vesicular location. Crippa et al. (2005) used synaptobrevin 2-expressing vesicles that were isolated from astrocytes to show that bafilomycin A$_1$ decreased the ability of vesicles to take up glutamate. Additionally, Ca^{2+}-dependent glutamate release from astrocytes was blocked by bafilomycin A$_1$ (Araque et al. 2000; Bezzi et al. 2001; Pasti et al. 2001; Montana et al. 2004; Zhang et al. 2004b; also see Figure 15.3). These studies present evidence for the existence of a functional V-ATPase in astrocytes.

To date, three isoforms of VGLUTs have been described in the CNS, annotated as VGLUTs 1, 2, and 3. VGLUTs 1 and 2 have a complementary pattern of expression in presynaptic terminals of the adult CNS. VGLUT 1 is expressed in the telencephalon, and VGLUT 2 in the diencephalon. However, VGLUT 3 expression differs from those of VGLUTs 1 and 2 because it is expressed within the cell bodies, somatodendritic, and axonal regions of many neurons that are traditionally considered not to release glutamate, but may release other transmitters (reviewed in Fremeau et al. 2002). Nonetheless, all three isoforms of VGLUTs have been detected in astrocytes (Fremeau et al. 2002; Bezzi et al. 2004; Kreft et al. 2004; Montana et al. 2004; Zhang et al. 2004b; Anlauf and Derouiche 2005). Ultrastructural examination has revealed VGLUTs in astrocytic processes adjacent to synaptic terminals *in situ* (Fremeau et al. 2002; Bezzi et al. 2004; Zhang et al. 2004b). Punctate labeling of VGLUTs with distributions throughout astrocytic bodies and processes (Bezzi et al. 2004; Kreft et al. 2004; Montana et al. 2004; Zhang et al. 2004b; Anlauf and Derouiche 2005), as well as colocalization with the SNARE proteins, cellubrevin and synaptobrevin 2 (Bezzi et al. 2004; Montana et al. 2004; Crippa et al. 2005), reveal the presence of a vesicle

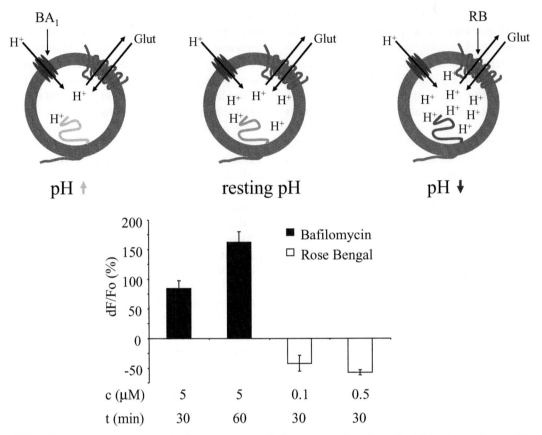

FIGURE 15.4. Bafilomycin A$_1$ alkalinizes the vesicular lumen, and Rose Bengal acidifies it, consistent with these agents' actions on V-ATPase (Top, brown) and VGLUTs (Top, pink), respectively. Synapto-pHluorin has a pH-sensitive fluorescent protein, pHluorin (Top; yellow, green, and blue), attached to the C-terminus of synaptobrevin 2 (Top, red). Since pHluorin is located in the lumen of the vesicle, it senses intravesicular pH. At resting intravesicular pH (Top, Middle, blue; Bottom, dF/Fo = 0) vesicular proton transport through VGLUTs [allowing glutamate (Glut) transport to the lumen] reduces the intravesicular proton concentration created by V-ATPase. Blockade of V-ATPase using bafilomycin A$_1$ (BA$_1$) can collapse the proton gradient leading to alkalinization of the intravesicular lumen and an increase in pHluorin's fluorescence (Top; left, yellow; Bottom, solid black bars). Rose Bengal (RB), at submicromolar concentrations, has a much higher affinity for VGLUTs than for V-ATPase (Ogita et al. 2001). By inhibiting VGLUTs and their proton transport activity, RB leads to vesicular acidification (Top, right, blue; Bottom, open bars). Changes in synapto-pHluorin fluorescence are expressed as dF/Fo (%) after background subtraction. Concentration (c) and exposure times (t) of pharmacological agents are given in μM and minutes, respectively. Modified from Montana et al. (2004).

population in astrocytes that is capable of accumulating and releasing glutamate. Indeed, isolated astrocytic vesicles can take up glutamate (Crippa et al. 2005). Additionally, when the vesicular proton transport of VGLUTs was inhibited by Rose Bengal, an allosteric modulator of VGLUTs, there was a reduction in intravesicular pH, as measured by the decrease of synap-to-pHluorin fluorescence in astrocytes [(Montana et al. 2004); also see Figure 15.4], indicating the presence of functional VGLUTs. Rose Bengal also caused a reduction in Ca^{2+}-dependent exocytic glutamate release in astrocytes (Montana et al. 2004; also see Figure 15.3).

VGLUTs are developmentally regulated in astrocytes and neurons. There is a gradual age-dependent decrease in the expression of all three VGLUT isoforms in astrocytes from the visual cortices of rats (Montana et al. 2004). In freshly-isolated astrocytes originating from 1- to 2-day-old rat pups, there is a high likelihood that VGLUTs 1, 2, and 3 are coexpressed in a single cell. Although there has been no direct demonstration of the presence of these proteins in single cells, the proportion of cells expressing individual proteins (86% for VGLUT 1, 99% for VGLUT 2, and 74% for VGLUT 3) supports this inference (see Figure 15.5A). However, the probability of dual or triple expression of VGLUTs in individual astrocytes is greatly reduced when astrocytes isolated from 55-day-old animals were studied, as is the proportion of astrocytes expressing individual VGLUTs. Since each VGLUT isoform is expressed in about a third of the astrocytic population, it is tempting to suggest that astrocytes could show highly complementary distributions of these three proteins, such as is found in adult neurons. An alternative extreme possibility is that only a third of astrocytes express VGLUTs, but express all three isoforms (see Figure 15.5B). Moreover, because freshly-isolated astrocytes are devoid of their processes or contain only their proximal parts, it is possible that in older animals the proteins already have been shipped to distal processes resulting in negative immunoreactivity of astrocytic bodies; similar redistribution of exocytic proteins in neurons has been described during synaptogenesis (Basarsky et al. 1994). However, slices from 35- to 70-day-old rats and confocal immunofluorescence microscopy detection reported that 32 and 8 percent of astrocytes in the dentate molecular layer of hippocampus were positive for VGLUT 1 and VGLUT 2, respectively (Bezzi et al. 2004).

VGLUT isoforms are coexpressed in neuronal populations transiently during development and in different brain regions (Boulland et al. 2004). For example, parallel fibers in the cerebellum of mice switch from VGLUT 2 to VGLUT 1 expression in the first month after birth (Miyazaki et al. 2003); during the midst of this period individual neurons express both isoforms. The subcellular localization of VGLUTs 1 and 2 coexpression during development in individual hippocampal neurons was recently addressed using genetic manipulation of mice, with two different outcomes (Fremeau et al. 2004; Wojcik et al. 2004; also see commentary by Schuske and Jorgensen 2004). The results from Fremeau et al. (2004) are most consistent with the segregation of VGLUT 1- or VGLUT 2-containing vesicles into different synaptic terminals of an individual neuron, and the data from Wojcik et al. (2004) are consistent with the notion that VGLUTs 1 and 2 are colocalized on the same synaptic vesicles. The functional significance of coexpression and switching of VGLUT isoforms remains unclear, although it might be related to different isoforms associating with different vesicular proteins (Pahner et al. 2003). The functional properties of VGLUT isoforms, including substrate specificity (glutamate) and affinity (K_m ~1–2 mM), do not differ substantially (Bellocchio et al. 2000; Takamori et al. 2000, 2001; Bai et al. 2001; Fremeau et al. 2001, 2002; Gras et al. 2002; Schafer et al. 2002; Varoqui et al. 2002). Thus, besides the number of VGLUTs on a vesicle and the magnitude of the proton gradient, the availability of cytoplasmic glutamate represents a significant parameter that should be considered relative to a role of VGLUTs in synaptic transmission.

The cytoplasmic glutamate concentrations in neurons and astrocytes are differentially maintained and regulated (for review see Danbolt 2001; Hertz and Zielke 2004); the cornerstone of this regulation is a metabolic interplay between neurons and astrocytes, that is, the glutamate-glutamine cycle. In astrocytes, but not in neurons, glutamate is converted to glutamine by the enzyme glutamine synthetase. Glutamine is transported out of astrocytes into the extracellular space and is taken up by

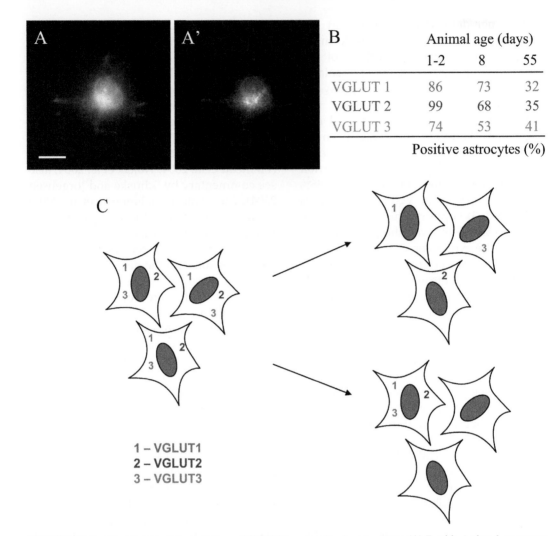

B	Animal age (days)		
	1-2	8	55
VGLUT 1	86	73	32
VGLUT 2	99	68	35
VGLUT 3	74	53	41

Positive astrocytes (%)

1 – VGLUT1
2 – VGLUT2
3 – VGLUT3

FIGURE 15.5.   **Developmental regulation of VGLUT expression in astrocytes.** (**A**) Freshly-isolated astrocytes were identified based on their ability to accumulate β-Ala-Lys-N_ε-AMCA, which selectively accumulates in astrocytes after its uptake, mediated by the PepT2 peptide transporter (Dieck et al. 1999). (**A'**) Double labeling using indirect immunocytochemistry against VGLUT 2 reveals punctate immunoreactivity. (**B**) Table showing expression of VGLUT 1, 2, and 3 in freshly isolated astrocytes originating from animals at various ages. (**C**) Possible mechanism of age-dependent distribution of VGLUTs in astrocytes (see Section II.C for details). AMCA, 7-amino-4-methylcoumarin-3-acetic acid. Scale bar, 10 µm. A and B are modified from Montana et al. (2004).

neurons/presynaptic terminals, where it is converted to glutamate for synaptic release. The glutamate released from neurons is taken up mainly by astrocytes via high affinity glutamate transporters expressed on their plasma membranes (K_m ~20 µM; for details see Danbolt 2001), but also by neurons (see Figure 15.6). Interestingly, the glial function to "chemically split or take up" transmitters was speculated by Lugaro almost 100 years ago (Lugaro 1907). In addition to serving as an excitatory neurotransmitter, glutamate is also a metabolic fuel that gets oxidatively degraded. Its degradation must be compensated for by net (*de novo*) glutamate synthesis from glucose, which is the only glutamate precursor that crosses the blood-brain barrier in sufficient

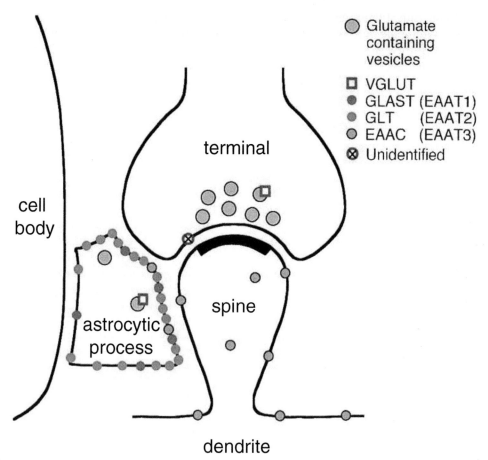

FIGURE 15.6. **Schematic representation of the localization of glutamate transporters at the hippocampal tripartite synapse.** Glutamatergic synapses in the hippocampus are contacted by astrocytes. GLAST and GLT are present in the astrocytic membranes at a high density of 2300 and 8500 molecules per μm^2, respectively, with highest concentration toward the neuropil. Quantitative information on EAAC and on molecularly unidentified transporter at presynaptic terminals is not available (for details see Section 4.2 of Danbolt (2001)). Neurons and astrocytes possess VGLUT-containing vesicles. The L-glutamate/L-aspartate transporter (GLAST-1) and the glial L-glutamate transporter (GLT-1) are also called excitatory amino acid transporters (EAAT1 and EAAT2, respectively); the excitatory amino acid carrier 1 (EAAC) is also called EAAT 3. GLT-1, GLAST, and EAAC are rodent forms of proteins that are highly homologous to human EAATs. Modified from Danbolt (2001), with permission from Elsevier.

quantities. This net synthesis of glutamate via the tricarboxylic acid cycle occurs in astrocytes, but not in neurons, via the enzyme pyruvate carboxylase. Therefore, in astrocytes the source of cytoplasmic glutamate is derived from *de novo* synthesis and from the uptake of extracellular glutamate via plasma membrane transporters. In neurons, the source of cytoplasmic glutamate is also dual, although from the conversion of glutamine (provided by astrocytes) to glutamate via the enzyme glutaminase and from the uptake of extracellular glutamate by glutamate transporters located in the plasma membrane (albeit the molecular identity of presynaptic transporters is debated; reviewed in Danbolt 2001). In synaptic terminals, the glutamate concentration reaches 10 to 15 mM (Attwell et al. 1993), which allows

VGLUTs to operate at their maximal rate to concentrate glutamate into synaptic vesicles, with the intravesicular concentration of glutamate reaching ~60 mM (Burger et al. 1989). When glutamate is released exocytically, the estimated concentration of glutamate in the synaptic cleft is ~1.1 mM (Clements et al. 1992). Astrocytes are responsible for removing glutamate from the synaptic cleft and for maintaining the extracellular glutamate concentrations of ~1 to 2 μM (Anderson and Swanson 2000). Cytoplasmic glutamate concentration in astrocytes is estimated to be 0.1 to 5 mM (Attwell et al. 1993), lower than in neurons, presumably due to the presence of glutamine synthetase. These low mM concentrations would just allow operation of VGLUTs to concentrate glutamate inside astrocytic vesicles. A recent study by Crippa et al. (2005) comparing glutamate uptake into astrocytic and synaptic vesicles, reported that the uptake of glutamate into astrocytic vesicles was three times lower than uptake into synaptic vesicles at a pH of 7.4. The diameters of astrocytic vesicles (majority 30–80 nm) are similar to those reported for neurons (40–60 nm) (de Camilli and Navone 1987), and assuming similar protein content in astrocytic and neuronal vesicles, the estimated intravesicluar concentration of glutamate in astrocytic vesicles is ~20 mM. Ca^{2+}-dependent glutamate release from astrocytes results in localized extracellular glutamate of 1 to 100 μM (Innocenti et al. 2000).

D. Other Indicators of Exocytosis in Astrocytes

In 1910, Nageotte described secretory granules in glia of gray matter and proposed that these cells may release substances into the blood, acting like an endocrine gland (Nageotte 1910). Only recently, however, has there been accumulating evidence of morphological and functional correlates of the exocytic process in astrocytes. The core SNARE complex proteins synaptobrevin 2, its homologue, cellubrevin (Parpura et al. 1995a), syntaxin 1 (Maienschein et al. 1999), and SNAP-23 (Wilhelm et al. 2004) are expressed in astrocytes, and display punctate immunoreactivity. Additionally, punctate immunoreactivity has been reported for the Ca^{2+} sensor Syt 4 (Zhang et al. 2004a), and for the proteins involved in filling astrocytic vesicles with glutamate: V-ATPase (Wilhelm et al. 2004) and VGLUTs 1, 2, and 3 (Bezzi et al. 2004; Montana et al. 2004; Zhang et al. 2004b; Anlauf and Derouiche 2005). Ultrastructural studies of astrocytes revealed that synaptobrevin 2 can be associated with electron-lucent (clear) and some dense core vesicular structures, with diameters ranging from 100 to 700 nm (Maienschein et al. 1999; also see Figure 15.7A), and most dense core (and some smaller and less dense) vesicles were associated with secretogranin 2 (Calegari et al. 1999; Coco et al. 2003). Cellubrevin and VGLUTs 1 and 2 were associated with small (mean diameter of ~30 nm) clear vesicles (Bezzi et al. 2004). Using a preloading technique that stimulated membrane recycling and trapping of styryl dyes (FM 1-43 or FM 2-10) in secretory organelles, astrocytes displayed a punctate pattern of FM fluorescence (Chen et al. 2005; also see Figure 15.7B). When this loading was followed by photoconversion and electron microscopy, Chen et al. (2005) showed that astrocytic vesicles had a mean diameter of 310 nm. Immuno-isolated synaptobrevin 2-containing vesicles are predominantly clear and heterogeneous in size, ranging from 30 nm to over 100 nm (the majority had diameters of 30–80 nm); in addition to clear vesicles, there were some dense core vesicles representing about 2 percent of the total number of vesicles isolated (Crippa et al. 2005). Synaptobrevin 2-containing vesicles also contained VGLUT 2 and Syt 4. Unlike neuronal synaptic vesicles that contain a large pool of SNAP-25 recycled from the plasma membrane, astrocytic vesicles do not contain detectable amounts of SNAP-23.

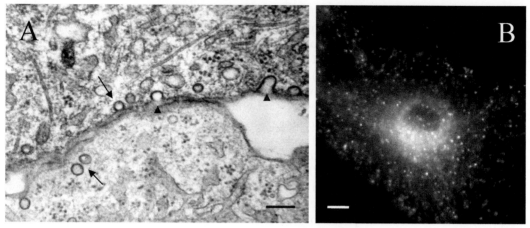

FIGURE 15.7. **Morphological correlates of exocytosis in astrocytes.** (**A**) An electron micrograph showing the contact between two astrocytes. There are several clear vesicles (arrows) close to the plasma membrane, which exhibits some omega shapes consistent with vesicles fusing with or pinching off the plasma membrane (arrowheads). (**B**) Astrocytes exposed to the coapplication of a recycling dye RH-414 (40 μM) and bradykinin (1 μM), known to cause Ca²⁺-dependent glutamate release, followed by a wash, accumulated RH-414, exhibiting a punctate pattern of fluorescence throughout entire cell body. Scale bars, 200 nm in **A**, and 10 μm in **B**. Micrograph courtesy of Dr. Srdija Jeftinija, Iowa State University.

During the process of exocytosis, there is a net addition of vesicular membrane to the plasma membrane, which is eventually counterbalanced by endocytosis involved in retrieval of plasma membrane (see Chapters 6 and 7). This process can be directly assessed by monitoring changes in C_m. Stimulation resulting in elevated $[Ca^{2+}]_i$, using photolysis of caged Ca^{2+} or t-ACPD, results in an increase in astrocyte surface area as assessed by capacitance measurements (Kreft et al. 2004; Zhang et al. 2004b). A delayed decrease in C_m caused by endocytotic retrieval of membrane has been observed and is consistent with the presence of endocytic system in astrocytes (Megias et al. 2000). The rate of endocytosis is dependent on the stimulation level similar to the exo-endocytic relationship in neurons (von Gersdorff and Matthews 1994; Neves and Lagnado 1999); higher increases in cytoplasmic $[Ca^{2+}]$ to 20–60 μM result in larger decreases in C_m, with estimated rates of 50 fF/s (Zhang et al. 2004b). The average increase in C_m, induced by application of t-ACPD, is 2.4 pF (Zhang et al. 2004b). Although individual vesicle

fusion events were not recorded due to technical limitations, one can estimate the number of full vesicular fusions. If we assume an astrocytic vesicular diameter to be 50 nm and the specific surface capacitance of 1 μF/cm² (Neher and Marty 1982), then each vesicle can contribute 79 aF of capacitance when fusing with the plasma membrane. Therefore, the total increase in C_m of 2.4 pF would correspond to the fusion of about 30,500 vesicles with the plasma membrane. However, kiss-and-run events would contribute little to the increase in C_m when compared to full fusions. It has been estimated that kiss-and-run represents approximately 60 percent, and full fusion represents approximately 40 percent of the total number of exocytic events in astrocytes stimulated with an mGluR agonist (Bezzi et al. 2004). Given these estimates, the number of astrocytic vesicles fusing upon stimulation would then correspond to about 76,000. Since the processes of a single astrocyte can make contacts with over 100,000 synapses (Bushong et al. 2002), assuming that astrocytic vesicles are distributed evenly over all their processes, there would be only a small number of vesicles available at each individual astrocytic process and some processes would probably be devoid of vesicles, consistent with ultrastructural data (Bezzi et al. 2004).

An alternative method of examining plasma membrane dynamics during exocytosis uses fluorescent styryl dyes to monitor cumulative exocytosis (see Chapter 6) (Betz et al. 1992; Smith and Betz 1996). Exocytic release of atrial natriuretic peptide (ANP) stimulated by the application of a Ca^{2+} ionophore, in the presence of extracellular Ca^{2+}, caused concomitant uptake of the membrane recycling dye, FM 4-64 (Krzan et al. 2003). This finding also confirmed the existence of previously described exocytosis via dense core vesicles in cultured astrocytes (Calegari et al. 1999). When a similar loading of astrocytes during regulated exocytosis is followed by extensive rinsing, styryl dye gets trapped in secretory organelles (see Figure 15.7). The resulting punctate FM fluorescence can be diminished, due to FM discharge to the extracellular space (referred to as destaining), when vesicles undergo fusion with the plasma membrane. Chen et al. (2005) used glutamate or laser photostimulation to produce increases in astrocytic $[Ca^{2+}]_i$ and induce regulated exocytosis. Styryl dye accumulation may reflect membrane retrieval during regulated exocytosis. These dyes can also accumulate during constitutive membrane cycling (Zhang et al. 2004a). Unlike neurons, astrocytes exhibit high levels of constitutive membrane retrieval (Zhang et al. 2004a), consistent with the reported release of glutamate under basal conditions (Parpura et al. 1994, 1995b) (see later). The process of exocytosis is characterized by quantal release of neurotransmitter, which has been demonstrated in astrocytes using three different technical approaches: biosensor cells (Pasti et al. 2001), TIRFM (Bezzi et al. 2004), and amperometry (Chen et al. 2005). Pasti et al. (2001) transfected human embryonic kidney (HEK) 293 cells with NMDAR subunits 1 and 2A, and

used them as glutamate biosensors/ reporters (NR cells). These NR cells were cocultured with astrocytes. Selective stimulation of astrocytes via activation of AMPARs/mGLURs caused oscillatory $[Ca^{2+}]_i$ elevations in astrocytes, some of which were followed by an NMDAR mediated $[Ca^{2+}]_i$ elevation in adjacent NR cells; whole-cell patch clamp recordings of NR cells indicated that the activation of NMDARs on NR cells caused inward currents. This glutamate-mediated signaling was tetanus toxin and bafilomycin A_1 sensitive. The current kinetics suggested pulsatile and quantal release of glutamate from astrocytes, since their rise times (as fast as 1.5 msec; with a mean of 20 msec) and decay time constants (with a mean of 563 msec) were in good agreement with those of NMDA excitatory PSCs (EPSCs) or those currents obtained with prolonged applications of NMDA on neurons (Collingridge et al. 1988; Lester et al. 1990; Lester and Jahr 1992). In addition to currents with fast kinetics, slower currents (with rise and decay times of >200 msec and 3 sec, respectively) also were recorded and most likely reflect the nature of the contact between astrocytes and NR cells. This study demonstrated quantal-like release events recorded from astrocytes consistent with vesicular exocytosis.

Further evidence for vesicular exocytosis from astrocytes was obtained using TIRFM (Bezzi et al. 2004). Cultured astrocytes were transfected with either VGLUT 1-EGFP or VGLUT 2-EGFP, which displayed punctate patterns of EGFP fluorescence throughout astrocytic bodies and processes. To examine spatio-temporal characteristics of the exocytosis of VGLUT positive vesicles, astrocytes were loaded with acridine orange (AO), a well-known vital stain for nucleic acids (Ranadive and Korgaonkar 1960; Schiffer and Vaharu 1962), which also accumulates in acidic compartments in a self-quenched and red-shifted state (Palmgren 1991). AO can be released into the extracellular space when an AO-loaded organelle fuses with the plasma membrane. Fusions can be detected as transient increases (flashes) of light due to dequenching, followed by the disappearance of fluorescent signal from the organelle due to diffusion of AO out of the organelle. Double (AO and VGLUT) positive organelles frequently were found in small groups that underwent slow Brownian motion (Bezzi et al. 2004). Stimulation of astrocytes with an mGluR agonist, dihydroxyphenylglycine (DHPG), known to cause Ca^{2+}-dependent release of glutamate from astrocytes (Bezzi et al. 1998), caused a slow exocytic burst (discussed earlier), recorded as AO flashes. This burst caused by 1 mM DHPG involved approximately 120 flashes, corresponding to about one-third of the VGLUT-positive puncta detected in the imaging plane and possibly representing a readily releasable pool. In the midst of the burst (~300 ms) EGFP fluorescence of double-labeled organelles either persisted in place (60%) or disappeared (40%), suggesting two modes of fusion: kiss-and-run and full fusion, respectively. When astrocytes were exposed to ionomycin or α-latrotoxin, there were almost three times as many flashes (320 and 340 flashes, respectively) elicited when compared to those induced by DHPG, and about 90 percent of the AO puncta disappeared. Assuming that each fluorescent dot represented a single vesicle, then the total number of vesicles is much smaller than that calculated from capacitance measurements. This should not come as a surprise, since TIRFM detects only the fusion of vesicles with the astrocytic surface within ~100 nm of the glass interface (Steyer and Almers 2001), but not fusion events occurring throughout the entire cell as detected in capacitance measurements. In a different study, Crippa et al. (2005) expressed a synaptobrevin 2-EGFP chimeric protein in astrocytes that displayed puncate fluorescence under confocal microscopy. When astrocytes were stimulated with ionomycin, about 60 percent of the fluorescent puncta disappeared; there was a "compensatory"

increase in the plasma membrane fluorescence consistent with full fusion of synaptobrevin 2-EGFP labeled vesicles. Similarly, the number of synaptobrevin 2-EGFP puncta was reduced by about 50 percent when astrocytes were stimulated with α-latrotoxin. Additionally, stimulation of astrocytes with the mGluR and the AMPAR agonist L-quisqualate resulted in a time-dependent reduction in the number of synaptobrevin 2-EGFP puncta that preferentially occurred in the processes rather than in the cell body. Together, these complementary studies provide further evidence for vesicular fusion with the plasma membrane, in the cell bodies and processes of astrocytes.

Amperometric measurement can provide information on different modes of exocytosis. When a vesicle fuses with the plasma membrane the initial opening of a fusion pore is detected as a prespike foot (reviewed in Mosharov and Sulzer 2005). This is followed by dilation of the pore and full fusion of the vesicle with the plasma membrane, which discharges its intravesicular content to generate a spike in the amperometric recording. In kiss-and-run events, however, fusion pores open and close without dilation, so that in these recordings the event appears as a stand-alone foot, without a spike. Amperometry uses carbon fibers to detect oxidizable secretory products or transmitters (Wightman et al. 1991) such as dopamine, but can also detect nonoxidizable transmitters, such as glutamate, when carbon fibers are enzymatically treated, albeit with much reduced temporal resolution (~300 ms) due to the speed of the enzymatic reaction (Pantano and Kuhr 1993). Hence, to obtain amperometric measurements from astrocytes, Chen et al. (2005) loaded the cells with dopamine. They found indirect evidence for dopamine storage in glutamatergic vesicles, and thus concluded dopamine could act as a surrogate transmitter for glutamate. Stimulation of astrocytes with glutamate (500 μM) resulted exclusively in kiss-and-run events, whereas more robust mechanical stimulation elicited both kiss-and-run and full fusion events. In neurons, kiss-and-run is characterized by transmitter release from a fusion pore that can open and close multiple times a second (Klyachko and Jackson 2002; Staal et al. 2004) in a process known as *flickering*. Flickering was rare in astrocytes, although these cells predominately exhibited just "single kisses" (Chen et al. 2005). Fusion pore (prespike foot) opening can be detected in neurons preceding about 10 percent of full fusion events (Klyachko and Jackson 2002), whereas during full fusion events in astrocytes such pores were detected only in ~0.4 percent of events (Chen et al. 2005). The comparison of kinetics of astrocytic amperometric spikes to those obtained from neurons (Bruns and Jahn 1995) indicates that the quantal events in astrocytes (Chen et al. 2005) are much slower, larger, and wider than those in neurons (see Table 15.1), perhaps consistent with the apparent large size of astrocytic vesicles (310 nm) reported in this study.

IV. VESICULAR TRAFFICKING

Delivery and transport of secretory vesicles to the plasma membrane exocytic sites has been investigated in astrocytes (Crippa et al. 2005; Giau et al. 2005; Potokar et al. 2005). To address the mobility of vesicles in astrocytes, Potokar et al. (2005) expressed chimeric pro-ANP fused with the emerald green fluorescent protein (ANP-emd; Han et al. 1999), which is localized in a punctate pattern (Krzan et al. 2003; Kreft et al. 2004; Potokar et al. 2005) and is stored in a population of vesicles distinct from glutamatergic vesicles (Kreft et al. 2004). All vesicles containing ANP-emd showed mobility that could be classified into two types. The majority of vesicles (65%) displayed nondirectional mobility (average directionality index 0.1, where 1 indicates a straight line) determined by free diffusion with an average speed of 0.3 μm/sec. The other vesicles displayed directional mobility (average directionality index 0.4, maximum at 0.7), with an average speed of 0.5 μm/sec (maximum speed of 3 μm/sec). Vesicles with directional mobility traveled 1.5 times longer and showed five times larger maximal displacement than nondirectional vesicles. Directional vesicles frequently changed: (1) the speed of their mobility, showing pauses and saltatory motion, the latter being associated with microtubules (Freed and Lebowitz 1970); and (2) their movement pattern, showing changes in direction after pauses, previously described in neurites of PC12 cells (Schutz et al. 2004). In neurons, the speed of fast transport of vesicles along microtubules is in the range of 0.8 to 3.5 μm/sec (Grafstein and Forman 1980; Goldstein and Yang 2000). Thus, directional vesicles in astrocytes have speed and characteristics similar to vesicular trafficking observed in neurons.

To study the mobility of small synaptic-like vesicles in astrocytes, Crippa et al. (2005) expressed synaptobrevin 2-EGFP resulting in a punctate pattern of fluorescence throughout astrocytic bodies and processes (Crippa et al. 2005; Giau et al. 2005; Potokar et al. 2005). The trafficking and dynamics of these vesicles were examined using confocal microscopy. The vesicles were either stationary (75%) or showed mobility (25%). Based on the type of motion displayed, the vesicles were classified as either long-range, directional (mean values of speed above 0.6 μm/s), or short-range, nondirectional (mean values of speed below 0.4 μm/s) vesicles. Directional vesicles displayed six times larger maximal displacements than nondirectional vesicles. Some of the directional vesicles displayed pauses and rapid changes in direction, including reversals. These findings agree with the behavior of ANP-emd containing vesicles in astrocytes, as descried earlier (Potokar et al. 2005). In astrocytic processes, however, the movement of synaptobrevin

2-EGFP vesicles was predominately directional. Vesicles were observed to rapidly travel toward the leading tip of an astrocytic process, or a growth cone. Once they reached the tip these vesicles were then stabilized/immobilized. As previously discussed, when astrocytes were stimulated to display regulated exocytosis, these vesicles were seen to fuse more readily at the processes than at the astrocytic bodies, suggesting that the vesicular release sites might be preferentially located in the processes.

Giau et al. (2005) studied constitutive secretion of protease nexin-1 (PN-1) in astrocytes and glioma C6 cells. They inserted EGFP between the signal peptide sequence (aa 1–20) and mature PN1 (aa 21–408) to make a chimeric protein, referred to as GFP/PN-1. When expressed in astrocytes and glioma C6 cells, GFP/PN-1 localized within the secretory pathway, in the *trans*-Golgi network (TGN) and transport vesicles, and localized in a punctate pattern, with each cell expressing an average of 320 puncta. PN-1 was constitutively secreted from both types of glial cells. In C6 glial cells, more than 95 percent of GPF/PN-1-containing vesicles were transported from the TGN along microtubule tracks, displaying discontinuous movement with an average speed of ~0.4μm/sec (maximal speed of ~1μm/sec), which exhibited fast runs, pauses, and changes in direction. Although some vesicles moved in reverse, the overall mobility of vesicles was directed from the TGN to the plasma membrane, where the fluorescent puncta disappeared due to discharge of GFP/PN-1. Some vesicles paused at the plasma membrane before their fluorescence disappeared. In these cases, the cortical F-actin network underneath the plasma membrane acted as a barrier that hindered the constitutive exocytosis of GFP/PN-1, since its disruption, via activation of GPCRs by vasointestinal/pituitary adenylate cyclase peptides (VPAC) or via direct activation of G-proteins by mastroparan, caused an increase in GFP/PN-1 release into the extra-cellular space. Receptor-mediated regulation of PN-1 constitutive secretion triggered by VPAC receptor activation was demonstrated not only for glioma C6 cells, but also for astrocytes. F-actin was previously demonstrated to constrain chromaffin cell granules in regulated exocytosis (Oheim and Stuhmer 2000). Thus, the F-actin cortical network can act as a barrier for constitutive and regulated exocytosis.

V. INPUT AND RELEASE SITES ON ASTROCYTES

It is clear that astrocytes can listen and talk to neurons via transmitter release (Araque et al. 1999; Haydon 2001). These data suggest the existence of localized input and release sites for the exchange of signals between neurons and astrocytes. As discussed previously, glutamatergic neuron-glia signaling occurs via spillover of synaptically released glutamate (Dani et al. 1992; Porter and McCarthy 1996). However, neuron-glia signaling can also occur through direct synapse-like connections, ectopic sites at presynaptic terminals, or even classical synaptic contacts. This indicates that glia are capable of responding to neuronal activity at levels of neurotransmitter release due to asynchronous quantal release and/or a single action potential.

There are direct inputs from neurons to glia through synapse-like, synaptoid contacts (see Figure 15.8A), where axonal projections end on pituicytes, specialized astrocytic cells, without notable presence of postsynaptic differentiation (PSD) on pituicytes (Wittkowski and Brinkmann 1974; van Leeuwen et al. 1983; Buijs et al. 1987). Stimulation of the pituitary stalk led to depolarization of pituicytes *in situ*, an effect mediated by GABA and dopamine receptors on this glial cell type (Mudrick-Donnon et al. 1993). Similar contacts exist on septo-hippocampal astrocytes that receive inputs from norepinephrine terminals (Milner et al. 1995).

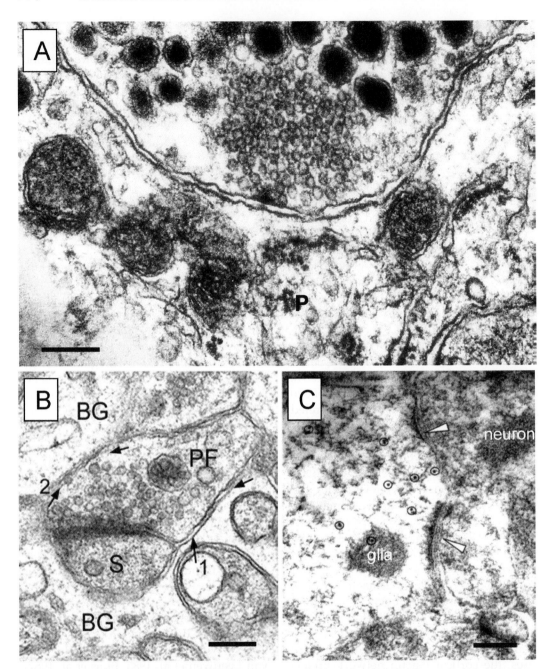

FIGURE 15.8. **Types of contacts between neuronal synaptic terminals and astrocytic (GFAP positive) cells.**
(**A**) Neurosecretory axon containing both dense core and clear secretroy vesicles making a synaptoid contact with a pituitary astrocyte (P). Note complete absence of postsynaptic membrane differentiations on the astrocyte apposing the presynaptic terminal. (**B**) Parallel fiber terminal (PF) is making a synapse on a dendritic spine (S) of a Purkinje neuron. Close appositions of Bergmann glia (BG) and PF membranes contain extracellular electron dense material (arrows, 1 and 2). Synaptic vesicles are morphologically docked to the presynaptic differentiations and close to the presynaptic membrane facing the BG. (**C**) Synaptic structures with typical neuronal synaptic terminals containing vesicles contact (arrowheads) the EGFP/GFAP-positive glial cells labeled using indirect immunocytochemistry with anti-GFP and secondary antibody conjugated to 12 nm gold particles (circles). Note postsynaptic densities in glial elements. Scale bars, 330 nm in **A**, 100 nm in **B**, and 250 nm in **C**. Micrograph in **A** courtesy of Dr. Glenn I. Hatton, University of California Riverside. **B** is reprinted from Matsui et al. (2005), with permission from the *Journal of Neuroscience* (Copyright 2005 by the Society for Neuroscience). **C** is reprinted from Jabs et al. (2005), with permission from the Company of Biologists Ltd.

In Bergmann glia, an astocytic cell found in the cerebellum, quantal responses to glutamate release from climbing and parallel fibers can be recorded *in situ* (Matsui and Jahr 2003, 2004; Matsui et al. 2005). The kinetics of these events are comparable to those recorded in the adjacent neurons, Purkinje cells, which are synaptic targets of climbing and parallel fibers. AMPARs expressed by Purkinje cells are activated by exocytosis occurring at active zones, and they are activated in Bergmann glia at ectopic release sites, defined as release sites outside of the active zones of the presynaptic terminals. Ectopic release activates low-affinity Ca^{2+}-permeable AMPARs in Bergmann glia (subunits GluR 1 and 4, but lacking GluR2; see Geiger et al. 1995), providing a rapid form of point-to-point information transfer from neuron to Bergmann glia (Matsui et al. 2005). At the EM level, parallel fiber-Purkinje cell synapses were surrounded by Bergmann glial processes (see Figure 15.8B). The distance between pre- and postsynaptic (neuronal) plasma membranes within the synaptic cleft was about 1.9 times wider than the distance between the plasma membranes of the presynaptic terminal and apposing Bergmann glia process. Ectopic sites have electron dense material within the extracellular space. The plasma membranes of Bergmann glia apposing the ectopic sites of parallel fibers were decorated with GluR1-associated gold particles, suggesting that they contain clusters of AMPARs. Such clustering was not evident for the plasma membranes of Bergmann glia surrounding climbing fibers. Nonetheless, the estimated density of AMPARs at the Bergmann glia membrane surrounding presynaptic terminals was, on average, 77 receptors/μm^2 (Matsui et al. 2005), much less than the 1280 receptors/μm^2 reported at the neuronal PSD (Tanaka et al. 2005). Additionally, the reported clustering of AMPARs at postsynaptic membranes in Bergmann glia was not noteworthy, since Bergmann glia have an average of approximately 47 receptors/ μm^2 on their somata, as compared to approximately 3 receptors/μm^2 on somata of Purkinje cells. Previous work has described glial microdomains at the Bergmann glial membranes ensheathing parallel fiber-Purkinje cell synapses (Grosche et al. 1999). Electrical stimulation of parallel fibers caused rapid and transient $[Ca^{2+}]_i$ elevations only in small compartments within Bergmann glial processes. Such Ca^{2+} microdomains seem to operate independently within the same process. It is if astrocytes can have point-to-point neuronal inputs at subsets of their processes, so that perhaps their processes may be involved in integrative functions similar to those of neuronal dendrites.

Direct glutamatergic and GABAergic synapses on glial cells expressing the proteoglycan NG2 (therefore, regarded as oligodendrocyte precursor cells) were observed in the hippocampus (Bergles et al. 2000; Lin and Bergles 2004). Similarly, direct glutamatergic and GABAergic synapses were observed on GluR cells in the hippocampus (Jabs et al. 2005). A subset of weakly EGFP expressing cells were observed in mice expressing EGFP under the control of the human GFAP promoter, and shown to coexpress ionotropic GluRs, presumably AMPARs. These cells were termed GluR cells. Since these GFAP positive cells not only coexpress S-100β, a common astrocytic marker, but also the proteoglycan NG2, a common OPC marker (but see Greenwood and Butt 2003; Aguirre et al. 2004) and have transcripts for the neuronal (but see Conti et al. 1998 and also Figure 15.6) excitatory amino acid carrier 1 (EAAC1), they do not match a definition of the classical astrocyte (Kimelberg 2004). Ultrastructurally, neuronal inputs onto GluR cells show all the morphological characteristics of chemical synapses, with the typical features of the neuronal presynaptic terminal and apposing PSDs in postsynaptic glial compartments (see Figure 15.8C). Although the molecular composition of glial PSDs has not been defined, it is reasonable to assume that they may contain mGluR 5 and AMPA receptors together with their respective interacting proteins, Homer and PICK 1, all found expressed in hypothalamic astrocytes (Ojeda et al. 2003). Direct synaptic connections onto classical astrocytes have not as yet been reported. However, the existence of multiple structures mediating the transfer of information from neurons to GFAP positive cells using several different routes—spillover, synaptoid, ectopic sites, and direct chemical synapses—is exciting.

The majority of synaptic communication occurs in neuronal processes. Molecularly and structurally, astrocytic processes also are equipped for receiving signals from neurons. Because synaptobrevin 2-containing vesicles in cultured astrocytes preferentially fuse to the plasma membrane at processes (Crippa et al. 2005), and since these vesicles contain the AMPAR subunits, GluR 2, 3 and to a lesser extent GluR 1, these GluRs should be located at various points along the plasma membrane of the processes, owing to vesicular trafficking. Such a scenario exists at neuronal presynaptic terminals, where there is a major pool of these AMPAR subunits residing in the membranes of synaptic vesicles; AMPARs were internalized after their activation and recruited to the surface from vesicular pools after depolarization (see Chapters 8 and 9) (Schenk et al. 2003). In addition to AMPARs, astrocytes express group I mGluRs (Bezzi et al. 1998) (Cai et al. 2000). Strong mGluR 5 immunostaining has been detected on some astrocytic processes surrounding asymmetrical synapses; the staining in astrocytes was continuous along the plasma membrane unlike the punctate immunoreactivity on neuronal plasma membranes (van den Pol et al. 1995). Metabotropic GluR5 is a GPCR that is known to stimulate phospolipase C resulting in production of IP_3 that mobilizes Ca^{2+} from endoplasmic reticulum Ca^{2+} stores. In addition to mGluR 5, astrocytic

bodies and processes also express bradykinin receptors (Hosli and Hosli 1993), another GPCR linked to IP_3 production. Astrocytes in all regions of the adult rat brain express IP_3R_2. Hippocampal astrocytes and Bergmann glia displayed fine hair-like branches that originated from the network of processes; the branches contained punctate immunorecativity for IP_3R_2 (Holtzclaw et al. 2002). This staining was colocalized with S-100β immunoreactivity, and juxtaposed to presynaptic nerve terminals and PSDs, identified by staining synaptophysin and PSD-95, respectively. Thus, this structure is reminiscent of astrocytic/Bergmann glial microdomains where glial processes enwrap synapses in the hippocampus (Spacek and Harris 1998; Ventura and Harris 1999) and cerebellum (Grosche et al. 1999). Additionally, astrocytic processes surrounding synapses contain smooth endoplasmic reticulum (Spacek and Harris 1998; Bezzi et al. 2004), which contains IP_3Rs and acts as a Ca^{2+} store. Therefore, astrocytic processes contain the molecular and structural components, as well as the strategic (perisynaptic) location, to receive signals from neurons. Indeed, the activation of astrocytic mGluRs in response to neuronal glutamatergic transmission was demonstrated as an increase in astrocytic $[Ca^{2+}]_i$ levels (Porter and McCarthy 1996).

Hippocampal astrocytes *in situ* have spontaneous $[Ca^{2+}]_i$ oscillations mediated by IP_3 receptor activation (Nett et al. 2002). These astrocytic $[Ca^{2+}]_i$ oscillations occur at different parts of a process with only moderate differences in pattern. However, the pattern of oscillations between different processes is very different. Similarly, Bergmann glial cells also have asynchronous spontaneous $[Ca^{2+}]_i$ elevations without neuronal input in different microdomains within the same process (Grosche et al. 1999). Thus, astrocytes of the cerebellum and hippocampus seem to be capable of displaying asynchronous $[Ca^{2+}]_i$

oscillations that are confined to distinct domains of cells, processes, and perhaps to portions of a process. Biochemical, ultra-structural, and electrophysiological evidence supports the notion that astrocytic release sites are found in processes. Vesicles expressing synaptobrevin 2 fuse more readily at the processes than at the astrocytic body (Crippa et al. 2005). Clear vesicles are detected in astrocytic perisynaptic processes (Bezzi et al. 2004). Unlike nerve terminals that contain tightly packed clusters of vesicles that are morphologically predocked with the plasma membrane, astrocytic vesicles in processes were more diffuse, often in smaller and less ordered groups than synaptic vesicles. Thus, astrocytes do not seem to posses classical active zones. Astrocytic vesicles were cellubrevin, and VGLUT 1 and 2 positive, and were located in close proximity to extrasynaptic NMDARs on nerve terminals or dendrites. Distances between the astrocytic plasma membrane and neuronal synaptic elements were similar to the distance spanning the synaptic cleft. VGLUT 1 and Syt 4 were found in hippocampal astrocytic processes adjacent to nerve terminals *in situ*, and synaptobrevin 2, complexin, munc-18, and SNAP-23 all were detected in processes of freshly isolated astrocytes (Zhang et al. 2004a,b). These data suggest astrocytic processes possess exocytic secretory machinery capable of glutamate release.

Electrophysiological studies support the notion that astrocytic processes can release glutamate. For example, Araque et al. (1998b) demonstrated that exocytoctic glutamate release from astrocytes increased the frequency of mPSCs in adjacent neurons, an effect mediated by extrasynaptic NMDARs. Similarly, intracellular photolysis of caged IP_3 to raise astrocyte $[Ca^{2+}]_i$ caused an increase in the frequency of spontaneous AMPA EPSCs in pyramidal neurons that was mediated by group I mGluRs (Fiacco and McCarthy 2004). Exocytic release of glutamate from astrocytes can also modu-late action-potential evoked synaptic transmission via neuronal presynaptic mGluRs (Araque et al. 1998a). Therefore, astrocytic processes appear to have input and release sites for bidirectional communication with neurons.

Astrocytic perisynaptic processes are dynamic structures. Distal astrocytic processes have small dimensions (often 50–100 nm wide), with scarce actin filaments as the only cytoskeletal element (intermediate filaments, i.e., GFAP, or microtubules are absent). They contain the actin-binding proteins ezrin and radixin, which are likely to contribute to the plasticity/motility of these perisynaptic processes (Derouiche and Frotscher 2001; Derouiche et al. 2002). Indeed, astrocytic perisynaptic processes display highly dynamic morphological changes near active synaptic terminals *in situ*, occurring with a time-course of minutes (Hirrlinger et al. 2004). Such mobility could affect the strength of glutamatergic synaptic transmission, since these distal processes contain plasma membrane glutamate transporters that can reduce the extracellular glutamate concentration. Additionally, localized expression of ephrin-A3 on astrocytic processes that enwrap spines can activate EphA4 receptors, enriched on dendritic spines, resulting in dendritic spine retraction (Murai et al. 2003). Thus, the motility of astrocytic processes may play a role in regulating the structure of neuronal synapses and in modulating synaptic strength by regulating extracellular glutamate concentrations.

VI. CONCLUDING REMARKS

The exocytosis of glutamate from astrocytes shares many similarities with neuronal exocytosis, although astrocytes lack electrical excitability and possess slightly different molecular machinery.

Morphologically, astrocytes appear as polarized cells with their processes contacting brain capillaries (see commentary by Peppiatt and Attwell 2004) and neuronal somata, as well as enwraping synapses (Hirrlinger et al. 2004). The emerging picture is that the locations where neuronal synapses are enwrapped by perisynaptic astrocytic processes (tripartite synapses) contain sites for the bilateral exchange of information between neurons and astrocytes. It appears that astrocytes can independently provide computation at such subcellular locations, by having asynchronous $[Ca^{2+}]_i$ dynamics, and allowing feedback at the level of a single synapse. Astrocytes could integrate asynchronous signals received from many synapses and provide a global output. When astrocytic $[Ca^{2+}]_i$ becomes globally elevated, an entire group of neurons possessing synaptic sites within the astrocytic domain may be affected, resulting in spatial coordination and synchronization of neuronal activity via astrocytic glutamate release (Angulo et al. 2004; Fellin et al. 2004). Thus, it appears that astrocytes can control synaptic networks and in such a capacity they represent an integral component of the computational power of the brain rather than being just brain "glue." Consequently, it is not surprising that in an ascending phylogeny, the glia (astrocytes) : neuron ratio increases and peaks within the human brain, having both relatively and absolutely the greatest number of glia (Reichenbach 1989; Nedergaard et al. 2003).

Acknowledgments

The authors' work is supported by a grant from the National Institute of Mental Health (R01 MH 69791) and a grant from Department of Defense/Defense Advanced Research Planning Agency/Defense Microelectronics Activity under Award No. DMEA90-02-2-0216. We thank Erik B. Malarkey for comments on the earlier versions of the chapter. Vladimir Parpura is an Institute for Complex Adaptive Matter Senior Fellow.

References

Adler, E.M., Augustine, G.J., Duffy, S.N., Charlton, M.P. (1991). Alien intracellular calcium chelators attenuate neurotransmitter release at the squid giant synapse. *J. Neurosci.* **11**, 1496–1507.

Aguado, F., Espinosa-Parrilla, J.F., Carmona, M.A., Soriano, E. (2002). Neuronal activity regulates correlated network properties of spontaneous calcium transients in astrocytes in situ. *J. Neurosci.* **22**, 9430–9444.

Aguirre, A.A., Chittajallu, R., Belachew, S., Gallo, V. (2004). NG2-expressing cells in the subventricular zone are type C-like cells and contribute to interneuron generation in the postnatal hippocampus. *J. Cell Biol.* **165**, 575–589.

Ales, E., Tabares, L., Poyato, J.M., Valero, V., Lindau, M., Alvarez de Toledo, G. (1999). High calcium concentrations shift the mode of exocytosis to the kiss-and-run mechanism. *Nat. Cell Biol.* **1**, 40–44.

Anderson, C.M., Swanson, R.A. (2000). Astrocyte glutamate transport: Review of properties, regulation, and physiological functions. *Glia* **32**, 1–14.

Angulo, M.C., Kozlov, A.S., Charpak, S., Audinat, E. (2004). Glutamate released from glial cells synchronizes neuronal activity in the hippocampus. *J. Neurosci.* **24**, 6920–6927.

Anlauf, E., Derouiche, A. (2005). Astrocytic exocytosis vesicles and glutamate: A high-resolution immunofluorescence study. *Glia* **49**, 96–106.

Araque, A., Parpura, V., Sanzgiri, R.P., Haydon, P.G. (1998a). Glutamate-dependent astrocyte modulation of synaptic transmission between cultured hippocampal neurons. *Eur. J. Neurosci.* **10**, 2129–2142.

Araque, A., Sanzgiri, R.P., Parpura, V., Haydon, P.G. (1998b). Calcium elevation in astrocytes causes an NMDA receptor-dependent increase in the frequency of miniature synaptic currents in cultured hippocampal neurons. *J. Neurosci.* **18**, 6822–6829.

Araque, A., Parpura, V., Sanzgiri, R.P., Haydon, P.G. (1999). Tripartite synapses: Glia, the unacknowledged partner. *Trends Neurosci.* **22**, 208–215.

Araque, A., Li, N., Doyle, R.T., Haydon, P.G. (2000). SNARE protein-dependent glutamate release from astrocytes. *J. Neurosci.* **20**, 666–673.

Aravanis, A.M., Pyle, J.L., Tsien, R.W. (2003a). Single synaptic vesicles fusing transiently and successively without loss of identity. *Nature* **423**, 643–647.

Aravanis, A.M., Pyle, J.L., Harata, N.C., Tsien, R.W. (2003b). Imaging single synaptic vesicles undergoing repeated fusion events: Kissing, running, and kissing again. *Neuropharmacology* **45**, 797–813.

Arcuino, G., Cotrina, M., Nedergaard, M. (2004). Mechanism and significance of asytocytic Ca2+

signaling. *In* Glial ⇔ neuronal signaling, G.I. Hatton, V. Parpura, eds., 349–363. Kluwer Academic Publishers, Boston, MA.

Arcuino, G., Lin, J.H., Takano, T., Liu, C., Jiang, L., Gao, Q. et al. (2002). Intercellular calcium signaling mediated by point-source burst release of ATP. *Proc. Natl. Acad. Sci. USA* **99**, 9840–9845.

Attwell, D., Barbour, B., Szatkowski, M. (1993). Nonvesicular release of neurotransmitter. *Neuron* **11**, 401–407.

Augustine, G.J., Eckert, R. (1984). Divalent cations differentially support transmitter release at the squid giant synapse. *J. Physiol.* **346**, 257–271.

Bai, L., Xu, H., Collins, J.F., Ghishan, F.K. (2001). Molecular and functional analysis of a novel neuronal vesicular glutamate transporter. *J. Biol. Chem.* **276**, 36764–36769.

Basarsky, T.A., Parpura, V., Haydon, P.G. (1994). Hippocampal synaptogenesis in cell culture: Developmental time course of synapse formation, calcium influx, and synaptic protein distribution. *J. Neurosci.* **14**, 6402–6411.

Bellocchio, E.E., Reimer, R.J., Fremeau, R.T., Jr., Edwards, R.H. (2000). Uptake of glutamate into synaptic vesicles by an inorganic phosphate transporter. *Science* **289**, 957–960.

Bergles, D.E., Roberts, J.D., Somogyi, P., Jahr, C.E. (2000). Glutamatergic synapses on oligodendrocyte precursor cells in the hippocampus. *Nature* **405**, 187–191.

Betz, W.J., Mao, F., Bewick, G.S. (1992). Activity-dependent fluorescent staining and destaining of living vertebrate motor nerve terminals. *J. Neurosci.* **12**, 363–375.

Bezzi, P., Gundersen, V., Galbete, J.L., Seifert, G., Steinhauser, C., Pilati, E., Volterra, A. (2004). Astrocytes contain a vesicular compartment that is competent for regulated exocytosis of glutamate. *Nat. Neurosci.* **7**, 613–620.

Bezzi, P., Carmignoto, G., Pasti, L., Vesce, S., Rossi, D., Rizzini, B.L. et al. (1998). Prostaglandins stimulate calcium-dependent glutamate release in astrocytes. *Nature* **391**, 281–285.

Bezzi, P., Domercq, M., Brambilla, L., Galli, R., Schols, D., De Clercq, E. et al. (2001). CXCR4-activated astrocyte glutamate release via TNF alpha: Amplification by microglia triggers neurotoxicity. *Nat. Neurosci.* **4**, 702–710.

Boulland, J.L., Qureshi, T., Seal, R.P., Rafiki, A., Gundersen, V., Bergersen, L.H. et al. (2004). Expression of the vesicular glutamate transporters during development indicates the widespread corelease of multiple neurotransmitters. *J. Comp. Neurol.* **480**, 264–280.

Bruns, D., Jahn, R. (1995). Real-time measurement of transmitter release from single synaptic vesicles. *Nature* **377**, 62–65.

Buijs, R.M., van Vulpen, E.H., Geffard, M. (1987). Ultrastructural localization of GABA in the supra-optic nucleus and neural lobe. *Neuroscience* **20**, 347–355.

Burger, P.M., Mehl, E., Cameron, P.L., Maycox, P.R., Baumert, M., Lottspeich, F. et al. (1989). Synaptic vesicles immunoisolated from rat cerebral cortex contain high levels of glutamate. *Neuron* **3**, 715–720.

Bushong, E.A., Martone, M.E., Jones, Y.Z., Ellisman, M.H. (2002). Protoplasmic astrocytes in CA1 stratum radiatum occupy separate anatomical domains. *J. Neurosci.* **22**, 183–192.

Cai, Z., Schools, G.P., Kimelberg, H.K. (2000). Metabotropic glutamate receptors in acutely isolated hippocampal astrocytes: developmental changes of mGluR5 mRNA and functional expression. *Glia* **29**, 70–80.

Calegari, F., Coco, S., Taverna, E., Bassetti, M., Verderio, C., Corradi, N. et al. (1999). A regulated secretory pathway in cultured hippocampal astrocytes. *J. Biol. Chem.* **274**, 22539–22547.

Capogna, M., Gahwiler, B.H., Thompson, S.M. (1995). Presynaptic enhancement of inhibitory synaptic transmission by protein kinases A and C in the rat hippocampus in vitro. *J. Neurosci.* **15**, 1249–1260.

Carmignoto, G. (2000). Reciprocal communication systems between astrocytes and neurones. *Prog. Neurobiol.* **62**, 561–581.

Carmignoto, G., Pasti, L., Pozzan, T. (1998). On the role of voltage-dependent calcium channels in calcium signaling of astrocytes in situ. *J. Neurosci.* **18**, 4637–4645.

Charles, A.C., Merrill, J.E., Dirksen, E.R., Sanderson, M.J. (1991). Intercellular signaling in glial cells: Calcium waves and oscillations in response to mechanical stimulation and glutamate. *Neuron* **6**, 983–992.

Chen, G., Ewing, A.G. (1995). Multiple classes of catecholamine vesicles observed during exocytosis from the Planorbis cell body. *Brain Res.* **701**, 167–174.

Chen, G., Gavin, P.F., Luo, G., Ewing, A.G. (1995). Observation and quantitation of exocytosis from the cell body of a fully developed neuron in Planorbis corneus. *J. Neurosci.* **15**, 7747–7755.

Chen, X., Wang, L., Zhou, Y., Zheng, L.H., Zhou, Z. (2005). "Kiss-and-run" glutamate secretion in cultured and freshly isolated rat hippocampal astrocytes. *J. Neurosci.* **25**, 9236–9243.

Chieregatti, E., Chicka, M.C., Chapman, E.R., Baldini, G. (2004). SNAP-23 functions in docking/fusion of granules at low Ca2+. *Mol. Biol. Cell* **15**, 1918–1930.

Chow, R.H., Klingauf, J., Neher, E. (1994). Time course of Ca2+ concentration triggering exocytosis in neuroendocrine cells. *Proc. Natl. Acad. Sci. USA* **91**, 12765–12769.

Clements, J.D., Lester, R.A., Tong, G., Jahr, C.E., Westbrook, G.L. (1992). The time course of glutamate in the synaptic cleft. *Science* **258**, 1498–1501.

Coco, S., Calegari, F., Pravettoni, E., Pozzi, D., Taverna, E., Rosa, P. et al. (2003). Storage and release of ATP from astrocytes in culture. *J. Biol. Chem.* **278**, 1354–1362.

Collingridge, G.L., Herron, C.E., Lester, R.A. (1988). Synaptic activation of N-methyl-D-aspartate receptors in the Schaffer collateral-commissural pathway of rat hippocampus. *J. Physiol.* **399**, 283–300.

Conti, F., DeBiasi, S., Minelli, A., Rothstein, J.D., Melone, M. (1998). EAAC1, a high-affinity glutamate tranporter, is localized to astrocytes and gabaergic neurons besides pyramidal cells in the rat cerebral cortex. *Cereb. Cortex* **8**, 108–116.

Cornell-Bell, A.H., Finkbeiner, S.M., Cooper, M.S., Smith, S.J. (1990). Glutamate induces calcium waves in cultured astrocytes: long-range glial signaling. *Science* **247**, 470–473.

Crippa, D., Schenk, U., Francolini, M., Rosa, P., Verderio, C., Zonta, M. et al. (2006). Synaptobrevin 2-expressing vesicles in astrocytes: Insights into molecular characterization, dynamics and exocytosis. *J. Physiol.* **570**, 567–582.

Csordas, G., Thomas, A.P., Hajnoczky, G. (1999). Quasi-synaptic calcium signal transmission between endoplasmic reticulum and mitochondria. *EMBO J.* **18**, 96–108.

Dai, H., Shin, O.H., Machius, M., Tomchick, D.R., Sudhof, T.C., Rizo, J. (2004). Structural basis for the evolutionary inactivation of Ca2+ binding to synaptotagmin 4. *Nat. Struct. Mol. Biol.* **11**, 844–849.

Danbolt, N.C. (2001). Glutamate uptake. *Prog. Neurobiol.* **65**, 1–105.

Dani, J.W., Chernjavsky, A., Smith, S.J. (1992). Neuronal activity triggers calcium waves in hippocampal astrocyte networks. *Neuron* **8**, 429–440.

de Camilli, P., Navone, F. (1987). Regulated secretory pathways of neurons and their relation to the regulated secretory pathway of endocrine cells. *Ann. NY Acad. Sci.* **493**, 461–479.

Delaney, K.R., Zucker, R.S. (1990). Calcium released by photolysis of DM-nitrophen stimulates transmitter release at squid giant synapse. *J. Physiol.* **426**, 473–498.

Derouiche, A., Frotscher, M. (2001). Peripheral astrocyte processes: Monitoring by selective immunostaining for the actin-binding ERM proteins. *Glia* **36**, 330–341.

Derouiche, A., Anlauf, E., Aumann, G., Muhlstadt, B., Lavialle, M. (2002). Anatomical aspects of gliasynapse interaction: The perisynaptic glial sheath consists of a specialized astrocyte compartment. *J. Physiol. Paris* **96**, 177–182.

Diamond, M.C., Scheibel, A.B., Murphy, G.M., Jr., Harvey, T. (1985). On the brain of a scientist: Albert Einstein. *Exp. Neurol.* **88**, 198–204.

Dieck, S.T., Heuer, H., Ehrchen, J., Otto, C., Bauer, K. (1999). The peptide transporter PepT2 is expressed in rat brain and mediates the accumulation of the fluorescent dipeptide derivative beta-Ala-Lys-Nepsilon-AMCA in astrocytes. *Glia* **25**, 10–20.

Dodge, F.A., Jr., Rahamimoff, R. (1967). Co-operative action a calcium ions in transmitter release at the neuromuscular junction. *J. Physiol.* **193**, 419–432.

Evanko, D.S., Zhang, Q., Zorec, R., Haydon, P.G. (2004). Defining pathways of loss and secretion of chemical messengers from astrocytes. *Glia* **47**, 233–240.

Fellin, T., Pascual, O., Gobbo, S., Pozzan, T., Haydon, P.G., Carmignoto, G. (2004). Neuronal synchrony mediated by astrocytic glutamate through activation of extrasynaptic NMDA receptors. *Neuron* **43**, 729–743.

Fiacco, T.A., McCarthy, K.D. (2004). Intracellular astrocyte calcium waves in situ increase the frequency of spontaneous AMPA receptor currents in CA1 pyramidal neurons. *J. Neurosci.* **24**, 722–732.

Finbow, M.E., Harrison, M.A. (1997). The vacuolar H+-ATPase: a universal proton pump of eukaryotes. *Bio. Chem. J.* **324 (Pt 3)**, 697–712.

Finkbeiner, S. (1992). Calcium waves in astrocytes-filling in the gaps. *Neuron* **8**, 1101–1108.

Freed, J.J., Lebowitz, M.M. (1970). The association of a class of saltatory movements with microtubules in cultured cells. *J. Cell Biol.* **45**, 334–354.

Fremeau, R.T., Jr., Kam, K., Qureshi, T., Johnson, J., Copenhagen, D.R., Storm-Mathisen, J. et al. (2004). Vesicular glutamate transporters 1 and 2 target to functionally distinct synaptic release sites. *Science* **304**, 1815–1819.

Fremeau, R.T., Jr., Troyer, M.D., Pahner, I., Nygaard, G.O., Tran, C.H., Reimer, R.J. et al. (2001). The expression of vesicular glutamate transporters defines two classes of excitatory synapse. *Neuron* **31**, 247–260.

Fremeau, R.T., Jr., Burman, J., Qureshi, T., Tran, C.H., Proctor, J., Johnson, J. et al. (2002). The identification of vesicular glutamate transporter 3 suggests novel modes of signaling by glutamate. *Proc. Natl. Acad. Sci. USA* **99**, 14488–14493.

Gandhi, S.P., Stevens, C.F. (2003). Three modes of synaptic vesicular recycling revealed by single-vesicle imaging. *Nature* **423**, 607–613.

Geiger, J.R., Melcher, T., Koh, D.S., Sakmann, B., Seeburg, P.H., Jonas, P., Monyer, H. (1995). Relative abundance of subunit mRNAs determines gating and Ca2+ permeability of AMPA receptors in principal neurons and interneurons in rat CNS. *Neuron* **15**, 193–204.

Giau, R., Carrette, J., Bockaert, J., Homburger, V. (2005). Constitutive secretion of protease nexin-1 by glial cells and its regulation by G-protein-coupled receptors. *J. Neurosci.* **25**, 8995–9004.

Giaume, C., Meme, W., Koulakoff, A. (2004). Astrocyte gap junctions and glutamate-induced neurotoxicity. *In* Glial ⇔ neuronal signaling, G.I. Hatton, V. Parpura, eds., 323–348. Kluwer Academic Publishers, Boston, MA.

Goldstein, L.S., Yang, Z. (2000). Microtubule-based transport systems in neurons: The roles of kinesins and dyneins. *Annu. Rev. Neurosci.* **23**, 39–71.

Golovina, V.A. (2005). Visualization of localized store-operated calcium entry in mouse astrocytes. Close proximity to the endoplasmic reticulum. *J. Physiol.* **564**, 737–749.

Grafstein, B., Forman, D.S. (1980). Intracellular transport in neurons. *Physiol. Rev.* **60**, 1167–1283.

Gras, C., Herzog, E., Bellenchi, G.C., Bernard, V., Ravassard, P., Pohl, M. et al. (2002). A third vesicular glutamate transporter expressed by cholinergic and serotoninergic neurons. *J. Neurosci.* **22**, 5442–5451.

Greenwood, K., Butt, A.M. (2003). Evidence that perinatal and adult NG2-glia are not conventional oligodendrocyte progenitors and do not depend on axons for their survival. *Mol. Cell Neurosci.* **23**, 544–558.

Grosche, J., Matyash, V., Moller, T., Verkhratsky, A., Reichenbach, A., Kettenmann, H. (1999). Microdomains for neuron-glia interaction: Parallel fiber signaling to Bergmann glial cells. *Nat. Neurosci.* **2**, 139–143.

Guthrie, P.B., Knappenberger, J., Segal, M., Bennett, M.V., Charles, A.C., Kater, S.B. (1999). ATP released from astrocytes mediates glial calcium waves. *J. Neurosci.* **19**, 520–528.

Han, W., Ng, Y.K., Axelrod, D., Levitan, E.S. (1999). Neuropeptide release by efficient recruitment of diffusing cytoplasmic secretory vesicles. *Proc. Natl. Acad. Sci. USA* **96**, 14577–14582.

Hassinger, T.D., Atkinson, P.B., Strecker, G.J., Whalen, L.R., Dudek, F.E., Kossel, A.H., Kater, S.B. (1995). Evidence for glutamate-mediated activation of hippocampal neurons by glial calcium waves. *J. Neurobiol.* **28**, 159–170.

Haydon, P.G. (2001). GLIA: Listening and talking to the synapse. *Nat. Rev. Neurosci.* **2**, 185–193.

Heidelberger, R., Heinemann, C., Neher, E., Matthews, G. (1994). Calcium dependence of the rate of exocytosis in a synaptic terminal. *Nature* **371**, 513–515.

Heinemann, C., Chow, R.H., Neher, E., Zucker, R.S. (1994). Kinetics of the secretory response in bovine chromaffin cells following flash photolysis of caged Ca2+. *Biophys. J.* **67**, 2546–2557.

Hepp, R., Perraut, M., Chasserot-Golaz, S., Galli, T., Aunis, D., Langley, K., Grant, N.J. (1999). Cultured glial cells express the SNAP-25 analogue SNAP-23. *Glia.* **27**, 181–187.

Hertz, L., Zielke, H.R. (2004). Astrocytic control of glutamatergic activity: Astrocytes as stars of the show. *Trends Neurosci.* **27**, 735–743.

Hirrlinger, J., Hulsmann, S., Kirchhoff, F. (2004). Astroglial processes show spontaneous motility at active synaptic terminals in situ. *Eur. J. Neurosci.* **20**, 2235–2239.

Holtzclaw, L.A., Pandhit, S., Bare, D.J., Mignery, G.A., Russell, J.T. (2002). Astrocytes in adult rat brain express type 2 inositol 1,4,5-trisphosphate receptors. *Glia* **39**, 69–84.

Hosli, E., Hosli, L. (1993). Autoradiographic localization of binding sites for neuropeptide Y and bradykinin on astrocytes. *Neuroreport* **4**, 159–162.

Hua, X., Malarkey, E.B., Sunjara, V., Rosenwald, S.E., Li, W.H., Parpura, V. (2004). Ca2+-dependent glutamate release involves two classes of endoplasmic reticulum Ca(2+) stores in astrocytes. *J. Neurosci. Res.* **76**, 86–97.

Huang, L.Y., Neher, E. (1996). Ca2+-dependent exocytosis in the somata of dorsal root ganglion neurons. *Neuron* **17**, 135–145.

Innocenti, B., Parpura, V., Haydon, P.G. (2000). Imaging extracellular waves of glutamate during calcium signaling in cultured astrocytes. *J. Neurosci.* **20**, 1800–1808.

Jabs, R., Pivneva, T., Huttmann, K., Wyczynski, A., Nolte, C., Kettenmann, H., Steinhauser, C. (2005). Synaptic transmission onto hippocampal glial cells with hGFAP promoter activity. *J. Cell Sci.* **118**, 3791–3803.

Jeftinija, S.D., Jeftinija, K.V., Stefanovic, G. (1997). Cultured astrocytes express proteins involved in vesicular glutamate release. *Brain Res.* **750**, 41–47.

Jeftinija, S.D., Jeftinija, K.V., Stefanovic, G., Liu, F. (1996). Neuroligand-evoked calcium-dependent release of excitatory amino acids from cultured astrocytes. *J. Neurochem.* **66**, 676–684.

Jensen, A.M., Chiu, S.Y. (1990). Fluorescence measurement of changes in intracellular calcium induced by excitatory amino acids in cultured cortical astrocytes. *J. Neurosci.* **10**, 1165–1175.

Kang, J., Jiang, L., Goldman, S.A., Nedergaard, M. (1998). Astrocyte-mediated potentiation of inhibitory synaptic transmission. *Nat. Neurosci.* **1**, 683–692.

Katz, B., Miledi, R. (1965). The measurement of synaptic delay, and the time course of acetylcholine release at the neuromuscular junction. *Proc. R. Soc. Lond. B. Biol. Sci.* **161**, 483–495.

Kettenmann, H., Schipke, C.G. (2004). Calcium signalling in astrocytes. *In* Glial ⇔ neuronal signaling, G.I. Hatton, V. Parpura, eds., 297–321. Kluwer Academic Publishers, Boston, MA.

Kimelberg, H.K. (2004). The problem of astrocyte identity. *Neurochem. Int.* **45**, 191–202.

Klyachko, V.A., Jackson, M.B. (2002). Capacitance steps and fusion pores of small and large-dense-core vesicles in nerve terminals. *Nature* **418**, 89–92.

Kreft, M., Stenovec, M., Rupnik, M., Grilc, S., Krzan, M., Potokar, M. et al. (2004). Properties of Ca(2+)-dependent exocytosis in cultured astrocytes. *Glia* **46**, 437–445.

Krzan, M., Stenovec, M., Kreft, M., Pangrsic, T., Grilc, S., Haydon, P.G., Zorec, R. (2003). Calcium-dependent exocytosis of atrial natriuretic peptide from astrocytes. *J. Neurosci.* **23**, 1580–1583.

Latour, I., Hamid, J., Beedle, A.M., Zamponi, G.W., Macvicar, B.A. (2003). Expression of voltage-gated Ca2+ channel subtypes in cultured astrocytes. *Glia* **41**, 347–353.

Lester, R.A., Jahr, C.E. (1992). NMDA channel behavior depends on agonist affinity. *J. Neurosci.* **12**, 635–643.

Lester, R.A., Clements, J.D., Westbrook, G.L., Jahr, C.E. (1990). Channel kinetics determine the time course of NMDA receptor-mediated synaptic currents. *Nature* **346**, 565–567.

Lin, S.C., Bergles, D.E. (2004). Synaptic signaling between GABAergic interneurons and oligodendrocyte precursor cells in the hippocampus. *Nat. Neurosci.* **7**, 24–32.

Liu, Q.S., Xu, Q., Arcuino, G., Kang, J., Nedergaard, M. (2004). Astrocyte-mediated activation of neuronal kainate receptors. *Proc. Natl. Acad. Sci. USA* **101**, 3172–3177.

Llinas, R., Steinberg, I.Z., Walton, K. (1981). Relationship between presynaptic calcium current and postsynaptic potential in squid giant synapse. *Biophys. J.* **33**, 323–351.

Llinas, R., Sugimori, M., Silver, R.B. (1992). Microdomains of high calcium concentration in a presynaptic terminal. *Science* **256**, 677–679.

Lugaro, E. (1907). Sulle funzioni della nevroglia. *Riv. Pat. Nerv. Ment.* **12**, 225–233.

MacVicar, B.A. (1984). Voltage-dependent calcium channels in glial cells. *Science* **226**, 1345–1347.

Maienschein, V., Marxen, M., Volknandt, W., Zimmermann, H. (1999). A plethora of presynaptic proteins associated with ATP-storing organelles in cultured astrocytes. *Glia* **26**, 233–244.

Malarkey, E.B., Parpura, V. (2005). The role of TRPC1 in internal Ca2+ regulation in astrocytes. *Biophysical Journal* **88**, 84A–85A.

Malarkey, E.B., Parpura, V. (2006). Glutamate release from astrocytes: Impact on neuronal function. *In* Hepatic encephalopathy and nitrogen metabolism, D. Häussinger, G. Hircheis, F. Schliess, eds., in press. Springer, London-Heidelberg.

Matsui, K., Jahr, C.E. (2003). Ectopic release of synaptic vesicles. *Neuron* **40**, 1173–1183.

Matsui, K., Jahr, C.E. (2004). Differential control of synaptic and ectopic vesicular release of glutamate. *J. Neurosci.* **24**, 8932–8939.

Matsui, K., Jahr, C.E., Rubio, M.E. (2005). High-concentration rapid transients of glutamate mediate neural-glial communication via ectopic release. *J. Neurosci.* **25**, 7538–7547.

Megias, L., Guerri, C., Fornas, E., Azorin, I., Bendala, E., Sancho-Tello, M. et al. (2000). Endocytosis and transcytosis in growing astrocytes in primary culture. Possible implications in neural development. *Int. J. Dev. Biol.* **44**, 209–221.

Miesenbock, G., De Angelis, D.A., Rothman, J.E. (1998). Visualizing secretion and synaptic transmission with pH-sensitive green fluorescent proteins. *Nature* **394**, 192–195.

Milner, T.A., Kurucz, O.S., Veznedaroglu, E., Pierce, J.P. (1995). Septohippocampal neurons in the rat septal complex have substantial glial coverage and receive direct contacts from noradrenaline terminals. *Brain Res.* **670**, 121–136.

Miyazaki, T., Fukaya, M., Shimizu, H., Watanabe, M. (2003). Subtype switching of vesicular glutamate transporters at parallel fibre-Purkinje cell synapses in developing mouse cerebellum. *Eur. J. Neurosci.* **17**, 2563–2572.

Montana, V., Ni, Y., Sunjara, V., Hua, X., Parpura, V. (2004). Vesicular glutamate transporter-dependent glutamate release from astrocytes. *J. Neurosci.* **24**, 2633–2642.

Mosharov, E.V., Sulzer, D. (2005). Analysis of exocytic events recorded by amperometry. *Nat. Methods* **2**, 651–658.

Mudrick-Donnon, L.A., Williams, P.J., Pittman, Q.J., MacVicar, B.A. (1993). Postsynaptic potentials mediated by GABA and dopamine evoked in stellate glial cells of the pituitary pars intermedia. *J. Neurosci.* **13**, 4660–4668.

Murai, K.K., Nguyen, L.N., Irie, F., Yamaguchi, Y., Pasquale, E.B. (2003). Control of hippocampal dendritic spine morphology through ephrin-A3/EphA4 signaling. *Nat. Neurosci.* **6**, 153–160.

Nageotte, J. (1910). *Phenomenes de secretion dans le protoplasma des cellules nevrogliques de la substance grise.* *C. R. Soc. Biol. (Paris)* **68**, 1068–1069.

Nedergaard, M. (1994). Direct signaling from astrocytes to neurons in cultures of mammalian brain cells. *Science* **263**, 1768–1771.

Nedergaard, M., Ransom, B., Goldman, S.A. (2003). New roles for astrocytes: Redefining the functional architecture of the brain. *Trends Neurosci.* **26**, 523–530.

Neher, E., Marty, A. (1982). Discrete changes of cell membrane capacitance observed under conditions of enhanced secretion in bovine adrenal chromaffin cells. *Proc. Natl. Acad. Sci. USA* **79**, 6712–6716.

Nett, W.J., Oloff, S.H., McCarthy, K.D. (2002). Hippocampal astrocytes in situ exhibit calcium oscillations that occur independent of neuronal activity. *J. Neurophysiol* **87**, 528–537.

Neves, G., Lagnado, L. (1999). The kinetics of exocytosis and endocytosis in the synaptic terminal of goldfish retinal bipolar cells. *J. Physiol.* **515 (Pt 1)**, 181–202.

Newman, E.A. (2003). New roles for astrocytes: Regulation of synaptic transmission. *Trends Neurosci.* **26**, 536–542.

Newman, E.A., Zahs, K.R. (1998). Modulation of neuronal activity by glial cells in the retina. *J. Neurosci.* **18**, 4022–4028.

Ni, Y., Parpura, V. (2002). Monitoring of exocytosis in astrocytes using synapto-phluorines. *Soc. Neurosci. Absrt.* Program No. 527.2.

Nolte, C., Matyash, M., Pivneva, T., Schipke, C.G., Ohlemeyer, C., Hanisch, U.K. et al. (2001). GFAP promoter-controlled EGFP-expressing transgenic mice: a tool to visualize astrocytes and astrogliosis in living brain tissue. *Glia* **33**, 72–86.

Ogata, K., Kosaka, T. (2002). Structural and quantitative analysis of astrocytes in the mouse hippocampus. *Neuroscience* **113**, 221–233.

Ogata, K., Hirata, K., Bole, D.G., Yoshida, S., Tamura, Y., Leckenby, A.M., Ueda, T. (2001). Inhibition of vesicular glutamate storage and exocytic release by Rose Bengal. *J. Neurochem.* **77**, 34–42.

Oheim, M., Stuhmer, W. (2000). Tracking chromaffin granules on their way through the actin cortex. *Eur. Biophys. J.* **29**, 67–89.

Ojeda, S.R., Prevot, V., Heger, S., Lomniczi, A., Dziedzic, B., Mungenast, A. (2003). Glia-to-neuron signaling and the neuroendocrine control of female puberty. *Ann. Med.* **35**, 244–255.

Pahner, I., Holtje, M., Winter, S., Takamori, S., Bellocchio, E.E., Spicher, K. et al. (2003). Functional G-protein heterotrimers are associated with vesicles of putative glutamatergic terminals: implications for regulation of transmitter uptake. *Mol. Cell. Neurosci.* **23**, 398–413.

Palmgren, M.G. (1991). Acridine orange as a probe for measuring pH gradients across membranes: Mechanism and limitations. *Anal. Biochem.* **192**, 316–321.

Pantano, P., Kuhr, W.G. (1993). Dehydrogenase-modified carbon-fiber microelectrodes for the measurement of neurotransmitter dynamics. 2. Covalent modification utilizing avidin-biotin technology. *Anal. Chem.* **65**, 623–630.

Parpura, V. (2004). Glutamate-mediated bi-directional signaling between neurons and astrocytes. *In* Glial neuronal signaling, G.I. Hatton, V. Parpura, eds., 365–395. Kluwer Academic Publishers, Boston, MA.

Parpura, V., Haydon, P.G. (2000). Physiological astrocytic calcium levels stimulate glutamate release to modulate adjacent neurons. *Proc. Natl. Acad. Sci. USA* **97**, 8629–8634.

Parpura, V., Scemes, E., Spray, D.C. (2004). Mechanisms of glutamate release from astrocytes: Gap junction "hemichannels," purinergic receptors and exocytic release. *Neurochem. Int.* **45**, 259–264.

Parpura, V., Tong, W., Yeung, E.S., Haydon, P.G. (1998). Laser-induced native fluorescence (LINF) imaging of serotonin depletion in depolarized neurons. *J. Neurosci. Methods* **82**, 151–158.

Parpura, V., Fang, Y., Basarsky, T., Jahn, R., Haydon, P.G. (1995a). Expression of synaptobrevin II, cellubrevin and syntaxin but not SNAP-25 in cultured astrocytes. *FEBS Lett.* **377**, 489–492.

Parpura, V., Basarsky, T.A., Liu, F., Jeftinija, K., Jeftinija, S., Haydon, P.G. (1994). Glutamate-mediated astrocyte-neuron signalling. *Nature* **369**, 744–747.

Parpura, V., Liu, F., Brethorst, S., Jeftinija, K., Jeftinija, S., Haydon, P.G. (1995b). Alpha-latrotoxin stimulates glutamate release from cortical astrocytes in cell culture. *FEBS Lett.* **360**, 266–270.

Parri, H.R., Crunelli, V. (2001). Pacemaker calcium oscillations in thalamic astrocytes in situ. *Neuroreport* **12**, 3897–3900.

Parri, H.R., Crunelli, V. (2003). The role of Ca2+ in the generation of spontaneous astrocytic Ca2+ oscillations. *Neuroscience* **120**, 979–992.

Parri, H.R., Gould, T.M., Crunelli, V. (2001). Spontaneous astrocytic Ca2+ oscillations in situ drive NMDAR-mediated neuronal excitation. *Nat. Neurosci.* **4**, 803–812.

Pasti, L., Pozzan, T., Carmignoto, G. (1995). Long-lasting changes of calcium oscillations in astrocytes. A new form of glutamate-mediated plasticity. *J. Biol. Chem.* **270**, 15203–15210.

Pasti, L., Volterra, A., Pozzan, T., Carmignoto, G. (1997). Intracellular calcium oscillations in astrocytes: A highly plastic, bidirectional form of communication between neurons and astrocytes in situ. *J. Neurosci.* **17**, 7817–7830.

Pasti, L., Zonta, M., Pozzan, T., Vicini, S., Carmignoto, G. (2001). Cytosolic calcium oscillations in astrocytes may regulate exocytic release of glutamate. *J. Neurosci.* **21**, 477–484.

Peppiatt, C., Attwell, D. (2004). Neurobiology: Feeding the brain. *Nature* **431**, 137–138.

Peters, A., Palay, S.L., and Webster, H.D. (1991). The fine structure of the nervous system. Oxford University Press, New York-Oxford.

Pizzo, P., Burgo, A., Pozzan, T., Fasolato, C. (2001). Role of capacitative calcium entry on glutamate-induced calcium influx in type-I rat cortical astrocytes. *J. Neurochem.* **79**, 98–109.

Porter, J.T., McCarthy, K.D. (1996). Hippocampal astrocytes in situ respond to glutamate released from synaptic terminals. *J. Neurosci.* **16**, 5073–5081.

Potokar, M., Kreft, M., Pangrsic, T., Zorec, R. (2005). Vesicle mobility studied in cultured astrocytes. *Biochem. Biophys. Res. Commun.* **329**, 678–683.

Ranadive, N.S., Korgaonkar, K.S. (1960). Spectrophotometric studies on the binding of acridine orange to ribonucleic acid and deoxyribonucleic acid. *Biochim. Biophys. Acta* **39**, 547–550.

Reichenbach, A. (1989). Glia:neuron index: Review and hypothesis to account for different values in various mammals. *Glia* **2**, 71–77.

Roberts, W.M., Jacobs, R.A., Hudspeth, A.J. (1990). Colocalization of ion channels involved in frequency selectivity and synaptic transmission at presynaptic active zones of hair cells. *J. Neurosci.* **10**, 3664–3684.

Robinson, I.M., Ranjan, R., Schwarz, T.L. (2002). Synaptotagmins I and IV promote transmitter release independently of Ca(2+) binding in the C(2)A domain. *Nature* **418**, 336–340.

Rupnik, M., Kreft, M., Sikdar, S.K., Grilc, S., Romih, R., Zupancic, G. et al. (2000). Rapid regulated dense-core vesicle exocytosis requires the CAPS protein. *Proc. Natl. Acad. Sci. USA* **97**, 5627–5632.

Sabatini, B.L., Regehr, W.G. (1999). Timing of synaptic transmission. *Annu. Rev. Physiol.* **61**, 521–542.

Sankaranarayanan, S., Ryan, T.A. (2001). Calcium accelerates endocytosis of vSNAREs at hippocampal synapses. *Nat. Neurosci.* **4**, 129–136.

Sankaranarayanan, S., De Angelis, D., Rothman, J.E., Ryan, T.A. (2000). The use of pHluorins for optical measurements of presynaptic activity. *Biophys. J.* **79**, 2199–2208.

Schafer, M.K., Varoqui, H., Defamie, N., Weihe, E., Erickson, J.D. (2002). Molecular cloning and functional identification of mouse vesicular glutamate transporter 3 and its expression in subsets of novel excitatory neurons. *J. Biol. Chem.* **277**, 50734–50748.

Schenk, U., Verderio, C., Benfenati, F., Matteoli, M. (2003). Regulated delivery of AMPA receptor subunits to the presynaptic membrane. *Embo. J.* **22**, 558–568.

Schiffer, L.M., Vaharu, T. (1962). Acridine orange as a useful chromosomal stain. *Am. J. Clin. Pathol.* **37**, 669–670.

Schuske, K., Jorgensen, E.M. (2004). Neuroscience. Vesicular glutamate transporter–shooting blanks. *Science* **304**, 1750–1752.

Schutz, G.J., Axmann, M., Freudenthaler, S., Schindler, H., Kandror, K., Roder, J.C., Jeromin, A. (2004). Visualization of vesicle transport along and between distinct pathways in neurites of living cells. *Microsc. Res. Tech.* **63**, 159–167.

Smith, C.B., Betz, W.J. (1996). Simultaneous independent measurement of endocytosis and exocytosis. *Nature* **380**, 531–534.

Smith, S.J., Augustine, G.J., Charlton, M.P. (1985). Transmission at voltage-clamped giant synapse of the squid: Evidence for cooperativity of presynaptic calcium action. *Proc. Natl. Acad. Sci. USA* **82**, 622–625.

Sorensen, J.B., Nagy, G., Varoqueaux, F., Nehring, R.B., Brose, N., Wilson, M.C., Neher, E. (2003). Differential control of the releasable vesicle pools by SNAP-25 splice variants and SNAP-23. *Cell* **114**, 75–86.

Spacek, J., Harris, K.M. (1998). Three-dimensional organization of cell adhesion junctions at synapses and dendritic spines in area CA1 of the rat hippocampus. *J. Comp. Neurol.* **393**, 58–68.

Staal, R.G., Mosharov, E.V., Sulzer, D. (2004). Dopamine neurons release transmitter via a flickering fusion pore. *Nat. Neurosci.* **7**, 341–346.

Stanley, E.F. (1986). Decline in calcium cooperativity as the basis of facilitation at the squid giant synapse. *J. Neurosci.* **6**, 782–789.

Steyer, J.A., Almers, W. (2001). A real-time view of life within 100 nm of the plasma membrane. *Nat. Rev. Mol. Cell Biol.* **2**, 268–275.

Sudhof, T.C. (2001). Alpha-Latrotoxin and its receptors: Neurexins and CIRL/latrophilins. *Annu. Rev. Neurosci.* **24**, 933–962.

Sudhof, T.C. (2004). The synaptic vesicle cycle. *Annu. Rev. Neurosci.* **27**, 509–547.

Takamori, S., Rhee, J.S., Rosenmund, C., Jahn, R. (2000). Identification of a vesicular glutamate transporter that defines a glutamatergic phenotype in neurons. *Nature* **407**, 189–194.

Takamori, S., Rhee, J.S., Rosenmund, C., Jahn, R. (2001). Identification of differentiation-associated brain-specific phosphate transporter as a second vesicular glutamate transporter (VGLUT2). *J. Neurosci.* **21**, RC182.

Tanaka, J., Matsuzaki, M., Tarusawa, E., Momiyama, A., Molnar, E., Kasai, H., Shigemoto, R. (2005). Number and density of AMPA receptors in single synapses in immature cerebellum. *J. Neurosci.* **25**, 799–807.

Thomas, D.M., Ferguson, G.D., Herschman, H.R., Elferink, L.A. (1999). Functional and biochemical analysis of the C2 domains of synaptotagmin IV. *Mol. Biol. Cell* **10**, 2285–2295.

Thomas, P., Wong, J.G., Almers, W. (1993a). Millisecond studies of secretion in single rat pituitary cells stimulated by flash photolysis of caged Ca2+. *EMBO J.* **12**, 303–306.

Thomas, P., Wong, J.G., Lee, A.K., Almers, W. (1993b). A low affinity Ca2+ receptor controls the final steps in peptide secretion from pituitary melanotrophs. *Neuron* **11**, 93–104.

Trudeau, L.E., Emery, D.G., Haydon, P.G. (1996). Direct modulation of the secretory machinery underlies PKA-dependent synaptic facilitation in hippocampal neurons. *Neuron* **17**, 789–797.

Trudeau, L.E., Fang, Y., Haydon, P.G. (1998). Modulation of an early step in the secretory machinery in hippocampal nerve terminals. *Proc. Natl. Acad. Sci. U S A* **95**, 7163–7168.

van den Pol, A.N., Romano, C., Ghosh, P. (1995). Metabotropic glutamate receptor mGluR5 subcellular distribution and developmental expression in hypothalamus. *J. Comp. Neurol.* **362**, 134–150.

van Leeuwen, F.W., Pool, C.W., Sluiter, A.A. (1983). Enkephalin immunoreactivity in synaptoid elements on glial cells in the rat neural lobe. *Neuroscience* **8**, 229–241.

Varoqui, H., Schafer, M.K., Zhu, H., Weihe, E., Erickson, J.D. (2002). Identification of the differentiation-associated Na+/PI transporter as a novel vesicular glutamate transporter expressed in a distinct set of glutamatergic synapses. *J. Neurosci.* **22**, 142–155.

Venance, L., Stella, N., Glowinski, J., Giaume, C. (1997). Mechanism involved in initiation and propagation of receptor-induced intercellular calcium signaling in cultured rat astrocytes. *J. Neurosci.* **17**, 1981–1992.

Ventura, R., Harris, K.M. (1999). Three-dimensional relationships between hippocampal synapses and astrocytes. *J. Neurosci.* **19**, 6897–6906.

Verderio, C., Pozzi, D., Pravettoni, E., Inverardi, F., Schenk, U., Coco, S. et al. (2004). SNAP-25 modulation of calcium dynamics underlies differences in

GABAergic and glutamatergic responsiveness to depolarization. *Neuron* **41**, 599–610.

Verkhratsky, A. (2002). The endoplasmic reticulum and neuronal calcium signalling. *Cell Calcium* **32**, 393–404.

Verkhratsky, A., Kettenmann, H. (1996). Calcium signalling in glial cells. *Trends Neurosci.* **19**, 346–352.

Verkhratsky, A., Steinhauser, C. (2000). Ion channels in glial cells. *Brain Res. Brain Res. Rev.* **32**, 380–412.

Verkhratsky, A., Orkand, R.K., Kettenmann, H. (1998). Glial calcium: Homeostasis and signaling function. *Physiol. Rev.* **78**, 99–141.

Virchow, R. (1856). *Gesammelte Abhandlungen zur wissenschaftlichen Medicin*. Verlag von Meidinger Sohn & Comp., Frankfurt.

Voets, T., Moser, T., Lund, P.E., Chow, R.H., Geppert, M., Sudhof, T.C., Neher, E. (2001). Intracellular calcium dependence of large dense-core vesicle exocytosis in the absence of synaptotagmin I. *Proc. Natl. Acad. Sci. U S A* **98**, 11680–11685.

Volterra, A., Meldolesi, J. (2005). Astrocytes, from brain glue to communication elements: The revolution continues. *Nat. Rev. Neurosci.* **6**, 626–640.

von Gersdorff, H., Matthews, G. (1994). Inhibition of endocytosis by elevated internal calcium in a synaptic terminal. *Nature* **370**, 652–655.

Wang, C.T., Lu, J.C., Bai, J., Chang, P.Y., Martin, T.F., Chapman, E.R., Jackson, M.B. (2003). Different domains of synaptotagmin control the choice between kiss-and-run and full fusion. *Nature* **424**, 943–947.

Wightman, R.M., Jankowski, J.A., Kennedy, R.T., Kawagoe, K.T., Schroeder, T.J., Leszczyszyn, D.J. et al. (1991). Temporally resolved catecholamine spikes correspond to single vesicle release from individual chromaffin cells. *Proc. Natl. Acad. Sci. U S A* **88**, 10754–10758.

Wilhelm, A., Volknandt, W., Langer, D., Nolte, C., Kettenmann, H., Zimmermann, H. (2004). Localization of SNARE proteins and secretory organelle proteins in astrocytes in vitro and in situ. *Neurosci. Res.* **48**, 249–257.

Wittkowski, W., Brinkmann, H. (1974). Changes of extent of neuro-vascular contacts and number of neuro-glial synaptoid contacts in the pituitary posterior lobe of dehydrated rats. *Anat. Embryol. (Berl.)* **146**, 157–165.

Wojcik, S.M., Rhee, J.S., Herzog, E., Sigler, A., Jahn, R., Takamori, S. et al. (2004). An essential role for vesicular glutamate transporter 1 (VGLUT1) in postnatal development and control of quantal size. *Proc. Natl. Acad. Sci. U S A* **101**, 7158–7163.

Yoshihara, M., Adolfsen, B., Galle, K.T., Littleton, J.T. (2005). Retrograde signaling by Syt 4 induces presynaptic release and synapse-specific growth. *Science* **310**, 858–863.

Zhang, J.M., Wang, H.K., Ye, C.Q., Ge, W., Chen, Y., Jiang, Z.L. et al. (2003). ATP released by astrocytes mediates glutamatergic activity-dependent heterosynaptic suppression. *Neuron.* **40**, 971–982.

Zhang, Q., Haydon, P.G. (2005). Roles for gliotransmission in the nervous system. *J. Neural. Transm.* **112**, 121–125.

Zhang, Q., Fukuda, M., Van Bockstaele, E., Pascual, O., Haydon, P.G. (2004a). Synaptotagmin IV regulates glial glutamate release. *Proc. Natl. Acad. Sci. U S A* **101**, 9441–9446.

Zhang, Q., Pangrsic, T., Kreft, M., Krzan, M., Li, N., Sul, J.Y. et al. (2004b). Fusion-related release of glutamate from astrocytes. *J. Biol. Chem.* **279**, 12724–12733.

Zhong, Y., Wu, C.F. (1991). Altered synaptic plasticity in Drosophila memory mutants with a defective cyclic AMP cascade. *Science* **251**, 198–201.

PROTEIN TRAFFICKING AND NEURONAL DISEASE

The endosomal system is a target in a number of neurodegenerative diseases. The pathology in this system may be causative or secondary to the disease process. Defining the specific endosomal compartments involved and the nature of the defects could help to account for certain distinctive features of disease such as age of onset, targeting of particular neuronal populations, and severity of symptoms. The selective vulnerability of neurons to impairments in a fundamental cellular process is curious, but may reflect the unique functions that endosomes serve in neurons that help to maintain the fidelity and efficiency of neuronal communication. This section contains three examples of neurodegenerative diseases whose etiology/pathology are thought to involve defects in trafficking.

16

Trafficking Defects in Huntington's Disease

E. TRUSHINA AND C. T. McMURRAY

I. GENERAL CONSIDERATIONS

The complement of adult neurons is largely maintained throughout development in healthy humans. Some neurogenesis can occur in the adult brain (Gage and McAllister 2005; Sanai et al. 2005). However, the inability to regenerate renders neurons particularly vulnerable to cellular malfunctions and to DNA damage (Charrin et al. 2005; Mattson 2002). The architecture of neurons differs from most other cell types. Long axons must carry out cellular functions at a distance, as much as 1.0m (Williamson et al. 1996; Bradke and Dotti 1998) from the cell body where the bulk of gene expression occurs. Consequently, survival and signal transmission critically depend on internalization, delivery, and recycling of proteins, lipids, and solutes between intracellular compartments and plasma membrane. All of these functions rely on the fidelity of axonal and membrane trafficking (Williamson et al. 1996; Bradke and Dotti 1998; Bonifacino and Glick 2004; Maxfield and McGraw 2004). Since trafficking is essential for neuronal maintenance, it is perhaps not surprising that malfunctions of the trafficking machinery are emerging as defects associated with a number of neurodegenerative diseases (McMurray 2000). One of the best-studied examples is Huntington's disease (HD).

HD is one of a growing list of hereditary and progressive neurodegenerative disorders in which the underlying mutation is a CAG expansion within the coding sequence of the gene (see Figure 16.1A) (Bates 2005; Kovtun and McMurray 2005). This

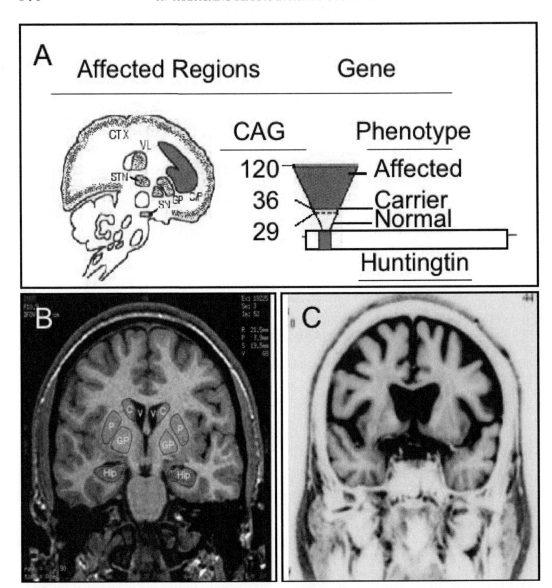

FIGURE 16.1. **Genetics and pathophysiology of Huntington's disease. (A)** The relationship between CAG repeat number and HD pathophysiology. Schematic representation of the HD gene: the small red bar indicates the position of the CAG repeat stretch within the N-terminal portion of the coding sequence. Upside-down triangle represents increasing number of CAG repeats. Base of triangle in white represents normal unaffected individuals with 6–29 CAG repeats. Dotted line indicates unaffected carriers for disease with 29–35 CAG repeats. Red part of the triangle indicates affected individuals with 36–120 CAG repeats (left). Regions of neuronal loss in HD. Red regions indicate the major areas of neuronal loss in HD patients with 36–120 CAG repeats. These brain regions control movement. C/P is the caudate/putamen; CTX is the cortex; GP is globus pallidus; STN is subthalamic nucleus; VL is ventrolateral thalamic nucleus; SN is substantia nigra. **(B)** Magnetic resonance image of a control human brain. Letters defined as in **A** except for hip, hippocampus. All regions of the brain are fully developed. **(C)** HD brain. Note the cortex has extensive invaginations and atrophy; C/P and parts of the GP are missing. These cells have died.

mutation is referred to as *trinucleotide expansion*, since the number of CAG triplets present in a mutated gene is greater than the number found in a normal gene. The number of CAG triplets in the disease gene continues to increase as the disease gene is inherited (see Figure 16.1A). HD is an autosomal dominant disease. Therefore, the expanded CAG repeat is usually present in only one of two alleles in the cell. Unaffected individuals may have 6 to 25 CAG triplets in both alleles; in HD patients, the disease allele typically contains 36 to 120 CAG triplets (Bates 2005). The HD protein, called huntingtin (htt), contains a polymorophic glutamine repeat at its N-terminus (Bates 2005; Hague et al. 2005). As the CAG repeat number grows, the growing polyglutamine tract produces htt with increasingly aberrant properties, and leads to selective neuronal death in the striatum (see Figure 16.1B, C). Neurons that are particularly affected in HD are located in the cerebral cortex and the caudate/putamen. Loss of brain cells in these regions causes memory deficits and uncontrolled muscle movements (chorea) that characterize the disease. Once symptoms begin, death usually occurs within 15 to 20 years (Hague et al. 2005). Although the molecular mechanism for cell death is unknown, emerging data suggest that expression of the mutant form of htt (mhtt) can cause sequestration of the trafficking machinery and/or alter motility of organelles and cellular cargo. Thus, there is intense interest in the relationship between trafficking defects and HD pathophysiology.

This chapter will review data that implicate defects in vesicle trafficking as a causative factor leading to toxicity in HD. We will specifically focus on three points: the interactions that implicate htt as a trafficking protein, how trafficking defects arise from expression of mhtt, and whether defective trafficking is a cause or consequence of toxicity. Finally, we will discuss common links among neurodegenerative disorders and defects in trafficking.

II. EVIDENCE THAT htt IS A TRAFFICKING PROTEIN

A role for htt in trafficking initially was suggested based on the isolation of interacting protein partners (Harjes and Wanker 2003) (see Figure 16.2). Yeast two-hybrid screens revealed that htt interacted with components of the trafficking machinery. Htt-associated protein (HAP1) was one of the first interacting proteins discovered and like htt, HAP1 is enriched in brain (Li and Li 2005; Li et al. 1998a,b) and associates with microtubules and vesicles. HAP1 partially colocalizes with htt in cells and on sucrose gradient *in vitro* (Block-Galarza et al. 1997). Both htt and HAP1 interact with the trafficking motors, kinesin heavy chain and the dynactin p150[glued], an accessory protein for the microtubule motor protein dynein (see Chapter 2) (Li and Li 2005; Engelender et al. 1997; Li et al. 1998a). In *Drosophila*, htt associates with Milton, a protein with homology to HAP1 that is also linked to kinesin-dependent axonal transport of mitochondria (Stowers et al. 2002).

Functional evidence demonstrating that htt plays a role in axonal trafficking has accumulated over the years. Early analysis revealed that htt and HAP1 accumulated on either side of a crushed rat sciatic nerve, suggesting htt was actively transported from distant cellular sites in both retrograde and anterograde directions (Block-Galarza et al. 1997). Whether htt was a cargo or a transport accessory protein was not known. Recent data obtained in mice (Trushina et al. 2004), *Drosophila* (Gunawardena et al. 2003) and isolated squid axoplasm (Szebenyi et al. 2003) have provided functional evidence that htt itself is a vesicular trafficking protein, or, at least, strongly influences the process. In squid axoplasm, expression of a truncated polyglutamine fragment from the androgen receptor inhibits axonal trafficking compared to wild-type protein. When the same peptide is expressed in dividing SYH-SY5Y cells, neurite outgrowth is prevented upon

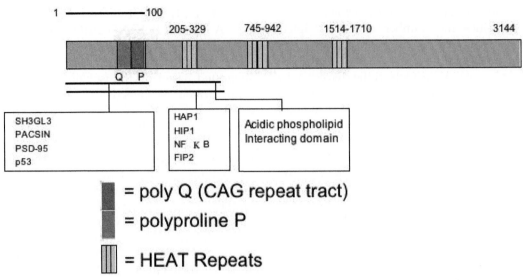

FIGURE 16.2. **Schematic diagram of the Huntington's disease gene product (Htt) and some of its interacting partners and their regions of interaction, if known.** Htt contains a glutamine (Q) and proline (P) rich regions at the N terminus, and three regions with HEAT repeats (after Htt, elongation factor 3, regulatory A subunit of protein phosphatase 2A and TOR1). Many interacting proteins associate at the N-terminal regions. Huntington's interacting protein 1, HIP1; Huntington's interacting protein 12, HIP12; protein kinase C and casein kinase substrate in neurons, PACSIN1; Src homology-containing Grb2-like protein 3, SH3GL3; postsynaptic density protein 95, PSD-95; Huntington's interacting protein 14, HIP14; Huntington's associating protein 1, HAP1; nuclear factor κb, NFκb. The polyQ region is encoded by the expanded CAG tract in the disease gene.

differentiation with retinoic acid and BDNF, suggesting that trafficking defects are microtubule-dependent (Szebenyi et al. 2003). In *Drosophila*, expression of mhtt suppresses axonal transport of GFP-EGF receptor (Gunawardena et al. 2003) or YFP-APP protein (Gauthier et al. 2004), and promotes vesicle and organelle accumulation in axons. These axonal trafficking defects appear to be a direct consequence of mhtt since expression of either truncated mhtt or expanded polyglutamine regions produces trafficking defects in isolated squid axoplasm where neither a nucleus nor protein synthesis is present (Szebenyi et al. 2003).

Expression of mhtt in mice destabilizes microtubule tracts in primary neurons and alters their morphology (Trushina et al. 2003), and results in defective axonal transport of vesicles and mitochondria very early in development (Trushina et al. 2004). Moreover, vesicle transport and uptake of the retrograde dye, fluorogold, is sup-pressed in animals expressing mhtt with 72 polyglutamins. Two days after injection into living animals, the uptake of fluoro-gold in the mhtt-expressing animals is suppressed compared to control animals (Trushina et al. 2004). Thus, expression of the full-length endogenous form of mhtt in mice causes aberrant trafficking both *in vivo* and *in vitro*. Reduction of htt expression compromises trafficking of mitochondria in mouse striatal neurons in both anterograde and retrograde directions (Trushina et al. 2004). Neurons with about 50 percent loss of htt display a more severe trafficking defect than those expressing mhtt, indicating that mhtt can support some of the normal functions of htt. These results are consistent with earlier genetic studies. Both White et al. (1997) and Nasir et al. (1995) demonstrated that mice lacking htt die early in embryogenesis but that behavioral and developmental defects can be partially rescued by one allele of mhtt. Consistent

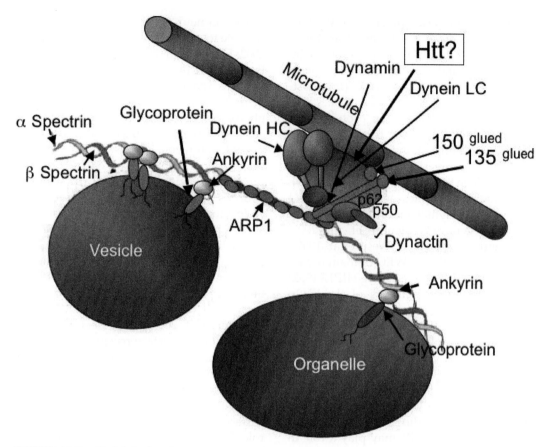

FIGURE 16.3. **Model for huntingtin (htt) function in axonal vesicle transport.** Model for interaction of dynactin/dynein motors with microtubules, membrane intermediates, and Htt. The dynactin complex powers retrograde translocation of pre-Golgi cargo (vesicle (red) or organelle (dark orange)) along microtubules (gray elongated) in axons. Dynein comprises two heavy chains (Dynein HC, gold) containing the minus-end directed motor domains, and light chains (Dynein LC, purple). The dynein motor complex consists of dynein HC and LC, p150[Glued] related protein 1 (red), dynactin (p50/dynamitin and p62, red balls), and the F-actin-like protein (Arp1) (magenta). The anchoring of the dynactin complex to membranes may be mediated by an ankyrin (light gray circles) and the spectrin meshwork (gray helix). Htt interacting partners include dynactin, microtubules, and p150[glued]. Thus, htt (orange) is positioned as a microtubule binding protein, which contacts these proteins. Cargo membranes interact with membrane glycoproteins, which bind to the spectrin filaments.

with these data, expression of mhtt impairs trafficking relative to control mice but partially restores the defect observed in the conditional knock-out animals. These data indicate that htt has a normal function in axonal trafficking, and impaired axonal trafficking appears to be a function gained by mhtt.

As with the axonal transport machinery, many of the htt interactions with endocytic proteins occur at the N-terminal portion of

htt containing the polyglutamine and polyproline regions (Qin et al. 2004). Thus, htt also has been implicated in endocytosis (Paschen and Mengesdorf 2005; Harjes and Wanker 2003) (see Figure 16.3). Htt associates with clathrin, colocalizes with mature clathrin-coated vesicles, and decorates clathrin-coated pits (Mattson 2002; Velier et al. 1998; Kim et al. 1999). Htt may form a complex with clathrin through Huntingtin interacting protein (HIP1) (Vecchi and Di

Fiore 2005; Engqvist-Goldstein et al. 1999; Wanker et al. 1997; Kalchman et al. 1997). HIP1 directly binds clathrin light chain, α-adaptin A and C, and colocalizes with the clathrin coat adaptor, AP2 (Engqvist-Goldstein et al. 1999; Waelter et al. 2001). HIP1 organizes these coat proteins and stimulates their assembly through its conserved C-terminal domain, known as the I/LWEQ module (Legendre-Guillemin et al. 2005; Chen et al. 2005; Wesp et al. 1997). This module promotes actin binding and prevents depolymerization of actin filaments (Legendre-Guillemin et al. 2005; Chen and Brodsky 2005; Wesp et al. 1997). The function of the I/LWEQ domain is consistent with HIP1 and its isolog HIP1R (Seki et al. 1998) being members of the Sla2p protein family of actin-organizing proteins required for the internalization step of endocytosis (Wanker et al. 1997; Kalchman et al. 1997; Engqvist-Goldstein et al. 1999). Additionally, the I/LWEQ module of HIP1 promotes both its dimerization with HIP1R, and its association with the plasma membrane via HIP1R binding to F-actin (Senetar et al. 2004). Thus, HIP1 may play a role in endocytosis by linking actin to the plasma membrane. Conceivably, plasma membrane association by HIP1R could be stabilized with the help of the HEAT domains in htt. HEAT domains are regions of low homology shared by htt with elongation factor 3, the p65 regulatory A subunit of protein phosphatase 2A, and TOR1, among others (Andrade and Bor 1995). HEAT repeats are leucine-rich domains, predicted to form superhelical channels in the plasma membrane (Groves et al. 1999). HEAT domains directly bind membranes through electrostatic interactions with acidic phospholipids such as phosphatidylinositol 4,5-bisphosphate (Kegel et al. 2005) (see Figure 16.2). Thus, the htt/HIP1 interaction may regulate endocytosis at the interface between actin filaments and the lipids of the plasma membrane.

Htt also interacts with HIP14, a protein predominantly expressed in medium spiny projection neurons, the cell type most susceptible to cell death in HD (Singaraja et al. 2002; Huang et al. 2004; Ducker et al. 2004). Further, HIP14 is associated with clathrin and has sequence homology with Ark1p, a protein essential for endocytosis of pheromone receptors in yeast. In fact, the ankyrin domain of HIP14 can restore both viability and endocytosis of an ark1(-/-) yeast strain (Singaraja et al. 2002). No direct functional evidence exists in striatal neurons to confirm a direct role for htt in endocytosis. However, impaired endocytosis has been observed in mice lacking htt-interacting proteins. For example, loss of HIP1 impairs endocytosis of the AMPA receptor, an important ion channel that binds glutamate and regulates calcium entry into cells (see Chapters 8 and 9) (Metzler et al. 2003). Htt associates with a number of cell surface receptors, including the EGF receptor (Lievens et al. 2005) and the dopamine receptor.

To date, the majority of data has focused on htt as a trafficking component at the cell surface and along the axons. However, a role for htt in trafficking is likely to be broader than previously suspected (see Figures 16.4 and 16.5). Early work suggested that htt may operate at multiple intermediate steps in vesicle maturation (Sharp et al. 1995; DiFiglia et al. 1995). For example, when a truncated HIP1 protein is overexpressed, large perinuclear vesicle-like structures containing HIP1, htt, clathrin, and internalized transferrin are observed (DiFiglia et al. 1995). Thus, htt through an interaction with HIP1 might be involved in trafficking steps subsequent to internalization. In differentiating neurons, htt localizes with uncoated vesicles (Velier et al. 1998). The identity of these vesicles is not yet clear. However, clathrin coats are lost during or shortly after transfer of vesicles from actin filaments to microtubules (Bonifacino and Glick 2004; Maxfield and McGraw 2004). In this regard, it is interesting to note that htt interacts with FIP-2 that acts with Rab8, a GTPase involved in regu-

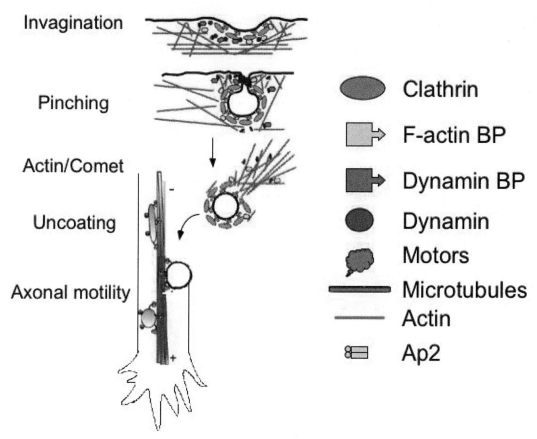

FIGURE 16.4. **Steps of clathrin dependent endocytosis.** Color-coding indicates proteins in each step. (1) Invagination and assembly of the clathrin coat on actin filaments. (2) Mature clathrin coated vesicle is pinched-off by dynamin constriction. (3) Release form the membrane together with actin "comet" tails. (4) Transfer to microtubules and uncoating. (5) Transport along axons. BP, binding proteins.

lation of endocytic and secretory transport (Hattula et al. 2000). The cellular function of FIP-2 is not well understood, but it appears to play a role in stabilizing the interface between actin and microtubule tracts. Consistent with this function, cells lacking Rab8 cannot maintain normal cell morphology (Hattula et al. 2000). Thus, it is possible that htt acts at the actin/microtubule interface to "hand-off" vesicles from the actin cytoskeleton to the microtubules for axonal trafficking via a FIP-2/Rab8 complex.

Golgi-derived vesicles shuttle cargo for exocytosis (Bonifacino and Glick 2004; Chernomordik and Kozlov 2005) (see Figure 16.5). Another htt interacting partner, HIP14, localizes to the Golgi with the Golgi marker GM130, which stabilizes the Golgi stacks and interacts with vesicle docking protein p115 (Singaraja et al. 2002). Thus, htt may be involved in trafficking in the secretory pathway. Ultrastructural and biochemical studies reveal that htt localizes with tubular cisternae and vesicles of the Golgi complex, and synaptic vesicles (Velier et al. 1998; Singaraja et al. 2002). Htt is also associated with synaptophysin, a synaptic vesicle membrane protein, consistent with the notion that htt is transported to presynaptic sites (DiFiglia et al. 1995). Loss of synaptic vesicle motility induced by mhtt may reduce the supply of vesicles in pools needed for synaptic transmission (see Chapter 6). Electrophysiological

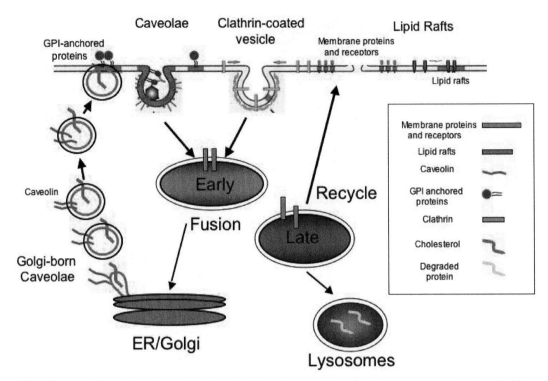

FIGURE 16.5. **Clathrin and nonclathrin endocytic pathways.** Components indicated in the color-coded key. Three pathways for endocytosis are indicated. Clathrin-independent caveolin-1 related pathway involves internalization through caveolae. Caveolae are flask-shaped vesicles enriched with cholesterol and lipids that are born in the Golgi or at the plasma membrane. Golgi-derived caveolae mediate exocytosis of lipids and deliver lipids to the plasma membrane. Lipids can also be brought into the cell through caveolae, although this function is carried on primarily through the clathrin pathway. Clathrin-coated vesicles (CCV) assemble a clathrin coat and accessory proteins (see Figure 16.4). CCVs carry receptors and membrane proteins into the cell. Caveolae and CCVs fuse at the microtubules to form early endosomes. Early endosomes can be recycled to the plasma membrane or mature to the lysosomes. Lipid rafts are cholesterol and lipid-rich regions of the membrane that could also participate in the endocytosis.

measurements confirm that synaptic transmission is inhibited in primary neurons from mice harboring mhtt with 72 repeats, and in animals and in *Drosophila* when mhtt of 128 repeats is expressed (Hodgson et al. 1999; Lee et al. 2004).

The role of htt has been inferred by the function of its interacting partners rather than htt directly. Thus, a role for htt in trafficking, in some cases, remains "guilt by association." For example, htt interacts with HIP1, which modulates endocytosis of GABA receptors and thereby regulates inhibitory neurotransmission (Metzler et al. 2003). However, loss of HIP1 (and presumably loss of the htt/HIP1 interaction) does

not influence transferrin entry into the cell (Metzler et al. 2003). Thus, htt may affect endocytosis, but only for a subset of receptors, and/or HIP1 may not always be associated with htt. It is also possible that htt mediates endocytosis through caveolae, an uncoated, lipid-containing vesicle (Stan 2005; Martin and Parton 2005). Caveolae transport cargo via microtubules in association with trafficking motors (Hommelgaard et al. 2005; Li et al. 2005). Ultrastructural studies indicate that htt colocalizes with caveolae in neurons (Peters et al. 2002) (see Figure 16.5), although none of the accumulating data clearly resolves whether htt is an integral vesicular component, or whether it functions as a motor accessory protein. Finally, the vast majority of studies linking endocytosis and htt function have been performed in fibroblasts, cells that do not manifest HD pathophysiology. Because cell death is largely limited to the neurons in disease, the expansion mutation may have different consequences in dividing cells and may not reflect neuronal physiology. Thus, data linking cell death directly with trafficking defects in dividing cells must be interpreted with caution. Nonetheless, the preponderance of data support the idea that htt has a role in both axonal trafficking and endocytosis, both of which are disrupted in HD. Whether defective axonal trafficking is a cause or consequence of toxicity, however, is poorly understood.

III. INHIBITION OF TRAFFICKING: A PRIMARY DEFECT OR A CONSEQUENCE OF AGGREGATION?

Long polyglutamine tracts are known to form hydrogen-bonded, β-sheets ("polar zippers") and are prone to aggregation (Perutz 1994; 1995). Indeed, the earliest observation in HD postmortem tissue was the presence of intracellular aggregates called inclusion bodies. Formation of inclusions marks disease pathophysiology, and is widely observed in vertebrate, invertebrate, and cell models for HD (Kim and Tanzi 1998; Sanchez et al. 2003). The presence of aggregated material characterizes HD as well as a number of other neurodegenerative diseases, including Alzheimer's disease (AD), Parkinson's disease (PD), and amyotrophic lateral sclerosis (ALS). The aggregates among neurodegenerative diseases do not have the same composition, nor do they reside in the same subcellular space as inclusions observed in HD. For example, in AD, plaques are extracellular rather than intracellular, and comprise Aβ peptide rather than htt (see Chapter 17).

However, inhibition of fast axonal trafficking is commonly observed among aggregation disorders, raising the possibility that the aggregates themselves are the underlying problem. Much circumstantial evidence supports the notion that aggregation is the primary toxicity and that trafficking defects are a consequence of toxicity.

In HD, mhtt accumulations have been found in axons of striatal projection neurons in mouse models for HD and in human patient brains. Dystrophic striatal and corticostriatal neurites in HD exhibit characteristics of blocked axons, including accumulations of vesicles and organelles that associate with htt aggregates in swollen axonal projections and terminals (DiFiglia et al. 1997; Sapp et al. 1999). Aggregates appear to block axonal trafficking. Consistent with a causative function, some candidate drugs that enhance survival in mhtt expressing mice or in cells can also reverse inclusion formation (Zhang et al. 2005). For example, the C2–8 compound suppresses aggregation and neurodegeneration *in vivo* in a *Drosophila* HD model (Zhang et al. 2005). The efficacy of this compound in enhancing survival of mice has not yet been demonstrated. However, these data suggest an essential role for polyQ aggregation in HD pathology. In human HD brains, and in the brains of mice expressing mhtt, inclusions appear to increase with disease progression. The inclusions present in human HD brain contain elements of trafficking motors and cytoskeletal components (Trushina et al. 2004), suggesting that expression of mhtt causes the sequestration of trafficking machinery, and providing a mechanism by which trafficking can be disrupted by aggregation. Indeed, redistribution of soluble motor proteins, including dynein heavy and light chains, p150^{Glued}, kinesin heavy and light chains, into insoluble pools also is observed in a *Drosophila* model for HD (Gunawardena et al. 2003). After induction of a truncated mhtt protein containing 128 glutamines and expressed under the control of an inducible promotor in *Drosophila*, aggregates containing mhtt were found to be transported along axons, and accumulated in growth cones (Lee et al. 2004). During this process, mhtt was able to trap and transport an unrelated, recombinant disheveled protein with 108 polyglutamines (Dsh-108). With time, axonal trafficking was inhibited, and only cytoplasmic aggregates caused the trafficking problem. In the same system, expression of the spinocerebellar ataxia-3 protein (SCA3) containing 78 polyglutamines caused aggregate formation exclusively in the nucleus, and did not inhibit trafficking. However, when coexpressed with mhtt-128, both proteins were found in cytoplasmic aggregates and axonal transport was inhibited. These data suggest that aggregation is the primary cause of toxicity and trafficking defects are a consequence of this toxicity.

However, despite a plethora of data suggesting a relationship between aggregates and trafficking defects, a direct cause-effect relationship has been nearly impossible to prove. Much of the existing data might easily be used to argue that inhibition of intracellular trafficking is a primary defect, and precedes aggregation. Although aggregates are observed in HD patients, many grade 1 patients display HD without visible signs of aggregates (Sapp et al. 1999). In humans, aggregation is not a requirement for toxicity, suggesting that aberrant processes occur prior to aggregate formation. In *Drosophila*, mutations in dynein or p150[Glued] cause axonal transport defects in larval nerves that resemble the neurodegenerative phenotype caused by htt expression but without aggregation (Gunawardena et al. 2003; Hafezparast et al. 2003). These data are suggestive of a connection between neuronal death and defective axonal trafficking, but not aggregation. Consistent with that idea, YAC72 and HD150 knock-in mice, both models for HD, do not develop obvious signs of motor defects or inclusions until late in life (Hodgson et al. 1999; Lin et al. 2001), yet axonal trafficking is altered in primary embryonic striatal neurons (E17) from these animals (Trushina et al. 2004). In these models, trafficking defects appear to precede aggregation. Finally, aggregation does not correlate with toxicity in cell models. In a fibroblast model for HD, cells expressing a truncated form of mhtt undergo cell death and aggregates are observed. However, treatment of these cells with caspase 3 inhibitors prevents apoptosis, but has no effect on aggregate formation (Kim et al. 1999). Similarly, expression of a truncated form of mhtt causes aggregation in cells, but the number of cells undergoing apoptosis does not correlate with the percentage of the neurons that form aggregates (Saudou et al. 1998).

Since htt is a microtubule binding protein, collapse of the cytoskeleton might impair the progression of motors and cargo in the cell, possibly making them susceptible to aggregation. Indeed, expression of full-length mhtt causes neurite retraction, collapse of cytoskeleton, and commitment to cell death, events that can be blocked by taxol (Trushina et al. 2003). Additionally, microtubule depolymerization affects clearance of mhtt from the cells increasing cellular toxicity (Ravikumar et al. 2005). Expression of a truncated, mutant androgen receptor prevents full neurite outgrowth in SH-SY5Y cells during differentiation to a neuronal phenotype that occurs when these cells are cultured with retinoic acid and brain-derived neurotrophic factor (Szebenyi et al. 2003). These events preceded signs of cellular toxicity, indicating that mhtt affects cytoskeleton integrity and causes the trafficking impairment. If this hypothesis is correct, then dendritic effects are likely to be primary events in pathogenesis. Mitochondria and vesicles are critical cargos that are transported along microtubule tracts. In both mouse (Trushina et al. 2004) and *Drosophila* (Guanawardena et al. 2003) models of expanded triple repeat disorders, trafficking of mitochondria is dramatically impaired. Defective mitochondria can lead to toxic consequences, such as reduction in ATP synthesis (Beal 2005; Sugars et al. 2004; Trushina et al. 2004; Gines et al. 2003), loss of calcium buffering (Panov et al. 2002, 2005), and/or oxidative damage (Browne and Beal 2004).

The discrepancies between aggregation and potentially toxic phenomenon have prompted a closer look at the relationship between aggregation and cell death. Many recent reports suggest that, at early stages of disease progression, formation of inclusions is likely to be neuroprotective due to active autophagy (Slow 2005; Webb et al. 2004). Early studies revealed that dopamine stimulation is toxic to postnatal striatal neurons from R6/2 mice, and correlated with oxidative damage. Thus, abnormal aggregates were observed in autophagic/lysosomal vesicles of mhtt-expressing mice relative to controls (Kegel et al. 2000; Petersen et al. 2001), and typically colocalized with oxidized and ubiquitinated proteins (Petersen et al. 2001). These results suggested that neuroprotective effects of autophagy were altered in mhtt expressing cells. Based on these results, autophagy has been widely investigated, and it is generally accepted that it is a neuroprotective response, reducing the level of cellular aggregation, and enhancing cell survival when polyglutamine and polyalanine proteins are expressed (Webb et al. 2004). In fact, the integrity of autophagy appears to be important for normal cell trafficking. Mutations of dynein motors impair autophagic clearance of aggregates and can enhance cell death (Ravikumar et al. 2005). Thus, nuclear inclusions, at least initially, are likely to be a consequence rather than a cause of disease. Thus, current data are most consistent with the idea that inclusions mark the progression of the disease and serve as a protective mechanism by which the cell attempts to rid itself of toxic moieties and decrease their devastating effect (Slow 2005). Although a clear cause-effect relationship has not been established, impaired trafficking could contribute to the cell death through one or several steps including destabilization of cytoskeleton, sequestration of motor proteins, or the derailing of their cargos.

IV. TRAFFICKING VERSUS NUCLEAR TOXICITY

If vesicular trafficking defects are a primary defect, then a major unresolved question is whether suppression of vesicle motility causes toxicity. In the fly, inactivation of trafficking motors using inhibitory antibodies to kinesin impaired trafficking but did not, on its own, cause cell death (Guanawrdena et al. 2003). Thus, mhtt toxicity must require additional functional disturbance beyond the impaired axonal trafficking. Interestingly, cell death did occur in fly when short truncated fragments containing the polyglutamine tract were expressed and found in the nucleus. Thus, data have raised the issue of whether nuclear events are most relevant to toxicity. In support of this idea, the N-terminal, truncated form of the mhtt can bind to and interfere with nuclear factors such as p53 (Steffan et al. 2000), CREB binding protein (CBP) (Steffan et al. 2000; Nucifora et al. 2001), corepressor (Kegel et al. 2002), and transcriptional activator Sp1 (Dunah et al. 2002). The mutation in full-length htt prevents its normal ability to bind to, and sequester, a repressor of brain-derived neurotrophic factor (BDNF) expression, reducing the availability of BDNF to striatal neurons (Zuccato et al. 2003). With respect to transcriptional dysfunction, the most successful strategy for treatment of HD has been the use of histone deacetylase (HDAC) inhibitors (Steffan et al. 2001). These compounds are known to improve survival in yeast (Hughes et al. 2001), affected neurons in cell culture (Hoshino et al. 2003), Drosophila (McCambell et al. 2001), and mouse (Hockly et al. 2003) models for HD. Exposure to HDAC inhibitors is thought to increase acetylation of histones resulting in open chromatin and beneficial effects on gene expression.

In contrast, several lines of evidence suggest that nuclear events may not be sufficient to account for the initial toxic effects of mhtt. In presymptomatic or grade 1 tissue from HD patients, nuclear inclusions are often absent, and cytoplasmic inclusions dominate (Sapp et al. 1999), suggesting that early stages of disease are associated with cytoplasmic events. HDACs are thought to mediate their therapeutic effects by offsetting transcriptional dysfunction via increased histone acetylation. However, microtubule-binding proteins also are acetylated and HDACs may produce beneficial effects by regulating trafficking. If the transcriptional activator Sp1 is a major target of mhtt, presumably mediating toxicity by repression of Sp1-dependent genes, then these genes should be suppressed. However, expression of many Sp1-dependent genes are unaffected by expression of mhtt (Krainc et al. 1998). Additionally, Sp1 specifically interacts with mhtt, and this interaction is stronger with the soluble mutant form than with the insoluble form associated with intranuclear aggregates (Li et al. 2002), suggesting that aggregation would prevent Sp1-dependent transcriptional dysfunction, despite the fact that toxicity is associated with increased aggregation. Finally, it should be mentioned that the idea of transcriptional dysfunction, as a primary event in toxicity, has been based on models of mice or flies expressing a short, truncated fragment of mhtt. Interestingly, short mhtt fragments can freely diffuse into the nucleus in fibroblasts (Trushina et al. 2003). Thus, deleterious transcriptional effects observed in these models for disease may take place because small fragments can enter the nucleus as in experimental systems. *In vivo*, fragments must arise from cleavage of the endogenous, full-length protein, yet neither the timing nor the extent of *in vivo* cleavage of the full-length protein is known. Recent observations from several models for HD

(Dyer and McMurray 2001) as well as SCA7 (Yoo et al. 2003) reveal that full-length mhtt may be resistant to proteolysis in human disease tissue and in mouse models for disease.

A sequential mechanism may help resolve some of these issues. Cytoplasmic dysfunction caused by expression of mhtt precedes nuclear dysfunction (Trushina et al. 2003). In individual striatal neurons transfected with a human cDNA encoding GFP-tagged full-length mhtt and monitored from the beginning of expression to cell death, toxicity appears to occur by a distinctly sequential mechanism. Loss of neurite integrity and collapse of the microtubule cytoskeleton were observed while full-length mhtt was present in the cytoplasm. Detection of GFP in the nucleus occurred very late in cell progression to death when the nuclear envelope became compromised, allowing mhtt entrance. Another export signal prevented mhtt nuclear accumulation but not cell death, strongly suggesting that mhtt initiates toxicity in the cytoplasm, consistent with the physical features of the mhtt protein. Htt is a cytoplasmic protein of 350 kD with no canonical nuclear localization signal (Hackam et al. 1999). Thus, a large endogenous mhtt protein might initiate toxicity in the cytoplasm, but once in the nucleus, truncated fragments of mhtt might exacerbate toxicity.

V. COMMON TIES BETWEEN HD AND OTHER NEURODEGENERATIVE DISORDERS

Defective axonal trafficking has been linked to neurodegeneration in polyglutamine disorders (Gunawardena and Goldstein 2005), HD (Trushina et al. 2004; Gunawardena et al. 2003; Szebenyi et al. 2003), AD (see Chapter 17) (Gunawardena and Goldstein 2001; Stokin et al. 2005), and ALS (Collard et al. 1995; Williamson and

Cleveland 1999; Breuer et al. 1997). Cellular dysfunctions associated with each of these diseases point to a common mechanism in which defective trafficking plays a key role.

VI. CYTOSKELETAL AND VESICULAR DYSFUNCTION

Early observations revealed that htt was a microtubule binding protein, and appeared to bind the polymerized microtubules *in vitro* rather then its monomeric subunits (Tukamoto et al. 1997), and to stabilize them. Microtubule abnormalities are observed in neurons when mhtt is expressed. A specific model for how alterations in microtubules might cause toxicity in HD has been reported recently. Htt, along with HAP1, enhances trafficking of BDNF growth factor along the microtubules, and expression of mhtt interferes with that process (Gauthier et al. 2004). Inhibition of trafficking arises from a competition between mhtt interaction with HAP1 and HAP1 interaction with microtubules (Gauthier et al. 2004). Specifically, the expanded polyglutamine tract in mhtt increases its interaction with HAP1 and dynactin p150[Glued] and reduces the association of the HAP1/dynactin p150[Glued] complex with microtubules. This toxic competition derails trafficking of BDNF, which is essential for growth/survival of striatal neurons.

Tau is a microtubule-binding protein essential for cytoskeletal stability (Drubin and Kirschner 1986). In AD, mutations identified in tau reduce the ability of tau proteins to stabilize microtubules and increases their propensity to assemble into abnormal filaments (Brandt and Leschik 2004). Consequently, neurofibrillary tangles (NFTs) are found in the axons of neurons from AD patients (Kuret et al. 2005). NFTs are composed of hyperphosphorylated tau, which promotes the sequestration of normal tau, MAP1A/MAP1B, and MAP2 (Kuret et al. 2005). This toxic gain of function of the pathological tau appears to be due to its abnormal hyperphosphorylation, because dephosphorylation converts it to a normal state. Posttranslational events also regulate mhtt. SUMOylation of the mhtt fragment by small ubiquitin-like modifier (SUMO)–1 stabilizes mhtt, reduces its ability to form aggregates, and promotes neurodegeneration, in part by preventing mhtt ubiquitination and degradation (Steffan et al. 2004).

Cytoskeletal abnormalities are also key features of human pathology in ALS, spinal-bulbar muscular atrophy (SBMA), and spinal muscular atrophy (SMA). Familial forms of ALS (FALS) have missense mutations in the Cu/Zn superoxide dismutase gene (SOD1) (Rosen et al. 1993). However, most of the sporadic mutations are associated with formation of axonal aggregates of neurofilaments in mouse models and in human patients. Some sporadic cases of ALS seem to arise from point mutations in neurofilament proteins (Cairns et al. 2004), and ultrastructural analysis reveals a paucity of essential proteins and organelles including mitochondria in the degenerating axons (Collard et al. 1995). In mouse models of ALS, slowing of axonal trafficking is detected in motor neurons at least six months prior to the onset of clinical symptoms. These data strongly implicate dysfunctional trafficking as the pathological event underlying ALS (Williamson and Cleveland 1999).

This viewpoint is reinforced by the discovery that two new forms of ALS, ALS8 and ALS2, arise from mutations in the genes encoding trafficking proteins. A new locus for ALS at 20q13.3 has been mapped in a large Brazilian family with 28 affected members distributed across four generations (Nishimura et al. 2004). In all patients of this family, as well as in six additional families with a different diagnosis, the missense mutation encoded a defective vesicle-associated membrane protein (VAMP), or synaptobrevin-associated membrane protein B (VAPB). Ultrastructural and biochemical analyses indicate that these VAP proteins can associate with microtubules and often are found at the junction between intracellular vesicles and cytoskeletal structures (Soussan et al. 1999; Skehel et al. 2000). In *Drosophila*, the DVAP-33A homologue can localize in the neuromuscular junctions. The VAPB acts during ER-Golgi transport and secretion (Soussan et al. 1999; Foster et al. 2000), and disruption of this function may lead to the accumulation of transport intermediates in cytosolic membranous aggregates.

Additional mutations in ALS have been identified in the *ALS2* gene, and lead to a rare recessive form that presents early in life and progresses much more slowly than the classical form (Hadano et al. 2001; Yang et al. 2001). Alsin, the protein encoded by *ALS2*, contains pleckstrin homology (PH) domain known to be involved in the regulation of the actin cytoskeleton in neurons (Hadano et al. 2001; Yang et al. 2001; Shaw 2001; Topp et al. 2004). Although the role of alsin is not well understood, two small deletions in *ALS2* severely truncate the predicted protein product, and such mutations recently have been associated with juvenile-onset primary lateral sclerosis (Yang et al. 2001) and infantile-onset hereditary spastic paralegia (Gros-Louis et al. 2003). Alsin appears to participate in guanine nucleotide exchange on Rab5 via its Vps9p domain located at its COOH terminus (Topp et al. 2004). Activation of Rab5 is essential for protein trafficking through the early stages of the endocytic pathway. Thus, it has been suggested that disease-causing mutation in alsin interferes with neuronal maintenance by altering membrane trafficking of signaling endosomes and synaptic vesicles (Topp et al. 2004).

Finally, the defective intracellular trafficking of caveolae has been implicated in Niemann-Pick disease type C (NPC) (Sturley et al. 2004). Caveolae are vesicular invaginations in the plasma membrane, and are specifically enriched with cholesterol and sphingolipids and are involved in trafficking of cholesterol (Kurzchalia and Parton 1999; Smart et al. 1999) (see Figure 16.5). NPC is characterized by the accumulation of unesterified cholesterol and sphingolipids within cells in the endosomes and lysosomes of various tissues (Sturley et al. 2004). The disease is autosomal recessive and is caused by mutations in one of two genetic loci, NPC1 and NPC2 (Sturley et al. 2004). NPC1 protein is predominantly located within the late endosomal membrane, but is also transiently associated with lysosomes and the TGN. NPC2 is a soluble lysosomal protein. The proposed mechanisms suggest that the mutant proteins could derail delivery of LDL-derived cholesterol to the proper cellular compartments, or transport of endogenously synthesized cholesterol to the PM (Chang et al. 2005). The inhibition of Rab 7 and Rab 5 observed in NPC leads to decreased motility of endosomes (Lebrand et al. 2002; Choudhury et al. 2004) (see Chapter 17). Overexpression of Rab 9 in NPC cells corrects their lipid trafficking defects, suggesting a possible therapeutic target (Choudhury et al. 2002; Walter et al. 2003). Growing evidence suggests that additional clathrin-independent and caveolin 1-related endocytosis could play a significant role in prion disease (Massimino et al. 2002), and if altered, could cause signaling malfunction contributing to neurodegeneration (see Chapter 18).

VII. MOTOR PROTEINS

Inclusions present in human HD brain contain components of soluble trafficking motors and cytoskeletal components (Trushina et al. 2004). Thus, the defect in HD might be considered a functional knockout of the trafficking machinery. Such functional defects link HD with other neurodegenerative disorders (Hirokawa and Takemura 2003). For example, KIF1B is a specific motor for anterograde transport of mitochondria (Nangaku et al. 1994) (see Figure 16.6), and mutations in this protein have been linked to Charcot-Marie-Tooth

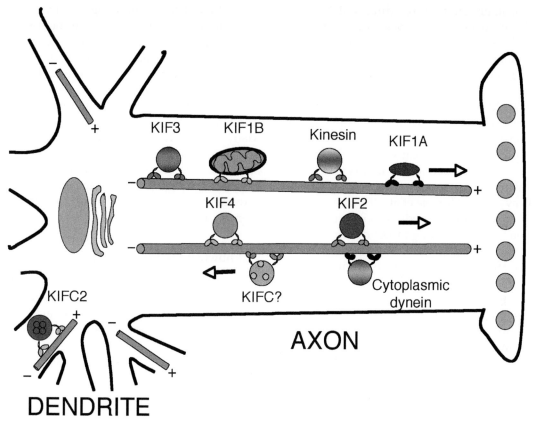

FIGURE 16.6. **Kinesin- and dynein-related trafficking within the cell.** Representation of the microtubule-based transport of vesicles and organelles within the axon (**a**) and the cell body (**b**). Kinesin family proteins (KIFs) participate primarily in the delivering of the cargo toward the plus end of the microtubules, and dynein delivers cargo toward the minus end. In neurons, KIF2 and KIF4 work mostly in juvenile stages. Neuron-specific KIFs and ubiquitous KIFs are drawn in the same cell. CGN, *cis*-Golgi network; TGN, *trans*-Golgi network; ECV, endosomal carrier vesicle; microtubules are shown in yellow; ER, endoplasmic reticulum. Black arrows indicate the direction of transport. Adapted from Hirokawa Science **279**: 519–526, 1998.

disease type 2A (CMT) (Zhao et al. 2001). CMT, the most common inherited peripheral neuropathy in humans, is clinically characterized by weakness and atrophy of distal muscles, depressed or absent deep-tendon reflexes, and mild sensory loss. KIF1B is a plus-end-directed motor that transports synaptic vesicle precursors in the axon from the cell body to the synapse. In CMT, both the number of synapses and the density of synaptic vesicles at the synapse are reduced, consistent with a defect in the transport of synaptic vesicle precursors. More recently, two additional mutations in members of kinesin superfamily, KIF21A

and KIF5A, have been linked to neuropathological disorder of the superior ocular nerve (Yamada et al. 2003), and to a subtype of hereditary spastic paraplegia (Reid et al. 2002) (see Figure 16.6). Although KIF5A expression is known to be neuron-specific (Xia et al. 2003), details of its cellular functions have only begun to be characterized. Although mice lacking KIF5A die at birth, postnatal removal of KIF5A results in an apparent defect in neurofilament transport (Xia et al. 2003).

In AD, APP-induced neuronal death exhibits striking and consistent genetic interactions with the molecular motor machinery, suggesting a direct link between the induction of neuronal death by APP and its function as a kinesin-I receptor (Gunawardena and Goldstein 2001). Alteration in vesicular trafficking appears to be a very early event in AD progression (Stokin et al. 2005), and overexpression of APP-like protein (APPL) in neurons disrupts axonal transport. In *Drosophila* model for AD, overexpression of APPL causes a cuticle tanning defect in enclosed adults that is genetically enhanced by reducing the dosage of kinesin heavy chain (Torroja et al. 1999).

In ALS animal models, functional inhibition of the trafficking motor dynein recapitulates the signs of human motor neurons disease. Disruption of the dynactin complex in motor neurons by the overexpression of the dynactin component, dynamitin, results in a progressive motor neuron neurodegeneration phenotype with features similar to those of amyotrophic lateral sclerosis (ALS) (LaMonte et al. 2002). Mutations in dynactin subunit p150 are linked to a slow and progressive motor neuron disease in humans (Puls et al. 2003). Additionally, disease loci were mapped to the cytoplasmic dynein heavy chain gene in two lines of mice showing progressive motor impairment and loss of motor neurons (Hafezparast et al. 2003). These studies suggest that impairment of transport components can result in disease. There are unmistakable links between HD and other neurodegenerative diseases including disruptions in the cytoskeleton, cytoskeletal motor protein function, and vesicular transport.

VIII. THERAPEUTICS AND FUTURE DIRECTIONS

In the coming decade, at least 2 percent of Americans will be afflicted with some form of Alzheimer's (4,000,000), Parkinson's (1,500,000), or Huntington's disease (200,000); related dominant ataxias; and amyotrophic lateral sclerosis (30,000). At present all of these neurodegenerative diseases are incurable. No effective therapeutic approach is currently available for these patients, yet the number of affected individuals is expected to grow exponentially. There is a growing consensus that common properties of trafficking may underlie a number of neurodegenerative disorders. Many unresolved questions must be answered in order to design effective therapeutics. However, a better understanding of basic trafficking mechanisms is likely to provide new avenues and hope for future therapeutics.

Acknowledgements

We thank I. Kovtun for careful reading of the manuscript. This work was supported by the Mayo Foundation, the National Institutes of Health grants NS40738 and GM 066359 (to CTM).

References

Andrade, M.A., Bor, P. (1995). HEAT repeats in the Huntington's disease protein. *Nat. Genet.* **11**, 115–116.

Bates, G.P. (2005). History of genetic disease: The molecular genetics of Huntington disease—A history. *Nat. Rev. Genet.* **6**, 766–773.

Beal, M.F. (2005). Mitochondria take center stage in aging and neurodegeneration. *Ann. Neurol.* **58**, 495–505.

Block-Galarza, J., Chase, K.O., Sapp, E., Vaughn, K.T., Vallee, R.B., DiFiglia, M., Aronin, N. (1997). Fast transport and retrograde movement of huntingtin and HAP 1 in axons. *Neuroreport* **8**, 2247–2251.

Bonifacino, J.S., Glick, B.S. (2004). The mechanisms of vesicle budding and fusion. *Cell* **116**, 153–166.

Bradke, F., Dotti, C.G. (1998). Membrane traffic in polarized neurons. *Biochim. Biophys. Acta* **1404**, 245–258.

Brandt, R., Leschik, J. (2004). Functional interactions of tau and their relevance for Alzheimer's disease. *Curr. Alzheimer Res.* **1**, 255–269.

Breuer, A.C., Lynn, M.P., Atkinson, M.B., Chou, S.M., Wilbourn, A.J., Marks, K.E. et al. (1997). Fast axonal transport in amyotrophic lateral sclerosis: An intra-axonal organelle traffic analysis. *Neurology* **37**, 738–748.

Browne, S.E., Beal, M.F. (2004). The energetics of Huntington's disease. *Neurochem. Res.* **29**, 531–546.

Cairns, N.J., Lee, V.M., and Trojanowski, J.Q. (2004). The cytoskeleton in neurodegenerative diseases. *J. Pathol.* **204**, 438–449.

Chang, T.Y., Reid, P.C., Sugii, S., Ohgami, N., Cruz, J.C., Chang, C.C. (2005). Niemann-Pick type C disease and intracellular cholesterol trafficking. *J. Biol. Chem.* **280**, 20917–20920.

Charrin, B.C., Saudou, F., Humbert, S. (2005). Axonal transport failure in neurodegenerative disorders: The case of Huntington's disease. *Pathol. Biol.* **53**, 189–192.

Chen, C.Y., Brodsky, F.M. (2005). Huntingtin-interacting protein 1 (Hip1) and Hip1-related protein (Hip1R) bind the conserved sequence of clathrin light chains and thereby influence clathrin assembly in vitro and actin distribution in vivo. *J. Biol. Chem.* **280**, 6109–6117.

Chernomordik, L.V., Kozlov, M.M. (2005). Membrane hemifusion: Crossing a chasm in two leaps. *Cell* **123**, 375–382.

Choudhury, A., Dominguez, M., Puri, V., Sharma, D.K., Narita, K., Wheatley, C.L. et al. (2002). Rab proteins mediate Golgi transport of caveola-internalized glycosphingolipids and correct lipid trafficking in Niemann-Pick C cells. *J. Clin. Invest.* **109**, 1541–1550.

Choudhury, A., Sharma, D.K., Marks, D.L., Pagano, R.E. (2004). Elevated endosomal cholesterol levels in Niemann-Pick cells inhibit rab4 and perturb membrane recycling. *Mol. Biol. Cell* **15**, 4500–4511.

Collard, J.F., Cote, F., Julien, J.P. (1995). Defective axonal transport in a transgenic mouse model of amyotrophic lateral sclerosis. *Nature* **375**, 61–64.

DiFiglia, M., Sapp, E., Chase, K., Schwarz, C., Meloni, A., Young, C. et al. (1995). Huntingtin is a cytoplasmic protein associated with vesicles in human and rat brain neurons. *Neuron* **14**, 1075–1081.

DiFiglia, M., Sapp, E., Chase, K.O., Davies, S.W., Bates, G.P., Vonsattel, J.P., Aronin, N. (1997). Aggregation of huntingtin in neuronal intranuclear inclusions and dystrophic neurites in brain. *Science* **277**, 1990–1993.

Drubin, D.G., Kirschner, M.W. (1986). Tau protein function in living cells. *J. Cell Biol.* **103**, 2739–2746.

Ducker, C.E., Stettler, E.M., French, K.J., Upson, J.J., Smith, C.D. (2004). Huntingtin interacting protein 14 is an oncogenic human protein: Palmitoyl acyltransferase. *Oncogene* **23**, 9230–9237.

Dunah, A.W., Jeong, H., Griffin, A., Kim, Y.M., Standaert, D.G., Hersch, S.M. et al. (2002). Sp1 and TAFII130 transcriptional activity disrupted in early Huntington's disease. *Science* **296**, 2238–2243.

Dyer, R.B., McMurray, C.T. (2001). Mutant protein in Huntington disease is resistant to proteolysis in affected brain. *Nat. Genet.* **29**, 270–278.

Engelender, S., Sharp, A.H., Colomer, V., Tokito, M.K., Lanahan, A., Worley, P. et al. (1997). Huntingtin-associated protein 1 (HAP1) interacts with the p150Glued subunit of dynactin. *Hum. Mol. Genet.* **13**, 2205–2212.

Engqvist-Goldstein, A.E., Kessels, M.M., Chopra, V.S., Hayden, M.R., Drubin, D.G. (1999). An actin-binding protein of the Sla2/Huntingtin interacting protein 1 family is a novel component of clathrin-coated pits and vesicles. *J. Cell. Biol.* **147**, 1503–1518.

Foster, L.J., Weir, M.L., Lim, D.Y., Liu, Z., Trimble, W.S., Klip, A. (2000). A functional role for VAP-33 in insulin-stimulated GLUT4 traffic. *Traffic* **1**, 512–521.

Gage, F.H., McAllister, A.K. (2005). Neuronal and glial cell biology. *Curr. Opin. Neurobiol.* **15**, 497–499.

Gauthier, L.R., Charrin, B.C., Borrell-Pages, M., Dompierre, J.P., Rangone, H., Cordelieres, F.P. et al. (2004). Huntingtin controls neurotrophic support and survival of neurons by enhancing BDNF vesicular transport along microtubules. *Cell* **118**, 127–138.

Gines, S., Seong, I.S., Fossale, E., Ivanova, E., Trettel, F., Gusella, J.F. et al. (2003). Specific progressive cAMP reduction implicates energy deficit in presymptomatic Huntington's disease knock-in mice. *Hum. Mol. Genet.* **12**, 497–508.

Gros-Louis, F., Meijar, I.A., Hand, C.K., Dube, M.-P., MacGregor, D.L., Seni, M.-H. et al. (2003). An ALS2 gene mutation causes hereditary spastic paraplegia in a Pakistani kindred. *Ann. Neurol.* **53**, 144–145.

Groves, M.R., Hanlon, N., Turowski, P., Hemmings, B.A., Barford, D. (1999). The structure of the protein phosphatase 2A PR65/A subunit reveals the conformation of its 15 tandemly repeated HEAT motifs. *Cell* **96**, 99–110.

Gunawardena, S., Goldstein, L.S. (2001). Disruption of axonal transport and neuronal viability by amyloid precursor protein mutations in Drosophila. *Neuron* **32**, 389–401.

Gunawardena, S., Goldstein, L.S. (2005). Polyglutamine diseases and transport problems: Deadly traffic jams on neuronal highways. *Arch. Neurol.* **62**, 46–51.

Gunawardena, S., Her, L.S., Brusch, R.G., Laymon, R.A., Niesman, I.R., Gordesky-Gold, B. et al. (2003). Disruption of axonal transport by loss of huntingtin or expression of pathogenic polyQ proteins in Drosophila. *Neuron* **40**, 25–40.

Hackam, A.S., Singaraja, R., Zhang, T., Gan, L., Hayden, M.R. (1999). In vitro evidence for both the nucleus and cytoplasm as subcellular sites of pathogenesis in Huntington's disease. *Hum. Mol. Genet.* **8**, 25–33.

Hadano, S., Hand, C.K., Osuga, H., Yanagisawa, Y., Otomo, A., Devon, R.S. et al. (2001). A gene encoding a putative GTPase regulator is mutated in familial amyotrophic lateral sclerosis 2. *Nat. Genet.* **29**, 166–173.

Hafezparast, M., Klocke, R., Ruhrberg, C., Marquardt, A., Ahmad-Annuar, A., Bowen, S. et al. (2003). Mutations in dynein link motor neuron degeneration to defects in retrograde transport. *Science* **300**, 808–812.

Hague, S.M., Klaffke, S., Bandmann, O. (2005) Neurodegenerative disorders: Parkinson's disease and

Huntington's disease. *J. Neurol. Neurosurg. Psychiatry* **76**, 1058–1063.

Harjes, P., Wanker, E.E. (2003) The hunt for huntingtin function: Interaction partners tell many different stories. *Trends Biochem. Sci.* **28**, 425–433.

Hattula, K., Peranen, J. (2000). FIP-2, a coiled-coil protein, links Huntingtin to Rab8 and modulates cellular morphogenesis. *Curr. Biol.* **10**, 1603–1606.

Hirokawa, N., Takemura, R. (2003). Biochemical and molecular characterization of diseases linked to motor proteins. *Trends Biochem. Sci.* **28**, 558–565.

Hockly, E., Richon, V.M., Woodman, B. et al. (2003). Suberoylanilide hydroxamic acid, a histone deacetylase inhibitor, ameliorates motor deficits in a mouse model of Huntington's disease. *Proc. Natl. Acad. Sci. USA* **100**, 2041–2046.

Hodgson, J.G., Agopyan, N., Gutekunst, C.A., Leavitt, B.R., LePiane, F., Singaraja, R. et al. (1999). A YAC mouse model for Huntington's disease with full-length mutant huntingtin, cytoplasmic toxicity, and selective striatal neurodegeneration. *Neuron* **23**, 181–192.

Hommelgaard, A.M., Roepstorff, K., Vilhardt, F., Torgersen, M.L., Sandvig, K., van Deurs, B. (2005). Caveolae: Stable membrane domains with a potential for internalization. *Traffic* **6**, 720–724.

Hoshino, M., Tagawa, K., Okuda, T., Murata, M., Oyanagi, K., Arai, N. et al. (2003). Histone deacetylase activity is retained in primary neurons expressing mutant huntingtin protein. *J. Neurochem.* **87**, 257–267.

Huang, K., Yanai, A., Kang, R., Arstikaitis, P., Singaraja, R.R., Metzler, M. et al. (2004). Huntingtin-interacting protein HIP14 is a palmitoyl transferase involved in palmitoylation and trafficking of multiple neuronal proteins. *Neuron* **44**, 977–986.

Hughes, R.E., Lo, R.S., Davis, C. et al. (2001). Altered transcription in yeast expressing expanded polyglutamine. *Proc. Natl. Acad. Sci. USA* **98**, 13201–13206.

Kalchman, M.A., Koide, H.B., McCutcheon, K., Graham, R.K., Nichol, K., Nishiyama, K. et al. (1997). HIP1, a human homologue of S. cerevisiae Sla2p, interacts with membrane-associated huntingtin in the brain. *Nat. Genet.* **16**, 44–53.

Kegel, K.B., Kim, M., Sapp, E., McIntyre, C., Castano, J.G., Aronin, N., DiFiglia, M. (2000). Huntingtin expression stimulates endosomal-lysosomal activity, endosome tubulation, and autophagy. *J. Neurosci.* **20**, 7268–7278.

Kegel, K.B., Meloni, A.R., Yi, Y., Kim, Y.J., Doyle, E., Cuiffo, B.G. et al. (2002). Huntingtin is present in the nucleus, interacts with the transcriptional corepressor C-terminal binding protein, and represses transcription. *J. Biol. Chem.* **277**, 7466.

Kegel, K.B., Sapp, E., Yoder, J., Cuiffo, B., Sobin, L., Kim, Y.J. et al. (2005). Huntingtin associates with acidic phospholipids at the plasma membrane. *J. Biol. Chem.* **280**, 36464–36473.

Kim, M., Velier, J., Chase, K., Laforet, G., Kalchman, M.A. et al. (1999). Forskolin and dopamine D1 receptor activation increase huntingtin's association with endosomes in immortalized neuronal cells of striatal origin. *Neurosci.* **89**, 1159–1167.

Kim, T, Tanzi, R. (1998). Neuronal intranuclear inclusions in polyglutamine diseases, nuclear weapons or nuclear fallout? *Neuron* **21**, 657–659.

Kittler, J.T., Thomas, P., Tretter, V., Bogdanov, Y.D., Haucke, V., Smart, T.G., Moss, S.J. (2004). Huntingtin-associated protein 1 regulates inhibitory synaptic transmission by modulating gamma-aminobutyric acid type A receptor membrane trafficking. *Proc. Natl. Acad. Sci. USA* **101**, 12736–12741.

Kovtun, I., McMurray, C.T. (2005). Mechanisms of DNA repair underlying DNA expansion. The genetic basis for neurological disease, R. Wells, ed. (in press).

Krainc, D., Bai, G., Okamoto, S., Carles, M., Kusiak, J.W., Brent, R.N., Lipton, S.A. (1998). Synergistic activation of the N-methyl-D-aspartate receptor subunit 1 promoter by myocyte enhancer factor 2C and Sp1. *J. Biol. Chem.* **273**, 26218–26224.

Kuret, J., Chirita, C., Congdon, E., Kannanayakal, T., Li, G., Necula, M. et al. (2005). Pathways of tau fibrillization. *Biochimica et Biophysica Acta* **1739**, 167–178.

Kurzchalia, T.V., Parton, R.G. (1999). Membrane microdomains and caveolae. *Curr. Opin. Cell Biol.* **11**, 424–431.

LaMonte, B.H., Wallace, K.E., Holloway, B.A., Shelly, S.S., Ascano, J., Tokito, M. et al. (2002). Disruption of dynein/dynactin inhibits axonal transport in motor neurons causing late-onset progressive degeneration. *Neuron* **34**, 715–727.

Lebrand, C., Corti, M., Goodson, H., Cosson, P., Cavalli, V., Mayran, N. et al. (2002). Late endosome motility depends on lipids via the small GTPase Rab7. *EMBO J.* **21**, 1289–1300.

Lee, W.M., Yoshihara, M., Littleton, J.T. (2004). Cytoplasmic aggregates trap polyglutamine-containing proteins and block axonal transport in a *Drosophila* model of Huntington's disease. *PNAS* **101**, 3224–3229.

Legendre-Guillemin, V., Metzler, M., Lemaire, J.F., Philie, J., Gan, L., Hayden, M.R., McPherson, P.S. (2005). Huntingtin interacting protein 1 (HIP1) regulates clathrin assembly through direct binding to the regulatory region of the clathrin light chain. *J. Biol. Chem.* **280**, 6101–6108.

Li, S.H., Gutekunst, C.A., Hersch, S.M., Li, X.J. (1998a). Interaction of huntingtin-associated protein with dynactin P150Glued. *J. Neurosci.* **18**, 1261–1269.

Li, S.H., Hosseini, S.H., Gutekunst, C.A., Hersch, S.M., Ferrante, R.J., Li, X.J. (1998b). A human HAP1 homologue. Cloning, expression, and interaction with huntingtin. *J. Biol. Chem.* **273**, 19220–19227.

Li, S-H., Cheng, A.L., Zhou, H., Lam, S., Rao, M., Li, H., Li, X-J. (2002). Interaction of Huntington disease protein with transcriptional activator Sp1. *Mol. Cell Biol.* **22**, 1277–1287.

Li, X.J., Li, S.H. (2005). HAP1 and intracellular trafficking. *Trends Pharmacol. Sci.* **26**, 1–3.

Li, X.A., Everson, W.V., Smart, E.J. (2005). Caveolae, lipid rafts, and vascular disease. *Trends Cardiovasc. Med.* **15**, 92–96.

Lievens, J.C., Rival, T., Iche, M., Cheniweiss, H., Birman, S. (2005). Expanded polyglutamine peptides disrupt EGF receptor signaling and glutamate transporter expression in Drosophila. *Hum. Mol. Genet.* **14**, 713–724.

Lin, C.H., Tallaksen-Greene, S., Chien, W.M., Cearley, J.A., Jackson, W.S., Crouse, A.B. et al. (2001). Neurological abnormalities in a knock-in mouse model of Huntington's disease. *Hum. Mol. Genet.* **10**, 137–144.

Martin, S., Parton, R.G. (2005). Caveolin, cholesterol, and lipid bodies. *Semin. Cell Dev. Biol.* **16**, 163–174.

Massimino, M.L., Griffoni, C., Spisni, E., Toni, M., Tomasi, V. (2002). Involvement of caveolae and caveolae-like domains in signalling, cell survival and angiogenesis. *Cell. Signal.* **14**, 93–98.

Mattson, M.P. (2002). Accomplices to neuronal death. *Nature* **415**, 377–379.

Maxfield, F.R., McGraw, T.E. (2004). Endocytic recycling. *Nat. Rev. Mol. Cell Biol.* **5**, 121–132.

McCampbell A., Taye A.A., Whitty L. et al. (2001). Histone deacetylase inhibitors reduce polyglutamine toxicity. *Proc. Natl. Acad. Sci. USA* **98**, 15179–15184.

McMurray, C.T. (2005). To die or not to die: DNA repair in neurons. *Mutat. Res.* **577**, 260–274.

Metzler, M., Li, B., Gan, L., Georgiou, J., Gutekunst, C.A., Wang, Y. et al. (2003). Disruption of the endocytic protein HIP1 results in neurological deficits and decreased AMPA receptor trafficking. *EMBO J.* **22**, 3254–3266.

Nangaku, M., Sato-Yoshitake, R., Okada, Y., Noda, Y., Takemura, R., Yamazaki, H., Hirokawa, N. (1994). KIF1B, a novel microtubule plus end-directed monomeric motor protein for transport of mitochondria. *Cell* **79**, 1209–1220.

Nasir, J., Floresco, S.B., O'Kusky, J.R., Diewert, V.M., Richman, J.M., Zeisler, J. et al. (1995). Targeted disruption of the Huntington's disease gene results in embryonic lethality and behavioral and morphological changes in heterozygotes. *Cell* **81**, 811–823.

Nishimura, A.L., Mitne-Neto, M., Silva, H.C., Richieri-Costa, A., Middleton, S., Cascio, D. et al. (2004). A mutation in the vesicle-trafficking protein VAPB causes late-onset spinal muscular atrophy and amyotrophic lateral sclerosis. *Am. J. Hum. Genet.* **75**, 822–831.

Nucifora, F.C. Jr., Sasaki, M., Peters, M.F., Huang, H., Cooper, J.K., Yamada, M. et al. (2001). Interference by huntingtin and atrophin-1 with CBP-mediated transcription leading to cellular toxicity. *Science* **291**, 2423–2428.

Panov, A.V., Gutekunst, C.A., Leavitt, B.R., Hayden, M.R., Burke, J.R., Strittmatter, W.J., Greenamyre, J.T. (2002). Early mitochondrial calcium defects in Huntington's disease are a direct effect of polyglutamines. *Nat. Neurosci.* **5**, 731–736.

Panov, A.V., Lund, S., Greenamyre, J.T. (2005). Ca2+-induced permeability transition in human lymphoblastoid cell mitochondria from normal and Huntington's disease individuals. *Mol. Cell. Biochem.* **269**, 143–152.

Paschen, W., Mengesdorf, T. (2005). Endoplasmic reticulum stress response and neurodegeneration. *Cell Calcium* **38**, 409–415.

Peters, P.J., Ning, K., Palacios, F., Boshans, R.L., Kazantsev, A., Thompson, L.M. et al. (2002). Arfaptin 2 regulates the aggregation of mutant huntingtin protein. *Nat. Cell Biol.* **4**, 240–245.

Perutz, M.F., Johnson, T., Suzuki, M., Finch, J.T. (1994). Glutamine repeats as polar zippers: Their possible role in inherited neurodegenerative diseases. *Proc. Natl. Acad. Sci. USA* **91**, 5355–5358.

Perutz, M.F. (1995). Glutamine repeats as polar zippers: their role in inherited neurodegenerative disease. *Mol. Med.* **1**, 718–721.

Petersen, A., Larsen, K.E., Behr, G.G., Romero, N., Przedborski, S., Brundin, P., Sulzer, D. (2001). Expanded CAG repeats in exon 1 of the Huntington's disease gene stimulate dopamine-mediated striatal neuron autophagy and degeneration. *Hum. Mol. Genet.* **10**, 1243–1254.

Pineda, J.R., Canals, J.M., Bosch, M., Adell, A., Mengod, G., Artigas, F., Ernfors, P., Alberch, J. (2005). Brain-derived neurotrophic factor modulates dopaminergic deficits in a transgenic mouse model of Huntington's disease. *J. Neurochem.* **93**, 1057–1068.

Puls, I., Jonnakuty, C., LaMonte, B.H., Holzbaur, E.L., Tokito, M., Mann, E. et al. (2003). Mutant dynactin in motor neuron disease. *Nat. Genet.* **33**, 455–456.

Qin, Z.H., Wang, Y., Sapp, E., Cuiffo, B., Wanker, E., Hayden, M.R. et al. (2004). Huntingtin bodies sequester vesicle-associated proteins by a polyproline-dependent interaction. *J. Neurosci.* **24**, 269–281.

Ravikumar, B., Acevedo-Arozena, A., Imarisio, S., Berger, Z., Vacher, C., O'Kane, C.J. et al. (2005). Dynein mutations impair autophagic clearance of aggregate-prone proteins. *Nat. Genet.* **37**, 771–776.

Reid, E., Kloos, M., Ashley-Koch, A., Hughes, L., Bevan, S., Svenson, I.K. et al. (2002). A kinesin heavy chain (KIF5A) mutation in hereditary spastic paraplegia (SPG10). *Am. J. Hum. Genet.* **71**, 1189–1194.

Rosen, D.R., Siddique, T., Patterson, D., Figlewicz, D.A., Sapp, P., Hentati, A. et al. (1993). Mutations in Cu/Zn superoxide dismutase gene are associated with familial amyotrophic lateral sclerosis. *Nature* **362**, 59–62.

Sanai, N., Alvarez-Buylla, A., Berger, M.S. (2005). Neural stem cells and the origin of gliomas. *N. Engl. J. Med.* **353**, 811–822.

Sanchez, I., Mahlke, C., Yuan, J. (2003). Pivotal role of oligomerization in expanded polyglutamine neurodegenerative disorders. *Nature* **421**, 373–379.

Sapp, E., Penney, J., Young, A., Aronin, N., Vonsattel, J.P., DiFiglia, M. (1999). Axonal transport of N-terminal huntingtin suggests early pathology of corticostriatal projections in Huntington disease. *J. Neuropathol. Exp. Neurol.* **58**, 165–173.

Saudou, F., Finkbeiner, S., Devys, D., Greenberg, M.E. (1998). Huntingtin acts in the nucleus to induce apoptosis but death does not correlate with the formation of intranuclear inclusions. *Cell* **95**, 55–66.

Seki, N., Muramatsu, M., Sugano, S., Suzuki, Y., Nakagawara, A., Ohhira, M. et al. (1998). Cloning, expression analysis, and chromosomal localization of HIP1R, an isolog of huntingtin interacting protein (HIP1). *J. Hum. Genet.* **43**, 268–271.

Senetar, M.A., Foster, S.J., McCann, R.O. (2004). Intrasteric inhibition mediates the interaction of the I/LWEQ module proteins Talin1, Talin2, Hip1, and Hip12 with actin. *Biochemistry* **43**, 15418–15428.

Sharp, A.H., Love, S.J., Schilling, G., Li, S.H., Li, X.J., Bao, J. et al. (1995). Widespread expression of Huntington's disease gene (IT15) protein product. *Neuron* **14**, 1065–1074.

Shaw, P.J. (2001). Genetic inroads in familial ALS. *Nat. Genet.* **29**, 103–104.

Singaraja, R.R., Hadano, S., Metzler, M., Givan, S., Wellington, C.L., Warby, S. et al. (2002). HIP14, a novel ankyrin domain-containing protein, links huntingtin to intracellular trafficking and endocytosis. *Hum. Mol. Genet.* **11**, 2815–2828.

Skehel, P.A., Fabian-Fine, R., Kandel, E.R. (2000). Mouse VAP33 is associated with the endoplasmic reticulum and microtubules. *Proc. Natl. Acad. Sci. USA* **97**, 1101–1106.

Slow, E.J. (2005). Inclusions to the rescue? Neuroprotective role for huntingtin inclusions in HD. *Clin. Genet.* **67**, 228–229.

Smart, E.J., Graf, G.A., McNiven, M.A., Sessa, W.C., Engelman, J.A., Scherer, P.E. et al. (1999). Caveolins, liquid-ordered domains, and signal transduction. *Mol. Cell Biol.* **19**, 7289–7304.

Soussan, L., Burakov, D., Daniels, M.P., Toister-Achituv, M., Porat, A., Yarden, Y., Elazar, Z. (1999). ERG30, a VAP-33-related protein, functions in protein transport mediated by COPI vesicles. *J. Cell Biol.* **146**, 301–311.

Stan, R.V. (2005). Structure of caveolae. *Biochim. Biophys. Acta* Oct 5 (online).

Steffan, J.S., Kazantsev, A, Spasic-Boskovic, O., Greenwald, M., Zhu, Y.Z., Gohler, H. et al. (2000). The Huntington's disease protein interacts with p53 and CREB-binding protein and represses transcription. *Proc. Natl. Acad. Sci. USA* **97**, 6763–6768.

Steffan, J.S., Bodai, L., Pallos, J., Poelman, M., McCampbell, A., Apostol, B.L. et al. (2001). Histone deacetylase inhibitors arrest polyglutamine-dependent neurodegeneration in Drosophila. *Nature* **413**, 7.

Steffan, J.S., Agrawal, N., Pallos, J., Rockabrand, E., Trotman, L.C., Slepko, N. et al. (2004). SUMO modification of Huntingtin and Huntington's disease pathology. *Science* **304**, 100–104.

Stokin, G.B., Lillo, C., Falzone, T.L., Brusch, R.G., Rockenstein, E., Mount, S.L. et al. (2005). Axonopathy and transport deficits early in the pathogenesis of Alzheimer's disease. *Science* **307**, 1282–1288.

Stowers, R.S., Megeath, L.J., Gorska-Andrzejak, J., Meinertzhagen, I.A., Schwarz, T.L. (2002). Axonal transport of mitochondria to synapses depends on milton, a novel Drosophila protein. *Neuron* **36**, 1063–1077.

Sturley, S.L., Patterson, M.C., Balch, W., Liscum, L. (2004). The pathophysiology and mechanisms of NP-C disease. *Biochim. Biophys. Acta* **1685**, 83–87.

Sugars, K.L., Brown, R., Cook, L.J., Swartz, J., Rubinsztein, D.C. (2004). Decreased cAMP response element-mediated transcription: An early event in exon 1 and full-length cell models of Huntington's disease that contributes to polyglutamine pathogenesis. *J. Biol. Chem.* **279**, 4988–4999.

Szebenyi, G., Morfini, G.A., Babcock, A., Gould, M., Selkoe, K., Stenoien, D.L. et al. (2003). Neuropathogenic forms of huntingtin and androgen receptor inhibit fast axonal transport. *Neuron* **40**, 41–52.

Topp, J.D., Gray, N.W., Gerard, R.D., Horazdovsky, B.F. (2004). Alsin is a Rab5 and Rac1 guanine nucleotide exchange factor. *J. Biol. Chem.* **279**, 24612–24623.

Torroja, L., Chu, H., Kotovsky, I., White, K. (1999). Neuronal overexpression of APPL, the Drosophila homologue of the amyloid precursor protein (APP), disrupts axonal transport. *Curr. Biol.* **9**, 489–492.

Trushina, E., Heldebrant, M.P., Perez-Terzic, C.M., Bortolon, R., Kovtun, I.V., Badger, J.D. 2nd et al. (2003). Microtubule destabilization and nuclear entry are sequential steps leading to toxicity in Huntington's disease. *Proc. Natl. Acad. Sci. USA* **100**, 12171–12216.

Trushina, E., Dyer, R.B., Badger, J.D. 2nd, Ure, D., Eide, L., Tran, D.D. et al. (2004). Mutant huntingtin impairs axonal trafficking in mammalian neurons in vivo and in vitro. *Mol. Cell Biol.* **24**, 8195–8209.

Tukamoto, T., Nukina, N., Ide, K., Kanazawa, I. (1997). Huntington's disease gene product, huntingtin, associates with microtubules in vitro. *Mol. Brain Res.* **51**, 8–14.

Valenza, M., Rigamonti, D., Goffredo, D., Zuccato, C., Fenu, S., Jamot, L. et al. (2005). Dysfunction of the cholesterol biosynthetic pathway in Huntington's disease. *J. Neurosci.* **25**, 9932–9939.

Vecchi, M., Di Fiore, P.P. (2005). It's HIP to be a hub: New trends for old-fashioned proteins. *J. Cell Biol.* **170**, 169–171.

Velier, J., Kim, M., Schwarz, C., Wan Kim, T., Sapp, E. et al. (1998). Wild-type and mutant huntingtin function in vesicle trafficking in the secretory and endo-cytic pathways. *Exper. Biol.* **152**, 34–40.

Waelter, S., Scherzinger, E., Hasenbank, R., Nordhoff, E., Lurz, R., Goehler, H. et al. (2001). The huntingtin interacting protein HIP1 is a clathrin and alpha-adaptin-binding protein involved in receptor-mediated endocytosis. *Hum. Mol. Genet.* **10**, 1807–1817.

Walter, M., Davies, J.P., Ioannou, Y.A. (2003). Telomerase immortalization upregulates Rab9 expression and restores LDL cholesterol egress from Niemann-Pick C1 late endosomes *J. Lipid Res.* **44**, 243–253.

Wanker, E.E., Rovira, C., Scherzinger, E., Hasenbank, R., Walter, S., Tait, D. et al. (2005). HIP-I: a huntingtin interacting protein isolated by the yeast two-hybrid system. *Hum. Mol. Genet.* **6**, 487–495.

Webb, J.L., Ravikumar, B., Rubinsztein, D.C. (2004). Microtubule disruption inhibits autophagosome-lysosome fusion: Implications for studying the roles of aggresomes in polyglutamine diseases. *Int. J. Biochem. Cell Biol.* **36**, 2541–2550.

Wesp, A., Hicke, L., Palecek, J., Lombardi, R., Aust, T., Munn, A.L., Riezman, H. (1997). End4p/Sla2p interacts with actin-associated proteins for endocytosis in Saccharomyces cerevisiae. *Mol. Biol. Cell* **8**, 2291–2306.

White, J.K., Auerbach, W., Duyao, M.P., Vonsattel, J.P., Gusella, J.F., Joyner, A.L., MacDonald, M.E. (1997). Huntingtin is required for neurogenesis and is not impaired by the Huntington's disease CAG expansion. *Nat. Genet.* **17**, 404–410.

Williamson, T.L., Marszalek, J.R., Vechio, J.D., Bruijn, L.I., Lee, M.K., Xu, Z. et al. (1996). Neurofilaments, radial growth of axons, and mechanisms of motor neuron disease. *Cold Spring Harb. Symp. Quant. Biol.* 709–723.

Williamson, T.L., Cleveland, D.W. (1999). Slowing of axonal transport is a very early event in the toxicity of ALS-linked SOD1 mutants to motor neurons. *Nat. Neurosci.* **2**, 50–56.

Xia, C.H., Roberts, E.A., Her, L.S., Liu, X., Williams, D.S., Cleveland, D.W., Goldstein, L.S. (2003). Abnormal neurofilament transport caused by targeted disruption of neuronal kinesin heavy chain KIF5A. *J. Cell Biol.* **161**, 55–66.

Yamada, K., Andrews, C., Chan, W.M., McKeown, C.A., Magli, A., de Berardinis, T. et al. (2003). Heterozygous mutations of the kinesin KIF21A in congenital fibrosis of the extraocular muscles type 1 (CFEOM1). *Nat. Genet.* **35**, 318–321.

Yang, Y., Hentati, A., Deng, H.X., Dabbagh, O., Sasaki, T., Hirano, M. et al. (2001). The gene encoding alsin, a protein with three guanine-nucleotide exchange factor domains, is mutated in a form of recessive amyotrophic lateral sclerosis. *Nat. Genet.* **29**, 160–165.

Yoo, S.Y., Pennesi, M.E., Weeber, E.J., Xu, B., Atkinson, R., Chen, S. et al. (2003). SCA7 knockin mice model human SCA7 and reveal gradual accumulation of mutant ataxin-7 in neurons and abnormalities in short-term plasticity. *Neuron* **37**, 383–401.

Zhao, C., Takita, J., Tanaka, Y., Setou, M., Nakagawa, T., Takeda, S. et al. (2001). Charcot-Marie-Tooth disease type 2A caused by mutation in a microtubule motor KIF1Bbeta. *Cell* **105**, 587–597.

Zhang, X., Smith, D.L., Meriin, A.B., Engemann, S., Russe, D.E., Roark, M. et al. (2005). A potent small molecule inhibits polyglutamine aggregation in Huntington's disease neurons and suppresses neurodegeneration in vivo. *Proc. Natl. Acad. Sci. USA* **102**, 892–897.

Zuccato, C., Ciammola, A., Rigamonti, D., Leavitt, B.R., Goffredo, D., Conti, L. et al. (2001). Loss of huntingtin-mediated BDNF gene transcription in Huntington's disease. *Science* **293**, 493–498.

Zuccato, C., Tartari, M., Crotti, A., Goffredo, D., Valenza, M., Conti, L. et al. (2003). Huntingtin interacts with REST/NRSF to modulate the transcription of NRSE-controlled neuronal genes. *Nat. Genet.* **35**, 76–83.

17

Neuronal Protein Trafficking in Alzheimer's Disease and Niemann-Pick Type C Disease

ANNE M. CATALDO AND RALPH A. NIXON

Sporadic Alzheimer's disease (AD), a chronic neurodegenerative disease of late age, is genetically and biochemically distinct from Niemann-Pick type C disease (NPC), an inherited storage disorder of early childhood characterized by cholesterol mistrafficking and accumulation. Despite these differences, AD and NPC share pathological features including neu-rodegeneration, neurofibrillary abnormalities, and, to some extent, β-amyloid accumulation. In both disorders, endocytic pathway function is prominently impaired. Moreover, disease onset is promoted by inheritance of the ε4 isoform of apolipo-protein-E, a major carrier protein for cholesterol, suggesting that cholesterol mishandling plays a role in AD as well as

NPC. In this chapter, we review the cell biology underlying the defects in these disorders with a special emphasis on dysfunction of the endocytic pathway and lysosomal system. We address how faulty lipid trafficking and alterations of Rab proteins associated with early and late stage endocytic compartments—that is, Rabs 4, 7, and 9—may disrupt vesicular transport leading to altered sorting and accumulation of endosomal cargoes and impaired net movement of late endosomal compartments. The redistribution of lysosomal proteases among different types of endocytic organelles and its relevance to the altered metabolism of amyloid-β precursor protein and amyloid-β formation are also considered.

I. INTRODUCTION

The endocytic pathway initially captured the interest of Alzheimer's disease (AD) researchers as a trafficking and metabolic pathway for amyloid-β precursor protein (APP). An ubiquitous membrane protein, APP is linked to the pathogenesis of AD by the existence of *App* gene mutations that cause rare familial forms of AD and also by the effect of *App* gene triplication, which contributes to the invariable development of AD in older individuals with Down syndrome (Trisomy 21). APP is the source of amyloid-β peptide (Aβ), which accumulates pathologically within neurons and extracellularly as oligomeric or fibrillar complexes, called β-amyloid. The neurotoxicity of Aβ, in its various forms, is regarded as a catalyst of the progressive degeneration and loss of neurons in AD, although additional APP metabolic products may also be pathogenic. With the discovery that organelles within the endocytic pathway generate Aβ, it also has been appreciated that these organelles already appear abnormal at the earliest stages of AD, suggesting not only a basis for Aβ overproduction, but also a basis for the disrup-

tion of key neuronal functions independent of Aβ toxicity. This notion is in accord with an emerging awareness in the Alzheimer field that Aβ may be necessary, but not sufficient to cause AD, and that additional factors are likely to be involved. The possibility that a primary disturbance of endocytic function may be critical to AD development is gaining strong support from evidence that a surprising number of neurodegenerative diseases are caused by mutations in genes encoding protein that regulate the endocytic pathway (Nixon 2005).

The inclusion of Niemann-Pick type C (NPC) disease with AD in this chapter, on its surface, may seem arbitrary. NPC begins at the start of life; the onset of AD is near the end of the life cycle. Other differences are also obvious. Infantile NPC, unlike AD, has prominent systemic effects and affects somewhat different neuronal populations. The cellular pathology of these two diseases, however, has intriguing commonalities and a consideration of these similarities, some of which involve the endocytic pathway, may be instructive about underlying mechanisms in both diseases. This chapter, in three main sections, will address first the pathobiology of endocytic trafficking in AD, then in NPC, and finally will consider abnormalities in both diseases that suggest common cellular mechanisms. Pathological features in AD and NPC discussed throughout the text are presented in Table 17.1.

II. ALZHEIMER'S DISEASE

A. Background

Alzheimer's disease (AD), the most prevalent human neurodegenerative disorder, involves progressive neuron loss leading to profoundly impaired memory and intellect. The neuropathology of AD is characterized by the presence of two hallmark lesions (Yankner 1996). One of these

TABLE 17.1. Pathological Features of Sporadic Alzheimer's Disease and Niemann-Pick Type C Disease

	AD	NPC
Neurodegeneration	✓	✓
Aβ Localization		
a) Intracellular	✓	?
b) Extracellular	✓	✓
Neurofibrillary Tangles	✓	✓
Endocytic Alterations		
a) Early stage organelles: early, recycling, sorting endosomes		
rab5	✓	✓
rab4	✓	✓
rab11	?	✓
EEA1	✓	✓
b) Late stage organelles: late endosomes, late endosomal hybrids		
rab7	✓	✓
rab9	?	✓
LBPA	✓	✓
Lysosomal system		
a) Mannose receptors:		
MPR 300 (215)	✓	✓
MPR46	✓	?
b) Hydrolases		
BACE	✓	?
Cathepsin D	✓	✓
Lipid Trafficking	?	✓
ApolipoproteinE Effects	✓	✓

? = unknown.

lesions involves the intraneuronal formation of neurofibrillary tangles, consisting mainly of aggregated forms of the microtubule-associated protein, tau. A second lesion is the "senile" or neuritic plaques composed of degenerating and dystrophic dendrites or axons in association with extracellular β-amyloid. Sporadic AD (SAD) has a complex etiology reflecting influences of genetic risk factors (Kang et al.

1997; Katzmann et al. 2001) and environmental factors (Sparks 1997; Cummings et al. 1998; Launer et al. 1999). The most common genetic influence is inheritance of *ApoE4*, one of three allelic forms encoding the lipid-carrier protein, apolipoprotein-E (APOE) (Corder et al. 1993). Fewer than 10 percent of AD cases are inherited as autosomal dominant diseases and have an early-onset before age 65. Mutations of two related genes encoding presenilin 1 (PS1) and presenilin 2 (PS2) cause more than half of these cases of early-onset familial AD (FAD) (Levy-Lehad et al. 1995; Sherrington et al. 1995; Price and Sisodia 1998). Mutations of *App* also cause early onset AD in less than 1 percent of cases (Levy et al. 1990; Goate 1998; Price and Sisodia 1998). All these mutations in *App* and *PS* appear to influence the proteolytic processing of APP, modifying the amount and/or length of the Aβ peptide generated (Borschelt et al. 1996; Citron et al. 1997; Sastre et al. 2001; Edbauer et al. 2003), but they may also potentially disturb other biological processes mediated by these proteins (Sisodia and St George-Hyslop 2002; Medina and Dotti 2003; Leyssen et al. 2005).

Newly synthesized APP in its mature glycosylated form is delivered from the *trans*-Golgi network to the plasma membrane, where it undergoes a fairly rapid turnover. Two aspartyl proteases at the cell surface, TACE and ADAM-10 (Buxbaum et al. 1998; Lopez-Perez et al. 2001), mediate α-cleavage of APP at a site within its lumenal/extracellular domain, liberating a large soluble amino-terminal fragment (sAPPα) having neurotrophic properties (Mattson 1997), and creating a membrane-associated carboxyl-terminal fragment (CTF). Alternative metabolism of cell surface APP occurs once it is internalized within early endosomes where the aspartyl protease, BACE (Vassar et al. 1999; Capell et al. 2000; Walter et al. 2001), carries out a β-cleavage more distally along the lumenal/extracellular domain. Beta-cleavage of APP releases a soluble fragment

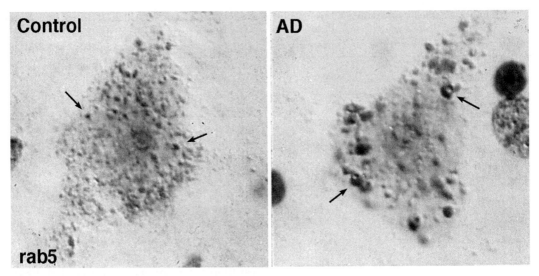

FIGURE 17.1.   **Early endosomes in neurons of AD brain are abnormally enlarged.** Early endosomal compartments (arrows) in representative neurons from lamina III of normal aged control and AD brains immunostained with the rab5 and enhanced by Normarski optics show the striking increase in the size of these organelles in the AD brain compared with the control.

(sAPPβ) and generates a membrane-associated CTF (βCTF) containing the whole Aβ peptide (Grbovic et al. 2003; Mathews and Nixon 2003). Aβ finally is generated from βCTF by an intramembrane γ-cleavage that yields predominantly a 40-mer peptide (Aβ40) and smaller amounts of a 42-mer peptide (Aβ42). The formation of Aβ liberates a cytoplasmic tail peptide that can form a multimeric complex that stimulates transcription (Cao and Sudhof 2001). The γ-cleavage of APP is mediated by a γ-secretase complex composed of the proteins presenilin, nicastrin, apH-1, and PEN (Edbauer et al. 2003; Go et al. 2004), which is present in many cellular locations, including at the plasma membrane and in early endosomes, late endosomes, autophagic vacuoles, and lysosomes (Cupers et al. 2001; Pasternak et al. 2003).

B. Alzheimer's Disease and the Endocytic Pathway

The appearance of the intraneuronal neurofibrillary tangles and senile plaques that define AD is preceded by an elevation in the levels of soluble and oligomeric Aβ peptide (McLean et al. 1999). At the earliest stage of disease when soluble Aβ levels are beginning to rise, many neurons in the cortex and hippocampus exhibit enlarged rab5-positive endosomes (Cataldo et al. 2004) (see Figure 17.1), most of which contain Aβ immunoreactivity (Cataldo et al. 2004). Effectors of Rab5 function, such as EEA-1 and rabaptin 5, also translocate to these enlarged endosomes and the expression of Rab4, a marker of endosome recycling, is increased, further suggesting that endocytic pathway function is altered (Cataldo et al. 1997; 2000). Gene profiling analyses of single neurons displaying enlarged endosomes (Elarova et al. 2004) show elevated expression of Rab5 and Rab4 at the mRNA level, consistent with increased neuronal endocytosis (Cataldo et al. 2000; 2004). Enlarged endosomes also contain abnormally high levels of lysosomal hydrolases, that are due in part to an elevated expression of MPR46 (Cataldo et al. 1997; Mathews et al. 2002b). This endosomal pathology seems neither to be a response to Aβ overproduction (Cataldo et al. 2000), nor a general feature of stressed or degenerating neurons in neurodegenerative disease (Cataldo et al. 2000). Various lines of evidence support the conclusion that endosome abnormalities lead to early rises in brain Aβ. For example, Aβ generation increases substantially when either the endosome enlargement or increased cathepsin delivery to early endosomes seen in AD brain is reproduced in cells by over-expressing Rab5 or MPR46, respectively (Mathews et al. 2002b; Grbovic et al. 2003). Moreover, endosomes contain β-site APP cleaving enzyme (BACE) (Wang et al. 1999; Huse et al. 2000), presenilin (Lah and Levey 2000), and APP (Haass et al. 1992a, 1993; Nordstedt et al. 1993). βCTF, the γ-secretase cleaved carboxyl terminal domain of APP, is generated directly in endosomes (Mathews et al. 2002a,b; Grbovic et al. 2003) and Aβ is detectable in enlarged endosomes of neurons from AD brains (Cataldo et al. 2004). Although Aβ generation may occur at several subcellular sites— ER (Chyung et al. 1997; Greenfield et al. 1999), medial Golgi saccules (Thinakaran et al. 1996), TGN (Xu et al. 1997; Vassar et al. 1999), autophagic vacuoles (Yu et al. 2004; 2005)—and along the endocytic pathway (Haass et al. 1992b; Koo and Squazzo 1994; Soriano et al. 1999), endosomes are the only one of these organelles that are detectably abnormal at a stage when intracellular Aβ first accumulates and before amyloid begins to be deposited extracellularly. Finally, Aβ production falls markedly when APP internalization is reduced by APP mutagenesis or when endocytosis is blocked (Koo and Squazzo 1994; Perez et al. 1999; Soriano et al. 1999), and rises substantially when endosome function is modified (Mathews et al. 2002b; Grbovic et al. 2003). Antibodies to the APP β-cleavage site, which target endosomal APP processing, significantly lower Aβ production and are being explored as potential immunization strategies for the treatment of AD (Arbel et al. 2005; Paganetti et al. 2005).

Nearly all individuals with Down syndrome (Trisomy 21) develop the neuropathologic lesions of AD by the age of 45, and nearly all these individuals who live into their sixties become demented (Lai and Williams 1989). Down syndrome (DS) develops in 0.1 to 0.2 percent of the population due to the triplication of chromosome 21, resulting in retarded physical and mental development. The age of onset of dementia, like that of AD, is influenced by APOE subtype (Lai et al. 1999; Deb et al. 2000). The early onset and high incidence of AD in DS have been ascribed to triplication of *App*, a gene located on chromosome 21, that, in mutant form, causes early onset AD (Price et al. 1998). Amyloid-β pathology in DS individuals is preceded by the same neuronal endosomal pathology that is observed in AD (Cataldo et al. 2000). Interestingly, endosomal pathology slowly begins to appear even before birth, becoming prominent in the years preceding the appearance of classical AD neuropathology. A similar endosomal phenotype also is observed in a genetic mouse model of DS (Ts65Dn), in which a critical segment of chromosome 16, the mouse homologue of chromosome 21, is triplicated (Galdzicki and Siarey 2003). This model reproduces key features of the DS dysmorphogenic phenotype, and adult mice develop atrophy and degeneration of basal forebrain cholinergic neurons and possibly hippocampal and neocortical neurons, in the absence of amyloid-β deposition or NFT formation (Holtzman et al. 1996). Endosomal pathology appears at early ages in neurons within these same brain regions. Significantly, these endocytic abnormalities are eliminated in Ts65Dn mice when *App* gene dosage is reduced from 3 to 2 (Cataldo et al. 2003), establishing a direct relationship between triplication of this AD-related gene and endosome dysfunction. APP or its metabolites must act in concert with products of one or more additional genes in the small trisomic segment of chromosome 16 to induce the endosomal and neurodegenerative phenotype, because it is not reproduced by APP overexpression alone. Interestingly, several of these trisomic genes on chromosome 16 function in endocytosis.

C. Possible Consequences of Endosome Dysfunction in AD

Neurons rely heavily on endocytosis to transduce extracellular signals necessary for survival, including hormones and growth factors (Sofroniew et al. 2001). Ligand receptor complexes (e.g., nerve growth factor (NGF)-TrkA) continue to signal after they are internalized in association with activated components of the Ras-MAPK pathway and other signaling protein effectors located on the endosome (Howe et al. 2001). Trophic signals are transduced while the endosome undergoes retrograde axonal transport. Recent evidence indicates that these signaling endosomes that convey retrograde NGF signals exhibit characteristics of early endosomes (Delcroix et al. 2003). A defect in this process may explain the failure of retrograde signaling by NGF in the Ts65Dn mouse, which results in degenerative changes in basal forebrain cholinergic neurons (Cooper et al. 2001). In these neurons, NGF seems to be normally internalized, but its retrograde transport is markedly impaired. The resulting reduced trophic support and atrophy of these neurons can be partly reversed by chronic NGF infusion into the brain (Cooper et al. 2001). Whether the defect reflects retrograde transport failure, premature degradation of signaling molecules, an endosome sorting defect, or a combination of these problems, is unclear. In this regard, early endosome dysfunction and elevated cathepsin levels are evident in Tg65Dn neurons, as in AD and DS (Cataldo et al. 2003). Moreover, growing evidence suggests that additional vesicular transport steps may be disturbed in neurons in AD and in AD animal models (Burke et al. 1990; Goldstein 2001; Terwel et al. 2002; Pigino

et al. 2003). In this regard, autophagic vacuoles and related vesicular compartments of the macroautophagy pathway accumulate in massive numbers in dystrophic dendrites of affected neurons in AD, reflecting not only impaired retrograde transport but also a delayed or defective maturation of autophagosomes to lysosomes (Nixon et al. 2005; Yu et al. 2005). It will be interesting to determine whether this abnormality of macroautophagy that occurs at a later stage of AD is related to the earlier disturbances in endocytic trafficking, since the autophagic and endocytic pathways converge at late endosomes. A single defect in endocytic trafficking or fusion could possibly result in dysfunction in both pathways.

D. The Cholesterol Connection

Inheritance of the *ApoE4* genotype, the most prevalent risk factor for AD development, exacerbates the endocytic pathology seen in sporadic AD, while also promoting an earlier onset of AD and more severe amyloid pathology (Corder et al. 1993; Strittmatter et al. 1993). APOE is the major cholesterol carrier protein in brain; the association between the *ApoE* genotype and AD development suggests that cholesterol may have pathogenic significance in AD. Indeed, other genetic factors linked to cholesterol metabolism, including polymorphisms of CYP46, appear to accelerate the risk of developing late-onset AD (Kolsch et al. 2002; Papassotiropoulos et al. 2003; Wolozin 2003). Epidemiological studies suggest that elevated plasma cholesterol levels have a facilitating effect, albeit a complex one, on AD development (Sparks 1997; reviewed in Friedland 2002; Chauhan 2003; Poirier 2005). In brain, as in other peripheral organs, cellular cholesterol levels are tightly regulated. Increased cholesterol in the membrane alters its fluidity and can affect the activity or function of lipid-anchored, integral membrane proteins, including APP. Nonamyloidogenic processing of membrane APP by α-secretase requires it to be cleaved at a fixed distance from the membrane (Sisodia et al. 1992). A change in lipid ordering that perturbs the spatial relationship of APP with the membrane could inhibit α-cleavage of APP. Growing evidence now supports this hypothesis and indicates that elevated cellular cholesterol levels promote Aβ generation and accumulation, and reduced sAPPα levels (Bodovitz and Klein 1996; Kojro et al. 2001; Puglielli et al. 2001, 2003). The subcellular distribution of cholesterol seems to play a significant role in regulating APP processing (Runz et al. 2002). When cholesterol accumulates within early and late endocytic compartments, the levels of γ-secretases within these same compartments rise. Alternatively, slowing cholesterol transport from endocytic compartments to the ER reduces reinternalization of surface APP and subsequently lowers Aβ generation. Our own preliminary studies indicate that, *in vitro* and *in vivo*, elevated cholesterol levels promote AD-like endocytic pathology and increase APP processing to Aβ. These observations suggest that cholesterol-induced endocytic pathway dysfunction could be a cellular mechanism for enhanced amyloidogenic APP processing within the endocytic pathway, and may explain the facilitating effect of *ApoE4* genotype on AD development.

III. NIEMANN-PICK DISEASE

A. Background

Niemann-Pick Disease type C (NPC) is an inherited autosomal storage disorder with an eventually fatal neurodegenerative course. The age of disease onset may vary from the perinatal period to adult although the late infantile and juvenile forms are most common (60 to 70% of cases) (Vanier and Suzuki 1998; Vincent and Erickson 2003; Vanier and Millat 2003; Sturley et al. 2004). The initial common clinical manifestations of NPC include progressive cerebellar ataxia, bulbar dysfunction, dysarthria, seizures, and cognitive decline (Vanier and Suzuki 1998; Vanier and Millat 2003; Vincent and Erickson 2003; Sturley et al. 2004). Classification of the NPC phenotypes is determined from the age of onset and neurological symptoms. Symptomatology of the adult onset form is similar to the juvenile type with an insidious onset and progressive dementia (Vanier and Suzuki 1998; Vanier and Millat 2003; Vincent and Erickson 2003; Sturley et al. 2004). Psychosis may be present in the early stages of disease and extrapyramidal movement disorders are more frequent in the adult compared to the juvenile form of the disease. The majority (95%) of NPC cases are caused by mutations in the *npc1* gene located on chromosome 18 (Carstea et al. 1997; Ioannou et al. 1997; reviewed in Vincent et al. 2003; Sturley et al. 2004); the remainder have been linked to mutations in the *HE1/npc2* gene (Naureckiene et al. 2000; reviewed in Vincent and Erickson 2003; Sturley et al. 2004) on chromosome 14. The classic pathological changes associated with NPC are both neural and visceral and include hepatosplenomegaly and the accumulation of abnormal amounts of cholesterol and other lipids such as sphingomyelin, lysobisphosphatidic acid (LBPA), glycosphingolipids, and phospholipids, within subcellular late endosomes and lysosomes in neural and nonneural cell types (Pentchev et al. 1994; Vanier and Suzuki 1998). The primary neuropathological features of NPC are progressive neuronal cell loss, axonal alterations, demyelination and, in some forms of NPC, the presence of neurofibrillary tangles (Suzuki et al. 1995), which are structurally and biochemically similar to those seen in AD brains, and intracellular and extracellular Aβ deposits (Yamazaki et al. 2001; Saito et al. 2002).

Niemann-Pick protein-1 (NPC1) is the gene product of *npc1*. NPC1 is a large, glycosylated, membrane-spanning protein (Pentchev et al. 1994; 2004) that is synthesized in the ER and undergoes posttranslational modification in the Golgi apparatus.

It contains several interesting domains, including a dileucine motif at its carboxy-terminal tail that is involved in clathrin-mediated internalization and delivery to late endosomes (Neufeld et al. 1999; Watari et al. 1999; Wiegand et al. 2003; Frolov et al. 2003; reviewed in Mukherjee and Maxfield 2004), and a sterol sensing domain that regulates the sorting and transport of internalized cholesterol esters from late endosomes and lysosomes to other subcellular compartments (Blanchette-Mackie et al. 1988; Neufeld et al. 1999; Coxey et al. 1993; Higgins et al. 1999; Garver et al. 2002; reviewed in Mukherjee and Maxfield 2004). NPC1 has been localized to late stage endocytic organelles that share characteristics consistent with both late endosomes and lysosomes: (1) Rab7-, LAMP2-positive compartments negative for cation-independent mannose 6-phosphate receptor (MPR); (2) Rab9-positive late endosomes, LAMP1-positive lysosomes, and Rab11-positive *trans*-Golgi saccules; and (3) LAMP1-, LAMP2-negative compartments. Although the precise role of NPC1 remains unclear, the association between NPC1 and cholesterol have reinforced its importance in the retroendocytic trafficking and recycling of cholesterol from late endosomes and lysosomes to the plasma membrane, the ER (Zhang et al. 2001; Ko et al. 2001; Garver et al. 2002; Lusa et al. 2001; Liscum and Munn 1999; Neufeld et al. 1999), and the Golgi apparatus. NPC2 is the gene product of *HE1/npc2* and a small soluble protein that is found in the lumen of lysosomes (Naureckiene et al. 2000; reviewed in Vincent and Erickson 2003; Sturley et al. 2004). Like NPC1, NPC2 possesses an N-terminal signal peptide, is synthesized in the ER, and is glycosylated in the Golgi apparatus (Naureckiene et al. 2000; reviewed in Vincent and Erickson 2003; Mukherjee and Maxfield 2004; Sturley et al. 2004). Similar to most lysosomal hydrolases, NPC2 binds to mannose 6-phosphate receptors, which deliver NPC2 from the *trans*-Golgi saccules to late endosomes and

lysosomes (Naureckiene et al. 2000; reviewed in Vincent and Erickson 2003; Sturley et al. 2004; reviewed in Mukherjee and Maxfield 2004).

Mutations in either NPC1 or NPC2 can cause alterations in lipid transport that lead to lipid storage and delayed lipid movement to the targeted cellular destinations (Blanchette-Mackie et al. 1988; Blom et al. 2003; Ikonen and Holtta-Vuori 2004; Liscum 2000; Vincent and Erickson 2003; Sturley et al. 2004). NPC1 and NPC2 proteins may act sequentially through a common pathway to achieve proper lipid sorting and transport (Pentchev et al. 1994; Liscum 2000; Blom et al. 2003; Vincent and Erickson 2003; Ikonen and Holtta-Vuori 2004; Sturley et al. 2004; Walkley and Suzuki 2004). Patients with mutations in either NPC1 or NPC2 are clinically indistinguishable (Pentchev et al. 1985, 1994; reviewed in Ory 2000), and fibroblasts obtained from individuals with NPC caused by defects in either NPC1 or NPC2 proteins display identical morphologies and biochemical profiles (Reid et al. 2003; reviewed in Paul et al. 2004; Pentchev et al. 2004). Consistent with the human data, studies involving mice with mutations in either NPC1 or NPC2 or both, have shown that these animals display a similar disease onset and course (Xie et al. 2000) and identical pathologies.

How does dysfunction of NPC1 or NPC2 lead to alterations in neuronal protein trafficking? Let us first briefly consider the normal trafficking itineraries with a specific emphasis on lipid trafficking.

B. Lipids and Compartments of the Endocytic Pathway

The configuration of the endocytic pathway, including its constituent organelles and their relationships, has been described in detail in previous chapters. In mammalian cells, vesicular trafficking along the endocytic pathway occurs through a series of budding and fusion events that is regulated by several groups of structural and functional proteins called SNARE proteins, and small GTPases of the Rab family (Zerial and McBride 2001) (see Chapters 6, 7, and 14). Following internalization, proteins and lipids from the cell surface initially are directed to early endosomes. These early stage endocytic vesicles can be divided into two groups: sorting endosomes and recycling endosomes. From these compartments, most internalized proteins and lipids, including plasma membrane lipids and proteins, are recycled back to the cell surface to be reused. Some internalized components, (e.g., receptors, transporters, and ion channels) are transported from early endosomes to late endosomes and lysosomes for terminal degradation.

In mammalian cells, protein-lipid interactions are critical for membrane stability. The distribution of proteins and lipids on discrete organelle membranes is distinct and the maintenance of these proportions is essential for biological functions (Puri et al. 1999; Maxfield and Wustner 2002; Reid et al. 2003; Riddell et al. 2001a,b). The lipid composition of the plasma membrane has the highest concentration of free sterols and is rich in cholesterol, glycosphingolipids, phosphatidylethanolamine, and phosphatidyl serine (reviewed in Brown and Goldstein 1986; Dietschy and Turley 2001). The plasma membrane contains about 60 to 80 percent of total cellular cholesterol, and as much as 30 to 40 percent of the plasma membrane is cholesterol (reviewed in Brown and Goldstein 1986; Dietschy and Turley 2001). Small changes in the cholesterol content of the plasma membrane can alter membrane fluidity (reviewed in Brown and Goldstein 1986; Dietschy and Turley 2001; Hao et al. 2004). Cholesterol and lipids within the plasma membrane are not distributed homogeneously, but are enriched within lateral domains called rafts that are resistant to solubilization at low temperatures by nonionic detergents and believed to play a role in cholesterol transport (Lange and Steck 1996; Simons and Ikonen 1997; Simons and Gruenberg 2000)

(see Chapters 1 and 13). Much of the plasma membrane in many cell types exists as raft-like domains. Rafts also have been reported in endosomes and the Golgi apparatus (Lange and Steck 1996; Simons and Ikonen 1997; Simons and Gruenberg 2000). Because the plasma membrane contributes membrane for the formation of early endosomes, it generally is accepted that the overall lipid composition of early endosomes and recycling endosomes is like that of the plasma membrane-containing high concentrations of cholesterol and glycosphingolipids (Lange and Steck 1996; Kobayashi et al. 1998; also see Hao et al. 2004 for additional references). In contrast to the plasma membrane and early stage endocytic organelles, late endosomes do not contain high amounts of cholesterol. Instead, these compartments are rich in neutral lipids, like triglycerides and cholesterol esters. Unlike other endocytic compartments, late endosomes contain an unusual glycophospholipid within their internal membrane, called lysobisphosphatidic acid (LPBA) (Kobayashi et al. 1998). Late endosomes are composed of extensive networks of these LBPA-rich membranes (Kobayashi et al. 1998). Lysosomes generally contain small amounts of sterols (Lange and Steck 1996; Zinser et al. 1993; see Pichler and Riezman 2004 for review and other references).

In normal cells, intracellular cholesterol levels and distribution are tightly regulated. Cells acquire cholesterol through exogenous uptake and endogenous *de novo* biosynthesis (Brown and Goldstein 1986; Lange and Steck 1996; Liscum and Munn 1999; also see Dietschy and Turley 2001; Maxfield and Wustner 2002; Sugii et al. 2003; Pichler and Riezman 2004 for review and additional references). The internalization of plasma membrane-derived cholesterol likely occurs during receptor mediated or constitutive endocytic vesicular trafficking or by less well-defined transport routes involving the ER. LDL carrying esterified cholesterol is internalized and transported to lysosomes where free cholesterol can be transported back to the plasma membrane. Cellular cholesterol obtained by *de novo* synthesis is transported from the ER to the plasma membrane bypassing the Golgi apparatus. However, some studies suggest that newly synthesized cholesterol can also be delivered to the plasma membrane by vesicular transport routes via the ER and *trans*-Golgi network. Cholesterol is necessary for caveolar uptake as well as internalization via clathrin-coated pits. In contrast, clathrin-independent internalization is insensitive to cholesterol loading or depletion. Recycling of GPI-anchored proteins from endosomes to the plasma membrane is slowed in the presence of cholesterol (Mayor et al. 1998), whereas reflux of GPI-anchored proteins to the plasma membrane is accelerated when cholesterol levels are low. Increased cholesterol levels may interfere with proper recycling, and trap materials destined for recycling in endosomes or could alter the membrane properties of endosomes such that organellar interactions and transport become abnormal. Cholesterol also modulates late-stage endocytic transport by forming sterol-rich anchoring networks in endosomes through its interactions with the acidic phospholipid-binding protein, annexin II (Johnstone et al. 1992; Waisman 1995).

C. Endocytic Trafficking in Niemann-Pick Disease

Most cholesterol in mammalian cells is internalized with low-density lipoproteins (LDL), which are delivered first to early endosomes and subsequently to late endosomes and lysosomes. In late endosomes cholesterol is hydrolyzed to unesterified free cholesterol, partitioned into the LBPA-rich membrane network and then released into the lumen. The network of LBPA-rich internal membranes within late endosomes collects cholesterol and regulates the distribution of free cholesterol throughout the cell. Under normal conditions, cholesterol does not accumulate in late endosomes.

Numerous studies suggest that cholesterol trafficking from the plasma membrane to intracellular compartments is mediated by tubulovesicular NPC1-sorting organelles that are closely associated with late endosomes. Interruption in the transport of these NPC1 sorting compartments from late endosomes is thought to be the primary defect in NPC (Wojtanik and Liscum 2003; Te Vruchte et al. 2004). The elegant studies of Kobayashi et al. (1998) support the notion that the LBPA-rich internal membrane within late endosomes are affected in NPC and produce the cholesterol-engorged late endosomes associated with the pathobiology. Cholesterol accumulation in late endosomes interferes with late endosomal sorting, trafficking from late endosomes to lysosomes, and the formation of late endosomal hybrid organelles (Mullock et al. 1998). Studies by Neufeld et al. (1999) have shown that the defective sorting and transport in NPC does not appear to be limited to sterols, but is a more generalized alteration in endosomal regulated transport that affects the bulk retrograde egress of vesicular compartments that recycle to the plasma membrane from late endosomes.

Another late endosomal sorting/trafficking defect associated with NPC involves the sorting of cation-independent mannose 6-phosphate targeted proteins within late endosomes. Studies using BSA-gold tracers have shown that a dynamic equilibrium exists between late endosomes and lysosomes (Griffiths 1996), and that both organelles contain lysosomal hydrolases. In contrast to lysosomes, which contain the bulk of the intracellular hydrolase pool, late endosomes are responsible for the majority of proteolysis, although they contain only 20 percent of the cell's total hydrolases (Tjelle et al. 1996; Bright et al. 1997; also reviewed in Pillay et al. 2002). Late endosomes play an important role in the sorting of cation-independent (CI) MPR/IGF2-tagged ligands that possess a mannose 6-phosphate recognition sequence, and include lysosomal hydrolases and insulin growth factor. The CIMPR/IGF2 receptor delivers nascent hydrolases, which are packaged in the *trans*-Golgi network, to late endosomes and lysosomes and then recycles back to the *trans*-Golgi for reuse (Kornfeld 1992). CIMPR/IGF2 receptor is concentrated predominantly in late endosomes in primary fibroblasts obtained from patients with NPC (Umeda et al. 2003; reviewed in Mukherjee and Maxfield 2004). Additionally, late endosomes are also severely immobile in NPC (Ko et al. 2001; Zhang et al. 2001; Le Brand et al. 2002). Late endosome motility is closely linked to its lipid content. Loading late endosomes with cholesterol or using the hydrophobic amine U18666A, which causes cholesterol accumulation like NPC, inhibits endosomal motility and stagnates late endosomal transport (LeBrand et al. 2002). Studies by LeBrand et al. (2002) have shown that the association between cholesterol and late endosomal mobility occurs via the small GTPase, Rab7. Membrane cholesterol content may contribute to the regulation of the Rab7 GTPase cycle that controls late endosomal movement. Like other Rab protein family members, Rab7 interacts with the membrane bilayer by cysteine geranylgeranylation (Zerial and McBride 2001). These small GTPases cycle between the active GTP-bound and the inactive GDP-bound forms. The GDP-bound form is a target for guanine nucleotide dissociation inhibitor (GDI). Rab proteins cycle between organellar membranes and the cytosol through the activity of GDI, which removes the GDP-Rab from the membrane and returns it to the cytosol (Dirac-Svejstrup et al. 1997; Luan et al. 1999; reviewed in Goody et al. 2005). Cholesterol accumulation alters membrane bilayer fluidity and increases the levels of membrane-bound Rab7. In late endosomes of NPC patients, cholesterol accumulation may interfere with the ability of GDI to extract Rab7 from the membrane, because cholesterol renders its membrane (prenyl) anchor less accessible. Cholesterol accumulation is also thought to interfere with the ability of late endosomes to switch from minus- to plus-ended motility. The organelles are able to retain cytoplasmic dynein but are unable to acquire kinesin activity, causing a loss in bidirectional movement and eventual stagnation. Rab7 and its effector, RILP, are thought to play an important role in the regulation of this process. Infection of normal fibroblasts treated with the hydrophobic amine, U18666A, which causes NPC-like cholesterol accumulation, with the dominant-negative Rab7 mutant, I125N, restores motility to cholesterol-loaded late endosomes (LeBrand et al. 2002). Late endosomes also retain Rab9, another Rab GTPase. Rab9 participates in protein sorting from late endosomes to lysosomes by stimulating the recycling of MPRs from late endosomes to the *trans*-Golgi network (Lombardi et al. 1993). Studies by Choudhury et al. (2002) and Walter et al. (2001) have shown that in NPC defective fibroblasts, the overexpression of Rab9 or Rab7 appears to partially rescue the late endosomal transport defects associated with NPC by reducing cholesterol accumulation in late endosomes and promoting the trafficking of BODIPY-labeled lactosylceramide to the *trans*-Golgi saccules. The results of these studies imply that in NPC, abnormalities in Rab-mediated vesicular trafficking are either the direct result of cholesterol accumulation within late endosomes or of NPC-related defects in late endosomal motility resulting in cholesterol accumulation in late endosomes. Recycling endosomes contain both Rab4 and Rab11 (Sheff et al. 1999; reviewed in Zerial and McBride 2001). In NPC, the defects in late stage endocytic trafficking linked to abnormal lipid accumulation are not limited to late-stage endocytic processes, but have been shown to impair early-stage endocytic events by altering Rab4-mediated recycling (Choudhury et al. 2004), possibly by disrupting Rab4-specific domains on early endosomes. The link between these perturbations and the excess cholesterol in NPC has been supported by findings showing that cholesterol depletion of NPC cells restored fast recycling of labeled lactosylceramide and the endosomal membrane distribution of Rab4 (Choudhury et al. 2004).

IV. NPC AND AD: COMMON PATHOLOGIES

Although major differences exist between NPC and AD, these diseases exhibit fascinating parallels, both in their neuropathology and cellular pathobiology, that we believe could be linked to the defects in cholesterol trafficking leading to endocytic dysfunction.

A. Neurodegeneration and Cytoskeletal Alterations

The alterations in cholesterol transport in NPC may play a central role in altered tau phosphorylation, NFT formation, and neurodegeneration (Fan et al. 2001; Zhang et al. 2001; Bu et al. 2002; Ohm et al. 2003; Walkey and Suzuki 2004). Cell atrophy and loss in NPC are most prominent within populations of cortical pyramidal neurons and cerebellar Purkinje cells. Other vulnerable neuronal populations include neurons in CA1 and CA2 fields of the hippocampus, the thalamus, basal ganglia, and substantia nigra (Ikonen and Holtta-Vuori 2004; Paul et al. 2004). Several of these neuronal populations, such as pyramidal cells in the hippocampus and cerebral cortex, also are severely affected in AD. Neurons in these brain regions exhibit swollen cell bodies and axonal hillocks, stunted dendritic arbors, and reduced elaboration of axons and dendrites (Karten et al. 2002; 2003). Finally, neurodegeneration in association with the presence of neurofibrillary tangles (NFTs) in NPC has been reported in several studies (Suzuki et al. 1995; Saito et al. 2002). The NFTs found in NPC brains are composed of hyperphosphorylated isoforms of tau in the form of paired helical filaments (PHFs), and are particularly abundant in the cell soma and swollen portions of dendrites and axons of neurons in cortex and hippocampus similar to AD pathology (Love et al. 1995).

Microtubules are major constituents of the neuronal cytoskeleton and play important roles in a number of neuronal functions including axonal transport (see Chapters 2 and 13). The structure and stability of microtubules is regulated by microtubule associated proteins (MAPs), including tau. Cholesterol levels are critical to the stabilization and organization of microtubules, dendritogenesis, spine formation myelination, and the formation of axon collaterals (Fan et al. 2001; Ohm et al. 2003; see Walkley and Suzuki 2004 for additional references). Distal portions of axons isolated from the brains of NPC mutant mice were found to have lower cholesterol levels compared to wild-type mice, suggesting that cholesterol loss may be interfering with proper axonal maintenance (Sawamura et al. 2003; Zhang et al. 2001; reviewed in Distl et al. 2003; Ohm et al. 2003; Treiber-Held et al. 2003; Walkley and Suzuki 2004). Decreased cholesterol levels can result in hyperphosphorylation of MAP2, a microtubule protein enriched in the cell soma and dendrites (Fan et al. 2001). Moreover, in NPC, hyperphosphorylated forms of tau, MAP2b, and MAP2c proteins aggregate in the soma and axons and are accompanied by increased levels of cdk5 kinase, a regulator of microtubule motor-associated transport (Zhang et al. 2001; Bu et al. 2002; reviewed in Distl et al. 2003; Ohm et al. 2003; Walkley and Suzuki 2004).

B. Amyloid

Circumstances that promote increased Aβ levels, such as the overexpression of normal APP in Down syndrome mutated forms of APP, or physiologic conditions, such as high cholesterol levels, impair efficient Aβ clearance. Hypercholesterolemia is thought to be an important risk factor for AD development (Sparks 1997; see Friedland 2002; Chauhan 2003; Poirier 2005 for review and additional references). Increased cholesterol perturbs the normal lipid-ordering of the membrane and increases membrane rigidity. Therefore, membrane-associated proteins like APP are likely to be affected by cholesterol-induced changes to membrane lipid composition. Cholesterol plays a major regulatory role in the α-secretase cleavage of APP. Alpha-secretase cleavage requires a very specific membrane-associated orientation between enzyme and substrate (Sparks 1997; see Friedland 2002; Chauhan 2003; Poirier 2005 for additional references). In the plasma membrane, a reduction in membrane fluidity resulting from increased cholesterol is thought to inhibit the lateral movement of APP and preclude its interaction with α-secretase (Bodovitz and Klein 1996; Kojro et al. 2001; see Runz et al. 2002 for additional references). In fact, increased cellular cholesterol reduces α-secretase activity and is synergisitic with Aβ generation. Numerous *in vitro* studies have shown that in a variety of cell types the addition of exogenous cholesterol causes increased Aβ production (Bodovitz and Klein 1996; Kojro et al. 2001; Kirsch et al. 2002; see Puglielli et al. 2003 for review and additional references). Similar results have been observed in APP transgenic mice fed diets high in cholesterol, which resulted in increases in soluble brain Aβ levels and the presence of Aβ-containing senile plaques in the brain parenchyma (Refolo et al. 2000). The consequences of altered cholesterol metabolism associated with NPC or with drugs that produce an NPC-like phenotype is accumulation of Aβ peptides and APP carboxy-terminal fragments such as occurs in NPC (Jin et al. 2004).

In addition to impairing normal vesicular sorting and transport, increased cellular cholesterol and its accumulation in endosomes and lysosomes has been shown to cause translocation and/or accumulation of β/γ-secretases in endosomes and lysosomes

(Runz et al. 2002; Jin et al. 2004). In neurons of NPC brain, the presence of Aβ peptides or carboxy-terminal β-stubs colocalized with elevated levels of cat D in endosomal compartments, a distribution that was most striking in the most vulnerable cell types (Jin et al. 2004).

C. Apolipoprotein-E

Apolipoprotein-E (ApoE) is a constituent of many lipoproteins and plays a significant role in cholesterol metabolism and distribution in brain where it mediates the exchange of cholesterol among neurons. In the CNS, ApoE is predominantly synthesized and secreted by astroglia and microglia and is internalized by neurons by receptor-mediated endocytic uptake through the LDL, LRP, and VLDL receptors, the ApoE receptor 2, or megalin (reviewed in Beffert et al. 1998). The human *ApoE* gene has three alleles—ε2, ε3, and ε4—that encode for three protein isoforms, Apo2, ApoE3, and ApoE4. Polymorphisms of the *ApoE* gene affect the risk for AD. Inheritance of one or both copies of the ε4 allele is a major risk factor for AD development (Corder et al. 1993; Strittmatter et al. 1993) and accelerates disease onset and the levels of the Aβ deposition and accumulation in brain. Numerous biochemical and epidemiological studies support a link between hypercholesterolemia and AD development. The mechanistic link between increased cholesterol and AD is thought to be regulation of Aβ metabolism (Bodovitz and Klein 1996; Kojro et al. 2001; Refolo et al. 2000). Both APP and APP secretases are transmembrane proteins. Alterations in membrane cholesterol or in the proportion of membrane to cytosolic cholesterol ester is thought to alter the subcellular distribution, activities, and processing sites of APP secretases, particularly β- and γ-secretases, toward amyloidogenic APP processing (Bodovitz et al. 1996; Kojro et al. 2001; Puglielli et al. 2001; Riddell et al. 2001a,b). Like AD, ApoE genotype appears to affect the severity of AD-like neuropathology in the brains of patients with NPC. Saito et al. (2002) showed that inheritance of the ε4 allele enhanced both neurofibrillary pathology and Aβ deposition in the brains of patients with NPC caused by multiple mutations in the *NPC1* gene (Saito et al. 2002). In contrast to NPC patients homozygous for the ε4 allele, NPC patients with ApoE3 or ApoE2 genotypes had fewer neurofibrillary tangles and no evidence of Aβ deposition.

D. Endosomal Pathway Dysfunction

In normal cells, molecules are internalized from the cell surface (see Figure 17.2) and, following sorting, either recycle back to the cell surface (Riederer et al. 1994; Ganley et al. 2004; Holtta-Vuori et al. 2002) or are delivered to late endosomes, and subsequently to lysosomes for terminal degradation. Enzymes required for degradation are generated in the biosynthetic pathway (ER and Golgi apparatus) and delivered by 215 and 46 kDa mannose 6-phosphate (MPR) receptors to late endosomes and lysosomes. In sporadic AD and DS, endocytic uptake and recycling are upregulated. This endocytic pathology is reflected by enlargement of Rab5 positive early endosomal and Rab7-positive late endosomal compartments, and increased numbers of Rab4-positive endocytic recycling compartments (Cataldo et al. 1997; 2000). MPR46-directed delivery of lysosomal hydrolases to early endosomes, which normally contain low levels, is enhanced. Abnormal recycling to the plasma membrane results from cholesterol-induced inhibition of Rab4-positive endocytic recycling compartments. MPR-tagged hydrolases are delivered to late endosomes and lysosomes, but the Rab9 directed return of MPR to the Golgi is reduced. Lysosomal hydrolase trafficking to early endosomes by MPR46 is increased (Jin et al. 2004). Dominant negative mutants of Rab7 and Rab9 reproduce the late endocytic pathology seen in NPC in

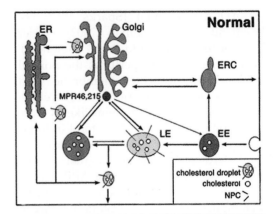

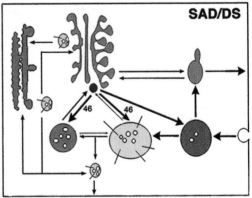

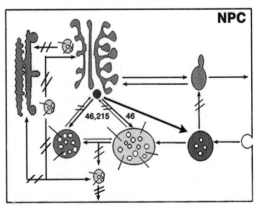

FIGURE 17.2. Schematic representation of normal and dysfunctional endocytic-lysosomal trafficking pathways and cholesterol movement in control, AD/DS, and NPC cells.

normal fibroblasts and enhance cholesterol accumulation, whereas overexpression of wild-type Rab7 and Rab9 in NPC fibroblasts appear to reduce the endocytic pathology (Choudhury et al. 2002; Narita et al. 2005).

V. CONCLUSION

The endosomal system is a target in a number of neurodegenerative diseases. The pathology in this system may be causative or secondary to the disease process. Defining the specific endosomal compartments involved and the nature of the defects could help to account for certain distinctive features of disease such as severity, age of onset, predilection for particular neuronal populations, and impact on neuron function. The particular vulnerability of the nervous system to impairments of a universal cellular process remains unclear although it is likely to reflect the unique functions that endosomes serve in neurons to maintain cellular communication. The presence of endosome dysfunction in distinct neurodegenerative disorders (AD and NPC) that share hallmark AD-like tauopathy, highlights an important relationship that may reveal mechanistic insight.

Acknowledgements

The authors wish to sincerely thank Ms. Corrinne Peterhoff for assistance with artwork and Ms. Heather Braunstein for secretarial assistance in the preparation of this manuscript.

References

Arbel, M., Yacoby, I., Solomon, B. (2005). Inhibition of amyloid precursor protein processing by β-secretase through site-directed antibodies. *Proc. Natl. Acad. Sci. U. S. A.* **102**, 7718–7723.

Beffert, U., Danik, M., Krzywkowski, P., Ramassamy, C., Berrada, F., Poirier, J. (1998). The neurobiology of apolipoproteins and their receptors in the CNS and Alzheimer's disease. *Brain Res. Brain Res. Rev.* **27**, 119–142.

Blanchette-Mackie, E.J., Dwyer, N.K., Amende, L.M., Kruth, H.S., Butler, J.D., Sokol, J. et al. (1988). Type-C Niemann-Pick disease: Low density lipoprotein uptake is associated with premature cholesterol accumulation in the Golgi complex and excessive cholesterol storage in lysosomes. *Proc. Natl. Acad. Sci. U. S. A.* **85**, 8022–8026.

Blom, T.S., Linder, M.D., Snow, K., Pihko, H., Hess, M.W., Jokitalo, E. et al. (2003). Defective endocytic trafficking of NPC1 and NPC2 underlying infantile Niemann-Pick type C disease. *Hum. Genet.* **12**, 257–272.

Bodovitz, S., Klein, W.L. (1996). Cholesterol modulates alpha-secretase cleavage of amyloid precursor protein. *J. Biol. Chem.* **271**, 4436–4440.

Borchelt, D.R., Thinakaran, G., Eckman, C.B., Lee, M.K., Davenport, F., Ratovitsky, T. et al. (1996). Familial Alzheimer's disease-linked presenilin 1 variants elevate Abeta1-42/1-40 ratio in vitro and in vivo. *Neuron* **17**, 1005–1013.

Bright, N.A., Reaves, B.J., Mullock, B.M., Luzio, J.P. (1997). Dense core lysosomes can fuse with late endosomes and are formed from the resultant hybrid organelles. *J. Cell Sci.* **110**, 2027–2040.

Brown, M.S., Goldstein, J.L. (1986). A receptor-mediated pathway for cholesterol homeostasis. *Science* **232**, 34–47.

Bu, B., Li, J., Davies, P., Vincent, I. (2002). Deregulation of cdk5, hyperphosphorylation, and cytoskeletal pathology in the Niemann-Pick type C murine model. *J. Neurosci.* **22**, 6515–6525.

Burke, W.J., Park, D.H., Chung, H.D., Marshall, G.L., Haring, J.H., Joh, T.H. (1990). Evidence for decreased transport of tryptophan hydroxylase in Alzheimer's disease. *Brain Res.* **537**, 83–87.

Buxbaum, J.D., Liu, K.N., Luo, Y., Slack, J.L., Stocking, K.L., Peschon, J.J. et al. (1998). Evidence that tumor necrosis factor alpha converting enzyme is involved in regulated alpha-secretase cleavage of the Alzheimer amyloid protein precursor. *J. Biol. Chem.* **273**, 27765–27767.

Cao, X., Sudhof, T.C. (2001). A transcriptionally [correction of transcriptively] active complex of APP with Fe65 and histone acetyltransferase Tip60. *Science* **293**, 115–120.

Capell, A., Steiner, H., Romig, H., Keck, S., Baader, M., Grim, M.G. et al. (2000). Presenilin-1 differentially facilitates endoproteolysis of the beta-amyloid precursor protein and Notch. *Nat. Cell Biol.* **2**, 205–211.

Capell, A., Steiner, H., Willem, M., Kaiser, H., Meyer, C., Walter, J. et al. (2000). Maturation and propeptide cleavage of beta-secretase. *J. Biol. Chem.* **273**, 30849–30854.

Carstea, E.D., Morris, J.A., Coleman, K.G., Loftus, S.K., Zhang, D., Cummings, C. et al. (1997). Niemann-Pick C1 disease gene: Homology to mediators of cholesterol homeostasis. *Science* **277**, 228–231.

Cataldo, A.M., Barnett, J.L., Pieroni, C., Nixon, R.A. (1997). Increased neuronal endocytosis and protease delivery to early endosomes in sporadic Alzheimer's disease: Neuropathologic evidence for a mechanism of increased beta-amyloidogenesis. *J. Neurosci.* **17**, 6142–6151.

Cataldo, A.M., Peterhoff, C.M., Troncoso, J.C., Gomez-Isla, T., Hyman, B.T., Nixon, R.A. (2000). Endocytic pathway abnormalities precede amyloid beta deposition in sporadic Alzheimer's disease and Down syndrome: Differential effects of APOE genotype and presenilin mutations. *Am. J. Pathol.* **157**, 277–286.

Cataldo, A.M., Petanceska, S., Peterhoff, C.M., Terio, N.B., Epstein, C.J., Villar, A. et al. (2003). App gene dosage modulates endosomal abnormalities of Alzheimer's disease in a segmental trisomy 16 mouse model of down syndrome. *J. Neurosci.* **23**, 6788–6792.

Cataldo, A.M., Petanceska, S., Terio, N.B., Peterhoff, C.M., Troncoso, J.C., Durham, R. et al. (2004). A-beta localization to abnormal endosomes coincides with early increases in soluble A-beta in Alzheimer's disease brain. *Neurobiol. Aging* **25**, 1263–1272.

Chauhan, N. (2003). Membrane dynamics, cholesterol homeostasis, and Alzheimer's disease. *J. Lipid Res.* **44**, 2019–2029.

Choudhury, A., Dominguez, M., Puri, V., Sharma, D.K., Narita, K., Wheatley, C.L. et al. (2002). Rab proteins mediate Golgi transport of caveola-internalized glycosphingolipids and correct lipid trafficking in Niemann-Pick C cells. *J. Clin. Invest.* **109**, 1541–1550.

Choudhury, A., Sharma, D.K., Marks, D.L., Pagano, R.E. (2004). Elevated endosomal cholesterol levels in Niemann-Pick cells inhibit rab4 and perturb membrane recycling. *Mol. Biol. Cell* **15**, 4500–4511.

Chyung, A.S.C., Greenberg, B.D., Cook, D.G., Doms, R.W., Lee, V.M. (1997). Novel beta-secretase cleavage of beta-amyloid precursor protein in the endoplasmic reticulum/intermediate compartment of NT2N cells. *J. Cell Biol.* **138**, 671–680.

Citron, M., Westaway, D., Xia, W., Carlson, G., Diehl, T., Levesque, G. et al. (1997). Mutant presenilins of Alzheimer's disease increase production of 42-residue amyloid beta-protein in both transfected cells and transgenic mice [see comments]. *Nat. Med.* **3**, 67–72.

Cooper, J.D., Salehi, A., Delcroix, J.D., Howe, C.L., Belichenko, P.V., Chua-Couzens, J. et al. (2001). Failed retrograde transport of NGF in a mouse model of Down's syndrome: Reversal of cholinergic neurodegenerative phenotypes following NGF infusion. *Proc. Natl. Acad. Sci. U. S. A.* **98**, 10439–10444.

Corder, E.H., Saunders, A.M., Strittmatter, W.J., Schmechel, D.E., Gaskell, P.C., Small, G.W. et al. (1993). Gene dose of apolipoprotein E type 4 allele and the risk of Alzheimer's disease in late onset families. *Science* **261**, 921–923.

Coxey, R.A., Pentchev, P.G., Campbell, G., Blanchette-Mackie, E.J. (1993). Differential accumulation of cholesterol in Golgi compartments of normal and Niemann-Pick type C fibroblasts incubated with LDL: A cytochemical freeze-fracture study. *J. Lipid Res.* **34**, 1165–1176.

Cummings, J.L., Vinters, H.V., Cole, G.M., Khachaturian, Z.S. (1998). Alzheimer's disease: Etiologies, pathophysiology, cognitive reserve, and treatment opportunities. *Neurology* **51**, S2–S17; discussion S65–S67.

Cupers, P., Bentahir, M., Craessaerts, K., Orlans, I., Vanderstichele, H., Saftig, P. et al. (2001). The discrepancy between presenilin subcellular localization and gamma-secretase processing of amyloid precursor protein. *J. Cell Biol.* **154**, 731–740.

Deb, S., Braganza, J., Norton, N., Williams, H., Kehoe, P.G., Williams, J., Owen, M.J. (2000). APOE epsilon 4 influences the manifestation of Alzheimer's disease in adults with Down's syndrome. *Br. J. Psychiatry* **176**, 468–472.

Delcroix, J.D., Valletta, J.S., Wu, C., Hunt, S.J., Kowal, A.S., Mobley, W.C. (2003). NGF signaling in sensory neurons: Evidence that early endosomes carry NGF retrograde signals. *Neuron* **39**, 69–84.

Dietschy, J.M., Turley, S.D. (2001). Cholesterol metabolism in the brain. *Curr. Opin. Lipidol.* **12**, 105–112.

Dirac-Svejstrup, A.B., Sumizawa, T., Pfeffer, S.R. (1997). Identification of a GDI displacement factor that releases endosomal Rab GTPases from Rab-GDI. *EMBO J.* **16**, 465–472.

Distl, R., Treiber-Held, S., Albert, F., Meske, V., Harzer, K., Ohm, T.G. (2003). Cholesterol storage and tau pathology in Niemann-Pick type C disease in the brain. *J. Pathol.* **200**, 104–111.

Edbauer, D., Winkler, E., Regula, J.T., Pesold, B., Steiner, H., Haass, C. (2003). Reconstitution of gamma-secretase activity. *Nat. Cell Biol.* **5**, 486–488.

Elarova, I., Che, S., Ruben, M.D., Nixon, R.A., Ginsberg, S.D. (2004). Expression profiling of hippocampal neurons in a mouse model of Down's syndrome (Ts65Dn). *Proc. Soc. Neurosci.* **30**, 335.

Fan, Q.-W., Yu, W., Senda, T., Yanagisawa, K., Michikawa, M. (2001). Cholesterol-dependent modulation of tau phosphorylation in cultured neurons. *J. Neurochem.* **76**, 391–491.

Friedland, R.P. (2002). Lipid metabolism, epidemiology, and the mechanisms of Alzheimer's disease. *Ann. N. Y. Acad. Sci.* **977**, 387–390.

Frolov, A., Zielinski, S.E., Crowley, J.R., Dudley-Rucker, N., Schaffer, J.E., Ory, D.S. (2003). NPC1 and NPC2 regulate cellular cholesterol homeostasis through generation of low density lipoprotein cholesterol-derived oxysterols. *J. Biol. Chem.* **278**, 25517–25525.

Galdzicki, Z., Siarey, R.J. (2003). Understanding mental retardation in Down's syndrome using trisomy 16 mouse models. *Genes Brain Behav.* **2**, 167–178.

Ganley, I.G., Carroll, K., Bittova, L., Pfeffer, S. (2004). Rab9 GTPase regulates late endosome size and requires effector interaction for its stability. *Mol. Biol. Cell* **15**, 5420–5430.

Garver, W.S., Heidenreich, R.A. (2002). The Niemann-Pick C proteins and trafficking of cholesterol through the late endosomal/lysosomal system. *Curr. Mol. Med.* **2**, 485–505.

Go, Y.M., Gipp, J.J., Mulcahy, R.T., Jones, D.P. (2004). H_2O_2-dependent activation of GCLC-ARE4 reporter occurs by mitogen-activated protein kinase pathways without oxidation of cellular glutathione or thioredoxin-1. *J. Biol. Chem.* **279**, 5837–5845.

Goate, A.M. (1998). Monogenetic determinants of Alzheimer's disease: APP mutations. *Cell Mol. Life Sci.* **54**, 897–901.

Goldstein, L.S. (2001). Transduction. When worlds collide—Trafficking in JNK. *Science* **291**, 2102–2103.

Goody, R.S., Rak, A., Alexandrov, K. (2005). The structural and mechanistic basis for recycling of Rab proteins between membrane compartments. *Cell Mol. Life Sci.* **62**, 1657–1670.

Grbovic, O.M., Mathews, P.M., Jiang, Y., Schmidt, S.D., Dinakar, R., Summers-Terio, N.B. et al. (2003). Rab5-stimulated up-regulation of the endocytic pathway increases intracellular beta-cleaved amyloid precursor protein carboxyl-terminal fragment levels and Abeta production. *J. Biol. Chem.* **278**, 31261–31268.

Greenfield, J.P., Tsai, J., Gouras, G.K., Hai, B., Thinakaran, G., Checler, F. et al. (1999). Endoplasmic reticulum and trans-Golgi network generate distinct populations of Alzheimer beta-amyloid peptides. *Proc. Natl. Acad. Sci. U. S. A.* **96**, 742–747.

Griffiths, G. (1996). On vesicles and membrane compartments. *Protoplasma* **195**, 37–58.

Haass, C., Koo, E.H., Mellon, A., Hung, A.Y., Selkoe, D.J. (1992a). Targeting of cell-surface beta-amyloid precursor protein to lysosomes: alternative processing into amyloid-bearing fragments. *Nature* **357**, 500–503.

Haass, C., Hung, A.Y., Schlossmacher, M.G., Teplow, D.B., Selkoe, D.J. (1993). Beta-amyloid peptide and a 3-kDa fragment are derived by distinct cellular mechanisms. *J. Biol. Chem.* **268**, 3021–3024.

Haass, C., Schlossmacher, M.G., Hung, A.Y., Vigo-Pelfrey, C., Mellon, A., Ostaszewski, B.L. et al. (1992b). Amyloid beta-peptide is produced by cultured cells during normal metabolism [see comments]. *Nature* **359**, 322–325.

Hao, M., Mukherjee, S., Sun, Y., Maxfield, F.R. (2004). Effects of cholesterol depletion and increased lipid unsaturation on the properties of endocytic membranes. *J. Biol. Chem.* **279**, 14171–14178.

Higgins, M.E., Davies, J.P., Chen, F.W., Ioannou, Y.A. (1999). Niemann-Pick C1 is a late endosome-resident protein that transiently associates with lysosomes and the trans-Golgi network. *Mol. Genet. Metab.* **68**, 1–13.

Holtta-Vuori, M., Tanhuanpaa, K., Mobius, W., Somerharju, P., Ikonen, E. (2002). Modulation of cellular cholesterol transport and homeostasis by rab11. *Molec. Biol. Cell* **13**, 3107–3122.

Holtzman, D.M., Santucci, D., Kilbridge, J., Chua-Couzens, J., Fontana, D.J., Daniels, S.E. et al. (1996). Developmental abnormalities and age-related neurodegeneration in a mouse model of Down syndrome. *Proc. Natl. Acad. Sci. U. S. A.* **93**, 13333–13338.

Howe, C.L., Valletta, J.S., Rusnak, A.S., Mobley, W.C. (2001). NGF signaling from clathrin-coated vesicles:

Evidence that signaling endosomes serve as a platform for the Ras-MAPK pathway. *Neuron* **32**, 801–814.

Huse, J.T., Pijak, D.S., Leslie, G.J., Lee, V.M., Doms, R.W. (2000). Maturation and endosomal targeting of beta-site amyloid precursor protein-cleaving enzyme. The Alzheimer's disease beta-secretase. *J. Biol. Chem.* **275**, 33729–33737.

Ikonen, E., Holtta-Vuori, M. (2004). Cellular pathology of Niemann-Pick type C disease. *Sem. Cell Dev. Biol.* **15**, 445–454.

Ioannou, Y.A., Higgins, M.E., Comly, M., Cooney, A., Brown, A., Kaneski, C.R. et al. (1997). Niemann-Pick C1 disease gene: Homology to mediators of cholesterol homeostasis. *Science* **277**, 228–231.

Jin, L.-W., Maezawa, I., Vincent, I., Bird, T. (2004). Intracellular accumulation of amyloidogenic fragments of amyloid-β precursor protein in neurons with Niemann-Pick type C defects is associated with endosomal abnormalities. *Am. J. Pathol.* **164**, 975–985.

Johnstone, S.S., Hubaishy, I., Waisman, D.M. (1992). Phosphorylation of annexin II tetramer by protein kinase C inhibits aggregation of lipid vesicles by the protein. *J. Biol. Chem.* **267**, 25976–25981.

Kang, D.E., Saitoh, T., Chen, X., Xia, Y., Masliah, E., Hansen, L.A. et al. (1997). Genetic association of the low-density lipoprotein receptor-related protein gene (LRP), an apolipoprotein E receptor, with late-onset Alzheimer's disease. *Neurology* **49**, 56–61.

Karten, B., Vance, D.E., Campenot, R.B., Vance, J.E. (2003). Trafficking of cholesterol from cell bodies to distal axons in Niemann-Pick C1-deficient neurons. *J. Biol. Chem.* **278**, 4168–4175.

Karten, B., Vance, D.E., Campenot, R.B., Vance, J.E. (2002). Cholesterol accumulates in cell bodies, but is decreased in distal axons, of Niemann-Pick C1-deficient neurons. *J. Neurochem.* **83**, 1154–1163.

Katzmann, D.J., Babst, M., Emr, S.D. (2001). Ubiquitin-dependent sorting into the multivesicular body pathway requires the function of a conserved endosomal protein sorting complex, ESCRT-I. *Cell* **106**, 145–155.

Kirsch, C., Eckert, G.P., Mueller, W.E. (2002). Cholesterol attenuates the membrane perturbing properties of beta-amyloid peptides. *Amyloid.* **9**, 149–159. PMID: 12408677 [PubMed—indexed for MEDLINE]

Ko, D.C., Gordon, M.D., Jin, J.Y., Scott, M.P. (2001). Dynamic movements of organelles containing Niemann-Pick C1 protein: NPC1 involvement in late endocytic events. *Mol. Biol. Cell* **12**, 601–614.

Kobayashi, T., Gu, F., Gruenberg, J. (1998). Lipids, lipid domains and lipid-protein interactions in endocytic membrane traffic. *Sem. Cell Dev. Biol.* **9**, 517–526.

Kojro, E., Gimpl, G., Lammich, S., Marz, W., Fahrenholtz, F. (2001). Low cholesterol stimulates the nonamyloidogenic pathway by its effect on the α-secretase ADAM 10. *Proc. Natl. Acad. Sci.* **98**, 5815–5820.

Kolsch, H., Lutjohann, D., Ludwig, M., Schulte, A., Ptok, U., Jessen, F. et al. (2002). Polymorphism in the cholesterol 24S-hydroxylase gene is associated with Alzheimer's disease. *Mol. Psychiatry*, **7**, 899–902.

Koo, E.H., Squazzo, S.L. (1994). Evidence that production and release of amyloid beta-protein involves the endocytic pathway. *J. Biol. Chem.* **269**, 17386–17389.

Kornfeld, S. (1992). Structure and function of the mannose 6-phosphate/insulinlike growth factor II receptors. *Annu. Rev. Biochem.* **61**, 307–330.

Lah, J.J., Levey, A.I. (2000). Endogenous presenilin-1 targets to endocytic rather than biosynthetic compartments. *Mol. Cell Neurosci.* **16**, 111–126.

Lai, F., Williams, R.S. (1989). A prospective study of Alzheimer disease in Down syndrome. *Arch. Neurol.* **46**, 849–853.

Lai, F., Kammann, E., Rebeck, G.W., Anderson, A., Chen, Y., Nixon, R.A. (1999). APOE genotype and gender effects on Alzheimer disease in 100 adults with Down syndrome. *Neurology* **53**, 331–336.

Lange, Y., Steck, T.L. (1996). The role of intracellular cholesterol transport in cholesterol homeostasis. *Trends Cell Biol.* **6**, 205–208.

Launer, L.J., Andersen, K., Dewey, M.E., Letenneur, L., Ott, A., Amaducci, L.A. et al. (1999). Rates and risk factors for dementia and Alzheimer's disease: Results from EURODEM pooled analyses. EURODEM Incidence Research Group and Work Groups. European Studies of Dementia. *Neurology* **52**, 78–84.

LeBrand, C., Corti, M., Goodson, H., Cosson, P., Cavalli, V., Mayran, N. et al. (2002). Late endosome motility depends on lipids via the small GTPase rab7. *EMBO J.* **21**, 1289–1300.

Levy, E., Carman, M.D., Fernandez-Madrid, I.J., Power, M.D., Lieberburg, I., van Duinen, S.G. et al. (1990). Mutation of the Alzheimer's disease amyloid gene in hereditary cerebral hemorrhage, Dutch type. *Science* **248**, 1124–1126.

Levy-Lahad, E., Wasco, W., Poorkaj, P., Romano, D.M., Oshima, J., Pettingell, W.H. et al. (1995). Candidate gene for the chromosome 1 familial Alzheimer's disease locus [see comments]. *Science* **269**, 973–977.

Leyssen, M., Ayaz, D., Hebert, S.S., Reeve, S., De Strooper, B., Hassan, B.A. (2005). Amyloid precursor protein promotes post-developmental neurite arborization in the Drosophila brain. *EMBO J.* **24**, 2944–2955.

Liscum, L. (2000). Niemann-Pick type C mutations cause lipid traffic jam. *Traffic* **1**, 218–225.

Liscum, L., Munn, J. (1999). Intracellular cholesterol transport. *Biochim. Biophys. Acta* **1438**, 19–37.

Lombardi, D., Soldati, T., Riederer, M.A., Goda, Y., Zerial, M., Pfeffer, S.R. (1993). Rab9 functions in transport between late endosomes and the trans Golgi network. *EMBO J.* **12**, 677–682.

Lopez-Perez, E., Zhang, Y., Frank, S.J., Creemers, J., Seidah, N., Checler, F. (2001). Constitutive alpha-secretase cleavage of the beta-amyloid precursor protein in the furin-deficient LoVo cell line: Involvement of the pro-hormone convertase 7 and the disintegrin metalloprotease ADAM10. *J. Neurochem.* **76**, 1532–1539.

Love, S., Bridges, L.R., Case, C.P. (1995). Neurofibrillary tangles in Niemann-Pick disease type C. *Brain* **118**, 119–129.

Luan, P., Balch, W.E., Emr, S.D., Burd, C.G. (1999). Molecular dissection of guanine nucleotide dissociation inhibitor function in vivo. Rab-independent binding to membranes and role of Rab recycling factors. *J. Biol. Chem.* **274**, 14806–14817.

Lusa, S., Blom, T.S., Eskelinen, E.L., Kuismanen, E., Mansson, J.E., Simons, K., Ikonen, E. (2001). Depletion of rafts in late endocytic membranes is controlled by NPC1-dependent recycling of cholesterol to the plasma membrane. *J. Cell Sci.* **114**, 1893–1900.

Mathews, P.M., Nixon, R.A. (2003). Setback for an Alzheimer's disease vaccine: Lessons learned. *Neurology* **61**, 7–8.

Mathews, P.M., Jiang, Y., Schmidt, S.D., Grbovic, O.M., Mercken, M., Nixon, R.A. (2002a). Calpain activity regulates the cell surface distribution of amyloid precursor protein. Inhibition of calpains enhances endosomal generation of beta-cleaved C-terminal APP fragments. *J. Biol. Chem.* **277**, 36415–36424.

Mathews, P.M., Guerra, C.B., Jiang, Y., Grbovic, O.M., Kao, B.H., Schmidt, S.D. et al. (2002b). Alzheimer's disease-related overexpression of the cation-dependent mannose 6-phosphate receptor increases Abeta secretion: Role for altered lysosomal hydrolase distribution in beta-amyloidogenesis. *J. Biol. Chem.* **277**, 5299–5307.

Mattson, M.P. (1997). Cellular actions of beta-amyloid precursor protein and its soluble and fibrillogenic derivatives. *Physiol. Rev.* **77**, 1081–1132.

Maxfield, F., Wustner, D. (2002). Intracellular cholesterol transport. *J. Clin. Invest.* **110**, 891–898.

Mayor, S., Sabharanjak, S., Maxfield, F.R. (1998). Cholesterol-dependent retention of GPI-anchored proteins in endosomes. *EMBO J.* **17**, 4626–4638.

McLean, C.A., Cherny, R.A., Fraser, F.W., Fuller, S.J., Smith, M.J., Beyreuther, K. et al. (1999). Soluble pool of Abeta amyloid as a determinant of severity of neurodegeneration in Alzheimer's disease. *Ann. Neurol.* **46**, 860–866.

Medina, M., Dotti, C.G. (2003). RIPped out by presenilin-dependent gamma-secretase. *Cell Signal* **15**, 829–841.

Mukherjee, S., Maxfield, F.R. (2004). Lipid and cholesterol trafficking in NPC. *Biochim. Biophys. Acta* **1685**, 28–37.

Mullock, B.M., Bright, N.A., Fearon, C.W., Gray, S.R., Luzio J.P. (1998). Fusion of lysosomes with late endosomes produces a hybrid organelle of intermediate density and is NSF dependent. *J. Cell Biol.* **140**, 591–601.

Narita, K., Choudhury, A., Dobrenis, K., Sharma, D.K., Holicky, E.L., Marks, D.L. et al. (2005). Protein transduction of rab9 in Niemann-Pick C cells reduces cholesterol storage. *FASEB J.* Epub ahead of print, June 22, 2005.

Naureckiene, S., Sleat, D.E., Lackland, H., Fensom, A., Vanier, M.T., Wattiaux, R. et al. (2000) Identification of HE1 as the second gene of Niemann-Pick C disease. *Science* **290**, 2298–2301.

Neufeld, E.B., Wastney, M., Patel, S., Suresh, S., Cooney, A.M., Dwyer, N.K. et al. (1999). The Niemann-Pick C1 protein resides in a vesicular compartment linked to retrograde transport of multiple lysosomal cargo. *J. Biol. Chem.* **274**, 9627–9635.

Nixon, R.A. (2005). Endosome function and dysfunction in Alzheimer's disease and other neurodegenerative diseases. *Neurobiol. Aging* **26**, 373–382.

Nordstedt, C., Caporaso, G.L., Thyberg, J., Gandy, S.E., Greengard, P. (1993). Identification of the Alzheimer beta/A4 amyloid precursor protein in clathrin-coated vesicles purified from PC12 cells. *J. Biol. Chem.* **268**, 608–612.

Ohm, T.G., Treiber-Held, S., Distl, R., Glockner, F., Schonheit, B., Tamanai, M., Meske, V. (2003). Cholesterol and tau protein—findings in Alzheimer's and Niemann Pick C's disease. *Pharmacopsychiatry* **36 (Suppl 2)**, S120–S126.

Ory, D.S. (2000). Niemann-Pick type C: A disorder of cellular cholesterol trafficking. *Biochim. Biophys. Acta* **1529**, 331–339.

Paganetti, P., Calanca, V., Galli, C., Stefani, M., Molinari, M. (2005). Beta-site specific intrabodies to decrease and prevent generation of Alzheimer's Abeta peptide. *J. Cell Biol.* **168**, 863–868.

Papassotiropoulos, A., Streffe, J.R., Tsolaki, M., Schmid, S., Thal, D., Nicosia, F. et al. (2003). Increased brain beta-amyloid load, phosphorylated tau, and risk of Alzheimer disease associated with an intronic CYP46 polymorphism. *Arch. Neurol.* **60**, 29–35.

Pasternak, S.H., Bagshaw, R.D., Guiral, M., Zhang, S., Ackerley, C.A., Pak, B.J. et al. (2003). Presenilin-1, nicastrin, amyloid precursor protein, and gamma-secretase activity are co-localized in the lysosomal membrane. *J. Biol. Chem.* **278**, 26687–26694.

Paul, C.A., Boegle, A.K., Maue, R.A. (2004). Before the loss: Neuronal dysfunction in Niemann-Pick type C disease. *Biochim. Biophys. Acta* **1685**, 63–76.

Pentchev, P.G. (2004). Niemann-Pick C research from mouse to gene. *Biochim. Biophys. Acta* **11**, 3–7.

Pentchev, P.G., Blanchette-Mackie, E.J., Dawidowicz, E.A. (1994). The NP-C gene: A key to pathways of intracellular cholesterol transport. *Trends Cell Biol.* **4**, 365–369.

Pentchev, P.G., Comly, M.E., Kruth, H.S., Vanier, M.T., Wenger, D.A., Patel, S., Brady, R.O. (1985). A defect in cholesterol esterification in Niemann-Pick disease (type C) patients. *Proc. Natl. Acad. Sci. U. S. A.* **82**, 8247–8251.

Perez, R.G., Soriano, S., Hayes, J.D., Ostaszewski, B., Xia, W., Selkoe, D.J. et al. (1999). Mutagenesis identifies new signals for beta-amyloid precursor protein endocytosis, turnover, and the generation of secreted fragments, including Abeta42. *J. Biol. Chem.* **274**, 18851–18856.

Pichler, H., Riezman, H. (2004). Where sterols are required for endocytosis. *Biochim. Biophys. Acta* **1666**, 51–61.

Pigino, G., Morfini, G., Pelsman, A., Mattson, M.P., Brady, S.T., Busciglio, J. (2003). Alzheimer's presenilin 1 mutations impair kinesin-based axonal transport. *J. Neurosci.* **23**, 4499–4508.

Pillay, C.S., Elliott, E., Dennison, C. (2002). Endolysosomal proteolysis and its regulation. *Biochem. J.* **363**, 417–429.

Poirier, J. (2005). Apolipoprotein E, cholesterol transport and synthesis in sporadic Alzheimer's disease. *Neurobiol. Aging* **26**, 355–361.

Price, D.L., Sisodia, S.S. (1998). Mutant genes in familial Alzheimer's disease and transgenic models. *Annu. Rev. Neurosci.* **21**, 479–505.

Price, D.L., Tanzi, R.E., Borchelt, D.R., Sisodia, S.S. (1998). Alzheimer's disease: Genetic studies and transgenic models. *Annu. Rev. Genet.* **32**, 461–493.

Puglielli, L., Konopka, G., Pack-Chung, E., Ingano, L.A., Berezovska, O., Hyman, B.T. et al. (2001). Acyl-coenzyme A: Cholesterol acyltransferase modulates the generation of the amyloid beta-peptide. *Nat. Cell Biol.* **3**, 905–912.

Puglielli, L., Tanzi, R.E., Kovacs, D.M. (2003). Alzheimer's disease: The cholesterol connection. *Nat. Neurosci.* **6**, 345–351.

Puri, V., Watanabe, R., Dominguez, M., Sun, X., Wheatley, C.L., Marks, D.L., Pagano, R.E. (1999). Cholesterol modulates membrane traffic along the endocytic pathway in sphingolipid-storage diseases. *Nat. Cell Biol.* **1**, 386–388.

Refolo, L.M., Malester, B., LaFrancois, J., Bryant-Thomas, T., Wang, R., Tint, G.S. et al. (2000). Hypercholesterolemia accelerates the Alzheimer's amyloid pathology in a transgenic mouse model. *Neurobiol. Dis.* **7**, 321–331.

Reid, P.C., Sugii, S., Chang, T.Y. (2003). Trafficking defects in endogenously synthesized cholesterol in fibroblasts, macrophages, hepatocytes, and glial cells from Niemann-Pick type C1 mice. *J. Lipid Res.* **44**, 1010–1019.

Riddell, D.R., Christie, G., Hussain, I., Dingwall, C. (2001a). Compartmentalization of beta-secretase (Asp2) into low-buoyant density, noncaveolar lipid rafts. *Curr. Biol.* **11**, 1288–1293.

Riddell, D.R., Sun, X.M., Stannard, A.K., Soutar, A.K., Owen, J.S. (2001b). Localization of apolipoprotein E receptor 2 to caveolae in the plasma membrane. *J. Lipid Res.* **42**, 998–1002.

Riederer, M.A., Soldati, T., Shapiro, A.D., Lin, J., Pfeffer, S.R. (1994). Lysosome biogenesis requires rab9 function and receptor recycling from endosomes to the *trans*-Golgi network. *J. Cell Biol.* **125**, 573–582.

Runz, H., Rietdorf, J., Tomic, I., De Bernard, M., Beyreuther, K., Pepperkok, R., Hartmann, T. (2002). Inhibition of intracellular cholesterol transport alters presenilin localization and amyloid precursor protein processing in neuronal cells. *J. Neurosci.* **22**, 1679–1689.

Saito, Y., Suzuki, K., Nanba, E., Yamamoto, T., Ohna, K., Murayama, S. (2002). Niemann-Pick type C disease: Accelerated neurofibrillary tangle formation and amyloid β deposition associated with apolipoprotein E ε4 homozygosity. *Ann. Neurol.* **52**, 351–355.

Sastre, M., Steiner, H., Fuchs, K., Capell, A., Multhaup, G., Condron, M.M. et al. (2001). Presenilin-dependent gamma-secretase processing of beta-amyloid precursor protein at a site corresponding to the S3 cleavage of Notch. *EMBO Rep.* **2**, 835–841.

Sawamura, N., Gong, J.S., Chang, T.Y., Yanagisawa, K., Michikawa, M. (2003). Promotion of tau phosphorylation by MAP kinase Erk1/2 is accompanied by reduced cholesterol level in detergent-insoluble membrane fraction in Niemann-Pick C1-deficient cells. *J. Neurochem.* **84**, 1086–1096.

Sheff, D.R., Daro, E.A., Hull, M., Mellman, I. (1999). The receptor recycling pathway contains two distinct populations of early endosomes with different sorting functions. *J. Cell Biol.* **145**, 123–139.

Sherrington, R., Rogaev, E.I., Liang, Y., Rogaeva, E.A., Levesque, G., Ikeda, M. et al. (1995). Cloning of a gene bearing missense mutations in early-onset familial Alzheimer's disease [see comments]. *Nature* **375**, 754–760.

Simons, K., Ikonen, E. (1997). Functional rafts in cell membranes. *Nature* **387**, 569–572.

Simons, K., Gruenberg, J. (2000). Jamming the endosomal system: Lipid rafts and lysosomal storage diseases. *Trends Cell Biol.* **10**, 459–462.

Sisodia, S.S. (1992). Beta-amyloid precursor protein cleavage by a membrane-bound protease. *Proc. Natl. Acad. Sci. U. S. A.* **89**, 6075–6079.

Sisodia, S.S., St George-Hyslop, P.H. (2002). Gamma-Secretase, Notch, Abeta and Alzheimer's disease: Where do the presenilins fit in? *Nat. Rev. Neurosci.* **3**, 281–290.

Sofroniew, M.V., Howe, C.L., Mobley, W.C. (2001). Nerve growth factor signaling, neuroprotection, and neural repair. *Annu. Rev. Neurosci.* **24**, 1217–1281.

Soriano, S., Chyung, A.S., Chen, X., Stokin, G.B., Lee, V.M., Koo, E.H. (1999). Expression of beta-amyloid precursor protein-CD3gamma chimeras to demonstrate the selective generation of amyloid beta(1-40) and amyloid beta(1-42) peptides within secretory and endocytic compartments. *J. Biol. Chem.* **274**, 32295–32300.

Sparks, D.L. (1997). Coronary artery disease, hypertension, ApoE, and cholesterol: A link to Alzheimer's disease? *Ann. N. Y. Acad. Sci.* **826**, 128–146.

Strittmatter, W.J., Weisgraber, K.H., Huang, D.Y., Dong, L.M., Salvesen, G.S., Pericak-Vance, M. et al. (1993). Binding of human apolipoprotein E to synthetic amyloid beta peptide: Isoform-specific effects and implications for late-onset Alzheimer disease. *Proc. Natl. Acad. Sci. U. S. A.* **90**, 8098–8102.

Sturley, S.L., Patterson, M.C., Balch, W., Liscum, L. (2004). The pathophysiology and mechanisms of NP-C disease. *Biochim. Biophys. Acta.* **1685**, 83–87.

Sugii, S., Reid, P.C., Ohgami, N., Du, H., Chang, T.Y. (2003). Distinct endosomal compartments in early trafficking of low density lipoprotein-derived cholesterol. *J. Biol. Chem.* **278**, 27180–27189.

Suzuki, K., Parker, C.C., Pentchev, P., Katz, D., Ghetti, B., D'Agostino, A.N., Carstea, E.D. (1995). Neurofibrillary tangles in Niemann-Pick disease type C. *Acta Neuropathol.* **89**, 227–238.

Terwel, D., Dewachter, L., Van Leuven, F. (2002). Axonal transport, tau protein, and neurodegeneration in Alzheimer's disease. *Neuromolecular Med.* **2**, 151–165.

Te Vruchte, D., Lloyd-Evans, E., Veldman, R.J., Neville, D.C.A., Dwek, R.A., Platt, F.M. (2004). Accumulation of glycosphingolipids in Niemann-Pick C disease disrupts endosomal transport. *J. Biol. Chem.* **279**, 26167–26175.

Thinakaran, G., Teplow, D.B., Siman, R., Greenberg, B., Sisodia, S.S. (1996). Metabolism of the "Swedish" amyloid precursor protein variant in neuro2a (N2a) cells. Evidence that cleavage at the "beta-secretase" site occurs in the golgi apparatus. *J. Biol. Chem.* **271**, 9390–9397.

Tjelle, T.E., Brech, A., Juvet, L.K., Griffiths, G., Berg, T. (1996). Isolation and characterization of early endosomes, late endosomes and terminal lysosomes: Their role in protein degradation. *J. Cell Sci.* **109**, 2905–2914.

Treiber-Held, S., Distl, R., Meske, V., Albert, F., Ohm, T.G. (2003). Spatial and temporal distribution of intracellular free cholesterol in brains of a Niemann-Pick type C mouse model showing hyperphosphorylated tau protein. Implications for Alzheimer's disease. *J. Pathol.* **200**, 95–103.

Umeda, A., Fujita, H., Kuronita, T., Hirosako, K., Himeno, M., Tanaka, Y. (2003). Distribution and trafficking of MPR300 is normal in cells with cholesterol accumulated in late endosomal compartments: Evidence for early endosome-to-TGN trafficking of MPR300. *J. Lipid Res.* **44**, 1821–1832.

Vanier, M.T., Millat, G. (2003). Niemann-Pick disease type C. *Clin. Genet.* **64**, 269–281.

Vanier, M.T., Suzuki, K. (1998). Recent advances in elucidating Niemann-Pick C disease. *Brain Pathol.* **8**, 163–174.

Vassar, R., Bennett, B.D., Babu-Khan, S., Kahn, S., Mendiaz, E.A., Denis, P. et al. (1999). Beta-Secretase cleavage of Alzheimer's amyloid precursor protein by the transmembrane aspartic protease BACE. *Science* **286**, 735–741.

Vincent, I., Bu, B., Erickson, R.P. (2003). Understanding Niemann-Pick type C disease: A fat problem. *Curr. Opin. Neurol.* **16**, 155–161.

Walkley, S.U., Suzuki, K. (2004). Consequences of NPC1 and NPC2 loss of function in mammalian neurons. *Biochim. Biophys. Acta* **1685**, 48–62.

Walter, J., Fluhrer, R., Hartung, B., Willem, M., Kaether, C., Capell, A. et al. (2001). Phosphorylation regulates intracellular trafficking of beta-secretase. *J. Biol. Chem.* **276**, 14634–14641.

Wang, J., Dickson, D.W., Trojanowski, J.Q., Lee, V.M. (1999). The levels of soluble versus insoluble brain Abeta distinguish Alzheimer's disease from normal and pathologic aging. *Exp. Neurol.* **158**, 328–337.

Waisman, D.M. (1995). Annexin II tetramer: structure and function. *Mol. Cell Biochem.* **149–150**, 301–322.

Watari, H., Blanchette-Mackie, E.J., Dwyer, N.K., Glick, J.M., Patel, S., Neufeld, E.B. et al. (1999). Niemann-Pick C1 protein: Obligatory roles for N-terminal domains and lysosomal targeting in cholesterol mobilization. *Proc. Natl. Acad. Sci. U. S. A.* **96**, 805–810.

Wiegand, V., Chang, T.-Y., Strauss, J.F., Fahrenholtz, F., Gimpl, G. (2003). Transport of plasma membrane-derived cholesterol and the function of Niemann-Pick C1 protein. *FASEB J.* **17**, 782–784.

Wojtanik, K.M., Liscum, L. (2003). The transport of low density lipoprotein-derived cholesterol to the plasma membrane is defective in NPC1 cells. *J. Biol. Chem.* **278**, 14850–14856.

Wolozin, B. (2003). Cyp46 (24S-cholesterol hydroxylase): A genetic risk factor for Alzheimer disease. *Arch. Neurol.* **60**, 16–18.

Xie, C., Burns, D.K., Turley, S.D., Dietschy, J.M. (2000). Cholesterol is sequestered in the brains of mice with Niemann-Pick type C disease but turnover is increased. *J. Neuropathol. Exp. Neurol.* **59**, 1106–1117.

Xu, H., Sweeney, D., Wang, R., Thinakaran, G., Lo, A.C., Sisodia, S.S. et al. (1997). Generation of Alzheimer beta-amyloid protein in the trans-Golgi network in the apparent absence of vesicle formation. *Proc. Natl. Acad. Sci. U. S. A.* **94**, 3748–3752.

Yamazaki, T., Chang, T.-Y., Haass, C., Ihara, Y. (2001). Accumulation and aggregation of amyloid β-protein in endosomes of Niemann-Pick type C cells. *J. Biol. Chem.* **276**, 4454–4460.

Yankner, B.A. (1996). New clues to Alzheimer's disease: Unraveling the roles of amyloid and tau. *Nat. Med.* **2**, 850–852.

Yu, W., Cuervo, A., Kumar, A., Peterhoff, C.M., Schmidt, S.D., Lee, J.H. et al. (2005). Macroautophagy—A novel amyloid-β (Aβ) peptide-generating pathway activated in Alzheimer's disease. *J. Cell Biol.* **171**, 87–98.

Yu, W.H., Kumar, A., Peterhoff, C., Shapiro Kulnane, L., Uchiyama, Y., Lamb, B.T. et al. (2004). Autophagic vacuoles are enriched in amyloid precursor protein-secretase activities: Implications for beta-amyloid peptide over-production and localization in Alzheimer's disease. *Int. J. Biochem. Cell Biol.* **36**, 2531–2540.

Zerial, M., McBride, H. (2001). Rab proteins as membrane organizers. *Nat. Rev. Mol. Cell Biol.* **2**, 107–117.

Zhang, M., Dwyer, N.K., Love, D.C., Cooney, A., Comly, M., Neufeld, E. et al. (2001). Cessation of rapid late endosomal tubulovesicular trafficking in Niemann-Pick type C1 disease. *Proc. Natl. Acad. Sci.* **98**, 4466–4471.

Zinser, E., Paltauf, F., Daum, G. (1993). Sterol composition of yeast organelle membranes and subcellular distribution of enzymes involved in sterol metabolism. *J. Bacteriol.* **175**, 2853–2858.

18

Trafficking of the Cellular Prion Protein and Its Role in Neurodegeneration

OISHEE CHAKRABARTI AND RAMANUJAN S. HEGDE

I. INTRODUCTION

Some of the most debilitating neuro-degenerative diseases can be viewed as a consequence of a seemingly unassuming protein, one among tens of thousands in a cell, that is present at the wrong place at the wrong time. In cell biological terms, the offending protein has deviated from the normal pathways of biosynthesis, trafficking, and degradation that characterize its lifetime. This leads to events and inter-actions that ordinarily would not have occurred. The ensuing consequences, which are not always immediately apparent or direct, manifest over remarkably long periods of time as complex molecular cas-cades that conclude with cellular dysfunc-tion and death. Given that the final outcome can be far removed from the initial event, it is exceedingly difficult to gain molecular insight into the inciting cause by simply examining the eventual dysfunction. Instead, it requires careful study of the normal events of biosynthesis, trafficking, and metabolism of a protein in order to then trace the downstream consequences of deviations that lead to disease. In this chapter, we will examine this concept for a particularly enigmatic set of neurodegener-ative diseases caused by the prion protein, PrP. Although the molecular basis of these diseases is far from clear, we aim to illus-trate, using the available information, how the key to interpreting these complex diseases lies in understanding protein trafficking pathways.

II. A BRIEF HISTORY OF PRION DISEASES

In the 1950s, Carleton Gajdusek and colleagues investigated the basis of Kuru, a slowly progressing and invariably fatal neurodegenerative disease endemic to the Fore tribes of New Guinea. These studies led to the discovery that Kuru was a transmissible disease involving the ritualistic cannibalism practiced by these tribes (Gajdusek and Zigas 1957). The unusual transmissible properties of this disease, in particular the long incubation times spanning decades, suggested an unconventional agent, which at that time was termed a slow virus. The pathological features of neurodegeneration in Kuru was noted by Hadlow (1959) to be remarkably similar in many ways to other unusually protracted neurodegenerative diseases that included scrapie, a disease of sheep known for centuries, and certain human familial diseases such as Creutzfeldt-Jakob disease (CJD) and Gerstmann-Straussler-Scheinker syndrome (GSS). The mechanistic links, if any, among these various diseases remained obscure for many years. Eventually, Gajdusek showed that Kuru and CJD, like scrapie, could be experimentally transmitted to animals with very long incubation times (Gajdusek et al. 1966, 1967; Gibbs et al. 1968). Although this established clear links between these various transmissible spongiform encephalopathies, or TSEs, the nature of the transmissible agent or the mechanisms underlying neurodegeneration remained unknown.

The strangely long incubation times for these infectious diseases drew immediate attention to the nature of the transmissible agent. Work on identifying this agent was slow and cumbersome until the discovery that hamsters were a particularly good model system (Gibbs and Gajdusek 1973; Manuelidis et al. 1978) due to a relatively short incubation time of the infectious agent (~100–150 days) in animals that are small and can be housed in large numbers. The transmissible agent displayed unusual properties, such as relative resistance to agents that usually damage nucleic acids (Alper et al. 1966, 1967). This raised the provocative notion that the transmissible agent was a proteinaceous particle lacking a nucleic acid, an idea that led to the coining of the term *prion* by Stanley Prusiner to distinguish it from conventional viruses or virions (Prusiner et al. 1982). Indeed, fractions of brain homogenate that were highly enriched for infectivity consistently contained a glycoprotein that was named the prion protein, or PrP (Bolton et al. 1982; McKinley et al. 1983; Prusiner et al. 1984). More remarkably, attempts to identify or even demonstrate the presence of a nucleic acid in these same preparations have consistently failed (Bellinger-Kawahara et al. 1986, 1987; Safar et al. 2005; reviewed in Soto and Castilla 2004), suggesting that PrP itself is the transmissible agent in prions. So what is this protein and where does it come from?

The answer to these questions came with the surprising finding that PrP is encoded by a normal gene in the host organism (Oesch et al. 1985; Basler et al. 1986.). Moreover, the corresponding protein is constitutively expressed in a wide range of tissues (most highly in the brain) at nearly all stages of development and throughout postnatal life (Lieberburg 1987; McKinley et al. 1988; Brown et al. 1990; Fournier 2001; Tichopad et al. 2003). However, the biochemical properties of the host-expressed PrP (which was termed PrP^C for cellular PrP) were distinct from the PrP found in association with the transmissible agent (this form of PrP became known as PrP^{Sc}, to designate its association with scrapie). Most notable among these differences was the relative insolubility and resistance to protease digestion of PrP^{Sc} relative to PrP^C

(Bolton et al. 1982; McKinley et al. 1983; Oesch et al. 1985). Conspicuously absent among the differences was any covalent modification that distinguished PrP^{Sc} and PrP^C (Hope et al. 1986; Stahl et al. 1993). Thus, they appeared to differ only in their folded structure, an idea that has subsequently been supported by numerous experimental approaches (Stahl et al. 1993; Pan et al. 1993). These properties of PrP^{Sc} seemed to explain two important features of the disease. First, the insolubility and aggregation-prone nature corresponded with the PrP-rich plaques observed histologically during the end stages of the disease (DeArmond et al. 1985; Kitamoto et al. 1986). Second, a model for protein-mediated disease transmission was suggested in which PrP^{Sc} would interact with and induce PrP^C to refold into additional copies of PrP^{Sc}. This mechanism of protein-only transmission of disease, articulated elegantly by Prusiner et al. (1982), has gained tremendous experimental support over the ensuing two decades and is now widely accepted as the explanation for PrP^{Sc} propagation in the absence of any nucleic acid (Prusiner 1998; Weissmann 1999; Collinge 2001; Aguzzi and Polymenidou 2004).

The cloning of the gene for PrP also seemed to explain another puzzling observation: that many cases of CJD and the phenotypically similar disease, GSS, were familial. It was discovered that in these instances, mutations in the PrP gene were to blame (Hsiao et al. 1989; Owen et al. 1989; Goldgaber et al. 1989; Dlouhy et al. 1992). The idea was that the mutation caused PrP^C to spontaneously misfold into the PrP^{Sc} isoform, thereby bypassing a need for the external acquisition of an infectious agent. Indeed, many of the approximately 20 mutations in PrP that are associated with familial neurodegenerative diseases have been biochemically shown to accumulate PrP^{Sc} (e.g., Telling et al. 1996; Mastrianni et al. 2001), and are successfully transmitted to animals (Chapman et al. 1992; Brown et al.

1994; Tateishi and Kitamoto 1995; Tateishi et al. 1995, 1996; Hoque et al. 1996; Telling et al. 1996; Mastrianni et al. 2001), supporting a mechanism of pathogenesis involving spontaneous PrP^{Sc} generation. Thus, by the mid-1990s, a plausible framework encompassing the major aspects of TSE pathogenesis could be constructed. In this view, the acquisition by a host organism of PrP^{Sc} molecules (either exogenously or by spontaneous conversion from PrP^C) would initiate a cascade in which more and more copies of host encoded PrP^C would be converted to PrP^{Sc} in a template-driven mechanism. Over time, the aggregation-prone and difficult to degrade PrP^{Sc} would accumulate, forming deposits in the brain that cause cellular dysfunction and neuronal death, culminating in massive spongiform degeneration observed at the late stages of disease. This framework was validated by several key observations: mice lacking PrP were immune to infection with PrP^{Sc} (Bueler et al. 1993; Sailer et al. 1994); the course of disease correlated with the accumulation of PrP^{Sc} (Jendroska et al. 1991); and inherited diseases caused by PrP mutations accumulated PrP^{Sc} and were transmissible (Chapman et al. 1992; Brown et al. 1994; Tateishi and Kitamoto 1995; Tateishi et al. 1995, 1996; Hoque et al. 1996; Telling et al. 1996; Mastrianni et al. 2001). Thus, though many of the molecular details remained to be described, the basic steps seemed clear.

However, certain observations continued to defy an obvious explanation within the framework that had placed PrP^{Sc} at the center of both disease transmission and neuronal pathology. The main confounding observation was that although some of the familial forms of PrP-mediated neurodegeneration were found to contain PrP^{Sc} (Telling et al. 1996; Mastrianni et al. 2001), others did not (Tateishi et al. 1990; Parchi et al. 1998; Yamada et al. 1999; Tagliavini et al. 2001; Piccardo et al. 2001). These non-PrP^{Sc} diseases also were not transmissible to experimental animals (Tateishi et al. 1990;

Tateishi and Kitamoto 1995). This further strengthened the correlation between PrP^{Sc} and the transmissible agent, but the relationship between PrP^{Sc} accumulation and neurodegeneration seemed more tenuous. Clearly, there were mutations in PrP that could lead to spongiform neurodegeneration without necessarily generating detectable levels of PrP^{Sc} or transmissible agent. This discordance between PrP^{Sc} levels and neurodegeneration also had been observed upon inoculation of mice homozygous or heterozygous for the PrP gene: although PrP$^{+/+}$ mice and PrP$^{-/+}$ mice accumulated PrP^{Sc} with the same time course and to the same levels after inoculation, they developed disease at different times (Bueler et al. 1994). The disease in the PrP$^{-/+}$ mice was delayed by several months despite high PrP^{Sc} levels, hinting that some aspect of PrP^{C} expression influenced neuropathology independently of PrP^{Sc} levels.

This idea of a key role for PrP^{C} metabolism in neurodegeneration gained significant support from brain grafting experiments carried out by Aguzzi and colleagues (Brandner et al. 1996). In these experiments, brain grafts from PrP$^{+/+}$ mice were placed into PrP$^{-/-}$ mice. Upon inoculation with PrP^{C}, the PrP$^{+/+}$ graft supported PrP^{Sc} replication and caused its marked accumulation to high levels throughout the brain. Remarkably, however, the only area where neuronal pathology was observed was in PrP$^{+/+}$ tissue. Even immediately adjacent PrP$^{-/-}$ cells, despite being exposed to high levels of PrP^{Sc}, were not damaged at the microscopic level. This, together with more recent experiments using conditional knockout methodology (Mallucci et al. 2003), established clearly that while PrP^{Sc} is a central player in the transmission of disease, its accumulation is neither necessary nor sufficient to cause neurodegeneration. It is not necessary, since there are clearly mutations in PrP that lead to neurodegeneration without generating PrP^{Sc} or transmissible agent. It is not sufficient because, in the context of neurons that are not actively expressing PrP^{C}, PrP^{Sc} deposits do not appear to be inherently toxic.

Thus, the transmissible phase of the disease can be uncoupled from the neurodegeneration phase of disease (the arguments for this conclusion are articulated in greater detail elsewhere; Hegde and Rane 2005). The absolute dependence on continuing PrP^{C} expression for neuronal death in these diseases directly implicates some feature of its biosynthesis, trafficking, and/or metabolism. Presumably, some yet to be defined deviation(s) from its normal biosynthetic and degradative pathways, which can be initiated either by a mutation or as an indirect consequence of PrP^{Sc} accumulation, leads to neuronal dysfunction and death. This relatively recent realization has renewed research interest in PrP cell biology, an area that has long been ignored in favor of studies on PrP^{Sc} formation and propagation.

III. THE CONCEPT OF A PrP-DERIVED PROTEOME

PrP cell biology is still in its infancy and there are many unknown or uncertain answers to even relatively basic questions. For this reason, it is more informative, and perhaps more productive, to map out a systematic plan of action for future studies than to overinterpret the heterogeneous and often contradictory studies thus far available. How can we decide on the best way to approach the molecular basis of PrP-mediated neurodegeneration? One approach would be to take cues from other similar diseases for which we arguably have more molecular insight. By looking at these related fields with the 20/20 acuity of hindsight, we can glean the key concepts and general approaches that have proven to be most instructive in understanding disease pathogenesis.

The study of neurodegeneration has shown that complex diseases of postmitotic neurons, such as Parkinson's, Huntington's, and Alzheimer's diseases, seem to share elements of a common process: protein misfolding, altered protein trafficking, and progressive polymerization (Aguzzi and Haass 2003; Dawson and Dawson 2003; Selkoe 2004). Although their molecular and cellular details differ, the tendency for normal neuronal proteins to develop altered conformations and aggregate as a function of time or genetic mutation precedes the earliest clinical signs of these diseases. The present understanding of these neurodegenerative diseases has relied heavily on cell biological concepts (Sherman and Goldberg 2001; Strooper and Annaert 2001; Kim et al. 2002b; Goldberg 2003; Selkoe 2004). In turn, basic research on these diseases also has revealed new mechanisms and pathways in cell biology.

In the case of Alzheimer's disease, attention was drawn to the major component of the aggregated plaques that characterize late stages of the disease, a peptide called aβ (Glenner and Wong 1984; Masters et al. 1985). This peptide turns out to be derived from a larger precursor, termed Alzheimer's precursor protein (APP). This realization led to many investigations into APP processing and trafficking in an attempt to identify the events leading to aβ production. This has led, thus far, to the characterization of a series of proteolytic processing events, identification of the associated proteases, and definition of a wide range of APP-derived molecular species that arise as a consequence of its normal (and abnormal) processing and metabolism (reviewed by Ling et al. 2003). Although the function and disease-relevance of each APP-derived species remains to be fully described, numerous viable and testable hypotheses and molecular tools have been generated from these analyses.

The cell biological investigations have continued into the realm of trafficking. This is clearly important for understanding where the substrate and various proteases can interact in the cell. The functional consequences of certain molecular species may rely on subtleties, such as the precise environment in which it is generated or resides (c.f., Kelly 1998; Lee et al. 2005). Thus, trafficking of the substrate, the proteases, and the metabolites are critical parameters of the disease. Indeed, precisely where the different molecular species reside remains an area of intense investigation and clearly influences cellular function and potentially, disease progression (Gao and Pimplikar 2001; Lustbader et al. 2004; Stokin et al. 2005). With many of the molecules in hand, and methods to manipulate their generation now emerging, a reasonable and logical path can be laid out along which future investigations should proceed.

It is clear that one of the most useful early approaches that helped advance our understanding of AD was the careful definition of the molecular species that arise as a consequence of APP expression: not only variants of the full length protein, but of all its various metabolites and fragments (including the critical aβ-42). This definition of an APP-derived proteome provided a focal point for framing important questions regarding the nature of the events that led to the generation of each molecular species within the proteome. Answers to these questions have driven the field toward the identification and characterization of many of the key molecular components that influence disease progression (such as the α-, β-, and γ-secretases). It therefore would seem prudent to similarly define a PrP-derived proteome that could form the basis on which to begin an investigation into its role in neurodegeneration.

The definition of this PrP-derived proteome would be important for several reasons. First, it would identify potential functional species that could help us understand its mysterious normal role in the cell. It is often implicitly assumed that the most abundant species (e.g., the full-length mature protein) is the functionally relevant form, but this is often not the case. For example, the minor or transiently generated fragment of numerous proteins is the active form, and the most abundant form is inactive. Notable examples include membrane-bound transcription factors (Hoppe et al. 2001) and caspases (Creagh et al. 2003). Thus, identifying all PrP species that are generated, even those that are minor, short-lived, or found only under certain conditions, may aid in delineating its function.

The second reason for defining the PrP-derived proteome would be to have a complete baseline with which to compare changes that occur during disease progression. Without carefully determining this baseline, relevant changes that may play a role in initiating disease would be difficult to identify. Third, characterizing the proteome will necessarily highlight questions that would have been difficult to frame otherwise, and which are likely to provide important insights into not only prion disease, but cell biology in general. For example, the realization that APP is processed by a protease in the center of its transmembrane domain posed a puzzle, that of intramembrane proteolysis, which led to the development of an entire new field in cell biology (Wolfe and Kopan 2004) that is now appreciated to be widely applicable to areas as diverse as intracellular signaling and cholesterol metabolism. Similarly, it is anticipated that features of PrP metabolism that at first seem unusual or unique to PrP will prove to be more generally applicable. Indeed, the initial puzzling observation that PrP biosynthesis at the ER membrane results in a heterogeneous population of topologically different forms (Hegde et al. 1998a; see later) appears to be a more widely applicable principle (Shaffer et al. 2005).

Diversity in the proteomic species of PrP could be generated in numerous qualitatively different ways (see Figure 18.1). The first would include different molecular species that are generated during its initial biosynthesis. For example, differentially modified, processed, folded or localized species all contribute to diversity in the proteome. The second way of generating

diversity is by differential trafficking. For example, the same molecule trafficking to two different cellular locations can be markedly different functionally, and hence represent different molecular species. And finally, yet additional molecular species can be generated by proteolytic processing events that are part of a protein's normal metabolism or degradation. Because any of these species might, if generated in inappropriate amounts, at the wrong locale, or at the wrong time, lead to cellular dysfunction, it is important to characterize their identity and pathways of generation. This task has been carried out in a relatively haphazard manner for PrP using different model systems, methodologies, and reagents. Only recently have systematic attempts been made to define more than just the most abundant PrP species (see Pan et al. 2002, and references therein). This analysis has revealed a far more complex PrP-derived proteome than generally assumed. Some important molecular species (with respect to neurodegeneration) are starting to emerge, arguing strongly for the value of a systematic approach to this problem. We will review next what is known about the generation, trafficking, degradation, and etiological role of the molecular species identified thus far. The large gaps in our knowledge of each of these facets of PrP metabolism will be highlighted.

IV. PrP BIOSYNTHESIS

The gene for the human prion protein contains an open reading frame of 253 codons located entirely within a single exon (Kretzschmar et al. 1986; Schatzl et al. 1995). PrP is expressed in nearly all tissues, with the highest levels in the central nervous system. It contains an N-terminal signal sequence that interacts with the signal recognition particle to mediate targeting to the endoplasmic reticulum (ER) (see Walter and Johnson 1994, for a review on protein

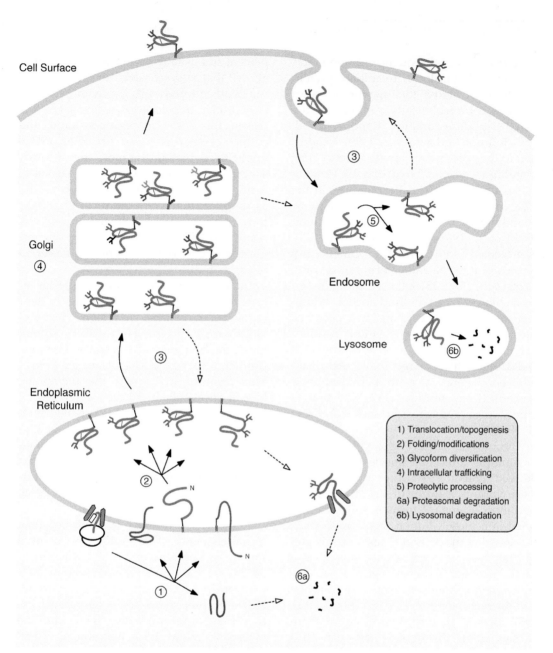

FIGURE 18.1. Diversification of the PrP-derived proteome during its biosynthesis, trafficking, and metabolism. This figure schematically depicts some of the main steps in cellular PrP metabolism, highlighting several points where multiple molecular variants are generated (summarized in the blue box). (1) During its synthesis, at least four topologically distinct forms of PrP can be generated: secPrP (blue), CtmPrP (red), NtmPrP (brown), and cyPrP (purple). (2) Some or all of these forms can be processed, folded, and modified in different ways, resulting in multiple variants for each topologic form of PrP. The example depicted here is variability in glycosylation site usage. (3) Intracellular trafficking of PrP within the secretory and endocytic pathways allows many of the PrP variants to reside in multiple compartments at any given time. (4) Within the Golgi, modifications to the glycans and the GPI anchor markedly increase the diversity of molecular species of PrP. (5) Endo-proteolytic processing of PrP generates truncated forms. The precise locations of such processing remain to be clearly established, but at least some processing is thought to occur in the endocytic pathway. (6) Ultimately, each of the numerous molecular species of PrP is degraded by one of two pathways. Most PrP is thought to be degraded by the lysosomal system, and some (usually forms that never exited the ER) are degraded by the cytosolic proteasomal system. Because all of these events are constantly occurring in the cell, there are dozens to hundreds of PrP-derived molecular species at steady state with their relative amounts varying in a cell-type–dependent manner. Alterations in the relative amounts of some of these molecular species can lead to cellular dysfunction and disease. For example, increased generation or persistence of CtmPrP or cyPrP can both cause neurodegeneration.

targeting). A signal sequence also interacts with the ER translocon to initiate protein translocation across the membrane, after which it is proteolytically removed by the signal peptidase complex (Rapoport et al. 1996; Johnson and van Waes 1999). In addition to the N-terminal signal sequence, PrP contains a C-terminal hydrophobic domain that is also proteolytically removed in the ER lumen at the time of glycosylphosphatidylinositol (GPI) anchor addition (Stahl et al. 1987). Thus, PrP that is fully translocated into the ER lumen and processed at the N- and C-terminus consists of residues 23 to 231 tethered to the membrane bilayer by a GPI lipid anchor. Within this sequence, PrP contains two potential sites for N-linked glycosylation (residues 181 and 197), two cysteines (at residues 179 and 214) that form an intramolecular disulfide bond, and a central hydrophobic domain (from residues 112–135) that is highly conserved across species (Schatzl et al. 1985).

The products of PrP biosynthesis result in multiple molecular species of the PrP-derived proteome that differ in the way that these various features of PrP are utilized. The most abundant form is fully translocated into the ER lumen (Hay et al. 1987), processed at the N- and C-terminus, is glycosylated at both consensus sites, has a single disulfide bond between the two cysteines, and contains a covalently attached GPI-anchor (Hope et al. 1986; Stahl et al. 1987; Turk et al. 1988; Pan et al. 1992). In studies that distinguish between this and other *topologically* different molecular species, this form is referred to as [sec]PrP (to indicate its complete translocation into the ER lumen, like a secretory protein; Hay et al. 1987; Hegde et al. 1998a). In most studies however, this form is assumed to be the *only* relevant cellular species of PrP, and is termed PrP[C]. Although most [sec]PrP is core glycosylated at both consensus N-linked sites, at least some fraction is glycosylated at only one of the two sites. Together, these [sec]PrP glycoforms represent more than 50

percent and usually more than 80 percent of all PrP that is synthesized in most studies utilizing cultured cells or mammalian *in vitro* protein translocation systems. For this reason, this often has been considered the correct cellular form of PrP (i.e., PrP[C]), implicitly depicting the more minor forms as aberrant species that represent mistakes, inefficiencies, or artifacts. However, the assumption that only the most abundant form of a protein is functionally relevant is risky, and must be cautiously considered given the plethora of counterexamples in biology.

In addition to this major PrP species, at least three other variants that differ in their topologies have been described (Hegde et al. 1998a; Stewart and Harris 2001). Among these, the best studied is a transmembrane form termed [Ctm]PrP (Hegde et al. 1998a). This form of PrP spans the membrane bilayer once at the highly conserved central hydrophobic domain and is oriented with the C-terminus in the ER. Because the glycosylation sites, disulfide bonded cysteines, and GPI-anchoring sequence are all in the C-terminal region, all of these modifications appear to occur on [Ctm]PrP (Hegde et al. 1998a; Stewart and Harris 2001).

Less clear is the fate of the N-terminal signal sequence. This domain is essential for the generation of [Ctm]PrP (Kim et al. 2002a), indicating that its interactions with the translocation machinery are important. However, precisely when during translocation the N-terminus slips back into the cytosol to allow the generation of [Ctm]PrP remains unclear. Studies with other proteins have shown that this can occur *after* signal sequence cleavage (Shaffer et al. 2005), suggesting that [Ctm]PrP could be signal cleaved. Indeed, we have found that although signal cleavage on [Ctm]PrP *in vitro* is relatively inefficient, it is nonetheless observed and may be more efficient *in vivo* (unpublished observations). Thus, studies based on signal sequence mutants concluding that [Ctm]PrP obligately contains an uncleaved signal may be premature

(Stewart et al. 2001). We instead favor the definition of [Ctm]PrP as a single-spanning membrane protein that is processed at both the N- and C-terminus, is glycosylated, GPI-anchored, and contains a disulfide bond. The available evidence also indicates that due to its topology, the C-terminus of [Ctm]PrP is folded differently than [sec]PrP as judged by differential susceptibility to limited protease digestion (c.f., Hegde et al. 1998a; 1999).

Insight into the role of [Ctm]PrP in neurodegenerative disease has come from the analysis of both naturally occurring and artificial mutants of PrP *in vitro* and in transgenic mice. Using *in vitro* systems, it was discovered that mutations within or near the membrane spanning region of [Ctm]PrP influenced the proportion of PrP that is synthesized in this form (Hegde et al. 1998a; Kim et al. 2001; Stewart and Harris 2001, 2003). Ordinarily, only about 10 percent of PrP synthesized *in vitro*, and even less made in most cultured cell systems, is [Ctm]PrP. However, mutations that increase or decrease the hydrophobicity of the transmembrane domain lead to a notable increase or decrease respectively, in this proportion of [Ctm]PrP. The expression of several such mutants in transgenic mice (in the PrP[-/-] background) revealed a striking correlation between the amount of [Ctm]PrP that was expressed and the susceptibility to neurodegeneration (Hegde et al. 1998a; 1999). In the most severe instances, the presence in brain of about 10 to 20 percent of total PrP in the [Ctm]PrP form led to neurodegenerative disease within two months after birth. Remarkably, even a single amino acid change from an alanine to valine (at position 117), resulting in barely detectable amounts of [Ctm]PrP in vivo (~2–5% of total) nonetheless caused neurodegeneration after about 18 months in aged animals (Hegde et al. 1998a). As expected, a version of PrP containing several mutations that together generate mostly [Ctm]PrP causes particularly severe neurodegeneration even at very low expression levels (Stewart et al.

2005). Importantly, in none of these instances has PrP[Sc] been detected, nor has it been possible to transmit the disease upon inoculation into other animals (Hegde et al. 1999). These studies illustrate that [Ctm]PrP can cause a neurodegenerative phenotype in experimental animals that resembles naturally occurring diseases caused by PrP in humans.

Interestingly, several of the naturally occurring disease-causing mutations in human PrP are within or near the central hydrophobic domain. Among these, the A117V mutation has been directly demonstrated to result in increased [Ctm]PrP *in vitro*, in mice, and in human tissue (Hegde et al. 1998a; 1999). Several other mutations described in humans (P105L, A118V, G114V, G131V) also increase the amount of [Ctm]PrP generated *in vitro* (Kim and Hegde 2002; unpublished results). Thus, [Ctm]PrP, although representing a minor proportion of the total cellular PrP, is clearly involved in causing neurodegenerative disease in several familial variants of GSS, and also is potentially involved in diseases caused by PrP[Sc] accumulation (Hegde et al. 1999; see later).

At the same time that [Ctm]PrP was identified, another transmembrane form of PrP, termed [Ntm]PrP, also was observed (Hegde et al. 1998a; Stewart and Harris 2001). This form also spans the membrane bilayer at the same central hydrophobic domain in PrP, but in the exact opposite orientation as [Ctm]PrP. Since the C-terminal domain of [Ntm]PrP is in the cytosol, it is not glycosylated, the two cysteines are presumably not disulfide bonded (since they are in the reducing environment of the cytosol), and the C-terminal hydrophobic signal is not replaced with the GPI anchor. It is currently unclear whether [Ntm]PrP is generated to a significant extent *in vivo*. In support of at least its transient generation *in vivo* is the observation in cultured cells of a nonglycosylated, signal sequence cleaved form of PrP seen during pulse labeling of newly synthesized PrP at the ER membrane (c.f., Drisaldi et al. 2003; Rane et al. 2004). At present, its role in neurodegenerative disease, if any, remains unknown.

Most recently, there has been increased interest in a form of PrP that is cytosolic. Although the pathways by which a cytosolic form of PrP (cyPrP) can be generated remain to be fully elucidated, there are at least two possibilities. One involves the retrotranslocation of PrP from the ER lumen to the cytosol (Yedidia et al. 2001; Ma and Lindquist 2001). It is presumed that due to incomplete maturation, a population of PrP may be recognized as misfolded and hence triaged into the ER-associated degradation pathway. Misfolded proteins in the ER lumen are thought to be first transported across the membrane back to the cytosol where they can be degraded by the proteasome system (Tsai et al. 2002; Meusser et al. 2005). If PrP follows this pathway, it could at least transiently reside in the cytosol. The other pathway for cyPrP generation involves molecules of nascent PrP that fail to either be targeted to or translocated across the ER membrane (Drisaldi et al. 2003; Rane et al. 2004). Such failed or aborted translocation products also would reside in the cytosol, and depending on the step at which they were aborted, may or may not contain the N-terminal signal sequence. Consistent with this idea, the signal sequence of PrP is not as efficient as certain other signal sequences in its interaction with the translocation apparatus *in vitro* (Rutkowski et al. 2001; Kim et al. 2002a) and *in vivo* (Levine et al. 2005). Furthermore, the generation of cyPrP could be markedly reduced by replacing the PrP signal sequence with a more efficient signal from another protein (Rane et al. 2004). The consequence of decreasing cyPrP in cultured cells was the reduced generation of toxic cytosolic PrP aggregates upon inhibition of proteasomal degradation (Rane et al. 2004). Therefore, at least one major pathway of cyPrP generation is a small but detectable inefficiency in PrP translocation into the ER lumen.

Whether, under normal conditions (i.e., a fully active proteasome) or in certain cell types, cyPrP plays a different role in the cell remains unclear. However, an emerging body of work suggests that under some circumstances, cyPrP is not necessarily aggregated (Mironov et al. 2003) and can be protective against cell death initiated by Bax, a pro-apoptotic protein (Bounhar et al. 2001; Roucou et al. 2003). It may be that cyPrP is harmful only when expressed inappropriately (i.e., at too high a level, or in a cell type that is unable to accommodate it), but ordinarily has a normal role in the appropriate context. It is also worth noting that at this point (and in contrast to the situation with [Ctm]PrP), naturally occurring mutations that lead to increased cyPrP by virtue of altered translocation into the ER have not been described. It has been speculated that certain disease-associated mutations in PrP lead to increased cyPrP via the retrotranslocation pathway (Ma and Lindquist 2001, 2002; Ma et al. 2002; Cohen and Taraboulos 2003), however this has been contested (Drisaldi et al. 2003). Thus, the functional or pathological roles of cyPrP remain to be investigated.

Why might a protein's biosynthesis be so heterogeneous as to produce four or more forms that differ in topology, modifications, localization, and folding? One clue comes from the observation that multiple *trans*-acting factors, in both the cytosol (Lopez et al. 1990) and ER membrane (Hegde et al. 1998b; Fons et al. 2003) apparently are involved in controlling the ratio of these forms. Analysis *in vivo* has revealed differences between cell types in how topogenic sequences are handled (Levine et al. 2005), suggesting that the relative amounts of the different PrP forms can vary in a cell type dependent manner. Indeed, dramatically skewed ratios of PrP topological isoforms, such as certain neurons expressing abundant cyPrP (Mironov et al. 2003; Barmada et al. 2004), have been observed *in vivo*. Thus, not only is PrP biosynthesis utilized to generate multiple molecular species that have distinct properties, but the ratios of these forms are subject to modulation in *trans* by factors that appear to be controlled in a cell type specific manner. These observations, together with the high degree of conservation of the elements in PrP that determine topology (Schatzl et al. 1995; Kim et al. 2001), suggest that all these molecular species may be functionally relevant in ways that remain to be elucidated. An important goal, from both a general cell biological perspective and for understanding PrP biology, would be to identify and elucidate the mechanism of the *trans*-acting factors that influence PrP translocation.

V. INTRACELLULAR TRAFFICKING OF PrP

Once the biosynthesis of PrP is completed at the ER, it is transported to other parts of the cell. Presumably, the different molecular species that are generated at the ER can be, and probably are sorted to different trafficking pathways. For example, extrapolation from the study of other model proteins would predict that species of PrP that are judged in a particular cell type to either be unnecessary or misfolded would be routed into degradation pathways. In contrast, forms that meet the requirements of cellular quality control would be transported to other cellular locales where the molecules are to serve their intended function. Since different cell types (or even the same cell type under different conditions) display both qualitative and quantitative differences in their quality control and trafficking machineries, it is likely that the triaging decisions would vary. Thus, what is true in one cell type (e.g., degradation of a particular molecular species) would not necessarily apply to another cell type. Such differences in trafficking are consistent with the observed disparities in localization and ratios of molecular species seen between cell types (for example, compare results in Liu et al. 2001; Laine et al. 2001; Kim et al. 2002a; Mironov et al. 2003; Galvan et al. 2005; Levine et al. 2005; Zhao et al. 2005). For these reasons, it is likely that viewing PrP trafficking as a single route followed by a single molecular species is too simplistic, and may explain the wide range of seemingly contradictory results obtained by different investigators using different model systems.

For the GPI-linked, glycosylated ^{sec}PrP that appears to be the most abundant form in most cell types, the trafficking follows the well-defined secretory pathway to the cell surface. In the generic tissue culture cell, this involves vesicular trafficking from the ER to the Golgi, intra-Golgi trafficking through the different cisternae, and transport from the *trans*-Golgi network to the cell surface (Borchelt et al. 1990; Taraboulos et al. 1992). During transit through the Golgi, PrP is further modified on its glycans and GPI anchor, especially with numerous sialic acids (Haraguchi et al. 1989; Stahl et al. 1992; Pan et al. 1992). Although this same ER $\rightarrow$ Golgi $\rightarrow$ cell surface paradigm is generally true in polarized cells such as neurons, finer distinctions such as localization to subdomains of the cell surface add further layers of complexity to PrP trafficking and function (Vey et al. 1996; Kaneko et al. 1997; Baron and Caughey 2003; Walmsley et al. 2003). In addition, potential interactions between PrP and other cellular proteins (c.f., Rieger et al. 1997; Spielhaupter and Schatzl 2001; Santuccione et al. 2005) may further influence important distinctions in localization (such as whether it is located within or excluded from rafts). For these reasons, caution should be exercised in overextrapolating the details (but not necessarily the general principles) obtained from simplified model systems to the highly specialized and diverse cell types in the central nervous system.

The general information available from multiple studies suggests that ^{sec}PrP has the capability of being routed selectively to one membrane surface of a polarized cell type. The parameters that control polarized localization remain largely speculative. Both intracellularly and at the cell surface, the GPI anchor and potentially other domains, can endow PrP with the potential to occupy detergent-resistant lipid microdomains or rafts (Taraboulos et al. 1995; Vey et al. 1996; Kaneko et al. 1997; Walmsley et al. 2003). Depending on the membrane surface at which PrP is located and the microdomain within which it resides, it is likely to be internalized via more than one pathway that includes clathrin-mediated endocytosis and caveolae (Shyng et al. 1993, 1994; Peters et al. 2003; Sunyach et al. 2003). Upon endocytosis, PrP either can be recycled to the cell surface (potentially being proteolytically cleaved in the process; Shyng et al. 1993) or routed to the lysosomal system for degradation. These observations provide the *potential* routes for PrP trafficking (reviewed extensively by Prado et al. 2004), but not necessarily the exact itinerary within any particular cell type.

Unfortunately, due to technical obstacles, relatively little is known about the precise trafficking pathways and subcellular destinations of PrP species in mature neurons *in situ*. Although some studies had suggested a focal distribution in synapses (Fournier et al. 1995; Haeberle et al. 2000), recent studies now appear to agree on a more uniform distribution on both axons and dendrites, being neither enriched nor excluded from any obvious region of the neuronal cell surface (Laine et al. 2001; Mironov et al. 2003; Galvan et al. 2005). Whether this applies to all neuronal subtypes remains to be thoroughly explored, but in general, secPrP goes through the standard secretory pathway by routes that are probably common to other GPI-anchored proteins to arrive at the surface of neurons. The most recent studies addressing this in cultured hippocampal neurons suggests axonal localization after, but not before complete differentiation (Galvan et al. 2005). Interest-

ingly, the polarized distribution occurs by a pathway that depends on the presence of rafts. As pointed out earlier, this is not necessarily extrapolatable to other model polarized cell types, or even to other types of neurons. However, the fact that the hippocampus consistently is affected in prion diseases makes these findings particularly valuable.

For the other forms of PrP, there is far less information. The first studies on CtmPrP, using transgenic mice expressing mutants that favor its production, indicated that it had left the ER and trafficked to at least the Golgi (Hegde et al. 1998a). Subsequently however, studies using overexpressed mutants in heterologous cell types suggested that CtmPrP is degraded in the ER (Stewart et al. 2001). This appears to have been a premature conclusion and potentially an artifact of either the cell type, the overexpression, or the mutant examined. This same mutant, when examined under its natural promoter in neurons of transgenic mice, showed a primarily Golgi distribution (Stewart and Harris 2005). Even though this is consistent with the original studies (Hegde et al. 1998a), it too may be a premature conclusion since the PrP mutant examined in this study contains an uncleaved signal sequence (Stewart et al. 2001), a property that is not necessarily a feature of CtmPrP. Thus, at this stage in our understanding, CtmPrP is clearly capable of passing ER quality control to travel via the secretory pathway to the Golgi, and possibly to the cell surface. A major limitation in gaining more precise information on CtmPrP trafficking and localization is the inability to distinguish it from the other form of PrP *in situ*, combined with the inability to generate animals that synthesize PrP exclusively in the CtmPrP form, due to its potent neurotoxicity.

The localization of NtmPrP remains totally obscure since it has not been observed in large amounts in most cell types examined even though it is often a prominent species when PrP is synthesized in cell-free systems. It is therefore possible that in certain cell types or under some conditions, NtmPrP is generated and either serves some functional role or influences disease progression. If so, the mechanisms to avoid its generation in other cell types remain unclear but may include a lack of production or its rapid degradation. One distinguishing feature of NtmPrP is that it would be signal cleaved, full length, and unglycosylated. These properties might be helpful in discriminating this form from the other topologic forms in future studies.

Cytoplasmic PrP in most cell types appears to be generated in small quantities and subject to rapid degradation, primarily if not exclusively by the ubiquitin-proteasome system. However, this is also potentially cell type specific since some cells appear to have particularly high levels of cyPrP that is neither in aggregates nor causative of obvious pathology (Mironov et al. 2003; Barmada et al. 2004). Thus, its trafficking does not seem to be as complex as the parameters that regulate its generation and degradation. One exception to this might be under conditions where proper isomerization of its proline residues is blocked by inhibition of peptidyl-prolyl isomerases using cyclosporine A. During chronic treatment (several days), a large fraction of PrP appears to be routed to aggresome-like structures (Cohen and Taraboulos 2003), a subcompartment of the cytosol that could be a depot for misfolded proteins (Kopito 2000). This pathway seems to be distinct from the route taken by cyPrP that accumulates during proteasome inhibition. Not only are the structures that accumulate different morphologically, but the cyclosporine-induced PrP accumulations arise even in cells containing fully active proteasomes.

VI. PrP DEGRADATION

The final stage in completely understanding the metabolism of PrP is to delineate the routes of degradation for each of the different forms of PrP, with particular attention to the products and intermediates that might be generated in the process. For cell surface secPrP, all the available data suggest that degradation occurs following endocytosis and trafficking into the endosomal-lysosomal pathway (Caughey et al. 1989; Taraboulos et al. 1992). Beyond this rather basic conclusion, additional details remain incomplete or controversial. For example, the route of endocytosis has been described to be via clathrin coated vesicles in some model systems (Shyng et al. 1993, 1994; Sunyach et al. 2003). This pathway is somewhat unusual for a GPI-anchored protein that does not have any portions exposed to the cytosol. Rather, internalization by a clathrin-independent pathway such as caveolae might have been expected, as has been suggested by recent studies (Peters et al. 2003). Consistent with caveolar internalization, PrP has been suggested to localize within detergent-resistant microdomains or rafts (Vey et al. 1996; Kaneko et al. 1997; Baron and Caughey 2003; Walmsley et al. 2003).

The fate or precise pathway after endocytosis remains to be studied carefully in most model systems. PrP appears to traffic to an endosomal compartment. From there, it can either be recycled to the plasma membrane (Shyng et al. 1993), or routed to the lysosomal pathway for degradation. The parameters that regulate recycling versus degradation remain to be clarified. However, it is clear that PrP does undergo proteolytic processing after reaching the cell surface, perhaps within the endosomal or lysosomal systems (Pan et al. 1992, 2002; Harris et al. 1993; Taraboulos et al. 1995). Two main processing events have been described. One involves cleavage after approximately amino acid 110, potentially by an ADAM family protease that generates a C-terminal fragment termed C1 (Vincent et al. 2001). The other cleavage occurs roughly near residue 80, may (at least indirectly) involve the activity of calpain

(Yadavalli et al. 2004), and generates a C-terminal fragment termed C2. The subsequent fates of C1 and C2 are not entirely clear, but at least a proportion of them may be recycled to the cell surface. Indeed, at steady-state in most cell types and many tissues, three prominent forms of PrP are observed that often have been assumed to be full length PrP that is glycosylated at two, one, or no sites. However, at least some of the lower molecular forms are indeed C1 and C2, and not different glycoforms, indicating that the proteolytic processing generates relatively abundant members of the PrP-derived proteome (Pan et al. 2002). The relative amounts of these forms seem to vary substantially in different brain regions, suggesting tissue-specific regulation of PrP processing (Liu et al. 2001; Zanusso et al. 2004).

The fate of the N-terminal fragments resulting from the proteolytic processing events has not been carefully characterized. They may either be shed into the medium (c.f., Zhao 2005) or degraded within the lysosomal system. There are yet additional proteolytic fragments that have been observed to varying extents in either normal and/or diseased brains (Salmona et al. 2003; Satoh et al. 2003; Zou et al. 2003; Roeber et al. 2005; Zanusso et al. 2005). For example, it is not clear whether the same PrP molecule could be processed at both the C1 and C2 sites. If so, it would generate a short peptide, a finding that is potentially of interest given that some peptide fragments of PrP (although not this one in particular) have been suggested to be neurotoxic. Additionally, several lower molecular weight fragments of PrP of variable abundance have been observed in brain tissue from various inherited forms of prion disease (Salmona et al. 2003; Satoh et al. 2003; Zou et al. 2003; Roeber et al. 2005; Zanusso et al. 2005). Because some of these fragments are enriched in diseased relative to normal brain, they have been suggested to play a causative role. How they are generated (i.e., which proteases, in what

compartment, and under what conditions) remains unknown, but is critical for eventually testing the hypothesis of their direct involvement in disease pathogenesis.

Finally, how these processed forms of PrP are subsequently degraded is not clear, but has been presumed to involve the lysosomal system. In addition, the half-life of these intermediate species is also not well characterized. Their functional role is also unclear given the lack of obvious function for PrP in general. It is worth noting that the N-terminally truncated species would not contain the octarepeat region proposed to bind copper (Hornshaw et al. 1995; Brown et al. 1997), which may affect its function and/or trafficking (Pauly and Harris 1998; Lee et al. 2001). By contrast, species that are either C-terminally truncated or have the GPI anchor cleaved would potentially be shed into the extracellular environment.

Another complicating factor is that the source of the products (i.e., their immediate or ultimate precursor) remains unknown for most species. It has largely been assumed that since [sec]PrP is the most abundant form of full length PrP, it is the precursor to all observed fragments. But this does not necessarily have to be the case, and whether some of them arise from other forms such as [Ctm]PrP, [Ntm]PrP, cyPrP, or yet other partially processed intermediates remains unknown. Reports that [Ctm]PrP may be degraded by the proteasome (Stewart et al. 2001) must be tempered by the fact that it has been demonstrated only with a mutant in a cell culture system that does not appear to faithfully recapitulate either wild-type [Ctm]PrP trafficking or processing. Thus, its degradation by the proteasome may be purely a consequence of its uncleaved signal sequence or its high level overexpression in a heterologous cell type.

CyPrP is degraded by cytosolic proteases, the most prominent of which is the ubiquitin-proteasome system. This is logical since this is the major pathway for degradation in the cellular compartment where cyPrP is located. Furthermore, both signal sequence cleaved and uncleaved versions of cyPrP have been shown to be stabilized by proteasome inhibition, and in some cases, found to be modified by ubiquitin (Yedidia et al. 2001; Ma and Lindquist 2001, 2002; Drisaldi et al. 2003; Cohen and Taraboulos 2003). However, the precise enzymes that mediate recognition and ubiquitination of PrP in the cytosol remain unknown. It is also unclear whether the different forms of cyPrP that arise through the different mechanisms (whether aborted translocation or retrotranslocation) are recognized and triaged for degradation by the same machinery.

Given that in some cell types, cyPrP is generated and accumulates to high levels (Mironov et al. 2003; Barmada et al. 2004), it may be that the recognition machinery for its degradation is differentially expressed. Furthermore, given that its inappropriate accumulation in the cell under some circumstances can lead to cell death (Ma et al. 2002; Rane et al. 2004), it is feasible that one mechanism of neuronal damage during prion disease is misregulation of the machinery for degradation of cyPrP, or by a similar logic, ^{Ctm}PrP. It is therefore attractive to postulate that the mechanism by which PrP^{Sc} accumulation results in neuronal death is via an indirect effect on the cellular pathways that metabolize cyPrP and/or ^{Ctm}PrP (Hegde and Rane 2003).

This would explain why PrP^{Sc} is not inherently toxic without simultaneous and active expression of endogenous PrP^{C} (Brandner et al. 1996). Consistent with this hypothesis, mutations that decrease the ability to generate ^{Ctm}PrP appear to make mice relatively refractory to the effects of PrP^{Sc} (Hegde et al. 1999). In contrast, mutations that predispose to the generation of ^{Ctm}PrP potentiate the neurodegenerative phenotype upon infection of mice with PrP^{Sc} (Hegde et al. 1999). Such an idea is imminently testable if we first identify the mechanisms and pathways of cyPrP and ^{Ctm}PrP generation and metabolism, and subsequently manipulate these events in *trans*

to determine the effects on PrP^{Sc}-mediated neurodegeneration. These directions represent important areas for future studies.

VII. THE IMPACT OF PrP^{Sc} ON PrP^{C} BIOSYNTHESIS, TRAFFICKING, AND DEGRADATION

The downstream consequences of PrP^{Sc} on PrP^{C} metabolism remain totally obscure. The main reason for this is the longtime focus on PrP^{C} only as a substrate for PrP^{Sc} replication. Thus, aside from PrP^{C} availability on the cell surface, it was not anticipated that some other facet of its metabolism could either be influenced by PrP^{Sc} or play a role in disease. This view was based on the widely held belief that PrP^{Sc} accumulation and aggregate deposition throughout the brain was directly neurotoxic. Since neurotoxicity proved to be more complex than this model and is clearly dependent on some yet undefined role for PrP^{C}, the relationships between their metabolic pathways necessitate careful scrutiny.

There are two general nonmutually exclusive ways in which neuronal death could be triggered by PrP^{Sc} accumulation such that active PrP^{C} expression is an absolute requirement. One is that a toxic intermediate or by-product is obligately generated during the PrP^{C} to PrP^{Sc} conversion, thereby necessitating ongoing conversion to cause neuronal death. In such a scenario, a similar by-product would be hypothesized to also arise (either directly or indirectly) as a consequence of certain mutations in PrP^{C}, thereby explaining the neurodegeneration seen in certain familial diseases that do not accumulate PrP^{Sc}. One candidate for such a toxic species might be a multimer or proto-fibril form of PrP that is neither the monomeric (or dimeric) PrP^{C} nor the much larger aggregates that characterize PrP^{Sc}. Such proto-fibrils have been hypothesized to be the toxic species in various neurodegenerative diseases that involve protein aggregates (Caughey and Lansbury 2003). In this model, proto-fibrils also would be generated by inherited PrP mutations to cause neurodegeneration, but only some mutants would further favor the conversion of these proto-fibrils to become PrP^{Sc}. This would explain the common features of both familial and transmissible prion diseases despite the fact that some of the familial forms do not accumulate PrP^{Sc}.

The second possibility is that PrP^{Sc} accumulation influences, by some indirect mechanism, the metabolism of PrP^{C} in a manner that generates or stabilizes a potentially harmful form. Given that several harmful forms of PrP have been described, their inappropriate generation in response to PrP^{Sc} accumulation is a feasible hypothesis (Hegde and Rane 2003). For example, if PrP^{Sc} accumulation results in increased accumulation of ubiquitin-conjugated proteins (as has been reported; Lowe et al. 1990; Ironside et al. 1993; Alves-Rodrigues et al. 1998; Kang et al. 2004), the cellular pools of free ubiquitin might be partially depleted. This could allow forms of PrP that ordinarily are degraded by the proteasome system (such as cyPrP or perhaps ^{Ctm}PrP) to have a longer half-life, effectively increasing their levels in the cell. The consequence of this could be to initiate neuronal cell death. In this scenario, the inherited disease causing mutations would have the same net consequence, but by another mechanism. This idea has been most well developed in the case of ^{Ctm}PrP. This form not only causes disease if its generation is increased directly (Hegde et al. 1998a; Stewart et al. 2005), but also seems to be elevated as an indirect consequence of PrP^{Sc} accumulation during transmissible prion disease (Hegde et al. 1999). Again, this general view would explain the need for ongoing PrP synthesis for neurodegeneration and the observation that PrP^{Sc} is not inherently toxic otherwise.

Although these models remain largely speculative, they are imminently testable provided that sufficient molecular insight is gained into PrP cell biology to experimentally manipulate its biosynthesis, trafficking, or degradation. For example, it has been suggested that ^{Ctm}PrP or cyPrP may need to be generated during disease pathogenesis to induce neurodegeneration (Hegde et al. 1999; Ma et al. 2002). The most convincing way to test such hypotheses is to elucidate the pathway of ^{Ctm}PrP or cyPrP production, identify a way to eliminate or substantially reduce its generation, and test whether the course of PrP^{Sc}-mediated neurodegeneration is altered. This indeed has been accomplished in a relatively crude manner for ^{Ctm}PrP (Hegde et al. 1999). In these experiments, mice were generated that expressed PrP mutants whose propensities to generate ^{Ctm}PrP differed. These were then all inoculated with PrP^{Sc} and the course of disease assessed. Remarkably, there was an inverse correlation between the ability to make ^{Ctm}PrP and susceptibility to PrP^{Sc}-mediated neurodegeneration (Hegde et al. 1999). In one particularly interesting example, mice expressing a mutant that generates very little ^{Ctm}PrP did not succumb to disease until PrP^{Sc} levels rose to severalfold higher levels than ordinarily seen at the time of clinical onset. Since this study, more precise methods to substantially reduce ^{Ctm}PrP (Kim and Hegde 2002) or cyPrP (Rane et al. 2004) have been identified. In particular, a critical step in the generation of both ^{Ctm}PrP and cyPrP appears to be the precise mode or strength of interaction between the signal sequence and the translocation channel (Rutkowski et al. 2001; Kim et al. 2002a; Kim and Hegde 2002; Fons et al. 2003; Rane et al. 2004). By changing the PrP signal sequence to one that is more efficient, ^{Ctm}PrP and cyPrP could be selectively reduced without altering any amino acids in the mature domain of PrP (Rutkowski et al. 2001; Kim et al. 2002a; Kim and Hegde 2002; Fons et al. 2003; Rane et al. 2004). Using these insights and tools, it therefore should be possible to ask whether the ability to generate these forms of PrP is important for

PrP^{Sc}-mediated neurodegeneration. Similarly, insight into other cell biological events, such as proteolytic processing of PrP, should eventually permit their role in disease etiology to be examined.

VIII. CONCLUDING REMARKS

PrP biosynthesis and metabolism constitute complex, variable, and regulated events that generate the wide range of molecular species of the PrP proteome. Alterations in the amounts, subcellular location, or temporal expression of certain members of this proteome are likely to play a critical role in the pathogenesis of PrP-associated diseases. One clear example of this is provided by studies on ^{Ctm}PrP, whose slight over-representation in the PrP proteome is sufficient to cause neurodegeneration in both humans and mice. Thus, it is apparent that a valuable immediate aim in the study of prion diseases would be to thoroughly define the proteome, delineate the pathways by which individual components within it are generated, and develop methods to manipulate and track particularly notable species.

Caution should be exerted in the extrapolation of results from one experimental system to another given the cell type specific differences in both biosynthesis and trafficking that already have been suggested. From the standpoint of neurodegeneration, it is logical to eventually analyze PrP biosynthesis and trafficking in the cell types (such as neurons of the hippocampus) that are most directly affected during the course of the disease. This has not been accomplished due to the difficulties in probing the molecular details of a protein's metabolism in differentiated and specialized postmitotic cells. Although model systems commonly used for cell biological studies are undoubtedly important for understanding the basic principles of PrP biology, it is likely that quirky variants on these principles will eventually explain why, despite widespread expression of PrP, disease pathology is restricted to certain cell types in the central nervous system. The development of tools to selectively and sensitively probe different molecular species of the PrP proteome *in situ* should be a priority for the future. We can then precisely analyze the immediate downstream consequences of disease-inciting events such as mutations within PrP or the accumulation of PrP^{Sc}. Such direct alterations to the PrP-derived proteome therefore would represent the most proximal causes of neuronal dysfunction and an attractive stage at which to intervene in preventing the development of disease.

Acknowledgements

We thank members of the Hegde Lab for fostering a creative, interactive, and productive environment to pursue research. We are particularly grateful to Neena Rane for stimulating discussions and Aarthi Ashok for carefully editing this manuscript. Work in the authors' laboratory is funded by the Intramural Program of the National Institute of Child Health and Human Development, National Institutes of Health.

References

Aguzzi, A., Haass, C. (2003). Games played by rogue proteins in prion disorders and Alzheimer's disease. *Science* **302**, 814–818.

Aguzzi, A., Polymenidou, M. (2004). Mammalian prion biology: One century of evolving concepts. *Cell* **116**, 313–327.

Alper, T., Cramp, W.A., Haig, D.A., Clarke, M.C. (1967). Does the agent of scrapie replicate without nucleic acid? *Nature* **214**, 764–766.

Alper, T., Haig, D.A., Clarke, M.C. (1966). The exceptionally small size of the scrapie agent. *Biochem. Biophys. Res. Commun.* **22**, 278–284.

Alves-Rodrigues, A., Gregori, L., Figueiredo-Pereira, M.E. (1998). Ubiquitin, cellular inclusions and their role in neurodegeneration. *Trends Neurosci.* **21**, 516–520.

Barmada, S., Piccardo, P., Yamaguchi, K., Ghetti, B., Harris, D.A. (2004). GFP-tagged prion protein is correctly localized and functionally active in the brains of transgenic mice. *Neurobiol. Dis.* **16**, 527–537.

Baron, G.S., Caughey, B. (2003). Effect of glyco-sylphosphatidylinositol anchor-dependent and -independent prion protein association with model raft membranes on conversion to the protease-resistant isoform. *J. Biol. Chem.* **278**, 14883–14892. Epub 12003 Feb 14819.

Basler, K., Oesch, B., Scott, M., Westaway, D., Walchli, M., Groth, D.F. et al. (1986). Scrapie and cellular PrP isoforms are encoded by the same chromosomal gene. *Cell* **46**, 417–428.

Bellinger-Kawahara, C., Cleaver, J.E., Diener, T.O., Prusiner, S.B. (1987). Purified scrapie prions resist inactivation by UV irradiation. *J. Virol.* **61**, 159–166.

Bellinger-Kawahara, C., Diener, T.O., McKinley, M.P., Groth, D.F., Smith, D.R., Prusiner, S.B. (1987). Purified scrapie prions resist inactivation by procedures that hydrolyze, modify, or shear nucleic acids. *Virology* **160**, 271–274.

Bolton, D.C., McKinley, M.P., Prusiner, S.B. (1982). Identification of a protein that purifies with the scrapie prion. *Science* **218**, 1309–1311.

Borchelt, D.R., Scott, M., Taraboulos, A., Stahl, N., Prusiner, S.B. (1990). Scrapie and cellular prion proteins differ in their kinetics of synthesis and topology in cultured cells. *J. Cell Biol.* **110**, 743–752.

Bounhar, Y., Zhang, Y., Goodyer, C.G., LeBlanc, A. (2001). Prion protein protects human neurons against Bax-mediated apoptosis. *J. Biol. Chem.* **276**, 39145–39149. Epub 32001 Aug 39124.

Brandner, S., Isenmann, S., Raeber, A., Fischer, M., Sailer, A., Kobayashi, Y. et al. (1996). Normal host prion protein necessary for scrapie-induced neurotoxicity. *Nature* **379**, 339–343.

Brown, D.R., Qin, K., Herms, J.W., Madlung, A., Manson, J., Strome, R. et al. (1997). The cellular prion protein binds copper in vivo. *Nature* **390**, 684–687.

Brown, H.R., Goller, N.L., Rudelli, R.D., Merz, G.S., Wolfe, G.C., Wisniewski, H.M., Robakis, N.K. (1990). The mRNA encoding the scrapie agent protein is present in a variety of non-neuronal cells. *Acta Neuropathol. (Berl.)* **80**, 1–6.

Brown, P., Gibbs, C.J., Jr., Rodgers-Johnson, P., Asher, D.M., Sulima, M.P., Bacote, A. et al. (1994). Human spongiform encephalopathy: The National Institutes of Health series of 300 cases of experimentally transmitted disease. *Ann. Neurol.* **35**, 513–529.

Bueler, H., Aguzzi, A., Sailer, A., Greiner, R.A., Autenried, P., Aguet, M., Weissmann, C. (1993). Mice devoid of PrP are resistant to scrapie. *Cell* **73**, 1339–1347.

Bueler, H., Raeber, A., Sailer, A., Fischer, M., Aguzzi, A., Weissmann, C. (1994). High prion and PrPSc levels but delayed onset of disease in scrapie-inoculated mice heterozygous for a disrupted PrP gene. *Mol. Med.* **1**, 19–30.

Caughey, B., Lansbury, P.T. (2003). Protofibrils, pores, fibrils, and neurodegeneration: Separating the responsible protein aggregates from the innocent

bystanders. *Annu. Rev. Neurosci.* **26**, 267–298. Epub 2003 Apr 2009.

Caughey, B., Race, R.E., Ernst, D., Buchmeier, M.J., Chesebro, B. (1989). Prion protein biosynthesis in scrapie-infected and uninfected neuroblastoma cells. *J. Virol.* **63**, 175–181.

Chapman, J., Brown, P., Rabey, J.M., Goldfarb, L.G., Inzelberg, R., Gibbs, C.J., Jr. et al. (1992). Transmission of spongiform encephalopathy from a familial Creutzfeldt-Jakob disease patient of Jewish Libyan origin carrying the PRNP codon 200 mutation. *Neurology* **42**, 1249–1250.

Cohen, E., Taraboulos, A. (2003). Scrapie-like prion protein accumulates in aggresomes of cyclosporin A-treated cells. *EMBO J.* **22**, 404–417.

Collinge, J. (2001). Prion diseases of humans and animals: their causes and molecular basis. *Annu. Rev. Neurosci.* **24**, 519–550.

Creagh, E.M., Conroy, H., Martin, S.J. (2003). Caspase-activation pathways in apoptosis and immunity. *Immunol. Rev.* **193**, 10–21.

Dawson, T.M., Dawson, V.L. (2003). Molecular pathways of neurodegeneration in Parkinson's disease. *Science* **302**, 819–822.

DeArmond, S.J., McKinley, M.P., Barry, R.A., Braunfeld, M.B., McColloch, J.R., Prusiner, S.B. (1985). Identification of prion amyloid filaments in scrapie-infected brain. *Cell* **41**, 221–235.

Dlouhy, S.R., Hsiao, K., Farlow, M.R., Foroud, T., Conneally, P.M., Johnson, P. et al. (1992). Linkage of the Indiana kindred of Gerstmann-Straussler-Scheinker disease to the prion protein gene. *Nat. Genet.* **1**, 64–67.

Drisaldi, B., Stewart, R.S., Adles, C., Stewart, L.R., Quaglio, E., Biasini, E. et al. (2003). Mutant PrP is delayed in its exit from the endoplasmic reticulum, but neither wild-type nor mutant PrP undergoes retrotranslocation prior to proteasomal degradation. *J. Biol. Chem.* **278**, 21732–21743. Epub 22003 Mar 21726.

Fons, R.D., Bogert, B.A., Hegde, R.S. (2003). Substrate-specific function of the translocon-associated protein complex during translocation across the ER membrane. *J. Cell Biol.* **160**, 529–539. Epub 2003 Feb 2010.

Fournier, J.G. (2001). Nonneuronal cellular prion protein. *Int. Rev. Cytol.* **208**, 121–160.

Fournier, J.G., Escaig-Haye, F., Billette de Villemeur, T., Robain, O. (1995). Ultrastructural localization of cellular prion protein (PrPc) in synaptic boutons of normal hamster hippocampus. *C. R. Acad. Sci. III* **318**, 339–344.

Gajdusek, D.C., Gibbs, C.J., Alpers, M. (1966). Experimental transmission of a Kuru-like syndrome to chimpanzees. *Nature* **209**, 794–796.

Gajdusek, D.C., Gibbs, C.J., Jr., Alpers, M. (1967). Transmission and passage of experimenal kuru to chimpanzees. *Science* **155**, 212–214.

Gajdusek, D.C., Zigas, V. (1957). Degenerative disease of the central nervous system in New Guinea: The endemic occurrence of kuru in the native population. *N. Engl. J. Med.* **257**, 974–978.

Galvan, C., Camoletto, P.G., Dotti, C.G., Aguzzi, A., Dolores Ledesma, M. (2005). Proper axonal distribution of PrP(C) depends on cholesterol-sphingomyelin-enriched membrane domains and is developmentally regulated in hippocampal neurons. *Mol. Cell. Neurosci.* **30**, 304–315.

Gao, Y., Pimplikar, S.W. (2001). The gamma-secretase-cleaved C-terminal fragment of amyloid precursor protein mediates signaling to the nucleus. *Proc. Natl. Acad. Sci. U. S. A.* **98**, 14979–14984. Epub 12001 Dec 14911.

Gibbs, C.J., Jr., Gajdusek, D.C. (1973). Experimental subacute spongiform virus encephalopathies in primates and other laboratory animals. *Science* **182**, 67–68.

Gibbs, C.J., Jr., Gajdusek, D.C., Asher, D.M., Alpers, M.P., Beck, E., Daniel, P.M., Matthews, W.B. (1968). Creutzfeldt-Jakob disease (spongiform encephalopathy): Transmission to the chimpanzee. *Science* **161**, 388–389.

Glenner, G.G., Wong, C.W. (1984). Alzheimer's disease: Initial report of the purification and characterization of a novel cerebrovascular amyloid protein. *Biochem. Biophys. Res. Commun.* **120**, 885–890.

Goldberg, A.L. (2003). Protein degradation and protection against misfolded or damaged proteins. *Nature* **426**, 895–899.

Goldgaber, D., Goldfarb, L.G., Brown, P., Asher, D.M., Brown, W.T., Lin, S. et al. (1989). Mutations in familial Creutzfeldt-Jakob disease and Gerstmann-Straussler-Scheinker's syndrome. *Exp. Neurol.* **106**, 204–206.

Hadlow, W.J. (1959). Scrapie and kuru. *Lancet* **2**, 289–290.

Haeberle, A.M., Ribaut-Barassin, C., Bombarde, G., Mariani, J., Hunsmann, G., Grassi, J., Bailly, Y. (2000). Synaptic prion protein immuno-reactivity in the rodent cerebellum. *Microsc. Res. Tech.* **50**, 66–75.

Haraguchi, T., Fisher, S., Olofsson, S., Endo, T., Groth, D., Tarentino, A. et al. (1989). Asparagine-linked glycosylation of the scrapie and cellular prion proteins. *Arch. Biochem. Biophys.* **274**, 1–13.

Harris, D.A., Huber, M.T., van Dijken, P., Shyng, S.L., Chait, B.T., Wang, R. (1993). Processing of a cellular prion protein: Identification of N- and C-terminal cleavage sites. *Biochemistry* **32**, 1009–1016.

Hay, B., Prusiner, S.B., Lingappa, V.R. (1987). Evidence for a secretory form of the cellular prion protein. *Biochemistry* **26**, 8110–8115.

Hegde, R.S., Mastrianni, J.A., Scott, M.R., DeFea, K.A., Tremblay, P., Torchia, M. et al. (1998a). A transmembrane form of the prion protein in neurodegenerative disease. *Science* **279**, 827–834.

Hegde, R.S., Rane, N.S. (2003). Prion protein trafficking and the development of neurodegeneration. *Trends Neurosci.* **26**, 337–339.

Hegde, R.S., Rane, N.S. (2005). The molecular basis of prion protein-mediated neuronal damage. *In* Neurodegeneration and prion disease, D.R. Brown, ed. Springer.

Hegde, R.S., Tremblay, P., Groth, D., DeArmond, S.J., Prusiner, S.B., Lingappa, V.R. (1999). Transmissible and genetic prion diseases share a common pathway of neurodegeneration. *Nature* **402**, 822–826.

Hegde, R.S., Voigt, S., Lingappa, V.R. (1998b). Regulation of protein topology by trans-acting factors at the endoplasmic reticulum. *Mol. Cell.* **2**, 85–91.

Hope, J., Morton, L.J., Farquhar, C.F., Multhaup, G., Beyreuther, K., Kimberlin, R.H. (1986). The major polypeptide of scrapie-associated fibrils (SAF) has the same size, charge distribution and N-terminal protein sequence as predicted for the normal brain protein (PrP). *EMBO J.* **5**, 2591–2597.

Hoppe, T., Rape, M., Jentsch, S. (2001). Membrane-bound transcription factors: Regulated release by RIP or RUP. *Curr. Opin. Cell Biol.* **13**, 344–348.

Hoque, M.Z., Kitamoto, T., Furukawa, H., Muramoto, T., Tateishi, J. (1996). Mutation in the prion protein gene at codon 232 in Japanese patients with Creutzfeldt-Jakob disease: A clinicopathological, immunohistochemical and transmission study. *Acta Neuropathol. (Berl.)* **92**, 441–446.

Hornshaw, M.P., McDermott, J.R., Candy, J.M. (1995). Copper binding to the N-terminal tandem repeat regions of mammalian and avian prion protein. *Biochem. Biophys. Res. Commun.* **207**, 621–629.

Hsiao, K., Baker, H.F., Crow, T.J., Poulter, M., Owen, F., Terwilliger, J.D. et al. (1989). Linkage of a prion protein missense variant to Gerstmann-Straussler syndrome. *Nature* **338**, 342–345.

Ironside, J.W., McCardle, L., Hayward, P.A., Bell, J.E. (1993). Ubiquitin immunocytochemistry in human spongiform encephalopathies. *Neuropathol. Appl. Neurobiol.* **19**, 134–140.

Jendroska, K., Heinzel, F.P., Torchia, M., Stowring, L., Kretzschmar, H.A., Kon, A. et al. (1991). Proteinase-resistant prion protein accumulation in Syrian hamster brain correlates with regional pathology and scrapie infectivity. *Neurology* **41**, 1482–1490.

Johnson, A.E., van Waes, M.A. (1999). The translocon: A dynamic gateway at the ER membrane. *Annu. Rev. Cell Dev. Biol.* **15**, 799–842.

Kaneko, K., Vey, M., Scott, M., Pilkuhn, S., Cohen, F.E., Prusiner, S.B. (1997). COOH-terminal sequence of the cellular prion protein directs subcellular trafficking and controls conversion into the scrapie isoform. *Proc. Natl. Acad. Sci. U. S. A.* **94**, 2333–2338.

Kang, S.C., Brown, D.R., Whiteman, M., Li, R., Pan, T., Perry, G. et al. (2004). Prion protein is ubiquitinated after developing protease resistance in the brains of scrapie-infected mice. *J. Pathol.* **203**, 603–608.

Kelly, J.W. (1998). The environmental dependency of protein folding best explains prion and amyloid diseases. *Proc. Natl. Acad. Sci. U. S. A.* **95**, 930–932.

Kim, S.J., Mitra, D., Salerno, J.R., Hegde, R.S. (2002a). Signal sequences control gating of the protein translocation channel in a substrate-specific manner. *Dev. Cell* 2, 207–217.

Kim, S., Nollen, E.A., Kitagawa, K., Bindokas, V.P., Morimoto, R.I. (2002b). Polyglutamine protein aggregates are dynamic. *Nat. Cell Biol.* 4, 826–831.

Kim, S.J., Hegde, R.S. (2002). Cotranslational partitioning of nascent prion protein into multiple populations at the translocation channel. *Mol. Biol. Cell* 13, 3775–3786.

Kim, S.J., Rahbar, R., Hegde, R.S. (2001). Combinatorial control of prion protein biogenesis by the signal sequence and transmembrane domain. *J. Biol. Chem.* 276, 26132–26140. Epub 22001 May 26118.

Kitamoto, T., Tateishi, J., Tashima, T., Takeshita, I., Barry, R.A., DeArmond, S.J., Prusiner, S.B. (1986). Amyloid plaques in Creutzfeldt-Jakob disease stain with prion protein antibodies. *Ann. Neurol.* 20, 204–208.

Kopito, R.R. (2000). Aggresomes, inclusion bodies and protein aggregation. *Trends Cell Biol.* 10, 524–530.

Kretzschmar, H.A., Stowring, L.E., Westaway, D., Stubblebine, W.H., Prusiner, S.B., Dearmond, S.J. (1986). Molecular cloning of a human prion protein cDNA. *DNA* 5, 315–324.

Laine, J., Marc, M.E., Sy, M.S., Axelrad, H. (2001). Cellular and subcellular morphological localization of normal prion protein in rodent cerebellum. *Eur. J. Neurosci.* 14, 47–56.

Lee, E.B., Zhang, B., Liu, K., Greenbaum, E.A., Doms, R.W., Trojanowski, J.Q., Lee, V.M. (2005). BACE overexpression alters the subcellular processing of APP and inhibits Abeta deposition in vivo. *J. Cell Biol.* 168, 291–302. Epub 2005 Jan 2010.

Lee, K.S., Magalhaes, A.C., Zanata, S.M., Brentani, R.R., Martins, V.R., Prado, M.A. (2001). Internalization of mammalian fluorescent cellular prion protein and N-terminal deletion mutants in living cells. *J. Neurochem.* 79, 79–87.

Levine, C.G., Mitra, D., Sharma, A., Smith, C.L., Hegde, R.S. (2005). The efficiency of protein compartmentalization into the secretory pathway. *Mol. Biol. Cell* 16, 279–291. Epub 2004 Oct 2020.

Lieberburg, I. (1987). Developmental expression and regional distribution of the scrapie-associated protein mRNA in the rat central nervous system. *Brain Res.* 417, 363–366.

Ling, Y., Morgan, K., Kalsheker, N. (2003). Amyloid precursor protein (APP) and the biology of proteolytic processing: Relevance to Alzheimer's disease. *Int. J. Biochem. Cell Biol.* 35, 1505–1535.

Liu, T., Zwingman, T., Li, R., Pan, T., Wong, B.S., Petersen, R.B. et al. (2001). Differential expression of cellular prion protein in mouse brain as detected with multiple anti-PrP monoclonal antibodies. *Brain Res.* 896, 118–129.

Lopez, C.D., Yost, C.S., Prusiner, S.B., Myers, R.M., Lingappa, V.R. (1990). Unusual topogenic sequence directs prion protein biogenesis. *Science* 248, 226–229.

Lowe, J., McDermott, H., Kenward, N., Landon, M., Mayer, R.J., Bruce, M. et al. (1990). Ubiquitin conjugate immunoreactivity in the brains of scrapie infected mice. *J. Pathol.* 162, 61–66.

Lustbader, J.W., Cirilli, M., Lin, C., Xu, H.W., Takuma, K., Wang, N. et al. (2004). ABAD directly links Abeta to mitochondrial toxicity in Alzheimer's disease. *Science* 304, 448–452.

Ma, J., Lindquist, S. (2001). Wild-type PrP and a mutant associated with prion disease are subject to retrograde transport and proteasome degradation. *Proc. Natl. Acad. Sci. U. S. A.* 98, 14955–14960. Epub 12001 Dec 14911.

Ma, J., Lindquist, S. (2002). Conversion of PrP to a self-perpetuating PrPSc-like conformation in the cytosol. *Science* 298, 1785–1788. Epub 2002 Oct 1717.

Ma, J., Wollmann, R., Lindquist, S. (2002). Neurotoxicity and neurodegeneration when PrP accumulates in the cytosol. *Science* 298, 1781–1785. Epub 2002 Oct 1717.

Mallucci, G., Dickinson, A., Linehan, J., Klohn, P.C., Brandner, S., Collinge, J. (2003). Depleting neuronal PrP in prion infection prevents disease and reverses spongiosis. *Science* 302, 871–874.

Manuelidis, E.E., Gorgacz, E.J., Manuelidis, L. (1978). Interspecies transmission of Creutzfeldt-Jakob disease to Syrian hamsters with reference to clinical syndromes and strains of agent. *Proc. Natl. Acad. Sci. U. S. A.* 75, 3432–3436.

Masters, C.L., Simms, G., Weinman, N.A., Multhaup, G., McDonald, B.L., Beyreuther, K. (1985). Amyloid plaque core protein in Alzheimer disease and Down syndrome. *Proc. Natl. Acad. Sci. U. S. A.* 82, 4245–4249.

Mastrianni, J.A., Capellari, S., Telling, G.C., Han, D., Bosque, P., Prusiner, S.B., DeArmond, S.J. (2001). Inherited prion disease caused by the V210I mutation: transmission to transgenic mice. *Neurology* 57, 2198–2205.

McKinley, M.P., Bolton, D.C., Prusiner, S.B. (1983). A protease-resistant protein is a structural component of the scrapie prion. *Cell* 35, 57–62.

McKinley, M.P., Lingappa, V.R., Prusiner, S.B. (1988). Developmental regulation of prion protein mRNA in brain. *Ciba Found. Symp.* 135, 101–116.

Meusser, B., Hirsch, C., Jarosch, E., Sommer, T. (2005). ERAD: The long road to destruction. *Nat. Cell Biol.* 7, 766–772.

Mironov, A., Jr., Latawiec, D., Wille, H., Bouzamondo-Bernstein, E., Legname, G., Williamson, R.A. et al. (2003). Cytosolic prion protein in neurons. *J. Neurosci.* 23, 7183–7193.

Oesch, B., Westaway, D., Walchli, M., McKinley, M.P., Kent, S.B., Aebersold, R. et al. (1985). A cellular gene encodes scrapie PrP 27–30 protein. *Cell* 40, 735–746.

Owen, F., Poulter, M., Lofthouse, R., Collinge, J., Crow, T.J., Risby, D. et al. (1989). Insertion in prion protein gene in familial Creutzfeldt-Jakob disease. *Lancet* **1**, 51–52.

Pan, K.M., Baldwin, M., Nguyen, J., Gasset, M., Serban, A., Groth, D. et al. (1993). Conversion of alpha-helices into beta-sheets features in the formation of the scrapie prion proteins. *Proc. Natl. Acad. Sci. U. S. A.* **90**, 10962–10966.

Pan, K.M., Stahl, N., Prusiner, S.B. (1992). Purification and properties of the cellular prion protein from Syrian hamster brain. *Protein Sci.* **1**, 1343–1352.

Pan, T., Li, R., Wong, B.S., Liu, T., Gambetti, P., Sy, M.S. (2002). Heterogeneity of normal prion protein in two-dimensional immunoblot: Presence of various glycosylated and truncated forms. *J. Neurochem.* **81**, 1092–1101.

Parchi, P., Chen, S.G., Brown, P., Zou, W., Capellari, S., Budka, H. et al. (1998). Different patterns of truncated prion protein fragments correlate with distinct phenotypes in P102L Gerstmann-Straussler-Scheinker disease. *Proc. Natl. Acad. Sci. U. S. A.* **95**, 8322–8327.

Pauly, P.C., Harris, D.A. (1998). Copper stimulates endocytosis of the prion protein. *J. Biol. Chem.* **273**, 33107–33110.

Peters, P.J., Mironov, A., Jr., Peretz, D., van Donselaar, E., Leclerc, E., Erpel, S. et al. (2003). Trafficking of prion proteins through a caveolae-mediated endosomal pathway. *J. Cell Biol.* **162**, 703–717.

Piccardo, P., Liepnieks, J.J., William, A., Dlouhy, S.R., Farlow, M.R., Young, K. et al. (2001). Prion proteins with different conformations accumulate in Gerstmann-Straussler-Scheinker disease caused by A117V and F198S mutations. *Am. J. Pathol.* **158**, 2201–2207.

Prado, M.A., Alves-Silva, J., Magalhaes, A.C., Prado, V.F., Linden, R., Martins, V.R., Brentani, R.R. (2004). PrPc on the road: trafficking of the cellular prion protein. *J. Neurochem.* **88**, 769–781.

Prusiner, S.B. (1982). Novel proteinaceous infectious particles cause scrapie. *Science* **216**, 136–144.

Prusiner, S.B. (1998). Prions. *Proc. Natl. Acad. Sci. U. S. A.* **95**, 13363–13383.

Prusiner, S.B., Groth, D.F., Bolton, D.C., Kent, S.B., Hood, L.E. (1984). Purification and structural studies of a major scrapie prion protein. *Cell* **38**, 127–134.

Rane, N.S., Yonkovich, J.L., Hegde, R.S. (2004). Protection from cytosolic prion protein toxicity by modulation of protein translocation. *EMBO J.* **23**, 4550–4559. Epub 2004 Nov 4554.

Rapoport, T.A., Jungnickel, B., Kutay, U. (1996). Protein transport across the eukaryotic endoplasmic reticulum and bacterial inner membranes. *Annu. Rev. Biochem.* **65**, 271–303.

Rieger, R., Edenhofer, F., Lasmezas, C.I., Weiss, S. (1997). The human 37-kDa laminin receptor precursor interacts with the prion protein in eukaryotic cells. *Nat. Med.* **3**, 1383–1388.

Roeber, S., Krebs, B., Neumann, M., Windl, O., Zerr, I., Grasbon-Frodl, E.M., Kretzschmar, H.A. (2005). Creutzfeldt-Jakob disease in a patient with an R208H mutation of the prion protein gene (PRNP) and a 17-kDa prion protein fragment. *Acta Neuropathol. (Berl.)* **109**, 443–448. Epub 2005 Mar 2001.

Roucou, X., Guo, Q., Zhang, Y., Goodyer, C.G., LeBlanc, A.C. (2003). Cytosolic prion protein is not toxic and protects against Bax-mediated cell death in human primary neurons. *J. Biol. Chem.* **278**, 40877–40881. Epub 42003 Aug 40812.

Rutkowski, D.T., Lingappa, V.R., Hegde, R.S. (2001). Substrate-specific regulation of the ribosome-translocon junction by N-terminal signal sequences. *Proc. Natl. Acad. Sci. U. S. A.* **98**, 7823–7828. Epub 2001 Jun 7819.

Safar, J.G., Kellings, K., Serban, A., Groth, D., Cleaver, J.E., Prusiner, S.B., Riesner, D. (2005). Search for a prion-specific nucleic acid. *J. Virol.* **79**, 10796–10806.

Sailer, A., Bueler, H., Fischer, M., Aguzzi, A., Weissmann, C. (1994). No propagation of prions in mice devoid of PrP. *Cell* **77**, 967–968.

Salmona, M., Morbin, M., Massignan, T., Colombo, L., Mazzoleni, G., Capobianco, R. et al. (2003). Structural properties of Gerstmann-Straussler-Scheinker disease amyloid protein. *J. Biol. Chem.* **278**, 48146–48153. Epub 42003 Sep 48111.

Santuccione, A., Sytnyk, V., Leshchyns'ka, I., Schachner, M. (2005). Prion protein recruits its neuronal receptor NCAM to lipid rafts to activate p59fyn and to enhance neurite outgrowth. *J. Cell Biol.* **169**, 341–354.

Satoh, K., Muramoto, T., Tanaka, T., Kitamoto, N., Ironside, J.W., Nagashima, K. et al. (2003). Association of an 11–12 kDa protease-resistant prion protein fragment with subtypes of dura graft-associated Creutzfeldt-Jakob disease and other prion diseases. *J. Gen. Virol.* **84**, 2885–2893.

Schatzl, H.M., Da Costa, M., Taylor, L., Cohen, F.E., Prusiner, S.B. (1995). Prion protein gene variation among primates. *J. Mol. Biol.* **245**, 362–374.

Selkoe, D.J. (2004). Cell biology of protein misfolding: The examples of Alzheimer's and Parkinson's diseases. *Nat. Cell Biol.* **6**, 1054–1061.

Shaffer, K.L., Sharma, A., Snapp, E.L., Hegde, R.S. (2005). Regulation of protein compartmentalization expands the diversity of protein function. *Dev. Cell*, in press.

Sherman, M.Y., Goldberg, A.L. (2001). Cellular defenses against unfolded proteins: A cell biologist thinks about neurodegenerative diseases. *Neuron* **29**, 15–32.

Shyng, S.L., Heuser, J.E., Harris, D.A. (1994). A glycolipid-anchored prion protein is endocytosed via clathrin-coated pits. *J. Cell Biol.* **125**, 1239–1250.

Shyng, S.L., Huber, M.T., Harris, D.A. (1993). A prion protein cycles between the cell surface and an endocytic compartment in cultured neuroblastoma cells. *J. Biol. Chem.* **268**, 15922–15928.

Soto, C., Castilla, J. (2004). The controversial protein-only hypothesis of prion propagation. *Nat. Med.* **10 Suppl**, S63–S67.

Spielhaupter, C., Schatzl, H.M. (2001). PrPC directly interacts with proteins involved in signaling pathways. *J. Biol. Chem.* **276**, 44604–44612. Epub 42001 Sep 44624.

Stahl, N., Baldwin, M.A., Hecker, R., Pan, K.M., Burlingame, A.L., Prusiner, S.B. (1992). Glycosylinositol phospholipid anchors of the scrapie and cellular prion proteins contain sialic acid. *Biochemistry* **31**, 5043–5053.

Stahl, N., Baldwin, M.A., Teplow, D.B., Hood, L., Gibson, B.W., Burlingame, A.L., Prusiner, S.B. (1993). Structural studies of the scrapie prion protein using mass spectrometry and amino acid sequencing. *Biochemistry* **32**, 1991–2002.

Stahl, N., Borchelt, D.R., Hsiao, K., Prusiner, S.B. (1987). Scrapie prion protein contains a phosphatidylinositol glycolipid. *Cell* **51**, 229–240.

Stewart, R.S., Drisaldi, B., Harris, D.A. (2001). A transmembrane form of the prion protein contains an uncleaved signal peptide and is retained in the endoplasmic reticulum. *Mol. Biol. Cell* **12**, 881–889.

Stewart, R.S., Harris, D.A. (2001). Most pathogenic mutations do not alter the membrane topology of the prion protein. *J. Biol. Chem.* **276**, 2212–2220. Epub 2000 Oct 2225.

Stewart, R.S., Harris, D.A. (2003). Mutational analysis of topological determinants in prion protein (PrP) and measurement of transmembrane and cytosolic PrP during prion infection. *J. Biol. Chem.* **278**, 45960–45968. Epub 42003 Aug 45921.

Stewart, R.S., Harris, D.A. (2005). A transmembrane form of the prion protein is localized in the Golgi apparatus of neurons. *J. Biol. Chem.* **280**, 15855–15864. Epub 12005 Jan 15825.

Stewart, R.S., Piccardo, P., Ghetti, B., Harris, D.A. (2005). Neurodegenerative illness in transgenic mice expressing a transmembrane form of the prion protein. *J. Neurosci.* **25**, 3469–3477.

Stokin, G.B., Lillo, C., Falzone, T.L., Brusch, R.G., Rockenstein, E., Mount, S.L. et al. (2005). Axonopathy and transport deficits early in the pathogenesis of Alzheimer's disease. *Science* **307**, 1282–1288.

Strooper, B.D., Annaert, W. (2001). Presenilins and the intramembrane proteolysis of proteins: Facts and fiction. *Nat. Cell Biol.* **3**, E221–E225.

Sunyach, C., Jen, A., Deng, J., Fitzgerald, K.T., Frobert, Y., Grassi, J. et al. (2003). The mechanism of internalization of glycosylphosphatidylinositol-anchored prion protein. *EMBO J.* **22**, 3591–3601.

Tagliavini, F., Lievens, P.M., Tranchant, C., Warter, J.M., Mohr, M., Giaccone, G. et al. (2001). A 7-kDa prion protein (PrP) fragment, an integral component of the PrP region required for infectivity, is the major amyloid protein in Gerstmann-Straussler-Scheinker disease A117V. *J. Biol. Chem.* **276**, 6009–6015. Epub 2000 Nov 6021.

Taraboulos, A., Raeber, A.J., Borchelt, D.R., Serban, D., Prusiner, S.B. (1992). Synthesis and trafficking of prion proteins in cultured cells. *Mol. Biol. Cell* **3**, 851–863.

Taraboulos, A., Scott, M., Semenov, A., Avrahami, D., Laszlo, L., Prusiner, S.B. (1995). Cholesterol depletion and modification of COOH-terminal targeting sequence of the prion protein inhibit formation of the scrapie isoform. *J. Cell Biol.* **129**, 121–132.

Tateishi, J., Brown, P., Kitamoto, T., Hoque, Z.M., Roos, R., Wollman, R. et al. (1995). First experimental transmission of fatal familial insomnia. *Nature* **376**, 434–435.

Tateishi, J., Kitamoto, T. (1995). Inherited prion diseases and transmission to rodents. *Brain Pathol.* **5**, 53–59.

Tateishi, J., Kitamoto, T., Doh-ura, K., Sakaki, Y., Steinmetz, G., Tranchant, C. et al. (1990). Immunochemical, molecular genetic, and transmission studies on a case of Gerstmann-Straussler-Scheinker syndrome. *Neurology* **40**, 1578–1581.

Tateishi, J., Kitamoto, T., Hoque, M.Z., Furukawa, H. (1996). Experimental transmission of Creutzfeldt-Jakob disease and related diseases to rodents. *Neurology* **46**, 532–537.

Telling, G.C., Parchi, P., DeArmond, S.J., Cortelli, P., Montagna, P., Gabizon, R. et al. (1996). Evidence for the conformation of the pathologic isoform of the prion protein enciphering and propagating prion diversity. *Science* **274**, 2079–2082.

Tichopad, A., Pfaffl, M.W., Didier, A. (2003). Tissue-specific expression pattern of bovine prion gene: Quantification using real-time RT-PCR. *Mol. Cell Probes* **17**, 5–10.

Tsai, B., Ye, Y., Rapoport, T.A. (2002). Retro-translocation of proteins from the endoplasmic reticulum into the cytosol. *Nat. Rev. Mol. Cell Biol.* **3**, 246–255.

Turk, E., Teplow, D.B., Hood, L.E., Prusiner, S.B. (1988). Purification and properties of the cellular and scrapie hamster prion proteins. *Eur. J. Biochem.* **176**, 21–30.

Vey, M., Pilkuhn, S., Wille, H., Nixon, R., DeArmond, S.J., Smart, E.J. et al. (1996). Subcellular colocalization of the cellular and scrapie prion proteins in caveolae-like membranous domains. *Proc. Natl. Acad. Sci. U. S. A.* **93**, 14945–14949.

Vincent, B., Paitel, E., Saftig, P., Frobert, Y., Hartmann, D., De Strooper, B. et al. (2001). The disintegrins ADAM10 and TACE contribute to the constitutive and phorbol ester-regulated normal cleavage of the cellular prion protein. *J. Biol. Chem.* **276**, 37743–37746. Epub 32001 Jul 37726.

Walmsley, A.R., Zeng, F., Hooper, N.M. (2003). The N-terminal region of the prion protein ectodomain contains a lipid raft targeting determinant. *J. Biol. Chem.* **278**, 37241–37248. Epub 32003 Jul 37214.

Walter, P., Johnson, A.E. (1994). Signal sequence recognition and protein targeting to the endoplasmic reticulum membrane. *Annu. Rev. Cell Biol.* **10**, 87–119.

Weissmann, C. (1999). Molecular genetics of transmissible spongiform encephalopathies. *J. Biol. Chem.* **274**, 3–6.

Wolfe, M.S., Kopan, R. (2004). Intramembrane proteolysis: theme and variations. *Science* **305**, 1119–1123.

Yadavalli, R., Guttmann, R.P., Seward, T., Centers, A.P., Williamson, R.A., Telling, G.C. (2004). Calpain-dependent endoproteolytic cleavage of PrPSc modulates scrapie prion propagation. *J. Biol. Chem.* **279**, 21948–21956. Epub 22004 Mar 21916.

Yamada, M., Itoh, Y., Inaba, A., Wada, Y., Takashima, M., Satoh, S. et al. (1999). An inherited prion disease with a PrP P105L mutation: Clinicopathologic and PrP heterogeneity. *Neurology* **53**, 181–188.

Yedidia, Y., Horonchik, L., Tzaban, S., Yanai, A., Taraboulos, A. (2001). Proteasomes and ubiquitin are involved in the turnover of the wild-type prion protein. *EMBO J.* **20**, 5383–5391.

Zanusso, G., Farinazzo, A., Prelli, F., Fiorini, M., Gelati, M., Ferrari, S. et al. (2004). Identification of distinct N-terminal truncated forms of prion protein in different Creutzfeldt-Jakob disease subtypes. *J. Biol. Chem.* **279**, 38936–38942. Epub 32004 Jul 38939.

Zhao, H., Klingeborn, M., Simonsson, M., Linne, T. (2006). Proteolytic cleavage and shedding of the bovine prion protein in two cell culture systems. *Virus Res.* **155**, 43–55.

Zou, W.Q., Capellari, S., Parchi, P., Sy, M.S., Gambetti, P., Chen, S.G. (2003). Identification of novel proteinase K-resistant C-terminal fragments of PrP in Creutzfeldt-Jakob disease. *J. Biol. Chem.* **278**, 40429–40436. Epub 42003 Aug 40412.

Index

Printed and bound by CPI Group (UK) Ltd, Croydon, CR0 4YY

08/05/2025

01865014-0002